TERRESTRIAL MAMMAL
CONSERVATION

Terrestrial Mammal Conservation

Global evidence for the effects of interventions for terrestrial mammals excluding bats and primates

Conservation Evidence Series Synopses
University of Cambridge, Cambridge, UK

Nick A. Littlewood, Ricardo Rocha, Rebecca K. Smith,
Philip A. Martin, Sarah L. Lockhart, Rebecca F. Schoonover,
Elspeth Wilman, Andrew J. Bladon, Katie A. Sainsbury,
Stuart Pimm and William J. Sutherland

OpenBook
Publishers

Contents

Advisory Board		xxi
About the authors		xxiii
Acknowledgements		xxv
1.	**About this book**	**1**
	The Conservation Evidence project	1
	The purpose of Conservation Evidence synopses	2
	Who this synopsis is for	2
	Background	3
	Scope of the Terrestrial Mammal Conservation synopsis	4
	Review subject	4
	Advisory board	5
	Creating the list of interventions	5
	Methods	6
	Literature searches	6
	Publication screening and inclusion criteria	10
	Study quality assessment & critical appraisal	14
	Data extraction	15
	Evidence synthesis	15
	Dissemination/communication of evidence synthesis	20
	How you can help to change conservation practice	20
	References	20
2.	**Threat: Residential and commercial development**	**23**
2.1.	Protect mammals close to development areas (e.g. by fencing)	24
2.2.	Keep cats indoors or in outside runs to reduce predation of wild mammals	24
2.3.	Use collar-mounted devices to reduce predation by domestic animals	25

2.4. Keep dogs indoors or in outside enclosures to reduce threats to wild mammals 29

2.5. Keep domestic cats and dogs well-fed to reduce predation of wild mammals 30

2.6. Translocate problem mammals away from residential areas (e.g. habituated bears) to reduce human-wildlife conflict 30

2.7. Issue enforcement notices to deter use of non-bear-proof garbage dumpsters to reduce human-wildlife conflict 38

2.8. Prevent mammals accessing potential wildlife food sources or denning sites to reduce nuisance behaviour and human-wildlife conflict 39

2.9. Provide diversionary feeding for mammals to reduce nuisance behaviour and human-wildlife conflict 41

2.10. Scare or otherwise deter mammals from human-occupied areas to reduce human-wildlife conflict 44

2.11. Retain wildlife corridors in residential areas 50

2.12. Install underpasses beneath ski runs 51

2.13. Provide woody debris in ski run area 53

3. Threat: Agriculture and aquaculture **55**
All farming systems 55

3.1. Establish wild flower areas on farmland 55

3.2. Create uncultivated margins around intensive arable or pasture fields 58

3.3. Provide or retain set-aside areas on farmland 64

3.4. Maintain/restore/create habitat connectivity on farmland 68

3.5. Manage hedgerows to benefit wildlife on farmland 68

3.6. Plant new or maintain existing hedgerows on farmland 69

3.7. Plant trees on farmland 71

3.8. Pay farmers to cover the costs of conservation measures 74

3.9. Provide refuges during crop harvesting or mowing 77

3.10. Use repellent on slug pellets to reduce non-target poisoning 78

3.11. Restrict use of rodent poisons on farmland with high secondary poisoning risk 79

Annual & Perennial Non-Timber Crops 80
3.12. Increase crop diversity for mammals 80
3.13. Create beetle banks on farmland 80
3.14. Plant crops to provide supplementary food for mammals 82
3.15. Change mowing regime (e.g. timing, frequency, height) 85
3.16. Leave areas of uncut ryegrass in silage field 85
3.17. Leave cut vegetation in field to provide cover 86
3.18. Establish long-term cover on erodible cropland 87
Livestock Farming & Ranching 88
3.19. Exclude livestock from semi-natural habitat (including woodland) 88
3.20. Reduce intensity of grazing by domestic livestock 94
3.21. Use livestock fences that are permeable to wildlife 102
3.22. Install mammal crossing points along fences on farmland 103
3.23. Use traditional breeds of livestock 106
3.24. Change type of livestock 108
Reduce human-wildlife conflict 110
3.25. Relocate local pastoralist communities to reduce human-wildlife conflict 110
3.26. Pay farmers to compensate for losses due to predators/wild herbivores to reduce human-wildlife conflict 111
3.27. Install non-electric fencing to exclude predators or herbivores and reduce human-wildlife conflict 114
3.28. Install electric fencing to reduce predation of livestock by mammals to reduce human-wildlife conflict 120
3.29. Exclude wild mammals using ditches, moats, walls or other barricades to reduce human-wildlife conflict 126
3.30. Use flags to reduce predation of livestock by mammals to reduce human-wildlife conflict 128
3.31. Use visual deterrents (e.g. scarecrows) to deter predation of livestock by mammals to reduce human-wildlife conflict 132
3.32. Use pheromones to deter predation of livestock by mammals to reduce human-wildlife conflict 134

3.33.	Use taste-aversion to reduce predation of livestock by mammals to deter human-wildlife conflict	134
3.34.	Dispose of livestock carcasses to deter predation of livestock by mammals to reduce human-wildlife conflict	140
3.35.	Use guardian animals (e.g. dogs, llamas, donkeys) bonded to livestock to deter predators to reduce human-wildlife conflict	141
3.36.	Use loud noises to deter predation of livestock by mammals to reduce human-wildlife conflict	148
3.37.	Translocate predators away from livestock to reduce human-wildlife conflict	151
3.38.	Provide diversionary feeding to reduce predation of livestock by mammals to reduce human-wildlife conflict	158
3.39.	Keep livestock in enclosures to reduce predation by mammals to reduce human-wildlife conflict	160
3.40.	Install electric fencing to protect crops from mammals to reduce human-wildlife conflict	161
3.41.	Install metal grids at field entrances to prevent mammals entering to reduce human-wildlife conflict	168
3.42.	Install automatically closing gates at field entrances to prevent mammals entering to reduce human-wildlife conflict	170
3.43.	Use tree nets to deter wild mammals from fruit crops to reduce human-wildlife conflict	171
3.44.	Deter predation of livestock by mammals by having people close by to reduce human-wildlife conflict	172
3.45.	Deter predation of livestock by herding livestock using adults instead of children to reduce human-wildlife conflict	173
3.46.	Deter predation of livestock by using shock/electronic dog-training collars to reduce human-wildlife conflict	174
3.47.	Fit livestock with protective collars to reduce risk of predation by mammals to reduce human-wildlife conflict	178
3.48.	Use lights and sound to deter predation of livestock by mammals to reduce human-wildlife conflict	180

3.49. Use scent to deter predation of livestock by   182
       mammals to reduce human-wildlife conflict

3.50. Use watchmen to deter crop damage by mammals   185
       to reduce human-wildlife conflict

3.51. Use mobile phone communications to warn farmers   185
       of problematic mammals (e.g. elephants)

3.52. Use fencing/netting to reduce predation of fish   186
       stock by mammals to reduce human-wildlife
       conflict

3.53. Establish deviation ponds in fish farms to reduce   187
       predation of fish stock by mammals to reduce
       human-wildlife conflict

3.54. Use lights and sound to deter crop damage by   188
       mammals to reduce human-wildlife conflict

3.55. Provide diversionary feeding to reduce crop   190
       damage by mammals to reduce human-wildlife
       conflict

3.56. Use scarecrows to deter crop damage by mammals   194
       to reduce human-wildlife conflict

3.57. Use loud noises to deter crop damage (e.g. banger   194
       sticks, drums, tins, iron sheets) by mammals to
       reduce human-wildlife conflict

3.58. Use noise aversive conditioning to deter crop   201
       damage by mammals to reduce human-wildlife
       conflict

3.59. Use ultrasonic noises to deter crop damage by   202
       mammals to reduce human-wildlife conflict

3.60. Use drones to deter crop damage by mammals to   204
       reduce human-wildlife conflict

3.61. Translocate crop raiders away from crops (e.g.   205
       elephants) to reduce human-wildlife conflict

3.62. Use negative stimuli to deter consumption of   207
       livestock feed by mammals to reduce human-
       wildlife conflict

3.63. Play predator calls to deter crop damage by   209
       mammals to reduce human-wildlife conflict

3.64. Use target species distress calls or signals to deter 209
       crop damage by mammals to reduce human-
       wildlife conflict

3.65. Use bees to deter crop damage by mammals (e.g.   213
       elephants) to reduce human-wildlife conflict

3.66.	Grow unattractive crop in buffer zone around crops (e.g. chili peppers) to reduce human-wildlife conflict	215
3.67.	Use chili to deter crop damage by mammals to reduce human-wildlife conflict	216
3.68.	Use light/lasers to deter crop damage by mammals to reduce human-wildlife conflict	221
3.69.	Use fire to deter crop damage by mammals to reduce human-wildlife conflict	223
3.70.	Use pheromones to deter crop damage by mammals to reduce human-wildlife conflict	225
3.71.	Use predator scent to deter crop damage by mammals to reduce human-wildlife conflict	225
3.72.	Use target species scent to deter crop damage by mammals to reduce human-wildlife conflict	228
3.73.	Use 'shock collars' to deter crop damage by mammals to reduce human-wildlife conflict	229
3.74.	Use repellents that taste bad ('contact repellents') to deter crop or property damage by mammals to reduce human-wildlife conflict	231
3.75.	Use repellents that smell bad ('area repellents') to deter crop or property damage by mammals to reduce human-wildlife conflict	238
3.76.	Use dogs to guard crops to reduce human-wildlife conflict	240
3.77.	Drive wild animals away using domestic animals of the same species to reduce human-wildlife conflict	241
4.	**Threat: Energy production and mining**	**243**
4.1.	Restore former mining sites	243
4.2.	Use electric fencing to deter mammals from energy installations or mines	251
4.3.	Use repellents to reduce cable gnawing	252
4.4.	Translocate mammals away from sites of proposed energy developments	254
5.	**Threat: Transportation and service corridors**	**257**
	Roads & Railroads	258
5.1.	Install tunnels/culverts/underpass under roads	258
5.2.	Install tunnels/culverts/underpass under railways	273

5.3.	Modify culverts to make them more accessible to mammals	278
5.4.	Install ledges in culverts under roads/railways	279
5.5.	Dig trenches around culverts under roads/railways	282
5.6.	Install fences around existing culverts or underpasses under roads/railways	282
5.7.	Install overpasses over roads/railways	287
5.8.	Install pole crossings for gliders/flying squirrels	299
5.9.	Install rope bridges between canopies	304
5.10.	Install one-way gates or other structures to allow wildlife to leave roadways	310
5.11.	Install barrier fencing along roads	315
5.12.	Install barrier fencing and underpasses along roads	323
5.13.	Install barrier fencing along railways	355
5.14.	Install wildlife warning reflectors along roads	356
5.15.	Install acoustic wildlife warnings along roads	365
5.16.	Install wildlife crosswalks	367
5.17.	Install wildlife exclusion grates/cattle grids	368
5.18.	Reduce legal speed limit	370
5.19.	Install traffic calming structures to reduce speeds	372
5.20.	Modify vegetation along roads to reduce collisions with mammals by enhancing visibility for drivers	373
5.21.	Modify the roadside environment to reduce collisions by reducing attractiveness of road verges to mammals	374
5.22.	Remove roadkill regularly to reduce kill rate of predators/scavengers	376
5.23.	Modify vegetation along railways to reduce collisions by reducing attractiveness to mammals	376
5.24.	Retain/maintain road verges as small mammal habitat	378
5.25.	Fit vehicles with ultrasonic warning devices	379
5.26.	Install signage to warn motorists about wildlife presence	382
5.27.	Use road lighting to reduce vehicle collisions with mammals	387
5.28.	Use chemical repellents along roads or railways	389
5.29.	Use alternative de-icers on roads	393
5.30.	Provide food/salt lick to divert mammals from roads or railways	394

5.31.   Use reflective collars or paint on mammals to   397
        reduce collisions with road vehicles
5.32.   Use wildlife decoy to reduce vehicle collisions with 397
        mammals
5.33.   Close roads in defined seasons   398
Utility & Service Lines   399
5.34.   Install crossings over/under pipelines   399
Shipping Lanes   402
5.35.   Install overpasses over waterways   402
5.36.   Install barrier fencing along waterways   404
5.37.   Provide mammals with escape routes from canals   404

6.   **Threat: Biological resource use**   **409**
Hunting & Collecting Terrestrial Animals   409
6.1.   Prohibit or restrict hunting of a species   409
6.2.   Ban private ownership of hunted mammals   412
6.3.   Site management for target mammal species carried   413
        out by field sport practitioners
6.4.   Set hunting quotas based on target species   414
        population trends
6.5.   Prohibit or restrict hunting of particular sex/   417
        breeding status/age animals
6.6.   Incentivise species protection through licensed   419
        trophy hunting
6.7.   Use selective trapping methods in hunting activities 420
6.8.   Use wildlife refuges to reduce hunting impacts   421
6.9.   Provide/increase anti-poaching patrols   423
6.10.   Make introduction of non-native mammals for   427
        sporting purposes illegal
6.11.   Commercially breed for the mammal production   428
        trade
6.12.   Promote sustainable alternative livelihoods   429
6.13.   Promote mammal-related ecotourism   429
6.14.   Ban exports of hunting trophies   430
Logging & Wood Harvesting   432
6.15.   Use selective harvesting instead of clearcutting   432
6.16.   Use patch retention harvesting instead of   437
        clearcutting
6.17.   Retain undisturbed patches during thinning   440
        operations

6.18.	Clear or open patches in forests	442
6.19.	Retain dead trees after uprooting	445
6.20.	Use thinning of forest instead of clearcutting	446
6.21.	Remove competing vegetation to allow tree establishment in clearcut areas	448
6.22.	Retain understorey vegetation within plantations	450
6.23.	Leave standing deadwood/snags in forests	452
6.24.	Leave coarse woody debris in forests	453
6.25.	Gather coarse woody debris into piles after felling	456
6.26.	Retain riparian buffer strips during timber harvest	458
6.27.	Retain wildlife corridors in logged areas	459
6.28.	Thin trees within forest	461
6.29.	Apply fertilizer to trees	469
6.30.	Fell trees in groups, leaving surrounding forest unharvested	472
6.31.	Coppice trees	474
6.32.	Allow forest to regenerate naturally following logging	475
6.33.	Harvest timber outside mammal reproduction period	477
6.34.	Control firewood collection in remnant native forest and woodland	477
6.35.	Plant trees following clearfelling	478
6.36.	Use tree tubes/small fences/cages to protect individual trees	479
6.37.	Provide supplementary feed to reduce tree damage	480
7.	**Threat: Human intrusions and disturbance**	**483**
7.1.	Use signs or access restrictions to reduce disturbance to mammals	483
7.2.	Set minimum distances for approaching mammals	484
7.3.	Set maximum number of people/vehicles approaching mammals	485
7.4.	Exclude or limit number of visitors to reserves or protected areas	486
7.5.	Provide paths to limit extent of disturbance to mammals	489
7.6.	Use voluntary agreements with locals to reduce disturbance	490
7.7.	Habituate mammals to visitors	491

7.8.	Translocate mammals that have habituated to humans (e.g. bears)	492
7.9.	Treat mammals to reduce conflict caused by disease transmission to humans	494
7.10.	Use conditioned taste aversion to reduce human-wildlife conflict in non-residential sites	496
7.11	Use non-lethal methods to deter carnivores from attacking humans	498

8.   **Threat: Natural system modifications**   **505**

8.1.	Use prescribed burning	505
8.2.	Burn at specific time of year	524
8.3.	Provide shelter structures after fire	526
8.4.	Thin trees to reduce wildfire risk	527
8.5.	Remove burnt trees and branches after wildfire	530
8.6.	Remove mid-storey vegetation in forest	531
8.7.	Remove understorey vegetation in forest	533
8.8.	Remove trees and shrubs to recreate open areas of land	535
8.9.	Provide artificial waterholes in dry season	537
8.10.	Use fencing to protect water sources for use by wild mammals	540
8.11.	Provide supplementary food after fire	540

9.   **Threat: Invasive alien and other problematic species**   **543**

9.1.	Use fencing to exclude grazers or other problematic species	543
9.2.	Use fencing to exclude predators or other problematic species	546
	Invasive Non-Native/Alien Species/Diseases	553
9.3.	Remove/control non-native amphibians (e.g. cane toads)	553
9.4.	Remove/control non-native invertebrates	554
9.5.	Remove/control non-native mammals	555
9.6.	Remove/control non-native mammals within a fenced area	571
9.7.	Remove/control non-native plants	573
9.8.	Control non-native/problematic plants to restore habitat	575

9.9. Reintroduce top predators to suppress and reduce the impacts of smaller non-native predator and prey species 575

9.10. Control non-native prey species to reduce populations and impacts of non-native predators 576

9.11. Provide artificial refuges for prey to evade/escape non-native predators 577

9.12. Remove/control non-native species that could interbreed with native species 578

9.13. Modify traps used in the control/eradication of non- native species to avoid injury of non-target mammal 579

9.14. Use conditioned taste aversion to prevent non-target species from entering traps 580

9.15. Use reward removal to prevent non-target species from entering traps 582

Problematic Native Species/Diseases 583

9.16. Remove or control predators 583

9.17. Sterilize predators 590

9.18. Remove or control competitors 591

9.19. Provide diversionary feeding for predators 593

9.20. Sterilise non-native domestic or feral species (e.g. cats and dogs) 593

9.21. Train mammals to avoid problematic species 596

9.22. Treat disease in wild mammals 598

9.23. Use vaccination programme 600

9.24. Eliminate highly virulent diseases early in an epidemic by culling all individuals (healthy and infected) in a defined area 605

9.25. Cull disease-infected animals 606

9.26. Use drugs to treat parasites 608

9.27. Establish populations isolated from disease 613

9.28. Control ticks/fleas/lice in wild mammal populations 615

10. Threat: Pollution 619

10.1. Reduce pesticide or fertilizer use 619

10.2. Leave headlands in fields unsprayed 622

10.3. Establish riparian buffers 623

10.4. Translocate mammals away from site contaminated by oil spill 624

11. Threat: Climate change and severe weather   **627**
       11.1.   Retain/provide migration corridors   627
       11.2.   Protect habitat along elevational gradients   628
       11.3   Translocate animals from source populations   629
               subject to similar climatic conditions
       11.4.   Provide dams/water holes during drought   630
       11.5.   Apply water to vegetation to increase food   631
               availability during drought
       11.6.   Remove flood water   632

12. Habitat protection   **635**
       12.1.   Legally protect habitat for mammals   635
       12.2.   Encourage habitat protection of privately-owned   640
               land
       12.3.   Build fences around protected areas   641
       12.4.   Retain buffer zones around core habitat   643
       12.5.   Increase size of protected area   644
       12.6.   Increase resources for managing protected areas   645

13. Habitat restoration and creation   **647**
       13.1.   Remove topsoil that has had fertilizer added to   647
               mimic low nutrient soil
       13.2.   Manage vegetation using livestock grazing   648
       13.3.   Manage vegetation using grazing by wild   652
               herbivores
       13.4.   Replant vegetation   654
       13.5.   Remove vegetation by hand/machine   654
       13.6.   Remove vegetation using herbicides   666
       13.7.   Restore or create grassland   670
       13.8.   Restore or create savannas   673
       13.9.   Restore or create shrubland   675
       13.10. Restore or create forest   677
       13.11. Restore or create wetlands   681
       13.12. Manage wetland water levels for mammal species   684
       13.13. Create or maintain corridors between habitat   685
               patches
       13.14. Apply fertilizer to vegetation to increase food   688
               availability
       13.15. Provide artificial refuges/breeding sites   689
       13.16. Provide artificial dens or nest boxes on trees   695

13.17. Provide more small artificial breeding sites rather than fewer large sites ... 710

14. **Species management** ... **713**
14.1. Cease/reduce payments to cull mammals ... 713
14.2. Temporarily hold females and offspring in fenced area to increase survival of young ... 715
14.3. Rehabilitate injured, sick or weak mammals ... 715
14.4. Hand-rear orphaned or abandoned young in captivity ... 723
14.5. Place orphaned or abandoned wild young with wild foster parents ... 727
14.6. Place orphaned or abandoned wild young with captive foster parents ... 730
14.7. Provide supplementary food to increase reproduction/survival ... 732
14.8. Provide supplementary water to increase reproduction/survival ... 747
14.9. Graze herbivores on pasture, instead of sustaining with artificial foods ... 752
Translocate Mammals ... 753
14.10. Translocate to re-establish or boost populations in native range ... 753
14.11. Translocate mammals to reduce overpopulation ... 788
14.12. Translocate predators for ecosystem restoration ... 790
14.13. Use holding pens at release site prior to release of translocated mammals ... 792
14.14. Hold translocated mammals in captivity before release ... 814
14.15. Use tranquillizers to reduce stress during translocation ... 824
14.16. Airborne translocation of mammals using parachutes ... 825
14.17. Release translocated mammals into fenced areas ... 826
14.18. Provide supplementary food during/after release of translocated mammals ... 842
Captive-breeding ... 854
14.19. Breed mammals in captivity ... 854
14.20. Place captive young with captive foster parents ... 857
14.21. Use artificial insemination ... 859

14.22.	Clone rare species	861
14.23.	Preserve genetic material for use in future captive breeding programs	863

Release captive-bred mammals — 865

14.24.	Release captive-bred individuals to re-establish or boost populations in native range	865
14.25.	Captive rear in large enclosures prior to release	882
14.26.	Use holding pens at release site prior to release of captive-bred mammals	885
14.27.	Provide live natural prey to captive mammals to foster hunting behaviour before release	904
14.28.	Train captive-bred mammals to avoid predators	907
14.29.	Release captive-bred mammals into fenced areas	909
14.30.	Provide supplementary food during/after release of captive-bred mammals	919

Release captive-bred/translocated mammals — 929

14.31.	Release translocated/captive-bred mammals in areas with invasive/problematic species eradication/control	929
14.32.	Release translocated/captive-bred mammals to islands without invasive predators	945
14.33.	Release translocated/captive-bred mammals in family/social groups	950
14.34.	Release translocated/captive-bred mammals in larger unrelated groups	967
14.35.	Release translocated/captive-bred mammals into area with artificial refuges/breeding sites	972
14.36.	Release translocated/captive-bred mammals at a specific time (e.g. season, day/night)	984
14.37.	Release translocated/captive-bred mammals to areas outside historical range	989

15. Education and awareness raising — **995**

15.1.	Encourage community-based participation in land management	995
15.2.	Use campaigns and public information to improve behaviour towards mammals and reduce threats	997
15.3.	Provide education programmes to improve behaviour towards mammals and reduce threats	999
15.4.	Provide science-based films, radio programmes, or books about mammals to improve behaviour towards mammals and reduce threats	1002

15.5.   Train and support local staff to help reduce   1003
        persecution of mammals

15.6.   Publish data on ranger performance to motivate   1004
        increased anti-poacher efforts

Appendix 1: Journals (and years) searched   **1007**

Index   **1017**

Advisory Board

We thank the following people for advising on the scope and content of this synopsis:

Esteban Brenes-Mora, Nai Conservation, Costa Rica

Steve Carter, Vincent Wildlife Trust, UK

Marco Festa-Bianchet, Université de Sherbrooke, Canada

José F. González-Maya, PROCAT Colombia/Internacional, Colombia

Clara Grilo, CDV — Transport Research Centre, Czech Republic

Menna Jones, Australian Mammal Society & University of Tasmania, Australia

Ros Kennerley, Durrell Conservation Trust,

Fiona Mathews, University of Sussex & The Mammal Society, UK

Nyumba Tobias Ochieng, University of Nairobi & African Conservation Centre, Kenya

Michela Pacifici, Sapienza University of Rome, Italy

Sara Roque, Iberian Wolf Research Team, Portugal

Miguel Rosalino, University of Lisbon, Portugal

Laurentiu Rozylowicz, University of Bucharest, Romania

About the authors

Nick Littlewood was a Research Associate in the Department of Zoology, University of Cambridge, UK and is now a Lecturer at SRUC (Scotland's Rural College), UK.

Ricardo Rocha was a Research Associate in the Department of Zoology, University of Cambridge, UK and is now a Post-doctoral Researcher at the University of Porto, Portugal.

Rebecca K. Smith is a Senior Research Associate in the Department of Zoology, University of Cambridge, UK.

Philip Martin is a Research Associate in the Department of Zoology, University of Cambridge, UK.

Sarah Lockhart was a Master's student at the Nicholas School of the Environment, Duke University, USA and is now a doctoral student and research assistant at the Center for Landscape Conservation Planning, University of Florida, USA.

Rebecca F. Schoonover was a Research Associate at Nicholas School of the Environment, Duke University, USA and is now a Brand Director.

Elspeth Wilman was a Master's student at Duke University, USA and is now a PhD Candidate at the University of Hong Kong.

Andrew J. Bladon is a Research Associate in the Department of Zoology, University of Cambridge, UK.

Katie A. Sainsbury is a Research Associate in the Department of Zoology, University of Cambridge, UK.

Stuart Pimm is the Doris Duke Chair of Conservation at the Nicholas School of the Environment, Duke University, USA.

William J. Sutherland is the Miriam Rothschild Professor of Conservation Biology at the University of Cambridge, UK.

Acknowledgements

This synopsis was made possible with funding from the MAVA Foundation, Arcadia and National Geographic Big Cats Initiative.

We would also like to thank Anna Berthinussen, William Morgan, Kate Willott and all others who contributed and provided help and advice.

1. About this book

The Conservation Evidence project

The Conservation Evidence project has four main parts:

1. The **synopses** of the evidence captured for the conservation of particular species groups or habitats, such as this synopsis. Synopses bring together the evidence for each possible intervention. They are freely available online and, in some cases, available to purchase in printed book form.

2. An ever-expanding **database of summaries** of previously published scientific papers, reports, reviews or systematic reviews that document the effects of interventions. This resource comprises over 6,616 pieces of evidence, all available in a searchable database on the website www.conservationevidence.com.

3. *What Works in Conservation*, which is an assessment of the effectiveness of interventions by expert panels, based on the collated evidence for each intervention for each species group or habitat covered by our synopses. This is available as part of the searchable database and is published as an updated book edition each year (www.conservationevidence.com/content/page/79).

4. An online, **open access journal:** *Conservation Evidence* publishes new pieces of research on the effects of conservation management interventions. All our papers are written by, or in conjunction with, those who carried out the conservation work and include some monitoring of its effects (http://conservationevidencejournal.com/).

 https://doi.org/10.11647/OBP.0234.01

The purpose of Conservation Evidence synopses

Conservation Evidence synopses **do**

- Bring together scientific evidence captured by the Conservation Evidence project (over 6,616 studies so far) on the effects of interventions to conserve biodiversity

- List all realistic interventions for the species group or habitat in question, regardless of how much evidence for their effects is available

- Describe each piece of evidence, including methods, as clearly as possible, allowing readers to assess the quality of evidence

- Work in partnership with conservation practitioners, policymakers and scientists to develop the list of interventions and ensure we have covered the most important literature

Conservation Evidence synopses **do not**

- Include evidence on the basic ecology of species or habitats, or threats to them

- Make any attempt to weight or prioritize interventions according to their importance or the size of their effects

- Weight or numerically evaluate the evidence according to its quality

- Provide recommendations for conservation problems, but instead provide scientific information to help with decision-making

Who this synopsis is for

If you are reading this, we hope you are someone who has to make decisions about how best to support or conserve biodiversity. You might be a land manager, a conservationist in the public or private sector, a farmer, a campaigner, an advisor or consultant, a policymaker, a researcher or someone taking action to protect your own local wildlife. Our synopses summarize scientific evidence relevant to your conservation objectives and the actions you could take to achieve them.

We do not aim to make your decisions for you, but to support your decision-making by telling you what evidence there is (or isn't) about the effects that your planned actions could have.

When decisions have to be made with particularly important consequences, we recommend carrying out a systematic review, as the latter is likely to be more comprehensive than the summary of evidence presented here. Guidance on how to carry out systematic reviews can be found from the Centre for Evidence-Based Conservation at the University of Bangor (www.cebc.bangor.ac.uk).

Background

At present, more than 6,300 extant mammal species are known to science (Burgin *et al.* 2018). They inhabit most of the planet's habitats and, following a commonly observed biogeographic pattern, increase in diversity with increasing proximity to the equator and peak in tropical regions (Schipper *et al.* 2008). Mammals are key providers of crucial ecosystem roles, such as herbivory, predation and seed dispersal, and they generate numerous benefits to human well-being (e.g. food, recreation and income; Schipper *et al.* 2008). Yet, over the last few decades, direct and indirect drivers of population decline, such as habitat loss, overexploitation, pollution and the impact of invasive species, have led to widespread declines in mammal population sizes and ranges (Ceballos *et al.* 2017; Ripple *et al.* 2017).

The fragile status of our mammalian fauna was reflected in the last complete IUCN assessment of the conservation status of the group, which revealed that at least one-fifth of all mammal species are currently at risk of extinction in the wild (Schipper *et al.* 2008). Extinction risks are particularly high in large-bodied species and, although the decline in mammal populations is a global pattern, the conservation status of mammal species in the Indomalayan and Australasian realms is deteriorating the fastest (Hoffmann *et al.* 2011). Conservation efforts have managed to counteract some of these population declines and, in some instances, even prevent species extinctions (Hoffmann *et al.* 2015). In fact, habitat protection and management, legal protection, and *ex-situ* conservation followed by reintroduction have contributed to the improvement of the conservation status of at least 24 species of mammal (Hoffmann *et al.* 2011). Furthermore, without conservation efforts at least 148 ungulates would have deteriorated in their IUCN red list category placement, including six species that would now likely be extinct in the wild (Hoffmann *et al.* 2015).

Evidence-based knowledge is key for planning successful conservation strategies and for the cost-effective allocation of scarce conservation resources. Targeted reviews have already collated evidence on the effects of particular interventions aimed at improving the conservation status of mammals. For example, a recent review of management practices for feral cats *Felis catus* in Australia has shown that the establishment of predator-free refuges on offshore islands, or within fenced mainland enclosures, has been crucial for the conservation of numerous threatened Australian mammals (Doherty *et al.* 2017). However, most conservation interventions targeting mammals have not yet been synthesised within a formal review and those that have could benefit from periodic update in light of new research.

Targeted reviews are labour-intensive and expensive. Furthermore, they are ill-suited for areas where the data are scarce and patchy. Here, we use a subject-wide evidence synthesis approach (Sutherland *et al.* 2019) to simultaneously summarize the evidence for the wide range of interventions dedicated to the conservation of all terrestrial mammals (excluding bats and primates). By simultaneously targeting the entire body of interventions, we are able to review the evidence for each intervention cost-effectively, and the resulting synopsis can be updated periodically and efficiently. The synopsis is freely available at www. conservationevidence.com and, alongside the Conservation Evidence online database, is a valuable asset to the toolkit of practitioners and policy makers seeking sound information to support mammal conservation. We aim to periodically update the synopsis to incorporate new research. The methods used to produce the Terrestrial Mammal Conservation Synopsis are outlined below.

Scope of the Terrestrial Mammal Conservation synopsis

Review subject

This synopsis focuses on the evidence for the effectiveness of global interventions for the conservation of terrestrial mammals, excluding bats and primates, each of which are covered in separate synopses (Berthinussen *et al.* 2019; Junker *et al.* 2017). It also excludes all species within mammal families comprised primarily of marine species, namely cetaceans (whales, dolphins and allies), pinnipeds (seals, sea lions and walruses) and sirenians (manatees and dugong). These are being covered in a separate synopsis. The Terrestrial Mammal Conservation synopsis was produced using a subject-wide evidence synthesis approach. This is defined as a systematic method of evidence synthesis that covers entire subjects at once (e.g. bird or forest conservation), including all review topics within that subject (e.g. the effects of each conservation intervention) at a fine scale and analysing results through study summary and expert assessment, or through meta-analysis; the term can also refer to any product arising from this process (Sutherland *et al.* 2019).

This synthesis covers evidence for the effects of conservation interventions for wild terrestrial mammals. We have not included evidence from the literature on husbandry of captive terrestrial mammals, such as those kept in zoos. However, where interventions carried out in captivity are relevant to the conservation of wild declining or threatened species, they were included, e.g. captive breeding for the purpose of reintroductions. For this synthesis, conservation interventions include management measures that aim to conserve wild terrestrial mammal populations and ameliorate the deleterious effects of threats. The output of the project is an authoritative, freely accessible evidence-base that will support mammal conservation objectives with the latest evidence and help to achieve conservation outcomes.

Advisory board

An advisory board made up of international conservationists and academics with expertise in terrestrial mammal conservation has been formed. These experts inputted into the synopsis at two key stages: a) producing the comprehensive list of conservation interventions for review, and b) reviewing the draft evidence synthesis. The advisory board is listed above and online (https://www.conservationevidence. com/content/page/119).

Creating the list of interventions

At the start of the project, a comprehensive list of interventions was developed by searching the literature and in partnership with the advisory board. The list was also checked by Conservation Evidence to ensure that it followed the standard structure.

The aim was to include all interventions that have been carried out or advised to support populations or communities of wild terrestrial mammals (excluding bats and primates), whether evidence for the effectiveness of an intervention is available or not. During the synthesis process further interventions were discovered and integrated into the synopsis structure.

The list of interventions was organized into categories based on the IUCN classifications of direct threats: (https://www.iucnredlist.org/resources/threat-classification-scheme) and conservation actions: (https://www.iucnredlist.org/resources/conservation-actions-classification-scheme).

In total, we found 294 conservation and/or management interventions that could be carried out to conserve terrestrial mammal (excluding bats and primates) populations. The evidence was reported as 1,261 summaries from 935 relevant publications found during our searches (see Methods below).

Methods

Literature searches

Literature was obtained from the Conservation Evidence discipline-wide literature database, and from searches of additional subject-specific literature sources (see Appendix 1). The Conservation Evidence discipline-wide literature database is compiled using systematic searches of journals (all titles and abstracts) and report series ('grey literature'); relevant publications describing studies of conservation interventions for all species groups and habitats were saved from each and were added to the database. The final list of evidence sources searched for this synopsis is published in this synopsis document (see Appendix 1), and the full list of journals and report series is published online (https://www.conservationevidence.com/journalsearcher/synopsis).

a) Global evidence

Evidence from all around the world was included.

b) Languages included

The following non-English journals published in Spanish and Portuguese were searched and relevant papers extracted.

- Therya Vol. 1, Issue 1 (2010) — Vol. 8, Issue 3 (2017)
- Galemys Vol. 1 (2011) — Vol. 7 (2017)
- Boletim da Sociedade Brasileira de Mastozoologia Vol. 66 (2013) — Vol. 78 (2017)
- Mastozoologia Neotropical Vol. 1, Issue 1 (1994) — Vol. 24, Issue 1 (2017)
- Mammalogy Notes Vol. 1, Issue 1 (2014) — Vol. 4, Issue 1 (2017)
- Revista Mexicana de Mastozoología Vol. 1 (1995) — Vol. 7, Issue 2 (2017)

All other journals searched are published in English or at least carry English summaries (see below). A recent study on the topic of language barriers in global science indicates that approximately 35% of conservation studies may be in non-English languages (Amano *et al.* 2016). While searching only a small number of non-English language journals may therefore potentially introduce some bias to the review process, project resources and time constraints determined the number of journals that could be searched within the project timeframe.

c) Journals searched

i) From Conservation Evidence discipline-wide literature database

All of the journals (and years) listed in Appendix 1 were searched prior to or during the completion of this project by authors of other synopses, and relevant papers added to the Conservation Evidence discipline-wide literature database. An asterisk indicates the journals most relevant to this synopsis. Others are less likely to include papers relevant to this synopsis, but if they did, those papers were summarised.

ii)   Update searches

The authors of this synopsis updated the search of the following journals:

- Hystrix, the Italian Journal of Mammalogy (2014–2016)
- Journal of Mammalogy (2013–2017)
- Mammal Review (2013–2017)
- Mammal Study (2013–2017)
- Mammalia (2013–2017)
- Mammalian Biology (2013–2017)

iii) New searches

Additional, focussed searches of journals most relevant to the conservation of terrestrial mammal populations listed in Appendix 1 were undertaken. These journals were identified through expert judgement by the project researchers and the advisory board.

- Acta Theriologica (1997–2014)
- Australian Mammalogy (2000–2017)
- Biotropica (1990–2017)
- Mammal Research (2001–2017)

d) Reports from specialist websites searched

i)   From Conservation Evidence discipline-wide literature database

All report series (and years) below have already been systematically searched for the Conservation Evidence project. An asterisk indicates the report series most relevant to this synopsis. Others are less likely to have included reports relevant to this synopsis but, if they did, they have been summarised.

- Amphibian Survival Alliance 1994–2012 Vol 9–Vol 104
- British Trust for Ornithology 1981–2016 Report 1–687
- IUCN Invasive Species Specialist Group 1995–2013 Vol 1–Vol 33

- Scottish Natural Heritage* 2004–2015 Reports 1–945

ii) *Update searches*

Updated searches of report series already searched as part of the wider Conservation Evidence project were not undertaken for this synopsis.

No new report searches were undertaken for this synopsis due to time constraints.

e) Other literature searches

The online database (www.conservationevidence.com) was searched for relevant publications that have already been summarised. If such summaries existed, they were extracted and added to this synopsis update.

Where a systematic review was found for an intervention, if the intervention had a small literature (<20 papers), all available English language publications including the systematic review were summarised. If the intervention had a large literature (≥20 papers), then only the systematic review was summarised. Where a non-systematic review (or editorial, synthesis, preface, introduction etc.) was found for an intervention, all relevant and accessible English language publications referenced within it were included, but the review itself was not summarised. However, if the review also provided new/ collective data, then the review itself was also included/summarised. Relevant publications cited in other publications summarised for the synopsis were not included (due to time restrictions).

f) Supplementary literature identified by advisory board or relevant stakeholders

Relevant papers or reports suggested by the advisory board or relevant stakeholders were also included, if relevant.

g) Search record database

A database was created of all relevant publications found during searches. Reasons for exclusion were recorded for all studies included during screening but not summarised for the synopsis.

Publication screening and inclusion criteria

a) Screening

We acknowledge that the literature search and screening method used by Conservation Evidence, as with any method, results in gaps in the evidence. The Conservation Evidence literature database currently includes relevant papers from over 270 English language journals as well as over 150 non-English journals. Additional journals are frequently added to those searched, and years searched are often updated. It is possible that searchers will have missed relevant papers from those journals searched. Publication bias, where studies reporting negative or non-significant findings are less likely to be written up and published in journals (e.g. Dwan *et al.* 2013), will not be taken into account, and it is likely that additional biases will result from the evidence that is available. For example, there are often geographic biases in study locations.

b) Inclusion criteria

The following Conservation Evidence inclusion criteria were used.

1) There has to be an intervention that conservationists would be likely to do.

2) Its effects on biodiversity or ecosystem services must have been monitored quantitatively.

If the intervention can be used for conservation purposes, but is being done for a different purpose in the study in question, it should be included, provided the details of the intervention are the same and the effects on biodiversity or ecosystem services have been monitored.

For example, methods to rear bumblebees in captivity for commercial pollination have been used to support conservation of rare bumblebees. All studies testing these methods were included in our bee synopsis. Another example is the construction of artificial wetlands for amphibian conservation. Studies that monitor amphibian numbers in wetlands constructed largely for recreational purposes were included.

Interventions for captive animals are only included if they are directly relevant to the conservation of native wild species, e.g. breeding animals in captivity for release into natural habitats, or trials of animals' responses to interventions designed to reduce human-wildlife conflict.

Modelling studies that do not actually test the intervention vs a control on the ground are not included.

c) *Relevant subject*

Studies relevant to the synopsis subject were those focused on the conservation of wild, native terrestrial mammals (excluding bats and primates). All mammals belonging to groups that are primarily comprised of marine species (cetaceans, pinnipeds and sirenians) were also excluded. For the remaining mammal groups, all species were deemed relevant for this synopsis, including those that may spend most of their time in water (e.g. sea otter *Enhydra lutris*).

d) *Relevant types of intervention*

An intervention has to be one that could be put in place by a manager, conservationist, policy maker, advisor or consultant to protect, manage, restore or reduce the impacts of threats to wild, native terrestrial mammals. Alternatively, interventions may aim to change human behaviour (actual or intentional), which is likely to protect, manage, restore or reduce threats to terrestrial mammal populations.

If the following two criteria were met, a combined intervention was created within the synopsis, rather than repeating evidence under all the separate interventions: a) there are five or more publications that use the same well-defined combination of interventions, with very clear description of what they were, without separating the effects of each individual intervention, and b) the combined set of interventions is a commonly used conservation strategy.

e) *Relevant types of comparator*

To determine the effectiveness of interventions, studies must include a comparison, i.e. monitoring change over time (typically before and after the intervention was implemented), or for example at treatment and control sites. Alternatively, a study could compare one specific intervention (or implementation method) against another. For example, this could be comparing the abundance of a mammal species before and after woodland is restored, or the reduction in mammal mortality at roads with different underpass designs.

Exceptions, which may not have a control but were still included, are for example the effectiveness of captive breeding or rehabilitation programmes or use made of nest boxes for arboreal mammals or of wildlife overpasses across roads.

f) Relevant types of outcome

Below we provide a list of included metrics:

- Community response
  - Community composition
  - Richness/diversity

- Population response
  - Abundance: mammal activity (relative abundance), number, presence/absence
  - Reproductive success: mating success, birth rate, infant survival
  - Survival: survival, mortality
  - Condition: body mass, weight, size, forearm length, disease symptoms

- Behaviour
  - Uptake
  - Use
  - Behaviour change: movement, range, timing (e.g. emergence, foraging period)

- Other
  - Human-wildlife conflict
  - Human behaviour change
  - Genetic diversity

g) *Relevant types of study design*

The table below lists the study designs included. The strongest evidence comes from randomized, replicated, controlled trials with paired-sites and before-and-after monitoring.

Table 1. Study designs

Term	Meaning
Replicated	The intervention was repeated on more than one individual or site. In conservation and ecology, the number of replicates is much smaller than it would be for medical trials (when thousands of individuals are often tested). If the replicates are sites, pragmatism dictates that between five and ten replicates is a reasonable amount of replication, although more would be preferable. We provide the number of replicates wherever possible. Replicates should reflect the number of times an intervention has been independently carried out, from the perspective of the study subject. For example, 10 plots within a mown field might be independent replicates from the perspective of plants with limited dispersal, but not independent replicates for larger motile animals such as birds. In the case of translocations/release of captive bred animals, replicates should be sites, not individuals.
Randomized	The intervention was allocated randomly to individuals or sites. This means that the initial condition of those given the intervention is less likely to bias the outcome.
Paired sites	Sites are considered in pairs, within which one was treated with the intervention and the other was not. Pairs, or blocks, of sites are selected with similar environmental conditions, such as soil type or surrounding landscape. This approach aims to reduce environmental variation and make it easier to detect a true effect of the intervention.
Controlled*	Individuals or sites treated with the intervention are compared with control individuals or sites not treated with the intervention. (The treatment is usually allocated by the investigators (randomly or not), such that the treatment or control groups/sites could have received the treatment).

Term	Meaning
Before-and-after	Monitoring of effects was carried out before and after the intervention was imposed.
Site comparison*	A study that considers the effects of interventions by comparing sites that historically had different interventions (e.g. intervention vs no intervention) or levels of intervention. Unlike controlled studies, it is not clear how the interventions were allocated to sites (i.e. the investigators did not allocate the treatment to some of the sites).
Review	A conventional review of literature. Generally, these have not used an agreed search protocol or quantitative assessments of the evidence.
Systematic review	A systematic review follows an agreed set of methods for identifying studies and carrying out a formal 'meta-analysis'. It will weight or evaluate studies according to the strength of evidence they offer, based on the size of each study and the rigour of its design. All environmental systematic reviews are available at: www.environmentalevidence.org/index.htm.
Study	If none of the above apply, for example a study measuring change over time in only one site and only after an intervention. Or a study measuring use of nest boxes at one site.

* Note that 'controlled' is mutually exclusive from 'site comparison'. A comparison cannot be both controlled and a site comparison. However, one study might contain both controlled and site comparison aspects e.g. study of fertilized grassland, compared to unfertilized plots (controlled) and natural, target grassland (site comparison).

Study quality assessment & critical appraisal

We did not quantitatively assess the evidence from each publication or weight it according to quality. However, to allow interpretation of the evidence, we made the sample size and design of each study we reported clear.

We critically appraised each potentially relevant study and excluded those that did not provide data for a comparison to the treatment, did not statistically analyse the results (or if included it was stated in the summary paragraph that statistical analysis was not carried out) or had obvious errors in their design or analysis. A record of the reason for excluding any of the publications included during screening was kept within the synopsis database.

Data extraction

Data on the effectiveness of the relevant intervention (e.g. mean species abundance inside or outside a protected area; reduction in mortality after installation of an overpass) were extracted from, and summarised for, publications that included the relevant subject, types of intervention, comparator and outcomes outlined above. The total number of publications included following data extraction is 931.

At the start of each month, authors swapped three summaries with another author to ensure that the correct type of data had been extracted and that the summary followed the Conservation Evidence standard format.

Evidence synthesis

a) Summary protocol

Each publication usually had just one paragraph for each intervention it tested describing the study. Summaries were in plain English and, where possible, were no more than 150 words long, though more complex studies required longer summaries. Each summary used the following format:

> A [TYPE OF STUDY] in [YEARS X-Y] in [HOW MANY SITES] in/of [HABITAT] in [REGION and COUNTRY] [REFERENCE] found that [INTERVENTION] [SUMMARY OF ALL KEY RESULTS] for [SPECIES/ HABITAT TYPE]. [DETAILS OF KEY RESULTS, INCLUDING DATA]. In addition, [EXTRA RESULTS, IMPLEMENTATION OPTIONS, CONFLICTING RESULTS]. The [DETAILS OF EXPERIMENTAL DESIGN, INTERVENTION METHODS and KEY DETAILS OF SITE CONTEXT]. Data was collected in [DETAILS OF SAMPLING METHODS].

Type of study — see terms and order in Table 1.

Site context — for the sake of brevity, only nuances essential to the interpretation of the results are included. The reader is always encouraged to read the original source to get a full understanding of the study site (e.g. history of management, physical conditions, landscape context).

For example:

A controlled study in 2008 of a grassland and woodland site in Nevada, USA (1) found that reducing grazing intensity by long-term exclusion of domestic livestock resulted in a higher species richness and abundance of small mammals. More small mammal species were recorded on ungrazed land (six) than on grazed land (four). Small mammal abundance on ungrazed land (0.08 animals/trap night) was higher than on grazed land (0.05 animals/trap night). Three species were caught in sufficient quantities for individual analyses. The Great Basin pocket mouse *Perognathus parvus* was more abundant on ungrazed than grazed land (0.05 vs 0.02 individuals/trap night) as was western jumping mouse *Zapus princeps* (0.02 vs 0.00 individuals/trap night). Deer mice *Peromyscus maniculatus* showed no preference (0.01 vs 0.01 individuals/trap night). Sampling occurred in a 10-ha enclosure, characterised by mixed shrubs and trees, from which domestic livestock were excluded at least 50 years previously and in a similar sized, adjacent cattle-grazed grassland. Small mammals were sampled using lines of snap-traps, over three or four nights, in July 2008.

(1) Rickart E.A., Bienek K.G. & Rowe R.J. (2013) Impact of livestock grazing on plant and small mammal communities in the Ruby Mountains, northeastern Nevada. *Western North American Naturalist*, 73, 505–515.

A replicated study in 1999–2004 in a wetland on an island in Catalonia, Spain (2) found that all 69 bat boxes of two different designs were used by soprano pipistrelles Pipistrellus pygmaeus with an average occupancy rate of 71%. During at least one of the four breeding seasons recorded, 96% of boxes were occupied and occupation rates by females with pups increased from 15% in 2000 to 53% in 2003. Bat box preferences were detected in the breeding season only, with higher abundance in east-facing bat boxes (average 22 bats/box) compared to west-facing boxes (12 bats/box), boxes with double compartments (average 25 bats/box) compared to single compartments (12 bats/box) and boxes placed on posts (average 18 bats/box) and houses (average 12 bats/box). Abundance was low in bat boxes on trees (average 2 bats/box). A total of 69 wooden bat boxes (10 cm deep × 19 cm wide × 20 cm high) of two types (44 single and 25 double compartment) were placed on three supports (10 trees, 29 buildings and 30 electricity posts) facing east and west. From July 2000 to February 2004, the boxes were checked on 16 occasions. Bats were counted in boxes or upon emergence when numbers were too numerous to count within the box.

(2) Flaquer C., Torre I. & Ruiz-Jarillo R. (2006) The value of bat-boxes in the conservation of *Pipistrellus pygmaeus* in wetland rice paddies. *Biological Conservation*, 128, 223–230.

b) *Terminology used to describe the evidence*

Unless specifically stated otherwise, results reflect statistical tests performed on the data, i.e. we only state that there was a difference if it was a significant difference or state that there was no difference if it was not significant. Table 1 above defines the terms used to describe the study designs.

c) *Dealing with multiple interventions within a publication*

When separate results are provided for the effects of each of the different interventions tested, separate summaries have been written under each intervention heading. However, when several interventions were carried out at the same time and only the combined effect reported, the results were described with a similar paragraph under all relevant interventions. The first sentence makes it clear that there was a combination of interventions carried out, i.e. '… (REF) found that [x intervention], along with [y] and [z interventions] resulted in [describe effects]'. Within the results section we also added a sentence such as: 'It is not clear whether these effects were a direct result of [x], [y] or [z] interventions', or 'The study does not distinguish between the effects of [x], and other interventions carried out at the same time: [y] and [z].'

d) Dealing with multiple publications reporting the same results

If two publications described results from the same intervention implemented in the same space and at the same time, we only included the most stringently peer-reviewed publication (i.e. journal of the highest impact factor). If one included initial results (e.g. after year one) of another (e.g. after 1–3 years), we only included the publication covering the longest time span. If two publications described at least partially different results, we included both but made it clear they were from the same project in the paragraph, e.g. 'A controlled study... (Gallagher et al. 1999; same experimental set-up as Oasis et al. 2001)...'.

e) Taxonomy

Taxonomy was not updated but followed that used in the original publication. Where possible, common names and Latin names were both given the first time each species was mentioned within each summary.

f) Key messages

Each intervention for which evidence is found has a set of concise, bulleted key messages at the top, which was written once all the literature had been summarised. These include information such as the number, design and location of studies included.

The first bullet point describes the total number of studies that tested the intervention and the locations of the studies, followed by key information on the relevant metrics presented under the headings and sub-headings shown below (with number of relevant studies in parentheses for each).

- **X studies** examined the effects of [INTERVENTION] on [TARGET POPULATION]. Y studies were in [LOCATION 1][1,2] and Z studies were in [LOCATION 2][3,4].

    - *Locations will usually be countries, ordered based on chronological order of studies rather than alphabetically, i.e. 'USA[1], Australia[2]' rather than 'Australia[2], USA[1]'. However, when more than 4–5 separate countries, they may be grouped into regions to make it clearer e.g. Europe, North America. The distribution of studies amongst habitat types may also be added here if relevant.*

COMMUNITY RESPONSE (x STUDIES)

- Community composition (x studies):
- Richness/diversity (x studies):

POPULATION RESPONSE (x STUDIES)

- Abundance (x studies):
- Reproductive success (x studies):
- Survival (x studies):
- Condition (x studies):

BEHAVIOUR (x STUDIES)

- Uptake (x studies):
- Use (x studies):
- Behaviour change (x studies):

OTHER (x STUDIES) (*Included only for interventions/chapters where relevant*)

- **[Sub-heading(s) for the metric(s) reported will be created] (x studies):** If no evidence was found for an intervention, the following text was added in place of the key messages above:
- We found no studies that evaluated the effects of [INTERVENTION] on [TARGET POPULATION].

'We found no studies' means that we have not yet found any studies that have directly evaluated this intervention during our systematic journal and report searches. Therefore, we have no evidence to indicate whether or not the intervention has any desirable or harmful effects.

g) Background information

Background information for an intervention is provided to describe the intervention and where we feel recent knowledge is required to interpret the evidence. This is presented before the key messages and relevant references included in the reference list at the end of the intervention section. In some cases, where a body of literature has strong implications for terrestrial mammal conservation, but does not directly test interventions for their effects, we may also refer the reader to this literature in the background sections.

Dissemination/communication of evidence synthesis

The information from this synopsis update will be available in three ways:

- This synopsis pdf, downloadable from www. conservationevidence.com, which contains the study summaries, key messages and background information on each intervention.

- The searchable database at www.conservationevidence.com, which contains all the summarized information from the synopsis, along with updated expert assessment scores.

- A chapter in *What Works in Conservation*, available as a pdf to download and a book from www.conservationevidence.com/ content/page/79, which contains the key messages from the synopsis as well as updated expert assessment scores on the effectiveness and certainty of the synopsis, with links to the online database.

How you can help to change conservation practice

If you know of evidence relating to terrestrial mammal conservation that is not included in this synopsis, we invite you to contact us via our website www.conservationevidence.com. You can submit a published study by clicking 'Submit additional evidence' on the right-hand side of an intervention page. If you have new, unpublished evidence, you can submit a paper to the *Conservation Evidence* journal. We particularly welcome papers submitted by conservation practitioners.

References

Amano T., González-Varo J.P. & Sutherland W.J. (2016) Languages are still a major barrier to global science. *PLoS Biology*, 14, e2000933, https://doi. org/10.1371/journal.pbio.2000933

Berthinussen A., Richardson O.C. & Altringham J.D. (2019) *Bat Conservation: Global Evidence for the Effects of Interventions*. Synopses of Conservation Evidence Series. University of Cambridge, Cambridge, UK.

Burgin C.J., Colella J.P., Kahn P.L. & Upham N.S. (2018) How many species of mammals are there? *Journal of Mammalogy*, 99, 1–11, https://doi.org/10.1093/jmammal/gyx147

Ceballos G., Ehrlich P.R. & Dirzo, R. (2017) Biological annihilation via the ongoing sixth mass extinction signaled by vertebrate population losses and declines. *Proceedings of the National Academy of Sciences*, 114, E6089–E6096, https://doi.org/10.1073/pnas.1704949114

Doherty T.S., Dickman C.R., Johnson C.N., Legge S.M., Ritchie E.G & Woinarski J.C.Z. (2017) Impacts and management of feral cats *Felis catus* in Australia. *Mammal Review*, 47, 83–97, https://doi.org/10.1111/mam.12080

Dwan, K., Gamble, C., Williamson, P.R., & Kirkham, J.J., (2013) Systematic review of the empirical evidence of study publication bias and outcome reporting bias—an updated review. *PloS ONE* 8(7), e66844, https://doi.org/10.1371/journal.pone.0066844

Hoffmann M., Belant J.L., Chanson J.S., Cox N.A., Lamoreux J., Rodrigues A.S., Schipper J. & Stuart S.N. (2011) The changing fates of the world's mammals. *Philosophical Transactions of the Royal Society B: Biological Sciences*, 366, 2598–2610, https://doi.org/10.1098/rstb.2011.0116

Hoffmann M., Duckworth J.W., Holmes K., Mallon D.P., Rodrigues A.S. & Stuart S.N. (2015) The difference conservation makes to extinction risk of the world's ungulates. *Conservation Biology*, 29, 1303–1313, https://doi.org/10.1111/cobi.12519

Junker J., Kühl H.S., Orth L., Smith R.K., Petrovan S.O. & Sutherland W.J. (2017) *Primate conservation: Global evidence for the effects of interventions*. University of Cambridge, Cambridge, UK.

Ripple W.J., Newsome T.M., Wolf C., Dirzo R., Everatt K.T., Galetti M., Hayward M.W., Kerley G.I., Levi T., Lindsey P.A. & Macdonald D.W. (2015) Collapse of the world's largest herbivores. *Science Advances*, 1, p.e1400103, https://doi.org/10.1126/sciadv.1400103

Schipper et al. (2008) The status of the world's land and marine mammals: diversity, threat, and knowledge. *Science*, 322, 225–230, https://doi.org/10.1126/science.1165115

Sutherland W.J., Taylor N.G., MacFarlane D., Amano T., Christie A.P., Dicks L.V., Lemasson A.J., Littlewood N.A., Martin P.A., Ockendon N., Petrovan S.O., Robertson R.J., Rocha R., Shackelford G.E., Smith R.K., Tyler E.H.M. & Wordley C.F.R. (2019) Building a tool to overcome barriers in the research-implementation space: the Conservation Evidence database. *Biological Conservation*, 238, 108199, https://doi.org/10.1016/j.biocon.2019.108199

2. Threat: Residential and commercial development

Background

Threats from residential and commercial development include the destruction of habitat, pollution and impacts from transportation and service corridors. Interventions in response to these threats are described in the following chapters: *Habitat protection*, *Habitat restoration and creation*, *Threat: Pollution* and *Threat: Transportation and service corridors*. The interventions that are more specific to development, including development of recreational facilities, are discussed in this section.

This section also includes interventions aimed at reducing human-wildlife conflict where continuation of this conflict can prompt calls for management actions including lethal control of the species involved.

Residential development can result in an increase in populations of domestic cats *Felis catus* and dogs *Canis lupus familiaris*, which can prey on wild mammals. Interventions that involve reducing predation by cats and dogs in residential areas are included here but see also interventions within *Invasive alien and other problematic species*.

    https://doi.org/10.11647/OBP.0234.02

2.1. Protect mammals close to development areas (e.g. by fencing)

https://www.conservationevidence.com/actions/2324

- We found no studies that evaluated the effects of protecting mammals close to development areas (e.g. by fencing).

'We found no studies' means that we have not yet found any studies that have directly evaluated this intervention during our systematic journal and report searches. Therefore, we have no evidence to indicate whether or not the intervention has any desirable or harmful effects.

Background

Mammals living at the edge of developed areas may face particular threats from predation by domestic animals, persecution, road traffic and disturbance. Fencing could be erected in some situations, to reduce exposure of wild mammals to such threats.

2.2. Keep cats indoors or in outside runs to reduce predation of wild mammals

https://www.conservationevidence.com/actions/2326

- **One study** evaluated the effects on potential prey mammals of keeping cats indoors or in outside runs. This study was in the UK[1].

COMMUNITY RESPONSE (0 STUDIES)

POPULATION RESPONSE (1 STUDY)

- **Survival (1 study):** One replicated study in the UK[1] found that keeping domestic cats indoors at night reduced the number of dead or injured mammals that were brought home.

BEHAVIOUR (0 STUDIES)

Background

Domestic cats *Felis catus* can be major predators on wild mammals. For example, one study estimated that domestic cats in the UK brought home 52–63 million mammals over a five-month period (Wood *et al.* 2003). Keeping them indoors, or in enclosed outdoor runs, may substantially reduce their impact on wild mammals.

See also: *Use collar-mounted devices to reduce predation by domestic animals.*

Woods M., Mcdonald R. & Harris S. (2003) Predation of wildlife by domestic cats *Felis catus* in Great Britain. *Mammal Review*, 33, 174–188, https://doi. org/10.1046/j.1365-2907.2003.00017.x

A replicated study in 1997 in urban and rural areas in the UK (1) found that domestic cats *Felis catus* that were kept indoors at night brought home fewer dead or injured mammals than cats that were allowed outside. The average number of mammals brought home by cats that were kept indoors at night (6.0) was less than the number delivered by those that were allowed outside (8.9). Between April and August, cat owners recorded the number of prey brought home by 90 cats which were kept inside at night and 192 cats which were allowed outside. Only cats living in households with no other cats were included in the study.

(1) Woods M., McDonald R.A. & Harris S. (2003) Predation of wildlife by domestic cats *Felis catus* in Great Britain. *Mammal Review*, 33, 174–188, https://doi.org/10.1046/j.1365-2907.2003.00017.x

2.3. Use collar-mounted devices to reduce predation by domestic animals

https://www.conservationevidence.com/actions/2332

- **Five studies** evaluated the effects on mammals of using collar-mounted devices to reduce predation by domestic animals. Three studies were in the UK[1,2,3], one was in Australia[4] and one was in the USA[5].

COMMUNITY RESPONSE (0 STUDIES)

POPULATION RESPONSE (5 STUDIES)

- **Survival (5 studies):** Five replicated studies (including four randomized, controlled studies), in the UK[1,2,3], Australia[4] and the USA[5], found that bells[1,2,3], a sonic device[3], and a neoprene flap (which inhibits pouncing)[4] mounted on collars, and a brightly coloured and patterned collar[5] all reduced the rate at which cats predated and returned home with mammals. In one of these studies, an effect was only found in autumn, and not in spring[5].

BEHAVIOUR (0 STUDIES)

Background

Domestic animals can predate a range of wild mammals, with cats *Felis catus* a potentially significant predator. For example, one study estimated that domestic cats in the UK brought home 52–63 million mammals over a five-month period (Woods *et al.* 2003). Various measures have been suggested, or are enacted, to try to reduce this predation, including a range of deterrents or warnings attached to collars that are worn by cats.

Woods M., Mcdonald R. & Harris S. (2003) Predation of wildlife by domestic cats *Felis catus* in Great Britain. *Mammal Review*, 33, 174–188, https://doi.org/10.1046/j.1365-2907.2003.00017.x

A replicated, randomized, controlled study in 1999 in urban and rural areas of Lancashire, UK (1) found that domestic cats *Felis catus* wearing a bell brought home fewer dead/injured mammals than did cats without a bell. Over an eight-week period, the total number of mammals brought home by cats when wearing bells (82) was less than half than that delivered during periods without a bell (167). The rate of delivery of items did not change over time, suggesting cats did not adapt to hunting with bells. Between July and October, a total of 41 cats were randomly allocated to either: four weeks without a bell followed by four weeks with a bell, four weeks with a bell followed by four weeks without, or alternate weeks with and without a bell, beginning with one week with a bell. Bells were fitted to a collar. Only cats that previously brought prey home and wore a collar were investigated. The number of prey delivered was recorded by cat owners.

A replicated study in 1997 in urban and rural areas in the UK (2) found that domestic cats *Felis catus* wearing a bell brought home fewer dead/injured mammals than cats without a bell. The average number of mammals brought home by cats with bells fitted to a collar (5.6) was smaller than the number delivered by cats not wearing a bell (9.9). Between April and August, cat owners recorded the number of prey brought home by 92 cats which wore bells and 190 cats which did not wear bells. Only cats living in households with no other cats were included in the study.

A replicated, randomized, controlled study in 2002–2003 in the UK (3) found that fewer mammals were brought home by domestic cats *Felis catus* fitted with a bell or a sonic device on their collar than by cats wearing a plain collar, but the type of device did not matter. In 2002, fewer mammals were returned by cats equipped with a bell (120) or a CatAlert™ sonic device (111) than by cats wearing a plain collar (181). In 2003, the average number of mammals returned was similar for cats equipped with one bell (0.07 mammals/cat/day), two bells (0.07 mammals/cat/day) or a CatAlert™ sonic device (0.05 mammals/cat/day). Between April and August 2002, 68 cats were fitted with each of the three types of collar (a bell, a sonic device or a plain collar) for one month at a time, in a random order. Owners recorded live prey items and collected dead items for identification. Between May and September 2003, 67 cats were fitted with a collar with either one bell, two bells or a sonic device. Owners recorded all prey items, and identified them to species wherever possible. Sonic devices were set to 'permanently on'.

A replicated, randomized, controlled study in 2005 in a residential area in Perth, Australia (4) found that domestic cats *Felis catus* wearing a collar with a CatBib™ 'pounce protector' (a neoprene flap that hangs from the collar) brought home fewer mammals than did cats without a CatBib™. When equipped with a CatBib™, cats brought home fewer mammals (total of 59) than when not wearing a collar (total of 105). Adding a bell to the CatBib™ did not further reduce the number of mammals returned (with bell: 26, without bell: 33). Wearing a CatBib™ stopped 45% of cats from catching mammals altogether. In

November–December 2005, in a random order, 56 cats underwent a period of three weeks wearing a CatBib™ and three weeks without a CatBib™. For the three weeks with a CatBib™, cats were randomly assigned either a CatBib™ only or a CatBib™ and bell. Only cats that frequently brought home intact prey were included in the study. Owners collected dead prey items and recorded live prey before release.

A replicated, randomized, controlled study in 2013–2014 in a residential area of New York state, USA (5) found that domestic cats *Felis catus* wearing collars with bright colours and patterns brought home fewer mammals than did cats with no collars in autumn, but not in spring. From September–November 2013, 54 cats brought home fewer mammals (0.6/cat) in six weeks spent wearing a Birdsbesafe® collar with bright colours and patterns than the same cats did during six weeks without a collar (1.2/cat). However, in a repeat experiment from April–June 2014, there was no difference (with collar: 1.1/cat; without collar: 1.1/cat). Cats were randomly allocated to one of two groups, beginning with or without a Birdsbesafe® collar, and the treatment on each cat was changed every two weeks throughout a 12-week period. Only cats that regularly brought home intact prey were included in the study. Owners collected dead prey items and recorded live prey before release.

(1) Ruxton G.D., Thomas S. & Wright J.W. (2002) Bells reduce predation of wildlife by domestic cats (*Felis catus*). *Journal of Zoology*, 256, 81–83, https://doi.org/10.1017/s0952836902000109

(2) Woods M., McDonald R.A. & Harris S. (2003) Predation of wildlife by domestic cats *Felis catus* in Great Britain. *Mammal Review*, 33, 174–188, https://doi.org/10.1046/j.1365-2907.2003.00017.x

(3) Nelson S.H., Evans A.D. & Bradbury R.B. (2005) The efficacy of collar-mounted devices in reducing the rate of predation of wildlife by domestic cats. *Applied Animal Behaviour Science*, 94, 273–285, https://doi.org/10.1016/j.applanim.2005.04.003

(4) Calver M., Thomas S., Bradley S. & McCutcheon H. (2007) Reducing the rate of predation on wildlife by pet cats: The efficacy and practicability of collar-mounted pounce protectors. *Biological Conservation*, 137, 341–348, https://doi.org/10.1016/j.biocon.2007.02.015

(5) Willson S.K., Okunlola I.A. & Novak J.A. (2015) Birds be safe: can a novel cat collar reduce avian mortality by domestic cats (*Felis catus*)? *Global Ecology and Conservation*, 3, 359–366, https://doi.org/10.1016/j.gecco.2015.01.004

2.4. Keep dogs indoors or in outside enclosures to reduce threats to wild mammals

https://www.conservationevidence.com/actions/2334

- We found no studies that evaluated the effects on mammals of keeping dogs indoors or in outside enclosures to reduce threats to wild mammals.

'We found no studies' means that we have not yet found any studies that have directly evaluated this intervention during our systematic journal and report searches. Therefore, we have no evidence to indicate whether or not the intervention has any desirable or harmful effects.

Background

Domestic dogs *Canis lupus familiaris* may have multiple negative impacts on wild mammals including through predation, disease transmission and disturbance (Hughes & Macdonald 2013). In some places, domestic dogs roam freely and are major predators of wild mammals. For example, Wierzbowska *et al.* (2016) estimated that over 33,000 wild animals (primarily mammals, especially brown hare *Lepus europaeus* and roe deer *Capreolus capreolus*) were killed by free-ranging dogs annually in Poland. Keeping dogs indoors or in outside enclosures may reduce their impacts, including predation, on wild mammals.

Hughes J. & Macdonald D.W. (2013) A review of the interactions between free-roaming domestic dogs and wildlife. *Biological Conservation*, 157, 341–351, https://doi.org/10.1016/j.biocon.2012.07.005

Wierzbowska I.A., Hędrzak M., Popczyk P., Okarma H. & Crooks K.R. (2016) Predation of wildlife by free-ranging domestic dogs in Polish hunting grounds and potential competition with the grey wolf. *Biological Conservation*, 201, 1–9, https://doi.org/10.1016/j.biocon.2016.06.016

2.5. Keep domestic cats and dogs well-fed to reduce predation of wild mammals

https://www.conservationevidence.com/actions/2335

- We found no studies that evaluated the effects on mammals of keeping domestic cats and dogs well-fed to reduce predation of wild mammals.

'We found no studies' means that we have not yet found any studies that have directly evaluated this intervention during our systematic journal and report searches. Therefore, we have no evidence to indicate whether or not the intervention has any desirable or harmful effects.

Background

Domestic pets can be major predators on wild mammals. For example, an estimated 57 million mammals are killed by domestic cats *Felis catus* in the UK each year (Wood *et al.* 2003) while negative impacts of domestic dogs *Canis lupus familiaris* on wild mammals include predation, disease transmission and disturbance (Hughes & Macdonald 2013). Keeping animals well fed might reduce their hunting activities and other interactions with wild mammals.

Woods M., Mcdonald R. & Harris S. (2003) Predation of wildlife by domestic cats *Felis catus* in Great Britain. *Mammal Review*, 33, 174–188, https://doi.org/10.1046/j.1365-2907.2003.00017.x

Hughes J. & Macdonald D.W. (2013) A review of the interactions between free-roaming domestic dogs and wildlife. *Biological Conservation*, 157, 341–351, https://doi.org/10.1016/j.biocon.2012.07.005

2.6. Translocate problem mammals away from residential areas (e.g. habituated bears) to reduce human-wildlife conflict

https://www.conservationevidence.com/actions/2336

- **Eleven studies** evaluated the effects of translocating problem mammals (such as bears) away from residential areas to reduce human-wildlife conflict. Six studies were in the USA[1–5,11], two were in Canada[7,8], one was Russia[6], one was in India[9] and one was in Romania[10].

COMMUNITY RESPONSE (0 STUDIES)

POPULATION RESPONSE (6 STUDIES)

- **Survival (6 studies):** A controlled study in the USA[3] found that grizzly bears translocated away from conflict situations had lower survival rates than did non-translocated bears. A replicated study in the USA[11] found that fewer than half of black bears translocated from conflict situations survived after one year. Two of three studies (two controlled), in the USA[2,4,5], found that after translocation away from urban sites, white-tailed deer survival was lower than that of non-translocated deer. The third study found that short-term survival was lower but long-term survival was higher than that of non-translocated deer. A study in Russia[6] found that most Amur tigers translocated after attacking dogs or people did not survive for a year after release.

BEHAVIOUR (0 STUDIES)

OTHER (6 STUDIES)

- **Human-wildlife conflict (6 studies):** Five studies (including one controlled and two replicated studies), in the USA[1,3,11] and Canada[7,8], of brown/grizzly[1,3] or black[7,8,11] bears translocated away from residential areas or human-related facilities, found that at least some returned to their original capture location[1,7,8,11] and/or continued to cause nuisance[3,8]. In two of the studies[1,8], most returned to their capture area and one black bear returned six times following translocation[7]. A before-and-after study in India[9] found that leopards translocated away from human-dominated areas, attacked more humans and livestock than before-translocation. A controlled study in Romania[10] found that translocated brown bears occurred less frequently inside high potential conflict areas than outside, the opposite to bears that had not been translocated.

Background

There is a variety of ways in which mammals in urban, residential or other human-occupied locations can come into conflict with people. Some species may raid garbage and create a mess while doing so, some may cause damage to gardens or parks, some may act aggressively towards humans and some mammals present substantial road traffic hazards. In many communities, there is a pressure to address these issues by focussing solutions on preventing or deterring mammals from accessing such areas. One such method is translocation, typically to an area away from habitation. This intervention can fail if translocated animals continue to cause problems at residential areas (including by returning to their capture site) or if survival of translocated animals is low. If the intervention is successful, it can reduce incentives for carrying out lethal control of such animals.

See also: *Species management — Translocate mammals.*

A study in 1979–1981 of a large boreal and subarctic forest area in Alaska, USA (1) found that translocated Alaskan brown bears *Ursus arctos* did not settle at their release site and most returned to their capture area. Twelve of 20 translocated adult bears returned to their capture area in 13–133 days. Returning bears had been released, on average, closer to their capture site (145–255 km) than had non-returning bears (168–286 km). No translocated female bears were known to have produced young in the following year. Forty-seven bears were caught between 22 May and 22 June 1979, marked and transported by vehicle or aircraft. Adults were radio-collared and relocation data were adequate for monitoring movements and survival of 20 of these. Bears were monitored by radio-tracking from an airplane in May–October 1979 and from other radio-tracking data and hunter kills in 1979–1981.

A controlled study in 1984–1988 at four woodland and grassland sites in Illinois, USA (2) found that following translocation away from urban sites to reduce human-wildlife conflict, white-tailed deer *Odocoileus virginianus*, had a lower survival rate that did deer that were not translocated. Annual survival of translocated adult female deer

(34%) was lower than that of resident adult female deer at one of the original capture sites (73%). Fifty deer (25 females, 25 males) were caught, mostly with rocket nets, between 18 December and 31 March in 1984–1988, at three largely urban sites. They were released at a rural site, ≤80 km from capture sites. Females were radio-collared and monitored every one to two weeks initially, then less frequently. Survival was compared with that of 12 additional females that were caught, radio-collared, and released at the capture site.

A controlled study in 1975–1993 in a forested national park in Wyoming, USA (3) found that grizzly bears *Ursus arctos* translocated away from bear-human conflict situations had lower survival rates than did non-translocated bears and over one third required multiple translocations. Translocated bears had a lower annual survival rate (83%) than that of non-translocated bears (89%). Of 81 translocated bears, 50 were moved once, 15 were moved twice, nine were moved three times, four were moved four times and three were moved five times. In a 20,000-km² study area, 81 bears were translocated 3–128 km away from human conflict situations, such as having entered residential areas. With recaptures, there were 138 bear translocations in total between 1975 and 1993. Survival was compared with that of 160 bears captured and released without translocation during the same period. Bears were monitored by radio-tracking from an aircraft.

A controlled study in 1995–1996 in a residential and forest area in South Carolina, USA (4) found that white-tailed deer *Odocoileus virginianus* translocated from a residential area to a nearby forest had lower short-term survival but higher long-term survival than did non-translocated deer. After three months, a lower proportion of translocated deer (52%) was alive, than of non-translocated deer (76%). After 12 months, a higher proportion of translocated deer was alive (39%) than of non-translocated deer (33%). Fifty percent of translocated deer dispersed from the release site whereas no non-translocated deer dispersed. Nineteen deer were caught with rocket nets in a residential area, in December 1995. Ten were moved 3 km and released in a forest preserve. Nine were released at the capture site. Deer were radio-collared and were monitored for up to 12 months.

A study in 1997–2000 of a residential area and a forest in Missouri, USA (5) found that after translocation away from a residential area, white-tailed deer *Odocoileus virginianus* had a lower survival rate than did deer that were not translocated. Annual survival after one year for translocated deer (30%) was lower than for non-translocated deer (69%). Among translocated deer, the largest causes of death were hunting (33%) and muscle weakness following capture ('capture myopathy'; 29%). Among non-translocated deer, roadkill (68%) and hunting (12%) were the largest causes of death. Eighty deer (51 male, 29 female) were caught in a residential area in January–February 1999, radio-collared, and released in a conservation area 160 km away. At the same capture site, additional deer (quantity not stated) were caught, radio-collared, and released at point of capture from December 1997 to March 1998.

A study in 2001–2004 in a mountainous protected area in eastern Russia (6) found that following translocation of Amur tigers *Panthera tigris altaica* that had attacked dogs *Canis lupus familiaris* or people around villages, most did not survive for a year after release. One of the four translocated tigers survived for at least 10 months. The other three were killed by people, between 20 days and one year after release. Two of the animals killed were suspected to have been poached, while one was killed after killing domestic dogs. In 2001–2003, four tigers that had been involved in attacks on domestic dogs (three tigers) or a human (one tiger) were translocated 150–350 km to a protected area. Before release, two tigers that were emaciated when caught were held in a 1-ha enclosure for 162–388 days. All tigers were fitted with radio-collars and released into areas known to be used by wild tigers. Animals were radio-tracked approximately weekly, over an unspecified period, by researchers on foot, in vehicles, or in a plane.

A study in 1994–1997 of extensive forest and a residential area in Ontario, Canada (7) found that repeated translocation of an adult female black bear *Ursus americanus* that habitually fed from garbage containers did not prevent it from returning and resuming nuisance behaviour at the capture site. The bear was translocated six times, over distances of 40–389 km (average 152 km), and returned each time to the initial capture area. On two of the returns to the capture area, the bear was accompanied by cubs. The maximum distance between any two capture sites was 10 km. The bear habitually foraged at unsecured garbage containers in residential areas. It was caught and translocated six times between June 1994 (when estimated to be nine years old) and

1997. It was ear-tagged at first capture and radio-collared at the time of the second capture and translocation.

A replicated study in 1982–1997 in three mainly forested areas in Ontario, Canada (8) found that translocating black bears *Ursus americanus* that caused nuisance around habitation or other human-related installations reduced their nuisance behaviour, though some animals continued to cause problems. Among translocated bears, ≥30% were involved in at least one further nuisance event. This occurred mostly in adult females (48%), followed by adult males (39%), juvenile females (26%) and juvenile males (18%). Seventy-three percent of translocated adult bears returned to their area of capture, compared to 29% of juveniles. Bears released further from their capture point were less likely to return (data presented as statistical model coefficients). In each of three regions, bear relocation and tag recovery data were obtained. In total, 123 bears were relocated after displaying nuisance behaviour, and were moved on average 70–80 km. Study periods in the three areas spanned three, four and 14 years.

A before-and-after study in 1993–2003 in a largely arable area in Maharashtra, India (9) found that after leopards *Panthera pardus fusca* were translocated away from human-dominated areas, the frequency and fatality of leopard attacks on humans increased and attacks on livestock increased. There were more leopard attacks on humans after translocations began (8–24/year) than before (1–7/year) and these resulted in more human fatalities (after: 3–11/year; before: 0–2/year). There were more leopard attacks on livestock after translocations began (average 166 attacks/year) than in the 12 months before translocations began (106 attacks). Authors reported that the attacks were by the translocated leopards. In a 4,275-km² study area, with a human population density of 185 people/km², 103 leopard translocations occurred between February 2001 and December 2003. Eighty-six leopards were caught in human-dominated areas, with 29 translocated <60 km to either of two natural forest sites and 56 moved >200 km to release sites elsewhere. Eleven leopards from outside the study area were also released at the natural forest sites. Location data were not available for six translocations. Human attack data during the translocation period were compared with those collated for 1993–2000.

A controlled study in 2008–2011 in a mixed landscape in the Eastern Romanian Carpathians, Romania (10) found that brown bears *Ursus arctos* translocated to reduce conflict with humans, some of which had been rehabilitated as orphans, occurred less frequently inside high potential conflict areas than outside. Bears were present less frequently inside high potential conflict areas than outside if they had been translocated (occurrences inside: 501; outside: 1,517) or rehabilitated (inside: 462; outside: 1,180) and particularly if they had been rehabilitated and translocated (inside: 245; outside: 963). Bears that had not been translocated or rehabilitated occurred inside the high potential conflict areas more than outside (inside: 2,166; outside: 1,067). Rehabilitated and translocated bears spent less time (9 hrs) in the conflict areas than those that had not been rehabilitated and translocated (14 hrs). Similar time was spent in those areas by bears that had just been translocated (4 hrs) or rehabilitated (6 hrs). Eight bears were radio-tracked for 3–17 months (541–1,869 locations/bear) in 2008–2011 across the 15,822 km^2 study site. There were two bears of each of four types: translocated but not rehabilitated, translocated and rehabilitated, not translocated but rehabilitated and not translocated or rehabilitated. The four bears (two male) were translocated >60–100 km from their capture site due to conflict with humans (damage and/or frequently visited settlements, e.g. waste disposal sites). Four bears (two male) were orphan bear cubs that were released after rehabilitation in relatively natural conditions for a maximum of two years. High potential conflict areas were those with human settlements, partially agricultural fields and woodlands.

A replicated study in 1995–1997 in an unspecified number of mountain sites in Colorado, USA (11) found that after translocation of black bears *Ursus americanus* that were involved in conflict with humans, fewer than half survived after one year and some returned to capture sites. One year after translocation, 50% of adult black bears and 28% of sub-adult bears had survived. Of 66 captured bears, 14 returned to capture sites and 16 repeated some form of problem behaviour. In May and October of 1995–1997, sixty-six bears that were considered a nuisance or threat to human safety were captured. All were individually marked with ear tags and lip tattoos and were fitted with radio-collars. Within two days of capture, bears were translocated to release sites. Bears were radio-tracked opportunistically, from the ground and from a plane, once a week, in May–October of 1995–1997.

(1) Miller S. & Ballard W. (1982) Homing of transplanted Alaskan brown bears. *The Journal of Wildlife Management*, 46, 869–876, http://doi.org/10.2307/3808219

(2) Jones J.M & Witham J.H. (1990) Post-translocation survival and movements of metropolitan white-tailed deer. *Wildlife Society Bulletin*, 18, 434–441.

(3) Blanchard B.M. & Knight R.R. (1995) Biological consequences of relocating grizzly bears in the Yellowstone Ecosystem. *The Journal of Wildlife Management*, 59, 560–565, http://doi.org/10.2307/3802463

(4) Cromwell J.A., Warren R.J. & Henderson D.W. (1999) Live-capture and small-scale relocation of urban deer on Hilton Head Island, South Carolina. *Wildlife Society Bulletin*, 27, 1025–1031.

(5) Beringer J., Hansen L.P., Demand J.A., Sartwell J., Wallendorf M. & Mange R. (2002) Efficacy of translocation to control urban deer in Missouri: costs, efficiency, and outcome. *Wildlife Society Bulletin*, 30, 767–774.

(6) Goodrich J.M. & Miquelle D.G. (2005) Translocation of problem Amur tigers *Panthera tigris altaica* to alleviate tiger-human conflicts. *Oryx*, 39, 454–457, https://doi.org/10.1017/s0030605305001146

(7) Landriault L., Hall M., Hamr J. & Mallory, F. (2006) Long-range homing by an adult female black bear, *Ursus americanus*. *The Canadian Field-Naturalist*, 120, 57–60, https://doi.org/10.22621/cfn.v120i1.246

(8) Landriault L.J., Brown G.S., Hamr J. & Mallory F.F. (2009) Age, sex and relocation distance as predictors of return for relocated nuisance black bears *Ursus americanus* in Ontario, Canada. *Wildlife Biology*, 15, 155–164, https://doi.org/10.2981/07-084

(9) Athreya V., Odden M., Linnel J. & Karanth U. (2011) Translocation as a tool for mitigating conflict with leopards in human dominated landscapes of India. *Conservation Biology*, 25, 133–141, https://doi.org/10.1111/j.1523-1739.2010.01599.x

(10) Pop, I.M., Sallay, A., Bereczky, L. & Chiriac, S. (2012) Land use and behavioral patterns of brown bears in the South-Eastern Romanian Carpathian Mountains: A case study of relocated and rehabilitated individuals. *Procedia Environmental Sciences*, 14, 111–122, https://doi.org/10.1016/j.proenv.2012.03.011

(11) Alldredge M.W., Walsh D.P., Sweanor L.L., Davies R.B. & Trujillo A. (2015) Evaluation of translocation of black bears involved in human–bear conflicts in South-central Colorado. *Wildlife Society Bulletin*, 39, 334–340, https://doi.org/10.1002/wsb.526

2.7. Issue enforcement notices to deter use of non-bear-proof garbage dumpsters to reduce human-wildlife conflict

https://www.conservationevidence.com/actions/2345

- **One study** evaluated the effects of issuing enforcement notices to deter use of non-bear-proof garbage dumpsters to reduce human-wildlife conflict. This study was in the USA[1].

COMMUNITY RESPONSE (0 STUDIES)

POPULATION RESPONSE (0 STUDIES)

BEHAVIOUR (0 STUDIES)

OTHER (1 STUDY)

- **Human-wildlife conflict (1 study):** A replicated, controlled, before-and-after study in the USA[1] found that issuing enforcement notices requiring appropriate dumpster use did not reduce garbage accessibility to black bears.

Background

Bears can be opportunistic feeders that sometimes raid sources of food left by humans. If food in garbage containers is not secured, this too can be targeted. As well as potentially causing mess, bears attracted to garbage containers may come to associate humans with sources of food and their behaviour may become problematic, through displays of aggression or boldness. Such animals may be translocated or lethally controlled. The issue could be reduced if food in garbage containers is made inaccessible to bears. Issuing enforcement notices is one way of attempting to increase compliance with legislation requiring proper use of bear-proof dumpsters.

See also: *Translocate problem mammals away from residential areas (e.g. habituated bears) to reduce human-wildlife conflict.*

A replicated, controlled, before-and-after study in 2008 of four alleyways in business and residential areas in Colorado, USA (1) found that issuing enforcement notices requiring appropriate dumpster use did not reduce garbage accessibility to black bears *Ursus americanus*. Changes in the proportion of dumpsters violating legislation in alleyways where enhanced enforcement occurred (after enforcement: 20% of dumpsters; before: 42%) did not significantly differ from those in alleyways without enhanced enforcements (after: 24% of dumpsters; before: 49%). Similarly, there was no significant difference in changes in legislation compliance between individual dumpsters issued with enforcement notices (after issuing: 36% of dumpsters; before: 72%) and those not (after: 17% of dumpsters; before: 36%). In treatment alleys (with 37 dumpsters) there were daily patrols. Twenty-two written notices were issued on 18 dumpsters and two verbal warnings were given. Two additional alleys (30 dumpsters) had continuing lower level of enforcement action. Pre- and post-treatment surveys took place between 1 July and 25 August 2008. Dumpsters were regarded as violating legislation if they were not bear-resistant or if food waste was otherwise accessible.

(1) Baruch-Mordo S., Breck S.W., Wilson K.R. & Broderick J. (2011) The carrot or the stick? Evaluation of education and enforcement as management tools for human-wildlife conflicts. *PLoS ONE*, 6, e15681, https://doi.org/10.1371/journal.pone.0015681

2.8. Prevent mammals accessing potential wildlife food sources or denning sites to reduce nuisance behaviour and human-wildlife conflict

https://www.conservationevidence.com/actions/2346

- **Two studies** evaluated the effects of preventing mammals accessing potential wildlife food sources or denning sites to reduce nuisance behaviour and human-wildlife conflict. One study was in the USA[1] and one was in Switzerland[2].

COMMUNITY RESPONSE (0 STUDIES)

POPULATION RESPONSE (0 STUDIES)

BEHAVIOUR (0 STUDIES)

OTHER (2 STUDIES)

- **Human-wildlife conflict (2 studies):** A replicated, controlled study in the USA[1] found that electric shock devices prevented American black bears from accessing or damaging bird feeders. A before-and-after study in Switzerland[2] found that electric fencing excluded stone martens from a building.

Background

Some mammals will utilize food, denning sites or other resources in human modified environments in such ways that risks them being regarded as exhibiting nuisance behaviour. Such behaviour might include damaging property, creating mess, causing noise disturbance or posing a perceived thrseat to humans. If mammals can be excluded from such situations, such as through electric fencing, this may reduce human-wildlife conflict and might, thus, reduce motivations for carrying out lethal control of such animals.

A replicated, controlled study in 2004 of 10 forest sites in Minnesota, USA (1) found that installing electric shock devices prevented American black bears *Ursus americanus* from accessing or damaging bird feeders. Bird feeders protected by electric shock devices suffered less bear damage (none of 10 was accessed or damaged) than did unprotected feeders (four of 10 accessed or destroyed). Two imitation bird feeders were installed at each of 10 sites, ≥30 km apart. One feeder was protected by an electric shock device, the 'Nuisance Bear Controller'. This device had two 6-volt batteries wired to an automobile vibrator coil/condenser, emitting 10,000–13,000 volts through a disk when contact was made by an animal. The other feeder was unprotected. Ground around each feeder was cleared to enable identification of bear signs. Feeders were in place from 1 July to 15 November 2004. They were monitored, and bait replenished, at least weekly.

A before-and-after study in 2006 on a building in Switzerland (2) found that electric fencing excluded stone martens *Martes foina* from the property. The rate of martens passing through gaps into the building's attic after electric fence installation was lower (0.1 martens/day) than before the fence was installed (1.9 martens/day). It was lower still (0 martens/day) after the fence was modified. The property, built in the 1950s, was used frequently by martens, resulting in serious damage.

Two electric fence types were deployed: wire mesh net for larger gaps and electric wire strands for small openings. Marten movements were monitored by video camera from 12 June to 27 July 2006. This covered nine nights before and seven nights after fence installation and 10 further nights after a crevice was modified by adding an extra electric wire strand. Checks were made for marten re-entry over a further 103 nights, by monitoring for bait removal and for faeces.

(1) Breck S., Lance N. & Callahan, P. (2006) A shocking device for protection of concentrated food sources from black bears. *Wildlife Society Bulletin*, 34, 23–26, https://doi.org/10.2193/0091-7648(2006)34[23:asdfpo]2.0.co;2

(2) Kistler C., Hegglin D., von Wattenwyl K. & Bontadina F. (2013) Is electric fencing an efficient and animal-friendly tool to prevent stone martens from entering buildings? *European Journal of Wildlife Research*, 59, 905–909, https://doi.org/10.1007/s10344-013-0752-5

2.9. Provide diversionary feeding for mammals to reduce nuisance behaviour and human-wildlife conflict

https://www.conservationevidence.com/actions/2323

- **Three studies** evaluated the effects of providing diversionary feeding for mammals to reduce nuisance behaviour and human-wildlife conflict. Two studies were in the USA[1,3] and one was in Slovenia[2].

COMMUNITY RESPONSE (0 STUDIES)

POPULATION RESPONSE (0 STUDIES)

BEHAVIOUR (1 STUDY)

- **Uptake (1 study):** A site comparison study in Slovenia[2] found that 22–63% of the estimated annual energy content of the diet of brown bears comprised provided diversionary food.

OTHER (2 STUDIES)

- **Human-wildlife conflict (2 studies):** Two before-and-after studies (one also a site comparison) in the USA[1,3] found that diversionary feeding reduced nuisance behaviour by black bears.

Background

Some mammals are attracted to residential or business areas by availability of food or other resources. Whilst many such mammals go unnoticed some, such as bears that raid garbage bins, can be perceived as a threat to humans or can cause damage to property or create a mess. Such animals are sometimes managed by being translocated to sites away from built-up areas whilst lethal control may be carried out in some situations. If diversionary feeding can reduce the extent to which animals exhibit nuisance behaviour, this may reduce motivations for carrying out lethal control or other intensive management.

See also: *Agriculture and aquaculture — Provide diversionary feeding to reduce crop damage by mammals to reduce human-wildlife conflict* and *Provide diversionary feeding to reduce predation of livestock by mammals to reduce human-wildlife conflict.*

A before-and-after study in 1981–1991 in an area of forest, residences and recreation facilities in Minnesota, USA (1) found that diversionary feeding reduced nuisance behaviour by black bears *Ursus americanus*. During eight years in which diversionary feeding was used, fewer bears (two bears) were removed for nuisance behaviour than in the three years before diversionary feeding started (six bears). Bears that visited the feeding site did not exhibit nuisance behaviour. A diversionary feeding site was operated during 1984–1991. This site was 0.25–3.4 km from a range of problem areas, including homes, a campground and a picnic site with unsecured bins and other food sources. The feeding location was stocked with beef fat and, sometimes, grapes. Bears were monitored using radio-tracking and direct observation and by ear tag returns from hunters.

A site comparison study in 1993–1998 in three regions comprising mainly forest and agricultural fields in Slovenia (2) found that providing diversionary feeding to reduce human-brown bear *Ursus arctos* conflict resulted in 22–63% of the estimated annual energy content of the diet of bears comprising supplementary food. Across the three regions, supplemental food was highest in the diet and was the most important food items in spring (maize: 27%; carrion: 26%), but not in summer (total 26%) and autumn (27%). The annual proportion of maize in the diet increased with the density of feeding sites (low density: 10–20%; high density: 52%). The proportion of all supplementary food in the diet followed a similar pattern (low density feeding sites: 22–33%; high density: 63%). In the three regions there was at least one carrion feeding site/60 km^2 of bear habitat (annual estimate: 33–146 kg/km^2) and maize feeding sites at average densities of one site/5.6 km^2 of bear habitat (annual estimate: 70–280 kg/km^2). Approximately two-thirds of feeding sites were supplied with food throughout the year. One region had a higher intensity of supplemental feeding (34 feeding sites/km^2) than the other two (16 feeding sites/km^2). A total of 714 brown bear scats were collected opportunistically (153–313/season, 220–260/region) from March to November 1993–1998 across the three regions and analysed.

A before-and-after and site comparison study in 2007 of 20 local communities in Lake Tahoe Basin, USA (3) found that diversionary feeding of black bears *Ursus americanus* during a drought reduced human-bear conflicts, particularly in communities closest to feeding sites. Overall, the total number of human-bear conflicts/month was lower three months after diversionary feeding commenced (834) compared to one month before (1,819), although the difference was not tested for statistical significance (data reported in Stringham & Bryant 2016). Average daily declines in conflicts during the three months of feeding were greater at seven communities located 1 km from feeding sites (1.2%) than at three communities located ≥8 km from feeding sites (0.6%). Diversionary feeding was carried out in September–November 2007 after human-bear conflicts increased during a drought. Fruit and nuts were scattered over a 100 m^2 area at 10 forest sites located 1–20 km from 20 communities. Human-bear conflicts (bears in yards, homes etc.) were reported to a telephone hotline in May–November 2007.

Stringham S. & Bryant, A. (2016) Commentary: Distance-dependent effectiveness of diversionary bear bait sites. *Human–Wildlife Interactions*, 10, 128–131, https://doi.org/10.26077/d5bv-c877

(1) Rogers L.L. (2011) Does diversionary feeding create nuisance bears and jeopardize public safety? *Human–Wildlife Interactions*, 5, 287–295.

(2) Kavčič, I., Adamič, M., Kaczensky, P., Krofel, M., Kobal, M. & Jerina, K. (2015) Fast food bears: brown bear diet in a human-dominated landscape with intensive supplemental feeding. *Wildlife Biology*, 21, 1–8, https://doi.org/10.2981/wlb.00013

(3) Stringham S.F. & Bryant, A. (2015) Distance-dependent effectiveness of diversionary bear bait sites. *Human–Wildlife Interactions*, 9, 229–235, https://doi.org/10.26077/5a9d-rk41

2.10. Scare or otherwise deter mammals from human-occupied areas to reduce human-wildlife conflict

https://www.conservationevidence.com/actions/2347

- **Ten studies** evaluated the effects of scaring or otherwise deterring mammals from residential areas to reduce human-wildlife conflict. Six studies were in the USA[3,4,5,7,8,9], three were in Canada[1,2,6] and one was in Tanzania[10].

COMMUNITY RESPONSE (0 STUDIES)

POPULATION RESPONSE (0 STUDIES)

BEHAVIOUR (0 STUDIES)

OTHER (10 STUDIES)

- **Human-wildlife conflict (10 studies):** Two of four studies (including one randomized and controlled study) in the USA[3,4,5,8], found that a range of noise and pain deterrents did not prevent black bears from returning to urban areas or other human-occupied sites[3,4]. The other two studies[5,8] found that such actions did deter them from seeking food at human-occupied sites. Two of three studies, in the USA[7,9] and Canada[6], found that chasing nuisance black bears with dogs[7] and chasing elk with people or dogs[6] caused them to stay away longer or remain further from human occupied areas. The other study found that attempts to scare coyotes did not cause them to avoid human occupied areas[9]. A before-and-after study in Canada[1] found that an electric fence prevented polar bear entry to a compound. A study in Canada[2] found that chemical and acoustic repellents did not deter polar bears from baits in most cases. A replicated study in Tanzania[10] found that drones caused African savanna elephants to quickly leave residential areas.

Background

There is a variety of ways in which mammals in urban, residential or other human-occupied locations can come into conflict with people. Some species may raid garbage and create a mess while doing so, some may cause damage to gardens or parks, some may act aggressively towards humans and some mammals present substantial road traffic hazards. In many communities, there is a pressure to address these issues by focussing solutions on preventing or deterring mammals from accessing such areas. If non-lethal means can be successfully deployed, this could reduce incentives for achieving this through carrying out lethal control of such species.

A before-and-after study in 1983–1985 at a research compound in Manitoba, Canada (1) found that after the area was enclosed with an electric fence, no polar bears *Ursus maritimus* entered it. Over a total of approximately five months over two summers with the fence installed, no polar bears entered the compound. However, before the fence was installed in those years and in the previous year before it was first installed, nine different bears visited the compound, some on multiple occasions. The study was conducted in a research compound where 10–15 biologists resided between May and September each year. In July–September 1984 and June–September 1985, a temporary two-strand electric fence was erected around the 300-m compound perimeter. The two strands of wire were 30 and 60 cm above the water or ground. The fence emitted 40 pulses/min of direct current (peak output of 8,000 volts). When the fence activated, two 110-decibel horns also sounded.

A study in 1978 at a shrubland and grassland site in Manitoba, Canada (2) found that acoustic deterrents and baits treated with chemical deterrents did not, in most cases, repel polar bears *Ursus maritimus*. Out of 55 visits, acoustic deterrents repelled bears on 17 visits and did not repel them on 38 visits. From 294 visits, chemical deterrent repelled bears five times but did not repel them during 289 visits. However, bears remained for shorter periods at chemical repellent-treated bait stations (average 98–317 s) than at baits without repellents (average 420 s). In October–November 1978, polar bears were attracted to 13 bait stations with sardines. Stations were all 100–500 m from a 6-m-high tower, from which bear responses were observed. At one bait station, a loudspeaker was placed 5m from the bait. Sounds played through the loudspeaker included bear sounds, human shouting, killer whale sounds, radio noise and human hissing and barking like a bear. Ten bait stations were sprayed with dog-repellents or household chemicals. Two bait stations had no repellents.

A study in 1990–1998 of a largely forested national park in North Carolina and Tennessee, USA (3) found that following capture and release back at capture sites, most black bears *Ursus americanus* did not subsequently repeat nuisance behaviour, such as entering picnic sites or campgrounds. For 50 out of 85 captures, bears were not subsequently sighted at capture locations during the remainder of that year. In four further cases, no management action was required that year, even if the bear was re-sighted at its capture location. In a 2,080-km² national park, 63 bears exhibiting nuisance behaviour (such as raiding bins) were captured by live-trapping or darting. Bears were immobilised, individually marked and had a tooth extracted (for aging) before release, after recovery from anaesthesia, <150 m from their capture site.

A randomized, controlled study in 1997–2002 in residential areas and adjacent forest across at least four mountain ranges in Nevada, USA (4) found that subjecting nuisance black bears *Ursus americanus* to deterrents intended to scare them, did not prevent their return to urban areas. The average time for bears to return to urban areas after treatments did not differ significantly between those chased by dogs *Canis lupus familiaris* in addition to noise and projectile deterrents (154 days), those subject to the same deterrents excluding chasing by dogs (88 days) or those not subject to deterrents (65 days). Fifty-seven of the 62 bears in the study returned to urban areas. Forty-four of these returned within 40 days. Nuisance bears (which raided garbage) were captured and radio-collared between July 1997 and April 2002. They were randomly assigned to deterrent treatments including chasing by dogs (20 bears), deterrent treatments excluding chasing by dogs (21 bears) or no deterrent (20 bears). Additional to chasing by dogs, deterrents entailed pepper spraying, firing 12-gauge rubber buckshot or rubber slugs, loud cracker shells and shouting. Deterrents were administered at release sites, 1–75 km from capture locations.

A replicated, controlled study in 2004 of ten forest sites in Minnesota, USA (5) found that installing electric shock devices prevented American black bears *Ursus americanus* from accessing or damaging bird feeders. Bird feeders protected by electric shock devices suffered less bear damage (none of ten accessed or damaged) than did unprotected feeders (four of ten accessed or destroyed). Two imitation bird feeders were installed at each of ten sites, ≥30 km apart. One feeder was protected by an electric shock device, the Nuisance Bear Controller. This device had two 6-volt batteries wired to an automobile vibrator coil/condenser, emitting 10,000–13,000 volts through a disk when contact is made by an animal. The other feeder was unprotected. Ground around each feeder was cleared to enable identification of bear signs. Feeders were in place from 1 July to 15 November 2004. They were monitored, and bait replenished, at least weekly.

A controlled study in 2001–2002 at a town and surrounding forest in Alberta, Canada (6) found that after being chased by humans, the average distance of elk *Cervus canadensis* from the town increased more than it did for elk chased by dogs *Canis lupus familiaris* or for elk that were not chased. The average distance of elk from the town boundary increased for all treatment groups but the increase was larger for elk chased by humans (after: 1,130 m; before: 184 m) than for elk chased by dogs (after: 1,041 m; before: 535 m) or for elk that were not chased (after: 881 m; before: 629 m). Twenty-four elk were radio-collared. Each was assigned to being chased by humans, chased by dogs or not chased, 10 times, from November 2001 to March 2002. Chases lasted 15 minutes and covered averages of 1,148 m when humans (shooting starter pistols) chased elk and 1,219 m when two border collie dogs chased elk. Non-chased elk moved an average of 49 m during 15 minutes. Capture and collar-fitting may have produced some aversive response though animal handling was uniform across groups. Displacement from the town boundary was calculated from daily sightings or radio-signals, from September 2001 to March 2002.

A study in 2005–2006 at a site comprising marsh, forest, farmland, and residential areas in Louisiana, USA (7) found that chasing nuisance black bears *Ursus americanus* with dogs *Canis lupus familiaris*, in addition to making noise and shooting with rubber buckshot, increased the amount of time until they next exhibited nuisance behaviour compared to solely making noise and shooting rubber buckshot. Black bears subjected to chasing by dogs, loud noise and shooting with rubber buckshot took longer to return to nuisance behaviour (58 days) than did bears that were subjected to loud noise and shooting with rubber buckshot but not chasing by dogs (48 days). Between April 2005 and July 2006, eleven bears reported to be exhibiting nuisance behaviour were live-trapped. All were immobilized and fitted with radio-collars. Upon release, six bears were subjected to loud noise, shooting with rubber buckshot and chasing with dogs and five were subjected to loud noise and shooting with rubber buckshot alone. Bears were monitored for recurring nuisance behaviour for up to 5 months after release.

A study in 2002–2005 in a national park in California, USA (8) found that aversive conditioning reduced the number of black bears *Ursus americanus* that were accustomed to seeking food at human-frequented locations revisiting. Of 29 bears accustomed to taking human-food, 17 ceased to do so, six required continued aversion conditioning and six 'persistent offenders' were removed or killed for safety reasons. Over 150 bears were subject to 1,050 aversive conditioning events. Of these, 729 events involved 36 individual food-conditioned or habituated bears (seven became habituated in the final year of the study, so their subsequent behaviour was not assessed). Five personnel drove bears from campsites and other human-occupied areas by throwing rocks and using sling shots, pepper spray, rubber slug projectiles and chasing. All actions were accompanied by shouting. Aversive conditioning actions were carried out each summer, from June 2002 to September 2005.

A replicated, controlled study in 2014 of four urban areas in Colorado, USA (9) found that attempts to scare away coyotes *Canis latrans* did not decrease their use of areas also frequently used by people. On trails frequently travelled by people, the overlap between coyote and human activity was similar where community-level programmes were run to scare coyotes and where programmes were not run (data presented as coefficients of overlap, incorporating frequency and timing of use). On trails with less human traffic, overlap between coyote and human activity was greater where programmes were run than where they were not run. These differences were not tested for statistical significance. Four urban park and open space areas were studied. In two, community-level programmes were run. These primarily involved shouting, throwing objects, and/or aggressively approaching coyotes. Activities were promoted by signs, social media, emailing to multiple recipients, education stations and an online video. Programmes were not run in the two control areas. Coyote and human use of trails were monitored using five camera traps in each area for a 3–4-week period, generating >50,000 independent records of people and coyotes.

A replicated study in 2016 in two savanna reserves in Tanzania (10) found that using drones to deter African savanna elephants *Loxodonta africana* from towns led to elephants leaving the sites quickly. On all 13 occasions, when drones were deployed, elephants began to flee within one minute. Elephants were typically herded to an area > 1 km from villages. Before using drones, rangers were trained during three 4-day workshops. In February–March and May–August 2015 and in March– April 2016, rangers deployed drones in 13 situations when elephants were found close to villages. Each drone was fitted with a flashlight, to locate elephants at night, and, during the day, a live video feed from a camera on the drone was used. Elephant responses were recorded over 60-second intervals for the first 10 minutes of the drone flight.

(1) Davies J.C. & Rockwell R.F. (1986) An electric fence to deter polar bears. *Wildlife Society Bulletin*, 14, 406–409.

(2) Miller G.D. (1987) Field tests of potential polar bear repellents. *Bears: Their Biology and Management*, 7, 383–390, https://doi.org/10.2307/3872649

(3) Clark J.E., van Manen F.T. & Pelton M.R. (2002) Correlates of success for on-site releases of nuisance black bears in Great Smoky Mountains National Park. *Wildlife Society Bulletin*, 30, 104–111.

(4) Beckmann J., Lackey C. & Berger J. (2004) Evaluation of deterrent techniques and dogs to alter behavior of 'nuisance' black bears. *Wildlife Society Bulletin*, 32, 1141–1146, https://doi.org/10.2193/0091-7648(2004)032[1141:eodtad]2.0.co;2

(5) Breck S., Lance N. & Callahan P. (2006) A shocking device for protection of concentrated food sources from black bears. *Wildlife Society Bulletin*, 34, 23–26, https://doi.org/10.2193/0091-7648(2006)34[23:asdfpo]2.0.co;2

(6) Kloppers E.L., St. Clair C. & Hurd T.E. (2005) Predator-resembling aversive conditioning for managing habituated wildlife. *Ecology and Society*, 10, 31, https://doi.org/10.5751/es-01293-100131

(7) Leigh J. & Chamberlain M.J. (2008) Effects of aversive conditioning on behavior of nuisance Louisiana black bears. *Human-Wildlife Conflicts*, 2, 175–182, https://doi.org/10.26077/frgt-yq55

(8) Mazur R.L. (2010) Does aversive conditioning reduce human–black bear conflict? *The Journal of Wildlife Management*, 74, 48–54, https://doi.org/10.2193/2008-163

(9) Breck S.W., Poessel S.A. & Bonnell M.A. (2017) Evaluating lethal and nonlethal management options for urban coyotes. *Human–Wildlife Interactions*, 11, 133–145, https://doi.org/10.5070/v427110686

(10) Hahn N., Mwakatobe A., Konuche J., de Souza N., Keyyu J., Goss M., Chang'a A., Palminteri S., Dinerstein E. & Olson D. (2017) Unmanned aerial vehicles mitigate human–elephant conflict on the borders of Tanzanian Parks: a case study. *Oryx*, 51, 513–516, https://doi.org/10.1017/s0030605316000946

2.11. Retain wildlife corridors in residential areas

https://www.conservationevidence.com/actions/2354

- **One study** evaluated the effects on mammals of retaining wildlife corridors in residential areas. This study was in Botswana[1].

COMMUNITY RESPONSE (0 STUDIES)

POPULATION RESPONSE (0 STUDIES)

BEHAVIOUR (1 STUDY)

- **Use (1 study):** A replicated study in Botswana[1] found that retained wildlife corridors in residential areas were used by 19 mammal species, including African elephants.

Background

Residential and commercial developments can fragment home ranges of mammal species, making access to some resources difficult or dangerous. Retention of wildlife corridors, such as undeveloped land, riversides, woodland strips or other habitat through which mammals can pass, may help to reduce or mitigate some of these impacts of development.

A replicated study in 2012–2014 in seven semi-arid residential and agricultural sites in northern Botswana (1) found that retained wildlife corridors in residential areas were used by African elephants *Locondonta africana* and 18 other mammal species. There were 2,619 camera-trap images of elephants captured, over 516 days. Elephant activity peaked in August, when 13 elephants/day were detected. Nineteen mammal species in total were recorded, including civet *Civettictis civetta* and buffalo *Syncerus caffer* (other species not named). Seven corridors that crossed urban and agricultural areas between a forest reserve and a major river were monitored using camera traps. The seven corridors were either fenced or otherwise ran between developed areas. They were 750–1,700 m long and 3–250 m wide. Camera traps were attached to trees or posts at 1.5–1.8 m high and operated for 24 hours/day from 1 November 2012 to 30 April 2014.

(1) Adams T.S., Chase M.J., Rogers T.L. & Leggett K.E. (2017) Taking the elephant out of the room and into the corridor: can urban corridors work? *Oryx*, 51, 347–353, https://doi.org/10.1017/s0030605315001246

2.12. Install underpasses beneath ski runs

https://www.conservationevidence.com/actions/2355

- **One study** evaluated the effects on mammals of installing underpasses beneath ski runs. This study was in Australia[1].

COMMUNITY RESPONSE (0 STUDIES)

POPULATION RESPONSE (0 STUDIES)

BEHAVIOUR (1 STUDY)

- **Use (1 study):** A replicated study in Australia[1] found that boulder-filled crossings beneath ski slopes were used by seven small mammal species.

Background

Infrastructure and land management associated with the ski industry has, on balance, a negative effect on mammals (Sato *et al.* 2013). One source of impact is habitat fragmentation, through construction of ski runs across previously forested slopes. Underpasses could facilitate mammal movements between habitat patches, especially if they mimic previous ground conditions across rocky slopes.

Sato C.F., Wood J.T. & Lindenmayer D.B. (2013) The effects of winter recreation on alpine and subalpine fauna: a systematic review and meta-analysis. PLoS ONE, 8, e64282, https://doi.org/10.1371/journal.pone.0064282

A replicated study in 2009–2013 in a woodland, heath, and grassland site in New South Wales, Australia (1) found that boulder-filled crossings beneath ski slopes were used by small mammals. Seven mammal species were detected using crossings. From 131 detections where mammals were identified to species, the most frequent were bush rat *Rattus fuscipes* (62 detections), broad-toothed rat *Mastacomys fuscus* (35 detection), dusky antechinus *Antechinus swainsonii* (21 detections) and black rat *Rattus rattus* (10 detections). Eight boulder-filled crossings were constructed under ski runs on grass slopes of a ski area that operated in June–September. Crossings linked remnant heath or woodland. Crossings comprised trenches, 0.4–2.4 m deep, 1–9 m wide, 12–79 m long and filled with rocks of 0.2–2 m diameter. Mammal passage was monitored using hair tubes every 3–6 m (4–13 tubes/crossing). Most crossings were surveyed biannually (7 days in each March–April and November–December) from March 2009 to April 2013.

(1) Schroder M. & Sato C.F. (2017) An evaluation of small-mammal use of constructed wildlife crossings in ski resorts. *Wildlife Research*, 44, 259–268, https://doi.org/10.1071/wr16102

2.13. Provide woody debris in ski run area

https://www.conservationevidence.com/actions/2356

- **One study** evaluated the effects on mammals of providing woody debris in ski run areas. This study was in the USA[1].

COMMUNITY RESPONSE (0 STUDIES)

POPULATION RESPONSE (1 STUDY)

- **Abundance (1 study):** A controlled study in the USA[1] found that placing woody debris on ski slopes did not affect overall small mammal abundance and had mixed effects on individual species abundances.

BEHAVIOUR (0 STUDIES)

Background

Ski-runs are traditionally created by removing trees and undergrowth along with removal of tree stumps and reshaping of topsoil by bulldozing (Ries 1996). As a result, they can present barriers to animal movement (Mansergh & Scotts 1989) and reduce animal abundance (Morrison *et al.* 1995). The provision of woody debris on ski runs may increase use by small mammals.

Mansergh I.M. & Scotts D.J. (1989) Habitat continuity and social organization of the mountain pygmy-possum restored by tunnel. *The Journal of Wildlife Management*, 53, 701–707, https://doi.org/10.2307/3809200

Morrison J.R., De Vergie W.J., Alldredge A.W. & Andree W.W. (1995) The effects of ski area expansion on elk. *Wildlife Society Bulletin*, 23, 481–489.

Ries J.B. (1996) Landscape damage by skiing at the Schauinsland in the Black Forest, Germany. *Mountain Research and Development*, 16, 27–40, https://doi.org/10.2307/3673893

A controlled study in 1999–2001 of coniferous forest and adjacent meadow in Colorado, USA (1) found that placing woody debris on ski slopes did not affect overall small mammal abundance and had mixed results on individual species. Differences in abundance between treatments were not tested for statistical significance. In the two years following ski run establishment, a similar number of small mammals was caught each year on a ski run with woody debris (76–77 individuals) and a run without (75–83 individuals). Red-backed voles *Clethrionomys gapperi* were more abundant where woody debris was added (23–43 individuals) than where no woody debris was added (1–23). Similar numbers of heather voles *Phenacomys intermedius* were caught in both areas (with debris: 10–16; without debris: 10–19) and there were fewer least chipmunk *Tamias minimus* in areas with woody debris (15–31 individual) than without (42–46 individuals). Ski runs were established in 1999. One run had one or more tree limbs placed end to end in rows across the run, with rows 3–9 m apart. The other did not contain woody debris. Small mammals were live-trapped over four consecutive days on three occasions in July–September 1999–2001.

(1) Hadley G.L. & Wilson K.R. (2004) Patterns of small mammal density and survival following ski-run development. *Journal of Mammalogy*, 85, 97–104, https://doi.org/10.1644/1545-1542(2004)085%3C0097:posmda%3E2.0.co;2

3. Threat: Agriculture and aquaculture

Background

In many parts of the world, much of the conservation effort is directed at reducing the impacts of agricultural intensification on biodiversity on farmland and in the wider countryside. A number of the interventions that we have captured reflect this. Further substantial threats from agriculture include loss of habitat and pollution (e.g. from fertilizer and pesticide use). Interventions in response to these threats are described in the following chapters: *Habitat restoration and creation, Threat: Natural system modifications* and *Threat: Pollution*.

All farming systems

3.1. Establish wild flower areas on farmland

https://www.conservationevidence.com/actions/2359

- **Four studies** evaluated the effects of establishing wild flower areas on farmland on small mammals. Two studies were in Switzerland[2,3], one in the UK[1] and one in Germany[4].

COMMUNITY RESPONSE (0 STUDIES)

POPULATION RESPONSE (4 STUDIES)

- **Abundance (4 studies):** Three of four site comparison studies (including three replicated studies), in Switzerland[2,3], the UK[1]

 https://doi.org/10.11647/OBP.0234.03

and Germany[4], found that sown wildflower areas contained more wood mice[1], small mammals[2,3] and common hamsters[4] compared to grass and clover set-aside[1], grasslands, crop and uncultivated margins[2], agricultural areas[3] and crop fields[4].

BEHAVIOUR (0 STUDIES)

Background

This intervention involves sowing areas with wild flowers, typically through agri-environment schemes. This includes set-aside areas, which are fields taken out of agricultural production and which may also enhance biodiversity within farmland.

See also *Provide or retain set-aside areas in farmland* for studies of set-aside under conventional management where no specific actions were taken to increase the wildflower content.

A site comparison study in 1996–1997 on two arable farms in southern UK (1) found that set-aside comprising a species-rich mix of grasses and native forbs was used more by wood mice *Apodemus sylvaticus* relative to availability, than was a simple grass and clover set-aside. Wood mice used species-rich set-aside proportionally to its availability within home ranges. Wood mice used grass/clover set-aside in lower proportion than its availability in home ranges. Data were presented as preference indices. Vegetation in the grass and forb set-aside was more species-rich than that in the grass and clover set-aside, though it was shorter and less dense. Grass and forb set-aside was established in 10-m strips adjacent to crops and hedgerows at one site. Grass and clover set-aside was established on 20-m margins and a 5-ha block at the second site. Nine wood mice were radio-tracked over three nights at each farm, in May–July of 1996 and 1997.

A replicated, site comparison study in 2003 on a farmed plain in Switzerland (2) found that sown wildflower strips contained more small mammals than did conventionally farmed grasslands, autumn-sown wheat fields and uncultivated herbaceous field margins. These comparisons were not tested for statistical significance. Small mammal densities varied greatly between sampling periods but peak densities were estimated at 1,047/ha in wildflower strips, 86/ha in farmed grasslands, 568/ha in wheat crops and 836/ha in herbaceous strips. Two small mammal species were caught in wildflower strips, with two each also in grassland and wheat and six in herbaceous margins. Wildflower strips (15 × 185 m) were sown with native species on fallow arable land. Grasslands (average 0.88 ha) were cut ≥5 times, each April–October and were fertilized. Autumn-sown wheat fields (average 1.3 ha) were harvested at the end of July. Herbaceous strips (5 × 320 m) comprised a range of herbaceous plant species along field margins. Small mammals were live-trapped on three fields of each treatment during 60-hour trapping sessions in March, May and July 2003. Densities were estimated using a capture-recapture method.

A replicated, site comparison study in 2005 in four agricultural areas in Switzerland (3) found that in most cases, following restoration, wildflower areas did not host more small mammals than nearby agricultural areas. In five of nine comparisons (between restored wildflower areas and wheat, maize and tobacco, over three sample seasons), there was no significant difference in the average abundance of small mammals in wildflower areas (458–1,285 animals/ha) and arable fields (34–682 animals/ha). In four of nine comparisons, small mammal abundance was significantly higher in restored wildflower areas (458–1,285 animals/ha) than in nearby arable fields (0–12 animals/ha). In four sites, live traps were placed in restored wildflower areas, wheat fields, maize fields, and tobacco fields. In each area, in May, July, and September 2005, three traps were placed every 5 m along two parallel 45-m-long transects, giving a total of 60 traps/area. Traps were operated over three nights and days at each area. Population sizes were estimated by mark-recapture techniques based on fur clipping of captured animals.

A replicated, site comparison study in 2013 on 28 fields in a mainly arable agricultural area in Bavaria, Germany (4) found that fields sown with wild flowers under an agri-environment scheme contained more common hamsters *Cricetus cricetus* than did crop fields. Hamster burrow density was higher in wildflower fields (3.2 hamster burrows/ha) than in crop fields (0.3 hamster burrows/ha). Fourteen wildflower fields were paired with similarly sized fields of maize, barley, oilseed rape, wheat or sugar beet. The study area measured approximately 50

× 20 km. Paired field were ≥200 m apart and wildflower fields were 440–21,500 m apart. Most wildflower fields were established on less-favoured arable land. They were sown, between 2008 and 2010, with annual and perennial wild and cultivated plants, and were unmanaged thereafter. Burrows, in which hamsters had overwintered and reopened the entrance on emergence in spring, were mapped in May–June 2013.

(1) Tattersall F.H., Fagiano A.L., Bembridge J.D., Edwards P., Macdonald D.W. & Hart B.J. (1999) Does the method of set-aside establishment affect its use by wood mice? *Journal of Zoology*, 249, 472–476, https://doi.org/10.1111/j.1469-7998.1999.tb01218.x

(2) Aschwanden J., Holzgang O. & Jenni L. (2007) Importance of ecological compensation areas for small mammals in intensively farmed areas. *Wildlife Biology*, 13, 150–158, https://doi.org/10.2981/0909-6396(2007)13[150:ioecaf]2.0.co;2

(3) Arlettaz R., Krähenbühl M., Almasi B., Roulin A. & Schaub M. (2010) Wildflower areas within revitalized agricultural matrices boost small mammal populations but not breeding barn owls. *Journal of Ornithology*, 151, 553–564, https://doi.org/10.1007/s10336-009-0485-0

(4) Fischer C. & Wagner C. (2016) Can agri-environmental schemes enhance non-target species? Effects of sown wildflower fields on the common hamster (*Cricetus cricetus*) at local and landscape scales. *Biological Conservation*, 194, 168–175, https://doi.org/10.1016/j.biocon.2015.12.021

3.2. Create uncultivated margins around intensive arable or pasture fields

https://www.conservationevidence.com/actions/2365

- **Nine studies** evaluated the effect of creating uncultivated margins around intensive arable, cropped grass or pasture fields on mammals. Six studies were in the UK[1,2,3,5,8,9], two were in Switzerland[4,6] and one was in the USA[7].

COMMUNITY RESPONSE (1 STUDY)

- **Richness/diversity (1 study):** One replicated, controlled study in the UK[2] found more small mammal species in uncultivated field margins than in blocks of set-aside.

POPULATION RESPONSE (9 STUDIES)

• **Abundance (9 studies):** One replicated, randomized, controlled study in the USA[7] found more small mammals in uncultivated and unmown field margins than in frequently mown margins. Three of seven replicated, site comparison studies (one randomized), in the UK[1,2,3,5,9] and Switzerland[4] found that uncultivated field margins had higher numbers of small mammals[1,2,4,5,9], bank voles[3] and brown hares[6] relative to crops (including grassland)[1,4] and set-aside[2]. The other four studies reported mixed or no effects on bank voles, wood mice and common shrews[3], small mammals[5,9] and brown hares[6]. One site comparison study in the UK[8] found that brown hares used grassy field margins more than expected based on their availability.

BEHAVIOUR (0 STUDIES)

Background

This intervention entails allowing field margin vegetation to regenerate naturally, typically without planting. It can involve some subsequent mowing. Field margins are not fertilized. This intervention includes field margins that run alongside waterways, where these are not otherwise managed, such as by planting trees (for which, see *Habitat Restoration and Creation -Restore or create riparian forest*).

A replicated, site comparison study in 1992–1998 on farms across southern UK (1) found that on uncultivated field margins, more small mammals were caught than in open crop fields. Results were not analysed for statistical significance. More small mammals were trapped in field margins (139 individuals) than in open fields (78 individuals) on conventional farms. The same pattern held on organic farms (margin: 142 individuals; field: 86). A higher proportion of individuals was trapped in margins at two primary study sites for wood mouse *Apodemus sylvaticus* (margin: 40–80%; field: 20–60%), bank vole *Myodes glareolus* (margin: 75–95%; field: 5–25%) and common shrew *Sorex araneus* (margin: 40–90%; field: 10–60%). Small mammals were sampled on two farms over 10 nights, four times/year, in 1992–1998. Live traps were set at 0, 1, 2, 3, 4, 5, 10, 20, 40 m into each field from the boundary. Sample areas included four each of conventional margins, organic margins, conventional crops and organic crops. An unspecified number ($\geq$12) of additional farms was also sampled, each in a single (unspecified) year. The study reports 54 sites were sampled. It is unclear if each of these was a different field. Further elements of the sampling design (such as margin dimensions and the proportion of traps that were in or outside of margins) are unclear.

A replicated, controlled study in 1996–1997 at two farms in Gloucestershire, UK (2) found that uncultivated field margins next to hedgerows hosted more small mammal individuals and species than did blocks of set-aside. Uncultivated margins had more small mammals (21 individuals, eight species/trap session) than did set-aside blocks (11 individuals, five species/trap session). Wood mice *Apodemus sylvaticus* comprised 76% of animals caught in margins and 50% of those caught in set-aside blocks. Species richness was higher in margins (2.6 species/ trap session) than in blocks (2.1 species/trap session). Diversity did not differ significantly between margins and blocks (result presented as indices). Margins (one/farm) comprised 20-m wide sections, covering 5 ha, adjacent to hedgerows. Blocks of set-aside (one/farm) also covered 5 ha. Set-aside was established by sowing a grass/clover mix in 1995. This was cut annually, in July or August. Grids of 49 live traps were set in the centre of set-aside blocks and spanning the margin and adjacent hedgerow and crop. Traps operated over five nights in March, June, September and December of 1996–1997.

A replicated, controlled study in 1999–2000 on an arable farm in North Yorkshire, UK (3) found that in uncultivated grassy field margins, more bank voles *Clethrionomys glareolus* were caught than in cultivated field edges in autumn, but not in spring, while numbers of wood mice *Apodemus sylvaticus* or common shrews *Sorex araneus* caught did not differ between uncultivated or cultivated margins. Total bank vole captures each autumn were higher in 3-m-wide grassy margins (13–14 individuals) and 6-m-wide grassy margins (26–38 individuals) than in cultivated field edges (1 individual) but differences between these treatments were not tested for statistical significance. There were no differences in spring (3-m margin: 9–10; 6-m margin: 2–7; cultivated:

0–18 individuals). Wood mouse catches did not differ significantly between field margin types (3-m margin: 1–29; 6-m margin: 0–18; cultivated: 7–22 individuals), nor did those of common shrew (3-m margin: 2–15; 6-m margin: 0–13; cultivated: 1–4 individuals). Grassy field margins were sown in autumn 1997. Small mammals were live-trapped in four 3-m grassy margins, four 6-m grassy margins and four cultivated field edges, over four weeks in spring (April–May) and four weeks in autumn (September–October) in each of 1999 and 2000.

A replicated, site comparison study in 2003 on a farmed plain in Switzerland (4) found that uncultivated herbaceous field margins contained more small mammals than did conventionally farmed grasslands and autumn-sown wheat fields, though fewer than did sown wildflower strips. These comparisons were not tested for statistical significance. Small mammal densities varied greatly between sampling periods but, at their peak, were estimated at 836/ha in herbaceous margins, 86/ha in farmed grasslands, 568/ha in wheat crops and 1047/ha in wildflower strips. Six small mammal species were caught in herbaceous margins compared to two in each of the other treatments. Herbaceous field margins (5 × 320 m) mainly comprised thistles *Cirsium* spp., common teasel *Dipsacus sylvestris*, St John's wort *Hypericum perforatum*, common mallow *Malva sylvestris* and mulleins *Verbascum* spp. Grasslands (average 0.88 ha) were cut ≥5 times each April–October and were fertilized. Autumn-sown wheat fields (average 1.3 ha) were harvested at the end of July. Wildflower strips (15 × 185 m) were sown with native species. Small mammals were live-trapped on three fields of each treatment during 60-hour trapping sessions in March, May and July 2003. Densities were estimated using a capture-recapture method.

A replicated, site comparison study in 2003–2004 in Yorkshire, UK (5) found that uncultivated field margins hosted similar numbers of small mammals compared to set-aside and farm woodland. There was no significant difference in the annual average numbers of small mammals caught in 2-m margins (2.9–4.4 individuals), 6-m margins (2.5–3.6), set-aside (1.6–2.0) and farm woodland (2.4–2.8). In the first year, more common shrews *Sorex araneus* were caught in 2-m margins (1.4 individuals) than in set-aside (0.6) or farm woodland (0.6) and more wood mice *Apodemus sylvaticus* were in 6-m margins (1.1) and farm woodland (1.4) than in set-aside (0.5). No other species differences between treatments were found. Field margins, sown with grass, were 2 m wide (cut every 2–3 years) or 6 m wide (cut every 1–3 years). Set-aside areas were fallow for ≥5 years, with ≥90% of the area cut annually. Farm woodland comprised young trees (age not stated), fenced and with grass generally uncut. Twelve small mammal traps were set in each of 20 plots/treatment (1 m from the habitat boundary) for four days in November–December in each of 2003 and 2004.

A replicated, site comparison study in 1992–2008 on 58 lowland arable and grassland sites in Switzerland (6) found that establishing uncultivated field margins, in the form of herbaceous strips alongside hedgerows, was associated with higher brown hares *Lepus europaeus* density in arable sites but not in grassland sites. Relative effects of herbaceous strips and hedgerows could not be separated. Hares density along herbaceous strips and adjacent hedgerows was higher than in the landscape as a whole in predominantly arable sites but there was no difference in densities in predominantly grassland sites (data presented as statistical models). Fifty-eight sites (40 mostly arable, 18 mostly grassland), of 71–1,950 ha extent (total area approximately 400 km²) were studied. Forty-three sites included areas managed under agri-environment funding. This entailed establishing 6-m-wide unfertilised herbaceous strips, cut once/year, alongside hedgerows, establishing set-aside areas and low-intensity management of meadows. Herbaceous strips and hedgerows covered 0.17% of arable sites and 0.13% of grassland sites. Vehicle-based spotlight surveys for hares were conducted twice in February–March. Ten sites were surveyed annually from 1992 to 2008 and 48 were, on average, surveyed biennially over that period.

A replicated, randomized, controlled study in 2009 of arable field margins at a site in North Carolina, USA (7) found that uncultivated and unmown field margins supported more small mammals than did frequently mown margins. There were more hispid cotton rats *Sigmodon hispidus* in margins planted with native grasses and flowers (average 8.8 animals/margin) or flowers only (7.5) and unmanaged fallow margins (3.3) than in unplanted mown margins (0). There were also more house mice *Mus musculus* in grass and flower margins (average 9.5 animals/margin), flower only margins (10.1) and unplanted fallow margins (8.8) than in unplanted mown margins (1.8). Three organic crop fields were each planted with soybeans, corn or hay crop and orchard grass. Four sections of margin (0.08 ha) within each of the three fields were assigned to the four treatments, of: planting native warm-season grasses and native prairie flowers, planting native prairie flowers only, leaving fallow without mowing and mowing 2–3 times/ month. Small mammals were live-trapped for three consecutive weeks in October and November 2009.

A site comparison study in 2009–2010 in a mixed farming area in North Yorkshire, UK (8) found that agri-environment grassy field margins had disproportionately high usage by brown hares *Lepus europaeus* during both feeding and resting periods, relative to available habitat areas. Hares spent 6.9% of time in grassy field margins during their main activity period and 13.0% during their inactive period, compared to margins covering of 3.5% of the study site. A total length of 10.8 km of grassy margins was established at field edges and along waterways within a 311-ha study area, through agri-environment funding. Margins comprised 2-m-wide strips and 6-m-wide 'conservation headlands'. They were seeded with a commercial field margin grass mixture, were not sprayed and were cut every two to three years. Fourteen adult hares were radio-tracked, for an average of 186 days each, between July 2009 and August 2010.

A replicated, controlled study in 2005–2011 on an arable farm in Buckinghamshire, UK (9) found that in wide grassy or grass and flower margins on arable fields, small mammal abundance in spring increased over the study period, but it remained stable in narrow, conventionally managed field margins. Small mammal abundance in spring rose by 140% on wide grassy margins and grass and flower margins over the first five years following establishment. There was no significant abundance change on conventional margins, nor any differences between margins in autumn population changes. Absolute counts are not presented in the paper. There were five replicates of three treatments, each on 43–70 ha of farmland. Treatments were conventional management (uncultivated, 2 m-wide field margins or 1 m margins alongside ditches), 6 m-wide grassy margins and 6 m-wide grass and wildflower margins. Margins were established in 2005. Small mammals were live-trapped, over three nights and two days, in November–December 2005, 2006, 2008 and 2010 and each following May.

(1) Brown R.W. (1999) Margin/field interfaces and small mammals. *Aspects of Applied Biology*, 54, 203–206.

(2) Tattersall F.H., Hart B.J., Manley W.J., Macdonald D.W. & Feber R.E. (1999) Small mammals on set-aside blocks and margins. *Aspects of Applied Biology*, 54, 131–138.

(3) Shore R.F., Meek W.R., Sparks T.H., Pywell R.F. & Nowakowski M. (2005) Will Environmental Stewardship enhance small mammal abundance on intensively managed farmland? *Mammal Review*, 35, 277–284, https://doi.org/10.1111/j.1365-2907.2005.00072.x

(4) Aschwanden J., Holzgang O. & Jenni L. (2007) Importance of ecological compensation areas for small mammals in intensively farmed areas. *Wildlife Biology*, 13, 150–158, https://doi.org/10.2981/0909-6396(2007)13[150:ioecaf]2.0.co;2

(5) Askew N.P., Searle J.B. & Moore N.P. (2007) Agri-environment schemes and foraging of barn owls *Tyto alba*. *Agriculture, Ecosystems & Environment*, 118, 109–114, https://doi.org/10.1016/j.agee.2006.05.003

(6) Zellweger-Fischer J., Kéry M. & Pasinelli G. (2011) Population trends of brown hares in Switzerland: The role of land-use and ecological compensation areas. *Biological Conservation*, 144, 1364–1373, https://doi.org/10.1016/j.biocon.2010.11.021

(7) Moorman C.E., Plush C.J., Orr D.B., Reberg-Horton C. & Gardner B. (2013) Small mammal use of field borders planted as beneficial insect habitat. *Wildlife Society Bulletin*, 37, 209–215, https://doi.org/10.1002/wsb.226

(8) Petrovan S.O., Ward A.I. & Wheeler P.M. (2013) Habitat selection guiding agri-environment schemes for a farmland specialist, the brown hare. *Animal Conservation*, 16, 344–352, https://doi.org/10.1111/acv.12002

(9) Broughton R.K., Shore R.F., Heard M.S., Amy S.R., Meek W.R., Redhead J.W., Turk A. & Pywell R.F. (2014) Agri-environment scheme enhances small mammal diversity and abundance at the farm-scale. *Agriculture, Ecosystems & Environment*, 192, 122–129, https://doi.org/10.1016/j.agee.2014.04.009

3.3. Provide or retain set-aside areas on farmland

https://www.conservationevidence.com/actions/2377

- **Four studies** evaluated the effects on mammals of providing or retaining set-aside areas on farmland. Three studies were in the UK[1,2,3] and one was in Switzerland[4].

COMMUNITY RESPONSE (0 STUDIES)

POPULATION RESPONSE (3 STUDIES)

- **Abundance** (**3 studies**): Three replicated studies (including two site comparison studies), in the UK[1,3] and Switzerland[4], found that set-aside did not enhance small mammal numbers relative to cropland[1] or to uncultivated field margins and farm woodland[3], or brown hare numbers relative to numbers on farms without set-aside areas[4].

BEHAVIOUR (1 STUDY)

- **Use** (**1 study**): A before-and-after study in the UK[2] found that use of uncut set-aside areas by wood mice increased after crop harvesting.

Background

Allocation of some farmland to set-aside (fields taken out of production) was compulsory under European Union agricultural policy from 1992 until 2008. The idea was to reduce production. However, set-aside has also been promoted as a method of enhancing biodiversity on farmland. Set-aside can be rotational (in a different place every year or two) or non-rotational (same place for 5–20 years) and fields can either be sown with fallow crops or left to naturally regenerate. Unlike fallow land, set-aside is not ploughed or harrowed except for the purpose of sowing. However, set-aside often is managed by cutting and/or spraying. In some cases, set-aside land has had wild flowers sown on it. Evidence for the effects of this management has been included under the intervention, *Establish wild flower areas on farmland*.

A replicated, controlled study in 1995 of set-aside on two farms in Gloucestershire, UK (1) found that establishing one-year set-aside areas on cropland did not increase small mammal abundance. Trapping success was lower in set-aside (0.6% of traps activated) than in the adjoining unharvested cereal crop (13% of traps activated) and hedgerow (30% of traps activated). Long-tailed field mouse *Apodemus sylvaticus* was the only species caught in set-aside. Sampling at two sites on each farm covered a hedgerow, a 20-m-wide strip of set-aside with adjacent cereal crop on one side of the hedge and a block of either set-aside (two sites) or cereal crop (two sites) on the other side. Set-aside was sown with a mix of wheat *Triticum aestivum* and oilseed rape *Brassica napus* (three sites) or left to regenerate naturally (one site). Fifty Longworth live traps were operated at each site for five nights/month in June–August 1995.

A before-and-after study in 1996–1997 on an arable farm in Wiltshire, UK (2) found that use of uncut set-aside areas by wood mice *Apodemus sylvaticus* increased after crop harvesting. After crop harvesting, uncut set-aside was used more than expected by chance, as were hedgerows. Cut set-aside was used less than expected by chance (results shown as preference indices). Use of cropped areas declined to an average 13% of wood mouse ranges after harvesting, from 54% before harvesting. Across two arable fields, a 3-ha block of set-aside and 3 km of 20-m-wide set-aside field margins were sown (grass/clover mix) in October 1995. In August 1996 and 1997, twenty-four alternate 50 × 6-m patches of cut and uncut set-aside were created alongside a hedge. The remaining 14-m width of set-aside was cut. Thirty-four wood mice were radio-tracked over ≥3 nights in June–July and September–November of 1996 and 1997.

A replicated, site comparison study in 2003–2004 in Yorkshire, UK (3) found that set-aside had similar numbers of small mammals compared to uncultivated field margins and farm woodland. There was no significant difference in the annual average numbers of small mammals caught in set-aside (1.6–2.0), 2-m margins (2.9–4.4 individuals), 6-m margins (2.5–3.6) and farm woodland (2.4–2.8). In the first year, fewer common shrews *Sorex araneus* were caught in set-aside (0.6) or farm woodland (0.6) than in 2-m margins (1.4 individuals) and fewer wood mice *Apodemus sylvaticus* were caught in set-aside (0.5) than in 6-m margins (1.1) and farm woodland (1.4). No other species differences between treatments were found. Set-aside areas were fallow for ≥5 years, with ≥90% of the area cut annually. Field margins, sown with grass, were 2 m wide (cut every 2–3 years) or 6 m wide (cut every 1–3 years). Farm woodland comprised young trees (age not stated), fenced and with grass generally uncut. Twelve small mammal traps were set in each of 20 plots/treatment (1 m from the habitat boundary) for four days in November–December in each of 2003 and 2004.

A replicated, site comparison study in 1992–2008 on 58 lowland arable and grassland sites in Switzerland (4) found that set-aside areas on farmland were not associated with higher brown hares *Lepus europaeus* densities. Set-aside areas were not associated with hare density in either predominantly arable or predominantly grassland areas (data presented as statistical models). Fifty-eight sites (40 mostly arable, 18 mostly grassland), of 71–1,950 ha extent (total area approximately 400 km²) were studied. Forty-three sites included areas managed under agri-environment funding. This entailed establishing set-aside areas (not mown or fertilized, usually sown with wildflower seeds and retained for 2–6 years), maintaining hedgerows (with adjacent herbaceous strips) and low intensity management of meadows. Set-aside covered 3.0% of arable sites and 4.6% of grassland sites. Vehicle-based spotlight surveys for hares were conducted twice in February–March. Ten sites were surveyed annually in 1992–2008 and 48 were, on average, surveyed biennially over that period.

(1) Tattersall F.H., Macdonald D.W., Manley W.J., Gates S., Feber R. & Hart B.J. (1997) Small mammals on one-year set-aside. *Acta Theriologica*, 42, 329–334, https://doi.org/10.4098/at.arch.97-33

(2) Tattersall F.H., Macdonald D.W., Hart B.J., Manley W.J. & Feber R.E. (2001) Habitat use by wood mice (*Apodemus sylvaticus*) in a changeable arable landscape. *Journal of Zoology*, 255, 487–494, https://doi.org/10.1017/s095283690100156x

(3) Askew N.P., Searle J.B. & Moore N.P. (2007) Agri-environment schemes and foraging of barn owls *Tyto alba*. *Agriculture, Ecosystems & Environment*, 118, 109–114, https://doi.org/10.1016/j.agee.2006.05.003

(4) Zellweger-Fischer J., Kéry M. & Pasinelli G. (2011) Population trends of brown hares in Switzerland: The role of land-use and ecological compensation areas. *Biological Conservation*, 144, 1364–1373, https://doi.org/10.1016/j.biocon.2010.11.021

3.4. Maintain/restore/create habitat connectivity on farmland

https://www.conservationevidence.com/actions/2381

- We found no studies that evaluated the effects on mammals of maintaining, restoring or creating habitat connectivity on farmland.

'We found no studies' means that we have not yet found any studies that have directly evaluated this intervention during our systematic journal and report searches. Therefore, we have no evidence to indicate whether or not the intervention has any desirable or harmful effects.

Background

Habitat destruction and fragmentation are important factors in the decline of some mammal populations. Small patches of habitat support smaller populations and if individuals are unable to move to other suitable areas, populations become isolated. This can make them more vulnerable to extinction. Maintaining, restoring or creating corridors of native vegetation between patches of suitable habitat in agricultural landscapes may help to maintain populations. Some specific actions that may encourage movements through farmland are covered in other interventions, including *Plant new or maintain existing hedgerows on farmland* and *Create uncultivated margins around intensive arable or pasture fields.*

3.5. Manage hedgerows to benefit wildlife on farmland

https://www.conservationevidence.com/actions/2382

- We found no studies that evaluated the effects on mammals of managing hedgerows to benefit wildlife on farmland.

'We found no studies' means that we have not yet found any studies that have directly evaluated this intervention during our systematic journal and report searches. Therefore, we have no evidence to indicate whether or not the intervention has any desirable or harmful effects.

Background

Hedgerows can be key habitats for farmland biodiversity, but they may need managing to maximize their value. Managing hedgerows to benefit wildlife involves one or more of the following management changes: reduce cutting frequency; reduce or avoid spraying; mow vegetation beneath hedgerows; fill gaps in hedges; coppice or lay to restore traditional hedge structure. See also *Plant new or maintain existing hedgerows on farmland*.

3.6. Plant new or maintain existing hedgerows on farmland

https://www.conservationevidence.com/actions/2383

- **Three studies** evaluated the effects on mammals of planting new or maintaining existing hedgerows on farmland. Two studies were in the UK[1,2] and one was in Switzerland[3].

COMMUNITY RESPONSE (0 STUDIES)

POPULATION RESPONSE (3 STUDIES)

- **Abundance (3 studies):** One of two replicated, site comparison studies, in the UK[2] and Switzerland[3], found that retaining and enhancing hedgerows along with other field boundary features was associated with higher brown hare density in arable sites but not in grassland sites[3] while the other study found that Irish hare numbers did not increase[2]. A replicated, site comparison study in the UK[1] found that establishing hedgerows alongside arable land increased small mammal abundance.

BEHAVIOUR (0 STUDIES)

Background

Agricultural intensification, including increases in field sizes and pesticides use, has resulted in a loss of field margin habitats, such as hedgerows. These features can provide a relatively undisturbed habitat for wildlife in intensively managed agricultural landscapes. Hedge planting and maintenance of existing hedges has, therefore, been proposed as a means of preserving and enhancing biodiversity. Such management is sometimes funded through agri-environmental schemes.

A replicated, site comparison study in 1999 on three primarily arable farms in Yorkshire, UK (1) found that establishing hedgerows alongside arable land increased small mammal abundance. Average small mammal abundance in hedgerows and adjacent rough margins (0.83 individuals/trap) was higher than on arable land (0.35 individuals/trap). Five species were caught in hedgerows and two in arable plots. Four hedgerows and ten 10 arable plots were surveyed. Hedgerow age and composition were not specified in the paper. Arable plots were sown with winter cereals and contained little cover. Small mammals were surveyed using Longworth live traps over four continuous days and nights, between 22 November and 4 December 1999.

A replicated, site comparison study in 2005 on 200 plots covering a range of agricultural habitats in Northern Ireland, UK (2) found that retaining and enhancing field boundaries, such as hedgerows and banks, as part of a wider suite of agri-environment measures, did not increase numbers of Irish hares *Lepus timidus hibernicus*. The effects of retaining and enhancing field boundaries cannot be separated from those of other agri-environment measures, which included reducing grazing intensity and managing nutrient systems. Hare abundance in agri-environment plots (0.45 hares/km transect) did not significantly differ from that in non-agri-environment plots (0.41 hares/km transect). One hundred and fifty 1-km² plots, on land enrolled into an agri-environment scheme 10–17 years previously, were selected along with 50 non-enrolled 1-km² plots, chosen to match enrolled plots for landscape characteristics. Hares were surveyed at night, in mid-winter, by spotlighting from a vehicle.

A replicated, site comparison study, in 1992–2008, on 58 lowland arable and grassland sites in Switzerland (3) found that maintenance of hedgerows (with adjacent herbaceous strips) on farmland was associated with higher brown hare *Lepus europaeus* density in arable sites but not in grassland sites. Relative effects of hedgerows and herbaceous strips could not be separated. Hare density along hedgerows and adjacent herbaceous strips was higher than in the landscape as a whole in predominantly arable sites but there was no difference in densities in predominantly grassland sites (data presented as statistical models). Fifty-eight sites (40 mostly arable, 18 mostly grassland), of 71–1,950 ha extent (total area approximately 400 km²) were studied. Forty-three sites included areas managed under agri-environment funding. This entailed maintaining hedgerows (unfertilized and unsprayed, with 6-m wide herbaceous strips), establishing set-aside areas and low-intensity management of meadows. Hedgerows and herbaceous strips covered 0.17% of arable sites and 0.13% of grassland sites. Vehicle-based spotlight surveys for hares were conducted twice in February–March. Ten sites were surveyed annually from 1992 to 2008 and 48 were, on average, surveyed biennially over that period.

(1) Moore N.P., Askew N. & Bishop J.D. (2003) Small mammals in new farm woodlands. *Mammal Review*, 33, 101–104, https://doi.org/10.1046/j.1365-2907.2003.00004.x

(2) Reid N., McDonald R.A. & Montgomery W.I. (2007) Mammals and agri-environment schemes: hare haven or pest paradise? *Journal of Applied Ecology*, 44, 1200–1208, https://doi.org/10.1111/j.1365-2664.2007.01336.x

(3) Zellweger-Fischer J., Kéry M. & Pasinelli G. (2011) Population trends of brown hares in Switzerland: The role of land-use and ecological compensation areas. *Biological Conservation*, 144, 1364–1373, https://doi.org/10.1016/j.biocon.2010.11.021

3.7. Plant trees on farmland

https://www.conservationevidence.com/actions/2386

- **Four studies** evaluated the effects on mammals of planting trees on farmland. Two studies were in the UK[1,2], one was in Italy[3] and one was in Australia[4].

COMMUNITY RESPONSE (0 STUDIES)

POPULATION RESPONSE (2 STUDIES)

- **Abundance (2 studies):** Two replicated studies (including one controlled, and one site comparison study), in the UK[1,2], found that farm woodland supported a higher small mammal abundance than on arable land[1] or similar abundance compared to uncultivated field margins and set-aside[2].

BEHAVIOUR (2 STUDIES)

- **Use (2 studies):** A study in Italy found that tree stands were used more by European hares compared to the wider farmed landscape[3]. A replicated study in Australia found that trees planted on farmland were used by koalas[4].

Background

Agricultural intensification, which includes increasing field size and pesticide use, has resulted in a loss of shelter and food resources for wildlife, such as that provided by areas of trees. These features can provide a relatively undisturbed habitat for wildlife in intensively managed agricultural landscapes. Tree planting may therefore diversify habitat availability and, in younger plantations, may also provide areas of longer uncut grass than is available elsewhere in the landscape.

A replicated, controlled study in 1999 on three mainly arable farms in Yorkshire, UK (1) found that establishing new woodland plantations on former arable land increased small mammal abundance. Average small mammal abundance in plantations (1.1 individuals/trap) was higher than on arable land (0.4 individuals/trap). Small mammal species richness in plantations (4–6 species/site) was also higher than on arable land (1–4 species/site), although this difference was not tested for statistical significance. Twelve plantations (0.17–2.0 ha), established in 1992–1997, were surveyed, along with arable plots adjacent to 10 of these. Plantations, predominantly of broad-leaved trees, were on ex-arable land. Dense grasses and other herbaceous plants dominated vegetation at time of surveys. Planted trees were ≤4 m high. Arable plots were sown with winter cereals and contained little cover. Small mammals were surveyed using Longworth live traps over four continuous days and nights, between 22 November and 4 December 1999.

A replicated, site comparison study in 2003–2004 in an agricultural area in Yorkshire, UK (2) found that farm woodland had similar numbers of small mammals compared to uncultivated field margins and set-aside. There was no significant difference in the annual average numbers of small mammals caught in farm woodland (2.4–2.8 individuals), 2-m-wide field margins (2.9–4.4), 6-m-wide field margins (2.5–3.6) and set-aside (1.6–2.0). In the first year, more wood mice *Apodemus sylvaticus* were caught in farm woodland (1.4 individuals) and in 6-m-wide margins (1.1) than in set-aside (0.5), but fewer common shrews *Sorex araneus* were in farm woodland (0.6 individuals) or set-aside (0.6) than in 2-m-wide margins (1.4). No other species differences between treatments were found. Farm woodland comprised young trees (age not stated), fenced and with grass generally uncut. Field margins, sown with grass, were 2 m wide (cut every 2–3 years) or 6 m wide (cut every 1–3 years). Set-aside areas were fallow for ≥5 years, with ≥90% of the area cut annually. Twelve small mammal traps were set in each of 20 plots/treatment (1 m from the habitat boundary) for four days in November–December in each of 2003 and 2004.

A study in 2005 in an area of arable farmland with scattered woodland cover in Lombardy Region, Italy (3) found that presence of tree stands increased the use of an area by European hares *Lepus europaeus*. Of plots where hare faecal pellets were present, 12% were in poplar groves, compared to 5% of plots where pellets were absent being in poplar groves. In addition, 16% of plots with pellets were in short rotation forestry compared to 6% of plots without pellets. Arboriculture comprised poplar groves and short-rotation (2–5 year) forestry. Habitat use was assessed by recording presence or absence of hare faecal pellets in 150 randomly located plots, of 1-m radius, across an 820-ha study area, in March–May 2005.

A replicated study in 2006 of 19 tree plots in New South Wales, Australia (4) found that trees planted on farmland were used by koalas *Phascolarctos cinereus*. Of the 19 plots surveyed, 14 had evidence of use by koalas. In eight plots, over 40% of trees inspected were used by koalas. Koala pellets were recorded under 16 of 25 tree species or species groups inspected. Trees closer to potential source populations and older trees were more likely to be used by koalas (results presented as statistical model). Nineteen plots (15 linear tree corridors and four patches of trees), aged 6–15 years (planted 1990–2001) were studied (plot sizes not stated). Plots were on 10 farms and in two roadside plantings. Every fifth tree (>2 m high), along pre-determined transects of up to 100 trees/plot, was assessed for presence of koala pellets within a 1-m radius of the tree base.

(1) Moore N.P., Askew N. & Bishop J.D. (2003) Small mammals in new farm woodlands. *Mammal Review*, 33, 101–104, https://doi.org/10.1046/j.1365-2907.2003.00004.x

(2) Askew N.P., Searle J.B. & Moore N.P. (2007) Agri-environment schemes and foraging of barn owls *Tyto alba*. *Agriculture, Ecosystems & Environment*, 118, 109–114, https://doi.org/10.1016/j.agee.2006.05.003

(3) Cardarelli E., Meriggi A., Brangi A. & Vidus-Rosin A. (2011) Effects of arboriculture stands on European hare *Lepus europaeus* spring habitat use in an agricultural area of northern Italy. *Acta Theriologica*, 56, 229–238, https://doi.org/10.1007/s13364-010-0019-4

(4) Rhind S.G., Ellis M.V., Smith M. & Lunney D. (2014) Do koalas Phascolarctos cinereus use trees planted on farms? A case study from north-west New South Wales, Australia. *Pacific Conservation Biology*, 20, 302–312, https://doi.org/10.1071/pc140302

3.8. Pay farmers to cover the costs of conservation measures

https://www.conservationevidence.com/actions/2387

- **Three studies** evaluated the effects on mammals of paying farmers to cover the costs of conservation measures. The three studies were in the UK[1,2,3].

COMMUNITY RESPONSE (0 STUDIES)

POPULATION RESPONSE (3 STUDIES)

- **Abundance (3 studies):** A replicated, controlled study in the UK[1] found that agri-environment scheme enrolment was associated with increased brown hare density in one of two regions studied. A replicated, site comparison study in Northern Ireland, UK[2] found that agri-environment scheme enrolment did not increase numbers of Irish hares. A replicated, controlled study in the UK (3) found that in field margins created through enrolment in an agri-environment scheme, small mammal abundance in spring increased, whereas it remained stable in conventionally managed margins.

BEHAVIOUR (0 STUDIES)

Background

Agri-environment schemes are government or inter-governmental schemes designed to compensate farmers financially for changing agricultural practice to be more favourable to biodiversity and the landscape. Agri-environment schemes represent many different specific interventions relevant to conservation. Where a study can be clearly assigned to a specific intervention, it appears in the appropriate section (e.g. *Create uncultivated margins around intensive arable or pasture fields* and *Establish wild flower areas on farmland*). This section includes broader evidence about the success of agri-environment policies, such as where specific actions are not clearly defined.

A replicated, controlled study, in 1998–2002, on 71 arable farms in two UK regions (1) found that increased semi-natural habitat cover through enrolment in an agri-environment scheme was associated with increases in brown hare *Lepus europaeus* density in one region but not another. In East Anglia, brown hare density on farms enrolled in the scheme increased by 35% from 1998–2003, compared to an 18% decline on non-enrolled farms. In the West Midlands, hare density changes from 1998–2003 did not differ significantly between farm types (enrolled farms: decline of 10.8%; non-enrolled farms: increase of 3.6%). Seventy-one farms were surveyed, 19 enrolled and 18 not enrolled in an agri-environment scheme in East Anglia and 19 enrolled and 15 not enrolled in West Midlands. The scheme (Arable Stewardship Pilot Scheme) incentivised a range of measures which are not specified in the study, but appear to include increasing woodland and set-aside areas. Enrolled farms operated under the scheme from 1998 onwards. Hares were surveyed from November–February in 1998–1999 and 2002–2003 by spotlighting after dark from a vehicle. Usually, ≥20 fields/farm were counted (≥30% of the farm area).

A replicated, site comparison study in 2005 on 200 plots covering a range of agricultural habitats in Northern Ireland, UK (2) found that retaining and enhancing field boundaries, reducing grazing intensity and managing nutrient systems through enrolment in an agri-environment scheme did not increase numbers of Irish hares *Lepus timidus hibernicus*. Hare abundance in agri-environment plots (0.45 hares/km transect) did not significantly differ from that in non-agri-environment plots (0.41 hares/km transect). One hundred and fifty 1-km² plots, on land that was enrolled into an agri-environment scheme 10–17 years previously, were selected along with 50 non-enrolled 1-km² plots, chosen to match enrolled plots for landscape characteristics. Hares were surveyed at night, in mid-winter, by spotlighting from a vehicle.

A replicated, controlled study in 2005–2011 on an arable farm in Buckinghamshire, UK (3) found that in wide grassy or grass and flower margins created on arable fields through enrolment in an agri-environment scheme, small mammal abundance in spring increased over the study period, but it remained stable in narrow, conventionally managed field margins. Small mammal abundance in spring rose by 140% on wide grassy margins and grass and flower margins over the first five years following establishment. There was no significant abundance change on conventional margins, nor any differences between margins in autumn population changes. Absolute counts are not presented in the paper. There were five replicates of three treatments, each on 43–70 ha of farmland. Treatments were 6 m-wide grassy margins ('Entry Level Scheme') and 6 m-wide grass and wildflower margins ('Entry Level Scheme Extra') both created as part of an agri-environment scheme, and conventional management (uncultivated, 2-m-wide field margins or 1 m margins alongside ditches). Margins were established in 2005. Small mammals were live-trapped, over three nights and two days, in November–December 2005, 2006, 2008 and 2010 and each following May.

(1) Browne S.J. & Aebischer N.J. (2003) *Arable Stewardship: impact of the pilot scheme on the brown hare and grey partridge after five years*. DEFRA contract ref. RMP1870vs3.

(2) Reid N., McDonald R.A. & Montgomery W.I. (2007) Mammals and agri-environment schemes: hare haven or pest paradise? *Journal of Applied Ecology*, 44, 1200–1208, https://doi.org/10.1111/j.1365-2664.2007.01336.x

(3) Broughton R.K., Shore R.F., Heard M.S., Amy S.R., Meek W.R., Redhead J.W., Turk A. & Pywell R.F. (2014) Agri-environment scheme enhances small mammal diversity and abundance at the farm-scale. Agriculture, Ecosystems & Environment, 192, 122–129, https://doi.org/10.1016/j.agee.2014.04.009

3.9. Provide refuges during crop harvesting or mowing

https://www.conservationevidence.com/actions/2389

- We found no studies that evaluated the effects on mammals of providing refuges during crop harvesting or mowing.

'We found no studies' means that we have not yet found any studies that have directly evaluated this intervention during our systematic journal and report searches. Therefore, we have no evidence to indicate whether or not the intervention has any desirable or harmful effects.

Background

During crop harvesting and mowing operations, mammals may move into adjacent areas of long grass or crops. If mowing/harvesting occurs from the outside of the field inwards, this behaviour can leave them trapped in the centre of the field and killed as the last patch is harvested. However, if unharvested refuges are left in fields then it is possible that mammals remain in them and survive.

3.10. Use repellent on slug pellets to reduce non-target poisoning

https://www.conservationevidence.com/actions/2390

- **One study** evaluated the effects on mammals of using repellent on slug pellets to reduce non-target poisoning. This study was in the UK[1].

COMMUNITY RESPONSE (0 STUDIES)

POPULATION RESPONSE (0 STUDIES)

BEHAVIOUR (1 STUDY)

- **Use (1 study):** A replicated, controlled study in the UK[1] found that, at some concentrations, food treated with a bitter substance was consumed less by wood mice but not by bank voles or common shrews.

Background

Poisons used to control slugs may also be ingested by non-target species, such as rodents. Such poisoning can lead to declines in rodent numbers (Shore et al. 1997). Substances that make slug pellets unattractive to small mammals, yet still effective on slugs, may help to reduce small mammal losses.

Shore R.F., Feber R.E., Firbank L.G., Fishwick S.K., Macdonald D.W. & Nøruma, U. (1997) The impacts of molluscicide pellets on spring and autumn populations of wood mice *Apodemus sylvaticus. Agriculture, Ecosystems & Environment*, 64, 211–217, https://doi.org/10.1016/s0167-8809(97)00039-x

A replicated, controlled study (year not stated) in an agricultural area in the UK (1) found that treating food with a bitter substance (Bitrex™; as a trial of its efficacy for deterring toxic slug pellet consumption) reduced consumption by wood mice *Apodemus sylvaticus* at some concentrations but did not change consumption rates of bank voles *Clethrionomys glareolus* or common shrews *Sorex aranaeus*. Wood mice avoided food treated with Bitrex at 100 ppm and 300 ppm but showed no avoidance at 50 ppm or 500–1,740 ppm (data not presented). Bank voles and common shrews showed no avoidance of food treated with Bitrex at 100 ppm or 300 ppm (data not presented). Wild small mammals were contained within small enclosures. Wood mice and bank voles were offered barley *Hordeum vulgare*. Common shrews were offered fly pupae. Food was sprayed with the Bitrex solution. Trails ran for eight hours overnight (wood mouse) or six hours night or day (bank vole and common shrew) with treated food only and with choices of treated and untreated food.

(1) Kleinkauf A., Macdonald D.W. & Tattersall F.H. (1999) A bitter attempt to prevent non-target poisoning of small mammals. *Mammal Review*, 29, 201–204, https://doi.org/10.1046/j.1365-2907.1999.00046.x

3.11. Restrict use of rodent poisons on farmland with high secondary poisoning risk

https://www.conservationevidence.com/actions/2391

• We found no studies that evaluated the effects on mammals of restricting use of rodent poisons on farmland that have secondary poisoning risks.

'We found no studies' means that we have not yet found any studies that have directly evaluated this intervention during our systematic journal and report searches. Therefore, we have no evidence to indicate whether or not the intervention has any desirable or harmful effects.

Background

Rodenticides are in common use around farms, houses and industrial sites. The most frequently used forms are anticoagulant rodenticides, which cause death in target animals by inhibiting blood clotting. Death can take several days after ingestion so poisoning may be passed on up the food chain both to predators and to scavengers. In some situations, a high proportion of predators may be exposed to secondary poisoning. For example, in one study 85% of fisher *Pekania pennanti* carcasses collected showed signs of exposure (Thompson *et al.* 2013) whilst another showed signs of exposure in 79% of invasive American Mink, with the risk of exposure being higher in areas with farms (Ruiz-Suárez *et al.* 2016). Restricting use of such poisons may reduce their ingestion by mammalian carnivores.

Thompson C., Sweitzer R., Gabriel M., Purcell K., Barrett R. & Poppenga R. (2013) Impacts of rodenticide and insecticide toxicants from marijuana cultivation sites on fisher survival rates in the Sierra National Forest, California. *Conservation Letters*, 7, 91–102, https://doi.org/10.1111/conl.12038

Ruiz-Suárez., Melero Y., Giela A., Henríquez-Hernández L.A., Sharp E., Boada L.D., Taylor M.J., Camacho M., Lambin X., Luzardo O.P. & Hartley G. (2016) Rate of exposure of a sentinel species, invasive American mink (*Neovison vison*) in Scotland, to anticoagulant rodenticides. *Science of the Total Environment*, 569–570, 1013–1021, https://doi.org/10.1016/j.scitotenv.2016.06.109

Annual & Perennial Non-Timber Crops

3.12. Increase crop diversity for mammals

https://www.conservationevidence.com/actions/2392

- We found no studies that evaluated the effects on mammals of increasing crop diversity.

'We found no studies' means that we have not yet found any studies that have directly evaluated this intervention during our systematic journal and report searches. Therefore, we have no evidence to indicate whether or not the intervention has any desirable or harmful effects.

Background

Some farmland heterogeneity is thought to be key in determining on-farm biodiversity (Benton *et al.* 2003). Therefore, increasing the range of different crops grown in a given year may increase the biological value of a farm.

Benton T.G., Vickery J.A. & Wilson J.D. (2003) Farmland biodiversity: is habitat heterogeneity the key? *Trends in Ecology & Evolution*, 18, 182–188, https://doi.org/10.1016/s0169-5347(03)00011-9

3.13. Create beetle banks on farmland

https://www.conservationevidence.com/actions/2393

- **One study** evaluated the effects on mammals of creating beetle banks on farmland. This study was in the UK[1].

COMMUNITY RESPONSE (0 STUDIES)

POPULATION RESPONSE (1 STUDY)

- **Abundance (1 study):** One replicated study in the UK[1] found that beetle banks had higher densities of harvest mouse nests than did field margins.

BEHAVIOUR (0 STUDIES)

Background

Beetle banks are raised strips which run through a field, typically planted with grasses. They primarily serve as an overwintering habitat for beetles, which provide pest control in the spring. By dividing the field, beetle banks reduce the distance that predators have to travel to reach the centre of the crop, a potential problem if overwintering habitat occurs only at the field edge. Beetle banks may also harbour other wildlife, such as small mammals.

A site comparison study in 1998 on an arable farm in Leicestershire, UK (1) found that beetle banks had higher densities of harvest mouse *Micromys minutus* nests than did field margins. The density of harvest mouse nests in beetle banks (117/ha) was higher than in field margins (14/ha). Beetle banks, created in 1992–1994, were 2–2.5 m wide, positioned down field centres and sown with tussock-forming grasses. They were cut during the first year but not thereafter. Field margins were ≥1 m wide, comprised perennial grasses and herbs and were mostly uncut. Harvest mouse nests were surveyed in September–November 1998 along 1,800 m length of beetle banks and 9,800 m length of field margins.

(1) Bence S.L., Stander K. & Griffiths M. (2003) Habitat characteristics of harvest mouse nests on arable farmland. *Agriculture, Ecosystems & Environment*, 99, 179–186, https://doi.org/10.1016/s0167-8809(03)00137-3

3.14. Plant crops to provide supplementary food for mammals

https://www.conservationevidence.com/actions/2394

- **Four studies** evaluated the effects on mammals of planting crops to provide supplementary food. Two studies were in the USA[1,2], one was in the UK[3] and one was in Spain[4].

COMMUNITY RESPONSE (0 STUDIES)

POPULATION RESPONSE (3 STUDIES)

- **Abundance (3 studies):** Two replicated, controlled studies (including one before-and-after study), in the UK[3] and Spain[4], found that crops grown to provide food for wildlife resulted in a higher abundance of small mammals in winter, but not in summer[3] and increased European rabbit abundance[4]. A replicated, randomized, controlled study in the USA[1] found that triticale (a cross between wheat and rye) held higher overwintering mule deer abundance relative to barley, annual ryegrass, winter wheat or rye.

BEHAVIOUR (2 STUDIES)

- **Use (2 studies):** A replicated, randomized, controlled study in the USA[1] found that mule deer consumed triticale (a cross between wheat and rye) more than they did barley, annual ryegrass, winter wheat or rye. A replicated, randomized, controlled study in the USA[2] found that supplementary food provided for game species was also consumed by lagomorphs and rodents.

Background

Crops may be planted to provide supplementary food for a range of mammal species, either of economic or conservation importance. The intervention includes also studies that measure the response of non-target mammals where the crop is nonetheless planted for a wildlife conservation purpose.

See also: *Species management — Provide supplementary food to increase reproduction/survival.*

A replicated, randomized, controlled study in 1979–1980 in a crop field in Texas, USA (1) found that on triticale (a cross between wheat and rye), overwintering mule deer *Odocoileus hemionus* abundance and crop consumption were higher than on barley, annual ryegrass, winter wheat or rye. The preference index (values >1 indicate selection for that grass and values <1 indicate avoidance) for the quantity of triticale removed by deer (1.37) was higher than for barley (0.90), annual ryegrass (0.99), wheat (0.87) and rye (0.66). Average deer abundance was also higher on triticale (12.8 deer/plot) compared to barley (7.0), annual ryegrass (10.1), wheat (5.8) and rye (9.0). In August 1979, five crop types were planted in five replicate blocks (four plots in each block were 0.125 ha, one was 0.063 ha). Grass species were randomly assigned to plots. Grass production and forage removal by deer were estimated monthly from November 1979 to March 1980 using paired caged and uncaged quadrats. Deer abundance was assessed by time lapse photography.

A replicated, randomized, controlled study in 1996–1997 of cropland on six ranches in Texas, USA (2) found that supplementary food provided for game species was also consumed by rodents and lagomorphs. Rodents ate 47% by biomass of winter oats *Avena sativa* grown for white-tailed deer *Odocoileus virginianus* that were consumed. Lagomorphs ate 10% and deer ate 44% of oats that were consumed. On each of six ranches, 2 ha of winter oats was grown. Twenty-four plots, each 1 m³, were established at each ranch from December 1996 to March 1997. Six plots were fenced using 10 × 10-cm mesh (to exclude deer), six using 2 × 3-cm mesh (to exclude deer and lagomorphs), six using 0.5 × 0.5-cm mesh (to exclude deer, lagomorphs and rodents) and six were unfenced. Consumption was assessed by comparing remaining oat biomass with that in the finest-mesh fenced plots.

A replicated, controlled study in 2004–2005 on four arable farms in southern UK (3) found that small mammals used plots sown with a wild bird seed mix more than wheat crop in winter but not in summer. In winter, more small mammals were caught on average in the wild bird mix (27 individuals/100 trap nights) than in adjacent crops (8 individuals/100 trap nights). However, in summer, fewer were caught in the wild bird mix (<1 individual/100 trap nights) than in adjacent crops (12 individuals/100 trap nights). A mix of white millet *Echinochloa esculenta*, linseed *Linum usitatissimum*, radish *Raphanus sativus* and quinoa *Chenopodium quinoa* was sown in a 150 × 30-m patch in the centre of a winter wheat crop on each of four farms, in April 2004 and 2005. Small mammals were live-trapped over three days and nights in November–December 2004 and again in May–June 2005.

A replicated, controlled, before-and-after study in 2004–2006 of forest, scrub and grassland mosaics on 14 estates in central Spain (4) found that sown grain crops were used more by, and had a higher abundance of, European rabbits *Oryctolagus cuniculus* relative to uncropped areas. Cropped plots had more rabbit latrines (52 latrines/km transect) than did uncropped plots (19 latrines/km transect). Rabbit relative abundance increased on sown areas (after sowing: 2.0 rabbits/km transect; before: 1.3) but not elsewhere on estates (after sowing: 3.0 rabbits/km transect; before: 3.3). Fourteen private estates in central Spain were studied. Across these, 125 plots were sown with barley and oat seed, at 150 kg/ha, in 2004–2006. There were 3–19 treatment plots/estate of 0.04–43.07 ha extent. For each treatment plot, an unsown control plot, ≥200 m away, with similar broad characteristics, was selected. Rabbit latrines were counted along transects in sown and unsown plots in late spring. Relative abundance was assessed by counting rabbits from transects in spring, before and after sowing.

(1) Wiggers E.P., Wilcox D.D. & Bryant F.C. (1984) Cultivated cereal grains as supplemental forages for mule deer in the Texas panhandle. *Wildlife Society Bulletin*, 12, 240–245.

(2) Donalty S., Henke S.E. & Kerr C.L. (2003) Use of winter food plots by nongame wildlife species. *Wildlife Society Bulletin*, 31, 774–778.

(3) Pywell R.F., Shaw L., Meek W., Turk A., Shore R.F. & Nowakowski M. (2007) Do wild bird seed mixtures benefit other taxa? *Aspects of Applied Biology*, 81, 69–76.

(4) Guil F., Fernández-Olallac M., Martínez-Jáuregui M., Moreno-Opoa R., Agudína S. & San Miguel-Ayanz A. (2014) Grain sowing aimed at wild rabbit *Oryctolagus cuniculus* L. enhancement in Mediterranean environments. *Journal for Nature Conservation*, 22, 552–558, https://doi.org/10.1016/j.jnc.2014.08.011

3.15. Change mowing regime (e.g. timing, frequency, height)

https://www.conservationevidence.com/actions/2399

- We found no studies that evaluated the effects of changing mowing regime (e.g. timing, frequency, height) on mammals.

'We found no studies' means that we have not yet found any studies that have directly evaluated this intervention during our systematic journal and report searches. Therefore, we have no evidence to indicate whether or not the intervention has any desirable or harmful effects.

Background

Numerous studies assess responses of grassland vegetation structure and composition to different mowing regimes. Responses of fauna are less frequently documented with invertebrate responses dominating among those that are published. Some mammalian herbivores may be sensitive to variations in grassland vegetation height and structure (Mero *et al.* 2015). An understanding of responses to changes in mowing regimes may assist with development of tailored management for particular species.

See also: *Habitat Restoration and Creation — Restore or create grassland.*

Mero, TO., Bocz R., Polyak L., Horvath G. & Lengyel S. (2015) Local habitat management and landscape-scale restoration influence small-mammal communities in grasslands. *Animal Conservation*, 18, 442–450, https://doi.org/10.1111/acv.12191

3.16. Leave areas of uncut ryegrass in silage field

https://www.conservationevidence.com/actions/2400

- We found no studies that evaluated the effects on mammals of leaving areas of uncut ryegrass in silage field.

'We found no studies' means that we have not yet found any studies that have directly evaluated this intervention during our systematic journal and

report searches. Therefore, we have no evidence to indicate whether or not the intervention has any desirable or harmful effects.

Background

This intervention involves leaving areas of uncut ryegrass *Lolium perenne* in silage fields. Ryegrass seeds are a potential food source for small mammals, but cutting ryegrass fields multiple times a year for silage removes seed heads before they can ripen and so reduces the food available the following winter. Leaving fields or plots uncut may provide overwinter food for small mammals and may also provide suitable habitat away from damaging harvesting machinery.

3.17. Leave cut vegetation in field to provide cover

https://www.conservationevidence.com/actions/2401

- **One study** evaluated the effects on mammals of leaving cut vegetation in field to provide cover. This study was in the USA[1.]

COMMUNITY RESPONSE (0 STUDIES)

POPULATION RESPONSE (1 STUDY)

- **Abundance (1 study):** A controlled, before-and-after study in the USA[1] found that increasing cover, by adding cut vegetation (hay), did not increase rodent abundance.

BEHAVIOUR (0 STUDIES)

Background

Leaving cut vegetation in a field, either following cutting or by adding hay from elsewhere, may increase ground-level shelter available to small mammals.

A controlled, before-and-after study in 1983–1984 on a prairie grassland in Kansas, USA (1) found that increasing cover, by adding cut vegetation (hay), did not increase rodent abundance. Rodent numbers were not significantly different after hay addition (19–28/census) compared to before hay addition (10–25/census). Rodent abundances in plots with no added hay likewise did not differ significantly over the same time periods (after: 14–45/census; before: 9–36/census). Three plots, 0.81 ha each, were established on brome grass *Bromus inermns* and prairie vegetation. One had 16 cm depth of hay added in January 1984. Two were left unmanaged. Small mammals were sampled using 100 Longworth live traps/plot. Trapping occurred over two nights, biweekly, from 12 weeks before hay addition (October 1983) until 26 weeks after hay addition (August 1984).

(1) Kotler B.P., Gaines M.S. & Danielson B.J. (1988) The effects of vegetative cover on the community structure of prairie rodents. *Acta Theriologica*, 33, 379–391, https://doi.org/10.4098/at.arch.88-32

3.18. Establish long-term cover on erodible cropland

https://www.conservationevidence.com/actions/2402

- **One study** evaluated the effects on mammals of establishing long-term cover on erodible cropland. This study was in the USA[1].

COMMUNITY RESPONSE (0 STUDIES)

POPULATION RESPONSE (1 STUDY)

- **Abundance (1 study):** A replicated, site comparison study in the USA[1], found that establishing long-term cover on erodible cropland did not increase the abundance of eastern cottontails.

BEHAVIOUR (0 STUDIES)

Background

Establishing long-term cover on cropland that is highly susceptible to erosion may be carried out for a number of reasons including conserving soil fertility, limiting carbon emissions and enhancing habitat for biodiversity. The provision of long-term cover has potential to benefit mammals that are able to exploit increased shelter and food resources.

A replicated, site comparison study in 1989–1990 on six areas of mostly arable farmland in Nebraska, USA (1) found that establishing long-term cover on erodible cropland was not associated with increased abundance of eastern cottontails *Sylvilagus floridanus*. The number of cottontails counted in areas with 18–21% long-term cover (2.1–6.7 cottontails/block) did not differ significantly from that in areas with 2–3% long-term cover (4.1–8.8 cottontails/block). Within six 23-km² farmland blocks, the proportion of land managed under an agri-environment scheme aimed at diversifying long-term cover types and reducing crop production was determined. In three blocks, 18–21% of cropland was in the scheme and in the other three, 2–3% was in the scheme. Long-term cover, established under 10-year contracts, included establishment of grasses and legumes. Live cottontails were counted from a vehicle while driving at 30–40 km/h, in May and June of 1989 and 1990.

(1) King J.W. & Savidge J.A. (1995) Effects of the Conservation Reserve Program on wildlife in southeast Nebraska. *Wildlife Society Bulletin*, 23, 377–385.

Livestock Farming & Ranching

3.19. Exclude livestock from semi-natural habitat (including woodland)

https://www.conservationevidence.com/actions/2407

- **Nine studies** evaluated the effects of excluding livestock from semi-natural habitat on mammals. Six studies were in the USA[1-5,9], two were in Spain[6,7] and one was in Australia[8].

COMMUNITY RESPONSE (2 STUDIES)

- **Richness/diversity (2 studies):** Two replicated, site comparison studies in the USA[2,4] found more small mammal species[2,4] on areas from which livestock were excluded.

POPULATION RESPONSE (9 STUDIES)

- **Abundance (9 studies):** Four out of eight studies (including four site comparisons and four controlled studies), in the

USA[1,2,3,4,5,9] and Spain[6,7], found that excluding grazing livestock led to higher abundances of mule deer[1], small mammals[4,6] and, when combined with provision of water, of European rabbits[7]. One study found higher densities of some but not all small mammals species[2] when livestock were excluded and the other three studies found that grazing exclusion did not lead to higher abundances of black-tailed hares[3], California ground squirrel burrows[5] or of five small mammal species[9]. A site comparison study in Australia[8] found more small mammals where cattle were excluded compared to high intensity cattle-grazing but not compared to medium or low cattle-grazing intensities.

BEHAVIOUR (0 STUDIES)

Background

This intervention involves preventing livestock from grazing certain semi-natural habitats, such as grasslands and woodland, to benefit wildlife. Mammal responses may be linked to reduction in competition from domestic herbivores or to changes in the vegetation structure.

See also *Reduce intensity of grazing by domestic livestock* for studies where livestock are removed from areas of permanent grassland.

A controlled study in 1982–1984 on a shrubland site in California, USA (1) found that inside a cattle-exclusion fence, there were more mule deer *Odocoileus hemionus* than there were outside it. This result was not tested for statistical significance. Over six sampling events, 192 faecal pellet clumps were counted inside the enclosure compared to 138 outside it. In June 1982, a prescribed burn was carried out across 4 ha of land. A 0.25-ha enclosure (cattle proof but not deer proof) was established on the burned area. Relative deer presence inside and outside the enclosure was assessed by counting pellet-groups in September 1982, February, August, and November 1983 and March and July 1984. Counts were made along 18 transects (5 m long) inside the enclosure and 18 outside the enclosure.

A replicated, site comparison study in 1990–1992 in a desert in south-central California, USA (2) found that excluding livestock led to more small mammal species, and higher densities of some small mammal species, compared to sheep-grazed areas. More species of small nocturnal rodents were found in ungrazed (3.7 species/sample) than in grazed areas (2.5 species/sample), and diversity was higher in ungrazed areas in all three years (data reported as diversity indices). The densities of three of five species were higher in ungrazed than in grazed plots (long-tailed pocket mouse *Chaetodipus formosus*: 26 vs 6 animals/ha; Merriam's kangaroo rat *Dipodomys merriami*: 31 vs 13; southern grasshopper mouse *Onychomys torridus*: 3 vs 0 respectively). The densities of the other two species did not differ significantly between grazed and ungrazed plots (little pocket mouse *Perognathus longimembris*: 29 vs 30 animals/ha; deer mouse *Peromyscus maniculatus*: 1 vs 0). Two pairs of 65-ha plots were established in 1990 with one plot inside an area fenced since 1978–1979 and one outside, in an area grazed by sheep (grazing intensity not stated). Over five periods of four to six nights, in May 1990–March 1992, mammals were caught in 64 Sherman traps/plot, 10 m apart.

A replicated, site comparison study in 1994 in a desert site in California, USA (3) found that in areas where livestock were excluded, there were fewer black-tailed hares *Lepus californicus*, compared to in sheep-grazed unfenced areas that were also driven over by off-road vehicles. Fewer black-tailed hares were found in fenced plots (0–1.5 hares/survey; 11 droppings/m^2) compared to in unfenced plots (1–4 hares/survey; 22–31 droppings/m^2). Two 2.25-ha plots that were fenced in 1980 were compared to two plots that were grazed by sheep (and driven over by off-road vehicles). Sites were matched for environmental variables. Hare numbers were estimated in May and July 1994 by counting the number of hares seen on four 1.25-km-long transects and the number of droppings in sixty 40 × 50-cm sampling units in each plot.

A replicated, site comparison study in 1998–1999 of a riparian grassland area in Pennsylvania, USA (4) found that stream margins, fenced to exclude grazing livestock, had a higher species richness and abundance of small mammals than did unfenced margins. There were more species in fenced stream margins (4.4 species/site) than in unfenced margins (2.6 species/site). More small mammals overall were caught in fenced (21.2/site) than in unfenced (9.7/site) margins. Three species were sufficiently abundant to analyse individually. There were more individuals in fenced than unfenced margins for meadow voles *Microtus pennsylvanicus* (fenced: 8.0; unfenced: 5.3 individuals) and meadow jumping mouse *Zapus hudsonius* (fenced: 9.1; unfenced: 3.5 individuals). No significant difference was found for short-tailed shrew *Blarina brevicauda* (fenced: 3.8; unfenced: 2.4 individuals). Nine 100-m-long riparian margins, fenced one to two years previously, were compared with nine 100-m-long unfenced (cattle-grazed) riparian margins. Three types of small-mammal trap were operated continually throughout April–July in 1998–1999.

A replicated, randomized, controlled study in 1991–1994 in grassland and savanna in California, USA (5) found that excluding grazing livestock did not increase the number of California ground squirrel *Spermophilus beecheyi* burrows. Changes in the number of active ground squirrel burrows, relative to pre-experiment numbers, did not differ between ungrazed and grazed plots (60–100% vs 40–100% of pre-experiment numbers). The spatial distribution of active burrow entrances did not differ between ungrazed and grazed plots (2.6–3.4 vs 2.2–4.1 m between nearest burrows). Three sites, each with four plots, were studied. Half of plots were in grassland, and half were in savanna. Half had cattle-exclusion fencing and half were cattle-grazed from spring to summer. Three ground squirrel colonies were mapped in each plot in autumn 1991 (pre-experiment). Fencing was erected late in 1991 and burrows were further mapped in autumns of 1992–1994.

A replicated, site comparison study in 1999–2001 of a grassland area in Castilla y Lyón, Spain (6) found more small mammals in plots from which cattle were excluded, compared to grazed plots. More individual small mammals were caught in grazing exclusion plots (0–16 individuals/plot) than in grazed plots (0–3 individuals/plot). Three species of mammal were found; white-toothed shrew *Crocidura russula* (61.6% of captures), common vole *Microtus arvalis* (31.9%), and wood mouse *Apodemus sylvaticus* (6.5%). Six grazing exclusion plots (2–10 ha) were established in reforestation areas in grasslands grazed by 2–10 cattle/ha. These areas were reforested in 1990, but few planted trees survived. Eight live traps were placed in each of 22 trapping plots (11 inside and 11 outside cattle exclosures). Traps were operated for three consecutive nights during September–October 1999 and 2000 and in June 2000 and 2001.

A controlled study in 2005–2007 in open forest and scrubland at a site in Córdoba province, Spain (7) found more European rabbits *Oryctolagus cuniculus* in a plot that was fenced to exclude large herbivores and with artificial warrens and water provided, than in an unmanaged area. Interventions were all carried out in the same plot, so their relative effects could not be separated. Average rabbit pellet counts were higher in the plot where the interventions were deployed (first year: 0.33 pellets/m^2/day; second year: 1.08 pellets/m^2/day) than in the unmanaged plot (first year: 0.02 pellets/m^2/day; second year: 0.03 pellets/m^2/day). A 2-ha plot was fenced to exclude large herbivores in March 2005. Rabbits and predators could pass through the fence. Five artificial warrens were installed and water was provided at one place. No management was carried out in an otherwise similar plot. Rabbit density was determined by monthly counts of pellets, from March 2005 to March 2007, in 0.5-m^2 circles every 100 m along a 1-km transect in each plot.

A site comparison study in 1993–2007 on a shrubland site in South Australia, Australia (8) found that excluding cattle increased abundances of small mammals compared to high intensity cattle grazing but not to medium or low grazing intensities. The average number of small mammals/sample at ungrazed points (3.6 individuals) was higher than with intensive cattle grazing (1.7 individuals) but not higher than the numbers with medium-(5.0) or low-intensity cattle grazing (7.7). Species richness followed a similar pattern (ungrazed: 1.7 species; intensive grazing: 1.2 species; medium grazing: 1.7, low intensity grazing: 2.2 species). Livestock were fenced out from an approximately 9 × 9-km area in 1986. Small mammals were sampled using pitfall traps for a 10-day period in either December or January 1993–1996 and again in 2007. Five points were sampled inside the enclosure (ungrazed) with 13 outside (grazed). Cattle grazing intensity was determined by dung counts. Low intensity grazing was <12 dung/ha, medium grazing was 12–100 dung/ha and intensive grazing was >120 dung/ha.

A replicated, controlled study in 1998–2006 in sagebrush shrubland previously affected by wildfire in California, USA (9) found that excluding livestock did not alter the abundance of five small mammal species. Over eight years, abundance of San Joaquin antelope squirrel

Ammospermophilus nelson did not differ significantly between areas where livestock were excluded (4–38 animals/plot) and grazed areas (2–29 animals/plot). The same pattern was true for short nosed kangaroo rat *Dipodomys nitratoides nitratoides* (1–55 vs 3–58 animals/plot), Heermann's kangaroo rat *Dipdomys heermanni* (0–4 vs 0–22), giant kangaroo rat *Dipodomys ingens* (0–4 vs 0–3), and San Joaquin pocket mouse *Perognathus inornatus inornatus* (1–10 vs 1–17). Four 2.6-km² areas were grazed by cattle and four 25-ha areas were fenced to exclude livestock. To estimate antelope squirrel abundance, 64 traps, baited with oats, at 40-m intervals, were established in each plot. To estimate abundance of other small mammals, 144 traps, baited with bird seed, were established in each plot at 10-m intervals. Traps were set for six consecutive days and nights in July–September 1998–2006.

(1) Roberts T.A. & Tiller R.L. (1985) Mule deer and cattle responses to a prescribed burn. *Wildlife Society Bulletin*, 13, 248–252.

(2) Brooks M.L. (1995) Benefits of protective fencing to plant and rodent communities of the western Mojave Desert, California. *Environmental Management*, 19, 65–74, https://doi.org/10.1007/bf02472004

(3) Brooks M. (1999) Effects of protective fencing on birds, lizards, and black-tailed hares in the western Mojave Desert. *Environmental Management*, 23, 387–400, https://doi.org/10.1007/s002679900194

(4) Giuliano W.M. & Homyack J.D. (2004) Short-term grazing exclusion effects on riparian small mammal communities. *Journal of Range Management*, 57, 346–350, https://doi.org/10.2111/1551-5028(2004)057[0346:sgeeor]2.0.co;2

(5) Fehmi J.S., Russo S.E. & Bartolome J.W. (2005) The effects of livestock on California ground squirrels (*Spermophilus beecheyi*). *Rangeland Ecology & Management*, 58, 352–359, https://doi.org/10.2458/azu_rangelands_v58i4_bartolome

(6) Torre I., Diaz M., Martínez-Padilla J., Bonal R., Vinuela J. & Fargallo J.A. (2007) Cattle grazing, raptor abundance and small mammal communities in Mediterranean grasslands. *Basic and Applied Ecology*, 8, 565–575, https://doi.org/10.1016/j.baae.2006.09.016

(7) Catalán I., Rodríguez-Hidalgo P. & Tortosa F.S. (2008) Is habitat management an effective tool for wild rabbit (Oryctolagus cuniculus) population reinforcement? *European Journal of Wildlife Research*, 54, 449–453, https://doi.org/10.1007/s10344-007-0169-0

(8) Read J.L. & Cunningham R. (2010) Relative impacts of cattle grazing and feral animals on an Australian arid zone reptile and small mammal assemblage. *Austral Ecology*, 35, 314–324, https://doi.org/10.1111/j.1442-9993.2009.02040.x

(9) Germano D.J., Rathbun G.B. & Saslaw L.R. (2012) Effects of grazing and invasive grasses on desert vertebrates in California. *The Journal of Wildlife Management*, 76, 670–682, https://doi.org/10.1002/jwmg.316

3.20. Reduce intensity of grazing by domestic livestock

https://www.conservationevidence.com/actions/2408

- **Thirteen studies** evaluated the effects on mammals of reducing the intensity of grazing by domestic livestock. Six studies were in the USA[1,2,3a,3b,9,10], six were in Europe[4,5,7,8,11,12] and one was in China[6].

COMMUNITY RESPONSE (3 STUDIES)

- **Richness/diversity (3 studies):** Two of three site comparison or controlled studies, in the USA[3a,3b] and Norway[12], found that reduced livestock grazing intensity was associated with increased species richness of small mammals[3b,12] whilst one study did not find an increase in species richness[3a].

POPULATION RESPONSE (13 STUDIES)

- **Abundance (13 studies):** Six of nine site comparison or controlled studies (including seven replicated studies), in the USA[2,3a,3b,9], Denmark[4], the UK[5], China[6], Netherlands[11] and Norway[12], found that reductions in livestock grazing intensity were associated with increases in abundances (or proxies of abundances) of small mammals[2,3b,4,5,9,11], whilst two studies showed no significant impact of reducing grazing intensity[3a,12] and one study showed mixed results for different species[6]. Two replicated studies (including one controlled and one site comparison study), in the UK[7] and in a range of European countries[8], found that reducing grazing intensity did not increase numbers of Irish hares[7] or European hares[8]. A controlled, before-and-after study, in the USA[1] found that exclusion of cattle grazing was associated with higher numbers of elk and mule deer. A replicated, site comparison study in the USA[10] found that an absence of cattle grazing was associated with higher numbers of North American beavers.

BEHAVIOUR (0 STUDIES)

Background

Overgrazing is responsible for the degradation of habitats across the world, being especially damaging in arid environments, where the removal of vegetation can quickly lead to soil erosion. Reducing grazing intensity may reduce the damage to vegetation and can also help reduce disturbance to mammals and accidental loss of nests of small mammal species.

A controlled, before-and-after study in 1981–1982 in a forest and meadow mosaic in Arizona, USA (1) found that an absence of cattle grazing was associated with higher numbers of elk *Cervus canadensis* and mule deer *Odocoileus hemionus*. There were 0.13 elk/km counted on transects in absence of cattle grazing and 0.01/km after grazing commenced whereas, concurrently, on a continually ungrazed pasture, 0.21 and 0.50 elk/km respectively were counted. The number of mule deer counted on transects fell from 0.07/km in absence of grazing to 0.00/km after grazing commenced whereas 0.02 mule deer/km were counted on a continually ungrazed pasture during both time periods. The 135 km²-study area was divided into two pastures. One was ungrazed in both years. The other was ungrazed in 1981 and stocked with cattle, at a rate of one animal unit (equivalent to a cow and suckling calf)/3 ha in May–July 1982. Elk and mule deer were counted in July and August, along a 48-km driving transect, 20 times in 1981 and 14 times in 1982.

A site comparison study in 1981–1983 on a grassland ranch in Arizona, USA (2) found that reducing grazing intensity by excluding livestock increased rodent abundance. More rodents were caught in an ungrazed area (428 individuals) than in a grazed area (328 individuals). This was the case for hispid pocket mouse *Perognathus hispidus* (38 vs 16 individuals), western harvest mouse *Reithrodonromys megalotis* (26 vs 4), white-footed mouse *Peromyscus leucopus* (45 vs 24), southern grasshopper mouse *Onychomys torridus* (42 vs 8) and hispid cotton rat *Sigmodon hispidus* (118 vs 49). Merriam's kangaroo rat *Dipodomys merriami* was less abundant in the ungrazed than the grazed area (5 vs 92 individuals). Silky pocket mouse *Perognathus flavus* abundance did not differ significantly between ungrazed and grazed areas (8 vs 5 individuals) and nor did deer mouse *Peromyscus maniculatus* abundance (146 vs 130). Livestock were fenced out of part of a 300-ha study area from 1968 onwards. The grazed part was stocked with approximately one cow/10 ha. Rodents were live-trapped, from two hours before sunset to two hours after sunrise, on 71 occasions, from July 1981 to January 1983.

A replicated, site comparison study in 1989–1991 of shrub grassland in a national park in Utah, USA (3a) found that reducing grazing intensity by excluding cattle from small enclosures did not increase small mammal abundance or species richness. Small mammal abundance in ungrazed enclosures (1.9 individuals/100 trap-nights) did not significantly differ from that in grazed areas (2.3 individuals/100 trap-nights). Small mammal species richness in enclosures (1.5 species/trap grid) did not significantly differ from that in grazed areas (1.6 species/trap grid). Cattle were excluded from four enclosures, three for six years prior to the study and one for 38 years. Enclosures measured 0.1–0.8 ha. Grazing outside enclosures was by 1,500 Animal Units (equivalent to a cow and suckling calf) across 35,499 ha in October–May. Small mammals were sampled in grids of Sherman live traps, one grid inside each enclosure. An identical grid was sampled simultaneously >500 m away from each enclosure. Grids were trapped for four consecutive days, between 1 May and 31 June. Three enclosures were sampled annually in 1989–1991, and one in 1990–1991.

A replicated, site comparison study in 1990 of shrub grassland at eight sites in two national parks in Utah, USA (3b) found that reducing grazing intensity by excluding cattle from areas of grassland increased small mammal abundance and species richness. Small mammal abundance in ungrazed sites (1.8 individuals/100 trap-nights) was higher than in grazed sites (1.0 individuals/100 trap-nights). Small mammal species richness in ungrazed sites (1.5 species/site) was higher than in grazed sites (1.0 species/site). Eight sites were sampled; four ungrazed for ≥30 years and four in a region grazed by 1,500 Animal Units (equivalent to a cow and suckling calf) across 35,499 ha in October–May. All sites were on large (≥ 100 ha) areas of shrub-grassland and were selected to match geological and soil characteristics. Each site was sampled using a grid of Sherman live traps, for four consecutive days, between 1 May and 31 June 1990.

A replicated, controlled study in 1998–2000 of pasture at a site in Denmark (4) found that in plots with reduced livestock grazing intensity, small mammal biomass was higher. Small mammal biomass peaks across the study in each of two plots/treatment were higher in ungrazed plots (287–959 g), intermediate in low-intensity sheep plots (251–801 g) and lowest in high-intensity cattle plots (64–195 g). The estimated population of field voles *Microtus agrestis* (the most abundant species recorded) was higher each year in ungrazed plots (29–94/plot) than in high-intensity cattle plots (3–27/plot), but was higher still in low-intensity sheep plots in two of three years (32–63/plot). In 1997, two meadows were divided into 70 × 300-m pens. One plot on each meadow was assigned to high-intensity cattle grazing (4.8 steers/ha), one to low intensity sheep grazing (4.5 ewes plus lambs/ha) and one was ungrazed. Grazing occurred from mid-May to mid-October, though was prevented on half of each pen until after hay cutting (late-June to early-July). The delayed grazing part was reversed the following year. Small mammals were live-trapped over three days and nights, every four weeks, over 31 trapping sessions, from June 1998 to October 2000.

A replicated, randomized, paired sites, controlled, before-and-after study in 2002–2004 on upland grassland in Scotland, UK (5) found that reducing sheep grazing intensity increased the abundance of field voles *Microtus agrestis*. In the first year of grazing treatments, the percentage of quadrats with vole signs was higher in ungrazed plots (20%), intermediate in lightly grazed plots (12%) and lowest in heavily grazed plots (4%). The same pattern held in the second year of treatments (ungrazed: 24%; lightly grazed: 11%; heavily grazed: 7%). Before grazing treatments were implemented, there was no significant difference in the frequency of vole signs between plots. Plots were all grazed similarly (stocking rate not stated) up to 2002. From spring 2003, there were six replicates (3.3 ha each) of no livestock grazing, light grazing (three ewes/plot) and heavy grazing (nine ewes/plot). Five 25 × 25-cm quadrats at each of five points/plot were searched for vole signs in April and October 2002–2004.

A replicated, site comparison study in 2001 and 2002 on two winter pasture areas in Sichuan, China (6) found that reduced livestock grazing intensity was associated with higher numbers of the tundra/lacustrine vole *Microtus oeconomus/limnophilus* complex but with lower numbers of Kam dwarf hamster *Cricetulus kamensis*. The numbers of tundra/lacustrine voles in low grazing intensity areas (7 individuals/100 trap nights) was higher than in medium (1/100 trap nights) or high grazing intensity areas (0/100 trap nights). The numbers of Kam dwarf hamster in low (0 individuals/100 trap night) and medium grazing intensity areas (0/100 trap nights) was lower than that in high grazing intensity areas (6/100 trap nights). Surveys were conducted in grassland and shrub areas in valley, wetland and slope habitats in winter pasture at 4,250 m altitude. Sites were grazed, in varying intensities, by yaks, sheep, goats, and horses, each October to early May. Small mammals were surveyed using back-break traps over three nights and days in July 2001 and July 2002.

A replicated, site comparison study in 2005 on 200 plots covering a range of agricultural habitats in Northern Ireland, UK (7) found that reducing grazing intensity as part of a wider suite of agri-environment measures did not increase numbers of Irish hares *Lepus timidus hibernicus*. The effects of reducing grazing intensity cannot be separated from those of other agri-environment measures, which included retaining and enhancing field boundary features and managing nutrient systems. Hare abundance in agri-environment plots (0.45 hares/km transect) did not significantly differ from that in non-agri-environment plots (0.41 hares/km transect). One hundred and fifty 1-km² plots, on land that was enrolled into an agri-environment scheme 10–17 years previously, were selected along with 50 non-enrolled 1-km² plots, chosen to match enrolled plots for landscape characteristics. Hares were surveyed at night, in mid-winter, by spotlighting from a vehicle.

A replicated, randomized, controlled study in 2002–2004 on grassland in France, Germany, Italy and the UK (8) found that areas with low livestock grazing intensities did not have more European hares *Lepus europaeus* than did areas with moderate livestock grazing intensities. Too few hares were recorded to enable statistical analyses. At the UK site, though, where most hares were recorded, numbers were similar between low intensity (14 hares) and moderate intensity (12 hares) grazing areas. Sites were grazed by the cattle Charolais × Fresian in the UK, Simmental in Germany and Charolais in France and by

Finnish Romanov sheep in Italy. Grazing rates differed, but low grazing intensity was 0.3–0.4 fewer animals/ha than moderate grazing intensity. There were three each of low and moderate intensity grazing paddocks (paddock size 0.4–3.6 ha) at one site in each of the four countries. Hares were counted every two weeks in early morning, from May to October, 2002–2004, during seven minutes of observation and whilst walking a transect in each paddock.

A controlled study in 2008 of a grassland and woodland site in Nevada, USA (9) found that reducing grazing intensity by long-term exclusion of domestic livestock resulted in a higher species richness and abundance of small mammals. More small mammal species were recorded on ungrazed land (six) than on grazed land (four). Small mammal abundance on ungrazed land (0.08 animals/trap night) was higher than on grazed land (0.05 animals/trap night). Three species were caught in sufficient quantities for individual analyses. The Great Basin pocket mouse *Perognathus parvus* was more abundant on ungrazed than grazed land (0.05 vs 0.02 individuals/trap night) as was western jumping mouse *Zapus princeps* (0.02 vs 0.00 individuals/trap night). Deer mice *Peromyscus maniculatus* showed no preference (0.01 vs 0.01 individuals/trap night). Sampling occurred in a 10-ha enclosure, characterised by mixed shrubs and trees, from which domestic livestock were excluded at least 50 years previously and in a similar sized, adjacent cattle-grazed grassland. Small mammals were sampled using lines of snap-traps, over three or four nights, in July 2008.

A replicated, site comparison study in 2013 in a forested area in New Mexico, USA (10) found that an absence of cattle grazing was associated with higher numbers of North American beavers *Castor canadensis*. The relative frequency of beaver dams was higher in the absence of cattle grazing than where cattle grazing was present (data presented as odds ratios). Data were collected along 57 sections of river, each 200 m long, of which 29 had beaver dams and 28 did not have beaver dams, though physical conditions were suitable for their construction. Field data were collected between 15 May and 15 August 2013. Livestock grazing was assessed by collating information on grazing consents and by surveying ungulate faeces.

A replicated, randomized, paired sites, controlled study in 2010–2013 on a coastal salt marsh in the Netherlands (11) found that plots grazed at lower intensity contained more signs of vole *Microtus* spp. presence than did plots grazed at higher intensity. After four years, a greater proportion of surveyed quadrats contained signs of vole presence in plots grazed at lower intensity than in plots grazed at high intensity (data not reported). Twelve plots were established (in three sets of four) on a historically grazed salt marsh. From 2010, six plots (two random plots/set) were grazed at each intensity: low (0.5 animals/ha) or high (1.0 animal/ha). Grazing occurred in summer (June–October) only. Half of the plots were grazed by cows and half by horses. In October 2013, sixty quadrats (2 m²) were surveyed in the higher elevations of each plot for signs of vole presence (runways, fresh plant fragments or faecal pellets). Some flooded quadrats were excluded from the analysis.

A randomized, replicated, controlled study in 2002–2005 at two heathland sites in Norway (12) found that excluding livestock with fences did not significantly change abundances of field voles *Microtus agrestis*. The number of animals trapped in plots that were fenced to exclude livestock did not differ significantly (6 animals/plot) from that in plots that were not fenced to exclude livestock (4 animals/plot). In 2002, at two sites, four 50 × 50-m plots were fenced to exclude livestock and four plots were not fenced. Sheep density prior to fencing was 32–48 sheep/ha. In June and August 2003–2005, thirty-six live traps baited with sunflower seeds and peanuts and with wool for bedding were placed in each plot and checked twice daily for five days. Captured animals were individually marked and released.

(1) Wallace M.C. & Krausman P.R. (1987) Elk, mule deer, and cattle habitats in Central Arizona. *Journal of Range Management*, 40, 80–83, https://doi. org/10.2307/3899367

(2) Bock C.E., Bock J.H., Kenney W.R. & Hawthorne V.M. (1984) Responses of birds, rodents, and vegetation to livestock exclosure in a semidesert grassland site. *Journal of Range Management*, 37, 239–242, https://doi. org/10.2307/3899146

(3) Rosenstock S.S. (1996) Shrub-grassland small mammal and vegetation responses to rest from grazing. *Journal of Range Management*, 49, 199–203.

(4) Schmidt N.M., Olsen H., Bildsøe M., Sluydts V. & Leirs H. (2005) Effects of grazing intensity on small mammal population ecology in wet meadows. *Basic and Applied Ecology*, 6, 57–66, https://doi.org/10.1016/j.baae.2004.09.009

(5) Evans D.M., Redpath S.M., Elston D.A., Evans S.A., Mitchell R.J. & Dennis P. (2006) To graze or not to graze? Sheep, voles, forestry and nature conservation in the British uplands. *Journal of Applied Ecology*, 43, 499–505, https://doi.org/10.1111/j.1365-2664.2006.01158.x

(6) Raoul F., Quére J-P., Rieffel D., Bernard N., Takahashi K., Scheifler R., Ito A., Wang Q., Qiu J., Yang W., Craig P.S. & Giraudoux P. (2006) Distribution of small mammals in a pastoral landscape of the Tibetan plateaus (Western Sichuan, China) and relationship with grazing practices. *Mammalia*, 70, 214–225, https://doi.org/10.1515/mamm.2006.042

(7) Reid N., McDonald R.A. & Montgomery W.I. (2007) Mammals and agri-environment schemes: hare haven or pest paradise? *Journal of Applied Ecology*, 44, 1200–1208, https://doi.org/10.1111/j.1365-2664.2007.01336.x

(8) Wallis De Vries M.F., Parkinson A.E., Dulphy J.P., Sayer M. & Diana E. (2007) Effects of livestock breed and grazing intensity on biodiversity and production in grazing systems. 4. Effects on animal diversity. *Grass and Forage Science*, 62, 185–197, https://doi.org/10.1111/j.1365-2494.2007.00568.x

(9) Rickart E.A., Bienek K.G. & Rowe R.J. (2013) Impact of livestock grazing on plant and small mammal communities in the Ruby Mountains, northeastern Nevada. *Western North American Naturalist*, 73, 505–515, https://doi.org/10.3398/064.073.0403

(10) Small B.A., Frey J.K. & Gard C.C. (2016) Livestock grazing limits beaver restoration in northern New Mexico. *Restoration Ecology*, 24, 646–655, https://doi.org/10.1111/rec.12364

(11) van Klink R., Nolte S., Mandema F.S., Lagendijk D.D.G., Wallis De Vries M.F., Bakker J.P., Esselink P. & Smit C. (2016) Effects of grazing management on biodiversity across trophic levels — the importance of livestock species and stocking density in salt marshes. *Agriculture, Ecosystems & Environment*, 235, 329–339, https://doi.org/10.1016/j.agee.2016.11.001

(12) Spirito F., Rowland M., Nielson R., Wisdom M. & Tabeni S. (2017) Influence of grazing management on resource selection by a small mammal in a temperate desert of South America. *Journal of Mammalogy*, 98, 1768–1779, https://doi.org/10.1093/jmammal/gyx106

3.21. Use livestock fences that are permeable to wildlife

https://www.conservationevidence.com/actions/2409

- **Two studies** evaluated the effects on target mammals of using livestock fences that are permeable to wildlife. Both studies were in the USA[1,2].

COMMUNITY RESPONSE (0 STUDIES)

POPULATION RESPONSE (0 STUDIES)

BEHAVIOUR (2 STUDIES)

- **Use (2 studies):** A study in the USA[1] found that wild ungulates crossed a triangular cross-section fence with varying success rates. A replicated, controlled study in the USA[2] found that fences with a lowered top wire were crossed more by elk than were conventional fences.

Background

Fences erected to retain domestic livestock or, in some cases, exclude wild herbivores or carnivores may also act as barriers to non-target species. Fence designs may be adapted to permit crossings and, thus, retain habitat connectivity for specific species. Fence designs are likely to vary between different situations, depending on the nature of the original fence and the species being targeted for continued access.

See also *Install mammal crossing points along fences on farmland.*

A study in 1988–1989 of shrubland and grassland along a national park boundary in Montana, USA (1) found that wild ungulates crossed a fence with a triangular cross-section (buck-and-pole fence) with varying success rates. Fence crossing success rates (away from gates) were mule deer *Odocoileus hemionus*: 85% of fence approaches, pronghorn *Antilocapra americana*: 72%, bison *Bison bison*: 46%, elk *Cervus canadensis*: 17%. Most bison crossings were achieved by damaging the fence. Other animals were generally able to pass through or below it. Some animals that did not cross the fence walked along until they found an open gate. The fence was 3.8 km long, had a width at the bottom of 165–175 cm and narrowed to a point at a height of 165–185 cm. Four rails were set on a slope on one side (the lowest being 25–59 cm above the ground). The other side comprised a single rail, 65–85 cm above the ground. Animal crossings were monitored by identifying tracks in snow, 10.5–109 hours after storms, on eight occasions from 5 January to 8 March 1988 and eight occasions from 16 November 1988 to 14 March 1989.

A replicated, controlled study in 1994 on a grassland site in New Mexico, USA (2) found that fences with a lowered top wire were crossed more by elk *Cersus elaphus* than were conventional fences. Of 10 fence designs trialled, two were crossed significantly more frequently than were conventional 100-cm high fences comprising four barbed wires. The two designs crossed most both involved lowering the top wire and fastening it to the second wire down, 80 cm above the ground. One also had the third wire attached to the bottom wire. These fences were crossed 4.6 and 4.3 times/day respectively. Conventional fences were crossed 2.3 times/day. No livestock escapes occurred during the trial. Fence sections, 15 m long, with 6–9 replicates of each design, were monitored for 21 days in late July–September 1994. Fence crossings were confirmed by presence of tracks and by breaks in a thread above the fence.

(1) Scott M.D. (1992) Buck-and-pole fence crossings by 4 ungulate species. *Wildlife Society Bulletin*, 20, 204–210.

(2) Knight J.E., Swensson E.J. & Sherwood H. (1997) Elk use of modified fence-crossing designs. *Wildlife Society Bulletin*, 25, 819–822.

3.22. Install mammal crossing points along fences on farmland

https://www.conservationevidence.com/actions/2410

- **Four studies** evaluated the effects on mammals of installing mammal crossing points along fences on farmland. Two studies were in Namibia[2,4] and one each was in the USA[1] and the UK[3].

COMMUNITY RESPONSE (0 STUDIES)

POPULATION RESPONSE (0 STUDIES)

BEHAVIOUR (4 STUDIES)

- **Use (4 studies):** A study in the USA[1] found that pronghorn antelopes crossed a modified cattle grid which prevented escape of domestic sheep and cows. A controlled, before-and-after study in Namibia[2] found installing swing gates through game fencing reduced the digging of holes by animals under the fence, whilst preventing large predator entry. A study in the UK[3] found that a vertical-sided ditch under an electric fence allowed access by otters. A before-and-after study in Namibia[4] found that tyres installed as crossings through fences were used by wild mammals and reduced fence maintenance requirements.

Background

Fences erected to retain domestic livestock or, in some cases, exclude wild herbivores or carnivores may also act as a barrier to non-target species. Crossings may be installed to retain habitat connectivity for specific species. Crossing designs vary between different situations depending on the nature of the original fence and the species being targeted for continued access.

For wildlife-permeable fencing (as opposed to specific crossing points) see *Use livestock fences that are permeable to wildlife.*

A study in 1965 of grassland at a site in Wyoming, USA (1) found that a modified pass based on a cattle grid design enabled passage by pronghorn antelopes *Antilocapra americana* whilst preventing escape of domestic sheep and cows. A total of 100 antelope were observed jumping across the grills, during five separate crossing events. Antelopes crossed grills at fence corners more than they crossed those along straight fences. A range of designs were trailed, the optimal being a 6-foot-long grill in a 5.5-foot-wide fence opening. The grill consisted of 13 bars at 6 inch-intervals. These were mounted on 10-inch-high timbers with earth ramps running up to both ends.

A controlled, before-and-after study in 2001–2002 on a game and livestock farm in Otjiwarongo district, Namibia (2) found that installing swing gates along animal routes in game fencing reduced the digging of holes by animals under the fence, whilst preventing large predator entry. Fewer holes were dug under a fence section with gates installed on animal routes (12.2 holes/survey) than on sections with evenly spaced gates (20.2 holes/survey) or no gates (19.1 holes/survey). Before gate installation, there was no significant difference in hole numbers between sections (animal route gates: 20.0 holes/survey; evenly spaced gates: 25.7 holes/survey; no gates: 21.7 holes/survey). Warthogs *Phacochoerus aethiopicus* were the most frequent gate users. Jackals *Canis mesomelas*, cheetahs *Acinonyx jubatus* and leopards *Panthera pardus* passed through holes but not the gates. A game fence (4,800 m long) was divided into three equal sections. One had six gates on established animal routes, one had eight evenly spaced gates and one had no gates. Swing gates comprised a metal frame (45 × 30 cm) covered with galvanised fencing (75-mm mesh). Holes were surveyed and filled at 3–15-day intervals, from August 2001 to April 2002. Animals were identified by signs and heat sensitive cameras.

A study in 2005 at a wetland reserve in Cambridgeshire, UK (3) found that a vertical-sided ditch under an electric fence allowed access to the site by otters *Lutra lutra*. Several otter spraints were found within the fenced area. Some were at the edge of the ditch under the fence, indicating probable otter use of that route. No evidence of red foxes *Vulpes vulpes* using the route was identified. The ditch, 1 m deep and 3 m wide, flowed under the boundary of the fenced reserve. Ditch sides were supported by wooden boards, to maintain the banks as vertical, so that entry could only be achieved by swimming. The fence, 1.3 m high and 2 km long, was electrified year-round. It was installed in 2005 to deter entry by foxes, for the purpose of reducing predation on nesting birds.

A before-and-after study in 2010 on a farm in Namibia (4) found that tyres installed as passageways through fences facilitated movements of wild mammals, especially carnivores, and reduced fence maintenance requirements. During 96 days, 11 mammal species, including nine carnivores, used one crossing. The most frequently recorded species were black-backed jackal *Canis mesomelas* (44 occasions), porcupine *Hystrix africaeaustralis* (21 occasions) and cheetah *Acinonyx jubatus* (nine occasions, seven different animals). Fewer fence holes needed mending after tyre installation (13.6 holes/day) than before (31.3 holes/day). Forty-nine discarded car tyres (37 cm radius opening) were installed at ground level into a 19.1-km-long, 2.4-m-high fence. Tyre locations, 35–907 m apart, were prioritised to areas of high warthog *Phacochoerus africanus* digging activity. One tyre was monitored with a camera trap for 96 days from August–December 2010. Holes needing maintenance were counted for 10 days before and 10 days after tyre installation.

(1) Mapston R.D., Zobell R.S., Winter K.B. & Dooley W.D. (1970) A pass for antelope in sheep-tight fences. *Journal of Range Management*, 23, 457–459, https://doi.org/10.2307/3896324

(2) Schumann M., Schumann B., Dickman A., Watson L.H. & Marker L. (2006) Assessing the use of swing gates in game fences as a potential non-lethal predator exclusion technique. *South African Journal of Wildlife Research*, 36, 173–181.

(3) Gulickx M.M.C., Beecroft R.C. & Green A.C. (2007) Creation of a 'water pathway' for otters *Lutra lutra*, under an electric fence at Kingfishers Bridge, Cambridgeshire, England. *Conservation Evidence*, 4, 28–29.

(4) Weise F.J., Wessels Q., Munro S. & Solberg M. (2014) Using artificial passageways to facilitate the movement of wildlife on Namibian farmland. *South African Journal of Wildlife Research*, 44, 161–166, https://doi.org/10.3957/056.044.0213

3.23. Use traditional breeds of livestock

https://www.conservationevidence.com/actions/2411

- **One study** evaluated the effects of using traditional breeds of livestock on wild mammals. This study was carried out in four European countries[1].

COMMUNITY RESPONSE (0 STUDIES)

POPULATION RESPONSE (0 STUDIES)

BEHAVIOUR (1 STUDY)

- **Use (1 study):** A replicated, randomized, controlled study in Europe[1] found that European hares did not use areas grazed by traditional livestock breeds more than they used areas grazed by commercial breeds.

Background

Traditional livestock breeds are often suggested to help enhance biodiversity, though motivations for doing so are often little studied and rely on anecdotal evidence (Rook *et al.* 2004).

Rook A.J., Dumont B., Isselstein J., Osoro K., Wallis De Vriese M.F. Parente G. & Mills J. (2004) Matching type of livestock to desired biodiversity outcomes in pastures — a review. *Biological Conservation*, 119, 137–150, https://doi.org/10.1016/j.biocon.2003.11.010

A replicated, randomized, controlled study in 2002–2004 on grassland in France, Germany, Italy and the UK (1) found that areas grazed by traditional livestock breeds did not have more European hares *Lepus europaeus* than did areas grazed by commercial breeds. Too few hares were recorded to enable statistical analyses. At the UK site, where most hares were recorded, numbers were similar between areas grazed by traditional breeds (15 hares) and commercial breeds (14 hares). Traditional cattle breeds were Devon, German Angus and Salers, compared with commercial Charolais × Fresian, Simmental and Charolais, in the UK, Germany and France respectively. In Italy traditional Karst sheep were compared with commercial Finnish Romanovs. There were three traditional breed paddocks and three commercial breed paddocks (paddock size 0.4–3.6 ha) at single sites in each of the four countries. Hares were counted every two weeks in early morning, from May to October of 2002–2004, during seven minutes of observation and by walking a transect in each paddock.

(1) Wallis De Vries M.F., Parkinson A.E., Dulphy J.P., Sayer M. & Diana E. (2007) Effects of livestock breed and grazing intensity on biodiversity and production in grazing systems. 4. Effects on animal diversity. *Grass and Forage Science*, 62, 185–197, https://doi.org/10.1111/j.1365-2494.2007.00571.x

3.24. Change type of livestock

https://www.conservationevidence.com/actions/2412

- **Two studies** evaluated the effect of changing type of livestock on mammals. One study was in the UK[1] and one was in the Netherlands[2].

COMMUNITY RESPONSE (0 STUDIES)

POPULATION RESPONSE (2 STUDIES)

- **Abundance (2 studies):** One replicated, randomized, paired sites, controlled, before-and-after study in the UK[1] found that sheep and cattle grazing increased field vole abundance relative to sheep-only grazing. One replicated, randomized, paired sites study in the Netherlands[2] found that cattle grazing increased vole abundance relative to horse grazing.

BEHAVIOUR (0 STUDIES)

Background

Domestic herbivores differ in the way that they graze. In particular, some species are more selective than others, and so will concentrate grazing in areas with highly palatable plant species. This may generate different effects on vegetation dynamics than does grazing by more generalist herbivores (Evans *et al.* 2015). Furthermore, large herbivores, such as cattle, may disturb the ground more through their footprints than is the case for smaller grazers, such as sheep. Such effects may produce a vegetation sward and structure than is more or less suited for wild mammals.

Evans D.M., Villar N., Littlewood N.A., Pakeman R.J., Evans S.A., Dennis P., Skartveit J. & Redpath S.M. (2015) The cascading impacts of livestock grazing in upland ecosystems: a 10-year experiment. *Ecosphere*, 6, article 42, https://doi.org/10.1890/ES14-00316.1

A replicated, randomized, paired sites, controlled, before-and-after study in 2002–2004 on an upland grassland site in Scotland, UK (1) found that, after two years, grazing with sheep and cattle increased field vole *Microtus agrestis* abundance relative to sheep-only grazing. In the first year of the experiment, a similar proportion of quadrats had signs of voles in sheep and cattle plots (11%) and sheep only plots (12%). In the second year, the proportion with vole signs was higher in sheep and cattle (16%) than sheep only plots (11%). Before the experiment began, there was no difference in the frequency of vole signs between plots. Plots were grazed similarly up to 2002 (rate not stated). From 2003, there were six replicates (each 3.3 ha) of sheep and cattle grazing (two ewes/plot and, for four weeks/year, two cattle each with a suckling calf) and sheep only grazing (three ewes/plot). Treatments were designed to have similar overall grazing intensity. Five 25 cm × 25 cm quadrats at each of five points in each plot were searched for vole signs in April and October of 2002–2004.

A replicated, randomized, paired sites, controlled study in 2010–2013 on a coastal salt marsh in the Netherlands (2) found that plots grazed by cattle contained more signs of vole *Microtus* spp. presence than did plots grazed by horses. After four years, a greater proportion of surveyed quadrats contained signs of vole presence in plots grazed by cattle than in plots grazed by horses (data not reported). Twelve plots were established (in three sets of four plots) on a grazed salt marsh. From 2010, six plots (two random plots/set) were grazed by each livestock type: cows (600 kg) or horses (700 kg). Grazing occurred in summer (June–October) only. Half of the plots were grazed at high intensity (1.0 animal/ha) and half were grazed at low intensity (0.5 animals/ha). In October 2013, sixty quadrats (2 m²) were surveyed in the higher elevations of each plot for signs of vole presence (runways, fresh plant fragments or faecal pellets). Some flooded quadrats were excluded from analyses.

(1) Evans D.M., Redpath S.M., Elston D.A., Evans S.A., Mitchell R.J. & Dennis P. (2006) To graze or not to graze? Sheep, voles, forestry and nature conservation in the British uplands. *Journal of Applied Ecology*, 43, 499–505, https://doi.org/10.1111/j.1365-2664.2006.01158.x

(2) van Klink R., Nolte S., Mandema F.S., Lagendijk D.D.G., Wallis De Vries M.F., Bakker J.P., Esselink P. & Smit C. (2016) Effects of grazing management on biodiversity across trophic levels — the importance of livestock species and stocking density in salt marshes. *Agriculture, Ecosystems & Environment*, 235, 329–339, https://doi.org/10.1016/j.agee.2016.11.001

Reduce human-wildlife conflict

3.25. Relocate local pastoralist communities to reduce human-wildlife conflict

https://www.conservationevidence.com/actions/2413

- **One study** evaluated the effects on mammals of relocating local pastoralists to reduce human-wildlife conflict. This study was in India[1].

COMMUNITY RESPONSE (0 STUDIES)

POPULATION RESPONSE (1 STUDY)

- **Abundance (1 study):** A study in India[1] found that after most pastoralists were relocated outside of an area, Asiatic lion numbers increased.

BEHAVIOUR (0 STUDIES)

Background

Species conservation can conflict with interests of local communities that own and manage grazing livestock. An intervention occasionally enacted is to relocate pastoralist communities to areas further away from the threatened species.

A study in 1974–2010 of forest and savanna in one area in Gujarat, India (1) found that after most pastoralists were relocated outside of the area, Asiatic lion *Panthera leo persica* numbers increased. The lion population increased during the study period from 180 in 1974 to 411 individuals 36 years later. This coincided with increased abundance of wild ungulates from 5,600 individuals prior to the start of the study, in 1969–1970, to 64,850 individuals in 2010. Scat analysis showed that domestic livestock formed 75% of lions' diets four years before the main study period which fell to 25% at the end of the study. A wildlife sanctuary was created in 1965 and was expanded and declared a National Park in 1975. Four further areas were protected between 1989 and 2007. Three core protected areas covered 1,452 km². Over two thirds of indigenous pastoral Maldharis and their livestock were relocated from the area, commencing in 1972. The number of domestic buffalo and cattle in the protected areas fell from 24,250 animals in the 1970s to 12,500 in the mid-1980s but then increased to 23,440 in 2010. Lions were visually surveyed at 5–6-year intervals, from 1974–2010.

(1) Singh H.S. & Gibson L. (2011) A conservation success story in the otherwise dire megafauna extinction crisis: The Asiatic lion (*Panthera leo persica*) of Gir forest. *Biological Conservation*, 144, 1753–1757, https://doi.org/10.1016/j.biocon.2011.02.009

3.26. Pay farmers to compensate for losses due to predators/wild herbivores to reduce human-wildlife conflict

https://www.conservationevidence.com/actions/2414

- **Five studies** evaluated the effects on mammals of paying farmers compensation for losses due to predators or wild herbivores to reduce human-wildlife conflict. Three studies were in Kenya[1,3,5] and one each was in Italy[2] and Sweden[4].

COMMUNITY RESPONSE (0 STUDIES)

POPULATION RESPONSE (5 STUDIES)

- **Abundance (2 studies):** Two studies, in Italy[2] and Sweden[4], found that compensating livestock owners for losses to predators led to increasing populations of wolves[2] and wolverines[4].

- **Survival (3 studies):** Three before-and-after studies (including two replicated studies), in Kenya[1,3,5], found that when pastoralists were compensated for livestock killings by predators, fewer lions were killed.

BEHAVIOUR (0 STUDIES)

Background

Where farmers suffer losses to wild mammals, either through predation of livestock or damage to crops, they may carry out lethal control of those mammals. Compensation schemes provide payments for losses to wild mammals and can have certain conditions, such as cessation of using lethal control or improving animal husbandry to reduce losses. The intervention includes schemes that make payments linked directly to losses (e.g. paying for each animal predated) and schemes that where payment is not linked directly to losses but instead to other mechanism that reduce incentives for killing wild mammals.

A before-and-after, site comparison study in 2001–2006 on a group ranch in Kajiado District, Kenya (1) found that compensating pastoralists for livestock predated by lions *Panthera leo* reduced the number of lions that pastoralists killed. Fewer lions were killed after the compensation fund commenced (five in 2003–2006) than before the fund commenced (24 in 2001–2002). Across five other group ranches, which lacked compensation funds, lion killings rose from nine in 2003 to 20 in 2004, 17 in 2005 and 32 in 2006. The lion population on the ranch where compensation was paid did not rise during the study period. The scheme was suspended from June 2003 to January 2004, April–June 2005 and in October 2005. At other times, pastoralists were compensated at market values for verified livestock losses to predators. Lower payments were made in cases of suboptimal animal husbandry. Fines were imposed for killing lions or other large predators.

A study in 1999–2009 of pasture and forest in Piedmont, Italy (2) found that when compensation was paid for livestock losses to wolves *Canis lupus* and dogs *Canis lupus familiaris*, an already expanding wolf population continued to grow. Over 11 years, the number of wolf packs increased from five to 20. Over the first five of these years, the annual number of attacks by wolves or dogs on livestock rose from 47 to 156. It then remained between 95 and 154 over the following six years. The scheme was established in 1999 to mitigate farmer-wolf conflict in a region with a recolonizing wolf population. Herders were compensated for livestock losses to wolves or dogs (as it is difficult to differentiate casualties due to these predators) and paid lump sums for indirect damages. From 2006, eligibility required using subsidised predation prevention measures, such as livestock guarding dogs, corrals and night confinement.

A replicated, before-and-after study in 2003–2011 of savanna grassland across three adjacent group ranches in southern Kenya (3) found that compensating for livestock predated by lions *Panthera leo* reduced lion killings by pastoralists. Prior to offering compensation, up to 25 lions/year were killed on two ranches and up to 10/year on the third. After introducing compensation payments, 2–15 lions/year were killed on two ranches and none was recorded killed on the third ranch. Compensating for loses was overall estimated to reduce lion killing by 87–91%. Compensation was paid for verified livestock losses to lions at the three group ranches between 2003 and 2008. Lion mortality data from 2003 to 2011 were collated primarily from community informants and direct interviews with lion hunters.

A study in 1996–2011 on tundra in northern Sweden (4) found that compensating reindeer herders for losses to wolverines *Gulo gulo* by paying for successful wolverine reproduction events was associated with an increase in wolverine abundance. The wolverine population grew at an annual rate of 4%. Male wolverines had a higher annual risk of being illegally killed (21%) than did female wolverines (8%), suggesting that payments were a greater disincentive to illegal killing of females. From 1996, payment rates to reindeer herders changed from being dependent on losses to predation to payment for documented wolverine reproductions (irrespective of predation levels). Population demography data were obtained from 95 wolverines (≥2 years old) radio-tracked in 1996–2011.

A before-and-after study in 2002–2013 in a savanna group ranch in the Amboseli–Tsavo ecosystem, Kenya (5) found that after introduction of a scheme to compensate for livestock killed by predators, fewer lions *Panthera leo* were killed or poisoned by pastoralists. Fewer lions were killed and poisoned during the six years after the scheme started (killed: 6; poisoned: 0) than the six years before (killed: 33; poisoned: 12). The number of livestock killed by lions did not differ significantly between the five years after the scheme commenced (cattle: 47–144/year; sheep and goats: 6–104/year) and the year before (cattle: 109; sheep and goats: 43). The study was conducted in a 1,133-km² group ranch, inhabited by 17,000 people and 20–30 lions. A compensation scheme for livestock killed by predators commenced in 2008. Livestock owners could claim between 35% and 70% of the market value of depredated livestock. The number of lions killed directly or poisoned was monitored between 2002 and 2013.

(1) Maclennan S.D., Groom R.J., Macdonald D.W. & Frank L.G. (2009) Evaluation of a compensation scheme to bring about pastoralist tolerance of lions. *Biological Conservation*, 142, 2149–2427, https://doi.org/10.1016/j.biocon.2008.12.003

(2) Dalmasso S., Vesco U., Orlando L., Tropini A. & Passalacqua C. (2012) An integrated program to prevent, mitigate and compensate Wolf (*Canis lupus*) damage in the Piedmont region (northern Italy). *Hystrix, the Italian Journal of Mammology*, 23, 54–61, https://doi.org/10.4404/hystrix-23.1-4560

(3) Hazzah L., Dolrenry S., Naughton L., Edwards C.T.T., Mwebi O., Kearney F. & Frank L. (2014) Efficacy of two lion conservation programs in Maasailand, Kenya. *Conservation Biology*, 28, 851–860, https://doi.org/10.1111/cobi.12244

(4) Persson J., Rauset G.R. & Chapron G. (2015) Paying for an endangered predator leads to population recovery. *Conservation Letters*, 8, 345–350, https://doi.org/10.1111/conl.12171

(5) Bauer H., Müller L., Van Der Goes D. & Sillero-Zubiri C. (2017) Financial compensation for damage to livestock by lions *Panthera leo* on community rangelands in Kenya. *Oryx*, 51, 106–114, https://doi.org/10.1017/s0030605 31500068x

3.27. Install non-electric fencing to exclude predators or herbivores and reduce human-wildlife conflict

https://www.conservationevidence.com/actions/2415

- **Eight studies** evaluated the effects on mammals of installing non-electric fencing to exclude predators or herbivores and reduce human-wildlife conflict. Two studies were in the USA[1,2] and one each was in Germany[3], the UK[4], Spain[5], China[6], Tanzania[7] and Kenya[8].

COMMUNITY RESPONSE (0 STUDIES)

POPULATION RESPONSE (0 STUDIES)

BEHAVIOUR (0 STUDIES)

OTHER (8 STUDIES)

- **Human-wildlife conflict (8 studies):** Four replicated studies (including three before-and-after studies), in USA[1], China[6], Tanzania[7] and Kenya[8], found that non-electric fencing reduced livestock predation by coyotes[1], Tibetan brown bears[6], and a range of mammalian predators[7,8]. A replicated, controlled study in USA[2] found that a high woven wire fence with small mesh, an overhang and an apron (to deter burrowing) was the most effective design at deterring crossings by coyotes. A replicated, controlled study in Germany[3] found that fencing with phosphorescent tape was more effective than fencing with normal yellow tape for deterring red deer and roe deer, but had no effect on crossings by wild boar or brown hare. Two studies (one replicated, before-and-after, site comparison and one controlled study) in the UK[4] and Spain[5] found that fences reduced European rabbit numbers[4] on or damage to[5] crops.

Background

Wild mammals can compete with domestic herbivores for food, can predate domestic herbivores or can damage crops. Human-wildlife conflict can be reduced if wild mammals can be effectively excluded from fields or other areas of crops or livestock. Non-electric fences are extensively used and can reduce the risk of wild mammal incursions into such sites. If successful, this could reduce incentives for carrying out lethal control of such mammals. Non-electric fences may be more suited to more extensive farming situations than are electric fences, as they may require less maintenance. This intervention also includes fortification of bomas (traditional livestock enclosures constructed by pastoralists) using conventional fencing materials such as fence wires.

A replicated, before-and-after, site comparison study in 1972–1977 in two pasture ranches in Oregon, USA (1) found that following erection of a fence to protect sheep, the number killed by coyotes *Canis latrans* was reduced to zero. Results were not tested for statistical significance. Over one year after fencing, no sheep were lost to coyotes in two fenced pastures. During the five years before fences were installed, 2% of sheep/pasture/year were killed by coyotes across one ranch and 24% across the other. On unfenced pastures on one of the ranches, 1% of sheep were lost to coyotes in the year that the fenced pasture was monitored, with 10% lost to coyotes on unfenced pastures on the other ranch. Two 5-ha pastures were fenced in November–December 1976. Fences were 1.8 m tall, made of wire, had a 41-cm overhang at a 60° angle from the fenced poles and an apron of old fence wire extending 61 cm out from the bottom, to inhibit digging under the fence. Ranchers monitored sheep kills by coyotes.

A replicated, controlled study in 1975–1976 in a captive facility in Oregon, USA (2) found that a high woven wire fence with small mesh, an overhang and an wire apron projecting out from the fence base (to deter burrowing) was the most effective of 34 fence designs at deterring crossings by coyotes *Canis latrans*. Fence performance varied from 0 to 71% of coyotes failing to cross fences. The best-performing non-electric fence prevented more crossings (14 of 15 trials) than did the best-performing electric fence (11 of 15 trials) or a standard sheep fence (6 of 15 trials). One of two coyotes, which had already crossed a standard sheep fence, crossed the best-performing fence during each of two tests whilst the other failed to cross it during four tests. Best-performing fence measurements were not stated explicitly but the paper recommends fences are ≥168 cm high, with mesh ≤15.2 × 10.2 cm and with an overhang and apron of ≥38 cm. Initial tests involved 10 coyotes, conditioned to walk a route, with 34 fence designs sequentially installed on the route. Subsequent trials, with five new coyotes, tested their ability to cross fences to reach a tethered rabbit. In final trials, coyotes that crossed a standard sheep fence and killed a tethered rabbit were tested using the best-performing fence design. Coyotes were wild caught. Trials were conducted from April 1975 to March 1976.

A replicated, controlled study in 1997 of four grassland fields and one cultivated field in central Germany (3) found that fencing with phosphorescent tape was more effective than fencing with normal yellow tape for deterring red deer *Cervus elaphus* and roe deer *Capreolus capreolus*, but had no effect on crossings by wild boar *Sus scrofa* or brown hare *Lepus europaeus*. At four grazing sites, areas surrounded by phosphorescent tape were avoided by red deer for four months and by roe deer for three weeks. Red deer entered areas fenced with yellow non-phosphorescent tape after one week and roe deer after one day. All deer species kept out of an area of willow fenced with phosphorescent strips for three weeks. After that, roe deer (but not red deer) tracks were found within the area. Wild boar and brown hare movements were not affected by tapes. PVC tape (4 cm wide) was attached 1 m high on 1.3-m iron posts. Four game grazing fields each had two 300-m² areas fenced off using phosphorescent strips and two with non-phosphorescent tape. After two months, all four areas were mown and the type of fencing was swapped. Mammal presence was assessed from droppings and tracks.

A replicated, before-and-after, site comparison study in 1980–1983 on 23 arable sites in southern UK (4) found that wire netting fences reduced European rabbit *Oryctolagus cuniculus* numbers on crops. Rabbit numbers on plots protected by fences with a buried fence base were lower 0–4 weeks after erection (7 rabbits/count) and 5–20 weeks after erection (7 rabbits/count) than before erection (41 rabbits/count). Numbers were also lower on plots protected by fences with the base folded horizontally along the ground 0–4 weeks after erection (11 rabbits/count) and 5–20 weeks after erection (7 rabbits/count) than they were before erection (45 rabbits/count). Rabbit numbers in unfenced plots remained constant throughout (0–4 weeks after erection: 16 rabbits/count; 5–20 weeks after erection: 13 rabbits/count; before erection: 14 rabbits/count). Fences (0.9 m high) were erected along one side of winter barley fields. Fences had bases buried 150 mm deep and then projecting horizontally underground for 150 mm (six sites), or laid out horizontally for 150 mm at ground level (seven sites). Ten unfenced sites were also monitored. Adult rabbits were counted using spotlights and binoculars in November–April between 1980 and 1983.

A controlled study in 2008 at three vineyards in Córdoba province, Spain (5) found that fencing reduced damage by European rabbits *Oryctolagus cuniculus* to common grape vines *Vitis vinifera* and resulted in greater grape vine yields. Grape vines within fenced plots had a lower percentage of buds and shoots removed by rabbits (0.5%) and greater yields (7 kg/vine) than unfenced plots (21%; 4.7 kg/vine). Each of three vineyard sites had a fenced plot and an unfenced plot. Fences were checked weekly. No details are provided about the fencing design. The proportion of buds and shoots removed by rabbits on 15–20 vines/ plot was recorded throughout the growing season in 2008. Grape vine yields were estimated during harvest from the number and size of grape clusters on each vine.

A replicated, before-and-after study in 2008–2009 of 19 households in Tibetan Autonomous Region, China (6) found that households fenced to exclude predators experienced fewer visits and lower rates of livestock predation by Tibetan brown bears *Ursus arctos pruinosus*. Results were not tested for statistical significance. In the year after fence installation, there were fewer bear visits (2.4/household) than in the year before (5.3/household). In the year after fence installation, fewer livestock were lost to bears (0.2/household) than in the year before (11.6/household). Fourteen fences were constructed around 19 households (some fences enclosed >1 household) and associated livestock in 2008. Fences were constructed of wire mesh (with mesh diagonal dimensions of ≤30 cm) and barbed wire, set on a steel frame. Each fence enclosed 120–1,000 sheep and goats. Bear visits and predation events were recorded by householders.

A replicated, before-and-after, site comparison study in 2003–2013 around two villages and associated pasture in Tanzania (7) found that fortifying bomas with trees and chain link fencing resulted in reduced predation of livestock by large mammalian predators. There was a lower rate of attacks by large predators on livestock in bomas after fortification (0.001 attacks/boma/month) than before (0.012 attacks/boma/month). Including bomas that remained unfortified throughout the study, the attack rate was lower overall on fortified bomas (0.001 attacks/boma/ month) than on unfortified bomas (0.009 attacks/boma/month). Between 2008 and 2013, 62 of 146 traditional bomas (built mainly from thorny branches) were fortified with 'living walls' (which combined fast-growing, thorny trees *Commiphora* sp. as fence posts at 0.5-m intervals, connected with chain link fencing). The average cost of the chain link was US$500/boma. Bomas were monitored for predator attacks from September 2003 to August 2013 (excluding January–February of 2006 and 2010).

A replicated, site comparison study in 2013–2015 of 308 savanna households in Narok County, Kenya (8) found that fewer livestock were lost to mammalian predators from fortified fenced areas than from traditional thorn-bush-fenced areas. Households holding their livestock in fortified fences lost fewer on average to predators (0.35 animal/month) than did households with livestock in traditional fenced areas (0.96 animals/month). The proportion of households not losing any livestock to mammalian predators over a year was higher for those using fortified fences (67%) than for those using traditional fences (15%). Mammalian predators included lions *Panthera leo*, leopards *Panthera pardus*, wild dogs *Lycaon pictus*, spotted hyenas *Crocuta crocuta*, honey badgers *Mellivora capensis*, cheetahs *Acinonyx jubatus* and baboons *Papio* sp. The study was based on 375 interviews, carried out from April 2013 to July 2015, with 308 Maasai households that housed livestock in fenced areas (bomas). Including some that were upgraded during the study, 179 households used fences fortified with posts, chain link wire and galvanized wire and 164 households used traditional fences made of thorny plants and branches during some or all of the period.

(1) de Calesta D.S. & Cropsey M.G. (1978) Field test of a coyote-proof fence. *Wildlife Society Bulletin*, 6, 256–259.

(2) Thompson B.C. (1979) Evaluation of wire fences for coyote control. *Journal of Range Management*, 32, 457–461, https://doi.org/10.2307/3898559

(3) Wölfel H. (1981) Testreihen zur Wirksamkeit von Leuchtbandfolien mit phosphoreszierenden Pigmenten bei der Wildschadensverhütung [Test trials on the effectiveness of strips of film with phosphorescent pigments in the prevention of damage by game]. *Zeitschrift für Jagdwissenschaft*, 27, 168–174, https://doi.org/10.1007/bf02243711

(4) McKillop I.G. & Wilson C.J. (1987) Effectiveness of fences to exclude European rabbits from crops. *Wildlife Society Bulletin*, 15, 394–401, https://doi.org/10.1016/0261-2194(92)90050-f

(5) Barrio I.C., Bueno C.G. & Tortosa F.S. (2010) Alternative food and rabbit damage in vineyards of southern Spain. *Agriculture, Ecosystems and Environment*, 138, 51–54, https://doi.org/10.1016/j.agee.2010.03.017

(6) Papworth S.K., Kang A., Rao M., Chin S.T., Zhao H., Zhao X. & Corrasco L.R. (2014) Bear-proof fences reduce livestock losses in the Tibetan Autonomous Region, China. *Conservation Evidence*, 11, 8–11.

(7) Lichtenfeld L.L., Trout C. & Kisimir E.L. (2015) Evidence-based conservation: predator-proof bomas protect livestock and lions. *Biodiversity and Conservation*, 24, 483–491, https://doi.org/10.1007/s10531-014-0828-x

(8) Sutton A.E., Downey M.G., Kamande E., Munyao F., Rinaldi M., Taylor A.K. & Pimm S. (2017) Boma fortification is cost-effective at reducing predation of livestock in a high-predation zone in the Western Mara region, Kenya. *Conservation Evidence*, 14, 32–38.

3.28. Install electric fencing to reduce predation of livestock by mammals to reduce human-wildlife conflict

https://www.conservationevidence.com/actions/2417

- **Eleven studies** evaluated the effects of installing electric fencing to reduce predation of livestock by mammals to reduce human-wildlife conflict. Six studies were in the USA[2,4a,4b,4c,6,7] (and a further one was presumed to be in the USA[1]) and one each was in Canada[3], South Africa[5], Brazil[8] and Spain[9].

COMMUNITY RESPONSE (0 STUDIES)

POPULATION RESPONSE (0 STUDIES)

BEHAVIOUR (0 STUDIES)

OTHER (11 STUDIES)

- **Human-wildlife conflict (11 studies):** Six out of 10 randomized and/or controlled or before-and-after studies (including eight replicated studies), in the USA[2,4a,4b,4c,6,7] (and a further one presumed to be in the USA[1]), Canada[3], Brazil[8] and Spain[9], found that electric fences reduced or prevented entry to livestock enclosures or predation of livestock by carnivores[1,3,4c,6,7,9]. Two studies[4a,4b] found that some designs of electric fencing prevented coyotes from entering enclosures and killing or wounding lambs. The other two studies found electric fencing did not reduce livestock predation or prevent fence crossings by carnivores[2,8]. A before-and-after study in South Africa[5] found that electrifying a fence reduced digging of burrows under the fence that black-backed jackals could pass through.

Background

Wild predatory mammals can come into conflict with humans if they predate domestic livestock. This conflict can be reduced if wild mammals can be effectively excluded from livestock enclosures. Electric fences are one means of doing this. If successful at reducing predation of livestock by carnivores, this could reduce incentives for carrying out lethal control of such species.

A replicated, controlled study (year not stated) of pasture at an undisclosed location, presumed to be in the USA (1) found that electric fencing prevented coyotes *Canis latrans* from entering an enclosure and killing lambs. During three trials, coyotes did not kill any of eight lambs in an enclosure surrounded by electric fencing but, in each trial, all eight lambs in an enclosure with conventional fencing were killed in 8–9 days. Two sheep enclosures (each 8,000 m²) were constructed within a coyote-proof 64-ha pasture. One enclosure had a 12-wire electric fence, 1.5 m high, with an additional electrified wire 20 cm outside the enclosure and 15 cm above the ground. The other enclosure had conventional wire fencing (81-cm woven wire with two strands of barbed wire, 15 cm apart, above the woven wire). For each of three trials, each lasting two weeks, a pair of wild-born captive coyotes was released into the pasture and eight lambs were placed in each of the two enclosures and observed daily. A different coyote pair was used for each trial.

A replicated, controlled study in 1975–1976 in a captive facility in Oregon, USA (2) found that most coyotes *Canis latrans* crossed electric fences and all 18 electric fence designs trialled were crossed by at least some coyotes. Coyotes crossed fences in 48–100% of the 20–30 tests/design. The most successful design (crossed in 13 of 27 tests) included three low-down electric wires laid out horizontally from the main vertical conventional fence (99-cm-high woven wire with two barbed wires above and one at the base). See paper for further details of fence designs. Tests involved 10 coyotes, conditioned to walk a route. Electric fences of 18 designs were sequentially placed along this route and

20–30 tests were conducted for each to see if coyotes would cross. The 18 designs represented modifications of standard fences used to house livestock in the study area, supplemented with wires charged by a 12-V battery. Trials were conducted from April 1975 to March 1976 and lasted each time for 10–15 minutes.

A replicated, before-and-after study in 1974–1978 on five farms in an area of boreal mixed-wood forest of Alberta, Canada (3) found that installing electric fences reduced the numbers of sheep killed by coyotes *Canis latrans*. These results were not tested for statistical significance. During the three years after electric fences were installed at five farms, fewer sheep were killed by coyotes (26) than during the three years before the electric fences were installed (147). The study was conducted in five farms, each covering 6–65 ha. An annual average of 44–550 sheep grazed at each farm in May–October. Between 0.8 and 3.2 km of electric fences were installed at each farm in 1976–1977. At two farms, fences had one or two strands of barbed wire spaced 15 cm apart above 81-cm-high woven wire, with a charged wire placed 15 cm above the ground and another 12 cm from the fence around the outside perimeter. At three farms, the fence was made of seven 2.7-mm wires alternating charged and grounded. Predation losses were reported by farmers.

A replicated, controlled study in 1977 at two sheep ranches in North Dakota, USA (4a) found that 12-wire electric fencing prevented coyotes *Canis latrans* from entering enclosures and killing lambs, but 6-wire electric fencing did not. At both ranches, 12-wire electric fencing prevented coyotes from killing lambs for at least 60 days, but 16–17 lambs were killed in 22–68 days in enclosures with conventional fencing. At one ranch, lambs were also killed in enclosures with 6–wire electric fencing (nine lambs killed in 20 days) and 6–wire electric fencing with a 'trip' wire (four lambs killed in four days). Two sheep ranches each had one enclosure with electric fencing (wires alternately charged) and one enclosure with conventional fencing (five strands of barbed wire, 104 cm high). Both ranches tested 12-wire electric fencing (168 cm high) for 60 days and conventional fencing for 22–68 days. One ranch tested 6-wire electric fencing (78 cm high) with and without an additional 'trip' wire (25 cm high, 51 cm from the fence) for four and 20 days respectively. All enclosures (1–1.5 ha) were kept stocked with 10 lambs and checked every other day for coyote kills during each of the six trials.

A replicated, before-and-after study in 1978 at two sheep ranches in Kansas, USA (4b) found that adding five electric wires to the outside of conventional fencing prevented coyotes *Canis latrans* from entering enclosures and killing or wounding lambs, but results varied when fewer wires were used. At one ranch, lambs were killed by coyotes in an enclosure with no electric wires (five lambs killed in 105 days) and four electric wires (one lamb killed in 17 days), but after adding a fifth wire no lambs were killed for at least 60 days. At the other ranch, lambs were killed or wounded in an enclosure with no electric wires (11 lambs killed in 11 days) and two electric wires (nine lambs killed or wounded in 14 days), but after adding two additional wires (total of four) no lambs were killed for at least 60 days. Two sheep ranches each had one enclosure (0.9–1.8 ha) with conventional fencing (woven wire and 1–2 strands of barbed wire, 110 cm high). At each ranch, enclosures were kept stocked with 10–20 lambs and checked for coyote kills during one trial (11–105 days) with conventional fencing only and two trials (11–60 days) with 2–5 electric wires added.

A replicated, before-and-after study in 1979 of 14 sheep producers in the USA (4c) found that installing electric fences or electric wires reduced predation of sheep by coyotes *Canis latrans*. Overall, the total number of sheep killed by coyotes was lower during a total of 228 months and 22 lambing seasons after electric fences or wires were installed (51 sheep) compared to during a total of 271 months and 27 lambing seasons before (1,064 sheep). However, the difference was not tested for statistical significance. In 1979, a total of 37 sheep producers using electric fencing or electric wires offset from existing conventional fencing were interviewed with a questionnaire. Fourteen responded with adequate information to compare sheep losses before and after electric fencing or wires were installed. Most respondents were reported to check their sheep at least once/day. Two-thirds answered questions from memory rather than written records.

A before-and-after study in 1983–1985 in a dry shrubland site in Cape Province, South Africa (5) found that electrifying a fence reduced digging of burrows under the fence that could then be used by black-backed jackals *Canis mesomelas* to enter and predate livestock. Fewer holes were dug under the fence after it was electrified (0–11 holes/week) than before (17–87 holes/week). Where the digger could be identified, holes were dug by black-backed jackals, warthogs *Phacochoerus africanus*, porcupines *Hystrix africaeaustralis*, bushpigs *Potamochoerus larvatus* and antbears *Orycteropus afer*. A 13.75-km-long game fence, that shared a boundary with five farms, was electrified by adding electric wires 250 mm away from both sides of the fence, 200 mm above the ground. The fence was monitored weekly for burrows for 33 weeks before electrification (September 1983 to May 1984) and for 44 weeks after (August 1984 to June 1985).

A replicated, before-and-after study in 1984–1985 of 51 sheep producers in Oregon, Washington and California, USA (6) found that installing electric fencing reduced predation of sheep by coyotes *Canis latrans*. The number of sheep killed by coyotes each year was lower during two or more years after electric fencing was installed (average 3.5 sheep/year; 0.3%) than during 1–7 years before (average 41 sheep/ year; 3.9%). Results were similar when sheep losses were included for producers that had electric fencing installed for one year only (before: 4.3% of sheep killed; after 0.7% killed; numbers not reported). More producers lost no sheep to coyotes after electric fencing was installed (28 of 51, 55%) than before (5 of 51, 10%). In 1984–1985, a total of 51 sheep producers that used electric fencing were interviewed. Electric fences enclosed areas of 1–1,550 ha containing 20–20,000 sheep. Sheep losses to coyotes were recorded during 1–7 years before electric fencing was installed and during one year (five producers) or two or more years (46 producers) after.

A randomized, replicated, controlled, before-and-after study in 2006 in a captive centre in Minnesota, USA and a replicated, controlled study in 2007 at 12 pastures in Montana, USA (7) found that electric fences with flags attached delayed grey wolf *Canis lupus* and red wolf *Canis rufus* entry. In the captive study, grey wolves and red wolves took longer (10 days) to cross electric fences with flags than non-electric fences with flags (1 day) or unfenced areas (<5 minutes). In the pasture study, wolves never entered pastures with electric fences and flags but twice entered pastures without electric fences and flags. The captive study ran for two weeks, using 45 wolves in 15 packs. Each pack (1–7 animals) was housed in a 105–925-m² enclosure. Five packs were offered food (white-tailed deer *Odocoileus virginianus*) positioned within an 18-m² electric fence (2,000 V) enclosure with red plastic flags (50 × 10 cm, 50 cm apart), five packs were offered food inside a non-electric fence with flags and five packs were offered food that was not protected by a fence or flags. Animals were monitored 24 hours/day with infra-red cameras. The pasture study was conducted in 12 cattle-grazed pastures (each 16–122 ha) enclosed with conventional barbed wire fences. Six pastures were further protected with electric fences with flags and six were not. Wolf tracks were monitored twice each week for three months.

A before-and-after study in 2006–2008 in a grassland-dominated cattle ranch in Mato Grosso do Sul, Brazil (8) found that after upgrading non-electric fences to become electric fences, a smaller percentage (but larger overall quantity) of cattle losses was due to killings by jaguars *Panthera onca*. These results were not tested for statistical significance. One year after upgrading fences to electric, 10% (50 of 504) of cattle losses were attributed to killings by jaguars. During the two years before non-electric fences were replaced by electric fences 24–85% (11 of 46 in one year and 24 of 28 in the other) of losses were attributed to killings by jaguars. The study was conducted on a 900-ha farm, fenced with five non-electrified wires at heights of 25, 50, 75, 100 and 125 cm. In February 2008, a 13,745-m perimeter fence was supplemented with two electrified wires (5,000–7,000 V), 25 and 50 cm above the ground. About 630 m of the fence was not electrified. Predation losses in the two years before the electric fence was installed were reported by farmers. After the electric fence was installed, losses were recorded by researchers.

A replicated, before-and-after study in 2012–2014 of two sheep flocks in Mediterranean forests and scrubland in Andalusia, Spain (9) found that electric fences prevented night-time predation by Iberian lynx *Lynx pardinus*. Over one winter and two spring lambing seasons following fence installation, no lynx or other predator attacks occurred inside fences. During the winter lambing season before fence installation, there were seven night-time predation events, involving 13 lambs. Electric fences (75 m perimeter, 106 cm high) were installed in early March 2013 (before the spring lambing season) for two sheep flocks. Fences contained a live braided plastic rope. Above the mesh were two 4-cm-wide conductor strips, giving a total height of 160 cm. Fences were powered from a solar rechargeable battery. Sheep were contained at night, but roamed freely, and suffered attacks, during daytime. All predator attacks on the two flocks were documented from December 2012 to May 2014.

(1) Gates N.L., Rich J.E., Godtel D.D. & Hulet, C.V. (1978) Development and evaluation of anti-coyote electric fencing. *Journal of Range Management*, 31, 151–153, https://doi.org/10.2307/3897668

(2) Thompson, B.C. (1979) Evaluation of wire fences for coyote control. *Journal of Range Management*, 32, 457–461, https://doi.org/10.2307/3898559

(3) Dorrance M.J. & Bourne J. (1980) An evaluation of anti-coyote electric fencing. *Journal of Range Management*, 33, 385–387, https://doi.org/10.2307/3897890

(4) Linhart S.B., Roberts J.D., & Dasch G.J. (1982). Electric Fencing Reduces Coyote Predation on Pastured Sheep. *Journal of Range Management*, 35, 276–281, https://doi.org/10.2307/3898301

(5) Heard H.W. & Stephenson A. (1987) Electrification of a fence to control the movements of black-backed jackals. *South African Journal of Wildlife Research*, 17, 20–24.

(6) Nass R.D. & Theade J. (1988) Electric fences for reducing sheep losses to predators. *Journal of Range Management* 41, 251–252, https://doi.org/10.2307/3899179

(7) Lance N.J., Breck S.W., Sime C., Callahan P. & Shivik J.A. (2010) Biological, technical, and social aspects of applying electrified fladry for livestock protection from wolves (*Canis lupus*). *Wildlife Research*, 37, 708–714, https://doi.org/10.1071/wr10022

(8) Cavalcanti S.M., Crawshaw P.G. & Tortato F.R. (2012) Use of electric fencing and associated measures as deterrents to jaguar predation on cattle in the Pantanal of Brazil. Pages 295–309 in: M.J. Somers and M.W. Hayward (eds.) *Fencing for Conservation. Restriction of Evolutionary Potential or a Riposte to Threatening Processes?* Springer, New York, NY, https://doi.org/10.1007/978-1-4614-0902-1_16

(9) Garrotea G., Lópeza G., Ruiza M., de Lilloa S., Buenoa J.F. & Simón M.A. (2015) Effectiveness of electric fences as a means to prevent Iberian lynx (*Lynx pardinus*) predation on lambs. *Hystrix, the Italian Journal of Mammalogy*, 26, 61–62, https://doi.org/10.4404/hystrix-26.1-10957

3.29. Exclude wild mammals using ditches, moats, walls or other barricades to reduce human-wildlife conflict

https://www.conservationevidence.com/actions/2420

- **Two studies** evaluated the effects of excluding wild mammals using ditches, moats, walls or other barricades to reduce human-wildlife conflict. One study was in Cameroon and Benin[1] and one was in Cameroon[2].

COMMUNITY RESPONSE (0 STUDIES)

POPULATION RESPONSE (0 STUDIES)

BEHAVIOUR (0 STUDIES)

OTHER (2 STUDIES)

- **Human-wildlife conflict (2 studies):** Two studies (including one before-and-after study and one site comparison), in Cameroon and Benin[1] and in Cameroon[2], found that fewer livestock were predated when they were kept in enclosures[2], especially when these were reinforced[1].

Background

This intervention includes the use of a range of barriers to prevent access to livestock by mammalian predators. If successful, this could reduce incentives for carrying out lethal control of predators.

A replicated, before-and-after study in 2004–2006 at a national park in Cameroon and a national park in Benin (1) found that when livestock enclosures were reinforced, fewer livestock were predated. In Cameroon, no cattle or pigs were predated from reinforced enclosures compared to six cattle predated (by lions *Panthera leo*) and 20 pigs predated (three by lions, 17 by hyenas *Crocuta crocuta*) from non-reinforced enclosures. In Benin, four cattle were predated (by lions) and 16 pigs (2 by lions, 14 by hyenas) from reinforced enclosures compared to 13 cattle predated (12 by lions, one by hyenas) and 53 pigs (28 by lions, 25 by hyenas) before reinforcements were added. In Cameroon, 75% of pastoralists across six villages in a national park buffer zone upgraded livestock enclosures. Enclosures comprised a thick layer of thorny shrubs and/or earth walls, with a safe gate (wood, or a complete tree *Acacia seyal* crown as a 'gate-plug'). Their performance was compared with that of non-reinforced enclosures over an unspecified period. In Benin, 13 enclosures were improved in 10 villages around a national park. The improved enclosures comprised sundried clay bricks covered with a clay/cement mixture ('banco'), similar to local houses. Livestock predation figures before (2004) and after (2005–2006) improvements were collated.

A site comparison study in 2008 of savanna around a national park in Cameroon (2) found that barricading livestock inside enclosures overnight reduced losses through predation by lions *Panthera leo*. Households owning enclosures lost an average of one animal/year to lion predation compared to two animals/year for households not owning enclosures. Owning enclosures did not reduce overall numbers of livestock predated by all mammalian predators (lions, spotted hyaenas *Crocuta crocuta* and jackals *Canis aureus*) (with enclosure: 4 animals predated/year; without enclosure: 5). However, fewer animals were lost by households that owned solid enclosures (2 animals/year) than those that owned enclosures made of thorny bushes (7 animals/year). In total, 207 resident pastoralists were interviewed for this study. Pastoralists reported the incidence of predation on livestock by large carnivores as well as whether their livestock were confined in enclosures at night. Villages were selected based on the tracking of movements of radio-collared lions.

(1) Bauer H., de Iongh H.H. & Sogbohossou E. (2010) Assessment and mitigation of human-lion conflict in West and Central Africa. *Mammalia*, 74, 363–367, https://doi.org/10.4404/hystrix-26.1-10957

(2) Tumenta P.N., de Iongh H.H., Funston P.J. & Udo de Haes H.A. (2013) Livestock depredation and mitigation methods practised by resident and nomadic pastoralists around Waza National Park, Cameroon. *Oryx*, 47, 237–242, https://doi.org/10.1017/s0030605311001621

3.30. Use flags to reduce predation of livestock by mammals to reduce human-wildlife conflict

https://www.conservationevidence.com/actions/2421

- **Five studies** evaluated the effects on mammals of using flags to reduce predation of livestock by mammals to reduce human-wildlife conflict. Three studies were in the USA[2,3,4], one was in Italy[1] and one was in Canada[2].

COMMUNITY RESPONSE (0 STUDIES)

POPULATION RESPONSE (0 STUDIES)

BEHAVIOUR (0 STUDIES)

OTHER (5 STUDIES)

- **Human-wildlife conflict (5 studies):** Three studies (including two before-and-after studies and a controlled study), in Italy[1], Canada[2] and the USA[4], found that flags hanging from fence lines (fladry) deterred crossings by wolves[1,2,4] but not by coyotes[4]. A further replicated, controlled study in the USA[5] found that electric fences with fladry were not crossed by wolves. A replicated, controlled, before-and-after study in the USA[3] found that fladry did not reduce total deer carcass consumption by a range of carnivores.

Background

Coloured flags (fladry) hung from fences are thought to deter crossings by wolves *Canis lupus* and potentially other predatory mammals. Thus, the intervention has potential for reducing predation on enclosed livestock. If successful, this could reduce incentives for carrying out lethal control of predatory mammals. The studies include both wild carnivores and captive wolves in experimental trials.

A before-and-after study in 1998 of captive animals in Italy (1) found that installing lines of flags (known as fladry) 50 cm high and ≤ 50 cm apart, deterred passage by gray wolves *Canis lupus*. Of 18 barrier designs trialled, four of five that were not crossed at all by two wolves involved lines of flags 50 cm high, with flags ≤50 cm apart. Three wolves in a larger enclosure made no crossings of a 50-cm-high flag line put in place to prevent access to one sixth, half and five sixths of the enclosure, even when the flag line split the enclosure in half with food placed at the opposite side. Flag lines comprised 50 × 10-cm red or grey flags. Two wolves, in a 120-m² enclosure, regularly paced along a fence line and barriers were set along this route. Three wolves, in an 850-m² enclosure, were excluded from varying proportions by flag lines. In all trials, wolves were observed for 30 minutes before and 30 minutes after each flag line was installed.

A replicated, before-and-after study in 2001–2002 on two pastures in Alberta, Canada (2) found that installing flags along fences (known as fladry) deterred wolves *Canis lupus* from entering pastures and predating livestock. Results were not tested for statistical significance. Before flags were installed, wolves approached pastures 2–7 times and predated livestock 2–5 times. With flags installed, wolves approached pastures 6–17 times but did not enter or predate livestock. After flags were removed, wolves approached twice and predated livestock 0–2 times. Plastic flags were placed at 50-cm intervals, suspended 50 cm above the ground on rope, 2 m out from the livestock fence. Two pastures (c.25 ha, 150 km apart) were studied. Each contained 100 cattle. Wolves were monitored by tracking signs in the snow, in winters of 2001 and 2002. Monitoring covered 60 days before flag installation, 60 days with flags installed and 60 days after flag removal.

A replicated, controlled, before-and-after study in 2002 of forest at six sites in Wisconsin, USA (3) found that installing lines of coloured flags (known as fladry) did not reduce overall deer carcass consumption by carnivores. Before installation, average consumption did not differ between carcasses assigned to treatments (flags: 2.0 kg/day; no flags: 1.6 kg/day). After flags were installed, consumption at these plots (2.5 kg/day) did not differ significantly from that at plots with no deterrent (3.3 kg/day). Wolves *Canis lupus*, black bears *Ursus americanus*, fishers *Martes pennanti* and foxes *Vulpes vulpes* visited plots. Study plots (30-m circumference) were established within territories of each of six wolf packs. A fresh deer carcass was placed in each plot. Plots were maintained for 9–35 days pre-treatment and 16–29 days during the treatment phase. The study ran during April–June 2002. Red flagging (100 × 7.5 cm) was suspended from perimeter ropes and was used at one plot in each territory and one plot had no deterrent. Carcasses were weighed every 2–3 days and replaced as required. Camera traps at three territories identified species visiting plots.

A replicated, randomized, controlled study in 2004–2005 in eight pasture and forest sites in Michigan, USA (4) found that tying coloured flags to a fence (known as fladry) reduced visits to pastures by gray wolves *Canis lupus* but not by coyotes *Canis latrans*. Fewer wolves were found in pastures where flags were used (0.3 visits/day) than outside pastures at the same sites (1.4 visits/day). There was no significant difference in wolf visitation rates where flags were not used (inside pasture: 0.7 visits/day; outside pasture: 0.3 visits/day). With flags, there was no significant difference in frequency of coyote visits in pastures (0.4 visits/day) and outside pastures at the same site (0.7 visits/day), and the same was true when flags were not used (inside pasture: 0 visits/day; outside pasture: 0.3 visits/day). In May 2004, red nylon flags were attached to fences at four randomly selected farms. At four other farms, no flags were used. One bait station, containing sand with sheep or cattle faeces, was placed inside each pasture and one outside each pasture fence. In May–August 2004 and 2005, each bait station was checked for wolf and coyote tracks.

A replicated, randomized, controlled study in 2007 at 12 pasture sites in Montana, USA (5) found that wolves *Canis lupus* did not visit sites with flags hanging from an electrified fence. The result was not tested for statistical significance. Relative effects of flags and electric fences cannot be separated in this study. Grey wolves *Canis lupus* did not visit any pastures with flags on electrified fences but twice visited pastures with conventional barbed wire fences. However, no livestock were killed by wolves in the pastures. The study was conducted in 12 pastures (16–122 ha), each with 40–200 cows. Pastures were contained within barbed wire fences. Six pastures (randomly selected) had electrified fences with red flags (50 × 10 cm) suspended from them, positioned outside existing fences and six did not. Wolf tracks were monitored twice weekly, for three months, in 2007.

(1) Musiani M. & Visalberghi E. (2001) Effectiveness of fladry on wolves in captivity. *Wildlife Society Bulletin*, 29, 91–98.

(2) Musiani M., Mamo C., Boitani L., Callaghan C., Gates C.C., Mattei L., Visalberghi E., Breck S. & Volpi G. (2003) Wolf depredation trends and the use of fladry barriers to protect livestock in western North America. *Conservation Biology*, 17, 1538–1547, https://doi.org/10.1111/j.1523-1739.2003.00063.x

(3) Shivik J.A., Treves A. & Callahan P. (2003) Nonlethal techniques for managing predation: primary and secondary repellents. *Conservation Biology*, 17, 1531–1537, https://doi.org/10.1111/j.1523-1739.2003.00062.x

(4) Davidson-Nelson S.J. & Gehring T.M. (2010) Testing fladry as a nonlethal management tool for wolves and coyotes in Michigan. *Human–Wildlife Interactions*, 4, 87–94, https://doi.org/10.26077/mdky-bs63

(5) Lance N.J., Breck S.W., Sime C., Callahan P. & Shivik J.A. (2010) Biological, technical, and social aspects of applying electrified fladry for livestock protection from wolves (*Canis lupus*). *Wildlife Research*, 37, 708–714, https://doi.org/10.1071/wr10022

3.31. Use visual deterrents (e.g. scarecrows) to deter predation of livestock by mammals to reduce human-wildlife conflict

https://www.conservationevidence.com/actions/2427

- **Two studies** evaluated the effects of using visual deterrents, such as scarecrows, to deter predation of livestock by mammals to reduce human-wildlife conflict. One study was in Kenya[1] and one was in Mexico[2].

COMMUNITY RESPONSE (0 STUDIES)

POPULATION RESPONSE (0 STUDIES)

BEHAVIOUR (0 STUDIES)

OTHER (2 STUDIES)

- **Human-wildlife conflict (2 studies):** A study in Kenya[1] recorded more livestock predation at bomas with scarecrows than those without scarecrows whereas a replicated, controlled study in Mexico[2] found that a combination of visual and sound deterrents reduced livestock predation.

Background

A range of visual deterrents, including scarecrows, may be used to deter carnivores from approaching livestock. If successful, such deterrents could reduce incentives for carrying out lethal control of carnivores.

A study in 2001–2005 of bushland and savanna in Laikipia and neighbouring districts, Kenya (1) found that at bomas with scarecrows positioned to deter predators, there were more, rather than fewer, carnivore attacks on livestock than at bomas without scarecrows. Scarecrows at bomas were associated with an increased risk of livestock attack by carnivores (results presented as odds ratio). The effect was strongest for leopards *Panthera pardus*. Scarecrows comprised cloth hung on trees or boma walls. They were present at 44% of 483 bomas (average 2.4/boma). Combining attacks on bomas with attacks on livestock herds grazing by day, the study documented 105 attacks by spotted hyenas *Crocuta crocuta*, 96 by leopards, 44 by African wild dogs *Lycaon pictus*, 35 by lions *Panthera leo* and 19 by cheetahs *Acinonyx jubatus*. From January 2001 to June 2005, eighteen local staff verified reports of livestock lost to predation and gathered data on animal husbandry practices used. Attacked bomas were compared to nearby bomas (median 323 m away) that had not been attacked.

A replicated, controlled study in 2010 of six farms in a forested area in central Mexico (2) found that visual and sound deterrents reduced predation of livestock on ranches. The relative effects of the two deterrent types were not assessed individually. No large predators (puma *Puma concolor* or jaguar *Panthera onca*) were detected on ranches that used deterrents compared with 2 detections/ranch and 2–4 livestock attacks/ranch where deterrents were not used. Out of six ranches (44–195 ha extent, ≥6 km apart), two cattle ranches and two goat ranches deployed deterrents whilst no deterrents were deployed on one cattle ranch and one goat ranch. Visual deterrents were shirts worn by livestock owners, hung around paddocks. Sound deterrents were recordings of voices, motors, pyrotechnics, barking dogs and bells, played twice daily for 40 min, between 06:00–08:00 and 20:00–22:00 h. Deterrents alternated weekly between visual and sound, through July–August 2010. Large predators were monitored using two camera traps/ranch and by searching for tracks and other signs.

(1) Woodroffe R., Frank L.G., Lindsey P.A., ole Ranah S.M.K. & Romañach S. (2007) Livestock husbandry as a tool for carnivore conservation in Africa's community rangelands: a case-control study. *Biodiversity and Conservation*, 16, 1245–1260, https://doi.org/10.1007/s10531-006-9124-8

(2) Zarco-González M.M. & Monroy-Vilchis O. (2014) Effectiveness of low-cost deterrents in decreasing livestock predation by felids: a case in Central Mexico. *Animal Conservation*, 17, 371–378, https://doi.org/10.1111/acv.12104

3.32. Use pheromones to deter predation of livestock by mammals to reduce human-wildlife conflict

https://www.conservationevidence.com/actions/2428

- We found no studies that evaluated the effects of using pheromones to deter predation of livestock by mammals to reduce human-wildlife conflict.

'We found no studies' means that we have not yet found any studies that have directly evaluated this intervention during our systematic journal and report searches. Therefore, we have no evidence to indicate whether or not the intervention has any desirable or harmful effects.

Background

Pheromones are chemical substances released into the environment by an animal that can affect the behaviour or physiology of other animals of the same species. If pheromones can be synthesised that deter wild mammalian predators from approaching and predating livestock, this could reduce the motivation among farmers for carrying out lethal control of such predators.

3.33. Use taste-aversion to reduce predation of livestock by mammals to deter human-wildlife conflict

https://www.conservationevidence.com/actions/2429

- **Nine studies** evaluated the effects of using taste-aversion to reduce predation of livestock by mammals to deter human-wildlife conflict. Six studies were in the USA[1,3,5,6,8a,8b], two were in Canada[4,7] and one was at an unnamed location[2].

COMMUNITY RESPONSE (0 STUDIES)

POPULATION RESPONSE (0 STUDIES)

BEHAVIOUR (0 STUDIES)

OTHER (9 STUDIES)

- **Human-wildlife conflict (9 studies):** Three of seven replicated studies (including three controlled studies), in the USA[1,3,5,6], Canada[4,7] and at an unnamed location[2], found that coyotes killed fewer sheep[1,3,7], rabbits[1] or turkeys[3] after taste-aversion treatment. The other four studies found that taste-aversion treatment did not reduce killing by coyotes of chickens[2], sheep[4,5] or rabbits[6]. A replicated, before-and-after study in the USA[8a] found that taste-aversion treatment reduced egg predation by mammalian predators whilst a replicated, controlled, paired sites study in the USA[8b] found no such effect.

Background

Wild mammalian predators can cause unacceptable levels of livestock losses. Human-wildlife conflict can be reduced if wild mammals can be effectively deterred from attacking livestock. This intervention covers the use of substances that cause unpleasant effects in mammals, such as gastrointestinal discomfort, but at a dose not intended to cause long-term harm to the animal. Most studies are trials using captive animals, especially coyotes *Canis latrans*. One study included here is a trial of using the same approach to deter predation of bird eggs. This would most likely find application in poultry or game rearing operations, and so is included here given that the intention could be to reduce economic losses caused by wild mammals. If the intervention is effective at reducing predation, it could reduce incentives for carrying out lethal control of mammalian predators.

A replicated, controlled, before-and-after study (year not stated) on captive animals in the USA (1) found that after conditioned taste-aversion treatment, coyotes *Canis latrans* did not catch and eat live lambs or rabbits. After one or two meals of lamb or rabbit meat containing lithium chloride (which causes gastrointestinal discomfort), six coyotes did not attack either lambs or rabbits. Three coyotes were held in individual pens. Over a 13-day period, coyotes alternated between being let into an enclosure with a live lamb or rabbit or with lamb meat containing lithium chloride. A similar experimental procedure was carried out with three different coyotes, which received rabbit meat containing lithium chloride.

A replicated study in 1975–1976 on captive animals (location not stated) (2) found that feeding dead chickens injected with lithium chloride to coyotes *Canis latrans* did not induce taste-aversive against taking live chickens. After eating dead chickens laced with lithium chloride (which causes gastrointestinal discomfort), two coyotes each killed and ate the single live chickens that they were offered. Three different coyotes between them killed and ate 25 of 31 live chickens offered. The five coyotes were offered 79 dead lithium chloride-laced chickens, from which 39 were uneaten, 23 were entirely eaten and 17 were partially eaten. Prior to lacing trials, each coyote was offered five live and five dead chickens (unlaced), all of which were eaten. Coyotes were then offered four to eight dead chickens, laced with lithium chloride. Following this, in daily trials, they were offered, in random order, a recently killed laced chicken or a live chicken. Two coyotes were offered single live chickens at this stage, and three were offered from three to nine live chickens each.

A replicated study in 1976–1977 of six livestock farms in a desert area of California, USA (3) found that after taste-aversion treatment, the number of sheep and turkeys killed by coyotes *Canis latrans* declined over time. In the second year that baits containing lithium chloride (which causes gastrointestinal discomfort) were used, the number of sheep killed by coyotes was lower (59 kills) than in the first year that baits were used (186 kills). The same pattern was true for the numbers of turkeys killed (data not presented). From August 1976 to April 1977, sheep carcasses containing lithium chloride were laid as bait, adjacent to areas where four sheep herds were grazing. Sheep herds were at least 12 km apart. From November 1976 to April 1977, turkey carcasses containing lithium chloride were laid as bait adjacent to two turkey farms. Turkey farms were 27 km apart. Methods used to monitor the numbers of animals killed were unclear.

A replicated, randomized, controlled, before-and-after study in 1978 on pastures in four areas in Alberta, Canada (4) found that lacing sheep meat baits with lithium chloride did not induce taste-aversive in coyotes *Canis latrans* against taking lambs. Average lamb predation rates on farms where baits were laced with lithium chloride (which causes gastrointestinal discomfort) (5.7/farm) did not significantly differ from those on farms without baits (7.5/farm). Over each of the previous two years, there was also no difference in predation rates between treatment farms (7.4 and 9.4/farm respectively) and control farms (6.1 and 9.5/ farm respectively). Four areas were studied, with five to eight sheep farms (≥8 km apart) in each. Half of farms had lithium chloride baits, half had baits without lithium chloride. Six to 10 baits (sheep meat, wrapped in sheep hide) were placed on each treatment farm in April 1978. Baits were replaced at least every three weeks. Baiting continued to September (to July on two farms). Few baits were consumed in one area, so predation data there were excluded from analyses. Predation rates were supplied by farmers for 1976–1978. Lethal control of coyotes was carried out when predation was confirmed.

A replicated, controlled study (year unspecified) in a research facility in Utah, USA (5) found that lithium chloride-injected bait did not induce taste aversion that prevented coyotes *Canis latrans* from killing lambs *Ovis aries*. Coyotes fed with baits containing lithium chloride (which causes gastrointestinal discomfort) took a similar length of time to kill a lamb after feeding (2.7 days) than did coyotes that had eaten bait without lithium chloride (2.7 days). Eight coyotes were held in separate kennels. At 08:00 each day, an individual animal was let into a 250-m² pen containing food. If a coyote consumed the food within 10 minutes on three consecutive days, then on the following day bait, in the form of sheep meat contained within sheep hide, was placed in the pen. For four coyotes, the baits contained lithium chloride (which induced gastrointestinal discomfort) and, for the other four, they did not. Coyotes were left in pens until they had eaten at least one bait. Following this, coyotes were let back into the pen along with a live lamb and the time it took for the coyote to kill the lamb was monitored.

A replicated, before-and-after study in 1983 in a research facility in Colorado, USA (6) found that feeding domestic European rabbits *Oryctolagus cuniculus* baited with an illness-inducing agent to coyotes *Canis latrans* did not change their predation rate on live rabbits. Coyotes killed all live rabbits presented to them both before and after being fed with rabbit meat and rabbit carcases baited with an illness-inducing agent. The study was conducted in a 6,400-m² enclosure of unspecified habitat. Three wild-caught adult coyotes were each presented with a series of live rabbits and made 10 consecutive kills. Each then received a control bait package (rabbit meat with an empty gelatin capsule) followed by five further live rabbits. Coyotes then received a bait package with a gelatin capsule containing lithium chloride, followed a day later by a live white rabbit. The next day, they received another lithium chloride-laced bait package followed by another live rabbit. Three days later, they received a lithium chloride-treated rabbit carcass and then live rabbits the following day. Bait packages were 227 g of rabbit meat containing 7 g of illness-inducing lithium chloride in a gelatin capsule. Baited rabbit carcasses were injected with 10 g of dissolved lithium chloride. No additional food was provided between trials.

A replicated, before-and-after study in 1975–1976 on 16 pastures in Saskatchewan, Canada (7) found that use of lithium chloride-treated baits to induce taste-aversion, was associated with reduced predation of sheep by coyotes *Canis latrans*. Losses of sheep and lambs to coyotes fell from 4% (892 predated out of 22,407 animals) in 1975 (before baits used) to 1.5% (301 predated out of 20,574 animals) in 1976. Factors such as animal husbandry and use of other coyote control methods were not controlled for. Sixteen sheep pastures (mix of private ownership and community cooperatives), holding 101–4,543 sheep, on which predation by coyotes was previously reported, were studied. Baseline predation data were collected in 1975. In 1976, lithium chloride baits (which induce gastrointestinal discomfort) were used at all sites (bait application methods not detailed in paper).

A replicated, before-and-after study in 1986 in three deciduous forest sites in Connecticut, USA (8a) found that dosing chicken eggs with emetine dihydrochloride reduced egg predation by inducing conditioned taste aversion in mammalian predators. The proportion of eggs predated daily was 85% at the end of the pre-treatment period (eggs not dosed), 10% at the end of the treatment period (eggs dosed with emetine) and remained low (17%) at the end of the post-treatment period (eggs not dosed). Mammals (mostly raccoons *Procyon lotor*, opossums *Didelphis virginia* and striped skunks *Mephitis mephitis*) predated 66% of eggs taken. At each of three sites (>4 km apart) 10 chicken eggs were placed >75 m apart. Pre-treatment, treatment and post-treatment each lasted three weeks. Eggs were placed for four days/week and checked (and replaced if predated) daily. During the treatment period, eggs were injected with 20–25 mg of emetine, which causes gastrointestinal discomfort. The study ran in June–September 1986.

A replicated, controlled, paired sites study in 1987 in eight deciduous forest sites in Connecticut, USA (8b) found that dosing chicken eggs with emetine dihydrochloride did not reduce egg predation by inducing conditioned taste aversion in mammalian predators. At treatment sites, the number of eggs predated that were dosed (5.0–8.7/week) or undosed (2.3–3.5/week) was not lower than the number predated at untreated sites (0.8–3.3). Racoons *Procyon lotor* were the main mammalian predator in this study. Four treatment sites each had 10 undosed eggs and 10 dosed eggs placed >75 m apart. Four further untreated sites each had 10 undosed eggs placed >75 m apart. Dosed eggs were injected with 20–25 mg of emetine, which causes gastrointestinal discomfort. Eggs were checked twice weekly in July–September 1987, and predated eggs were replaced.

(1) Gustavson C.R., Garcia J., Hankins W.G. & Rusiniak K.W. (1974) Coyote predation control by aversive conditioning. *Science*, 184, 581–583, https://doi.org/10.1126/science.184.4136.581

(2) Conover M.R., Francik J.G. & Miller D.E. (1977) An experimental evaluation of aversive conditioning for controlling coyote predation. *The Journal of Wildlife Management*, 41, 775–779, https://doi.org/10.2307/3800006

(3) Ellins S.R. & Catalano S.M. (1980) Field application of the conditioned taste aversion paradigm to the control of coyote predation on sheep and turkeys. *Behavioral and Neural Biology*, 29, 532–536, https://doi.org/10.1016/s0163-1047(80)92882-4

(4) Bourne J. (1982) A field test of lithium chloride aversion to reduce coyote predation on domestic sheep. *The Journal of Wildlife Management*, 46, 235–239, https://doi.org/10.2307/3808426

(5) Burns, R. J. (1983) Microencapsulated lithium chloride bait aversion did not stop coyote predation on sheep. *The Journal of Wildlife Management*, 47, 1010–1017, https://doi.org/10.2307/3808159

(6) Horn S.W. (1983) An evaluation of predatory suppression in coyotes using lithium chloride-induced illness. *The Journal of Wildlife Management*, 47, 999–1009, https://doi.org/10.2307/3808158

(7) Jelinski D.E., Rounds R.C. & Jowsey J.R. (1983) Coyote predation on sheep, and control by aversive conditioning in Saskatchewan. *Journal of Range Management*, 36, 16–19, https://doi.org/10.2307/3897972

(8) Conover M.R. (1990) Reducing mammalian predation on eggs by using a conditioned taste aversion to deceive predators. *The Journal of Wildlife Management*, 54, 360–365, https://doi.org/10.2307/3809055

3.34. Dispose of livestock carcasses to deter predation of livestock by mammals to reduce human-wildlife conflict

https://www.conservationevidence.com/actions/2432

- **One study** evaluated the effects of disposing of livestock carcasses to deter predation of livestock by mammals to reduce human-wildlife conflict. This study was in the USA[1].

COMMUNITY RESPONSE (0 STUDIES)

POPULATION RESPONSE (0 STUDIES)

BEHAVIOUR (0 STUDIES)

OTHER (1 STUDY)

- **Human-wildlife conflict (1 study):** One site comparison study in the USA[1] found that burying or removing sheep carcasses reduced predation on livestock by coyotes, but burning carcasses did not alter livestock predation rates.

Background

Leaving livestock carcasses in place on farms after death may attract mammalian carnivores that may also attack live farm animals. Carcasses can be removed to eliminate this form of attraction for predators. If this results in fewer predators being attracted to farms and, consequently, less predation on livestock, this could reduce incentives for carrying out lethal control of such predators.

A site comparison study in 1975–1976 of 97 sheep farms in Kansas, USA (1) found that when sheep carcasses were buried or removed, sheep losses to coyotes *Canis latrans* and dogs *Canis lupus familiaris* were reduced compared to leaving them on the pasture, but burning carcasses did not reduce predation. The proportion of sheep lost to coyotes or dogs each month was lower when carcasses were buried (0.05%) or removed (0.08%) than when they were left in place (0.14%). The rate when carcasses were burned (0.17%) did not differ from that of leaving them in place. Ninety-seven farms were studied, on which total sheep numbers varied through the study period from 14,578 to 17,023. Farmers recorded monthly sheep losses and husbandry methods for 15 months.

(1) Robel R.J., Dayton A.D., Henderson F.R., Meduna, R.L. & Spaeth, C.W. (1981) Relationships between husbandry methods and sheep losses to canine predators. *The Journal of Wildlife Management*, 45, 894–911, https://doi.org/10.2307/3808098

3.35. Use guardian animals (e.g. dogs, llamas, donkeys) bonded to livestock to deter predators to reduce human-wildlife conflict

https://www.conservationevidence.com/actions/2433

- **Twelve studies** evaluated the effects of using guardian animals (e.g. dogs, llamas, donkeys) bonded to livestock to deter mammals from predating these livestock to reduce human-wildlife conflict. Four studies were in the USA[1,2,3,6], two were in Kenya[4,5] and one each was in Solvakia[7], Argentina[8], Australia[9], Cameroon[10], South Africa[11], and Namibia[12].

COMMUNITY RESPONSE (0 STUDIES)

POPULATION RESPONSE (0 STUDIES)

BEHAVIOUR (0 STUDIES)

OTHER (12 STUDIES)

- **Human-wildlife conflict (12 studies):** Four of seven studies, (including four site comparison studies), in the USA[1,2],

Kenya[4,5], Solvakia[7], Australia[9] and Cameroon[10], found that guardian animals reduced attacks on livestock by predators. The other three studies reported mixed results with reductions in attacks on some but not all age groups[2] or livestock species[4] and reductions for nomadic but not resident pastoralists[10]. Two studies, (including one site comparison study and one before-and-after study), in Argentina[8] and Namibia[12], found that using dogs to guard livestock reduced the killing of predators by farmers[8,12] but the number of black-backed jackals killed by farmers and dogs combined increased[12]. A replicated, controlled study in the USA[3] found that fewer sheep guarded by llamas were predated by carnivores in one of two summers whilst a replicated, before-and-after study in South Africa[11] found that using dogs or alpacas to guard livestock reduced attacks by predators. A randomized, replicated, controlled study in USA[6] found that dogs bonded with livestock reduced contact between white-tailed deer and domestic cattle.

Background

Using animals to guard livestock is a long-established practice. Usually dogs *Canis lupus familiaris* are used but occasionally other animals (e.g. llamas *Lama glama*) may be used. In most cases, guardian animals are raised among livestock and bond to them. If guardian animals can reduce losses of livestock to predators, this may reduce motivations for lethal control of such predators.

A replicated study in 1981 of 36 ranches in North Dakota, USA (1) found that guard dogs *Canis lupus familiaris* reduced sheep losses to predation by coyotes *Canis latrans*. The average annual predation rate after commencing use of guard dogs (0.4% of the sheep flock) was lower than that before guard-dog use commenced (6%). In 1981, thirty-six ranchers were interviewed about livestock management and losses to predation in the 1976–1981 period. Between them, ranchers had 52 great Pyrenees dogs (44 working and eight training) and two working komondor dogs. All ranchers commenced using guardian dogs during the period. Guarded pastures were 4–486 ha in extent and guarded sheep flocks contained 10–1,300 animals. Dogs were raised with the sheep flock and remained with them most of the time.

A replicated, site comparison study in 1986 of 134 sheep producers in Colorado, USA (2) found that using livestock-guarding dogs *Canis lupus familiaris* reduced coyote *Canis latrans* predation of lambs in fenced pastures and some open ranges, but predation of ewes was not reduced in either. A lower percentage of lambs was killed by coyotes in fenced pastures with livestock-guarding dogs (0%) than without dogs (2–5%). In open ranges, a lower percentage of lambs was killed compared to 20 of 25 producers without dogs (with dogs: 1.2%; without dogs: 16%), this was not the case compared to the five producers without dogs that responded by telephone rather than post (without dogs: 3%). The percentage of ewes killed by coyotes did not differ significantly with dogs (fenced pastures: 0%; open ranges: 0.4%) or without dogs (fences pastures: 0.5–1%; open ranges: 1.1–1.5%). Sheep producers kept ewes and lambs with or without livestock-guarding dogs in fenced pastures (with dogs: 6–7 producers; without dogs: 87–92 producers) or open ranges (with dogs: 10 producers; without dogs: 25 producers). Average flock sizes were 90–321 lambs or ewes in fenced pastures and 910–2,440 lambs or ewes in open ranges. Seven breeds (or mixed breeds) of livestock-guarding dog were used (see original paper for details). The 134 sheep producers responded to postal or telephone surveys in 1986.

A replicated, controlled study in 1996–1997 on pasture in Utah, USA (3) found that using llamas *Lama glama* to guard sheep flocks reduced canine predation on lambs in one of two summers. Sheep flocks guarded by a llama lost a lower proportion of lambs to predators in the first summer season than did flocks without llamas. There was no significant difference in losses during the second summer season. Actual loss rates were not presented. Predation rates of ewes and predation in the winter season were very low across all flocks. Coyotes *Canis latrans*, domestic dogs *Canis lupus familiaris* and red foxes *Vulpes vulpes* accounted for 92% of losses to predators. Flocks with llamas averaged 301 sheep (including lambs). Flocks without llamas averaged 333 sheep and lambs. Twenty flocks were each guarded by a single llama. The number of flocks without llamas varied through the study, due to splitting and merging of flocks, from 8 to 29. Sheep producers reported fortnightly, from May 1996 to December 1997, on predation events and flock sizes.

A replicated, site comparison study in 1999–2000 of savanna across 10 ranches in Laikipia District, Kenya (4) found that at bomas with domestic dogs *Canis lupus familiaris* in attendance, fewer cattle were killed by predators, though there was no effect on predation of sheep or goats. Fewer cattle were killed by lions *Panthera leo*, leopards *Panthera pardus* and hyenas *Crocuta crocuta* and *Hyaena hyaena* combined when dogs were present at bomas (0.03 cattle/month) than at bomas without dogs (0.28 cattle/month). There was no significant relationship between dog presence and predation on sheep or goats (data not presented). Livestock were housed in bomas overnight, when 75% of recorded kills occurred. Data on livestock predation and predator deterrence activities at 84 bomas on 10 ranches (nine commercial ranches, one community area) were gathered from ranch managers. Ranches were monitored for 2–17 months, between January 1999 and May 2000.

A study in 2001–2005 of bushland and savanna in Laikipia and neighbouring districts of Kenya (5) found that when livestock were accompanied by one or more domestic dogs *Canis lupus familiaris*, fewer were attacked by carnivores. Livestock herds grazing by day and those held overnight in thornbrush bomas were less likely to be attacked by carnivores if accompanied by domestic dogs (results presented as odds ratios). Of 502 grazing herds, 24% were accompanied by one or more dogs (average 1.3 dogs/accompanied herd). Of 491 bomas, dogs were present at 71% (average 2.0 dogs/boma). The study documented 105 attacks by spotted hyenas *Crocuta crocuta*, 96 by leopards *Panthera pardus*, 44 by African wild dogs *Lycaon pictus*, 35 by lions *Panthera leo* and 19 by cheetahs *Acinonyx jubatus*. From January 2001 to June 2005, eighteen local staff verified reports of livestock lost to predation and gathered data on animal husbandry practices used. Attacked herds or bomas were compared to nearby herds (median 656 m away) or bomas (median 323 m away) that had not been attacked.

A randomized, replicated, controlled study in 2003 at two forest sites in Michigan, USA (6) found that dogs *Canis lupus familiaris* bonded with livestock reduced levels of contact (and potential for disease transmission) between white-tailed deer *Odocoileus virginianus* and domestic cattle. In dog-guarded pastures, deer came within 5 m of cattle fewer times (three instances) than in non-guarded pastures (79 instances). No deer were within 5 m of cattle when dogs were present, while 114 events occurred with dogs absent. Deer consumed hay less frequently in dog-guarded pastures (two instances) compared to pastures without dogs (303 instances). At each site, four 1.2-ha pastures, >200 m apart, were enclosed by electric fencing. Deer were baited into pastures with corn and alfalfa. Each pasture contained four calves while two pastures at each site also had a dog. Livestock guarding dogs were great Pyrenees, raised from eight-week-old pups, following standard training procedures. Visits of deer into pastures were monitored by direct observation and video surveillance, in March–August 2003.

A replicated, site comparison study in 2002 on 58 farms in Solvakia (7) found that farms using livestock-guarding dogs *Canis lupus familiaris* lost fewer livestock to predation than did farms without dogs. The number of livestock lost to predators (mainly grey wolf *Canis lupus*) in flocks with livestock-guarding dogs (1.1 sheep/flock) was not significantly different to that in unguarded flocks (3.3 sheep/flock). However, dog placement was prioritised at flocks with previously high predation rates. On farms where predation occurred, fewer livestock were lost in guarded (1.5 sheep/flock) than in unguarded flocks (5.0 sheep/flock). Pups (Slovenský čuvač and Caucasian shepherd dog) were reared alongside livestock. Of 34 pups placed on farms in 2000–2004, seventeen were successfully integrated into livestock flocks during the first full grazing season. Reported losses for 2002 were compared between 13 flocks with successfully integrated 1–2-year-old livestock-guarding dogs and 45 farms in the same regions without dogs.

A replicated, site comparison study in 2005–2011 on a grass-shrub steppe area in Patagonia, Argentina (8) found that use of dogs *Canis lupus familiaris* by goat herders to guard livestock reduced the killing of predators by herders. Results were not tested for statistical significance. Six of eight herders with working guard dogs reported that they no longer killed predators, one had never done so and one did so less frequently than previously. Nine herders who did not have working dogs all continued to kill predators. Most reported predation was by cougar *Puma concolor* and culpeo fox *Lycalopex culpaeus*. Thirty-seven puppies were placed with herders, of which 11 became successful livestock guarding dogs. Herders were interviewed monthly or bimonthly during the dog training period. Nine neighbouring herders without dogs were also interviewed. Interviews included questions about predator control activities carried out by the herders.

A before-and-after study in 1997–2010 on a grassland-dominated ranch in Queensland, Australia (9) found that when guardian dogs *Canis lupus familiaris* were used to protect livestock from dingoes *Canis dingo* and other predators, sheep mortality declined. By three years after the guardian dog programme commenced, annual sheep losses had fallen to 4% of the flock and remained at 4–7% over the following five years. In the six years before the programme commenced, there was 7–15% annual mortality of the sheep flock. Sheep mortality figures included all causes of death, not only predation. The study was conducted on a 47,000-ha ranch, hosting approximately 12,000–22,000 sheep and 4,000 cattle. Dingoes and feral dogs were the main livestock predators in the area. In 2002, twenty-four Maremma sheepdogs were integrated with the sheep. The sheepdogs worked unsupervised in groups of 1–4. They had access to self-feeders with dry dog food. Dingoes and wild dogs were also baited with poison and wild dogs were shot opportunistically.

A site comparison study in 2008 of savanna around a national park in Cameroon (10) found that using dogs *Canis lupus familiaris* to guard livestock reduced losses through predation among nomadic pastoralists but not among resident pastoralists. Among nomadic pastoralists that owned dogs (53% of all nomadic pastoralists), fewer livestock were lost to carnivores (six animals/year) than among those that did not own dogs (10 animals/year). Among resident pastoralists that owned dogs (33% of all resident pastoralists), there was no significant difference in the number lost to predators (five animals/year) compared to those that did not own dogs (four animals/year). Two hundred and seven resident pastoralists and 174 nomadic pastoralists were interviewed. Subjects reported the incidence of predation on livestock by large carnivores and details of animal husbandry techniques used. Villages were selected based on the tracking of movements of radio-collared lions.

A replicated, before-and-after study in 2007–2009 of four livestock farms in savanna and shrubland in Eastern Cape, South Africa (11) found that using dogs *Canis lupus familiaris* and alpacas *Vicugna pacos* to guard livestock reduced attacks by carnivores on livestock, compared to using lethal control of predators. Results were not tested for statistical significance. When guard animals were used, 0–15% of livestock were killed each year by predators, but when lethal predator-control methods were used 5–45% of livestock were killed. Costs of using non-lethal control were lower (0.73–6.02 USD/livestock animal) than were those of lethal control (0.95–7.94 USD/livestock animal). In August 2006–August 2007, all four farms used lethal methods, including trapping and shooting, to control black-backed jackals *Canis mesomelas*, caracals *Caracal caracal* and leopards *Panthera pardus*. In September 2007–September 2009, farms either used guard dogs (three farms) or alpacas (one farm) to protect animals. Farmers reported the number of livestock killed by predators and associated costs, each September, in 2007–2009.

A before-and-after study in 2009–2010 of 73 livestock farms in Namibia (12) found that placing dogs *Canis lupus familiaris* with farmers to guard livestock reduced the overall number of farmers that killed predators, but increased the numbers of black-backed jackals *Canis mesomelas* killed by farmers and dogs combined. Eighteen percent of farmers killed livestock predators in the year after dog placement compared to 31% in the previous year. The reduction was larger among subsistence farmers (0% after dog placement; 30% before) than commercial farmers (26% after dog placement; 32% before). However, the number of black-backed jackals killed by farmers and dogs combined in the year following dog placement (3.4/farm) was greater than the number killed by farmers alone the previous year (1.7/farm). There were no significant differences for killings of caracal *Caracal caracal* (farmer and dog: 0.19; farmer: 0.10), cheetah *Acinonyx jubatus* (farmer and dog: 0.02; farmer: 0.11) or leopard *Panthera pardus* (farmer and dog: 0; farmer: 0.02). Anatolian shepherd dogs were placed on 53 commercial farms and 20 subsistence farms. Farmers were interviewed between March 2009 and September 2010. Dogs were placed with a livestock flock at eight weeks old and averaged 39 months old at time of the study.

(1) Pfeifer W.K. & Goos M.W. (1982) Guard dogs and gas exploders as coyote depredation control tools in North Dakota. *Proceedings of the Tenth Vertebrate Pest Conference*, 55–61.

(2) Andelt W.F. (1992) Effectiveness of livestock guarding dogs for reducing predation on domestic sheep. *Wildlife Society Bulletin*, 20, 55–62.

(3) Meadows L.E. & Knowlton F.K. (2000) Efficacy of guard llamas to reduce canine predation on domestic sheep. *Wildlife Society Bulletin*, 28, 614–622.

(4) Ogada M.O., Woodroffe R., Oguge N.O. & Frank L.G. (2003) Limiting depredation by African carnivores: the role of livestock husbandry. *Conservation Biology*, 17, 1521–1530, https://doi.org/10.1111/j.1523-1739.2003.00061.x

(5) Woodroffe R., Frank L.G., Lindsey P.A., ole Ranah S.M.K. & Romañach S. (2007) Livestock husbandry as a tool for carnivore conservation in Africa's community rangelands: a case-control study. *Biodiversity and Conservation*, 16, 1245–1260.

(6) VerCauteren K.C., Lavelle M.J. & Phillips G.E. (2008) Livestock protection dogs for deterring deer from cattle and feed. *The Journal of Wildlife Management*, 72, 1443–1448, https://doi.org/10.2193/2007-372

(7) Rigg R., Finďo S., Wechselberger M., Gorman M.L., Sillero-Zubiri C. & Macdonald D.W. (2011) Mitigating carnivore–livestock conflict in Europe: lessons from Slovakia. *Oryx*, 45, 272–280, https://doi.org/10.1017/s0030605310000074

(8) González A., Novaro A., Funes M., Pailacura O., Bolgeri M.J. & Walker S. (2012) Mixed-breed guarding dogs reduce conflict between goat herders and native carnivores in Patagonia. *Human-Wildlife Interactions*, 6, 327–334.

(9) Van Bommel L., & Johnson C. N. (2012) Good dog! Using livestock guardian dogs to protect livestock from predators in Australia's extensive grazing systems. *Wildlife Research*, 39, 220–229, https://doi.org/10.1071/wr11135

(10) Tumenta P.N., de Iongh H.H., Funston P.J. & Udo de Haes H.A. (2013) Livestock depredation and mitigation methods practised by resident and nomadic pastoralists around Waza National Park, Cameroon. *Oryx*, 47, 237–242, https://doi.org/10.1017/S0030605311001621

(11) McManus J.S., Dickman A.J., Gaynor D., Smuts B.H. & Macdonald B.W. (2015) Dead or alive? Comparing costs and benefits of lethal and non-lethal human-wildlife conflict mitigation on livestock farms. *Oryx*, 49, 687–695, https://doi.org/10.1017/S0030605313001610

(12) Potgieter G.C., Kerley G.I.H. & Marker L.L. (2016) More bark than bite? The role of livestock guarding dogs in predator control on Namibian farmlands. *Oryx*, 50, 514–522, https://doi.org/10.1017/S0030605315000113

3.36. Use loud noises to deter predation of livestock by mammals to reduce human-wildlife conflict

https://www.conservationevidence.com/actions/2435

- **Three studies** evaluated the effects of using loud noises to deter predation of livestock by mammals to reduce human-wildlife conflict. Two studies were in the USA[1,2] and one was in Mexico[3].

COMMUNITY RESPONSE (0 STUDIES)

POPULATION RESPONSE (0 STUDIES)

BEHAVIOUR (0 STUDIES)

OTHER (3 STUDIES)

- **Human-wildlife conflict (3 studies):** Three replicated studies (including two controlled studies), in the USA[1,2] and Mexico[3], found that loud noises at least temporarily deterred sheep predation[1] or food consumption[2] by coyotes and (combined with visual deterrents) deterred livestock predation by large predators[3].

Background

This intervention specifically refers to use of sound, from various sources, to deter predation on livestock by wild mammalian carnivores. If successful, such an intervention could reduce livestock losses and, thus, reduce motivation for carrying out lethal control of predators.

A replicated study in 1979–1980 of three ranches in North Dakota, USA (1) found that gas exploders temporarily deterred sheep predation by coyotes *Canis latrans*. Installation and use of gas exploders stopped predation for 17–102 days. Sites selected for the study had suffered ≥5 sheep losses to predation by coyotes in the previous two weeks. Following this, propane gas exploders were installed in the pastures. Exploders were operated until the grazing season was over or until ≥2 verified coyote kills occurred. Two to three exploders/site fired at 8–20-minute intervals overnight and were moved every 4–5 days. Sheep farmers were compensated for losses to coyotes provided that exploders were used as the sole means of control. The trial operated on three sites, with pastures extending over 56–255 ha, and containing 190–1,000 sheep.

A replicated, controlled study on captive animals in Utah, USA (2) found that playing loud noises deterred consumption of food by coyotes *Canis latrans*. Six of 14 coyote pairs did not eat food while loud noises were playing repeatedly, whilst all seven coyote pairs not played loud noises ate their food. Food consumption was reduced if loud noises were activated solely when coyotes approached food. Twenty-one pairs of coyotes were held in 0.1-ha pens. An alarm was suspended 2 m above the door to the pen, where 100 g of food was positioned. For seven coyote pairs, the alarm sounded every 7–9 seconds for 1 hour. For seven more pairs, it activated solely when they approached the food. For seven further coyote pairs, it was not activated. Behaviour of coyotes was observed for 1 hour.

A replicated, controlled study in 2010 of six farms in a forested area in central Mexico (3) found that sound and visual deterrents reduced predation of livestock on ranches. The relative effects of the two deterrent types were not assessed individually. No large predators (puma *Puma concolor* or jaguar *Panthera onca*) were detected on ranches that used deterrents compared with 2 detections/ranch and 2–4 livestock attacks/ranch where deterrents were not used. Out of six ranches (44–195 ha extent, ≥6 km apart), two cattle ranches and two goat ranches deployed deterrents, whilst no deterrents were deployed on one cattle ranch and one goat ranch. Sound deterrents were recordings of voices, motors, pyrotechnics, barking dogs and bells, played twice daily for 40 minutes, between 06:00–08:00 h and 20:00–22:00 h. Visual deterrents were shirts worn by livestock owners, hung around paddocks. Deterrents alternated weekly between sound and visual, through July–August 2010. Large predators were monitored using two camera traps/ranch and by searching for tracks and other signs.

(1) Pfeifer W.K. & Goos M.W. (1982) Guard dogs and gas exploders as coyote depredation control tools in North Dakota. *Proceedings of the Tenth Vertebrate Pest Conference*, Monterey, California, USA, 55–61.

(2) Shivik J.A. & Martin D.J. (2000) Aversive and disruptive stimulus applications for managing predation. *Proceedings -Wildlife Damage Management Conferences*, Pennsylvania, USA, 9, 111–119.

(3) Zarco-González M.M. & Monroy-Vilchis O. (2014) Effectiveness of low-cost deterrents in decreasing livestock predation by felids: a case in Central Mexico. *Animal Conservation*, 17, 371–378, https://doi.org/10.1111/acv.12104

3.37. Translocate predators away from livestock to reduce human-wildlife conflict

https://www.conservationevidence.com/actions/2436

- **Eleven studies** evaluated the effects on mammals of translocating predators away from livestock to reduce human-wildlife conflict. Four studies were in the USA[1,2,3,7] two were in Botswana[9,11], one each was in Canada[4], Zimbabwe[6] and Namibia[10], one was in Venezuela and Brazil[8] and one covered multiple locations in North and Central America and Africa[5].

COMMUNITY RESPONSE (0 STUDIES)

POPULATION RESPONSE (8 STUDIES)

- **Reproductive success (2 studies):** Two studies, in Zimbabwe[6] and Namibia[10], found that predators translocated away from livestock bred in the wild after release.

- **Survival (8 studies):** Four of eight studies (including three replicated studies and a systematic review), in the USA[2,7], Canada[4], Zimbabwe[6], South America[8], Botswana[9,11] and Namibia[10], found that translocating predators reduced their survival[7] or that most did not survive more than 6–12 months after release[4,9,11]. Three studies found that translocated predators had similar survival to that of established animals[2,10] or persisted in the wild[6] and one study could not determine the effect of translocation on survival[8].

BEHAVIOUR (0 STUDIES)

OTHER (6 STUDIES)

- **Human-wildlife conflict (6 studies):** Four of six studies (including a review and a systematic review), in the USA[1,2,3,7], South America[8] and in North and Central America and Africa[5], found that some translocated predators continued to predate livestock or returned to their capture sites[1,2,5,7]. One study found that translocated predators were not subsequently involved in livestock predation[3] and one study could not determine the effect of translocation on livestock predation[5].

Background

Where mammalian predators cause unacceptable losses to farmers, through predation on livestock, they may be translocated from their point of capture and released some distance away. The release site may be an area away from where livestock are kept. The intervention can fail if translocated animals continue to predate livestock or if survival of translocated animals is low. If the intervention is successful, it can reduce incentives for carrying out lethal control of such animals. Several other interventions cover translocations that are primarily for conservation of rare or threatened species, such as *Translocate to re-establish or boost populations in native range.*

A study in 1975–1978 of an extensive primarily forested area in Minnesota, USA (1; same experimental set-up as 2) found that gray wolves *Canis lupus* translocated away from sites of livestock predation or harassment were less likely to return to capture sites if moved when younger or across greater distances. Of 15 translocations of <64 km, nine endpoints (sites of mortality, recapture or last radiolocation) were at original capture sites. Of 20 translocations of >64 km, no endpoints were at original capture sites. None of nine pups, whose endpoints were determined (following translocation of 64 km (two pups) or 111–321 km (seven pups), returned to original capture locations. Between February 1975 and May 1978, 62 adult wolves and 45 four-to seven-month-old pups were caught in an area of livestock predation and harassment by wolves. Wolves were ear-tagged and released into forests, 50–331 km from capture sites. Forty-one wolves were released individually. Sixty-six were released in groups of 2–6. Fifteen adults and four pups were fitted with radio-collars. Seventeen of these were tracked from an aircraft for 1–588 days. Thirty-five endpoints in total were determined from 32 wolves (23 adults and nine pups — second endpoints were determined for three recaptured wolves that were translocated twice).

A study in 1975–1978 of an extensive primarily forested area in Minnesota, USA (2; same experimental set-up as 1) found that gray wolves *Canis lupus* translocated away from sites of livestock predation or harassment had similar survival to that of established wolves. Annual survival for 17 radio-collared wolves (60%) was similar to survival in three studies of established wolves in the region (65%, 66% and 21–100%). Between February 1975 and May 1978, sixty-two adult wolves and 45 four-to seven-month-old pups were caught in an area of livestock predation or harassment by wolves. Wolves were ear-tagged and released into forests, 50–331 km from capture sites. Forty-one wolves were released individually. Sixty-six were released in groups of 2–6. Fifteen adults and four pups were fitted with radio-collars. Seventeen of these were tracked from an aircraft for 1–588 days.

A study in 1989–1992 of forest and meadow in an area of Oregon, USA (3) found that black bears *Ursus americanus* translocated away from areas with histories of bear attacks on sheep were not subsequently involved in livestock predation. None of five radio-collared, translocated bears was involved in sheep predation during the monitoring period ($\leq$1 year). However, four of the bears died during that period (three were shot and one found dead) and one either moved away or its radio-collar malfunctioned. Sixteen bears were translocated in 1990 and five in 1991 from areas where five bears had been killed in 1989 to protect livestock. Bears were released $\leq$20 miles from capture sites. Bears translocated in 1991 were radio-collared. One was monitored for approximately one year. The others were monitored for shorter, unspecified, periods.

A replicated study in 1988–1990 across parts of Alberta, Canada (4) found that three cougars *Felis concolor* translocated following predation of livestock survived for between 3.5 months and at least one year after release. An adult female (4.3 years old) was translocated 51 km following sheep predation. She was found dead, from a bacterial infection, 3.5 months later. A 20-month-old male was translocated 51 km. One year later he was recaptured, 79 km from the release site, following reports of goat killings. He was released 43 km away but not subsequently monitored. A 15-month-old male was translocated 63 km after having killed a dog *Canis lupus familiaris*, and was shot by a licensed hunter, 20 km from the release site, nine months later. All three cougars had been previously caught and either ear-tagged or radio-collared for monitoring and research. In this study, the adult female was radio-tracked from an airplane.

A review published in 1997 of translocation studies in North and Central America and southern Africa (5) found that many carnivores translocated to prevent livestock conflict or 'nuisance' behaviours returned to capture sites and/or resumed predation or nuisance behaviour. Ten of 11 studies of brown bears *Ursus arctos* and black bears *Ursus americanus* found that 45–100% of translocated bears returned up to 229 km to their capture site. Eight leopards *Panthera pardus* translocated to a national park immediately left the park and some (number not specified) resumed livestock predation. A further animal returned and resumed livestock predation following an 80-km translocation. Two further animals did likewise following translocation over an unspecified distance. Of 25 lions *Panthera leo* translocated 5–300 km (pooled from two studies), at least six resumed livestock killing. Of two jaguars *Panthera onca* translocated 160 km, at least one resumed livestock killing. Relevant studies on translocations to reduce livestock predation or nuisance behaviours were gathered for black bear (seven studies), brown bear (four studies), leopard (three studies), lion (two studies) and jaguar (two studies).

A study in 1994–1998 in a woodland savanna protected area in northern Zimbabwe (6) found that a population of cheetahs *Acinonyx jubatus* translocated to reduce livestock losses, persisted over four years and that translocated animals reproduced in the wild. At least 13 adult cheetahs and four cubs, were alive four years after the translocation of 17 individuals. Translocated cheetahs bred at least five times and at least two cubs survived to adulthood. In 1993–1994, fourteen adult cheetahs and three cubs were released into Matusadona National Park. Cheetahs had been captured in commercial ranches where they were causing livestock losses. At the time of release, the park had no resident cheetahs but had a high density of lions (0.31/ km^2) and hyenas (0.13/ km^2). Cheetah numbers were estimated until July 1998, from sightings by visitors and park workers.

A study in 1982–2002 in 25 temperate forest sites in Montana, Idaho, and Wyoming, USA (7) found that some wolves *Canis lupus* translocated away from areas of livestock predation continued to prey on livestock, some returned to their capture location and that translocation reduced wolf survival. Out of 63 translocated individual wolves and nine wolf groups, 19 wolves preyed on livestock following release. Of 81 wolves or wolf groups, 16 returned to their capture site, from 74–316 km away. Annual survival of translocated wolves (60%) was lower than that of non-translocated, resident wolves (73%). Eighty-eight individual wolves were translocated 74–515 km in 1989–2001, in response to livestock predation (75 wolves) or pre-emptively to avoid such conflict (13 wolves). Seven translocated wolves were moved twice and five were moved three times. Translocated wolves were radio-collared, and were monitored to the end of 2002. Survival data were also compiled over 1982–2002 from 399 non-translocated, resident wolves in the same general area.

A systematic review published in 2010 of studies in forest and savanna areas in Venezuela and Brazil (8) found insufficient evidence to determine whether or not translocating jaguars *Panthera onca* reduced livestock predation by jaguars, or hunting of jaguars or whether it increased survival of translocated individuals. Ten studies met review criteria. Of these, seven provided only qualitative data, whilst the three quantitative studies had methodological limitations. No evidence was identified for effectiveness of translocation in reducing livestock predation by jaguars or reducing hunting of jaguars. Of 14 translocated jaguars, four survived translocation and the follow-up monitoring period of three weeks to eight months, four died during capture or post-release monitoring and six further animals were insufficiently monitored to determine post-release survival. Keyword and database searches were used to collect 3,200 articles evaluating jaguar translocation. Of these, 10 met pre-defined criteria for inclusion in the review.

A replicated study in 2001–2008 on two savanna game reserves in Botswana (9) found that following translocation of four leopards *Panthera pardus* involved in livestock predation, three did not survive more than six months after release. Of four stock-raiding leopards translocated to a protected area, three were shot within six months, having left the release area and resumed livestock predation. The fourth animal returned to, and settled back within, its initial capture area. By comparison, four leopards resident within the protected area had stable home ranges. Four leopards (three male and one female), which were suspected of predating livestock, were released in a protected area, 33–158 km from capture sites. These animals, and four leopards resident in the protected area (one male, three female), were monitored by a combination of radio-and satellite-tracking between April 2001 and March 2008, for between 23 days and 53 months.

A controlled study in 2004–2014 across five regions of Namibia (10) found that following translocation (mostly of animals moved from sites of livestock predation), survival rates and home range sizes of leopards *Panthera pardus* did not differ significantly from those of resident leopards and that translocated females reproduced in the wild. The average annual survival rate of the six translocated leopards (93%) was not significantly different to that of 12 resident leopards (85%). The same applied for home range sizes (translocated: 54–481 km²; resident: 36–580 km²). Two of three translocated females reproduced in the wild, with conception occurring from eight months post-release. Livestock predation ceased for 16–29 months or entirely at pre-translocation capture sites, and was then lower (1–3 calves/year) than before translocation (5 calves in one year). Only one of six translocated leopards killed livestock (herded into range) at release sites. Eighteen leopards were trapped and fitted with GPS (14) or VHF (5) transmitter collars. Twelve were released at or close to their capture sites and six (4 'problem' animals) were released at an average distance of 403 km (47–754 km) from their capture site. Translocated animals spent an average of 203 days in captivity before release. VHF-tagged leopards were monitored at least weekly and GPS-tagged individuals were monitored daily, for an average of 718 days for translocated animals and 465 days for resident animals.

A replicated study in 2003–2011 of savanna and farmland at several sites across Botswana (11) found that nine of 11 cheetahs *Acinonyx jubatus* translocated away from farms, for livestock protection reasons, survived for less than one year. Eight translocated male cheetahs survived for 46 to at least 981 days (average 106) after release. Three females survived for 21–95 days (average 31) after release. Nine of the 11 cheetahs were known to have died (three were shot and for six, the cause of death was unknown). On one animal, the GPS-collar failed after 981 days and the outcome for one animal was unknown. Twenty-one cheetah social groups, involving 39 animals, were translocated. They were held for 0–16 days and then released 28–278 km from capture sites. Eleven translocated animals were monitored using satellite-or GPS-collars.

(1) Fritts S.H., Paul W.J. & Mech L.D. (1984) Movements of translocated wolves in Minnesota. The Journal of Wildlife Management, 48, 709–721, https://doi.org/10.2307/3801418

(2) Fritts S.H., Paul W.J. & Mech L.D. (1985) Can relocated wolves survive? *Wildlife Society Bulletin*, 13, 459–463.

(3) Armistead A.R., Mitchell K. & Connolly G.E. (1994) Bear relocations to avoid bear/sheep conflicts. *Proceedings of the 16th Vertebrate Pest Conference*, 31–35.

(4) Ross P.I. & Jalkotzy M.G. (1995) Fates of translocated cougars, *Felis concolor*, in Alberta. *The Canadian Field-Naturalist*, 109, 475–476.

(5) Linnell J.D.C., Aanes R., Swenson J.E., Odden J. & Smith M.E. (1997) Translocation of carnivores as a method for managing problem animals: a review. *Biodiversity and Conservation*, 6, 1245–1257.

(6) Purchase G.K. (1998) The Matusadona cheetah project: lessons from a wild-to-wild translocation. *Proceedings of a Symposium on Cheetahs as Game Ranch Animals*, Onderstepoort, 83–89.

(7) Bradley E.H., Pletscher D.H., Bangs E.E., Kunkel K.E., Smith D.W., Mack C.M., Meier T.J., Fontaine J.A., Niemeyer C.C. & Jimenez M.D. (2005) Evaluating wolf translocation as a nonlethal method to reduce livestock conflicts in the Northwestern United States. *Conservation Biology*, 19, 1498–1508.

(8) Isasi-Catala E. (2010) Is translocation of problematic jaguars (*Panthera onca*) an effective strategy to resolve human-predator conflicts? *CEE review* 8-018, SR55.

(9) Weilenmann M., Gusset M., Mills D.R., Gabanapelo T. & Schiess-Meier M. (2010) Is translocation of stock-raiding leopards into a protected area with resident conspecifics an effective management tool? *Wildlife Research*, 37, 702–707, https://doi.org/10.1071/WR10013

(10) Weise F.J., Lemeris J., Stratford K.J., van Vuuren R.J., Munro S.J., Crawford S.J., Marker L.L. & Stein A.B. (2015) A home away from home: insights from successful leopard (*Panthera pardus*) translocations. *Biodiversity and Conservation*, 24, 1755–1774.

(11) Boast L.K., Good L. & Klein R. (2016) Translocation of problem predators: is it an effective way to mitigate conflict between farmers and cheetahs *Acinonyx jubatus* in Botswana? *Oryx*, 50, 537–544, https://doi.org/10.1017/S0030605315000241

3.38. Provide diversionary feeding to reduce predation of livestock by mammals to reduce human-wildlife conflict

https://www.conservationevidence.com/actions/2437

- **Two studies** evaluated the effects of providing diversionary feeding to reduce predation of livestock by mammals to reduce human-wildlife conflict. One study was in the USA[1] and one was in Canada[2].

COMMUNITY RESPONSE (0 STUDIES)

POPULATION RESPONSE (1 STUDY)

- **Reproductive success (1 study):** A controlled study in the USA[1] found that diversionary feeding of predators did not increase overall nest success rates for ducks.

BEHAVIOUR (0 STUDIES)

OTHER (2 STUDIES)

- **Human-wildlife conflict (2 studies):** One of two studies (one controlled, one before-and-after study) in the USA[1] and Canada[2] found that diversionary feeding reduced striped skunk predation on duck nests. The other study found that diversionary feeding of grizzly bears did not reduce predation on livestock[2].

Background

Mammalian predators can cause unacceptable losses to farmers, through predation on livestock. If diversionary feeding can reduce the extent to which animals exhibit nuisance behaviour, this may reduce motivations for carrying out lethal control or other intensive management. See also: *Provide diversionary feeding to reduce crop damage by mammals to reduce human-wildlife conflict* and *Residential and commercial development — Provide diversionary feeding for mammals to reduce nuisance behaviour and human-wildlife conflict.*

A controlled study in 1993–1994 of 24 upland prairie areas in North Dakota, USA (1) found that diversionary feeding of predators reduced striped skunk *Mephitis mephitis* predation on duck *Anas* spp. nests, but overall nest success rates did not increase significantly. The proportion of predation events on large-clutch duck nests by striped skunks was lower in areas with diversionary feeding (11%) than in areas without feeding (24%). However, the proportion of duck nests in which at least one egg hatched did not differ significantly between feeding areas (41%) and areas without food provision (29%). In April–July 1993 and 1994, supplementary food (90–100 kg of fish offal and sunflower seeds) was distributed within 1–2 plots (50 x 200–300 m) in each of 12 areas every 3–4 days. Twelve control areas had no supplementary food. Each area contained 33–83 ha of upland nesting cover and was managed for duck production. In May–July 1993 and 1994, three searches for duck nests were conducted in each of the 24 areas using a vehicle-towed chain drag. A total of 1,008 nests (609 in feeding areas; 399 in areas without supplementary food) were marked and checked every 6–21 days or until abandoned/destroyed.

A before-and-after study in 1982–2013 in a forested and agricultural area of southwestern Alberta, Canada (2) found that diversionary feeding of grizzly bears *Ursus arctos* did not reduce predation on livestock. The frequency of grizzly bear-livestock incidents during the spring did not differ significantly during 14 years before (average 0.8 incidents/year) and 15 years after (average 3.3 incidents/year) diversionary feeding commenced. Road-killed ungulate carcasses were dropped by helicopter at sites close to grizzly bear dens each spring during 1998–2013. In 2012 and 2013, 149–160 carcasses were dropped at 14–15 sites in March–April (details for earlier years are not reported). All sites were within a 3,600-km² area comprising forested mountains adjacent to agricultural land. Remote trail cameras at feeding sites recorded grizzly bears. Complaint data (reports of grizzly bears harassing, mauling or killing livestock) were analysed for March–June in each year before (1982–1995) and after (1998–2013) diversionary feeding commenced.

(1) Greenwood R.J., Pietruszewski D.G. & Crawford R.D. (1998) Effects of food supplementation on depredation of duck nests in upland habitat. *Wildlife Society Bulletin*, 26, 219–226.

(2) Morehouse A.T. & Boyce M.S. (2017) Evaluation of intercept feeding to reduce livestock depredation by grizzly bears. *Ursus*, 28, 66–80, https://doi. org/10.2192/URSU-D-16-00026.1

3.39. Keep livestock in enclosures to reduce predation by mammals to reduce human-wildlife conflict

https://www.conservationevidence.com/actions/2438

- **One study** evaluated the effects of keeping livestock in enclosures to reduce predation by mammals to reduce human-wildlife conflict. This study was in Portugal[1].

COMMUNITY RESPONSE (0 STUDIES)

POPULATION RESPONSE (0 STUDIES)

BEHAVIOUR (0 STUDIES)

OTHER (1 STUDY)

- **Human-wildlife conflict (1 study):** A replicated study in Portugal[1] found fewer wolf attacks on cattle on farms where cattle were confined for at least some of the time compared to those with free-ranging cattle.

Background

Free-ranging livestock may be more vulnerable to attacks by predators than those contained indoors or in enclosures close to farm buildings. Here we consider the effectiveness of such methods of animal husbandry. If successful, this intervention could reduce incentives for carrying out lethal control of predators.

See also *Exclude wild mammals using ditches, moats, walls or other barricades to reduce human-wildlife conflict.*

A replicated, site comparison study in 2012–2014 of 68 cattle farms in a mountainous region dominated by agricultural land, forests and shrubs in northern Portugal (1) found that farms that often kept cattle in barns or enclosures suffered fewer wolf *Canis lupus* attacks than did farms with free-ranging cattle. The average annual number of wolf attacks was lower on farms that often confined cattle (2.4 attacks/year) than on farms with free-ranging cattle (9.0 attacks/year). Eighteen farms suffered no wolf attacks, 42 had 1–9 wolf attacks and eight had >9 wolf attacks. The study was conducted in an area of approximately 20,000 km². Semi-structured interviews were conducted in 2013–2014 with 68 cattle farmers reporting high or low levels of wolf-attacks during 2012–2013. Interview responses were used to classify farms as those that often confined cattle within fences or in barns year-round, or those using a free-ranging system, in which animals were rarely confined with fences or in barns (except at night during winter).

(1) Pimenta V., Barros I., Boitani L. & Beja P. (2017) Wolf predation on cattle in Portugal: Assessing the effects of husbandry systems. *Biological Conservation*, 207, 17–26, https://doi.org/10.1016/j.biocon.2017.01.008

3.40. Install electric fencing to protect crops from mammals to reduce human-wildlife conflict

https://www.conservationevidence.com/actions/2439

- **Eleven studies** evaluated the effects of installing electric fencing to protect crops from mammals to reduce human-wildlife conflict. Three studies were in Japan[4,7,9], three were in the USA[1,6,10], two were in the UK[2,3] and one each was in Namibia[5], India[8] and Guinea-Bissau[11].

COMMUNITY RESPONSE (0 STUDIES)

POPULATION RESPONSE (0 STUDIES)

BEHAVIOUR (0 STUDIES)

OTHER (11 studies)

- **Human-wildlife conflict (11 studies):** Nine of 11 studies (including three before-and-after studies and three controlled studies), in the USA[1,6,10], the UK[2,3], Japan[4,7,9], Namibia[5], India[8] and Guinea-Bissau[11], found that electric fences deterred crossings by mammals, ranging in size from European rabbits[2] to elephants[8]. Two studies had mixed results, with some fence designs deterring elephants[5] and black bears[10].

> **Background**
>
> Wild mammals can compete with domestic herbivores for food, can predate domestic herbivores or can damage crops. Human-wildlife conflict can be reduced if wild mammals can be effectively excluded from fields. Electric fences are extensively used and can reduce the risk of wild mammal incursions into such fields. If successful, they may reduce incentives for carrying out lethal control of such mammals.

A before-and-after study in 1961–1965 in a forest in New York State, USA (1) found that an electric fence reduced browsing on hardwood trees by white-tailed deer *Odocoileus virginanus*. Three years after fence erection, there were more unbrowsed stems inside the fence (43 unbrowsed stem/plot) than outside (16 unbrowsed stems/plot). There had been no difference in browsing rates before fence erection (inside fence line: 22 unbrowsed stems/plot; outside fence line: 22 unbrowsed stems/plot). The fence (2.5 miles perimeter) consisted of five wires, with the lower three electrified from November 1961. Browsing intensity was measured in plots measuring one rod-square (approximately 25 m²). Twenty plots inside and 20 outside the fence were surveyed in 1961 and 1964.

A replicated, before-and-after, site comparison study in 1980–1983 on 24 arable sites in southern UK (2) found that electric fences reduced European rabbit *Oryctolagus cuniculus* numbers on crops. Rabbit numbers fell on plots protected by a Flexinet® fence (0–4 weeks after erection: 6.7 rabbits/count; 5–20 weeks after erection: 7.6 rabbits/count; before erection: 42.7 rabbits/count) and a Livestok® fence (0–4 weeks after erection: 10.1 rabbits/count; 5–20 weeks after erection: 17.6 rabbits/count; before erection: 48.0 rabbits/count). Rabbit numbers in unfenced plots remained constant throughout (0–4 weeks after erection: 15.9 rabbits/count; 5–20 weeks after erection: 13.3 rabbits/count; before erection: 13.6 rabbits/count). Electric fences (0.5 m high) were erected along one side of winter barley fields. Flexinet® (seven sites) had 80 ×

80-mm mesh and Livestok® (seven sites) had 500 × 50-mm mesh. Ten unfenced sites were also monitored. Adult rabbits were counted using spotlights and binoculars in November–April between 1980 and 1983.

A controlled study in 1988–1989 on an arable farm in Devon, UK (3) found that electric fencing reduced damage to an oat *Avena sativa* crop by badgers *Meles meles* in one of two years. Results were not tested for statistical significance. In the first year, 1.8–2.6% of crop area in fields protected by electric fencing was damaged by badgers, compared to 9.6% in an unfenced field. In the second year, 2.2–4.3% of fenced crop was damaged compared to 1% of unfenced crop. Electric fences around two fields had parallel wires at 10 cm and 20 cm above the ground. Wires were connected to a fence energiser, powered from a 12-volt battery. A third field was unfenced. Vegetation short circuited the fence, especially in 1988. In 1989, dry conditions may have reduced soil conductivity, thus reducing fence voltage. Damage (mostly flattened stalks) was assessed by walking crops in August 1988 and 1989. Additionally, 1988 data were verified using aerial photographs.

A replicated study in 1997–1998 of 24 crop fields and two areas of beehives adjacent to woodlands in Nagano prefecture, Japan (4) found that electric fences prevented raids by Asiatic black bears *Ursus thibetanus*. No bears got through any of the electric fences. Bear activity near fences was documented 23 times, including three bears departing after touching the fence, one trying unsuccessfully to dig under the fence and eight raids on unprotected fields within 13–120 m of fences. In July–October of 1997 and 1998, twenty-four sweetcorn fields and two areas of beehives (area enclosed 0.001–0.75 ha) with recent history of bear-raids were fenced using Gallagher power fence systems for 2–65 nights/fence. Fences comprised four wires at 24 cm intervals with a further wire 30 cm outside the fence and 30 cm above the ground.

A replicated, before-and-after study in 1991–1995 on farmland and grassland at four sites in East Caprivi, Namibia (5) found that some electric fences reduced crop losses to elephants *Loxodonta africana*. At one village, where 31 farms were enclosed within a 9.5-km-long permanent electric fence, there were no compensation claims for losses to elephants over two years following installation, compared to 30 claims over the previous three years. A 4-km-long permanent electric fence at another site was unsuccessful, due to inadequate installation or maintenance. At a third site, temporary electric fences kept out elephants at one village in one year. In the second year, the fence was effective but elephants were able to walk around the side. At a fourth temporary fence site, no elephants returned after electric fence installation, so its effectiveness was untested. The two, 2 m-high, permanent steel wire electric fences comprised two strands of 2-mm steel wire attached to trees or poles. The temporary fences (<2 km long) at two villages comprised polyurethane cords which were threaded with wire strands and strung between trees. Fences were powered by 12-volt batteries. Data were collated from questionnaire surveys in 1991–1995.

A randomized, replicated, controlled study in 2002–2004 at a woodland and grassland site in Ohio, USA (6) found that electric fencing deterred white-tailed deer *Odocoileus virginianus* when turned on. Significantly fewer deer entered enclosures with electric fencing (0–1 deer/day) than entered enclosures without fencing (72–86 deer/day). When power was applied to fencing in week two, deer entries decreased 88–99%. When power was delayed 10 weeks, entries decreased 90%. When power was turned on and off within a 4-week period, entries decreased 57%. Corn consumption was lower in powered (<2–6.4 kg/day) than in unpowered sites (15–32 kg/day). Ten sites (> 1 km apart) each had two 5 × 5 m enclosures (9 m apart), fenced on three sides, each containing a feed trough that measured food (corn) consumption. Infra-red cameras monitored enclosures. In February 2002, 1.3-m-high electric fencing (7 kV; ElectroBraid™) was installed around one enclosure in each pair. After one week, the treatment and control were swapped. In March 2002, one feed trough was removed from each pair, leaving five sites with troughs, surrounded by electric fencing and five unfenced troughs, for three weeks. In December 2002, all sites had electric fencing but five had it turned on and five off for one week. Power was then off for two weeks and then the same repeated. Treatment and control sites were then swapped (10 weeks since start) with the power on for three weeks at treatment sites. In January 2004, five were fenced and five were controls without fencing, for six weeks. Before each trial there was a week with no treatments.

A study in 2007–2008 of three fences in Japan (7) found that electric fencing was effective at excluding a range of large and medium-sized wild mammals. No mammals were recorded inside any fences. Outside the lowest fence, there were 157 occurrences of eight species. Outside the intermediate-height fence, there were 96 occurrences of eight species. Outside the highest fence, there were 117 occurrences of three species. Japanese macaques *Macaca fuscata*, which can climb non-electrified fences, were among animals excluded at the highest fence. Fences enclosed areas of 100–930 m². They comprised metallic 15 × 29 mm mesh in 0.6-m-high × 1.8-m-wide sections. The lowest fence (0.6 m high) was a single section high. The intermediate fence (1.6 m high) comprised a single wire between two mesh sections. The highest fence (1.8 m high) comprised three wires and nylon netting between two mesh sections, with two ground wires above. A current (2,000–6,500 V) ran through metallic parts. A corrugated polyvinyl chloride sheet insulated the fence bottom from the ground.

A study in 2006–2009 in two areas of Assam, India (8) found that electric or chili fences reduced the probability of Asian elephants *Elephas maximus* damaging crops. The effectiveness specifically of electric fences was not analysed. The chance of crop damage occurring was lower when fences provided a barrier to crop-raiding elephants, compared to a range of other interventions or to no intervention (results presented as statistic model coefficients). However, loud noises alongside fences reduced their effectiveness. Within two study areas, 33 community members trained as monitors recorded 1,761 crop-raiding incidents, from 1 March 2006 to 28 February 2009. A range of deterrent methods, used singly or in combination, included two-strand electric fences, chili fencing (engine grease and ground chili paste, on a jute or coconut rope), chili smoke (from burning dried chilies, tobacco, and straw), spotlights, elephant drives (repelling wild elephants using domesticated elephants), fire and noise.

A replicated study in 2010 at four arable sites in Japan (9) found that a modified electric fence design was effective at excluding large and medium-sized mammals from crops. Fewer animals were recorded inside fences (0–3) than outside fences (60–327). Raccoon dog *Nyctereutes procyonoides* (one occurrence), sika deer *Cervus nippon* (two) and wild boar *Sus scrofa* (one) crossed fences. The most frequently recorded mammals outside fences were wild boar (112 occurrences), sika deer (373) and Japanese macaque *Macaca fuscata* (117). Four fences enclosed cops covering 100–1,700 m². They comprised insulated fiberglass poles

(8.5 mm diameter, 2.1 m long) at 2.5-m intervals. Nine electrified wires (0.9 mm diameter) were attached, up to 1.7 m high. Nylon net (45-mm mesh) was attached to the full fence height. Poles were flexible, so animals attempting to climb would retain ground contact and hence be shocked. Measured voltages were 3,600–6,800 V. Fences were checked at least weekly. Animals were monitored inside and outside fences using infrared-triggered cameras for ≥5 months from April–November 2010.

A site comparison study in 2010 in a forested area in Michigan, USA (10) found that two of four electric fence designs successfully excluded black bears *Ursus americanus*. Two of four electric fence designs excluded 100% of black bears from accessing bait within fenced enclosures during a total of 30–38 fence interactions. Bears breached the other two fence designs and accessed bait on three occasions during a total of 48–52 fence interactions. Each of four electric fence designs was tested at 2–3 baited sites within a 17-km² forested area. The fences enclosed a 13-m² area filled with 4–13 l of bait/day (including bread, cookies, trail mix, honey, bacon, sardines etc.). Fences were constructed with 2–3 rows of white polytape (1.3 cm) at different spacings (23–58 cm from the ground) and charged with 5,000 V (see original paper for details). Each site was baited for an average of three nights prior to fencing and was visited by bears during this time. Infrared cameras recorded bears interacting with the fences during 2–5 nights/site in June–August 2010.

A replicated, controlled study in 2008–2012 of 100 rice fields in the Bijagos archipelago and Oio and Gabau regions, Guinea Bissau (11) found that electric fences deterred hippopotamus *Hippopotamus amphibius* entry into fields. The proportion of fenced fields where hippopotamuses were detected (1.3%) was lower that of unfenced fields (80.0%). Hippopotamuses were monitored in 100 rice fields in 2008–2011 in Orango Islands National Park and Uno Island and, in 2012–2013, in Cacheu National Park. Seventy-five rice fields had electric fences and 25 were unfenced. Fences were 80 cm high, were made out of 2.5-mm-diameter aluminium wire, connected to an energizer unit. Fences also comprised rope between wooden stakes, with strips of red and white striped plastic at 1-m intervals. Vegetation was cut from within 2–3 m around the wires twice each week. Fenced and unfenced fields were surveyed every 3–4 days for hippopotamus footprints.

(1) Tierson W.C. (1969) Controlling deer use of forest vegetation with electric fences. *The Journal of Wildlife Management*, 33, 922–926, https://www.jstor.org/10.2307/3799326

(2) McKillop I.G. & Wilson C.J. (1987) Effectiveness of fences to exclude European rabbits from crops. *Wildlife Society Bulletin*, 15, 394–401.

(3) Wilson C.J. (1993) Badger damage to growing oats and an assessment of electric fencing as a means of its reduction. *Journal of Zoology*, 231, 668–675, https://doi.org/10.1111/j.1469-7998.1993.tb01949.x

(4) Huygens O. & Hayashi H. (1999) Using electric fences to reduce Asiatic black bear depredation in Nagano prefecture, central Japan. *Wildlife Society Bulletin*, 27, 959–964.

(5) O'Connell-Rodwell C.E., Rodwell T., Rice M. & Hart L.A. (2000) Living with the modern conservation paradigm: can agricultural communities co-exist with elephants? A five-year case study in East Caprivi, Namibia. *Biological Conservation*, 93, 381–391, https://doi.org/10.1016/S0006-3207(99)00108-1

(6) Seamans T.W. & VerCauteren K.C. (2006) Evaluation of ElectroBraide™ fencing as a white-tailed deer barrier. *Wildlife Society Bulletin*, 34, 8–15, https://doi.org/10.2193/0091-7648(2006)34[8:EOEFAA]2.0.CO;2

(7) Honda T., Miyagawa Y., Ueda H. & Inoue M. (2009) Effectiveness of newly-designed electric fences in reducing crop damage by medium and large mammals. *Mammal Study*, 34, 13–17, https://doi.org/10.3106/041.034.0103

(8) Davies T.E., Wilson S., Hazarika N., Chakrabarty J., Das D., Hodgson D.J. & Zimmermann A. (2011) Effectiveness of intervention methods against crop-raiding elephants. *Conservation Letters*, 4, 346–354, https://doi.org/10.1111/j.1755-263X.2011.00182.x

(9) Honda T., Kuwata H., Yamasaki S. & Miyagawa Y. (2011) A low-cost, low-labor-intensity electric fence effective against wild boar, sika deer, Japanese macaque and medium-sized mammals. *Mammal Study*, 36, 113–117, https://doi.org/10.3106/041.036.0203

(10) Otto T.E. & Roloff G.J. (2015) Black bear exclusion fences to protect mobile apiaries. *Human–Wildlife Interactions*, 9, 78–86, https://doi.org/10.26077/dn8a-3941

(11) González L.M., Montoto F.G., Mereck T., Alves J., Pereira J., de Larrinoa P.F., Maroto A., Bolonio L. & El-Kadhir N. (2017) Preventing crop raiding by the Vulnerable common hippopotamus *Hippopotamus amphibius* in Guinea-Bissau. *Oryx*, 51, 222–229, https://doi.org/10.1017/S003060531500109X

3.41. Install metal grids at field entrances to prevent mammals entering to reduce human-wildlife conflict

https://www.conservationevidence.com/actions/2440

- **Two studies** evaluated the effects on mammal incursions of installing metal grids at field entrances to reduce human-wildlife conflict. Both of these studies were in the USA[1,2].

COMMUNITY RESPONSE (0 STUDIES)

POPULATION RESPONSE (0 STUDIES)

BEHAVIOUR (0 STUDIES)

OTHER (2 STUDIES)

- **Human-wildlife conflict (2 studies):** One of two replicated studies (including one controlled study), in the USA[1,2], found that deer guards (horizontal, ground-level metal grids) reduced entry into enclosures by white-tailed deer[2] whilst the other found that they did not prevent crossings by mule deer or elk[1].

Background

Wild herbivores can compete with domestic herbivores for food and can damage crops. Fencing can exclude wild herbivores from fields but entranceways remain vulnerable to incursions, especially were regular vehicle access is required. Metal grids (sometimes known as cattle grids) fitted across field entrances may be used to exclude wild herbivores. If successful, this could reduce incentives for carrying out lethal control of such species.

See also *Install wildlife exclusion grates/cattle grids* for studies where the intention is to exclude herbivore access to roads rather than into fields.

A replicated study in 1972–1973 of two fences in Colorado, USA (1) found that steel rail deer guards did not prevent crossings through vehicle openings by mule deer *Odocoileus hemionus hemionus* or elk *Cervus canadensis*. In test conditions, 16 of 18 mule deer released adjacent to 12, 18 or 24-foot-wide guards, crossed the guards, in an average time of 173 s. During natural encounters, 11 mule deer and one elk crossed a 24-ft-long guard and four mule deer crossed a 12-ft-long guard. There were at least 11 approaches by mule deer and three by elk in which animals did not then cross. Guards, at vehicle openings in 8-ft-high fences, comprised flat steel rails, 0.5-inch-wide, 4 inches high and 120 inches long, set 4 inches apart. Rails were perpendicular to the traffic direction. Eighteen deer were released in situations where guard crossing providing the only exit. Deer and elk tracks, from natural encounters with two guards, were examined periodically, between 29 June 1972 and 19 April 1973.

A replicated, controlled study in 2006–2007, in three forest and grassland sites in Ohio, Iowa and Wisconsin, USA (2) found that deer guards (ground-level roller grids) reduced white-tailed deer *Odocoileus virginianus* entry into enclosures. Deer guards at two sites excluded more deer than did open enclosures (data not presented). At the third site, deer did not cross one deer guard but there were 2.5 incursions/day at the other compared to 0.4 incursions/day in open enclosures at that site. Deer-resistant enclosures (6 m × 6 m, baited with alfalfa cubes) were constructed at three sites. At each site, two enclosures (one each in forest and grassland) had a deer guard (a grid of rollers over a 1.5 × 3 m pit) and two (one each in forest and grassland) had open gateways. Deer incursions into enclosures were monitored using camera traps from December 2006 to April 2007.

(1) Reed D.F., Pojar T.M. & Woodard T.N. (1974) Mule deer responses to deer guards. *Journal of Range Management*, 27, 111–113.

(2) VerCauteren K.C., Seward N.W., Lavelle M.J., Fischer J.W. &Phillips G.E. (2009) Deer guards and bump gates for excluding white-tailed deer from fenced resources. *Human-Wildlife Conflicts*, 3, 145–153, https://doi.org/10.26077/sb9r-sh17

3.42. Install automatically closing gates at field entrances to prevent mammals entering to reduce human-wildlife conflict

https://www.conservationevidence.com/actions/2441

- **One study** evaluated the effects on mammal movements of installing automatically closing gates at field entrances to reduce human-wildlife conflict. This study was in USA[1].

COMMUNITY RESPONSE (0 STUDIES)

POPULATION RESPONSE (0 STUDIES)

BEHAVIOUR (0 STUDIES)

OTHER (1 STUDY)

- **Human-wildlife conflict (1 study):** A replicated, controlled study, in the USA[1] found that vehicle-activated bump gates prevented white-tailed deer from entering enclosures.

Background

Wild mammals can compete with domestic herbivores for food, can predate domestic herbivores or can damage crops. Human-wildlife conflict can be reduced if wild mammals can be effectively excluded from fields. Gates through fences can provide crossing points if there is a risk of the gate being left open. Gates that close automatically may reduce the risk of wild mammals entering such fields. If successful, this may reduce incentives for carrying out lethal control of such mammals.

A replicated, controlled study, in 2006–2007, in three forest and grassland sites in Ohio, Iowa and Wisconsin, USA (1) found that vehicle-activated bump gates prevented white-tailed deer *Odocoileus virginianus* entry into enclosures. Bump gates excluded deer from all enclosures. At enclosures without bump gates, there were averages across the three sites of 0.4, 33.0 and 49.0 deer entries/day. However, supplementary tests on a separate bump gate revealed that it did not always close securely following vehicle passage. Deer-resistant enclosures (6 × 6 m, baited with alfalfa cubes) were constructed at three sites. At each site, two enclosures (one each in forest and grassland) had bump gates installed (designed to open upon low-speed vehicle contact and close after vehicle passage) and two (one each in forest and grassland) had open gateways. Deer movements into enclosures were monitored using camera traps from December 2006 to April 2007.

(1) VerCauteren K.C., Seward N.W., Lavelle M.J., Fischer J.W. & Phillips G.E. (2009) Deer guards and bump gates for excluding white-tailed deer from fenced resources. *Human-Wildlife Conflicts*, 3, 145–153, https://doi.org/10.26077/sb9r-sh17

3.43. Use tree nets to deter wild mammals from fruit crops to reduce human-wildlife conflict

https://www.conservationevidence.com/actions/2442

• We found no studies that evaluated the effects of using tree nets to deter mammals from fruit crops to reduce human-wildlife conflict.

'We found no studies' means that we have not yet found any studies that have directly evaluated this intervention during our systematic journal and report searches. Therefore, we have no evidence to indicate whether or not the intervention has any desirable or harmful effects.

Background

Tree nets can be used to close off tree canopy pathways or other access in order to protect fruit crops from being accessed by mammals. Netting is cheap to install but can be labour intensive for subsistence farmers. If successful in protecting fruit crops, use of nets could reduce incentives for carrying out lethal control of mammals.

3.44. Deter predation of livestock by mammals by having people close by to reduce human-wildlife conflict

https://www.conservationevidence.com/actions/2444

- **One study** evaluated the effects of deterring predation of livestock by mammals by having people close by to reduce human-wildlife conflict. This study was in Kenya[1].

COMMUNITY RESPONSE (0 STUDIES)

POPULATION RESPONSE (0 STUDIES)

BEHAVIOUR (0 STUDIES)

OTHER (1 STUDY)

- **Human-wildlife conflict (1 study)**: One study in Kenya[1] recorded fewer attacks by predators on livestock in bomas when people were also present but the presence of people did not reduce predator attacks on grazing herds.

Background

Domestic livestock may be vulnerable to mammalian predators. Livestock can be guarded by animals, especially dogs *Canis lupus familiaris*, or by people (or both). This intervention involves people remaining close to livestock, either actively guarding or simply as a passive deterrent, such as by bringing livestock in at night to an area adjacent to human habitation. If the intervention results in fewer livestock being predated, this could reduce incentives for carrying out lethal control of predators.

A study in 2001–2005 of bushland and savanna across Laikipia and neighbouring districts, Kenya (1) found that when livestock in bomas were accompanied by people, fewer animals were attacked by carnivores, but there was no similar effect for grazing herds. Livestock kept in bomas overnight were less likely to be attacked when more herders were present. Presence of herders did not reduce the risk of attack for herds grazing away from bomas in the daytime (results presented as odds ratios). The 502 grazing herds were accompanied by an average of 2.1 herders. At 491 bomas, an average of 11.3 people were present. The study documented 105 attacks by spotted hyenas *Crocuta crocuta*, 96 by leopards *Panthera pardus*, 44 by African wild dogs *Lycaon pictus*, 35 by lions *Panthera leo* and 19 by cheetahs *Acinonyx jubatus*. From January 2001 to June 2005, eighteen local staff verified reports of livestock lost to predation and gathered data on animal husbandry practices used. Attacked herds or bomas were compared to nearby herds (median 656 m away) or bomas (median 323 m away) that had not been attacked.

(1) Woodroffe R., Frank L.G., Lindsey P.A., ole Ranah S.M.K. & Romañach S. (2007) Livestock husbandry as a tool for carnivore conservation in Africa's community rangelands: a case-control study. *Biodiversity and Conservation*, 16, 1245–1260, https://doi.org/10.1007/978-1-4020-6320-6_28

3.45. Deter predation of livestock by herding livestock using adults instead of children to reduce human-wildlife conflict

https://www.conservationevidence.com/actions/2445

• **One study** evaluated the effects on predatory mammal activities of herding livestock using adults instead of children to reduce human-wildlife conflict. This study was in Cameroon[1].

COMMUNITY RESPONSE (0 STUDIES)

POPULATION RESPONSE (0 STUDIES)

BEHAVIOUR (0 STUDIES)

OTHER (1 STUDY)

• **Human-wildlife conflict (1 study):** A site comparison study in Cameroon[1] found that using adults to herd livestock reduced losses through predation relative to that of livestock herded solely by children.

Background

Domestic livestock may be vulnerable to mammalian predators. Livestock may be guarded by people to deter predators. In some areas, guarding is routinely carried out by children. This intervention refers to guarding by adults instead of children.

A site comparison study in 2008 of savanna around a national park in Cameroon (1) found that using adults to herd livestock reduced losses through predation relative to livestock herded by children. Among resident pastoralist households, fewer livestock were lost to carnivores when the livestock were herded by adults (two animals/year) than by children (eight animals/year). Among nomadic pastoralist households, there were also fewer livestock lost to carnivores when herded by adults (five animals/year) than by children (16 animals/year). Among resident pastoralists that herded livestock, 42% of herders (60 herders) were adults. Among nomadic pastoralists that herded livestock, 72% (124 herders) were adults. Two hundred and seven resident pastoralists and 174 nomadic pastoralists were interviewed. Pastoralists reported the incidence of predation of livestock by large carnivores and details of animal husbandry techniques used. Villages studied were selected based on tracked movements of radio-collared lions.

(1) Tumenta P.N., de Iongh H.H., Funston P.J. & Udo de Haes H.A. (2013) Livestock depredation and mitigation methods practised by resident and nomadic pastoralists around Waza National Park, Cameroon. *Oryx*, 47, 237–242, https://doi.org/10.1017/s0030605311001621

3.46. Deter predation of livestock by using shock/ electronic dog-training collars to reduce human-wildlife conflict

https://www.conservationevidence.com/actions/2446

- **Five studies** evaluated the effects of using shock/electronic dog-training collars to deter predation of livestock to reduce human-wildlife conflict. All five studies were in the USA[1,2,3,4,5].

COMMUNITY RESPONSE (0 STUDIES)

POPULATION RESPONSE (0 STUDIES)

BEHAVIOUR (0 STUDIES)

OTHER (5 STUDIES)

- **Human-wildlife conflict (5 studies):** Three of four replicated studies (including two controlled studies), in the USA[2,3,4,5], found that electric shock collars reduced livestock predation or bait consumption by wolves, whilst one found that they did not reduce wolf bait consumption. One replicated, controlled study in the USA[1] found that electric shock collars reduced the frequency of attacks by captive coyotes on lambs[1].

Background

Electric shock collars may be used on mammalian predators as a form of aversive conditioning. A shock is administered if the animal approaches or attacks livestock. Some studies summarized below test the potential for aversive conditioning to work on captive animals using non-live food and some others studies look at wild mammals, but using artificial food. Whilst not directly assessing the effectiveness of the intervention in reducing livestock predation, these studies provide evidence as to the potential for shock collars to alter animals' behaviour in a way that could potentially be applied to wild predators in livestock production areas. If using shock collars can reduce livestock predation, this could reduce incentives for carrying out lethal control of predators.

A replicated study in 1997 on pasture at a site in Utah, USA (1) found that electric shock collars reduced the frequency of attacks by captive coyotes *Canis latrans* on lambs. During week 1 (five coyotes each spending 4–6 hours with lambs) there was a total of 10 attempted lamb attacks. During week 2 (five coyotes each spending two hours with lambs) there was one attempted attack. There were no attempted attacks in week 4, one in week 7 and none in weeks 11, 16 or 22 (five coyotes each spending two hours with lambs during each study week). All attempted attacks ceased upon electric shock administration. Five captive male coyotes (aged 5–9 years), which killed lambs in trials, were studied. Each was fitted with a Model 100 Lite electronic dog-training collar, set at maximum shock intensity. During each trial, one coyote and one lamb were held in a 679 m² enclosure. Shocks were administered when the coyote actively pursued the lamb.

A replicated, controlled study in 2002 of captive wolves *Canis lupus* in Minnesota, USA (2) found that electronic dog-training collars did not reduce the amount of food consumed by wolves *Canis lupus*. Wolves fitted with dog-training collars, which activated when close to the food, consumed 43% of food offered. This was not significantly different to the 84% of food eaten by wolves where no deterrent was used. Four groups of 1–4 captive wolves were each offered 1 kg of sled-dog chow for 1 hour during June or July 2002. The wolves wore electronic dog-training collars, which emitted an electric shock when ≤2 m from the food. Four further groups of 1–4 wolves were offered the same food, without any deterrent.

A replicated study in 1998–2001 on a cattle farm in Wisconsin, USA (3) found that electric shock collars deterred gray wolves *Canis lupis* from predating livestock. In the first year, one calf was killed (possibly by non-collared wolves) after the alpha-female wolf was fitted with a shock collar, compared to nine killed earlier that year. Two were killed over the following two years (by non-collared wolves). A second wolf, collared in the fourth study year and thought to be the new alpha female of the pack, appeared to stay off the farm while the collars were operational. Other pack members continued predating calves, and the pack was subsequently translocated. A female wolf was fitted with an electric shock-collar on 14 May 1998. This activated when she was ≤300 m from cattle pasture. A replacement collar, operating from 26 April to 15 August 1999, beeped and shocked when she came within 0.4 km. In 2000, the collar operated from 26 April–August with beeping only (no shock). The second female wolf's shock-collar operated from 31 May to 13 August 2001.

A replicated, randomized, controlled, before-and-after study in 2003–2004 in a forested area in Michigan, USA (4) found that wolves *Canis lupus* wearing electric shock collars avoided baited areas where shocks were administered, but aversion did not persist. Shocked wolves made fewer visits to the detection zone when shocked (treatment period: 9 visits/wolf) relative to pre-treatment (19 visits/wolf) and post-treatment (16 visits/wolf) periods. There was no corresponding decrease for non-shocked wolves (treatment: 18 visits/wolf; pre-treatment: 21; post-treatment: 19). Shocked wolves spent less time/visit in detection zones during the treatment period (13 minutes/wolf) relative to pre-treatment (77 minutes/wolf) and post-treatment (20 minutes/wolf) periods. No decrease was detected for non-shocked wolves (treatment: 63 minutes/wolf; pre-treatment: 76; post-treatment: 47). Ten wolves (one per pack) were radio-collared in 2003–2004. Five wolves (randomly selected) also received electric shock collars (Innotek Training Shock Collar). A dead deer was placed in each pack's territory every two to three days. Collared wolves ≤75 m from baits were detected and logged over two weeks (pre-treatment). Treatment wolves, ≤30 m from baits, were shocked (for 13 seconds) over the following two weeks (treatment). For two further weeks (post-treatment), collared wolf visits to the 75 m detection zone were logged.

A replicated study in 2005–2006 in a mostly forested area of Wisconsin, USA (5) found that electric shock collars reduced visits by gray wolves *Canis lupus* to baited zones. Shock-collared wolves spent less time in shock zones when collars were active than did wolves without shock collars (with shock collar: 1 min/day in baited zone; no shock collar: 14 min/day). The pattern continued post-treatment when collars were not activated (shock collar: 1 min/day; no shock collar: 21 min/day). Fourteen adult wolves (one in each pack) were caught. Ten had a radio collar and shock unit fitted. Four had a radio collar only fitted. Each pack was baited with a dead deer every three days. The shock zone was a 70-m radius from the bait. Shock collars were automatically activated within this zone during a 40-day shock period. Bait placement and monitoring continued for a further 40-day non-shock period. Radio data loggers recorded wolf visits to bait sites between May and September of 2005 and 2006.

(1) Andelt W.E., Phillips R.L., Gruver K.S. & Guthrie J.W. (1999) Coyote predation on domestic sheep deterred with electronic dog-training collar. *Wildlife Society Bulletin*, 27, 12–18.

(2) Shivik J.A., Treves A. & Callahan P. (2003) Nonlethal techniques for managing predation: primary and secondary repellents. *Conservation Biology*, 17, 1531–1537, https://doi.org/10.1111/j.1523-1739.2003.00062.x

(3) Schultz R.N., Jonas K.W., Skuldt L.H. & Wydeven A.P. (2005) Experimental use of dog-training shock collars to deter depredation by gray wolves. *Wildlife Society Bulletin*, 33, 142–148, https://doi.org/10.2193/0091-7648(2005)33[142:euodsc]2.0.co;2

(4) Hawley J.E., Gehring T.M., Schultz R.N., Rossler S.T. & Wydeven A.P. (2009) Assessment of shock collars as nonlethal management for wolves in Wisconsin. *The Journal of Wildlife Management*, 73, 518–525, https://doi.org/10.2193/2007-066

(5) Rossler S.T., Gehring T.M., Schultz R.N., Rossler M.T., Wydeven A.P. & Hawley J.E. (2012) Shock collars as a site-aversive conditioning tool for wolves. *Wildlife Society Bulletin*, 36, 176–184, https://doi.org/10.1002/wsb.93

3.47. Fit livestock with protective collars to reduce risk of predation by mammals to reduce human-wildlife conflict

https://www.conservationevidence.com/actions/2448

- **One study** evaluated the effects of fitting livestock with protective collars to reduce human-wildlife conflict on rates of livestock killings by predators. This study was in South Africa[1].

COMMUNITY RESPONSE (0 STUDIES)

POPULATION RESPONSE (0 STUDIES)

BEHAVIOUR (0 STUDIES)

OTHER (1 STUDY)

- **Human-wildlife conflict (1 study):** A replicated, before-and-after study in South Africa[1] found that livestock protection collars reduced predation on livestock by carnivores.

Background

Carnivores typically kill their prey by a fatal bite to the neck. Hard collars can protect animals' necks. This may increase the effort needed by predators to kill livestock and, thus, reduce the likelihood of a fatal bite. If the intervention results in fewer livestock predated, this could reduce incentives for carrying out lethal control of predators.

A replicated, before-and-after study in 2006–2009 of seven livestock farms in savanna and shrubland in Eastern Cape, South Africa (1) found that using livestock protection collars reduced livestock fatalities caused by predators, compared to the rate when predators were controlled by lethal means. Results were not tested for statistical significance. When livestock collars were used, 1–12% of livestock were killed each year by predators. When not using livestock collars but, instead, carrying out lethal predator control, 6–31% of livestock were killed. Costs of using livestock collars (3.5 USD/livestock animal) were comparable to those of lethal control (0.7–6.0 USD/livestock animal). In August 2006–August 2007, all seven farms used lethal methods, including trapping and shooting, to control black-backed jackals *Canis mesomelas*, caracals *Caracal caracal* and leopards *Panthera pardus*. In September 2007–September 2009, all farms fitted animals with epoxy–metal mesh collars that protected the animal's neck from predator bites. Farmers reported numbers of livestock killed by predators, and associated costs, in September in 2007–2009.

(1) McManus J.S., Dickman A.J., Gaynor D., Smuts B.H. & Macdonald, B.W. (2015) Dead or alive? Comparing costs and benefits of lethal and non-lethal human-wildlife conflict mitigation on livestock farms. *Oryx*, 49, 687–695, https://doi.org/10.1017/s0030605313001610

3.48. Use lights and sound to deter predation of livestock by mammals to reduce human-wildlife conflict

https://www.conservationevidence.com/actions/2449

- **Three studies** evaluated the effects of using lights and sound to deter predation of livestock by mammals to reduce human-wildlife conflict. All three studies were in the USA[1,2,3].

COMMUNITY RESPONSE (0 STUDIES)

POPULATION RESPONSE (0 STUDIES)

BEHAVIOUR (0 STUDIES)

OTHER (3 STUDIES)

- **Human-wildlife conflict (3 studies):** Three replicated studies (including one controlled study), in the USA[1,2,3], found that devices emitting sounds and lights deterred predators from predating sheep[1] or consuming bait[2,3].

Background

This intervention specifically refers to use of light and sound in combination, often delivered via a commercially-purchased frightening devise, designed to repel wild mammals. If successful, such an intervention could reduce predation of livestock by predators and thus reduce motivations for carrying out lethal control of carnivores. For different applications of similar devices, see *Use lights and sound to deter crop damage by mammals to reduce human-wildlife conflict.*

A replicated study in 1979–1983 on pasture at 20 sites in Colorado, Idaho, South Dakota, and Oregon, USA (1) found that strobe light and siren devices reduced predation of sheep by coyotes *Canis latrans*. Ten trials, using 1–2 strobe light and siren devices per pasture, provided an average 53 nights of protection (≤2 sheep losses) from coyotes. Five trials, using 3–6 devices per pasture, protected sheep for an average 91 nights. Predation rates prior to trials were not stated. During five trials on unfenced range with two siren and two strobe light devices on each site, sheep losses to coyotes were 44–95% lower than those during the previous year. Sheep on pasture were protected by units containing a commercial strobe light or a warbling siren or both. Trials occurred in 1979–1982. On rangeland, sheep were protected, from June/July to late September of 1982–1983, by two warbling-type siren units and two with strobe lights, active at night and operating at intervals of 7 or 13 minutes. Other coyote control ceased during this time.

A replicated, controlled study in 2002 in a captive facility in Minnesota and a replicated, controlled, before-and-after study in 2002 at six forest sites in Wisconsin, USA (2) found that movement-activated guard (MAG) devices (emitting sound and light deterrents) reduced food consumption by carnivores. Captive wolves *Canis lupus* ate less of food protected with MAG devices (14% of available food consumed) than of unprotected food (84% consumed). Wild carnivores consumed less of MAG-protected deer carcasses (1.1 kg/day) than of unprotected carcasses (3.3 kg/day). At the same time in sites with no device, there was no difference in consumption between the later period (1.8 kg/day) and the earlier period (1.6 kg/day). Wolves, black bears *Ursus americanus*, fishers *Martes pennanti* and foxes *Vulpes vulpes* visited plots. Six groups of 1–7 captive wolves were each offered 1 kg of sled-dog chow for 1 hour during June or July 2002. A MAG device activated when animals were ≤2 m from the food. Four groups of 1–4 wolves were offered the same food, without deterrent. Study plots (30-m circumference) were established within territories of six wild wolf packs. A fresh deer carcass was placed in each plot. The study ran during April–June 2002 for 9–35 days (pre-treatment) and 16–29 days (treatment phase). A MAG device was used at one plot in each territory and one plot had no deterrent. Carcasses were weighed every 2–3 days and replaced as required. Camera traps at three territories identified species visiting plots.

A replicated, randomized study in 2005 in a captive facility in Utah, USA (3) found that combined light and sound or using light alone deterred coyotes *Canis latrans* from eating bait more than did sound alone. Fewer coyotes consumed bait with both light and sound deterrents used (none, from five pairs) or with light alone used (one coyote from five pairs) than with sound alone used (four coyotes from five pairs).

Fifteen captive coyote pairs were housed separately in 0.1-ha outdoor pens, each with a frightening device. Devices produced noise (100 dB at 2 m), strobe light (400 cd) or noise and light combined, when motion was detected ≤2 m away. Stimuli lasted 20 s. Five coyote pairs were randomly assigned to each of the three treatments. Pork bait was placed 1 m from the frightening device. For eight days' acclimation, devices were inactive. Then one trial, lasting 1.5 h, was run each evening, over 10 evenings. Trials were conducted from 17 July to 31 August 2005.

(1) Linhart S.B. (1984) Strobe light and siren devices for protecting fenced-pasture and range sheep from coyote predation. *Proceedings of the Eleventh Vertebrate Pest Conference*, 154–156.

(2) Shivik J.A., Treves A. & Callahan P. (2003) Nonlethal techniques for managing predation: primary and secondary repellents. *Conservation Biology*, 17, 1531–1537, https://doi.org/10.1111/j.1523-1739.2003.00062.x

(3) Darrow P.A. & Shivik J.A. (2009) Bold, shy, and persistent: Variable coyote response to light and sound stimuli. *Applied Animal Behaviour Science*, 116, 82–87, https://doi.org/10.1016/j.applanim.2008.06.013

3.49. Use scent to deter predation of livestock by mammals to reduce human-wildlife conflict

https://www.conservationevidence.com/actions/2450

- **Three studies** evaluated the effects of using scent to deter predation of livestock by mammals to reduce human-wildlife conflict. Two studies were in the USA[1,3] and one was in Botswana[2].

COMMUNITY RESPONSE (0 STUDIES)

POPULATION RESPONSE (0 STUDIES)

BEHAVIOUR (0 STUDIES)

OTHER (3 STUDIES)

- **Human-wildlife conflict (3 studies):** Two of three studies (including one replicated, before-and-after study), in the USA[1,3] and Botswana[2], found that applying scent marks from unfamiliar African wild dogs[2] and grey wolves[3] restricted movements of these species. The other study found that applying scent marks from coyotes[1] did not restrict their movements.

Background

Predatory mammals often mark their home ranges with scent, especially by selecting sites for depositing faeces and urine. If artificially placing such scent marks can constrain predators to particular areas and, in particular, to avoid areas where livestock are kept, this might reduce predation of livestock. If effective, this could reduce incentives for carrying out lethal control of these predators.

A study in 2007–2009 of a shrubland and grassland wildlife refuge and a replicated, randomized study in 2006 at a captive facility in Utah, USA (1) found that applying coyote *Canis latrans* scent as a trial of its use in deterring livestock predation did not reduce visits by coyotes. In the wildlife refuge study, wild coyotes visited areas marked with other coyotes' scent more often (average 36 visits/coyote) than they visited non-marked areas (average 11 visits/coyote). In the captive study, coyotes visited areas marked with other coyotes' scent more often than they visited non-marked areas both at territory boundaries (marked: 17 visits; not marked: 6 visits) and within territories (marked: 13 visits; not marked: 7 visits). In the wildlife refuge, GPS-collar data were obtained from three coyotes that had been followed for >10 weeks to define home-ranges. Within each home range, 1–2 clearings (2 ha), >100 m apart, were randomly selected and either marked with coyote urine (1–2 ml every 1–2 m) or left unmarked. Coyotes were monitored for four weeks. The captive study was conducted over two 13–14-day periods in October–November 2006. Two from four coyote pairs, housed in 1-ha pens, were randomly selected to have the boundary of 7% of their pen area marked with urine and scats from other coyotes. Two pairs did not have their pens marked. The behaviour of each coyote was monitored for eight hours through direct observation.

A study in 2008–2010 at a savanna reserve in Botswana (2) found that applying scent marks from other African wild dogs *Lycaon pictus* at the reserve boundary caused resident wild dogs to return towards the centre of their range. Seven of eight scent mark applications were followed by wild dogs moving closer to the centre of their range within the reserve. An additional application, 24 h after initial applying scents, generated the same response on the eighth occasion. Wild dogs moved further in the day after application (average 7.2 km) than when no marks were applied (3.4 km). This response reduced movements onto neighbouring farmland and potential livestock depredation. Eighteen wild dogs were translocated to the reserve and released in April 2008. When they moved to the reserve boundary, 3–26 wild dog urine and faeces marks, brought from a different site, were applied 50–200 m from the pack. The pack was monitored, using GPS collars or visual observation, from September 2008 to February 2010.

A replicated, before-and-after study in 2008–2011 in three forest-dominated sites in Idaho, USA (3) found that marking grey wolf *Canis lupus* territories with lines of scent from other wolf packs restricted wolf movements in some but not all cases. Results were not tested for statistical significance. Overall, the proportion of location fixes indicating that wolves had crossed scent lines was variable after scents were deployed (0–23%) and before scent deployment (1–12%). No incursions across scent lines were recorded in single years for two wolf packs (out of five pack/year combinations). In other cases, there was less evidence of scent lines reducing incursions. Two parallel 10–36-km lines were marked across wolf pack territories in 2010 (two packs) and 2011 (three packs). Lines were marked with 3 ml of urine from a different wolf pack, every 500 m and with 6 ml of urine every 750 m, and scats every km. Scent marks were refreshed every 10–14 days in June–August. Wolf packs (8–14 wolves) were monitored by satellite tracking of 2–4 wolves in each pack for 3–4 years during May–September of 2008–2011.

(1) Shivik J.A., Wilson R.R. & Gilbert-Norton L. (2011) Will an artificial scent boundary prevent coyote intrusion? *Wildlife Society Bulletin*, 35, 494–497, https://doi.org/10.1002/wsb.68

(2) Jackson, C.R., McNutt, J.W. & Apps, P.J. (2012) Managing the ranging behaviour of African wild dogs (*Lycaon pictus*) using translocated scent marks. *Wildlife Research*, 39, 31–34, https://doi.org/10.1071/WR11070

(3) Ausband D.E., Mitchell M.S., Bassing, S.B. & White, C. (2013) No trespassing: using a biofence to manipulate wolf movements. *Wildlife Research*, 40, 207–216, https://doi.org/10.1071/WR12176

3.50. Use watchmen to deter crop damage by mammals to reduce human-wildlife conflict

https://www.conservationevidence.com/actions/2451

- We found no studies that evaluated the effects of using watchmen to deter crop damage by mammals to reduce human-wildlife conflict.

'We found no studies' means that we have not yet found any studies that have directly evaluated this intervention during our systematic journal and report searches. Therefore, we have no evidence to indicate whether or not the intervention has any desirable or harmful effects.

Background

Damage to agricultural crops by mammalian herbivores may cause substantial losses for some farmers. Although labour-intensive, farmers in some areas may directly guard crops. If this can reduce crop losses to mammals, it could reduce incentive for carrying out lethal control of such species.

3.51. Use mobile phone communications to warn farmers of problematic mammals (e.g. elephants)

https://www.conservationevidence.com/actions/2452

- We found no studies that evaluated the effects of using mobile phone communications to warn farmers of problematic mammals (e.g. elephants).

'We found no studies' means that we have not yet found any studies that have directly evaluated this intervention during our systematic journal and report searches. Therefore, we have no evidence to indicate whether or not the intervention has any desirable or harmful effects.

Background

Farmers may be vulnerable to loss of crops from raids by wild herbivores or to loss of livestock to mammalian predators. The large growth in use of mobile phones makes it easier for farmers to communicate the presence of problem animals to others in the general area. This may allow faster responses in deployment of prevention measures (Lewis *et al.* 2016). If this reduces crop damage or livestock predation, it might also reduce incentives for lethal control of wild herbivores or predators.

Lewis A.L., Baird T.D. & Sorice M.G. (2016) Mobile phone use and human-wildlife conflict in Northern Tanzania. *Environmental Management*, 58, 117–129, https://doi.org/10.1007/s00267-016-0694-2

3.52. Use fencing/netting to reduce predation of fish stock by mammals to reduce human-wildlife conflict

https://www.conservationevidence.com/actions/2454

- We found no studies that evaluated the effects of using fencing or netting to reduce predation of fish stock by mammals to reduce human-wildlife conflict.

'We found no studies' means that we have not yet found any studies that have directly evaluated this intervention during our systematic journal and report searches. Therefore, we have no evidence to indicate whether or not the intervention has any desirable or harmful effects.

Background

Fish farms can attract a range of mammalian predators, causing human-wildlife conflict. For example, questionnaire respondents from among fish farm operators and anglers in the Czech Republic reported between 7% and 17% of fish losses being due to predation by Eurasian otters *Lutra lutra* (Václavíková *et al.* 2011). If barriers, such as netting or fencing, can keep predators from accessing fish, this may reduce incentives for carrying out lethal control of such animals.

Václavíková M., Václavík T & Kostkan V. (2011) Otters vs. fishermen: Stakeholders' perceptions of otter predation and damage compensation in the Czech Republic. *Journal for Nature Conservation*, 19, 95–102, https://doi.org/10.1016/j.jnc.2010.07.001

3.53. Establish deviation ponds in fish farms to reduce predation of fish stock by mammals to reduce human-wildlife conflict

https://www.conservationevidence.com/actions/2455

- We found no studies that evaluated the effects on mammals of establishing deviation ponds in fish farms to reduce predation of fish stock by mammals to reduce human-wildlife conflict.

'We found no studies' means that we have not yet found any studies that have directly evaluated this intervention during our systematic journal and report searches. Therefore, we have no evidence to indicate whether or not the intervention has any desirable or harmful effects.

Background

Some mammals can become significant predators of fish being reared in fish farms. For example, one study found that rainbow trout *Onchorhynchus mykiss* from a fish farm formed 87% of biomass of prey consumed by otters *Lutra lutra* in the vicinity (Marques *et al.* 2007). Deviation ponds are sites where fish are made easily accessible to predators in order to keep them away from other, more valuable, fish kept elsewhere on the site. If effective, this intervention could reduce incentives for carrying out lethal control of mammalian predators of fish.

Marques C., Rosalino LM. & Santos-Reis M. (2007) Otter predation in a trout fish farm of Central-east Portugal: Preference for 'fast-food'? *River Research and Applications*, 23, 1147–1153, https://doi.org/10.1002/rra.1037

3.54. Use lights and sound to deter crop damage by mammals to reduce human-wildlife conflict

https://www.conservationevidence.com/actions/2456

- **Two studies** evaluated the effects of using both lights and sound to deter crop damage by mammals to reduce human-wildlife conflict. Both studies were in the USA[1,2].

COMMUNITY RESPONSE (0 STUDIES)

POPULATION RESPONSE (0 STUDIES)

BEHAVIOUR (0 STUDIES)

OTHER (2 STUDIES)

- **Human-wildlife conflict (2 studies):** Two replicated paired sites, controlled studies (one also randomized), in the USA[1,2], found that frightening devices, emitting lights and sound, did not reduce crop intrusions by white-tailed deer[1] or food consumption by elk and mule deer[2].

Background

This intervention specifically refers to use of light and sound in combination, typically delivered via a commercially-produced product designed to deter visits by wild mammals. If successful, such an intervention could reduce crop damage and, thus, reduce motivation for carrying out lethal control of herbivores.

See also: *Use light/lasers to deter crop damage by mammals to reduce human-wildlife conflict, Use loud noises to deter crop damage (e.g. banger sticks, drums, tins, iron sheets) by mammals to reduce human-wildlife conflict and Use noise aversive conditioning to deter crop damage by mammals to reduce human-wildlife conflict.*

A replicated, paired sites, controlled study in 1999 of corn fields at two sites in Nebraska, USA (1) found that a device emitting lights and sound (Electronic Guard) did not reduce crop visits by white-tailed deer

Odocoileus virginianus. The number of deer visits/km of field boundary did not differ between treatment fields protected by Electronic Guards (38–46/day) and unprotected control fields (40–56/day). Similarly, there was no difference between fields before devices operated (treatment fields: 24 visits/km/day; control fields: 21 visits/km/day) or after operations ceased (treatment fields: 47 visits/km/day; control field: 53 visits/km/day). Four groups of fields were studied at each of two sites. Fields were 0.5–2.5 km apart and separated by woodland. In each group, one field was protected by two Electronic Guard devices and one field was unguarded. Electronic Guards comprised a strobe light (60 flashes/minute) and siren (116 dB at 1 m). They operated at night, from when corn crops became susceptible to damage (13 July 1999 at one site and 25 July 1999 at the second site), for 18 days. Deer activity was assessed by counting tracks twice while devices operated, once during the two weeks before devices operated and once during the week after they operated.

A replicated, randomized, paired sites, controlled study in 2001 of pastures on a ranch in Colorado, USA (2) found that a device emitting lights and sound (Critter Gitter™) did not reduce combined elk *Cervus canadensis* and mule deer *Odocoileus hemionus* food consumption. Daily alfalfa consumption at bales protected by Critter Gitters™ (3.1–6.0 kg/day) did not differ from that at unprotected bales (2.8–7.3 kg/day). The Critter Gitter™ activated when infrared sensors detected movement and heat. When activated, an alarm (approaching 120 decibels) sounded for five seconds and a pair of red LEDs flashed. Five sites (>300 m apart) on private ranchland, adjacent to residential areas, were studied. Each site had two alfalfa bales, 60 m apart. One or two devices were positioned by one bale (selected randomly). The other bale was unprotected. Devices detected animals ≤2 m away. Alfalfa consumption was estimated visually, every two or three days, on 10 occasions.

(1) Gilsdorf J.M., Hygnstrom S.E., VerCauteren K.C. Blankenship E.E. & Engeman R.M. (2004) Propane exploders and Electronic Guards were ineffective at reducing deer damage in cornfields. *Wildlife Society Bulletin,* 32, 524–531, https://doi.org/10.2193/0091-7648(2004)32[524:PEAEGW]2.0 .CO;2

(2) VerCauteren K.C., Shivik J.A. & Lavelle M.J. (2005) Efficacy of an animal-activated frightening device on urban elk and mule deer. *Wildlife Society Bulletin,* 33, 1282–1287, https://doi.org/10.2193/0091-7648(2005)33[1282:eoa afd]2.0.co;2

3.55. Provide diversionary feeding to reduce crop damage by mammals to reduce human-wildlife conflict

https://www.conservationevidence.com/actions/2457

- **Six studies** evaluated the effects of providing diversionary feeding to reduce crop damage by mammals to reduce human-wildlife conflict. Three studies were in Canada[1a,1b,2] and one was in each of France[3], Spain[4] and Austria[5].

COMMUNITY RESPONSE (0 STUDIES)

POPULATION RESPONSE (0 STUDIES)

BEHAVIOUR (0 STUDIES)

OTHER (6 STUDIES)

- **Human-wildlife conflict (6 studies):** Three of six studies (including four controlled and one before-and-after study) in Canada[1a,1b,2], France[3], Spain[4] and Austria[5] found that diversionary feeding reduced damage by red squirrels[2] to pine trees and European rabbits[4] to grape vines, and resulted in fewer red deer[5] using vulnerable forest stands. Two studies found that diversionary feeding did not reduce damage by voles[1a] to apple trees or wild boar[3] to grape vines. One study[1b] found mixed results on damage by voles to crabapple trees depending on the food provided.

Background

Mammals can cause unacceptable losses to farmers, through feeding on crops. If diversionary feeding can reduce the extent to which animals exhibit nuisance behaviour, this may reduce motivations for carrying out lethal control or other intensive management.

See also: *Provide diversionary feeding to reduce predation of livestock by mammals to reduce human-wildlife conflict* and *Residential and commercial development — Provide diversionary feeding for mammals to reduce nuisance behaviour and human-wildlife conflict.*

A randomized, controlled study in 1983–1984 at an orchard in British Columbia, Canada (1a) found that diversionary feeding with treated plywood sticks did not reduce damage by voles *Microtus* spp. to spartan apple *Malus domestica* trees. The percentage of apple trees damaged by voles did not differ significantly in orchard blocks with treated plywood sticks (32%) or those without sticks (36%). Trees with treated plywood sticks around them had more bark and tissues removed by voles (average 20–27 cm²/tree) than trees without sticks (5 cm²/tree), although the difference was not tested for statistical significance. In November 1983, three treatments (plywood sticks treated with sucrose, soybean oil or sorbitol) were randomly assigned to each of three orchard blocks of 100 spartan apple trees (15 and 30 years old). Three plywood sticks (5 x 37.5 cm, 9 mm thick kiln-dried Douglas fir *Pseudotsuga menziesii*) were placed in a triangle around each tree, 1–2 cm from the base. One control orchard block had no plywood sticks. The area of bark and vascular tissues removed by voles was measured on each of the 400 trees in March 1984.

A randomized, controlled study in 1984–1985 at a newly planted orchard in British Columbia, Canada (1b) found that diversionary feeding with bark-mulch logs treated with soybean oil reduced damage by montane voles *Microtus montanus* to crabapple *Malus* spp. trees, but logs treated with apple or apple and soybean oil did not. Orchard blocks with logs treated with soybean oil had a lower percentage of trees damaged by voles (25%) and trees with stem or root girdling (4%) than those without logs (63% damaged; 25% girdling). The difference was not significant between orchards with logs treated with apple (46% damaged; 17% with girdling) or apple and soybean oil (58% damaged; 33% with girdling) and those without logs. In November 1984, logs made from sifted Douglas fir *Pseudotsuga menziesii* bark mulch mixed with wax and one of three treatments (soybean oil, apple powder or apple powder and soybean oil mixed together) were randomly assigned to each of three orchard blocks of 24 one-year-old crabapple trees. Three logs were placed around each tree, 8–10 cm from the base. Additional logs were added as required in December 1984–February 1985. One control orchard block had no logs. Numbers of trees with vole damage and stem or root girdling in each of the four orchard blocks were recorded in March 1985.

A controlled study in 1989–1990 of managed forest in British Columbia, Canada (2) found that diversionary feeding reduced damage by red squirrels *Tamiasciurus hudsonicus* to lodgepole pine *Pinus contorta* crop trees. In each of three years, lodgepole pine blocks with diversionary feeding had a lower percentage of trees damaged by squirrels (average 5–11%) and fewer damage wounds (average 0.02–0.13 wounds/tree) than control blocks without diversionary feeding (average 26–61% of trees damaged; 0.5–2 wounds/tree). In May and June 1989, sunflower seeds were manually distributed in piles (45 kg/ha) within a 20-ha lodgepole pine block, and one 20-ha control block had no seeds. In 1990, two 15-ha blocks had seeds manually distributed in piles (22.7 kg/ha), two 20-ha blocks had seeds distributed by helicopter (22.7 kg/ha), and two 15-ha control blocks had no seeds. In 1991, seeds were distributed across three areas of 131–200 ha by helicopter (20 kg/ha), and three control areas had no seeds. Squirrel damage was recorded within 16–24 circular plots located every 50 or 100 m in a grid pattern within each treatment and control block or area in 1989, 1990 and 1991.

A before-and-after study in 1990–1993 of 283 vineyards in Puechabon, France (3) found that diversionary feeding did not reduce damage by wild boar *Sus scrofa* to grape vines. Average grape vine losses caused by wild boar did not differ significantly during two years before diversionary feeding (193 kg/ha) and one year with diversionary feeding (151 kg/ha). In July–September 1993, a total of 4.7 tons of grain maize (25 kg/day) was distributed along a 4.5 km trail through woodland located 500–1,000 m from 283 vineyards. The 50 owners of the vineyards were questioned on the estimated amount of damage to grape vines caused by wild boar in 1990–1992 (before diversionary feeding) and 1993 (with diversionary feeding).

A controlled study in 2008 at three vineyards in Córdoba province, Spain (4) found that diversionary feeding reduced damage by European rabbits *Oryctolagus cuniculus* to common grape vines *Vitis vinifera*. Grape vines within plots with diversionary feeding had a lower percentage of buds and shoots removed by rabbits (11%) than those without diversionary feeding (21%). However, grape vine yield did not differ between vineyard plots with or without diversionary feeding (both 4.7 kg/vine). At each of three vineyard sites, one plot had diversionary feeding (50 kg fresh alfalfa placed in strips along the edge of the plot each week during the growing season), and a second plot did not. All plots were unfenced. The proportion of buds and shoots removed by rabbits on 15–20 vines/plot was recorded throughout the growing season in 2008. Grape vine yields were estimated during harvest from the number and size of grape clusters on each vine.

A study in 2009–2011 in a mixed timber forest in Austria (5) found that diversionary feeding of red deer *Cervus elaphus* resulted in fewer deer using forest stands vulnerable to deer damage. Forest stands vulnerable to deer browsing and bark-stripping (young and mid-aged stands) were used less by red deer in areas 1.3–1.5 km from winter feeding stations compared to areas further away (data reported as statistical model results). Supplementary food (mainly apple pomace and hay) was provided during winter (October–May) at seven feeding stations (1 station/19 km^2) within a 131-km^2 area of mixed forest managed for production of Norway spruce *Picea abies* and European larch *Larix decidua*. In 2009–2011, eleven red deer (seven males, four females) were radio-tracked to a total of 29,799 locations within the forest. Deer damage was not directly measured.

(1) Sullivan T.P. & Sullivan D.S. (1988) Influence of alternative foods on vole population and damage in apple orchards. *Wildlife Society Bulletin*, 16, 170–175.

(2) Sullivan T.P. & Klenner W. (1993) Influence of diversionary food on red squirrel population and damage to crop trees in young lodgepole pine forests. *Ecological Applications*, 3, 708–718, https://doi.org/10.2307/1942102

(3) Calenge C., Maillard D., Fournier P. & Fouque C. (2004) Efficiency of spreading maize in the garrigues to reduce wild boar (*Sus scrofa*) damage to Mediterranean vineyards. *European Journal of Wildlife Research*, 50, 112–120, https://doi.org/10.1007/s10344-004-0047-y

(4) Barrio I.C., Bueno C.G. & Tortosa F.S. (2010) Alternative food and rabbit damage in vineyards of southern Spain. *Agriculture, Ecosystems and Environment*, 138, 51–54, https://doi.org/10.1016/j.agee.2010.03.017

(5) Arnold J.M., Gerhardt P., Steyaert S., Hackländer K. & Hochbichler E. (2018) Diversionary feeding can reduce red deer habitat selection pressure on vulnerable forest stands, but is not a panacea for red deer damage. *Forest Ecology and Management*, 407, 166–173, https://doi.org/10.1016/j.foreco.2017.10.050

3.56. Use scarecrows to deter crop damage by mammals to reduce human-wildlife conflict

https://www.conservationevidence.com/actions/2459

- We found no studies that evaluated the effects of using scarecrows to deter crop damage by mammals to reduce human-wildlife conflict.

'We found no studies' means that we have not yet found any studies that have directly evaluated this intervention during our systematic journal and report searches. Therefore, we have no evidence to indicate whether or not the intervention has any desirable or harmful effects.

Background

Scarecrows are generally life-sized models of people that come in various designs, including static scarecrows and those that move, or inflate at intervals, to increase their impact. They are placed in crop fields, usually to deter visits by birds, but they could also be used to deter mammalian crop-raiders. If successful, this could reduce incentives for carrying out lethal control of such mammals.

3.57. Use loud noises to deter crop damage (e.g. banger sticks, drums, tins, iron sheets) by mammals to reduce human-wildlife conflict

https://www.conservationevidence.com/actions/2460

- **Ten studies** evaluated the effects of using loud noises to deter crop damage by mammals to reduce human-wildlife conflict. Three studies were in the USA[2,6,7], two were in Zimbabwe[4,5] and Kenya[8a,8b] and one each was in the UK[1], Namibia[3], and India[9].

COMMUNITY RESPONSE (0 STUDIES)

POPULATION RESPONSE (0 STUDIES)

BEHAVIOUR (0 STUDIES)

OTHER (10 STUDIES)

- **Human-wildlife conflict (10 studies):** Five of six studies (including two controlled, one replicated and two before-and-after studies), in the USA[2,6], Namibia[3], Kenya[8a,8b] and India[9], found that loud noises activated when an animal was in the vicinity reduced or partially reduced crop damage or crop visits by white-tailed deer[2], black-tailed deer (when combined with using electric shock collars)[6] and elephants[3,8a,9]. The other study[8b] found that using loud noises (along with chili fences and chili smoke) did not reduce crop-raiding by African elephants. Three studies (including two controlled studies), in the UK[1] and the USA[2,7], found that regularly sounding loud noises did not repel European rabbits[1] or white-tailed deer[2,7]. Two replicated studies, in Zimbabwe[4,5], found that, from among a range of deterrents, African elephants were repelled faster from crop fields when scared by firecrackers[5] or by a combination of deterrents that included drums[4].

Background

This intervention specifically refers to use of sound, from various sources, to deter visits by wild mammals into crops. If successful, such an intervention could reduce crop damage and, thus, reduce motivation for carrying out lethal control of herbivores.

See also: *Use lights and sound to deter crop damage by mammals to reduce human-wildlife conflict, Use noise aversive conditioning to deter crop damage by mammals to reduce human-wildlife conflict and Use ultrasonic noises to deter crop damage by mammals to reduce human-wildlife conflict.*

A before-and-after study in 1984 on grassland in Surrey, UK (1) found that an acoustic scaring device did not deter European rabbits *Oryctolagus cuniculus* from consuming bait. Bait consumption after the device was activated (2–361 g/bait pile/day), did not differ from that before the device was activated (7–368 g/bait pile/day). Five wild, adult rabbits were placed in a 50 × 40-m grass enclosure, with wooden hutches at one end. The opposite end housed the scaring device and 400-g piles of chopped carrots at 3, 6, 9, 12 and 15 m from the device. The device emitted 5-s bursts of rapidly pulsed sound, separated by 4-s silences. Bait was deposited on four days/week. Remaining carrots were removed and weighed to establish quantity consumed. Similar bait, in rabbit-proof cages, was used to correct weights for moisture changes. The enclosure contained sufficient grass to sustain rabbits without their need to eat carrots. The trial lasted four weeks, in March 1984, with the scaring device switched on midway through.

A randomized, controlled, before-and-after study in 1994–1995 on a grassland site in Ohio, USA (2) found that motion-activated propane exploders temporarily reduced white-tailed deer *Odocoileus virginianus* visits but regularly firing exploders did not. There were fewer deer visits in the week following deployment of motion-activated exploders, in two out of three seasons (23–94 visits/week) compared to the pre-treatment period (159–313 visits/week). In spring/early-summer and late-summer, visit rates returned to pre-treatment levels after 2–6 weeks. In autumn, exploders did not reduce deer visits. Regularly firing exploders did not reduce deer visit rates compared to pre-treatment levels in any weeks studied and neither did non-functioning exploders. The experiment used different combinations of three out of six feeding sites, during 9 August–12 September 1994, 20 September–24 October 1994 and 27 April–12 July 1995. Each time, a two-week pre-treatment period preceded a 3–9-week treatment period. Feeding sites (>1 km apart) were semi-circular fences around whole kernel corn. Treatments were propane exploders firing eight times in two minutes when motion was detected, exploders firing every 8–10 minutes and non-functioning exploders. Deer visits were monitored with electronic detecting devices.

A replicated study in 1993–1995 of farmland and grassland at 10 villages in East Caprivi, Namibia (3) found that car sirens connected to trip wires around crops were partially successful in reducing crop raiding by elephants *Loxodonta africana*. Sirens at three villages in the first year were all reported to have positive effects of reducing crop-raiding by elephants (actual crop-raiding frequencies not reported). In the second year, a positive effect of sirens was reported from one village, whilst elephants did not approach at three villages (so the system was untested) and at two further villages, the crop area was too large to protect using the system. In the third year, three villages reported positive effects whilst at a fourth, battery failure rendered the system ineffective. Sirens each protected 1–7 farms at 10 villages during one or two years of the trial. Each system comprised a car siren, a 12-V battery and a 10-s timer. Polyethylene cords were mounted on fences or trees to enclose fields. The siren activated for 10 s when the cord was pulled. Data were collated from questionnaire surveys in 1993–1995.

A replicated study in 1995–1996 in crop fields at a site surrounded by savanna in Sebungwe, Zimbabwe (4) found that African elephants *Loxodonta africana* were repelled faster from agricultural fields by groups of people banging drums (alongside a range of other deterrents) than by one person making less noise. Specific effects of banging on drums cannot be separated from those of other scaring tactics. Elephants were repelled faster when scared by people with drums, dogs *Canis lupus familiaris*, whips and large fires (4 minutes) or with drums, dogs, slingshots and burning sticks (10 minutes) than by one person sometimes with a dog and chasing elephants while banging on tins and yelling (14 minutes). When scared by actions that included drums, elephants charged at defenders 12 times out of 26 trials, though only charged two out of nine times when scared by a single person without drums. Elephants raiding crops were scared 15 times by 4–7 people with drums, dogs, whips and large fires, 11 times by 2–3 people with drums, dogs, slingshots, and burning sticks and 15 times by one person (sometimes with a dog, and sometimes hitting tins and yelling to deter elephants). Behavioural responses were monitored through a monocular. Distance between elephants and farmers was 20–40 m. Tests were conducted between 18:30 and 06:30 h. The number of fields was not specified.

A replicated study in 2001 of arable land in seven villages in Guruve District, Zimbabwe (5) found that using loud noises, by throwing firecrackers at crop-raiding elephants *Loxodonta africana*, repelled them faster than did traditional deterrents such as beating drums and throwing rocks. Elephants left faster when firecrackers were activated (average 6 minutes) than they did when traditional repellent methods alone were used (average 65 minutes). Seven villages were studied. At three villages, on 35 occasions, farmers threw locally made firecrackers at elephants that were attempting to raid crops. On 27 occasions, farmers at four villages used traditional methods to ward off elephants that attempted to raid crops, namely banging drums and throwing rocks with catapults. The study was conducted from 1 January to 30 June 2001 and data were collected by a team of observers.

A replicated, controlled study in two pastures in Washington, USA (6) found that playing loud noise, along with using shock collars, reduced damage by black-tailed deer *Odocoileus hemionus* to tree seedlings. The loud noise and electric shock were part of the same treatment, so their relative effects could not be separated. In areas where playing of loud noise was triggered, damage to tree seedlings was lower (0–1 bites) than in areas where loud noises were not triggered (0–25 bites). Three deer, fitted with shock collars, were placed in each of two 1.5-ha pastures. Within each pasture, four 20 × 20 m plots were established. In each plot, 16 red cedar *Thuja plicata* seedlings were planted at 1-m intervals. When deer entered two of the plots, a loud noise was played through a speaker and deer received an electric shock. When they entered the other two plots, no noise was played and they received no shock. Deer activity was measured by counting the number of bites taken from seedlings over a 21-day period.

A replicated, paired sites, controlled study in 1999 of corn fields at two sites in Nebraska, USA (7) found that loud noises from propane exploders did not reduce visits to crops by white-tailed deer *Odocoileus virginianus*. The number of deer visits/km of field boundary was similar in fields protected by propane exploders (31–36/day) and unprotected fields (40–56/day). Similarly, there were no significant difference between fields before devices operated (exploders: 17 visits/km/day; unprotected: 21 visits/km/day) or after (exploders: 37 visits/km/day; unprotected: 53 visits/km/day). Four groups of fields (0.5–2.5 km apart, separated by woodland) were studied at each of two sites. At each site, one field had propane exploders (two/field) and one was unguarded. Propane exploders fired at 15-minute intervals. They operated at night, from when corn crops became susceptible to damage (13 July 1999 at one site and 25 July 1999 at the second site), for 18 days. Deer activity was assessed by counting tracks twice while devices operated and once each in ≤2 weeks before and after this time.

A before-and-after and site comparison study in 2003–2004 of two farming areas in Laikipia, Kenya (8a) found that using loud noises, along with chili fences and chili smoke, reduced raiding and crop damage by African elephants *Loxodonta africana*. The study does not distinguish between the effects of loud noises and chilli deterrents. After farmers began using loud noises, along with chili fences and smoke, the total number of crop-raiding incidents (26) and the average area of crop damage (375 m²/incident) was lower than before deterrents were used (92 incidents; 585 m²/incident). However, the difference was not tested for statistical significance. At a control site without deterrents, crop-raiding increased (total 17–166 incidents) as did crop damage (average 328 m²–421 m²/incident) during the same time period. A group of farmers within a 0.03-km² area were provided with training and materials to deter crop-raiding elephants. Deterrents included loud noises (bangers, banger sticks, cow bells), chili fences (rope and cloth fences with chili and engine grease applied) and chili smoke (chili and dung briquettes burned at night). Some farmers also used watchtowers and torches. A second control area, of equal size and within 1 km, used no deterrents. Crop-raiding incidents and crop damage were recorded in each of the two areas before (June–December 2003) and after (June–December 2004) deterrents were introduced.

A replicated, before-and-after and site comparison study in 2004–2005 at 40 farms in Laikipia, Kenya (8b) found that using loud noises, along with chili fences and chili smoke, did not result in an overall reduction in crop-raiding by African elephants *Loxodonta africana*. The study does not distinguish between the effects of chilli deterrents and loud noises. After farmers began using loud noises, along with chili fences and chili smoke, the average number of crop-raiding incidents across all farms (2) was similar to before deterrents were used (2.5). At 10 control farms without deterrents, crop-raiding decreased (from an average of three incidents to one) during the same time period. Ten farmers in each of two areas were provided with training and materials to deter crop-raiding elephants. Deterrents included loud noises (bangers, banger sticks, cow bells), chili fences (rope and cloth fences with chili and engine grease applied) and chili smoke (chili and dung briquettes burned at night). Some farmers also used watchtowers and torches. Uptake of deterrent types varied between farms (see original paper for details). Ten control farms within each of the two areas used no deterrents. Crop-raiding incidents were recorded at all 40 farms before (February–November 2004) and after (February–November 2005) deterrents were introduced.

A study in 2006–2009 in two areas of Assam, India (9) found that using loud noises to scare Asian elephants *Elephas maximus* reduced the probability of elephants damaging crops. The chance of crop damage occurring was lower when noise was used to deter elephants compared to a range of other interventions or to no intervention (results presented as statistic model coefficients). Only fences and spotlights reduced crop raiding to a greater extent. Within two study areas, 33 community members, trained as monitors, recorded 1,761 crop-raiding incidents, from 1 March 2006 to 28 February 2009. A range of deterrent methods was used, singly or in combination, including noise (shouting, crackers or drums), chili smoke (from burning dried chilies, tobacco, and straw), spotlights, two-strand electric fences, chili fencing (engine grease and ground chili paste, on a jute or coconut rope), elephant drives (repelling wild elephants using domesticated elephants) and fire.

(1) Wilson C.J. & McKillop I.G. (1986) An acoustic scaring device tested against European rabbits. *Wildlife Society Bulletin*, 14, 409–411.

(2) Belant J.L., Seamans T.W. & Dwyer C.P. (1996) Evaluation of propane exploders as white-tailed deer deterrents. *Crop Protection*, 15, 575–578, https://doi.org/10.1016/0261-2194(96)00027-0

(3) O'Connell-Rodwell C.E., Rodwell T., Rice M. & Hart L.A. (2000) Living with the modern conservation paradigm: can agricultural communities co-exist with elephants? A five-year case study in East Caprivi, Namibia. *Biological Conservation*, 93, 381–391, https://doi.org/10.1016/S0006-3207(99)00108-1

(4) Osborn F.V. (2002) Capsicum oleoresin as an elephant repellent: field trials in the communal lands of Zimbabwe. *The Journal of Wildlife Management*, 66, 674–677, https://doi.org/10.2307/3803133

(5) Osborn F.V. & Parker G.E. (2002) Community-based methods to reduce crop loss to elephants: experiments in the communal lands of Zimbabwe. *Pachyderm*, 33, 32–38.

(6) Nolte D.L., VerCauteren K.C., Perry K.R. & Adams S.E. (2003) *Training deer to avoid sites through negative reinforcement*. USDA National Wildlife Research Center-Staff Publications, 264.

(7) Gilsdorf J.M., Hygnstrom S.E., VerCauteren K.C. Blankenship E.E. & Engeman R.M. (2004) Propane exploders and Electronic Guards were ineffective at reducing deer damage in cornfields. *Wildlife Society Bulletin*, 32, 524–531, https://doi.org/10.2193/0091-7648(2004)32[524:peaegw]2.0.co;2

(8) Graham M. & Ochieng T. (2008) Uptake and performance of farm-based measures for reducing crop raiding by elephants *Loxodonta africana* among

smallholder farms in Laikipia District, Kenya. *Oryx*, 42, 76–82, https://doi.org/10.1017/S0030605308000677

(9) Davies T.E., Wilson S., Hazarika N., Chakrabarty J., Das D., Hodgson D.J. & Zimmermann A. (2011) Effectiveness of intervention methods against crop-raiding elephants. *Conservation Letters*, 4, 346–354, https://doi.org/10.1111/j.1755-263x.2011.00182.x

3.58. Use noise aversive conditioning to deter crop damage by mammals to reduce human-wildlife conflict

https://www.conservationevidence.com/actions/2461

- **One study** evaluated the effects of using noise aversive conditioning to deter crop damage by mammals to reduce human-wildlife conflict. This study was in the USA[1].

COMMUNITY RESPONSE (0 STUDIES)

POPULATION RESPONSE (0 STUDIES)

BEHAVIOUR (0 STUDIES)

OTHER (1 STUDY)

- **Human-wildlife conflict (1 study):** A replicated, controlled study in USA[1] found that noise aversive conditioning reduced bait consumption by white-tailed deer.

Background

Aversive conditioning is the process of associating a negative stimulus with a secondary behaviour or outcome. In the case of this intervention, it involves associating a negative stimulus with a neutral one (noise) when carrying out undesirable behaviour (feeding on crops) to the extent that the neutral stimuli alone deters this behaviour. If this reduces crop damage, it may reduce motivations for carrying out lethal control of wild mammalian herbivores.

A replicated, controlled study in 2001 on a pasture site in Georgia, USA (1) found that attempts to condition white-tailed deer *Odocoileus virginianus* to avoid food when a metronome was played, by initially playing the sound alongside an electric wire deterrent, reduced, but did not eliminate, consumption of the food. With the metronome active but the electric wire deactivated, corn consumption (1.4–2.0 kg/day) was generally lower than at unprotected feeders (2.2 kg/day) but was higher than when both the metronome and electric wire deterrent were active (0–0.1 kg/day). Deer were studied in three 13-ha pasture plots, each containing two feeders, 6.5 m apart. Feeders comprised a plastic tray on a toolbox. At one feeder in each plot, the box housed an electric fence charger and an electronic metronome. An electric fence wire on each tray was likely to be touched by deer accessing corn. Each feeder was supplied with 2.3 kg/day of whole corn. Unconsumed corn was weighed and removed. Data were collected during six 5-day periods in April–May 2001. During the first, third and fifth periods, electric chargers and metronomes were activated. In alternate periods, only metronomes remained active.

(1) Gallagher G.R. & Prince R.H. (2003) Negative operant conditioning fails to deter white-tailed deer foraging activity. *Crop Protection*, 22, 893–895, https://doi.org/10.1016/s0261-2194(03)00048-6

3.59. Use ultrasonic noises to deter crop damage by mammals to reduce human-wildlife conflict

https://www.conservationevidence.com/actions/2479

- **One study** evaluated the effects of using ultrasonic noises to deter crop damage by mammals to reduce human-wildlife conflict. This study was in Australia[1].

COMMUNITY RESPONSE (0 STUDIES)

POPULATION RESPONSE (0 STUDIES)

BEHAVIOUR (0 STUDIES)

OTHER (1 STUDY)

- **Human-wildlife conflict (1 study):** A replicated, controlled, paired sites study in Australia[1] found that ultrasonic devices did not repel eastern grey kangaroos.

Background

Ultrasonic noise is sound waves at higher frequencies than those audible to humans. Different mammal species can detect sound at different ranges of frequencies, so some ultrasonic noises may be audible to a range of mammal species. If ultrasonic noises can deter animals from damaging crops, this could reduce motivation for carrying out lethal control of such species.

See also: *Use lights and sound to deter crop damage by mammals to reduce human-wildlife conflict, Use noise aversive conditioning to deter crop damage by mammals to reduce human-wildlife conflict and Use loud noises to deter crop damage (e.g. banger sticks, drums, tins, iron sheets) by mammals to reduce human-wildlife conflict.*

A replicated, controlled, paired sites study in 1995–1996 on a grassland site in Victoria, Australia (1) found that ultrasonic devices (ROO-Guard) did not repel eastern grey kangaroos *Macropus giganteus*. The number of kangaroo faecal pellets counted with the devices running (0.36–0.38 pellets/m²/day) was not significantly different from the number counted in the presence of dummy devices (0.17–0.20 pellets/m²/day). ROO-Guards were reported by the manufacturer to emit high frequency noise that is inaudible to humans but which deters kangaroos by masking their ability to hear predators. ROO-Guard Mk II devices were operated in December 1995–January 1996 in five open grassy areas of ≥100 m diameter. Each was paired with a similar area ≥850 m away, where an inactive device was simultaneously placed. Kangaroo use of each area was assessed by counting faecal pellets after 5–10 days.

(1) Bender H. (2003) Deterrence of kangaroos from agricultural areas using ultrasonic frequencies: efficacy of a commercial device. *Wildlife Society Bulletin*, 31, 1037–1046.

3.60. Use drones to deter crop damage by mammals to reduce human-wildlife conflict

https://www.conservationevidence.com/actions/2481

- **One study** evaluated the effects on mammals of using drones to deter crop damage by mammals to reduce human-wildlife conflict. This study was in Tanzania[1].

COMMUNITY RESPONSE (0 STUDIES)

POPULATION RESPONSE (0 STUDIES)

BEHAVIOUR (0 STUDIES)

OTHER (1 STUDY)

- **Human-wildlife conflict (1 study):** A replicated study in Tanzania[1] found that drones repelled African savanna elephants from crops within one minute.

Background

Wild herbivores can cause substantial damage to agricultural crops. Various methods may be used to deter animals from accessing crops or to scare away animals in the area. This intervention covers use of drones for scaring animals away from crop areas. If successful, the intervention could reduce incentives for carrying out lethal control of crop-raiding mammal species.

A replicated study in 2015–2016 in two savanna reserves in Tanzania (1) found that using drones to deter crop damage led to African savanna elephants *Loxodonta africana* leaving sites within one minute on all occasions. On all 38 occasions when drones were deployed to intercept elephants, the animals began to flee within one minute. Elephants were typically herded to an area > 1 km from croplands. Before drone use, rangers were trained during three 4-day workshops. In February–March and May–August 2015, and in March–April 2016, rangers deployed drones in 38 situations when elephants were found close to croplands or villages. Each drone was fitted with a flashlight, to locate elephants at night and, during the day, a live video feed from a camera on the drone was used. Elephant responses were recorded over 60-second intervals, during the first 10 minutes of the drone flight.

(1) Hahn N., Mwakatobe A., Konuche J., de Souza N., Keyyu J., Goss M., Chang'a A., Palminteri S., Dinerstein E. & Olson D. (2017) Unmanned aerial vehicles mitigate human–elephant conflict on the borders of Tanzanian Parks: a case study. *Oryx*, 51, 513–516, https://doi.org/10.1017/s0030605316000946

3.61. Translocate crop raiders away from crops (e.g. elephants) to reduce human-wildlife conflict

https://www.conservationevidence.com/actions/2485

- **Two studies** evaluated the effects on mammals of translocating crop-raiding animals away from crops to reduce human-wildlife conflict. One study was in Kenya[1] and one was in the USA[2].

COMMUNITY RESPONSE (0 STUDIES)

POPULATION RESPONSE (1 STUDY)

- **Survival (1 study)**: A controlled study in Kenya[1] found that translocated crop-raiding African elephants had a lower survival rate after release than did non-translocated elephants at the same site.

BEHAVIOUR (0 STUDIES)

OTHER (1 STUDY)

- **Human-wildlife conflict (1 study)**: A study in the USA[2] found that most American black bears translocated from sites of crop damage were not subsequently recaptured at sites of crop damage.

Background

Where wild mammals cause unacceptable damage to crops, they may be translocated from their point of capture and released some distance away. The release site may be an area away from where agricultural crops are grown. The intervention can fail if translocated animals continue to raid crops or if survival of translocated animals is low. If the intervention succeeds, it may reduce incentives for carrying out lethal control of such animals. Several other interventions cover translocations that are primarily for conservation of rare or threatened species, such as *Translocate to re-establish or boost populations in native range*.

A controlled study in 2005–2006 of savanna in and around a national park in Kenya (1) found that translocated crop-raiding African elephants *Loxodonta africana* had a lower survival rate than non-translocated elephants at the same site. Twenty-four of 150 translocated elephants died within 55 days of translocation; from dying during translocation (six elephants), poaching (one), shooting by problem animal control officers (two) and unknown causes (three), whilst 12 calves went missing and were presumed to have died. Out of 103 elephants that survived this period and were successfully monitored, four (4%) died over year following release, compared to 77 out of 6,395 (1%) during the same time period from the non-translocated population in the same park. One hundred and fifty elephants were translocated 160 km to a national park, in September 2005, to reduce human-elephant conflicts related to crop damage at the source location. Locations of translocated elephants and resident elephants were monitored 4–5 times/week at the receptor site from road transects and 2–3 times/week by aerial surveys.

A study in 2006–2007 across a large portion of northern Wisconsin, USA (2) found that most American black bears *Ursus americanus* translocated away from sites of damage to corn crops were not subsequently recaptured at sites of crop damage. Out of 520 translocated bears, 20 (4%) were recaptured during subsequent capture activities at sites of crop damage (including the original capture site). Average time to recapture was 45 days. Recaptured bears had been moved 40–64 km following initial capture. Of the total of 21 recaptures of 20 recaptured bears (one was recaptured twice), nine (43%) were at the original capture site and 15 (71%) were within 10 km of the original capture site. Bears were captured on 55 farms from 11 August to 9 October 2006 and 50 farms from 3 August to 12 October 2007. Skin samples were taken using a biopsy dart and 541 out of 567 samples produced genetic material that enabled identification of 520 individuals.

(1) Pinter-Wollman N., Isbell L.A. & Hart L.A. (2009) Assessing translocation outcome: Comparing behavioral and physiological aspects of translocated and resident African elephants (*Loxodonta africana*). *Biological Conservation*, 142, 1116–1124, https://doi.org/10.1016/j.biocon.2009.01.027

(2) Shivik J.A., Ruid D., Willging R.C. & Mock K.E. (2011) Are the same bears repeatedly translocated from corn crops in Wisconsin? *Ursus*, 22, 114–119, https://doi.org/10.2192/URSUS-D-10-00031.1

3.62. Use negative stimuli to deter consumption of livestock feed by mammals to reduce human-wildlife conflict

https://www.conservationevidence.com/actions/2486

- **One study** evaluated the effects of using negative stimuli to deter consumption of livestock feed by mammals to reduce human-wildlife conflict. This study was in the USA[1].

COMMUNITY RESPONSE (0 STUDIES)

POPULATION RESPONSE (0 STUDIES)

BEHAVIOUR (0 STUDIES)

OTHER (1 STUDY)

- **Human-wildlife conflict (1 study):** A replicated, controlled study in the USA[1] found that white-tailed deer presence at cattle feeders was usually reduced by a device that produced a negative stimulus.

Background

Livestock feed might also attract wild herbivores. This could produce a financial cost to farmers, through added feed costs and through transmission of disease, such as bovine tuberculosis, between wild and domestic herbivores (Phillips *et al.* 2003). Disease transmission may be greater where animals share foodstuffs. Hence, if wild herbivores can be effectively deterred from accessing livestock feed, this may reduce motivations for carrying out lethal control of wild herbivores.

Phillips C.J., Foster C.R., Morris P.A. & Teverson R. (2003) The transmission of *Mycobacterium bovis* infection to cattle. *Research in Veterinary Science*, 74, 1–15, https://doi.org/10.1016/s0034-5288(02)00145-5

A replicated, controlled study in 2005 of captive deer on a farm in Michigan, USA (1) found that a deer-resistant cattle feeder device reduced white-tailed deer *Odocoileus virginianus* presence at feeders for the first five of six weeks. Fewer deer were recorded on camera traps within 1 m of feeders with active devices (0–0.2 deer/activation) than of feeders without devices (0.7–1.9 deer/activation) during the first five treatment weeks. There was no significant difference during the sixth week (active device: 0.4 deer/activation; no device: 1.2 deer/activation). During four weeks before device activation, deer number recorded on camera traps were similar between feeders with (2.3–2.9 deer/ activation) and without (2.1–2.7 deer/activation) devices. Three feeders each were protected and unprotected by devices. Devices entailed a 3.4-m horizontal bar with a 1.6-m arm hanging on chains at each end, down to 45 cm above the ground. The rig rotated on a central pivot for 45 s, when an animal entered an infra-red-surveillance zone. Hanging arms struck animals within 1 m of feeders, startling, but not hurting, them. Monitoring, using camera traps, spanned 10 February to 10 March 2005 (devices inactive) and 13 May to 23 June 2005 (devices active).

(1) Seward N.W., Phillips G.E., Duquette J.F. & VerCauteren K.C. (2007) A frightening device for deterring deer use of cattle feeders. *The Journal of Wildlife Management*, 71, 271–276, https://doi.org/10.2193/2006-265

3.63. Play predator calls to deter crop damage by mammals to reduce human-wildlife conflict

https://www.conservationevidence.com/actions/2487

- We found no studies that evaluated the effects on mammals of playing predator calls to deter crop damage to reduce human-wildlife conflict.

'We found no studies' means that we have not yet found any studies that have directly evaluated this intervention during our systematic journal and report searches. Therefore, we have no evidence to indicate whether or not the intervention has any desirable or harmful effects.

Background

Wild herbivores can cause damage to crops. Calls of predators of these animals can be played in an attempt to deter wild herbivores from the area.

3.64. Use target species distress calls or signals to deter crop damage by mammals to reduce human-wildlife conflict

https://www.conservationevidence.com/actions/2488

- **Five studies** evaluated the effects of using target species distress calls or signals to deter crop damage by these species to reduce human-wildlife conflict. Two studies were in the USA[2,4] and one each was in Namibia[1], Australia[3] and Sri Lanka[5].

COMMUNITY RESPONSE (0 STUDIES)

POPULATION RESPONSE (0 STUDIES)

BEHAVIOUR (0 STUDIES)

OTHER (5 STUDIES)

- **Human-wildlife conflict (5 studies):** Two of five replicated studies (including four controlled studies), in the USA[2,4],

Namibia[1], Australia[3] and Sri Lanka[5], found that white-tailed deer[4] and Asian elephants[5] were deterred or repelled from areas by playing their respective distress calls. Two studies found that, in most cases, elephants[1] and white-tailed deer[2] were not deterred from entering or remaining at sites when distress calls were played. The fifth study found mixed results but, overall, eastern grey kangaroo foot-thumping noises did not increase numbers leaving a site[3].

Background

Some animals, especially species that routinely form social groups, produce calls or other audible signals when they detect danger. If artificially playing calls or signals from the same species can restrict movements of animals, this may assist in reducing damage to crops. If effective, the intervention could reduce incentives for carrying out lethal control of such species.

A replicated study in 1994 at three water holes in a grassland area in East Caprivi, Namibia (1) found that playing warning calls of elephants *Loxodonta africana* did not, in most cases, deter elephants from remaining at a site. In eight trials at three sites, groups of elephants (5–30 animals) were deterred from the site during three trials and undeterred during five. In six further trials involving 1–3 bull elephants, the animals were not deterred. Trail groups were not independent and some involved the same animals. Elephant warning calls, produced during times of apparent natural distress events, were recorded. They were played back on a portable cassette player at approximately 15-m distance from each herd as they visited water holes. Playback was activated when elephants pushed a tripwire.

A replicated, paired sites, controlled study in 2001 on arable fields alongside woodland at a site in Nebraska, USA (2) found that playing white-tailed deer *Odocoileus virginianus* distress calls did not affect deer intrusions into corn crops or subsequent corn yields. The rate of deer entries into fields was similar at fields protected by frightening devices (48–57 entries/km boundary/day) and unprotected fields (48–52 entries/km boundary/day). Similarly, there was no difference between fields before devices operated (device fields: 69 entries/km/ day; unprotected: 56 entries/km/day) or after devices were turned off (device fields: 23–46 entries/km/day; unprotected: 20–47 entries/km/ day). Average corn yields did not differ between fields with frightening devices (6,381 kg/ha) and unprotected fields (5,614 kg/ha). Six pairs of fields (6–20 ha, ≥0.5 km apart, matched for size, shape and location) were studied. Frightening devices played deer distress noises for 30 s when activated by deer breaking 50–200-m-long infrared beams. Two devices at each protected field covered 21–48% of the perimeter. Devices operated from 6–24 July 2001, when corn was most vulnerable to deer-damage. Deer activity was assessed by counting tracks twice during the device operating period, once five days before this and three times during 18 days after this time.

A replicated, randomized, controlled study in 1997–1998 at a shrubland site in Victoria, Australia (3) found that playing recordings of foot-thumping kangaroos increased vigilance in eastern grey kangaroos *Macropus giganteus* and caused more kangaroos to flee in the first few second, but did not cause more overall to flee. Where the foot-thumping noise was played, kangaroos increased vigilance more than did those played a background recording (data presented as indices). A higher proportion of kangaroos fled within the first 3 s of hearing foot-thumping (26%) than of hearing background noise (0%). However, in total, 63% of kangaroos fled, and there was no significant difference in the overall average time to fleeing between noise types (combined average time to fleeing of 25 s). Kangaroos were observed from hides alongside three perimeter fence holes (≥850 m apart). Foot-thumping or a background noise were played for 8 s (noise type selected randomly). Responses were assessed from videos of 236 kangaroos, on 15 nights (20.00 to 21.15 hrs), from 11 December 1997 to 5 February 1998. Fleeing time was measured in 112 adult kangaroos, 64 exposed to foot-thumping and 48 with background noise. Individual kangaroos were tested once/session.

A replicated, randomized, controlled, before-and-after study in 2010 in a deciduous forest in Utah, USA (4) found that devices playing deer distress calls reduced white-tailed deer *Odocoileus virginianus* visits and food uptake. Sites with devices had 0 deer visits/day when devices were active (treatment period) compared to 273 visits/day with devices inactive (pre-treatment). Concurrently, sites without devices had 122 visits/day (treatment period) and 169 visits/day (pre-treatment). Food consumption by deer was lower at sites with devices during treatment (0 litres) than pre-treatment phases (2,175 1). At sites without devices, consumption during treatment (1,100 1) and pre-treatment phases (1,585 1) was similar. Six sites, >0.6 km apart, were each enclosed in a U-shaped fence, 18.3 m long. Three sites, selected randomly, had a deer-activated frightening device installed. This played deer distress calls when an infra-red beam was broken. Sites were baited with >38 1 of alfalfa cubes in February 2010. Bait was topped up every second day. Deer visits were monitored using camera traps. Pre-treatment (device inactive) ran during 10–22 March 2010 while the treatment phase (device active) ran from 23 March to 4 April 2010.

A replicated, randomized, controlled study (year not stated) in a protected area containing forest, grassland, and wetland in Sri Lanka (5) found that playing recordings of elephant family groups to Asian elephants *Elephas maximus* led to more elephants fleeing the area compared to playing of other sounds. After playing the sound of elephant family groups, 11 of 17 elephants (65%) fled, compared to three of 31 (10%) when other sounds were played. Randomly selected elephants in the protected area were provided with a sugarcane, banana and palm frond mixture. Speakers were placed approximately 15 m from elephants. Sounds were played in a random order for one minute each, with a five-minute interval between sounds. Sounds played were: elephant group vocalizations (17 occasions), Sri Lankan hornets *Vespa affinis affinis* (12 occasions), lone female elephant vocalizations (8 occasions) and a chainsaw (11 occasions). Behaviour of animals was recorded during and after each playback.

(1) O'Connell-Rodwell C.E., Rodwell T., Rice M. & Hart L.A. (2000) Living with the modern conservation paradigm: can agricultural communities co-exist with elephants? A five-year case study in East Caprivi, Namibia. *Biological Conservation*, 93, 381–391, https://doi.org/10.1016/s0006-3207(99)00108-1

(2) Gilsdorf J.M., Hygnstrom S.E., VerCauteren K.C., Clements G.M., Blankenship E.E. & Engeman R.M. (2004) Evaluation of a deer-activated bioacoustic frightening device for reducing deer damage in cornfields. *Wildlife Society Bulletin*, 32, 515–523, https://doi.org/10.2193/0091-7648(2004)32[515:eoadbf]2.0.co;2

(3) Bender H. (2005) Effectiveness of the eastern grey kangaroo foot thump for deterring conspecifics. *Wildlife Research*, 32, 649–655, https://doi.org/10.1071/wr04091

(4) Hildreth A.M., Hygnstrom S.E. & VerCauteren K.C. (2013) Deer-activated bioacoustic frightening device deters white-tailed deer. *Human–Wildlife Interactions* 7, 107–113, https://doi.org/10.26077/12mz-1p38

(5) Wijayagunawardane M.P., Short R.V., Samarakone T.S., Nishany K.B., Harrington H., Perera B.V., Rassool R. & Bittner E.P. (2016) The use of audio playback to deter crop-raiding Asian elephants. *Wildlife Society Bulletin*, 40, 375–379, https://doi.org/10.1002/wsb.652

3.65. Use bees to deter crop damage by mammals (e.g. elephants) to reduce human-wildlife conflict

https://www.conservationevidence.com/actions/2489

- **Three studies** evaluated the effects on elephants of using bees to deter crop damage to reduce human-wildlife conflict. All three studies were in Kenya[1,2,3].

COMMUNITY RESPONSE (0 STUDIES)

POPULATION RESPONSE (0 STUDIES)

BEHAVIOUR (0 STUDIES)

OTHER (3 STUDIES)

- **Human-wildlife conflict (3 studies):** Three replicated studies (including one controlled study), in Kenya[1,2,3], found that beehive fences reduced crop raiding by African elephants.

Background

Conflicts between farmers and free-ranging elephants occur in parts of Africa. Farmers on small plots may lose large proportions of their crops to raids by elephants. Some elephants are said to be wary of foraging near African honeybees *Apis mellifera scutellata* (Vollrath & Douglas-Hamilton 2002). Thus, fences comprising bee hives linked by wires may deter entry to fields by elephants, as well as providing a further potential crop (honey) for farmers. If successful, the intervention could reduce incentives for carrying out lethal control of elephants.

Vollrath F. & Douglas-Hamilton I. (2002) African bees to control African elephants. *Naturwiss*, 89, 508–511, https://doi.org/10.1007/s00114-002-0375-2

A controlled study in 2007 on two farms in Laikipia, Kenya (1) found that a beehive fence (without resident bees) reduced crop-raiding by African elephants *Loxodonta africana*. Results were not tested for statistical significance. There were fewer successful crop raids on the farm protected by the beehive fence (7 raids) than on the unprotected farm (13 raids). Fewer individual elephants raided the protected farm (38) than the unprotected farm (95). The two farms, 466 m apart, each approximately 2 acres, grew similar mixes of maize *Zea mays*, potatoes *Solanum tuberosum*, sorghum *Sorghum* sp. and beans. On one farm, nine hives were suspended under thatch roofs, along a 90-m boundary. A wire between hives connected to the wires suspending hives, so an elephant pushing against it caused the hives to shake, and bees to emerge. However, hives were unoccupied during the trial. The second farm was unprotected. Elephant raids were documented by farmers over six weeks in August–September 2007.

A replicated, controlled study in 2008–2010 on agricultural land around two villages in Kenya (2) found that beehive fences reduced entry onto farmland by elephants *Loxodonta africana*. Elephants entered farmland through a beehive fence less often (1 occasion) than they did through traditional thorn bush barriers (31 occasions). Following entry to farmland, elephants also left less frequently through beehive fences (six occasions) than they did through thorn bush barriers (26 occasions). Thirty-four farms were studied, of which 17 were protected along parts of their perimeters by beehive fences and 17 were protected solely by traditional thorn bush barriers. Beehive fences comprised a total of 149 beehives deployed in June–August 2008 and 21 deployed in April 2009. Hives were positioned 10 m apart. Farms were monitored over three crop seasons, from June 2008 until June 2010.

A replicated study in 2012–2015 of 10 crop fields in an agricultural community in Kenya (3) found that beehive fences deterred crop raiding by African elephants *Loxodonta africana*. Of 238 elephants that approached farms with beehive fences, more turned away (190 elephants) than broke through to raid crops (48). On 65 occasions, elephant groups approached to ≤10 m from beehive fences. Of these, 39 groups (114 elephants) turned back at the fence and 26 groups (50 elephants) broke through fences. Eight farm plots, each 0.4 ha extent, were enclosed by beehive fences, built in June 2012 to February 2013. Fences comprised 12 beehives and 12 two-dimensional plywood dummy hives suspended from a wire running continuously between fence posts. Pushing the wire caused hives to rock and bees to emerge. Elephant movements around fences were recorded by farmers.

(1) King L.E., Lawrence A., Douglas-Hamilton I. & Vollrath F. (2009) Beehive fence deters crop-raiding elephants. *African Journal of Ecology*, 47, 131–137, https://doi.org/10.1111/j.1365-2028.2009.01114.x

(2) King L.E., Douglas-Hamilton I. & Vollrath F. (2011) Beehive fences as effective deterrents for crop-raiding elephants: field trials in northern Kenya. *African Journal of Ecology*, 49, 431–439. https://doi.org/10.1111/j.1365-2028.2011.01275.x

(3) King L.E., Lala F., Nzumu H., Mwambingu E. & Douglas-Hamilton I. (2017) Beehive fences as a multidimensional conflict-mitigation tool for farmers coexisting with elephants. *Conservation Biology*, 31, 743–752, https://doi.org/10.1111/cobi.12898

3.66. Grow unattractive crop in buffer zone around crops (e.g. chili peppers) to reduce human-wildlife conflict

https://www.conservationevidence.com/actions/2491

- We found no studies that evaluated the effects on mammals of growing unattractive crops (such as chili peppers) in buffer zones around crops to reduce human-wildlife conflict.

'We found no studies' means that we have not yet found any studies that have directly evaluated this intervention during our systematic journal and report searches. Therefore, we have no evidence to indicate whether or not the intervention has any desirable or harmful effects.

Background

Some crops are vulnerable to wild herbivores, such as elephants. Some other crops, such as chilli, may have a repellent effect for wild herbivores. Planting them around the perimeter of the main crop may act as a deterrent to approach by such wild herbivores. If successful, this may reduce the incentives for carrying out lethal control of such herbivores.

3.67. Use chili to deter crop damage by mammals to reduce human-wildlife conflict

https://www.conservationevidence.com/actions/2492

- **Seven studies** evaluated the effects on elephants of using chili to deter crop damage to reduce human-wildlife conflict. Four studies were in Zimbabwe[1,2,3,5], two were in Kenya[4a,4b] and one was in India[6].

COMMUNITY RESPONSE (0 STUDIES)

POPULATION RESPONSE (0 STUDIES)

BEHAVIOUR (0 STUDIES)

OTHER (7 STUDIES)

- **Human-wildlife conflict (7 studies):** Five of seven studies (including four replicated and two before-and-after studies), in Zimbabwe[1,2,3,5], Kenya[4a,4b] and India[6], found that chill-based deterrents (chili-spray, chili smoke, chili fences and chili extract in a projectile, in some cases along with other deterrents) repelled elephants at least initially[1,2,3,4a,5], whist two studies found that chili smoke (and in one case chili fences) did not reduce crop raiding[4b,6].

Background

This intervention covers use of chili in various forms for deterring crop damage. All studies are of its effectiveness against elephants *Loxodonta africana* and *Elephas maximus*. In some cases, trials were of deterrent effects of chili against elephants that were not actively crop-raiding. Studies in this intervention are all of situations where chili repellents are targeted specifically at potential crop raiding animal, using smoke, aerosol or projectile. If successful, the intervention could reduce incentives for carrying out lethal control of elephants.

See also *Use repellents that taste bad ('contact repellents') to deter crop or property damage by mammals to reduce human-wildlife conflict*, which includes use of Hot Sauce® and other chili-based repellents that are applied directly to crops.

A replicated study in 1993–1994 of savanna and farmland at two sites in Zimbabwe (1) found that a chili-based capsicum spray repelled elephants *Loxodonta africana*. In 19 of 22 tests in a national park, elephants retreated when sprayed with the capsicum aerosol. In three successful tests, elephants reacted to the sound of the spray discharging. Elephants also retreated in 16 of 18 tests carried out on farmland. In two tests, elephants appeared not to inhale the spray. Twenty-two tests were conducted in a national park from 16–22 July 1993, thirteen on bulls and nine on family groups. Capsicum sprays were discharged on foot or from vehicles (average 40 m from elephants) or by remote-control, 250 m from a watering hole. Eighteen tests were conducted on 1–14 elephants on farmland, on moonlit nights, from February–May 1994. Capsicum sprays were administered on foot or by remote-control. In all tests, elephants were settled for 5–20 mins, with staff in place, before testing. This helped to ensure that elephants' responses were not simply a reaction to human presence. A 10% capsicum oleoresin solution was then discharged from an aerosol can, upwind of elephants.

A replicated study in 1995–1996 in crop fields at a site surrounded by savanna in Sebungwe, Zimbabwe (2) found that a chili-based capsicum spray repelled crop-raiding African elephants *Loxodonta africana* faster than did scaring by combinations of people, dogs *Canis lupus familiaris*, slingshots, drums, whips, burning sticks large fires. Elephants were repelled faster when sprayed with capsicum aerosol (2 minutes) than when scared by one person with a small fire (and sometimes with a dog) (14 minutes), by two to three people with dogs and slingshots, drums and burning sticks (10 minutes) or by four to seven people with dogs, drums, whips and large fires (4 minutes). No elephants charged at defenders when sprayed with the capsicum aerosol but defenders were charged on 13–60% of occasions when elephants were scared by other means. Elephants raiding crops were scared 18 times using 10% capsicum oleoresin spray, 15 times by one person with a small fire (and sometimes with a dog), 11 times by 2–3 people with dogs, slingshots, drums and burning sticks and 15 times by 4–7 people with dogs, drums, whips and large fires. Behavioural responses were monitored by watching through a monocular. Distance between elephants and farmers was 20–40 m. Tests were conducted between 18:30 and 06:30 h. The number of fields studied was not specified.

A replicated study in 2001 of arable land in seven villages in Guruve District, Zimbabwe (3) found that burning chilies mixed with elephant *Loxodonta africana* dung, repelled crop-raiding elephants faster than did traditional deterrents of beating drums and throwing rocks. Elephants left faster (average 9 minutes) when chili mixed with dung was burned than they did when traditional repellent methods alone were used (average 65 minutes). Seven villages were studied. At three villages, farmers set fire to bricks made of elephant dung mixed with chili, to deter elephants that were attempting to raid crops, on 34 occasions. Farmers at four villages used traditional methods to scare off elephants that attempted to raid crops, namely banging drums and throwing rocks with catapults, on 27 occasions. The study was conducted from 1 January to 30 June 2001 and data were collected by a team of observers.

A before-and-after and site comparison study in 2003–2004 of two farming areas in Laikipia, Kenya (4a) found that using chili fences and chili smoke, along with loud noises, reduced raiding and crop damage by African elephants *Loxodonta africana*. The study does not distinguish between the effects of chilli deterrents and loud noises. After farmers began using chili fences and chili smoke, along with loud noises, the total number of crop-raiding incidents (26) and the average area of crop damage (375 m²/incident) was lower than before deterrents were used (92 incidents; 585 m²/incident). However, the difference was not tested for statistical significance. At a control site without deterrents, crop-raiding increased (total 17–166 incidents) as did crop damage (average 328 m²–421 m²/incident) during the same time period. A group of farmers within a 0.03-km² area were provided with training and materials to deter crop-raiding elephants. Deterrents included chili fences (rope and cloth fences with chili and engine grease applied), chili smoke (chili and dung briquettes burned at night) and loud noises (bangers, banger sticks, cow bells). Some farmers also used watchtowers and torches. A second control area, of equal size and within 1 km, used no deterrents. Crop-raiding incidents and crop damage were recorded in each of the two areas before (June–December 2003) and after (June–December 2004) deterrents were introduced.

A replicated, before-and-after and site comparison study in 2004–2005 at 40 farms in Laikipia, Kenya (4b) found that using chili fences and chili smoke, along with loud noises, did not result in an overall reduction in crop-raiding by African elephants *Loxodonta africana*. The study does not distinguish between the effects of chilli deterrents and loud noises. After farmers began using chili fences and chili smoke, along with loud noises, the average number of crop-raiding incidents across all farms (2) was similar to before deterrents were used (2.5). At 10 control farms without deterrents, crop-raiding decreased (from an average of three incidents to one) during the same time period. Ten farmers in each of two areas were provided with training and materials to deter crop-raiding elephants. Deterrents included chili fences (rope and cloth fences with chili and engine grease applied), chili smoke (chili and dung briquettes burned at night) and loud noises (bangers, banger sticks, cow bells). Some farmers also used watchtowers and torches. Uptake of deterrent types varied between farms (see original paper for details). Ten control farms within each of the two areas used no deterrents. Crop-raiding incidents were recorded at all 40 farms before (February–November 2004) and after (February–November 2005) deterrents were introduced.

A study in 2007 of grassland, thicket, woodland and water holes in a national park in Zimbabwe (5) found that, after being shot at with chili oil extract, most savanna elephants *Loxodonta africana* either ran away or backed up, but most soon resumed normal behaviour. When shot at, 11 (46%) of 24 elephants ran away, seven (29%) changed their behaviour and walked away and six (25%) did not change their behaviour. After 1 minute, seven (29%) were still running away, one (4%) was walking away and 16 (67%) had resumed normal behaviour. The study was conducted in a remote area of Hwange National Park in October 2007. Between 09:30 and 18:00 h, a professional hunter shot a ping-pong ball filled with chili oil extract at 24 elephants from 15–110 m using a gas-dispenser. Only eight elephants were hit by the balls, of which seven then released chili oil.

A study in 2006–2009, in two areas of Assam, India (6) found that using chili smoke to deter Asian elephants *Elephas maximus* did not reduce the probability of elephants raiding crops. The chance of crop damage occurring was not lower when chili smoke was used to deter crop-raiding elephants compared to a range of other interventions or to no intervention (results presented as statistic model). Within two study areas, 33 community members were trained as monitors to record the 1,761 crop-raiding incidents, from 1 March 2006 to 28 February 2009. A range of deterrents were used, singly or in combination. These included chili smoke (from burning dried chilies, tobacco, and straw), spotlights, two-strand electric fences, chili fencing (engine grease and ground chili paste, on a jute or coconut rope), elephant drives (using domesticated elephants to repel wild elephants), fire and noise.

(1) Osborn F.V. & Rasmussen L.E.L. (1995) Evidence for the effectiveness of an oleo-resin capsicum aerosol as a repellent against wild elephants in Zimbabwe. *Pachyderm*, 20, 55–64.

(2) Osborn F.V. (2002) Capsicum oleoresin as an elephant repellent: field trials in the communal lands of Zimbabwe. *The Journal of Wildlife Management*, 66, 674–677, https://doi.org/10.2307/3803133

(3) Osborn F.V. & Parker G.E. (2002) Community-based methods to reduce crop loss to elephants: experiments in the communal lands of Zimbabwe. *Pachyderm*, 33, 32–38.

(4) Graham M. & Ochieng T. (2008) Uptake and performance of farm-based measures for reducing crop raiding by elephants Loxodonta africana among smallholder farms in Laikipia District, Kenya. *Oryx*, 42, 76–82, https://doi.org/10.1017/S0030605308000677

(5) Le Bel S., Taylor R., Lagrange M., Ndoro O., Barra M. & Madzikanda H. (2010) An easy-to-use capsicum delivery system for crop-raiding elephants

in Zimbabwe: preliminary results of a field test in Hwange National Park. *Pachyderm*, 47, 80–89.

(6) Davies T.E., Wilson S., Hazarika N., Chakrabarty J., Das D., Hodgson D.J. & Zimmermann A. (2011) Effectiveness of intervention methods against crop-raiding elephants. *Conservation Letters*, 4, 346–354, https://doi. org/10.1111/j.1755-263x.2011.00182.x

3.68. Use light/lasers to deter crop damage by mammals to reduce human-wildlife conflict

https://www.conservationevidence.com/actions/2496

- **Two studies** evaluated the effects of using light or lasers to deter crop damage by mammals to reduce human-wildlife conflict. Both studies were in the USA[1,2].

COMMUNITY RESPONSE (0 STUDIES)

POPULATION RESPONSE (0 STUDIES)

BEHAVIOUR (0 STUDIES)

OTHER (2 STUDIES)

- **Human-wildlife conflict (2 studies):** A replicated, randomized, controlled study in the USA[1] found that red lasers did not disperse white-tailed deer from fields at night whilst a study in India[2] found that spotlights directed at the eyes of Asian elephants did reduce the probability of crop damage.

Background

This intervention specifically refers to use of directional light or lasers aimed at animals. If such lights can reduce crop damage by mammals, this may reduce incentives for carrying out lethal control of such species.

See also *Use lights and sound to deter crop damage to reduce human-wildlife conflict*.

A replicated, randomized, controlled study in 2001 in arable fields on two adjacent wildlife refuges straddling Nebraska and Iowa, USA (1) found that red lasers did not disperse white-tailed deer *Odocoileus virginianus* from fields at night. No differences were found in flight response between two different lasers (deer fled in 2–3% of encounters) or between these lasers and the control without lasers (3% fled). Thirty-two crop fields were randomly assigned one of two lasers, shone from a vehicle, or as the control (vehicle without laser). The two red lasers were the Desman® (633 nm, 5 mW, 12 mm beam) and Dissuader™ (650 nm, 68 mW, variable beam). Deer behaviour was monitored using night-vision binoculars on eight consecutive nights in July 2001 (total 177 deer encounters). Deer were initially located with a spotlight. Lasers were used for 2 minutes/deer, first on adjacent vegetation, then in a zig-zag manner, then on the body.

A study in 2006–2009 in two areas of Assam, India (2) found that using spotlights directed at the eyes of Asian elephants *Elephas maximus* reduced the probability of elephants causing crop damage. The chance of crop damage occurring was lower when spotlights were used to deter crop-raiding elephants compared to a range of other interventions or no intervention (results presented as statistical model coefficients). Only installing fences reduced crop raiding to a greater extent. Using loud noises alongside spotlighting reduced its effectiveness. Within two study areas, 33 community members were trained as monitors to record the 1,761 crop-raiding incidents, from 1 March 2006 to 28 February 2009. A range of deterrents were used, singly or in combination, including spotlights, chili smoke (from burning dried chilies, tobacco, and straw), two-strand electric fences, chili fencing (engine grease and ground chili paste, on a jute or coconut rope), elephant drives (using domesticated elephants to repel wild elephants), fire and noise.

(1) VerCauteren K.C., Hygnstrom S.E., Pipas M.J., Fioranelli P.B., Werner S.J. & Blackwell B.F. (2003) Red lasers are ineffective for dispersing deer at night. *Wildlife Society Bulletin*, 31, 247–252.

(2) Davies T.E., Wilson S., Hazarika N., Chakrabarty J., Das D., Hodgson D.J. & Zimmermann A. (2011) Effectiveness of intervention methods against crop-raiding elephants. *Conservation Letters*, 4, 346–354, https://doi.org/10.1111/j.1755-263X.2011.00182.x

3.69. Use fire to deter crop damage by mammals to reduce human-wildlife conflict

https://www.conservationevidence.com/actions/2499

- **Two studies** evaluated the effects on mammals of using fire to deter crop damage by mammals to reduce human-wildlife conflict. One study was in Zimbabwe[1] and one was in India[2].

COMMUNITY RESPONSE (0 STUDIES)

POPULATION RESPONSE (0 STUDIES)

BEHAVIOUR (0 STUDIES)

OTHER (2 STUDIES)

- **Human-wildlife conflict (2 studies):** A replicated study in Zimbabwe[1] found that a combination of large fires and people with drums and dogs repelled African elephants from crops faster than did a combination of people with dogs and slingshots, drums and burning sticks. A study in India[2] found that fire reduced the chance of Asian elephants damaging crops.

Background

Wild herbivores can cause substantial damage to agricultural crops. Various methods may be used to deter animals from accessing crops or to scare away animals in the area. This intervention covers use of fire for scaring animals away from crop areas. If successful, the intervention could reduce incentives for carrying out lethal control of crop-raiding mammals.

A replicated study in 1995–1996 in crop fields at a site surrounded by savanna in Sebungwe, Zimbabwe (1) found that when scared by a combination of large fires and people with dogs *Canis lupus familiaris*, whips and drums, African elephants *Loxodonta africana* were repelled faster from fields than by a combination of people with dogs, slingshots, drums and burning sticks. Elephants were repelled faster when scared with by large fires and people with dogs, whips and drums (4 minutes) than when scared by people with dogs, slingshots, drums and burning sticks (10 minutes). However, when scared by large fires and people with dogs, whips and drums, elephants charged at defenders during 60% of scaring attempts (9 of 15). Elephants raiding crops were scared 15 times by 4–7 people with multiple large fires, several dogs, whips and drums and 11 times by 2–3 people with dogs, slingshots, drums and burning sticks. Behavioural responses were monitored through a monocular. Elephants and farmers were 20–40 m apart. Tests were conducted between 18:30 and 06:30 h. The number of fields was not specified.

A study in 2006–2009, in two areas of Assam, India (2) found that using fire to deter crop-raiding Asian elephants *Elephas maximus* reduced the chance of crop damage occurring. The chance of crop damage occurring was lower when fire was used to deter crop-raiding elephants compared to a range of other interventions or no intervention (results presented as statistic model coefficients). Loud noise, fences and spotlights reduced crop raiding to a greater extent. Using loud noises alongside fire was less effective than using fire alone. Within two study areas, 33 community members trained as monitors, recorded 1,761 crop-raiding incidents, from 1 March 2006 to 28 February 2009. A range of deterrent methods was used, singly or in combination. These were fire (in pits or on hand-held fire torches), chili smoke (from burning dried chilies, tobacco, and straw), spotlights, two-strand electric fences, chili fencing (engine grease and ground chili paste, on a jute or coconut rope), elephant drives (using domesticated elephants to repel wild elephants) and noise.

(1) Osborn F.V. (2002) Capsicum oleoresin as an elephant repellent: field trials in the communal lands of Zimbabwe. *The Journal of Wildlife Management*, 66, 674–677, https://doi.org/10.2307/3803133

(2) Davies T.E., Wilson S., Hazarika N., Chakrabarty J., Das D., Hodgson D.J. & Zimmermann A. (2011) Effectiveness of intervention methods against crop-raiding elephants. *Conservation Letters*, 4, 346–354, https://doi.org/10.1111/j.1755-263X.2011.00182.x

3.70. Use pheromones to deter crop damage by mammals to reduce human-wildlife conflict

https://www.conservationevidence.com/actions/2503

- We found no studies that evaluated the effects of using pheromones to deter crop damage by mammals to reduce human-wildlife conflict.

'We found no studies' means that we have not yet found any studies that have directly evaluated this intervention during our systematic journal and report searches. Therefore, we have no evidence to indicate whether or not the intervention has any desirable or harmful effects.

Background

Pheromones are chemical substances released into the environment by an animal that can affect the behaviour or physiology of other animals of the same species. If pheromones can be synthesised that deter entry to crops by wild herbivores, this could reduce the motivation among farmers for carrying out lethal control of wild herbivores.

3.71. Use predator scent to deter crop damage by mammals to reduce human-wildlife conflict

https://www.conservationevidence.com/actions/2505

- **Three studies** evaluated the effects of using predator scent to deter crop damage by mammals to reduce human-wildlife conflict. All three studies were in the USA[1,2a,2b].

COMMUNITY RESPONSE (0 STUDIES)

POPULATION RESPONSE (0 STUDIES)

BEHAVIOUR (0 STUDIES)

OTHER (3 STUDIES)

- **Human-wildlife conflict (3 studies):** Two of three replicated, randomized, controlled studies (including two before-and-after studies), in the USA[1,2a,2b], found that coyote scent reduced food consumption by mountain beavers[1] and white-tailed deer[2a]. The third study found that it did not reduce trail use by white-tailed deer[2b].

Background

Wild herbivores may be sensitive to scents from predators and may alter their behaviour or visitation rates to a site accordingly (Wikenros *et al.* 2015). If scents can be deployed artificially, they could reduce crop damage caused by wild herbivores and, hence, reduce motivations for carrying out lethal control of these animals.

Wikenros C., Kuijper D.P.J., Behnke R. & Schmidt K. (2015) Behavioural responses of ungulates to indirect cues of an ambush predator. *Behaviour*, 152, 1019–1040, https://doi.org/10.1163/1568539X-00003266

A replicated, randomized, controlled study (year not stated) on captive animals from Washington State, USA (1) found that coyote *Canis latrans* urine was more effective at deterring food consumption by mountain beavers *Aplodontia rufa* than were four synthetic compounds. In two-choice feeding trials, the quantity of coyote urine-soaked food removed by male beavers (7 g) was lower than that of water-soaked food removed (14 g). The same pattern held for females (coyote urine: 1 g; water: 7 g). A3-Isopentenyl methyl sulfide (IMS) did not affect food choice when compared to an untreated 'blank' (IMS: 8–11 g; blank: 7 g), nor did 2,2-dimethylthietane (DMT) (DMT: 7–13 g; blank: 10–14 g). A mix of 2-propylthietane and 3-propyl-l,2-dithiolane (PT/PDT) reduced food retrieval (PT/PDT: 14 g; blank: 18 g) but the response was not apparent during longer (5 day) exposure (PT/PDT: 31 g; blank: 35 g). Twelve wild-caught mountain beavers (six male and six female) were held in captivity for several months prior to the experiment. Trials were run as choice tests between bowls 25 cm apart. Food remaining after one or two hours was weighed. Each beaver was used twice for each choice experiment.

A replicated, randomized, controlled, before-and-after study in 2000–2001 in a forest in Ohio, USA (2a) found that coyote *Canis latrans* hair reduced feeding at troughs by white-tailed deer *Odocoileus virginianus*. With one bag of coyote hair/trough, deer consumed less corn (103 kg) than before bag placement (246 kg). With three bags of coyote hair/trough, deer consumed less corn (46–108 kg/week) than in the week before bag placement (323 kg). At control toughs with empty bags, operated concurrently to experimental troughs, consumption (284–425 kg/week) did not differ to that in the week before bag placement (247–265 kg/week). Ten troughs (≥1 km apart) were fenced on three sides and stocked with whole kernel corn. Five were treatment troughs and five were controls. Stage I (January–February 2000) entailed one week with unprotected troughs. The following week, a nylon mesh bag containing 17 g of coyote hair was placed touching the back of treatment troughs. An empty bag was placed at control troughs. Stage II (January–March 2001) had a similar pre-treatment week, then five weeks with three bags, each containing 16 g of coyote hair, in front of each treatment trough. Three empty bags were placed at each control trough.

A replicated, randomized, controlled, before-and-after study in 2000 in a forest in Ohio, USA (2b) found that hanging bags of coyote *Canis latrans* hair did not reduce use of established trails by white-tailed deer *Odocoileus virginianus*. The number of deer using treatment trails did not differ significantly before (2.6 deer/day) or after (3.1 deer/day) placement of coyote hair bags. Similarly, the number of deer using non-treatment trails was not significantly different before (3.4 deer/day) or after (5.1 deer/day) placement of empty bags. Deer passes along 10 active trails (around 1 km apart) were recorded for three weeks (18 August to 8 September 2000) using infra-red monitors. A nylon mesh bag containing 16 g of coyote hair, was then suspended 2 m high from a tree along five randomly selected trails. Empty bags were hung at the other five trails. Monitoring continued for three further weeks (8–29 September 2000).

(1) Epple G., Mason J.R., Aronov E., Nolte D.L., Hartz R.A., Kaloostian R., Campbell D. & Smith A.B. (1995) Feeding responses to predator-based repellents in the mountain beaver (*Aplodontia rufa*). Ecological Applications, 5, 1163–1170.

(2) Seamans T.W., Blackwell B.F. & Cepek J.D. (2002) Coyote hair as an area repellent for white-tailed deer. *International Journal of Pest Management*, 48, 301–306, https://doi.org/10.1080/09670870210149853

3.72. Use target species scent to deter crop damage by mammals to reduce human-wildlife conflict

https://www.conservationevidence.com/actions/2506

- **One study** evaluated the effects on mammals of using target species scent to deter crop damage to reduce human-wildlife conflict. This study was in South Africa[1].

COMMUNITY RESPONSE (0 STUDIES)

POPULATION RESPONSE (0 STUDIES)

BEHAVIOUR (0 STUDIES)

OTHER (1 STUDY)

- **Human-wildlife conflict (1 study)**: A replicated, controlled study in South Africa[1] found that African elephants were not deterred from feeding by the presence of secretions from elephant temporal glands.

Background

Mammals often mark their territories with scent. If artificially placed scents from the same species can restrict movements of animals, this may assist in reducing damage to crops. If successful, this could reduce incentives for carrying out lethal control of such animals.

A replicated, controlled study in 1985 of shrubland in Limpopo, South Africa (1) found that compounds mimicking secretions from African elephant *Loxodonta africana* temporal glands did not deter feeding or otherwise change elephant behaviour. The rate of sniffing by captive elephants of hardboard pieces into which five scent compounds were absorbed (1–18 times/elephant/hour) did not differ from that for hardboards treated with carboxylic acids (2–15 times/elephant/ hour). The rates fell for all boards over the 10-day study. Boards hung directly over feeding troughs did not deter elephants from feeding. Wild elephants exposed to aerosols containing scent compounds or carboxylic acids did not change behaviour. Seven captive elephants, 9–12 months old, held in three pens, were exposed to secretions or carboxylic acid absorbed into hardboards fastened to the sides of pens. Boards were re-treated every two days. Lone wild bull elephants were exposed to scent compounds (18 times) or carboxylic acid (nine times) mixed with water and administered as aerosols. The study was conducted in July– August 1985.

(1) Gorman M.L. (1986) The secretion of the temporal gland of the African elephant *Loxodonta africana* as an elephant repellent. *Journal of Tropical Ecology*, 2, 187–190, https://doi.org/10.1017/S0266467400000766

3.73. Use 'shock collars' to deter crop damage by mammals to reduce human-wildlife conflict

https://www.conservationevidence.com/actions/2508

- **One study** evaluated the effects on mammals of using 'shock collars' to deter crop damage to reduce human-wildlife conflict. This study was in the USA[1].

COMMUNITY RESPONSE (0 STUDIES)

POPULATION RESPONSE (0 STUDIES)

BEHAVIOUR (0 STUDIES)

OTHER (1 STUDY)

- **Human-wildlife conflict (1 study):** A replicated, controlled study in the USA[1] found that electric shock collars (combined with loud noise) reduced damage caused by black-tailed deer to tree seedlings.

Background

Using electric shock collars on mammalian herbivores is a form of aversive conditioning. A shock is administered if the animal wearing a 'shock collar' approaches a pre-determined area, containing a crop. The potential for the technique to be effective may be assessed using captive animals in controlled experimental settings. Whilst not directly assessing the effectiveness of the intervention in reducing crop damage, such studies may provide evidence as to the potential for shock collars to alter animals' behaviour in a way that could potentially be applied to wild herbivores in crop production areas. If the intervention is successful, it may reduce incentives for carrying out lethal control of such animals.

A replicated, controlled study (year not stated) on two pastures in Washington, USA (1) found that using electric shock collars, along with playing loud noise, reduced damage by black-tailed deer *Odocoileus hemionus* to tree seedlings. As the loud noise and electric shock were part of the same treatment, their relative effects could not be separated. In areas where shock collars were triggered, damage to tree seedlings was lower (0–1 bites) than in areas where shock collars were not triggered (0–25 bites). Three deer, fitted with shock collars, were placed in each of two 1.5-ha pastures. Within each pasture, four 20 × 20-m plots were established. In each plot, 16 red cedar *Thuja plicata* seedlings were planted at 1-m intervals. When deer entered two of the plots, they received an electric shock and a loud noise was played through a speaker. When they entered the other two plots, they received no shock and no noise was played. Deer activity was measured by counting the number of bites taken from seedlings over a 21-day period.

(1) Nolte D.L., VerCauteren K.C., Perry K.R. & Adams S.E. (2003) *Training deer to avoid sites through negative reinforcement*. USDA National Wildlife Research Center-Staff Publications, 264.

3.74. Use repellents that taste bad ('contact repellents') to deter crop or property damage by mammals to reduce human-wildlife conflict

https://www.conservationevidence.com/actions/2509

- **Twelve studies** evaluated the effects of using repellents that taste bad ('contact repellents') to deter crop or property damage by mammals to reduce human-wildlife conflict. Nine studies were in the USA[1–4,5a,5b,5c,9,10], two were in the UK[7,8] and one was in Italy[6].

COMMUNITY RESPONSE (0 STUDIES)

POPULATION RESPONSE (0 STUDIES)

BEHAVIOUR (0 STUDIES)

OTHER (12 STUDIES)

- **Human-wildlife conflict (12 studies):** Five of 11 controlled studies (including 10 replicated studies), in the USA[1–4,5a,5b,5c,9], Italy[6] and the UK[7,8], of a range of contact repellents, found that they reduced herbivory or consumption of baits. The other six studies reported mixed results with at least some repellents at some concentrations deterring herbivory, sometimes for limited periods. A replicated, controlled study in the USA[10] found that a repellent did not prevent chewing damage by coyotes.

Background

This intervention considers specifically studies that assess effectiveness of repellents that are intended to be distasteful to wild mammals. Although some may produce some element of repellent odour, the main effect is generally when they are tasted, such as through licking or biting off vegetation to which it has been applied. Included here are tests of several repellents that are marketed commercially, especially to reduce browsing by herbivores on planted trees. The intervention also covers use of these repellents to deter damage to property.

See also: *Use repellents that smell bad ('area repellents') to deter crop or property damage by mammals to reduce human-wildlife conflict.*

A replicated, controlled study, in 1962–1964, on shrubland and a forest area of South Dakota, USA (1) found that applying repellents to trees reduced browsing by white-tailed deer *Odocoileus virginianus* and mule deer *Odocoileus hemionus*. Treated aspen *Populus tremuloides* shoots suffered less browsing than untreated shoots (zinc dimethyldithiocarbamate cyclohexylamine (ZAC)-treated: 3% removed; tetramethylthiuram disulfide (TMTD)-treated: 3%; untreated: 12%). The same pattern applied for wild chokeberry *Prunus virginiana* shrubs (ZAC-treated: 0.7% removed; TMDT-treated: 6.8%; untreated: 28.9%). On trees transplanted from nurseries, there was less browsing on ZAC-treated than untreated chokecherry (ZAC-treated: 0.1% removed; untreated: 6%), American plum *Prunus americana* (ZAC-treated removed: 0.1%; untreated: 19.8%) and caragana *Caragana arborescens* (ZAC-treated: 0.8% removed; untreated: 4.5%). Herbivory on naturally growing Aspen and chokeberry was compared between groups of ZAC-treated, TMTD-treated and untreated trees (10 trees in each case). Chokecherry, American plum and caragana were transplanted from nurseries to two sites where they were either treated with ZAC or were untreated (total ≤64 trees/species). Herbivory was assessed as the proportion of shoot lengths removed. Aspen and wild chokeberry trees were assessed over winters of 1962–1963 and 1963–1964. Transplanted chokecherry, American plum and caragana were assessed in winter of 1963–1964.

A replicated, randomized, controlled study in 1982–1985 at three tree nursery sites in Connecticut, USA (2) found that treating Japanese yew trees *Taxus cuspidata* with commercially available repellents reduced subsequent losses to herbivory by white-tailed deer *Odocoileus virginianus*. Results were not tested for statistical significance. The proportion of shoots browsed by white-tailed deer on trees treated with repellents (23%) was lower than the proportion browsed on untreated trees (41%). Over the three winters from 1982 to 1985, a total of 16 blocks of Japanese yew across three sites were studied. Each block was split into three plots (0.2–0.3 ha), which were randomly assigned to Big Game Repellent, Hinder® repellent or no treatment. Repellent was applied once annually, in November, following manufacturer instructions. Herbivory was assessed the following March, by inspecting 500–1,000 branch terminals in each plot.

A replicated, randomized, controlled study in 1989 on captive animals in Colorado, USA (3) found that chicken eggs, MGK® Big Game Repellent and coyote urine, used as repellents on foodstuffs, reduced consumption of that food by mule deer *Odocoileus hemionus* more than did treatment with thiram, Hinder®, soap and Ro·pel®. Deer consumed less food treated with chicken eggs (89 g/day), MGK® Big Game Repellent (94 g/day) and coyote urine (98 g/day) than food treated with thiram (212 g/day), Hinder® (223 g/day), soap (308 g/day) and Ro·pel® (399 g/day). It was not possible to assess which of these feeding rates differed significantly from consumption of food treated just with water (500 g/day). Three female and eight castrated male mule deer were held in individual pens. Repellents and a control (water) were sprayed daily on commercial deer pellets at a rate of 10 ml/500 g. Pellets were dried for 24 hours. The soap treatment involved hanging a bar of soap above the feed container. Food from each treatment was offered in different containers (500 g in each), which were randomized daily, for four days, in May and June 1989.

A replicated, controlled study in 1997 in a forest in Colorado, USA (4) found that aspens *Populus tremuloides* treated with the repellents Deer Away® and the highest concentration of Hot Sauce® were browsed less by elk *Cervus canadensis* than were untreated trees. There was less browsing on aspens treated with Deer Away® (42% of sprouts and terminal leaders browsed) and 6.2% Hot Sauce® (56% browsed) than on untreated aspens (77% browsed). Browsing rates on aspens treated with 0.62% Hot Sauce® (65%) and 0.062% Hot Sauce® (72%) did not differ significantly from those on untreated aspens. Four fenced pasture blocks (each 0.41 ha) each contained 10 strips (1 × 23 m) of sprouting aspen. Treatments were Deer Away® and Hot Sauce® at three concentrations (0.062%, 0.62%, 6.2%). Each treatment was applied to one strip in each pasture, five weeks before exposure to elk and to a further strip two weeks before exposure. Two strips remained untreated. Two captive elk were placed in each pasture block, from 3 August to 5 September 1997. Proportional browsing rates were assessed by examining all aspen sprouts in each pasture.

A replicated, randomized, controlled study in 1997 on captive animals in a forested site in Washington, USA (5a) found that Hot Sauce® repellent reduced most measures of tree browsing by black-tailed deer *Odocoileus hemionus columbi* for four weeks, but not subsequently. There were fewer damaged trees in treated than in untreated plots during the first two weeks but not during the third and fourth weeks. There were fewer damaged terminal buds and lateral bites in treated than in untreated plots across all four weeks. There was no difference in the number of trees stripped of all leaves in treated and untreated plots on day one, but fewer trees were stripped of all leaves in treated than untreated plots through to and including the fourth week. During weeks five and six, there were no differences in these measures between treated and untreated plots. Data were not presented. Three to four deer were held in each of four pens (0.75–2 ha). Two plots (>25 m apart) in each pen each contained three western red cedar *Thuja plicata* trees (0.5–1 m tall, 1 m apart). Plots were randomly assigned to a single application of 6.2% Hot Sauce® or were untreated. Tree damage was assessed between 4 February and 16 March 1997.

A replicated, controlled study (year not stated) on captive animals in Washington, USA (5b) found that treating food with Hot Sauce® repellent (as a trial of its effectiveness at reducing crop consumption) reduced consumption by porcupines *Erethizon dorsatum*, reduced consumption by pocket gophers *Thomomys mazama* at two of four concentrations and did not reduce consumption by mountain beavers *Aplodontia rufa*.

Porcupines consumed fewer treated than untreated apple pieces at all four Hot Sauce® concentrations. Pocket gopher consumption of apple pieces did not differ between treated and untreated food at 0.062% concentration. At 0.62%, fewer treated than untreated pieces were eaten on two of four days. At 3.1% and 6.2%, fewer treated than untreated pieces were eaten. Mountain beaver consumption of apple pieces did not differ between treated and untreated food at any of the four repellent concentrations. See paper for full details of results. Trials were carried out on four porcupines, 12 pocket gophers and 10 mountain beavers. All were held in enclosures and were offered two-choice tests between apple pieces treated with Hot Sauce®, a repellent containing capsaicin, and untreated apple pieces. Solutions containing 0.062%, 0.62%, 3.1% and 6.2% of Hot Sauce® were used. Each concentration was tested for four days with each animal. Tests ran consecutively, from lowest to highest concentrations of Hot Sauce® solution.

A replicated, randomized, controlled study (year not stated) on captive animals in Washington, USA (5c) found that treating cottonwood *Populus* spp. stems with Hot Sauce® repellent reduced the extent to which they were chewed by beavers *Castor canadensis*. At all three Hot Sauce® concentrations applied, chewing damage was lower in treated stems than in untreated stems (results expressed as damage indices). Eight adult beavers were housed in pens that contained 1-m-long cottonwood stems of 7–10 cm diameter. Adjacent pairs of stems were randomly assigned for treatment by Hot Sauce® at 0.062%, 0.62% and 6.2% concentrations and untreated stems were available. Beavers also had free access to apples, carrots, pelleted food and water. The test was run for six days, then repeated. Damage to cottonwood stems was assessed at the end of each six-day period.

A replicated, controlled study in 2001 on a site in Italy (6) found that two of three repellents significantly reduced browsing of olive trees *Olea europaea* by fallow deer *Dama dama* for three weeks following application. A lower proportion of plants treated with Eutrofit® was browsed, relative to untreated plants, at one, two and three weeks after application (reductions relative to untreated plants of 100%, 71% and 41% respectively). Tree Guard® similarly reduced the proportions of plants browsed relative to untreated plants (by 82%, 82% and 55% after one, two and three weeks respectively). Reductions in the proportions of plants treated with Hot Sauce® that were browsed relative to untreated plants (64%, 12% and 9% after one, two and three weeks respectively) were not significant. From four weeks onwards, no repellent reduced browsing relative to untreated trees. Olive cuttings, 1 year old and about 20 cm high, were planted in five blocks of 20 plants. In each block, five plants each were treated each with the commercially available repellents, Eutrofit®, Tree Guard® and Hot Sauce®, following manufacturer instructions. Browsing damage was assessed weekly, for eight weeks.

A controlled, before-and-after study in 1996 in a woodland in Oxfordshire, UK (7) found that European badgers *Meles meles* ate less food treated with the repellent, ziram, than untreated food, but cinnamamide and capsaicin treatments did not affect consumption rates. Badgers consumed 31–100% of ziran-treated bait over the first eight treatment nights, 0–10% over the ninth to sixteenth treatment nights and 0–3% from the seventeenth to twenty-eighth treatment nights. All untreated baits, and baits treated with cinnamamide and capsaicin, were consumed throughout the trial. A hexagon of paving slabs, each separated into four quadrants, was established. Each quadrant was supplied nightly with 20 g of Beta Puppy 1–6 months™ pelleted food. Untreated baits were used for 68 nights, followed by 56 nights during which treatment nights and control nights (untreated food) alternated. On treatment nights, the four quadrants on each slab each received one from pellets treated with ziram in the form of AAprotect™, cinnamamide with methanol, capsaicin with diethyl ether or untreated bait. Uneaten bait was weighed to determine consumption. The study ran from 19 July to 19 November 1996.

A replicated, randomized, controlled study (year not stated) in a woodland in Oxfordshire, UK (8) found that treating corn cobs with the repellent, ziram, reduced the rate of its consumption by European badgers *Meles meles*. Fewer corn cobs treated with ziram were damaged by badgers (39–63% of cobs) than were untreated cobs (82% of cobs). Among badgers that were repeat visitors to feeding stations, treated cobs were fed on (as opposed to rejected) on a lower proportion of occasions (10–34%) than were untreated cobs (60%). At two sites, 450 m apart, feeding stations were established, each offering 12 corn cobs and water. Sites were pre-baited, to encourage attendance, and the experiment ran for five nights. Cobs were treated, in equal numbers, with 5%, 10%, 20%

or 40% ziram in water or with water alone (as an untreated control). Treatments were assigned randomly across cobs.

A replicated, randomized, controlled study in 2006–2008 in two agricultural sites in Connecticut, USA (9) found that 10 commercially available repellents varied in effectiveness at reducing white-tailed deer *Odocoileus virginianus* herbivory on trees. At one site, trees treated with Chew-Not®, Deer-Away® Big Game Repellent, Bobbex®, Liquid Fence® and Hinder® had greater needle mass (140–234 g) than did untreated trees (14 g). Needle mass of trees treated with five other repellents (Repellex®, Deer Solution®, coyote urine, Plantskydd® and Deer-Off ®) (23–81 g) did not differ from that of untreated trees. Trees treated with Bobbex®, and Hinder® were taller (35–36 cm) than untreated trees (25 cm). Tree height when treated with the eight other repellents (23–31 cm) did not differ significantly from that of untreated trees. At the second site, where herbivory was light, there were no significant differences in tree heights and needle mass was not measured. At each of two sites, two blocks were established in May 2006, each with 12 groups of six yew *Taxus cuspidata* trees. Each treatment was applied randomly to one tree group in each block. Additionally, one group was untreated and one fenced. Repellent application followed manufacturer instructions. Trees were harvested in April 2008.

A replicated, controlled study (year not stated) on captive animals in Utah, USA (10) found that applying the repellent, Ropel®, to nylon items similar to those used on military airstrips did not reduce chewing damage caused by coyotes *Canis latrans*. Coyotes repeatedly tasted a lower proportion of Ropel®-treated items (67–75%) than of untreated items (58–83%). However, there was no difference in the proportion destroyed within 24 hours between treated (58–75%) and untreated items (58–83%). Twelve mated coyote pairs each had access to 1-m lengths of nylon strapping (3 cm wide, 3 mm thick) with three 0.2-m loops. Latex stickers aided adhesion of Ropel® and of water (as an untreated control solution) to nylon strapping. Solutions were applied four and one days before one treated and one untreated item were placed in each coyote pen. Coyote behaviour was monitored using camera traps.

(1) Dietz D.R. & Tigner J.R. (1968) Evaluation of two mammal repellents applied to browse species in the Black Hills. *The Journal of Wildlife Management*, 32, 109–114, https://doi.org/10.2307/3798244

(2) Conover M.R. (1987) Comparison of two repellents for reducing deer damage to Japanese yews during winter. *Wildlife Society Bulletin*, 15, 265–268.

(3) Andelt W.F., Burnham K.P. & Manning J.A. (1991) Relative effectiveness of repellents for reducing mule deer damage. *The Journal of Wildlife Management*, 55, 341–347, https://doi.org/10.2307/3809161

(4) Baker D.L., Andelt W.F., Burnham K.P. & Shepperd W.D. (1999) Effectiveness of Hot Sauce® and Deer Away® repellents for deterring elk browsing of aspen sprouts. *The Journal of Wildlife Management*, 63, 1327–1336, https://doi.org/10.2307/3802851

(5) Wagner K.K. & Nolte D.L. (2000) Evaluation of Hot Sauce® as a repellent for forest mammals. *Wildlife Society Bulletin*, 28, 76–83.

(6) Santilli F., Mori L. & Galardi L. (2004) Evaluation of three repellents for the prevention of damage to olive seedlings by deer. *European Journal of Wildlife Research*, 50, 85–89, https://doi.org/10.1007/s10344-004-0036-1

(7) Baker S.E., Ellwood S.A., Watkins R. & Macdonald D.W. (2005) Non-lethal control of wildlife: using chemical repellents as feeding deterrents for the European badger *Meles meles*. *Journal of Applied Ecology*, 42, 921–931, https://doi.org/10.1111/j.1365-2664.2005.01069.x

(8) Baker S.E., Ellwood S.A., Watkins R.W. & Macdonald D.W. (2005) A dose-response trial with ziram-treated maize and free-ranging European badgers *Meles meles*. *Applied Animal Behaviour Science*, 93, 309–321, https://doi.org/10.1016/j.applanim.2004.11.022

(9) Ward J.S. & Williams S.C. (2010) Effectiveness of deer repellents in Connecticut. *Human–Wildlife Interactions*, 4, 56–66, https://doi.org/10.26077/v0bn-9k23

(10) Miller E.A., Young J.K., Stelting S. & Kimball B.A. (2014) Efficacy of Ropel® as a coyote repellent. *Human-Wildlife Interactions*, 8, 271–278.

3.75. Use repellents that smell bad ('area repellents') to deter crop or property damage by mammals to reduce human-wildlife conflict

https://www.conservationevidence.com/actions/2511

• **One study** evaluated the effects of using repellents that smell bad ('area repellents') to deter crop or property damage by mammals to reduce human-wildlife conflict. This study was in the UK[1].

COMMUNITY RESPONSE (0 STUDIES)

POPULATION RESPONSE (0 STUDIES)

BEHAVIOUR (0 STUDIES)

OTHER (1 STUDY)

- **Human-wildlife conflict (1 study):** A randomized, replicated, controlled study in the UK[1] found that a repellent reduced use of treated areas by moles.

Background

This intervention covers use of manufactured repellents that emit a smell that is designed to repel animals from areas of crops or other property that is vulnerable to damage. If such repellents can prevent or reduce crop or property damage by wild mammals, this could reduce motivations for carrying out lethal control of these animals.

See also: *Use predator scent to deter crop damage by mammals to reduce human-wildlife conflict* and *Use pheromones to deter crop damage by mammals to reduce human-wildlife conflict.*

Randomized, replicated, controlled studies in 1989–1990 on three farms in Oxfordshire, UK (1) each found that a bone-oil based repellent (Renardine) reduced use of treated areas by moles *Talpa europaea*. Moles avoided the 25% of their home range that was treated with the repellent for 9–27 days (moles' home ranges treated similarly, but with water, were not avoided). With close to 100% of their home ranges treated, moles avoided reoccupying treated areas for 42 hours to at least nine days. Moles took longer to cross a repellent-treated slit, cut across their home ranges (26 days) than a similar water-treated slit (four hours). The repellent, Renardine [use of which is prohibited in some countries], was soaked into rolled toilet paper and pushed into one mole tunnel/ m² in the 25% most heavily used part of home ranges (three moles) in spring 1989 or into all identified tunnels in the home range (four moles) in late summer 1989. One site was used in each case. Water-soaked toilet paper acted as a control at the 25% site (two moles). At a third site, 0.5 l/m of Renardine was poured into a 50-cm-deep slit across six home ranges in autumn/winter 1990. The slit was filled with peat, and a further 0.5 l/m of Renardine poured on top. One further home range was treated similarly, but with water. Mole movements were monitored by radio-tracking.

(1) Atkinson R.P.D. & MacDonald D.W. (1994) Can repellents function as a non-lethal means of controlling moles (*Talpa europaea*)? *Journal of Applied Ecology*, 31, 731–736, https://doi.org/10.2307/2404163

3.76. Use dogs to guard crops to reduce human-wildlife conflict

https://www.conservationevidence.com/actions/2512

- **One study** evaluated the effects on mammals of using dogs to guard crops to reduce human-wildlife conflict. This study was in Zimbabwe[1].

COMMUNITY RESPONSE (0 STUDIES)

POPULATION RESPONSE (0 STUDIES)

BEHAVIOUR (0 STUDIES)

OTHER (1 STUDY)

- **Human-wildlife conflict (1 study):** A replicated study in Zimbabwe[1] found that people with dogs took longer to repel African elephants from crops compared to scaring them by using combinations of people, dogs, slingshots, drums, burning sticks, large fires and spraying with capsicum.

Background

Dogs *Canis lupus familiaris* are frequently used to guard livestock but this intervention covers the use of dogs to deter herbivores from damaging crops. If successful, this could reduce incentives for carrying out lethal control on crop-raiding mammal species.

A replicated study in 1995–1996 in agricultural fields surrounded by savanna in Sebungwe, Zimbabwe (1) found that African elephants *Loxodonta africana* took longer to be repelled from agricultural fields when scared only by people with dogs *Canis lupus familiaris* than by combinations of people, dogs, slingshots, drums, burning sticks, large fires and when sprayed with capsicum. Relative effects of the individual deterrents cannot be separated. Elephants were repelled more slowly when scared by one person with dogs (14 minutes) than when scared by people with dogs and slingshots, drums and burning sticks (10 minutes), by people with dogs, drums and large fires (4 minutes) or when sprayed with capsicum oleoresin (2 minutes). The study was conducted in communal lands surrounding a research area. Attempts were made to deter elephants raiding crops, 15 times by one person with dogs, 11 times by 4–7 people with dogs, drums and large fires, 11 times by 2–3 people with dogs and slingshots, drums and burning sticks and 18 times using a spray with 10% capsicum oleoresin. Behavioural responses were monitored using a monocular. Distance between elephants and farmers was 20–40 m. Tests were conducted between 18:30 and 06:30 h. The number of fields was not reported.

(1) Osborn F.V. (2002) Capsicum oleoresin as an elephant repellent: field trials in the communal lands of Zimbabwe. *The Journal of Wildlife Management*, 66, 674–677, https://doi.org/10.2307/3803133

3.77. Drive wild animals away using domestic animals of the same species to reduce human-wildlife conflict

https://www.conservationevidence.com/actions/2513

- **One study** evaluated the effects of using domestic animals to drive away wild mammals to reduce human-wildlife conflict. This study was in India[1].

COMMUNITY RESPONSE (0 STUDIES)

POPULATION RESPONSE (0 STUDIES)

BEHAVIOUR (0 STUDIES)

OTHER (1 STUDY)

- **Human-wildlife conflict (1 study):** One study in India[1] found that using domestic elephants to drive wild Asian elephants away from villages did not reduce the probability of elephants damaging crops.

Background

Domestic mammals may be used in attempts to repel wild mammals of the same species that are causing nuisance, such as be crop-raiding. This intervention is likely to be especially relevant where the wild animal presents a potential threat to people such that simply chasing animals away may not always be a viable or effective option. If the intervention is effective, this could reduce incentives for carrying out lethal control of the focal species.

A study in 2006–2009, in two areas of Assam, India (1) found that using domestic elephants to drive wild Asian elephants *Elephas maximus* away from villages did not reduce the probability of elephants damaging crops. The chance of crop damage occurring was not lower when domestic elephants were used to deter crop-raiding wild elephants, in comparison with a range of other interventions or no intervention (results presented as statistical model coefficients). Within two study areas, 33 community members trained as monitors recorded 1,761 crop-raiding incidents, from 1 March 2006 to 28 February 2009. A range of deterrence methods was used, singly or in combination, including using domesticated elephants to repel wild elephants, chili smoke (from burning dried chilies, tobacco, and straw), spotlights, two-strand electric fences, chili fencing (engine grease and ground chili paste, on a jute or coconut rope), fire and noise.

(1) Davies T.E., Wilson S., Hazarika N., Chakrabarty J., Das D., Hodgson D.J. & Zimmermann A. (2011) Effectiveness

4. Threat: Energy production and mining

Background

Energy production (renewable and non-renewable) and mining can have substantial impacts on terrestrial mammal populations through the destruction and pollution of habitats. Most interventions involve restoration of previously mined land, which may be hampered by contamination of the ground water or soil resulting from mining operations. Several other interventions consider actions to reduce human-wildlife conflict in order that motivations to carry out lethal control of these species will also be reduced.

For more general actions that relate to habitat restoration or addressing impacts of pollution, see chapters *Habitat restoration and creation* and *Threat: Pollution*.

4.1. Restore former mining sites

https://www.conservationevidence.com/actions/2490

- **Twelve studies** evaluated the effects of restoring former mining sites on mammals. Eleven studies were in Australia[2–12] and one was in the USA[1].

COMMUNITY RESPONSE (8 STUDIES)

- **Species richness (8 studies):** A review in Australia[10] found that seven of 11 studies indicated that rehabilitated areas had

 https://doi.org/10.11647/OBP.0234.04

lower mammal species richness compared to in unmined areas. Four of five replicated, site comparison studies, in Australia[2-4,6,9], found that mammal species richness was similar in restored mine areas compared to unmined areas[3,4,6] or higher in restored areas (but similar when considering only native species)[9]. One study found that species richness was lower in restored compared to in unmined areas[2]. A replicated, controlled study in Australia[8] found that thinning trees and burning vegetation as part of mine restoration did not increase small mammal species richness. A replicated, site comparison study in Australia[5] found that restored mine areas were recolonized by a range of mammal species within 10 years.

POPULATION RESPONSE (5 STUDIES)

- **Abundance (5 studies):** A review of rehabilitated mine sites in Australia[10] found that only two of eight studies indicated that rehabilitated areas had equal or higher mammal densities compared to those in unmined areas. One of three replicated, site comparison studies, in the USA[1] and Australia[2,9], found that small mammal density was similar on restored mines compared to on unmined land[1]. One study found that for three of four species (including all three native species studied) abundance was lower in restored compared to unmined sites[2] and one study found mixed results, including that abundances of two out of three focal native species were lower in restored compared to unmined sites[9]. A replicated, controlled study in Australia[8] found that thinning trees and burning vegetation as part of mine restoration did not increase small mammal abundance.

BEHAVIOUR (2 STUDIES)

- **Use (2 studies):** A replicated, site comparison study in Australia[7] found that most restored former mine areas were not used by koalas while another replicated site comparison study in Australia[11] found quokka activity to be similar in revegetated mined sites compared to in unmined forest.

OTHER (1 STUDY)

- **Genetic diversity (1 study):** A site comparison study in Australia[12] found that in forest on restored mine areas, genetic diversity of yellow-footed antechinus was similar to that in unmined forest.

Background

Restoration of former mining sites usually involves establishing native or non-native plants, often with the main aim of reducing erosion or reducing the concentration of pollutants (Wong 2003). However, this restoration may also benefit mammal species found in and around former mining sites by creating habitat conditions similar to those found prior to mining operations.

Wong M.H. (2003) Ecological restoration of mine degraded soils, with emphasis on metal contaminated soils. *Chemosphere*, 50, 775–780, https://doi. org/10.1016/s0045-6535(02)00232-1

A replicated, site comparison study in 1980–1981 of four restored areas of a mine and an adjacent unmined grassland in Wyoming, USA (1) found that on restored mine plots, small mammal density was similar to that found on unmined land. Average mammal density on two-year-old restored plots (14–16 individuals/ha) and 3–5-year-old restored plots (16–23 individuals/ha) were not significantly different to those on unmined plots (12–14 individuals/ha). More deer mice *Peromyscus maniculatus* were found in restored plots (13–18/ha) than in unmined plots (6–8/ha). The reverse was true for thirteen-lined ground squirrels *Spermophilus tridecemlineatus* (restored: 0.6–1.5/ha; unmined: 4.5–5.0/ha). Plots were restored by replacing mine deposits with topsoil followed by adding seed and fertilizer. Two restored areas were studied in 1980 and four (including the original two) in 1981. A nearby area of unmined rangeland was sampled both years. Small mammals in restored plots were live-trapped for 4–7 days/month in June–August 1980 and May–September 1981. On the unmined rangeland, mammals were live-trapped for 4–7 days in July both years.

A replicated, site comparison study in 1987–1988 in five heath and scrubland sites in Western Australia, Australia (2) found that after restoring natural vegetation on former sand mines, mammal species richness and abundance for most species was lower than found in undisturbed. Three species were recorded in each restored site and four in each undisturbed site. Fewer honey possums *Tarsipes rostratus* were recorded in restored sites (0.6–0.7/trap night) than in undisturbed sites (2.5–5.2/trap night). The same was true for ash-grey mouse *Pseudomys albocinereus* (0.1 vs 1.6–5.6/trap night) and white-tailed dunnart *Sminthopsis granulipes* (0 vs 0.4–2.3/trap night). Numbers of house mice *Mus musculus* did not differ between restored and undisturbed sites (3.6–5.0 vs 4.0–8.7/trap night). Two sites were restored following sand mining. Three sites were unmined. Restoration (starting in 1977 and 1982) involved reprofiling and reseeding. At one site, original topsoil was returned. Mammals were surveyed using pitfall and box traps, twice each month, from July 1987 to September 1988, for seven consecutive nights (three nights in July and September 1988).

A replicated, site comparison study in 1992–1998 of forest at two sites in Western Australia, Australia (3) found similar mammal species richness in forest restored on former bauxite mines compared with unmined jarrah forest. Results were not tested for statistical significance. The number of mammal species recorded in restored forest (10) was similar to that in unmined forest (9). Short-beaked echidna *Tachyglossus aculeatus* and the introduced feral cat *Felis catus* and European rabbit *Oryctolagus cuniculus* were found in restored but not in unmined forest. Common brushtail possum *Trichosurus vulpecula* and western brush wallaby *Macropus irma* were found in unmined but not restored forest. At each of two mines, one survey plot was established in restored forest and one in unmined forest. Restoration, commencing in 1990, involved disturbing and reprofiling the mine surface, to reverse compaction, and replacing topsoil and associated aggregate. Tree and understorey plant seeds were added. Mammals were surveyed, using three trap types, over four successive nights, in July–August 1992, 1995 and 1998. Native mammals were released and feral mammals were euthanized.

A replicated, site comparison study in 2000–2002 of woodland and scrub at five mines in Western Australia, Australia (4) found that restored sites had a similar mammal species richness compared to unmined sites. The average number of species/site/month in restored sites (2–4) was similar to that in unmined sites (2–5). The overall number of mammal species recorded/site was also similar (restored: 5–8; unmined: 4–7). Five former mine site waste dumps, where restoration had started 3–9 years previously, and an unmined area adjacent to each dump were sampled. At four mines, pit-traps and drift fencing were used to sample sites over a seven-day period, on 10 occasions, from spring 2000 to winter 2002. At one mine, sampling was carried out five times, from spring 2001 to winter 2002.

A replicated, site comparison study in 1978–2005 of former mines in jarrah forests in Western Australia, Australia (5) found that restored mined areas were recolonized by a range of mammal species within 10 years. Western grey kangaroo *Macropus fuliginosus*, mardo *Antechinus flavipes* and chuditch *Dasyurus geoffroii* were all first reported in restored mines 0–2 years after restoration, whereas common brushtail possum *Trichosurus vulpecula* was first reported after eight years and brush-tailed phascogale *Phascogale tapoatafa* after ten years. Mardo capture rates increased at restored sites (caught in 1% of traps 10 years after restoration) but remained lower than in adjacent undisturbed forest (2–11% of traps). Mined areas were revegetated using various techniques including topsoil return, deep ripping, understorey seeding of many local species and establishment of local eucalypt species. Wildlife corridors and specific microhabitats (e.g. hollow logs, stumps) were created. In 1993–1994, mammal nest boxes were placed in a range of sites (number not stated). Non-native red fox *Vulpes vulpes* control was carried out for several years from 1994. Mammals in restored areas (of varying ages and restoration techniques) and undisturbed forest were monitored using wire cage traps, large and medium aluminium box traps and pit traps.

A replicated, paired sites, site comparison study in 2000–2004 of five former mines and adjacent scrubland vegetation in Western Australia, Australia (6) found that mines undergoing restoration contained all small mammal species recorded on adjacent unmined land and higher overall abundance of small mammals. Results were not tested for statistical significance. Seven species were recorded in both restored mines and in adjacent unmined land. Three other species were only recorded in restored mines. In total, 211–493 mammals/site were caught in restored mines and 91–131 mammals/site were caught on unmined land. Five mines, which had been under restoration management for three to nine years, were studied along with adjacent unmined land. From June 2000 to January 2004, sampling was carried out 12 times on each of four sites and seven times on the fifth. Animals were sampled using pitfall traps or funnels along drift fences, for seven days (14 days on the final sample visit).

A replicated, site comparison study in 2005–2006 in woodland in Queensland, Australia (7) found that four of five restored mines were not used by koalas *Phascloarctos cinereus*, but that koala diet did not differ between those in restored and unmined sites. In four of five restored sites, koalas were not found, but they were found in two of three nearby unmined sites. There was no significant difference between diets of koalas in the occupied restored area and those in the two occupied unmined areas. In 1976–1977, areas mined for mineral sands were recontoured and trees, including *Eucalyptus* species, were planted. Eight koalas were radio-collared and located once/week for 12 months to determine the tree species they were using. To investigate diet and koala presence, dung was collected from study animals once, from five 50 × 50 m plots in restored sites and three in unmined areas.

A replicated, controlled study in 2002–2006 of forest at a site in Western Australia, Australia (8) found that thinning trees and burning vegetation, as part of mine restoration, did not increase small mammal species richness or abundance. Thinning and burning were carried out in the same plots, so their individual effects cannot be determined. Small mammal abundance in thinned and burned plots (4.0–4.2 individuals/ grid) did not differ significantly from that in plots that were not thinned and burned (2.5–4.7 individuals/grid). There was also no difference in species richness (thinned and burned: 2.0–2.8 species/grid; not thinned and burned: 1.5–2.0 species/grid). In 1984–1992, areas of a former bauxite mine were either planted with non-local tree species or sown with the seed of local tree species. Eight plots were thinned between December 2002 and July 2003 and then burned in November 2003. Eight different plots were not thinned or burned. Small mammals were monitored for four nights each in October and November–December 2005 and March and May 2006, using pitfall traps with drift fencing and live cage and box traps.

A replicated, site comparison study in 2005–2006 of two former mines in jarrah forests in Western Australia, Australia (9) found that in restored areas, overall mammal species richness was higher, native mammal species richness was similar, and differences in mammal abundances were mixed compared to unmined sites. Overall mammal species richness was higher in restored sites (2.4 species/site) than in unmined sites (0.4 species/site), but native species richness did not differ (data not reported). In three of four restoration age comparisons, there were more individuals in restored sites than in unmined sites for both house mice *Mus musculus* (1.7–4.0 vs 0 animals/grid) and western pygmy possum *Cercartetus concinnus* (0.9–1.0 vs 0.3 animals/grid). In three of four restoration age comparisons, there were fewer individuals in restoration sites than in unmined sites for common brushtailed possums *Trichosurus vulpecula* (0–0.8 vs 1 animals/grid) and yellow-footed antechinus *Altechinus flavipes* (0.8–1.8 grid vs 2 animals/grid). Small mammals were surveyed across two mine areas at sites where restoration commenced 4, 8, 12 and 17 years earlier (total six sites for each age class) and in six unmined forest sites. Mammals were trapped using grids with nine pitfall traps, four Elliott traps and Sheffield cage-traps, set along drift-fencing at each site. Traps were set for four nights/season, totalling 1,728 trap nights/treatment.

A review of rehabilitated mine sites in Australia (10) found that 62% of 13 studies indicated that rehabilitated areas had lower densities and/or species richness of mammals compared to in unmined areas. Seven of 11 studies found that rehabilitated areas had lower mammal species richness than unmined areas, while the other four found rehabilitated and unmined areas had equal or higher mammal species richness. Only two of eight studies found that rehabilitated areas had equal or higher mammal densities compared to unmined areas. Data for individual studies were not reported. Methods combining the use of fresh topsoil with planting seeds and seedlings were most successful for animal recolonization. Studies investigating faunal recolonization of rehabilitated mines in Australia were obtained from the literature, of which 13 of 71 monitored mammals. Studies often compared plots in rehabilitated areas (1–30 plots/study) with plots in unmined areas (1–22/study). Rehabilitated sites were up to 20 years old.

A replicated site comparison in 2012 in four revegetated mine sites and eight forest sites in Western Australia, Australia (11) found that after revegetating mined sites, quokka *Setonix brachyurus* activity did not differ in restored compared to in unmined forest sites. Quokka activity did not differ significantly between areas where forest had been revegetated after mining (detected on 4.7 nights/site) and forest that had never been mined (0–8.2 nights/site). Between 16 and 21 years before the study, part of the study landscape was sown with a seed mixture containing 76–111 plant species. In August–September 2012, a motion-sensitive-camera was strapped to a tree at a height of 0.3 m and was left active for 21 nights, in each of four restored sites, and eight unmined forests. Cameras were baited with apples, oats, honey and peanut butter. The number of nights on which quokkas were detected was recorded.

A site comparison study in 2005–2012 of jarrah forest at a site in Western Australia, Australia (12) found that in areas of forest restored following mining, genetic diversity of yellow-footed antechinus *Antechinus flavipes* was similar to that in unmined forest. Allelic richness (a measure of genetic diversity) was similar in restored forest (9.1) to that in unmined forest (9.1). Genetic analysis was based on 24 samples from restored forest and 33 from unmined forest. DNA samples were extracted from antechinus caught in pit and cage traps in 17 trapping grids in restored mine areas (3–21 years post-mining) and 22 grids in unmined forest areas. Grids were, on average, 1,095 m apart. Traps were operated for three or four periods of two weeks, each year, in 2005–2012.

(1) Hingtgen T.M. & Clark W.R. (1984) Small mammal recolonization of reclaimed coal surface-mined land in Wyoming. *The Journal of Wildlife Management*, 48, 1255–1261, https://doi.org/10.2307/3801786

(2) McNee S.A. & Collins B.G. (1995) Population ecology of vertebrates in undisturbed and rehabilitated habitats on the Northern Sandplain of Western Australia. *Bulletin No. 16*. Curtin University of Technology School of Environmental Biology

(3) Nichols O.G. & Nichols F.M. (2003) Long-term trends in faunal recolonization after bauxite mining in the jarrah forest of southwestern Australia. *Restoration Ecology*, 11, 261–272, https://doi.org/10.1046/j.1526-100x.2003.00190.x

(4) Thompson G.G. & Thompson S.A. (2005) Mammals or reptiles, as surveyed by pit-traps, as bio-indicators of rehabilitation success for mine sites in the goldfields region of Western Australia? Pacific *Conservation Biology*, 11, 268–286, https://doi.org/10.1071/pc050268

(5) Nichols O.G. & Grant C.D. (2007) Vertebrate fauna recolonization of restored bauxite mines — key findings from almost 30 years of

monitoring and research. *Restoration Ecology*, 15, S116–S126, https://doi. org/10.1111/j.1526-100x.2007.00299.x

(6) Thompson G.G. & Thompson S.A. (2007) Early and late colonizers in mine site rehabilitated waste dumps in the Goldfields of Western Australia. *Pacific Conservation Biology*, 13, 235–243, https://doi.org/10.1071/pc070235

(7) Woodward W., Ellis W.A., Carrick F.N., Tanizaki M., Bowen D. & Smith P. (2008) Koalas on North Stradbroke Island: diet, tree use and reconstructed landscapes. *Wildlife Research*, 35, 606–611, https://doi.org/10.1071/wr07172

(8) Craig M.D., Hobbs R.J., Grigg A.H., Garkaklis M.J., Grant C.D., Fleming P.A. & Hardy G.E.S.J. (2010) Do thinning and burning sites revegetated after bauxite mining improve habitat for terrestrial vertebrates? *Restoration Ecology*, 18, 300–310, https://doi.org/10.1111/j.1526-100x.2009.00526.x

(9) Craig M.D., Hardy G.E.S.J., Fontaine J.B., Garkakalis M.J., Grigg A.H., Grant C.D., Fleming P.A. & Hobbs R.J. (2012) Identifying unidirectional and dynamic habitat filters to faunal recolonisation in restored mine-pits. *Journal of Applied Ecology*, 49, 919–928, https://doi.org/10.1111/j.1365-2664.2012.02152.x

(10) Cristescu R.H., Frère C. & Banks P.B. (2012) A review of fauna in mine rehabilitation in Australia: current state and future directions. *Biological Conservation*, 149, 60–72, https://doi.org/10.1016/j.biocon.2012.02.003

(11) Craig M.D., White D.A., Stokes V.L. & Prince J. (2017) Can postmining revegetation create habitat for a threatened mammal? *Ecological Management & Restoration*, 18, 149–155, https://doi.org/10.1111/emr.12258

(12) Mijangos J.L., Pacioni C., Spencer P.B.S., Hillyer M. & Craig M.D. (2017) Characterizing the post-recolonization of Antechinus flavipes and its genetic implications in a production forest landscape. *Restoration Ecology*, 25, 738–748, https://doi.org/10.1111/rec.12493

4.2. Use electric fencing to deter mammals from energy installations or mines

https://www.conservationevidence.com/actions/2500

- We found no studies that evaluated the effects of using electric fencing to deter mammals from energy installations or mines.

'We found no studies' means that we have not yet found any studies that have directly evaluated this intervention during our systematic journal and report searches. Therefore, we have no evidence to indicate whether or not the intervention has any desirable or harmful effects.

Background

Mammals may cause damage to equipment if they enter energy installations or mines. There is also a direct risk to mammals from becoming trapped, falling into pits or being electrocuted. Electric fencing may be use around such sites to deter mammal entry. As well as reducing direct risks to mammals, if successful the intervention may also reduce the need to carry out lethal control of mammals on such sites.

See also: *Agriculture and aquaculture — Install electric fencing to protect crops from mammals to reduce human-wildlife conflict* and *Agriculture and aquaculture — Install electric fencing to reduce predation of livestock by mammals to reduce human-wildlife conflict.*

4.3. Use repellents to reduce cable gnawing

https://www.conservationevidence.com/actions/2502

- **One study** evaluated the effects of using repellents to reduce cable gnawing. This study was in the USA[1].

COMMUNITY RESPONSE (0 STUDIES)

POPULATION RESPONSE (0 STUDIES)

BEHAVIOUR (0 STUDIES)

OTHER (1 STUDY)

- **Human-wildlife conflict (1 study):** A randomized, replicated, controlled study in the USA[1] found that repellents only deterred cable gnawing by northern pocket gophers when encased in shrink-tubing.

Background

Human-wildlife conflict can arise where animals cause damage to equipment or installations. Damage, such as that caused by gophers to underground cables, can represent substantial financial losses (Ramey & McCann 1997). If repellents can reduce or prevent damage to cables, this might reduce incentives for carrying out lethal control of such animals.

Ramey C.A. & McCann G.R. (1997) Evaluating cable resistance to pocket gopher damage-a review. *Great Plains Wildlife Damage Control Workshop*, 13, 107–113.

A randomized, replicated, controlled study (year not stated) in a captive facility in Colorado, USA (1) found that repellents only deterred cable gnawing by northern pocket gophers *Thomomys talpoides* when encased in shrink-tubing. When repellents were contained within shrink-tubing, there were reductions in all four damage measures (mass loss, chewing depth, chewing width and volume of chewed area — see paper for details) for capsaicin-treated cables but just for two of the measures (mass loss and chewing depth) for denatonium benzoate-treated cables, when compared to cables treated with a non-deterrent substance. However, when applied to cables without shrink tubing, there was no reduction in the four damage measures for either capsaicin or denatonium benzoate-treated cables, compared to cables treated with a non-deterrent substance. Gophers were live-trapped in the wild and transferred to individual enclosures in captivity. Enclosures each had a 1.2-cm-diameter coaxial cable across an opening. Cables were sponged with capsaicin (six gophers) or denatonium benzoate (six gophers), each in solution with Indopol®, or with Indopol® alone (three gophers). The same treatments were applied to cables then encased in a shrink-tube coating (which adhered to the cable upon exposure to heat) with six gophers each offered cables treated with capsaicin, denatonium benzoate or Indopol® alone. In each case, after seven days, cables were assessed for weight and volume loss and for depth and width of gnawing damage.

(1) Shumake S.A., Sterner R.T. & Gaddis S.E. (1999) Repellents to reduce cable gnawing by northern pocket gophers. *The Journal of Wildlife Management*, 63, 1344–1349, https://doi.org/10.2307/3802853

4.4. Translocate mammals away from sites of proposed energy developments

https://www.conservationevidence.com/actions/2517

- **Two studies** evaluated the effects of translocating mammals away from sites of proposed energy developments. One study was in Brazil[1] and one was in Australia[2].

COMMUNITY RESPONSE (0 STUDIES)

POPULATION RESPONSE (0 STUDIES)

BEHAVIOUR (2 STUDIES)

- **Behaviour change (2 studies):** A study in Brazil[1] found that lesser anteaters translocated away from a hydroelectric development site remained close to release sites while a study in Australia[2] found that at least one out of eight chuditchs translocated from a site to be mined returned to its site of capture.

Background

Mammals may be vulnerable to habitat destruction at sites of developments such as energy generation installations or mines. If permission is granted for such developments to go ahead, translocating mammals away from the site may be a way of attempting to mitigate the effects of the development.

For related studies, see interventions within *Species Management-Translocate Mammals*.

A study in 1996–1998 of savanna at a hydroelectric development scheme in Goiás, Brazil (1) found that translocated lesser anteaters *Tamandua tetradactyla* remained close to release sites up to at least nine months after release. Anteaters moved 0.3–2.2 km from release sites during tracking periods. The greatest distances between recorded points in each anteater's range were 0.3–2.6 km. Eight adult lesser anteaters were moved from an area being flooded for a reservoir and were released at the edge of the reservoir (distances from capture to release sites not stated). They were monitored by radio-tracking, over two weeks each month. Animals were monitored for between four days and nine months and were located between two and thirty times in total, between December 1996 and February 1998.

A study in 2016 in a forest site in Western Australia, Australia (2) found that following translocation away from an area being cleared for mining, at least one out of eight chuditchs *Dasyurus geoffroii* returned to its area of capture. Out of eight translocated chuditchs, one was recaptured, 12 days after release, close to the initial capture site. Its recapture site was 13.5 km from the release point and 1 km from the original capture location. Between first capture and recapture, the individual had lost 13% of its body weight but was otherwise in good condition. In January–March 2016, eight chuditchs were live-trapped across four 53–73-ha woodland plots about to be cleared for mining. Chuditchs were marked with PIT-tags and released in a forest area, approximately 14 km away (linear distance). No details are provided about the release procedures or about post-release monitoring.

(1) Rodrigues F.H.G., Marinho-Filho J. & dos Santos H.G. (2001) Home ranges of translocated lesser anteaters *Tamandua tetradactyla* in the cerrado of Brazil. *Oryx*, 35, 166–169, https://doi.org/10.1017/s0030605300031732

(2) Cannella E.G. & Henry J. (2017) A case of homing after translocation of chuditch, *Dasyurus geoffroii* (Marsupialia: Dasyuridae). *Australian Mammalogy*, 39, 118–120, https://doi.org/10.1071/am16023

5. Threat: Transportation and service corridors

Background

The greatest threats from transportation and service corridors tend to be from the destruction of habitat and pollution. Interventions in response to these threats are described in the chapters *Habitat restoration and creation* and *Threat: Pollution*. However, often a more visible impact is that of mortality of mammals in collisions with road vehicles or trains (e.g. Rytwinski & Fahrig 2015). Substantial efforts can be put into reducing this threat, through actions such as providing underpasses or overpasses. The motivation is often to reduce risks to drivers though studies reported on here are those that describe the effectiveness in terms of wild mammal conservation. However, monitoring frequently just considers use of these structures rather than the overall effect on population status of target species. Some related interventions for waterways and pipelines are also included.

Rytwinski, T. & Fahrig, L. (2015) The impacts of roads and traffic on terrestrial animal populations. Pages 237–246, in: R. van der Ree, D. J. Smith & C. Grilo (eds) *Handbook of Road Ecology*. John Wiley & Sons, Ltd, UK, https://doi.org/10.1002/9781118568170.ch28

https://doi.org/10.11647/OBP.0234.05

Roads & Railroads

5.1. Install tunnels/culverts/underpass under roads

https://www.conservationevidence.com/actions/2514

- **Twenty-five studies** evaluated the effects on mammals of installing tunnels, culverts or underpass under roads. Eight studies were in the USA[7,11–14,18,19,24,25], four were in Australia[1,5,15,22], four were in Canada[8,9,16,23], two were in Spain[3,4], one each was in Germany[2], the Netherlands[6] and South Korea[17] and three were reviews with wide geographic coverage[10,20,21].

COMMUNITY RESPONSE (0 STUDIES)

POPULATION RESPONSE (3 STUDIES)

- **Survival (3 studies):** A study in South Korea[17] found that road sections with higher underpass density did not have fewer wildlife-vehicle collisions. A review[10] found that most studies recorded no evidence of predation of mammals using crossings under roads. A controlled, before-and-after, site comparison study in Australia[1] found that overwinter survival of mountain pygmy-possums increased after an artificial rocky corridor, which included two underpasses, was installed.

BEHAVIOUR (23 STUDIES)

- **Use (23 studies):** Seventeen of 20 studies (including seven replicated studies and two reviews), in the USA[7,11,13,14,18,19,24,25], Canada[8,9,16,23], Australia[5,15,22], Spain[3,4], the Netherlands[6], and across multiple continents[20,21], found that crossing structures beneath roads were used by mammals[3–9,11,14–16,19–22,24,25] whilst two studies found mixed results depending on species[18,23] and one study found that culverts were rarely used as crossings by mammals[13]. One of the studies[24] found that crossing structures were used by two of four species more than expected compared to their movements through adjacent habitats. A controlled, before-and-after, site comparison study in Australia[1] found that an artificial rocky corridor, which included two underpasses, was used by mountain pygmy-possums. A replicated study in Germany[2] found that use of tunnels by fallow deer was affected by tunnel colour and design. A study in the USA[12] found that a range of mammals used culverts, including those with shelves fastened to the sides.

- **Behaviour change (1 study):** A controlled, before-and-after, site comparison study in Australia[1] found that after an artificial rocky corridor, which included two underpasses, was installed, dispersal of mountain pygmy-possums increased.

Background

Tunnels, culverts and underpasses may provide safe road crossing opportunities for mammals. A range of different tunnels can be used, including purpose-built wildlife tunnels, culverts that assist with drainage and which can also be used by wildlife, and large passages beneath elevated road section which may sometimes also be used for local vehicle access.

Underpasses are frequently installed in conjunction with wildlife barrier fencing which funnels animals towards the tunnel and prevents them from accessing the road. For this combined intervention, see *Install barrier fencing and underpasses along roads.* See also *Install tunnels/culverts/underpass under railways.*

Studies included here are those where barrier fencing is not installed or not explicitly referred to in the study methods or where at least some underpasses were in unfenced areas. Most studies here report solely on the use of these structures, such as the number of crossings made. There is an absence of studies reporting on wider population-level effects of the presence of these structures.

A controlled, before-and-after, site comparison study in 1982–1986 of rock screes and boulder fields on a mountain in Victoria, Australia (1) found that an artificial rocky corridor, which included two underpasses, was used by mountain pygmy-possums *Burramys parvus* and female overwinter survival and male dispersal increased. Over 28 days, mountain pygmy-possum were recorded in a monitored underpass 60 times, bush rats *Rattus fuscipes* 21 times and dusky antechinus *Antechinus swainsonii* three times. The overwinter survival of female pygmy-possums was 96% of that at an undisturbed site after corridor construction, compared to 21% before. Before construction, sex ratios at the two sites differed, with males not dispersing at the developed site. After construction, both adult and juvenile males dispersed (population before: 25% male; after: 10% male). In 1985, a 60-m-long corridor, connecting a fragmented breeding area, was created. This included two adjacent tunnels (1 m diameter) under a road. The corridor and tunnels were filled with rocks to imitate scree. A remotely activated camera monitored one tunnel over 18 days in February–April and 10 days in October–November 1986. Possums were live-trapped in 1982–1986. Population composition was compared at the developed (ski resort) site and one undisturbed site.

A replicated study in 1994 of tunnels in enclosures in Germany (2) found that use of tunnels by fallow deer *Dama dama* was affected by tunnel colour and design. Deer used one tunnel significantly more in four of six paired trials. A white-painted tunnel was used more than a grey-painted tunnel (732 vs 425 passages) and also more than a black-painted tunnel (294 vs 153 passages). A black base was used more than one without a base (747 vs 584 passage). An unlit tunnel was used more than an indirectly-lit tunnel (581 vs 242). There was no significant difference in the use of tunnels with and without tree stumps within them. Two tunnels were erected in a 0.7-ha enclosure, each 2 m high, 2 m wide and 8 m long. Twenty deer accessed food through the tunnels. Tunnel use was registered by a photo-electric sensor. Trials were run with six tunnel design combinations: both tunnels unpainted; white vs grey; white vs black; black base (and 80 cm up sides) vs no base; indirect light on ceiling vs unlit; tree stumps in tunnel vs no stumps. Tunnels were painted off-white for base, lighting and tree stump trials.

A replicated study in 1994 of roads and railways in Madrid province, Spain (3) found that all 17 culverts under roads were used by mammals. The highest frequencies of tracks were from wood mice *Apodemus sylvaticus* (2.5 tracks/day), shrews *Sorex* spp. (0.5/day) and European rabbits *Oryctolagus cuniculus* (0.3/day). Rats *Rattus* sp. (0.1 tracks/day), hedgehogs *Erinaceus europaeus* (0.01/day), cats (mostly wild cat *Felis silvestris* — 0.04/day), red fox *Vulpes vulpes* (0.03/day), genet *Genetta genetta* (0.02/day) and weasel *Mustela nivalis* (0.01/day) were also detected. Small mammal use of culverts decreased with increased road width and culvert length and increased with increased culvert height, width and openness. Use by rabbits and carnivores decreased with increasing highway or railway width. Rabbit use also declined with increased boundary fence height (fences ran across culvert entrances, rather than funnelling animals towards them). Vegetation complexity had little influence. Five culverts were monitored under railways, two under a motorway and 10 under local roads. Structural, vegetation and traffic variables were recorded at each culvert. Use was monitored using marble (rock) dust over culvert floors to record tracks. Sampling was undertaken in 1994, over four days each in spring, summer, autumn and winter. Sampling extended to eight days at four culverts when deer were nearby.

A replicated study in 1993–1994 along four roads in Catalonia, Spain (4) found that underpasses were used by several mammal species. Small mammals used all rectangular culverts and 94% of circular culverts. Hares *Lepus* spp. and rabbits *Oryctolagus cuniculus* used 83% and 23% of rectangular and circular culverts respectively whilst carnivores used 88% and 75% respectively. Carnivores recorded were weasel *Mustela nivalis*, beech marten *Martes foina*, badger *Meles meles*, genet *Genetta genetta* and fox *Vulpes vulpes*. Wild boar *Sus scrofa* and roe deer *Capreolus capreolus* also used underpasses. Use was greater by small mammals for underpasses at the same level as the surroundings and those with natural substrate on the floor. Those with water were used less frequently. Rabbits did not use narrow structures (<1.5 m), whereas wild boar used underpasses >7 m wide. A total of 39 circular (1–3 m diameter) and 17 rectangular drains (4–12 m wide) and other underpasses were surveyed along four 10-km sections of road. Underpasses were monitored for four days/ season over a year, in 1993–1994. Animal tracks were monitored using marble power (50 cm wide) across the centre of each structure. Infra-red and photographic cameras were used at entrances.

A study in 1996–1997 along a highway in New South Wales, Australia (5) found that mammals used three underpasses. Between three and nine native mammal species used each of the tunnels. Common wombat *Vombatus ursinus*, swamp wallaby *Wallabia bicolor*, rats (*Rattus fuscipes,*

Rattus lutreolus) and bandicoots (*Perameles nasuta, Isoodon macrourus*) were the most frequently recorded. Four non-native species also used underpasses. The greatest number of species was recorded in the largest underpass, but the smallest underpass had the greatest frequency of use. A total of 43 native and 57 introduced mammals were killed on the road during the survey. Three underpasses (diameters: 1.5–10 m) were monitored from August 1996 to June 1997. Infra-red camera traps, track counts (sand 2 m inside entrances), trapping and nocturnal searches were used. Road-kill data were also collected.

A replicated study in 1997–1998 of 53 wildlife passages along waterways under roads at over 20 sites in the Netherlands (6) found that all passages were used by mammals. At least 16 mammal species used passages. Waterside banks extending under bridges were used by 14 species and other types of passageways by 10 species. Brown rats *Rattus norvegicus*, mice and voles were the most frequently recorded mammals (see original publication for details). For all mammals, frequency of use increased with increasing passage diameter and width, but was not affected by substrate. Culverts and bridges were adapted for wildlife, in the 1990s. In 1997, thirty-one passages (0.4–3.5 m wide) were monitored. These included extended banks (unpaved or paved), planks along bridge or culvert walls, planks floating on the water, concrete passageways and plastic gutters covered with sand. In 1998, twenty-two passages were monitored for the effect of width and substrate. These were wooden passageways along bridge or culvert walls (0.2–0.6 m wide). Monitoring involved weekly checks of tracks on sandbeds (for 4–7 weeks) and ink pads (12 weeks in 1997, four weeks in 1998) across passageways.

A study in 2000 along a highway in Vermont, USA (7) found that a concrete underpass was used by four mammal species to cross the road. Infra-red monitors recorded 190 confirmed or unconfirmed instances of animals using the tunnel. Where a species was identified, 58% of occurrences were racoon *Procyon lotor*, 27% were mink *Neovison vison*, 11% were weasel *Mustela frenata* and 4% were skunk *Mephitis mephitis*. The total number of passages by these species was not stated. The underpass was a concrete block structure, split along the middle by a concrete support. It was 97 m long, 3 m wide and 4 m high. A stream flowed through one tunnel and, at times of high water, through both tunnels, though a sloping floor ensured at least some dry passage. The underpass was monitored discontinuously from June–November 2000, using infrared monitors, cameras and footprint pads.

A replicated study in 1999–2000 along two highways in Alberta, Canada (8) found that drainage culverts were used by at least nine mammal species. A total of 618 crossings were recorded. Species recorded were coyote *Canis latrans* (1% of crossings), American marten *Martes americana* (12%), weasel *Mustela ermine* and *Mustela frenata* (28%), snowshoe hare *Lepus americanus* (3%), red squirrel *Tamiasciurus hudsonicus* (4%), bushy-tailed wood rat *Neotoma cinerea* (15%), shrew spp. *Sorex* spp. (8%), deer mouse *Peromyscus maniculatus* (28%) and vole spp. *Arvicolinae* (0.5%). Culvert use was positively correlated with traffic volume (for hare, squirrel and marten), culvert openness (marten), culvert height (weasel), through-culvert visibility (hare) and adjacent shrub cover (hare). A range of factors negatively affected culvert use by mammals (see paper for details). Thirty-six drainage culverts were monitored along a 55-km section of the Trans-Canada highway (two-and four-lane sections, with and without central reservation) and a 24-km section of highway 1A (two lanes, no central reservation). Crossings were determined from sooted track-plates (75 × 30 cm) in each culvert, checked weekly in January–April of 1999–2000 (≥ 12 times/culvert) and tracks in adjacent snow indicating culvert use.

A replicated study in 2000 along highways through two wetlands in British Columbia, Canada (9) found that culverts were used by small-to medium-sized mammals. Mammals used most of the eight dry culverts. In particular, there were frequent records of racoons *Procyon lotor* (on 11% of track plates) and species from the weasel family (on 32% of track plates — species not stated). Mice, voles and shrews combined were recorded on 31% of track plates. Racoons also used wet culverts on all nine occasions when tracks were not obscured by water. In 1995, twelve dry corrugated steel pipe culverts (average 35 long, 1 m diameter) were installed at 50-m intervals under a four-lane highway at one wetland. Eight were monitored. At another wetland, two wet cross-drainage corrugated steel pipe culverts (31 m long, 0.6 m diameter) were monitored. Aluminium track-plates, covered with soot, were installed 1–2 m inside each culvert and monitored over nine weekly intervals, in July–October 2000.

A review in 2000 of studies investigating whether mammalian predators use wildlife passages under roads and railways as 'prey-traps' (10) found that most studies recorded no evidence of predation in or around passages. Evidence suggested that predator species used different passages to their prey. Only one study, in Australia, suggested that tunnels increased predation risk and that study recorded only one predator in tunnels. However, no studies specifically investigated predator activity, densities or predation rates, or predator-induced prey mortality at passage sites relative to control sites away from passages, or before-and-after passage construction. A literature survey was carried out in July 2000 using BIOSIS (Biological Abstracts) and Proceedings of the First, Second and Third International Conference on Wildlife Ecology and Transportation.

A study in 1998–1999 in a fragmented urban area in California, USA (11) found that bobcats *Felis rufus* and coyotes *Canis latrans* used underpasses to cross a road. Nine road crossings (two by bobcats and seven by coyotes) out of 24 crossings where culverts were available within 100 m were through culverts and 15 (five by bobcats and 10 by coyotes) were over the road. Traffic levels were higher during crossings through culverts (2.1 cars/minute) than during crossings over the road (0.8 cars/minute). Results were not tested for statistical significance. The study was conducted northwest of Los Angeles from July 1998 to October 1999. Movements of 13 bobcats and nine coyotes were determined from 53 radio-tracking sessions (32 focussed on bobcats, 21 on coyotes). Locations were obtained every 30 minutes for 2–12 hours and road crossings were observed directly when possible.

A study in 2001–2003 along a highway through wetlands in Montana, USA (12) found that a range of mammals used culverts, including those with shelves fastened to sides. Twenty-three mammal species used culverts. These included six of the seven small mammal species that were recorded by trapping outside tunnel entrances; meadow vole *Microtus pennsylvanicus*, deer mouse *Peromyscus maniculatus*, vagrant shrew *Sorex vagrans*, Columbian ground squirrel *Spermophilus columbianus*, short-tailed weasel Mustela erminea and striped skunk *Mephitis mephitis*. Other mammals recorded using culverts included white-tailed deer *Odocoileus virginianus*, muskrat *Ondatra zibethicus*, raccoon *Procyon lotor*, coyote *Canis latrans* and red fox *Vulpes vulpes*. When water covered culvert floors, deer mice, short-tailed weasels, striped skunks and raccoons travelled along shelves in culverts. Meadow voles used tubes along culvert shelves. At least ten culverts (total number not clear) were monitored along a 6-mile section of Highway 93. Five had 25-inch-wide shelves installed. Culverts included some of 3–4 feet diameter and may have included others up to 10 feet wide. Monitoring was conducted from October 2001 to 2003 using heat-and motion-triggered cameras. Each month (March–October), small mammal populations adjacent to culverts were censused using 25 live traps, over three days.

A study in 2002 of mixed habitats including forest, swamp and farmland, along a highway in New York, USA (13) found that 19 culverts were rarely used as crossing points by mammals. The only crossings documented were five by northern racoons *Procyon lotor* at a single drainage culvert. Nineteen culverts were studied, along 141 km of highway, from 14 March to 29 April 2002. Culverts were categorised according to primary use: drainage (seven culverts), pedestrian underpass (nine), truck use (two) or bridge (one, where a river flowed beneath the road). Enabling white-tailed deer *Odocoileus virginianus* passage was also thought to be a motivation in installing at least some culverts. Animal passage was recorded using one camera trap at each culvert (average 40 days/site) and opportunistic snow-tracking when conditions permitted.

A replicated study in 1999–2000 along three major highways in California, USA (14) found that tunnels, culverts and underpasses were used by mammals. Fourteen of the 15 passages were used by racoons *Procyon lotor* (making 207 crossings), eight by opossums *Didelphis virginianus* (24 crossings), seven by coyotes *Canis latrans* (59 crossings), seven by bobcats *Lynx rufus* (36 crossings), five by striped skunks *Mephitis mephitis* (23 crossings), three by mule deer *Odocoileus hemionus* (26 crossings), one by spotted skunks *Spilogale putorius* (five crossings) and one by a mountain lion *Puma concolor* (one crossing). Crossing numbers include both verified and probable crossings. Rodents and cottontail rabbits *Sylvilagus audubonii* were also recorded. Six square livestock tunnels, five drainage culverts and four underpasses (surface roads or wide stream crossings) were studied. Passages were 44–218 m long and 2–238 m^2 in cross-section. Camera traps were used in four passages and powder stations to detect animal footprints in 12 passages.

One passage was monitored using both methods. Monitoring occurred over four consecutive days/month between July 1999 and June 2000.

A study in 2001–2003 on a road through rainforest in Queensland, Australia (15) found that underpasses beneath the road were used by a range of mammals. There were 237 crossings recorded by brown bandicoots *Isoodon obesulus*, 233 by red-legged pademelons *Thylogale stigmatica*, 230 by coppery brushtail possums *Trichosurus vulpecula johstoni*, two by Lumholtz's tree-kangaroos *Dendrolagus lumholtzi*, 53 by rodents and 13 by dogs *Canis lupus familiaris* or dingoes *Canis dingo*. Three underpasses (3.4 m high, 3.7 m wide), installed in 2001 below an upgraded two-lane road, were studied. Habitat enhancement features were added to each, such a soil, leaf and branch litter, rocks and logs and also vertical tree branches, to enable escape off the tunnel floor. Underpass use was monitored by weekly checks, over three years, for animal tracks in 1-m-wide strips of sand. Infrared-triggered cameras were used occasionally to confirm identifications.

A study in 2003 of a highway and railway in British Columbia, Canada (16) found that at least two of three crossing structures were used by mammals. Mule deer *Odocoileus hemionus* were detected using one small culvert (2.1 m wide, 1.5 m high, 30 m long) six times. They were not recorded using a larger (7 m wide, 5 m high, 40 m long) cattle underpass though signs of their presence were noted nearby. Black bears were detected 20 times passing through the smaller culvert and four times through the cattle underpass. Raccoons were detected twice at the cattle underpass. The smaller culvert had a soil substrate, was surrounded by vegetation and was relatively far from human activity. The cattle underpass had limited surrounding natural vegetation. No mammals were recorded using a third culvert (1.2 m wide and high, 30 m long), possibly due to camera malfunction. Culverts and the underpass ran under both the Trans-Canada Highway and Canadian Pacific Railway. They were monitored using infrared sensor cameras during August–November 2003. Animal tracks or signs around camera stations were also recorded.

A study in 2004–2006 in an area of rice fields and scattered forest in Jeollanamdo province, South Korea (17) found that highway underpasses were used by a range of mammals, though road sections with higher underpass density did not have fewer wildlife-vehicle collisions. Eleven wild mammal species were recorded using underpasses. The most frequent were raccoon dog *Nyctereutes procyonoides* (865 images), brown rat *Rattus norvegicus* (455), leopard cat *Prionailurus benalensis* (253), striped field mouse *Apodemus agrarius* (229), Siberian weasel *Mustela sibirica* (166), Eurasian otter *Lutra lutra* (35) and water deer *Hydropotes inermis* (32). Ninety-three roadkill mammals of 12 species were recorded. The most frequent were rodents (24 casualties), leopard cat (17), Siberian weasel (13) and water deer (12). Most mammals used all underpass types frequently, except water deer, which rarely used small passages. Use of seven circular culverts (0.8–1.2 m diameter), two box culverts (2.5 m wide and high) and five human underpasses (2.0–4.3 m wide and high), selected from 31 underpasses along a 6.6-km section of four-lane highway, were monitored from September 2005–August 2006. One or two infrared-operated cameras were installed 1–2 m inside each underpass for an average of 239 days/underpass. Wildlife-vehicle collisions were recorded daily from September 2004–August 2006.

A study in 2004–2005 at seven sites along roads through forest in Virginia, USA (18) found that white-tailed deer *Odocoileus virginianus* used underpasses to cross the road but black bears *Ursus americanus* did not. White-tailed deer crossed through four of seven underpasses monitored, with a total of 1,107 crossings detected. Black bears approached one underpass entrance three times, but did not cross through. Other mammals recorded in underpasses included opossums *Didelphis virginiana*, bobcats *Lynx rufus*, red foxes *Vulpes vulpes*, coyotes *Canis latrans*, raccoons *Procyon lotor* and groundhogs *Marmota monax* as well as squirrels and mice (see paper for details). Seven underpasses were monitored. Five were culverts (1.8–6.1 m wide, 1.8–4.6 m high and 21–79 m long). Two were crossings under bridges (13–94 m wide, 5–14 m high and 10–18 m long). Underpasses were not fenced and most had a narrow water section. Underpasses were monitored from June 2004 to May 2005, using one or two camera traps at each entrance.

A study in 2003–2005 along a highway through deciduous woodland in North Carolina, USA (19) found that mammals used a wildlife underpass. An estimated 299 mammal crossings of at least 10 species occurred (based on 126 crossings observed on a sample of video surveillance). Of these, an estimated 185 were white-tailed deer *Odocoileus virginianus* crossings. At least 17 deer approached the underpass but retreated without crossing. Other mammals crossing included red or grey fox *Vulpes vulpes* or *Urocyon cinereoargenteus*, raccoon *Procyon lotor*, woodchuck *Marmota monax*, gray squirrel *Sciurus carolinensis* and chipmunk *Tamias striatus*. Only four incidences of mammals killed by vehicles were recorded from December 2003 to June 2005. Two digital ultra-low-light video cameras and infrared spotlights monitored underpass use below a four-lane highway between December 2003 and May 2005. A sample of videos was viewed from 458 days of continual video recordings. The underpass was constructed in 1955, encompassing a 6-m width either side of a stream. It was 2–3 m high and 41 m long. Weekly surveys of vehicle-killed animals were undertaken on a 1.8-km section of road encompassing the underpass.

A global review in 2007 of 123 studies investigating the use of 1,864 wildlife crossings (20) found that all studies reported that the majority of underpasses and overpasses were used by wildlife. Of the 1,864 structures reported on, most were underpasses (83%), including culverts (742 examples), bridges (130), tunnels (340) and unknown types (333). Structures provided crossings over or under roads (113 studies), railways (5 studies), both (1 study), canals (2 studies) and a pipeline (1 study). Studies were from Europe (55 studies), the USA (30 studies), Canada (nine studies), South America (one study) and Australia (29 studies).

A review of 30 studies reporting on monitoring of 329 crossing structures in Australia, Europe and North America (21) found that mammals used most culverts and underpasses. Small mammals used pipes (demonstrated by 6/7 relevant studies), drainage culverts (5/5 studies), adapted culverts (5/5 studies), wildlife underpasses (3/4 studies) and bridge underpasses (2/3 studies). Arboreal mammals used pipes (1/1 studies), drainage culverts (4/4 studies), adapted culverts (4/4 studies) and bridge underpasses (1/1 studies). Medium-sized mammals used pipes (8/11 studies), drainage culverts (12/13 studies), adapted culverts (8/8 studies), wildlife underpasses (6/8 studies) and bridge underpasses (6/7 studies). Large mammals used pipes (6/9 studies), drainage culverts (11/12 studies), adapted culverts (11/11 studies), wildlife underpasses (24/24 studies) and bridge underpasses (14/15 studies). Larger mammals tended to use more open underpasses. Small and medium-sized mammals used underpasses with funnel-fencing or adjoining walls and those with vegetation cover close to entrances. Those with vegetation cover tended to be avoided by some ungulates. Thirty papers reporting monitoring of 329 crossing structures were reviewed. Fourteen papers investigated multiple structure types, resulting in a total of 52 studies of different structure types. Underpasses, from small drainage pipes to dry passage bridges, comprised 82% of crossings.

A study in 2010 of a road through forest and pastureland in New South Wales, Australia (22) found that bare-nosed wombats *Vombatus ursinus* used culverts to cross the road. Bare-nosed wombats used eight out of 19 monitored culverts. Wombats were recorded using culverts on 16 out of 190 camera-trap nights. One culvert was used three times in one night and three were used twice in one night. Other culverts were not used more than once in a night. The study was conducted along 8 km of a two-lane road. Nineteen concrete pipe culverts (40–60 cm diameter and 13–25 m long) were monitored between April and August 2010. A camera trap was set 1 m from each culvert entrance for 10 days. Five culverts were dry with earth substrate, nine were dry without earth substrate and five had constant water flow. Culverts were 40–2,200 m apart.

A study in 2009 at 10 sites along a highway through forest in Alberta, Canada (23) found that North American deer mice *Peromyscus maniculatus* used underpasses to cross a road but meadow voles *Microtus pennsylvanicus* and southern red-backed voles *Myodes gapperi* did not. Tracks of deer mice were recorded in 90% of track tubes in elliptical culverts, in 87% of track tubes in box culverts and in 75% of track tubes on open-span bridge underpasses. No tracks of meadow vole or southern red-backed vole were detected, despite their use of overpasses in the area. Over two weeks in September–October 2010, small mammals were surveyed in three elliptical metal culverts (4 m high, 7 m wide), five concrete box culverts (2.6 m high, 3.2 m wide) and two open-span bridge underpasses (3 m high, 11 m wide). Underpasses were unvegetated and entrances were characterized by roadside grasslands. Two parallel sample lines, each of five 30 × 10 cm track tubes with sooted metal sheet as a floor, were placed in the centre of each underpass. Mammals were identified from their footprints.

A study in 2015 along a highway in Montana, USA (24) found that underpasses were used by white-tailed deer *Odocoileus virginianus* and mule deer *Odocoileus hemionus* more than expected compared to their movements through adjacent habitats, but no difference was found for black bear *Ursus americanus* or coyote *Canis latrans*. Overall, white-tailed deer (recorded at all 15 underpasses) and mule deer (at five of 15 underpasses) had an average of 88% and 472% more movements/day respectively through underpasses than adjacent habitats. Black bear (recorded at seven of 15 underpasses) and coyote (at 13 of 15 underpasses) had an average of 112% and 75% more movements/day respectively through underpasses than adjacent habitats, but the difference was not significant. Fifteen elliptical underpasses were installed in 2006–2011 along a 91 km stretch of highway. Underpasses (7–8 m wide, 4–6 m high, 15–40 m long) were constructed from corrugated metal with a soil substrate and retaining walls extending 10 m from the roadside. Twelve of the 15 underpasses had 2.4-m high wildlife exclusion fencing. Infrared cameras recorded large mammal movements through each underpass (one camera/entrance) and at random locations within an adjacent 300 m² plot on each side (five cameras/plot) for 12–20 days in April–November 2015.

A replicated study in 2008–2011 of 265 culverts throughout Maryland, USA (25) found that culverts were used by a range of mammal species to cross roads. Crossings were made by northern raccoons *Procyon lotor* (0.79/culvert/day), Virginia opossums *Didelphis virginiana* (0.03/culvert/day), woodchucks *Marmota monax* (0.03/culvert/day), red foxes *Vulpes vulpes* (0.03/culvert/day), gray squirrels *Sciurus carolinensis* (0.02/culvert/day) and both common grey foxes *Urocyon cinereoargenteus* and white-footed mice *Peromyscus* spp (0.01/culvert/day). Between August 2008 and January 2011, a total of 265 randomly selected culverts were monitored using camera traps for a total of 31,317 camera-trap days. Culverts were located under paved roads and contained either a waterway, a route for water flow, or other depression. Culverts averaged 2.4 m wide, 1.9 m high and 46.4 m long. Each culvert was sampled at least nine times in 2008–2011, for 10–36 days each time, using one camera trap. The camera was placed at the approximate midpoint of the culvert or near the entrance.

(1) Mansergh I.M. & Scotts D.J. (1989) Habitat continuity and social organisation of the mountain pygmy-possum restored by tunnel. *The Journal of Wildlife Management*, 53, 701–707, https://doi.org/10.2307/3809200

(2) Woelfel H. & Krueger H.H. (1995) Zur Gestaltung von Wilddurchlässen an Autobahnen [On the design of game passages across highways]. *Zeitschrift für Jagdwissenschaft*, 41, 209–216, https://doi.org/10.1007/bf02239950

(3) Yanes M., Velasco J. & Suarez F. (1995) Permeability of roads and railways to vertebrates: the importance of culverts. *Biological Conservation*, 71, 217–222, https://doi.org/10.1016/0006-3207(94)00028-o

(4) Rosell C., Parpal J., Campeny R., Jove S., Pasquina A. & Velasco J.M. (1997) Mitigation of barrier effect on linear infrastructures on wildlife. Pages 367–372 in: *Habitat Fragmentation & Infrastructure*. Ministry of Transport, Public Works and Water Management, Delft, Netherlands.

(5) Norman T., Finegan A. & Lean B. (1998) *The role of fauna underpasses in New South Wales.* Proceedings -International Conference on Wildlife Ecology and Transportation, Florida Department of Transportation, Tallahassee, Florida, USA, 195–208.

(6) Veenbaas G. & Brandjes J. (1999) *Use of fauna passages along waterways under highways.* Proceedings -International Conference on Wildlife Ecology and Transportation, Florida Department of Transportation, Tallahassee, Florida, USA, 253–258.

(7) Austin J.M. & Garland L. (2001) *Evaluation of a wildlife underpass on Vermont State Highway 289 in Essex, Vermont.* Proceedings -2001 International Conference on Ecology and Transportation, Center for Transportation and the Environment, North Carolina State University, Raleigh NC, USA, 616–624.

(8) Clevenger A.P., Chruszcz B. & Gunson K. (2001) Drainage culverts as habitat linkages and factors affecting passage by mammals. *Journal of Applied Ecology*, 38, 1340–1349, https://doi.org/10.1046/j.0021-8901.2001.00678.x

(9) Fitzgibbon K. (2001) An evaluation of corrugated steel culverts as transit corridors for amphibians and small mammals at two Vancouver Island wetlands and comparative culvert trials. MA thesis. Royal Roads University, Vancouver, Canada.

(10) Little S.J., Harcourt R.G. & Clevenger A.P. (2002) Do wildlife passages act as prey-traps? *Biological Conservation*, 107, 135–145, https://doi.org/10.1016/S0006-3207(02)00059-9

(11) Tigas L.A., Van Vuren D.H. & Sauvajot R.M. (2002) Behavioral responses of bobcats and coyotes to habitat fragmentation and corridors in an urban environment. *Biological Conservation*, 108, 299–306, https://doi.org/10.1016/s0006-3207(02)00120-9

(12) Foresman K.R. (2003) *Small mammal use of modified culverts on the Lolo South project of western Montana — an update.* Proceedings -International

Conference on Ecology and Transportation, Center for Transportation and the Environment, North Carolina State University, Raleigh NC, USA, 342–343

(13) LaPoint S., Keys R.W. & Ray J.C. (2003) Animals crossing the Northway: are existing culverts useful? *Adirondack Journal of Environmental Studies*, 10, 11–17.

(14) Ng S.J., Dole J.W., Sauvajot R.M., Riley S.P.D. & Valone T.J. (2004) Use of highway undercrossings by wildlife in southern California. *Biological Conservation*, 115, 499–507, https://doi.org/10.1016/S0006-3207(03)00166-6

(15) Goosem M., Weston N. & Bushnell S. (2005) *Effectiveness of rope bridge arboreal overpasses and faunal underpasses in providing connectivity for rainforest fauna.* Proceedings -International Conference on Ecology and Transportation, Center for Transportation and the Environment, North Carolina State University, Raleigh NC, USA, 304–318.

(16) Krawchuk A., Larsen K.W., Weir R.D. & Davis H. (2005) Passage through a small drainage culvert by mule deer, *Odocoileus hemionus*, and other mammals. *The Canadian Field Naturalist*, 119, 296–298, https://doi.org/10.22621/cfn.v119i2.119

(17) Choi T.-Y. & Park C.H. (2007) *Can wildlife vehicle collision be decreased by increasing the number of wildlife passages in Korea?* Proceedings — International Conference on Ecology and Transportation, Center for Transportation and the Environment, North Carolina State University, Raleigh NC, USA, 392–400.

(18) Donaldson B. (2007) Use of highway underpasses by large mammals and other wildlife in Virginia: factors influencing their effectiveness. *Transportation Research Record: Journal of the Transportation Research Board*, 2011, 157–164.

(19) Kleist A.M., Lancia R.A. & Doerr P.D. (2007) Using video surveillance to estimate wildlife use of a highway underpass. *The Journal of Wildlife Management*, 71, 2792–2800, https://doi.org/10.3141%2F2011-17

(20) van der Ree R., van der Grift E., Mata C. & Suarez F. (2007) *Overcoming the barrier effect of roads — how effective are mitigation strategies? An international review of the use and effectiveness of underpasses and overpasses designed to increase the permeability of roads for wildlife.* Proceedings — International Conference on Ecology and Transportation, Center for Transportation and the Environment, North Carolina State University, Raleigh NC, USA, 423–431.

(21) Taylor B.D. & Goldingay R.L. (2010) Roads and wildlife: impacts, mitigation and implications for wildlife management in Australia. *Wildlife Research*, 37, 320–331, https://doi.org/10.1071/WR09171

(22) Crook N., Cairns S.C. & Vernes K. (2013) Bare-nosed wombats (*Vombatus ursinus*) use drainage culverts to cross roads. *Australian Mammalogy*, 35, 23–29, https://doi.org/10.1071/am11042

(23) D'Amico M., Clevenger A.P., Román J. & Revilla, E. (2015) General versus specific surveys: Estimating the suitability of different road-crossing structures for small mammals. *The Journal of Wildlife Management*, 79, 854–860, https://doi.org/10.1002/jwmg.900

(24) Andis A.Z., Huijser M.P. & Broberg L. (2017) Performance of arch-style road crossing structures from relative movement rates of large mammals. *Frontiers in Ecology and Evolution*, 5, 122, https://doi.org/10.3389/fevo.2017.00122

(25) Sparks J.L. & Gates J.E. (2017) Seasonal and regional animal use of drainage structures to cross under roadways. *Human–Wildlife Interactions*, 11, 182–191, https://doi.org/10.26077/x2b9-nk15

5.2. Install tunnels/culverts/underpass under railways

https://www.conservationevidence.com/actions/2519

- **Six studies** evaluated the effects on mammals of installing tunnels, culverts or underpass under railways. Two studies were in Spain[2,3], one was in each of Australia[1], Canada[5] and the Netherlands[6] and one reviewed literature from a range of countries[4].

COMMUNITY RESPONSE (0 STUDIES)

POPULATION RESPONSE (1 STUDY)

- **Survival (1 study)**: A review[4] found that most studies recorded no evidence of predation in or around passages under railways or roads of mammals using those passages.

BEHAVIOUR (5 STUDIES)

- **Use (5 studies)**: Five studies, in Spain[2,3], Australia[1], Canada[5] and the Netherlands[6], found that tunnels, culverts and underpasses beneath railways were used by a range of mammals including rodents[1,2,3,6], rabbits and hares[2,3,6], carnivores[2,3,5,6], marsupials[1], deer[5] and bears[5]. One of these studies found that existing culverts were used more than specifically designed wildlife tunnels[1].

Background

Tunnels, culverts and underpasses may provide safe railway crossing opportunities for wildlife. A range of different tunnels can be used, often in combination with wildlife barrier fencing which funnels animals towards the tunnel and prevents them from accessing the railway (see Install barrier fencing along railways). Studies summarised within this intervention cover both tunnels created specifically for wildlife and those that were created for other purposes (e.g. drainage or farm access) but where information about use of such structures by mammals is included. Studies mostly report on the use of these structures, such as the number of crossings made, rather than on wider population-level effects of their presence.

See also: *Install tunnels/culverts/underpass under roads* and *Install overpasses over roads/railways*.

A site comparison study in 1984–1985 in New South Wales, Australia (1) found that small and medium-sized mammals used established drainage culverts, but rarely used new wildlife tunnels. All five existing culverts were used by mammals. Bush rat *Rattus fuscipes* was recorded in all culverts (1–6 captures and/or tracks/culvert) and long-nosed bandicoot *Perameles nasuta* in one. Few signs of use were recorded in wildlife tunnels. Swamp wallaby *Wallabia bicolor* tracks were recorded in one tunnel in October 1984. No indication of tunnel use was found in January 1985. Five long-established drainage culverts (0.2 × 0.9 to 2.4 × 3.0 m) with dense surrounding vegetation and three of seven newly constructed wildlife tunnels (3 m diameter, 15–20 m long) with sandy floors and little vegetation, under a 35-km-long section of railway line, were monitored. Small mammal traps were set in all underpasses and cage traps in tunnels and one culvert. Tracks were recorded in sand and on soot-coated paper across passages. Culverts were surveyed for eight nights in September–October 1984 and tunnels for seven nights in October 1984 and five nights in January 1985 (15–242 trap nights/structure).

A replicated study in 1994 of 17 culverts under roads and railways in Madrid province, Spain (2) found that mammals used all 17 culverts studied. The highest frequencies of tracks were from wood mice *Apodemus sylvaticus* (2.5 tracks/culvert/day), shrews *Sorex* spp. (0.5/culvert/day) and European rabbit *Oryctolagus cuniculus* (0.3/culvert/day). Rats *Rattus* sp. (0.1 tracks/culvert/day), hedgehogs *Erinaceus europaeus* (0.01/culvert/day), cats (mostly wild cat *Felis silvestris* -0.04/culvert/day), red fox *Vulpes vulpes* (0.03/culvert/day), genet *Genetta genetta* (0.02/culvert/day) and weasel *Mustela nivalis* (0.01/culvert/day) were also detected. Small mammal use of culverts decreased with increased culvert length and increased with increasing culvert height, width and openness. Use by rabbits and carnivores decreased with increasing width of the railway or highway. Rabbit use also declined with increased boundary fence height. Vegetation complexity had little influence. Five culverts were monitored under railways, two under a motorway and 10 under local roads. Structural, vegetation and traffic variables were recorded at each culvert. Use was monitored using marble (rock) dust over culvert floors to record tracks. Sampling was undertaken in 1994, over four days each in spring, summer, autumn and winter. Sampling of four culverts extended to eight days when deer were in the vicinity.

A study in 1991–1992 along a high-speed railway through agricultural land in Castilla La Mancha, Spain (3) found that culverts and underpasses not specifically designed for wildlife were used as crossings under the railway by a range of mammals. Small mammals were recorded in culverts/underpasses (and two overpasses) 582 times (37 crossings/100 passage-days) and brown hare *Lepus granatensis* and European rabbit *Oryctolagus cuniculus* 89 times (5 crossings/100 passage-days). Tracks of four carnivore species, red fox *Vulpes vulpes*, wild cat *Felis silvestris*, common genet *Genetta genetta* and Iberian lynx *Lynx pardinus*, were recorded. No deer or wild boar *Sus scrofa* used passages. Rabbit and hare crossing rates were not affected by underpass design, vegetation cover at entrances or distance from scrubland. Small mammals preferred culverts ≤2 m wide. Fencing did not significantly affect relative crossing rates. Fifteen dry culverts and passages (e.g. small roads and two flyovers, 13–64 m long, 1.2–6.0 m wide, 1.2–3.5 m high) along a 25-km section of high-speed railway, were monitored.

Tracks in sand were monitored at each passage for 15–22 days/month between September 1991 and July 1992. The railway was fenced with 2-m-high wire netting in July 1991–March 1992.

A review in 2000 of studies investigating whether mammalian predators use wildlife passages under railways and roads as 'prey-traps' (4) found that most studies recorded no evidence of predation in or around passages. Evidence suggested that predator species used different passages to their prey. Only one study, in Australia, suggested that tunnels increased predation risk and that recorded only one predator in tunnels. However, no studies specifically investigated predator activity, densities or predation rates, or predator-induced prey mortality at passage sites relative to control sites away from passages, or before-and-after passage construction. A literature survey was carried out in July 2000 using BIOSIS (Biological Abstracts) and Proceedings of the First, Second and Third International Conference on Wildlife Ecology and Transportation.

A study in 2003 of culverts under a railway and highway in British Columbia, Canada (5) found that at least two of three underpasses were used by mammals. Mule deer *Odocoileus hemionus* were detected using one small culvert (2.1 m wide, 1.5 m high, 30 m long) six times. They were not recorded using a larger (7 m wide, 5 m high, 40 m long) cattle underpass though signs of their presence were noted nearby. Black bears were detected 20 times passing through the smaller culvert and four times through the cattle underpass. Raccoons were detected twice at the cattle underpass. The smaller culvert had a soil substrate, was surrounded by vegetation and was relatively far from human activity. The cattle underpass had limited surrounding natural vegetation. No mammals were recorded using a third culvert (1.2 m wide and high, 30 m long), possibly due to camera malfunction. Culverts and the underpass ran under both the Canadian Pacific Railway and Trans-Canada Highway. They were monitored using infrared sensor cameras during August–November 2003. Animal tracks or signs around camera stations were also recorded.

A study in 2003 at 14 underpasses beneath a railway through suburban and rural habitat in the Netherlands (6) found that several species of small-and medium-sized mammals used underpasses to cross the railway. Tracks identified in the monitored underpasses were from western hedgehog *Erinaceus europaeus* (recorded at two of the 14 underpasses), rabbit *Oryctolagus cuniculus* (two underpasses), brown rat *Rattus norvegicus* (4–5 underpasses), western polecat *Mustela putorius* (0–1 underpasses), red fox *Vulpes vulpes* (one underpass), mice, voles and shrews (13 underpasses), weasel *Mustela nivalis* and stoat *Mustela erminea* (11 underpasses) and pine *Martes martes* and stone marten *Martes foina* (one underpass). Ranges in the number of underpasses used reflect uncertainties in track identification. Fourteen underpasses (0.6 m wide, 0.3 m high and 19–32 m long), were installed beneath a 12-km stretch of railway in 1998–2003. Eleven underpasses were topped with grates (2–9 m long) between entrances and railway tracks. Mammal use was monitored between August and October 2003, using ink track-plates (0.6 × 2.4 m). Track-plates were checked on average at eight-day intervals.

(1) Hunt A., Dickens H.J. & Whelan R.J. (1987) Movement of mammals through tunnels under railway lines. *Australian Journal of Zoology*, 24, 89–93, https://doi.org/10.7882/AZ.1987.008

(2) Yanes M., Velasco J.M. & Suarez F. (1995) Permeability of roads and railways to vertebrates: the importance of culverts. *Biological Conservation*, 71, 217–222, https://doi.org/10.1016/0006-3207(94)00028-O

(3) Rodriguez A., Crema G. & Delibes M. (1996) Use of non-wildlife passages across a high speed railway by terrestrial vertebrates. *Journal of Applied Ecology*, 33, 1527–1540, https://doi.org/10.2307/2404791

(4) Little S.J., Harcourt R.G. & Clevenger A.P. (2002) Do wildlife passages act as prey-traps? *Biological Conservation*, 107, 135–145, https://doi.org/10.1016/s0006-3207(02)00059-9

(5) Krawchuk A., Larsen K.W., Weir R.D. & Davis H. (2005) Passage through a small drainage culvert by mule deer, *Odocoileus hemionus*, and other mammals. *The Canadian Field Naturalist*, 119, 296–298, https://doi.org/10.22621/cfn.v119i2.119

(6) van Vuurde M.R. & van der Grift E.A. (2005) The effects of landscape attributes on the use of small wildlife underpasses by weasel (*Mustela nivalis*) and stoat (*Mustela erminea*). *Lutra*, 48, 91–108.

5.3. Modify culverts to make them more accessible to mammals

https://www.conservationevidence.com/actions/2522

- **One study** evaluated the effects of modifying culverts to make them more accessible to mammals. This study was in the USA[1].

COMMUNITY RESPONSE (0 STUDIES)

POPULATION RESPONSE (0 STUDIES)

BEHAVIOUR (1 STUDY)

- **Use (1 study):** A replicated, site comparison study in the USA[1] found that modified culverts (with a dry walkway, open-air central section and enlarged entrances) were used more by bobcats to make crossings than were unmodified culverts.

Background

Culverts under roads may be used as crossing routes by mammals. This use reduces collision-associated risks to mammals and to motorists compared with crossings over the road surface. Some culverts may be less suited as crossing routes than others. For example, culverts with water flowing across their entire width may not be used by some mammals whilst tunnel length may also be a barrier to their use. A range of modifications can be made to try to increase culvert suitability for use by wild mammals.

A replicated, site comparison study in 1997–1999 in dry shrubland along a highway in Texas, USA (1) found that modified culverts were used more by bobcats *Lynx rufus* than were unmodified culverts. Use of crossings by cat spp. was higher at modified culverts (2.6 visits/month) than at unmodified culverts (0.5 visits/month). The rate of crossings at bridges (2.2 visits/month) was similar to that at modified culverts. Most cats recorded were bobcats, which accounted for 371 of 471 camera-trap images obtained at culvert entrances. Remaining images were of feral cats *Felis catus*. Five modified culverts, nine unmodified culverts and four bridges were monitored. Modified culverts had elevated central catwalks (to facilitate a dry crossing even when water was flowing through), open-air sections at the road centre (but fenced, to prevent escape at this part) and enlarged entrances. Crossings were checked two times/week from 1 July 1997 to 31 May 1999 for tracks. Remote cameras were used at seven crossings at a time, from 1 August 1997 to 31 May 1999, and were rotated among all crossings.

(1) Cain A.T., Tuovila V.R., Hewitt D.G. & Tewes M.E. (2003) Effects of a highway and mitigation projects on bobcats in Southern Texas. *Biological Conservation*, 114, 189–197, https://doi.org/10.1016/s0006-3207(03)00023-5

5.4. Install ledges in culverts under roads/railways

https://www.conservationevidence.com/actions/2523

- **Three studies** evaluated the effects on mammals of installing ledges in culverts under roads or railways. Two studies were in the USA[1,3] and one was in Portugal[2].

COMMUNITY RESPONSE (0 STUDIES)

POPULATION RESPONSE (0 STUDIES)

BEHAVIOUR (3 STUDIES)

- **Use (3 studies):** A replicated, controlled study in Portugal[2] found that under-road culverts with ledges were used more than culverts without ledges by two of five mammal species. A before-and-after study in the USA[3] found that installing ledges within under-road culverts did not increase the number or diversity of small mammal species crossing through them, and only one of six species used ledges. A study in the USA[1] found that ledges in under-road culverts were used by nine of 12 small mammal species and ledges with access ramps were used more often than those without.

Background

Culverts may be installed under roads to enable drainage. They are sometimes also used by mammals to cross under the road and, in some cases, roadside fencing will be designed to funnel mammals towards culvert entrances. However, some mammals are resistant to passing through tunnels that have water at their base (Serronha *et al.* 2013). Ledges may be installed on the sides of culverts, above the usual water level, to assist animal passage.

See also: *Install tunnels/culverts/underpasses under roads* and *Install tunnels/culverts/underpasses under railways*.

Serronha A.M., Mateus A.R.A., Eaton F., Santos-Reis M. & Grilo C. (2013) Towards effective culvert design: monitoring seasonal use and behavior by Mediterranean mesocarnivores. *Environmental Monitoring and Assessment*, 185, 6235–6246.

A study in 2005–2006 at six road sites in Colorado, USA (1) found that ledges in under-road culverts were used by nine of 12 small mammal species and ledges with access ramps were used more often than ledges without access ramps. Nine of 12 small mammal species that passed through the culverts used ledges (see original paper for details). Overall, a greater number of small mammal crossings were recorded along ledges with access ramps installed (total 443 crossings) than along those without (total 262 crossings). Temporary wooden ledges (15 cm wide) were installed in six concrete culverts (1–5 m wide, 1–1.3 m high, 9–48 m long) containing water. At each of the six culverts, access ramps were alternately attached or removed for 8–10 two-week periods in May–September 2005 and 2006. Motion-sensor cameras recorded small mammal movements through the culverts during a total of 16–20 weeks in May–September 2005 and 2006.

A replicated, controlled study in 2008–2009 of 32 culverts under roads in southern Portugal (2) found that under-road culverts with ledges were used more by two mammal species, less by two species and to a similar extent by one species compared to culverts without ledges. Culverts with ledges were used more by stone marten *Marte foina* and genet *Genetta genetta* (data reported as model results). However, red fox *Vulpes vulpes* and badger *Meles meles* used culverts with ledges less than they used those without ledges (data reported as model results). The use of culverts by European otter *Lutra lutra* was not altered by the presence of ledges (data reported as model results). In January–March 2008, wooden ledges, 50 cm wide, were installed in 15 culverts and no ledges were installed in 17 culverts. Two video cameras with movement and heat sensors were placed at one entrance of each culvert. Marble dust was spread covering the width of the culvert for monitoring footprints. Each culvert was monitored for seven consecutive nights, in each season, for a year after ledge installation.

A before-and-after study in 2012–2013 at seven road sites in New York, USA (3) found that installing ledges within under-road culverts did not increase the number or diversity of small mammal species crossing through them, and only one of six species used ledges. Overall, a similar number of small mammal crossings of six species were recorded in the seven culverts before (total 55 crossings) and after (total 58 crossings) ledges were installed, although no statistical tests were carried out. Racoons *Procyon lotor* were the only species recorded using ledges and did so during 58% of crossings, but similar numbers were recorded before (total 47 crossings) and after (total 41 crossings) ledge installation. In May–June 2013, plywood ledges (14 cm wide) and access ramps were installed through seven under-road culverts (1–3 m wide, 1–2 m high, 6–25 m long) containing water. Cat food was placed on ledges and ramps once after installation. A motion-sensor camera monitored each of the seven culverts for 12 weeks in June–September before (2012) and after (2013) ledges were installed.

(1) Meaney C.A., Bakeman M., Reed-Eckert M. & Wostl E. (2007) *Effectiveness of ledges in culverts for small mammal passage*. Report No. CDOT-2007-9. Colorado Department of Transportation Research Branch, USA.

(2) Villalva P., Reto D., Santos-Reis M., Revilla E., & Grilo C. (2013) Do dry ledges reduce the barrier effect of roads? *Ecological Engineering*, 57, 143–148, https://doi.org/10.1016/j.ecoleng.2013.04.005

(3) Kelley, A. (2014) *A test of simple ledges for facilitating mammal passage through inundated culverts*. Thesis. Union College, New York.

5.5. Dig trenches around culverts under roads/railways

https://www.conservationevidence.com/actions/2524

- **One study** evaluated the effects on mammals of digging trenches around culverts under roads and/or railways. This study was in South Africa[1].

COMMUNITY RESPONSE (0 STUDIES)

POPULATION RESPONSE (1 STUDY)

- **Survival (1 study):** A replicated, randomized, controlled, before-and-after study in South Africa[1] found that digging trenches alongside culverts did not reduce mammal mortality on roads.

BEHAVIOUR (0 STUDIES)

Background

Collisions with vehicles can be a large cause of mortality for mammal species (e.g. Forman & Alexander 1998). Underpasses installed beneath roads or drainage culverts may be made accessible to mammals with the intention of increasing connectivity of habitats and reducing the animal-vehicle collision risk associated with crossing the road. A range of means may be employed to help funnel animals towards such crossing points. These are usually fences or similar barriers to prevent animal crossings. However, trenches may be dug at some sites with the intention of inhibiting crossings, especially of small mammals.

See also: *Transportation and Service Corridors: Install barrier fencing along roads.*

Forman R.T.T & Alexander L.E. (1998) Roads and their major ecological effects. Annual Review of Ecology and Systematics, 29, 207–231.

A replicated, randomized, controlled, before-and-after study in 2015 along a road through dry savanna in Limpopo, South Africa (1) found that digging trenches alongside culverts did not reduce the number of mammals killed on roads. Results were not tested for statistical significance. One mammal (a South African pouched mouse *Saccostomus campestris*) was detected as a roadkill near culverts after trenches were dug and one (a red veld rat *Aethomys chrysophilus*) was found before they were dug. Over the same period, near culverts where no trenches were dug, two multimammate rats *Mastomys* spp. were detected as roadkills after trenches were dug at treatment sites and one was found before trenches were dug. The study was conducted in January–February 2015 along 400-m-long road sections with 2-m-wide culverts. In three sections, a 30-cm-deep trench, 2 m from the road verge, was dug for 200 m on either side of the culvert. Three road sections had no trench. Roadkills were counted at all sites over 20 days before the trench was dug and 20 days afterwards, by an observer in a car moving at 40–50 km/h.

(1) Collinson W.J., Davies-Mostert H.T. & Davies-Mostert W. (2017) Effects of culverts and roadside fencing on the rate of roadkill of small terrestrial vertebrates in northern Limpopo, South Africa. *Conservation Evidence*, 14, 39–43.

5.6. Install fences around existing culverts or underpasses under roads/railways

https://www.conservationevidence.com/actions/2525

- **Four studies** evaluated the effects on mammals of installing fences around existing culverts under roads/railways. Two studies were in the USA[1,2] one was in Portugal[3] and one was in South Africa[4].

COMMUNITY RESPONSE (0 STUDIES)

POPULATION RESPONSE (3 STUDIES)

- **Survival (3 studies):** Two out of three before-and-after studies (including a controlled and a site comparison study), in the USA[1], Portugal[3] and South Africa[4], found that installing or enhancing roadside fencing alongside existing culverts reduced mammal road mortality whilst one study found that such fences did not alter mammal road mortality.

BEHAVIOUR (1 STUDY)

- **Use (1 study):** A replicated, randomized, controlled, before-and-after study in the USA[2] found that fences installed to funnel animals to existing culverts did not increase culvert use by bobcats.

Background

Culverts are often installed under roads to aid or enable drainage whilst underpasses enable movement of traffic or apparatus such as farming machinery. Such passages are sometimes used by animals to make road crossings but many animals may nonetheless cross over the road surface and are then at risk of collision with vehicles. This intervention includes studies where fences are installed or extended specifically in a way designed to encourage animals to use existing passages rather than crossing over the road surface. It includes only studies that specifically assess the effectiveness of fencing in a way that can be separated from that of underpasses. For situations where roadside fencing is installed specifically to prevent animal access to roads, in some cases along with underpasses as part of an integrated road casualty reduction scheme, see Install barrier fencing along roads. See also *Install barrier fencing and underpasses along roads* for studies that assess the combined effectiveness of installing fending and underpasses.

See also: *Install tunnels/culverts/underpass under railways* and *Install tunnels/culverts/underpass under roads.*

A before-and-after study in 1976–1981 along a highway through shrubland in Wyoming, USA (1) found that after a fence alongside the highway that was connected to underpasses was made taller, fewer mule deer *Odocoileus hemionus* were killed. Results were not tested for statistical significance. In six migration seasons (three springs, three autumn–winters) after increasing the height of the fence, only one deer-vehicle accident occurred in the fenced area. In three migration seasons before fence construction (two spring and one autumn–winter), 53 deer–vehicle accidents occurred within the area to be fenced. The study was conducted along a stretch of highway constructed in late 1970. In 1977–1978, the height of a fence along the highway was increased from 4 ft to 8 ft along both sides of 7.8 miles of road. The fence allowed deer to access seven underpasses (length: 110–393 feet; width: 10–50 feet; height: 10–17 feet). Deer movement was monitored before (1976–1977) and after (1978–1981) fence heightening by direct observation, track counts, radio-tracking and automatic cameras. The highway was located across a migration route of 1,600–2,000 mule deer.

A replicated, randomized, controlled, before-and-after study in 1997–1999 in dry shrubland along a highway in Texas, USA (2) found that installing fences to funnel animals to existing culvert entrances did not increase culvert use by bobcats *Lynx rufus*. Fences did not significantly increase cat spp. use of culverts (data not presented). However, among four culverts most used by bobcats, two fenced culverts saw a rise in use after fence installation (after 7.2; before: 3.9 track sets/month) while two unfenced culverts saw a fall over this same time (after: 2.2; before: 2.9 track sets/month). Most cats (371 of 471 camera-trap images) were bobcats. The remainder were feral cats *Felis catus*. At six culverts, randomly selected from 12, wire net fences (1.6 m high) were erected at entrances, extending 100 m to each side, parallel to the road. Culverts were checked two times/week from 1 July 1997 to 31 May 1999 for cat spp. tracks. Remote cameras were used at culverts from 1 August 1997 to 31 May 1999. Fences were erected after the first year of monitoring.

A replicated, before-and-after, site comparison study in 2008–2009 of 64 culverts under roads in southern Portugal (3) found that fences connecting to existing under-road culverts did not alter mammal road mortality. After fence installation, there was a similar number of mammals killed by traffic (19 road-kills) compared to before (20 road-kills). There was also no significant difference in mammal road-kills between road sections where fences were installed (19 road-kills) and those that were not fenced (13 road-kills). In April 2008, 100-m-long fences with 2.5-cm mesh, buried to 50 cm deep and extending 50 cm above ground, were installed alongside the road at each side of 32 under-road culverts. These were in addition to existing livestock fencing. Another 32 culverts in the same area that were unfenced were selected for comparison. The number of mammals killed by traffic was recorded by highway maintenance staff for 10 months before and 10 months after fence installation.

A randomized, replicated, controlled, before-and-after study in 2015 along a road through dry savanna in Limpopo, South Africa (4) found that installing fences around existing culverts reduced mammal road casualties. Results were not tested for statistical significance. One scrub hare *Lepus saxatilis* was detected as a roadkill near fenced culverts compared to two bushveld gerbils *Tatera leucogaster* detected as roadkills before fencing was installed. Concurrently, two multimammate rats *Mastomys* sp. were detected as roadkills near unfenced culverts after fence installation at treatment sites compared to one before fence installation. The study was conducted along six 400-m-long road segments with culverts. In three segments, a 70-cm-high fence was erected extended 200 m along both sides of the road on either side of the culvert. The fence was approximately 2 m from the road verge, sloped at 45° away from the road and extended 30 cm below ground. Three segments remained unfenced. Roadkills were counted in all sites during a 20-day period before fences were installed (January 2015) and a 20-day period after (February 2015). Roadkills were counted by an observer in a car moving at 40–50 km/h.

(1) Ward A.L. (1982) Mule deer behavior in relation to fencing and underpasses on Interstate 80 in Wyoming. *Transportation Research Record*, 859, 8–13.

(2) Cain A.T., Tuovila V.R., Hewitt D.G. & Tewes M.E. (2003) Effects of a highway and mitigation projects on bobcats in Southern Texas. *Biological Conservation*, 114, 189–197, https://doi.org/10.1016/S0006-3207(03)00023-5

(3) Villalva P., Reto D., Santos-Reis M., Revilla E. & Grilo C. (2013) Do dry ledges reduce the barrier effect of roads? *Ecological Engineering*, 57, 143–148, https://doi.org/10.1016/j.ecoleng.2013.04.005

(4) Collinson W.J., Davies-Mostert H.T. & Davies-Mostert W. (2017) Effects of culverts and roadside fencing on the rate of roadkill of small terrestrial vertebrates in northern Limpopo, South Africa. *Conservation Evidence*, 14, 39–43.

5.7. Install overpasses over roads/railways

https://www.conservationevidence.com/actions/2526

- **Twenty-two studies** evaluated the effects on mammals of installing overpasses over roads or railways. Seven studies were in Canada[1,4,6,7,18,20,22], three were in Spain[2,8,11], three were in Australia[10,14,19], two were in Sweden[12,13], one each was in the Netherlands[5], Germany[15], Croatia[16] and the USA[21], and three (including two reviews) were conducted across multiple countries[3,9,17].

COMMUNITY RESPONSE (0 STUDIES)

POPULATION RESPONSE (4 STUDIES)

- **Survival (4 studies):** Four studies (including three before-and-after studies), in Canada[4], Sweden[12,13] and Australia[14], found that overpasses (in combination with roadside fencing) reduced collisions between vehicles and mammals. In two of these studies, data from overpasses and underpasses were combined for analysis[4,14].

BEHAVIOUR (21 STUDIES)

- **Use (21 studies):** Nineteen studies, in North America[1,6,7,18,20,21,22], Europe[2,3,5,8,11,12,13,15,16] and Australia[10,14,19], found that overpasses were used by mammals. A wide range of mammals was reported using overpasses, including rodents and shrews[1,5,6,8,11,20], rabbits and hares[2,8,11,16], carnivores[2,5,7,8,11,15,15], ungulates[3,5,7,8,10,11,12,13,16,21], bears[7,16,18,22], marsupials[10,14,19] and short-beaked echidna[10]. A review of crossing structures in Australia, Europe and North America[17] found that overpasses were used by a range of mammals, particularly larger mammal species. A global review of crossing structures (including overpasses)[9] found that all studies reported that the majority of crossings were used by wildlife.

Background

Wildlife overpasses are constructed to provide safe road and rail crossing opportunities for wildlife. A range of different structures can be used as overpasses including purpose-built 'green bridges', on which natural vegetation is established, through to multi-use crossings that are accessible to wildlife. Overpasses are often used in combination with wildlife barrier fences that prevent animals accessing the road and which funnel animals toward the overpasses (see *Install barrier fencing along roads* and *Install barrier fencing along railways*). Studies summarised within this intervention cover both overpasses created specifically for wildlife and those that were created for other purposes but where information about use of such structures by mammals is included. Studies mostly report on the use of such structures, such as the number of crossings made, rather than on wider population-level effects of their presence.

See also: *Install tunnels/culverts/underpass under railways* and *Install tunnels/culverts/underpass under roads*.

A replicated study in 1971–1973 of 21 highway overpasses constructed for wildlife use in Québec and Ontario, Canada (1) found that they were extensively used by woodchucks *Marmota monax*. Woodchucks or their burrows were recorded on 18 of 21 overpasses surveyed. Across four surveys on overpasses, minimum total woodchuck numbers were 16–22. On average, underpasses had 45 woodchucks/100 acres, a high figure compared to those reported by other authors in open flat ground. Twenty-one highway overpasses were built up with rubble and sand and covered with topsoil. Four overpasses had an average area of 72,000 square feet. Overpasses were surveyed once in 1971, twice in 1972 and once in 1973. Surveys were conducted in May, when grass (mainly *Agropyron repens*) was short. Animals and burrows on overpasses were counted from a vehicle (first two surveys) and on foot (last two surveys).

A study in 1991–1992 along a high-speed railway within agricultural land in Castilla La Mancha, Spain (2) found that two flyovers not designed for wildlife were used to cross the railway by small mammals, but not by deer or wild boar *Sus scrofa*. Small mammals were recorded, with data combined between two overpasses and 15 underpasses, 582 times (37/100 passage-days) and brown hare *Lepus granatensis* and European rabbit *Oryctolagus cuniculus*, 89 times (5/100 passage-days). Tracks of four carnivore species, red fox *Vulpes vulpes*, wild cat *Felis silvestris*, common genet *Genetta genetta* and Iberian lynx *Lynx pardinus*, were recorded. No deer or wild boar *Sus scrofa* were recorded using overpasses or underpasses. Two flyovers (small roads) crossing a 25-km section of a high-speed railway were monitored. Sand, 3 cm thick and 1 m wide, was put at one entrance to each. Animal tracks were monitored for 15–22 days/month between September 1991 and July 1992.

A replicated study in 1996 of roads in Germany, Switzerland, France and the Netherlands (3) found that mammals used flyovers as bridges/overpasses across roads, and frequency of their use tended to increase with overpass width. For all mammal species, frequency of use of the seven narrow overpasses (<15 m wide) was very low. Roe deer *Capreolus capreolus* used the nine medium-sized (15–50 m wide) and five wide overpasses (>50 m wide) significantly more frequently than they used narrow overpasses. Twenty-one wildlife flyovers/overpasses, in Germany (eight), Switzerland (six), France (four) and the Netherlands (three), were monitored using infra-red video equipment. Flyover widths were 3.4–186 m. Video surveys were carried out during a total of 223 nights.

A replicated, before-and-after study in 1981–1999 in temperate mixed woodland and grassland in Alberta, Canada (4) found that wildlife overpasses, underpasses and roadside barrier fencing reduced road deaths of large mammals. Species recorded as road casualties included coyote *Canis latrans*, black bear *Ursus americanus*, wolf *Canis lupus*, bighorn sheep *Ovis canadensis*, moose *Alces alces*, deer *Odocoileus* spp. and elk *Cervus canadensis*. Mammal-vehicle collisions were significantly lower during the two years after fencing (5–28/year) compared to the two years before (18–93/year) for all three road sections, despite an increase in traffic flow. Ungulate casualties declined by 80%. Most road deaths were within 1 km of the end of the fences. Deaths also occurred close to drainage structures. The Trans-Canada highway was expanded to four lanes and had 2.4-m-high wildlife exclusion fence installed in three phased sections, completed in 1984 (10 km), 1987 (16 km) and 1997 (18 km). In addition, 22 wildlife underpasses and two overpasses were constructed. Wildlife-vehicle collisions were monitored from May 1981 to December 1999.

A study in 1989 and 1994–1995 along a motorway between Arnhem and Apeldoorn in the Netherlands (5) found that a wildlife overpass was used by deer, wild boar *Sus scrofa*, rodents and carnivores. The overpass was used most frequently by red deer *Cervus elaphus* (1989: 0.1–9 crossings/night; 1994–1995: 4–21) and wild boar (1989: 0.5–21; 1994–1995: 0.5–8.5). It was used less often by roe deer *Capreolus capreolus* (1989: 2.0 crossings/night; 1995–1994: 0.5) and fallow deer *Dama dama* (data not presented). Twenty-five rodents and shrews, of three species, wood mouse *Apodemus sylvaticus*, common vole *Microtus arvalis* and common shrew *Sorex araneaus*, were caught on the overpass. Overpasses were also used by badger *Meles meles* and red fox *Vulpes vulpes*. Overall numbers of crossings was greater in 1994–1995 than 1989 (16 vs 12 crossings/night). The overpass was constructed in the late 1980s. It was 50 m wide, 95 m long and planted with trees. Large mammal tracks were recorded on a 5-m-wide sand strip across the overpass, on 93 occasions in 1989 and 114 occasions in May 1994–April 1995. Small mammals were caught during five nights in summer 1995 using 20 live traps at each end and 32 mouse-traps between.

A replicated study in 1999–2000 in Alberta, Canada (6) found that deer mice *Peromyscus maniculatus*, but not red-backed voles *Clethrionomys gapperi* or meadow voles *Microtus pennsylvanicus*, crossed wildlife overpasses. Forty percent of deer mice translocated across roads crossed back over when released alongside overpasses, but no voles did. More animals successfully returned through overpasses (and underpasses) with 100% vegetation cover at entrances (55–100% of animals) compared to those with 50% cover (20–76% of animals) or no cover (0–66% of animals). Those animals that crossed did so in 1–4 days. Two sparsely vegetated wildlife overpasses (75–79 m long, 15 m wide) were used. Territorial mice and voles were caught using Longworth live traps (166 caught in total), ear-tagged, coated with fluorescent powder, translocated across the road, released 2 m from overpasses (or underpasses) and followed as they returned. The amount of ground cover 2 m inside and outside entrances was manipulated to 100%, 50% and no cover, using spruce branches. Traps at original capture sites were monitored for four days after translocation. Animals that did not return were returned by hand. Monitoring was undertaken in July–October 1999 and 2000.

A study in 1997–2000 in Alberta, Canada (7) found that large herbivores and carnivores used two wildlife overpasses. A total of 640 visits to overpasses by elk *Cervus canadensis*, 1,086 by deer *Odocoileus* spp., 10 by black bear *Ursus americanus*, nine by grizzly bear *Ursus arctos*, eight by wolf *Canis lupus* and 12 by cougar *Puma concolor* were recorded, with the majority involving animals crossing the structures. Features that positively influenced use of crossings (two overpasses and 11 underpasses) included increased width, height and openness. Black bears and cougars, though, favoured more constricted crossing structures. Increased length and noise negatively influenced use of crossing structures for some species. Two 50-m-wide overpasses were monitored along an 18-km-stretch of the four-lane Trans-Canada Highway. Barrier fencing, 2.4-m-high, ran alongside the highway. Tracks were monitored at each end of each overpass (in 2 × 4 m of sand/clay), every 3–4 days, from November 1997 to August 2000. Infra-red activated cameras were also used. Information about structure, landscape and human activity were recorded for each overpass.

A study in 2002 in along a road in Zamora, Spain (8) found that wildlife overpasses were used by mammals. Overpasses were used by red deer *Cervus elaphus* (detected at wildlife overpasses on average of 2/10 days), small mammals (shrews, mice and voles; detected 1.0/10 days) and rabbits and hares (detected 4.5/10 days). Other overpasses, such as rural tracks, were used by small mammals (detected 6.4/10 days), rabbits and hares (3.3/10) and foxes *Vulpes vulpes* (1.4/10), but not by red deer. Two wildlife overpasses (16 m wide, 60 m long) and 16 general overpasses (rural tracks, 7–8 m wide, 58–62 m long) were monitored along a 72-km section of the A-52 motorway. The motorway had barrier fencing along its length. Marble dust (1 m wide cross) was used to record animal tracks for 10 days in June–September 2002. Camera traps were installed on some overpasses.

A global review in 2007 of 123 studies investigating the use of wildlife crossings (9) found that all studies reported that the majority of underpasses and overpasses were used by wildlife. A total of 1,864 structures were reported on, mainly underpasses (83%; including culverts (742 examples), bridges (130), tunnels (340) and unknown types (333)). Overpasses included land bridges (68), overpasses with small roads (112), canopy bridges (8), glider poles (1) and others (35). Structures provided crossings over or under roads (113 studies), railways (5 studies), both (1 study), canals (2 studies) and a pipeline (1 study). Studies were from Europe (55 studies), the USA (30 studies), Canada (nine studies), South America (one study) and Australia (29 studies).

A study in 2004–2007 in eucalypt woodland in Queensland, Australia (10) found that a wildlife bridge was used by mammals. A total of 1,240 herbivore scats were recorded on the bridge. Brown hare *Lepus capensis* scats were the most common (78%), followed by red-necked wallaby *Macropus rufogriseus* (15%), eastern grey kangaroo *Macropus giganteus* (5%), swamp wallaby *Wallabia bicolor* (1%), possum (1%) and short-beaked echidna *Tachyglossus aculeatus* (1%). Six mammals were killed on the road before construction and one afterwards. In 2004, a 1.3-km section of highway was upgraded to four lanes and a variety of wildlife crossings constructed, with barrier fencing (2.5 m high) between. Use of a large overpass (15–20 m wide, 70 m long, planted with grass, shrubs and trees) was monitored from six months after completion. Scats were recorded weekly from August 2005–February 2006 and for two weeks in June 2007. Road-kill was monitored twice weekly before construction (April–July 2004) and weekly afterwards, until June 2007.

A replicated study in 2001 in Zamora province, Spain (11) found that overpasses were used by mammals. Wildlife overpasses were used by red fox *Vulpes vulpes* (detected on average per overpass on 3.5/10 days), wild boar *Sus scrofa* (2.3/10 days), small mammals (shrews, mice and voles; 0.3/10 days) and rabbits and hares (3.0/10 days). Other overpasses, such as rural tracks, were also used by wild boar (detected on average per crossing on 0.7/10 days), small mammals (1.0/10 days), rabbits and hares (1.8/10 days), red deer *Cervus elaphus* (0.2/10 days), rats *Rattus* sp. (1.3/10 days), western hedgehogs *Erinaceus europaeus* (0.2/10 days), European badger *Meles meles* (0.2/10 days) and red fox (3.0/10 days). Cat and dog prints were also detected but could not be determined as being from either wild or domestic species. Overall, overpasses (not including wildlife overpasses) were used disproportionately more than were other crossings (which included underpasses and culverts — data presented as indices). Four wildlife overpasses (15–20 m wide, 60–62 m long) and six general overpasses (rural tracks, 7–8 m wide, 58–65 m long) were monitored along the A-52 motorway. The motorway had barrier fencing along its length. Marble dust (1-m-wide cross) was used to record animal tracks daily for 10 days in March–June 2001.

A before-and-after study in 2002–2004 in mixed forest and farmland in southwestern Sweden (12, same experimental study site as 13) found that following installation of two wildlife overpasses and barrier fencing, moose *Alces alces* used overpasses and collisions with vehicles decreased, but fencing created a barrier to movements. There were fewer moose-vehicle collisions after overpass and fence construction (zero/year) than before (2.7/year). During construction, 1.8 collisions/year were recorded. Moose were recorded crossing the highway 12 times after overpass and fence installation (during 18 months) and 47 times before installation (eight months). All crossings after construction were via the two wildlife overpasses. Home-range locations changed significantly, with ranges intersected by the highway decreasing to five out of 38 monitored ranges (13%) after fencing from 10 out of 38 (26%) before. Two 6-km sections of the European highway 6 were converted to a fenced four-lane highway in 2000–2004. A third section remained unfenced (3 km). The sections contained two wildlife overpasses, one wildlife underpass, three conventional road tunnels and two conventional bridges that could be crossed. Twenty-four moose were radio-collared. Locations were recorded every two hours before construction (February–September 2002), during construction (October 2002–May 2004) and after construction (June 2004–December 2005).

A before-and-after study in 2000–2005 in forest and farmland in southwestern Sweden (13, same experimental study site as 12) found that a wildlife overpass was used by moose *Alces alces* and roe deer *Capreolus capreolus* and, along with barrier fencing, it reduced road-kills. Deaths were reduced 70% from the 12-year pre-construction averages of 2.7 moose killed/year and 5.3 roe deer killed/year. From March 2002–June 2005, the overpass was crossed 437 times by roe deer and 95 times by moose (mainly at night). Roe deer, but not moose crossings, increased over the six-year study. Five to seven individual moose/year used the overpass. Overpass use declined with increased traffic flow. In 2000–2004, a 12-km section of the European Highway 6 was converted from two to four lanes and 2.2-m-high exclusion fencing was installed.

Two overpasses and one underpass were constructed. One hourglass-shaped overpass (29–17 m wide, 80 m long, 2 m high, with grey glass-shields to reduce incursion of highway noise and light) was monitored. Tracks were counted in sand beds twice/week and two infrared remote cameras were set overnight. Twenty-four moose were tracked using GPS collars for 22 months.

A site comparison study in 2006 along a highway in New South Wales, Australia (14) found that two wildlife overpasses were used by mammals and presence of crossing-structures along with roadside fencing reduced road-kills. There were fewer road-kills over seven weeks along the section with crossing-structures (0.02/km) than along a section without crossings (0.07/km). The most frequently recorded road casualties along both sections combined were bandicoots (16 casualties) and kangaroo and wallabies (nine casualties). Kangaroos and wallabies used the two overpasses more than they used two underpasses (104 vs 36 tracks). However, the overpasses were used less than were underpasses by bandicoots (28 vs 87) and rodents (15 vs 82). Use was similar for possums (overpasses: 9; underpasses: 14). There were two wildlife bridges (9–37 m wide, with vegetation) and two concrete box culverts (3 × 3 m, 42–63 m long), with 5 km of exclusion fencing, along a 12-km section of dual-carriageway highway. Tracks were monitored on sand plots across each crossing. Road-kill surveys were conducted along the 12-km section and along a 51-km two-lane section without crossings or fencing. Track and road-kill surveys were conducted up to three times/week over seven weeks in August–September 2006.

A study in 2001–2005 along a motorway through forest and agricultural land in Germany (15) found that most overpasses, viaducts and underpasses were used by wildcats *Felis silvestris* to cross roads. Wildcats used crossing structures on 18 of 21 (85%) of the occasions in which they were recorded <50 m from the motorway. Open-span viaducts were used by the highest proportion of cats (five out of seven for which viaducts fell within their home ranges). Forest road overpass were used by one out of eight cats for which road overpasses fell within their home ranges. Two open-span viaducts (335–660 m wide, 29 m long), two forest road overpasses (6 m wide, 46–61 m long) and three underpasses were monitored in 2002–2005. Twelve wildcats were radio-collared between January 2001 and February 2005. Animals were tracked at night for 3–30 months each, to monitor their road crossings.

A study in 1999–2003 along a road through beech and fir forest in Gorski kotar, Croatia (16) found that medium-large mammals used a wildlife overpass (a green bridge) and two other overpasses not specifically designed for wildlife. Monitoring of the green bridge revealed tracks of hare *Lepus europaeus* (49 tracks), wild boar *Sus scrofa* (66), roe deer *Capreolus capreolus* (166), red deer *Cervus elaphus* (103), fox *Vulpes vulpes* (83), badger *Meles meles* (2), brown bear *Ursus arctos* (39), grey wolf *Canis lupus* (4) and Eurasian lynx *Lynx lynx* (1). A similar range of species was recorded on the two other overpasses that were not designed as green bridges (see paper for data). A new highway was constructed in 1998–2004, with 2.1-m barrier fencing. Along a 9-km section, a 100-m-wide green bridge and two overpasses (742 and 835 m wide) above road tunnels, were monitored. Tracks (in snow, mud or sand) and other animal signs were counted 64 times at the green bridge and eight and 23 times at the two other overpasses, in January 1999–January 2001. One of the overpasses was also monitored using a camera trap.

A review of 30 papers monitoring 329 crossing structures in Australia, Europe and North America (17) found that overpasses were used by a range of mammals, particularly larger mammal species. Small mammals used conventional bridge overpasses (demonstrated by 2/4 relevant studies) and wildlife overpasses (4/7 studies). Arboreal mammals used wildlife overpasses (1/1 study). Medium-sized mammals used conventional bridge overpasses (4/5 studies) and wildlife overpasses (5/7 studies). Large mammals used conventional bridge overpasses (9/11 studies) and wildlife overpasses (23/23 studies). Studies suggested that ungulates used overpasses more when they were close to vegetation cover and a river or stream and less when they were in a cropland area. Narrow overpasses (<6 m wide) were not used by deer. Thirty papers, monitoring 329 crossing structures, were reviewed. Fourteen papers investigated multiple structure types, resulting in a total of 52 studies of different structure types. Overpasses included land bridges, wildlife overpasses with grass, trees or other vegetation, combined wildlife and vehicle overpasses, pole bridges and rope bridges.

A replicated study in 2006–2008 of two overpasses over a highway in a Natural Park in Alberta, Canada (18) found that American black bears *Ursus americanus* and grizzly bears *Ursus arctos* used the overpasses.

Over three years, a total of eight passages of American black bears (by one individual at each overpass) and 210 of grizzly bears (by 10 individuals at each overpass) were detected. Bear crossings were monitored at two overpasses (dimensions not stated) in Bow Valley, Banff National Park. Overpasses were built in the 1980s and 1990s, and cost >US$2 million each to construct. Bear tracks were counted in May–October 2006, April–October 2007 and April–October 2008 using track pads comprising 1.5–2 m of sandy loam. Track pads were checked every two days and the species, direction of travel, and number of animals was recorded. Individuals were identified by DNA analysis of hairs caught on barbed wires on overpasses.

A review of two studies in 2006–2008 in Australia (19) found that overpasses installed over roads were used by eastern grey kangaroos *Macropus giganteus*, red-necked wallabies *Macropus rufogriseus* and swamp wallabies *Wallabia bicolor*. All road overpasses used fencing to reduce likelihood of animals crossing roads rather than using overpasses. Overpasses in the review were 70 m long and 15 m wide.

A replicated study in 2009 at two sites along a highway through forest in Alberta, Canada (20) found that North American deer mice *Peromyscus maniculatus*, southern red-backed voles *Myodes gapperi* and meadow voles *Microtus pennsylvanicus* used overpasses to cross a road. Deer mouse tracks were recorded in 75% of track tubes established on overpasses. Southern red-backed vole tracks were detected in 15% and meadow vole in 5% of track tubes. Over two weeks in September–October 2010, small mammals were surveyed on two 50-m-wide wildlife overpasses above the Trans-Canada Highway. Overpasses consisted of sparse young trees, shrubs and open grassland. Two parallel sample lines, each with five 30 cm long × 10 cm diameter track tubes, with sooted metal sheet as a floor, were placed in the centre of each overpass. Mammals were identified from their footprints.

A replicated study in 2010–2014 of five crossing structures at two sites along a highway in Nevada, USA (21) found that more migratory mule deer *Odocoileus hemionus* used overpasses than underpasses to cross a road. More mule deer crossed the road across two overpasses (234–4,007 deer crossings/overpass/season) than through three underpasses (44–629 deer crossings/underpass/season). Crossing structures, 1.5–2.0 km apart, were located at important crossings for migratory deer. One site had one overpass and two underpasses. The other had one of each structure. Overpasses, made of concrete arches, were 31–49 m wide and 8–20 m long. Cylindrical underpasses were 8 m wide, 28 m long and 6 m tall. All structures had soil bases. Fencing, 2.4 m high, deterred deer from accessing the highway between crossings and extended 0.8–1.6 km beyond crossings at each site. Crossings were monitored, during six to eight mule deer migratory periods (between autumn 2010 and spring 2014) using camera traps, over 10 weeks in each migration (15 September to 1 December and 1 March to 15 May). Cameras were positioned 12 m apart along crossing structures.

A study in 1996–2014 of 18 overpasses and 19 culverts crossing a major highway in Alberta, Canada (22) found that overpasses were used by grizzly bears *Ursus arctos*, particularly in family groups. Over an 18-year period, grizzly bears used overpasses more often (241 crossings/structure) than they used culverts (122 crossings/structure). Over an eight-year period, bear family groups used overpasses more often (1.4 family groups/year/structure) than they used culverts (0.0–0.3 family groups/year/structure). In 1996–2006, 2-m-wide pads, were covered in sandy-loam soil to survey bear movements at 23 crossing structures. From 2008, remote cameras were installed at all crossing structures. As more crossing structures were built in the area, they were added to the survey, up to a maximum of 18 overpasses and 19 culverts. It is not clear when these structures were built.

(1) Doucet G.J., Sarrazin J. & Bider J-P.R. & Bider R. (1974) Use of highway overpass embankments by the woodchuck, *Marmota monax*. *The Canadian Field-Naturalist*, 88, 187–190.

(2) Rodriguez A., Crema G. & Delibes M. (1996) Use of non-wildlife passages across a high speed railway by terrestrial vertebrates. *Journal of Applied Ecology*, 33, 1527–1540, https://doi.org/10.2307/2404791

(3) Keller V. (1999) *The use of wildlife overpasses by mammals: results from infrared video surveys in Switzerland, Germany, France, and the Netherlands.* Proceedings of the 5th Infra Eco Network Europe Conference, Budapest, Hungary, 27–28.

(4) Clevenger A.P., Chruszcz B. & Gunson K.E. (2001) Highway mitigation fencing reduces wildlife-vehicle collisions. *Wildlife Society Bulletin*, 29, 646–653, https://doi.org/10.2307/3784191

(5) van Wieren S.E. & Worm P.B. (2001) The use of a motorway wildlife overpass by large mammals. *Netherlands Journal of Zoology*, 51, 97–105, https://doi.org/10.1163/156854201X00071

(6) McDonald W. & St Clair C.C. (2004) Elements that promote highway crossing structure use by small mammals in Banff National Park. *Journal of Applied Ecology*, 41, 82–93, https://doi.org/10.1111/j.1365-2664.2004.00877.x

(7) Clevenger A.P. & Waltho N. (2005) Performance indices to identify attributes of highway crossing structures facilitating movement of large mammals. *Biodiversity and Conservation*, 121, 453–464, https://doi.org/10.1016/j.biocon.2004.04.025

(8) Mata C., Hervàs I., Herranz J., Suàrez F. & Malo J.E. (2005) Complementary use by vertebrates of crossing structures along a fenced Spanish motorway. *Biological Conservation*, 124, 397–405, https://doi.org/10.1016/j.biocon.2005.01.044

(9) van der Ree R., van der Grift E., Mata C. & Suarez F. (2007) *Overcoming the barrier effect of roads — how effective are mitigation strategies? An international review of the use and effectiveness of underpasses and overpasses designed to increase the permeability of roads for wildlife.* Proceedings of the International Conference on Ecology and Transportation, Center for Transportation and the Environment, North Carolina State University, Raleigh NC, USA, 423–431.

(10) Bond A.R. & Jones N.J. (2008) Temporal trends in use of fauna-friendly underpasses and overpasses. *Wildlife Research*, 35, 103–112, https://doi.org/10.1071/wr07027

(11) Mata C., Hervàs I., Herranz J., Suàrez F. & Malo J.E. (2008) Are motorway passages worth building? Vertebrate use of road-crossing structures on a Spanish motorway. *Journal of Environmental Management*, 88, 407–415, https://doi.org/10.1016/j.jenvman.2007.03.014

(12) Olsson M.P.O. & Widen P. (2008) Effects of highway fencing and wildlife crossings on moose *Alces alces* movements and space use in southwestern Sweden. *Wildlife Biology*, 14, 111–117, https://doi.org/10.2981/0909-6396(2008)14[111:eohfaw]2.0.co;2

(13) Olsson M.P.O., Widen P. & Larkin J.L. (2008) Effectiveness of a highway overpass to promote landscape connectivity and movement of moose and roe deer in Sweden. *Landscape and Urban Planning*, 85, 133–139, https://doi.org/10.1016/j.landurbplan.2007.10.006

(14) Hayes I. & Goldingay R.L. (2009) Use of fauna road-crossing structures in north-eastern New South Wales. *Australian Mammalogy*, 31, 89–95, https://doi.org/10.1071/am09007

(15) Klar N., Herrmann M. & Kramer-Schadt S. (2009) Effects and mitigation of road impacts on individual movement behavior of wildcats. *The Journal of Wildlife Management*, 73, 631–638, https://doi.org/10.2193/2007-574

(16) Kusak J., Huber D., Gomerčić T., Schwaderer G. & Gužvica G. (2009) The permeability of highway in Gorski Kotar (Croatia) for large mammals.

European Journal of Wildlife Research, 55, 7–21, https://doi.org/10.1007/s10344-008-0208-5

(17) Taylor B.D. & Goldingay R.L. (2010) Roads and wildlife: impacts, mitigation and implications for wildlife management in Australia. *Wildlife Research*, 37, 320–331, https://doi.org/10.1071/wr09171

(18) Sawaya M.A., Clevenger A.P. & Kalinowski S.T. (2013) Demographic connectivity for ursid populations at wildlife crossing structures in Banff National Park. *Conservation Biology*, 27, 721–730, https://doi.org/10.1111/cobi.12075

(19) Bond A.R. & Jones D.N. (2014) Roads and macropods: interactions and implications. *Australian Mammalogy*, 36, 1–14, https://doi.org/10.1071/am13005

(20) D'Amico M., Clevenger A.P., Román J. & Revilla E. (2015) General versus specific surveys: Estimating the suitability of different road-crossing structures for small mammals. *The Journal of Wildlife Management*, 79, 854–860, https://doi.org/10.1002/jwmg.900

(21) Simpson N.O., Stewart K.M., Schroeder C., Cox M., Huebner K. & Wasley, T. (2016) Overpasses and underpasses: Effectiveness of crossing structures for migratory ungulates. *The Journal of Wildlife Management*, 80, 1370–1378, https://doi.org/10.1002/jwmg.21132

(22) Ford A.T., Barrueto M. & Clevenger A.P. (2017) Road mitigation is a demographic filter for grizzly bears. *Wildlife Society Bulletin*, 41, 712–719, https://doi.org/10.1002/wsb.828

5.8. Install pole crossings for gliders/flying squirrels

https://www.conservationevidence.com/actions/2546

- **Seven studies** evaluated the effects on gliders/flying squirrels of installing pole crossings. Six studies were in Australia[1,2,4–7] and one was in the USA[3].

COMMUNITY RESPONSE (0 STUDIES)

POPULATION RESPONSE (1 STUDY)

- **Survival (1 study):** A study in Australia[7] found that arboreal marsupials using artificial road crossing structures did not suffer high predation rates when doing so.

BEHAVIOUR (6 STUDIES)

- **Use (6 studies):** Six studies (five replicated), in Australia[1,2,4,5,6] and the USA[3], found that poles were used for crossing roads by squirrel gliders[1,2,4,5,6], sugar gliders[6] and Carolina northern flying squirrels[3].

Background

Wildlife crossings over or under roads may be installed to reduce the impact of the road on animal mortality and on habitat fragmentation. They usually take the form of tunnels or bridges of a range of designs. These may not be suitable for use by mammals that move by gliding from tree to tree. Glide poles have been trialled, especially in Australia (e.g. Ball & Goldingay 2008), to provide a means of reconnecting habitat and reducing road mortality for gliding mammal species. Monitoring typically takes the form of documenting use of poles rather than looking at population level effects or impacts on road mortality.

See also: *Install rope bridges between canopies.*

Ball T.M. & Goldingay R.L. (2008) Can wooden poles be used to reconnect habitat for a gliding mammal? *Landscape and Urban Planning*, 87, 140–146, https://doi.org/10.1016/j.landurbplan.2008.05.007

A replicated study in 2006–2010 of a pasture and two highways through a woodland in Queensland, Australia (1) found that lines of poles were used by squirrel gliders *Petaurus norfolcensis* to cross the gaps between trees. At the pasture site, squirrel gliders were detected on all five surveys of poles. At the highway crossing sites, gliders were detected on 25 out of 30 and 11 out of 16 surveys of poles. Summing records for each pole in each monitoring session, gliders were recorded on 13/20 poles at the pasture site and 130/240 and 32/114 poles at highway sites. Canopy gaps of 50–70 m were spanned by 5–8 poles, 5–12 m high and 5–22 m apart. One pole line was across a pasture and two were over existing wildlife bridges across highways. Poles had crossbars attached close to the top. Squirrel glider usage of poles was assessed using hair tube surveys between October 2006 and April 2010.

A replicated, site comparison study in 2006–2010 at four sites along two roads through forests in New South Wales and Queensland, Australia (2) found that glider poles along overpasses were used by squirrel gliders *Petaurus norfolcensis* for crossing roads. Gliders used glider poles along both overpasses where they were installed (detected on 30–66% of sample sessions). No gliders were detected in the middle of either overpass that did not have glider poles. Two overpasses (36–70 m long, 10–15 m wide, constructed in 2005–2008), each had eight glider poles installed. Poles were 6.5 m high and 5–12 m apart. Two further overpasses (62–66 m long, 19–37 m wide, constructed in 2002) had no poles. Between September 2006 and December 2010, gliders were surveyed 23–35 times at each site with poles, using hair-traps attached 1.8 m high on each pole. Overpasses without poles were surveyed 10 times, for 2–4 weeks each time, between May 2010 and June 2011, using six hair-traps/overpass, mounted 1.8 m high on trees or shrubs.

A replicated study in 2008–2010 at three sites along a road through forest in North Carolina, USA (3) found that crossing poles were used by Carolina northern flying squirrels *Glaucomys sabrinus coloratus* to cross the road. All three radio-tagged flying squirrels crossed the road with at least one using a crossing pole. Out of 25 videos of flying squirrels at crossing poles, 14 (56%) showed crossing attempts (landing on the opposite pole was not confirmed). In June 2008, six wooden poles (32 cm diameter) were set in three pairs on opposite sides of a two-lane road. Poles, 15 m apart, were buried 2.4 m into the ground and extended 14.3 m above ground. Each pole was fitted with a 3-m-long, 10 × 19-cm horizontal wooden launch beam at the top. In March 2009, three flying squirrels were fitted with radio-transmitters and released onto a crossing pole on the opposite side of the road from their capture location. They were tracked at least monthly between March–June 2009. Infrared motion detection cameras were used at each pole between March 2009 and June 2010 to detected crossings.

A replicated, site comparison study in 2007–2011 along a highway in Victoria, Australia (4) found that glider poles, along with canopy rope bridges across highways, were used occasionally by squirrel gliders *Petaurus norfolcensis*. Just one of seven radio-tracked squirrel gliders crossed the road where a glider pole was present compared to three of seven crossing canopy road bridges. Seven of 10 crossed a narrow single-lane-road without crossing structures but none of 12 crossed a wider highway with no crossing structures. Camera traps recorded 13 crossings by squirrel gliders at glider poles over 146 camera-trap nights. In July 2007, three glider poles and two rope bridges were installed along a 70-km-long section of four-lane divided highway. Poles (13 m high, 45 cm diameter) were installed in the centre of the highway to reduce glide distances required for road crossings. Camera traps monitored pole use (December 2009–March 2011; 22–87 nights/pole) and rope-bridge use (August 2007–May 2011; 787–873 nights/bridge). In 2010–2011, 42 gliders were radio-tracked at sites with and without crossings and at a narrow (<10 m wide) single-lane road.

A study in 2011–2012 at a site on a highway through woodland in Queensland, Australia (5) found that roadside glide poles were used by squirrel gliders *Petaurus norfolcensis* to cross the highway. Squirrel gliders were recorded on poles on 60 out of 310 nights monitored. Road crossings were confirmed on 16 nights of 125 when both sides were monitored. Three poles were installed across a 61-m-wide canopy gap. One pole was on each roadside. A third bridged a 35-m gap between the roadside and forest. The two poles at each side of the gap were thus 6 and 14 m from tree canopies. Poles, made from hardwood, were 30 cm diameter and 12 m high. Wooden crossbars were attached at 20 and 40 cm below the top. Squirrel gliders were monitored using a camera trap on the middle pole from 1 August 2011 to 30 June 2012 and an additional camera trap on the pole across the road from 27 February to 30 June 2012.

A replicated study in 2012–2014 at 15 sites along a highway though eucalyptus forest in Victoria, Australia (6) found that squirrel gliders *Petaurus norfolcensis* and sugar gliders *Petaurus breviceps* used glider poles to cross the road. Remote cameras detected 842 road crossings by squirrel gliders and 258 by sugar gliders using glider poles. The study was conducted in two sections of the Hume Freeway, located 200 km apart. In 2007–2009, fifteen pole crossings (≤5 poles/site) were erected spanning roads of 56–382 m wide. Poles were 13–18 m tall, 40–50 cm diameter and made of hardwood timber. A timber cross-beam (10 cm × 10 cm × 2.4 m) was fixed horizontally 0.5 m from the top of each pole (oriented parallel to the road edge). The number and height of poles used in each array varied with gap width and the height of roadside trees. Wildlife crossings were monitored from between April and June 2012 to February 2013, using motion-triggered cameras.

A study in 2007–2015 at five points along a highway through woodland in Victoria, Australia (7) found that arboreal marsupials using artificial road crossing structures did not suffer high predation rates when doing so. Among 13,488 detections of arboreal marsupials using glider pole crossings and rope bridges combined (separate figures not given in paper), there were no recorded instances of attempted predation of those using glider poles. One unsuccessful predation attempt was recorded from a rope bridge. In July 2007, five crossing structures were installed along 70 km of highway. Three were poles for gliders (one or two poles/crossing, 12–14 m tall) and two were rope mesh canopy bridges (70 m long, 5 m wide). Crossings were monitored with motion and heat activated cameras from July 2007 to February 2015. Cameras recorded 5–10 images, 3 s apart (2007–2011) or a 10–20 s video (2011–2015). Predation attempts were detectable when animals were ≤1 m from the top of each glider pole or ≤5 m from each end of a canopy bridge.

(1) Goldingay R.L., Taylor B.D. & Ball T. (2011) Wooden poles can provide habitat connectivity for a gliding mammal. *Australian Mammalogy*, 33, 36–43, https://doi.org/10.1071/am10023

(2) Taylor B.D. & Goldingay R.L. (2012) Restoring connectivity in landscapes fragmented by major roads: a case study using wooden poles as 'stepping stones' for gliding mammals. *Restoration Ecology*, 20, 671–678, https://doi.org/10.1111/j.1526-100x.2011.00847.x

(3) Kelly C.A., Diggins C.A. & Lawrence A.J. (2013) Crossing structures reconnect federally endangered flying squirrel populations divided for 20 years by road barrier. *Wildlife Society Bulletin*, 37, 375–379, https://doi.org/10.1002/wsb.249

(4) Soanes K., Lobo M.C., Vesk P.A., McCarthy M.A., Moore J.L. & van der Ree R. (2013) Movement re-established but not restored: Inferring the effectiveness of road-crossing mitigation for a gliding mammal by monitoring use. *Biological Conservation*, 159, 434–441, https://doi.org/10.1016/j.biocon.2012.10.016

(5) Taylor B.D. & Goldingay R.L. (2013) Squirrel gliders use roadside glide poles to cross a road gap. *Australian Mammalogy*, 35, 119–122, https://doi.org/10.1071/am12013

(6) Soanes K., Vesk P.A. & van der Ree R. (2015) Monitoring the use of road-crossing structures by arboreal marsupials: insights gained from

motion-triggered cameras and passive integrated transponder (PIT) tags. *Wildlife Research*, 42, 241–256, https://doi.org/10.1071/wr14067

(7) Soanes K., Mitchell B. & van der Ree R. (2017) Quantifying predation attempts on arboreal marsupials using wildlife crossing structures above a major road. *Australian Mammalogy*, 39, 254–257, https://doi.org/10.1071/am16044

5.9. Install rope bridges between canopies

https://www.conservationevidence.com/actions/2556

- **Ten studies** evaluated the effects on mammals of install rope bridges between canopies. Eight studies were in Australia[1–5,7,8,10], one was in Brazil[6] and one in Peru[9].

COMMUNITY RESPONSE (0 STUDIES)

POPULATION RESPONSE (1 STUDY)

- **Survival (1 study):** A study in Australia[10] found that arboreal marsupials using rope bridges did not suffer high predation rates when doing so.

BEHAVIOUR (9 STUDIES)

- **Use (9 studies):** Nine studies (including three replicated studies and a site comparison), in Australia[1–5,7,8], Brazil[6] and Peru[9] found that rope bridges were used by a range of mammals. Seven of these studies found between three and 25 species using rope bridges[1–4,7], one found that that they were used by squirrel gliders[5] and one that they were used by mountain brushtail possums and common ringtail possums but not by koalas and squirrel gliders[8]. One of the studies[9] found that crossing rates were higher over the canopy bridges than at ground level.

Background

Wildlife crossings over or under roads may be installed to reduce the impact of the road on animal mortality and on habitat fragmentation. They usually take the form of tunnels or bridges of a range of designs. These may not be suitable for use by mammals that spend most of their time higher up within trees. Rope bridges have been trialled, especially in Australia, to provide a means of reconnecting habitat and reducing road mortality for arboreal mammal species. Monitoring typically takes the form of documenting use of crossings rather than looking at population level effects or impacts on road mortality.

See also: *Install pole crossings for gliders/flying squirrels*.

A study in 2000–2002 along a road through highland rainforest in Queensland, Australia (1) found that all three rope bridges across the road were used by arboreal marsupials. Across the three rope bridges, six species of possums, Lumholtz's tree kangaroos *Dendrolagus lumholtzi* and fawn-footed melomys *Melomys cervinipes* were recorded, with 5–7 species/crossing recorded. The number of crossings was not documented. In 1995, a canopy bridge tunnel was erected 7 m above a 7-m-wide tree gap over a low-traffic road (4 vehicles/day). The bridge comprised a 50 × 50-cm rope tunnel, 14 m long, made of 10-mm silver rope attached to wooden poles, erected amongst trees on the roadside. In 2000, a 10-m-long, 50-cm-wide rope-bridge was erected 7 m high, spanning a 5-m gap over a forestry track. Additionally, a 25-cm-wide rope ladder was placed initially over the same track, then lengthened and moved in 2001 to span a 14-m-wide gap over a road carrying 150 vehicles/day. Mammal crossings were monitored in 2000–2002, through scat and hair analysis, remote photography and spotlighting surveys.

A study in 2000–2010 of four roads through rainforest in Queensland, Australia (2) found that all seven rope bridges connecting trees at each side of the road were used and nine mammal species in total were recorded. Of these, five species were directly observed crossing bridges. The remaining four were detected solely by other monitoring methods. Totals of 2–7 species/rope bridge were recorded. No mammals were found dead on roads in the vicinity of rope bridges (though details of searches for casualties are not stated). Seven rope bridges in total were erected at four sites in 1995–2005. Two were rope tunnels, with a square cross-section. The remainder were rope ladders, 0.25–0.5 m wide. Mammal use of bridges was monitored by direct observation by spotlight, faeces collected in nets or funnels below bridges, motion-and heat-sensitive cameras and hair collection using sticky tape.

A site comparison study in 2010–2011 at three overpasses along a road through forest in Queensland, Australia (3) found that squirrel gliders *Petaurus norfolcensis*, a brushtail possum *Trichosurus vulpecula* and a ringtail possum *Pseudocheirus perigrinus* used a rope bridge that connected between glider poles across the overpass. Squirrel gliders were detected using the rope bridge on 33 occasions during 27 of 166 survey nights. Over the same period, one brushtail possum and one ringtail possum were detected. No gliders crossed two overpasses that did not have glider poles or rope bridges. The study was conducted on an overpass (36 × 15 m, constructed in 2008) with eight glider poles, 6.5 m high, connected by a single rope (40 mm diameter). Two overpasses without poles or a rope bridge (62–66 m long, 19–37 m wide) were also monitored. Mammal crossings were surveyed using camera traps between September 2010 and April 2011. A camera was placed near the top of one end pole and directed along the connecting rope. Cameras were also placed in the middle of overpasses without poles.

A replicated study in 2008–2011 of five rope bridges at four sites along a highway through woodlands in New South Wales, Australia (4) found that rope bridges were used by six mammal species. Bridges were used by squirrel gliders *Petaurus norfolcensis* (44 records at two bridges), feathertail gliders *Acrobates pygmaeus* (nine records at three bridges), common ringtail possums *Pseudocheirus peregrinus* (seven records at one bridge), common brushtail possums *Trichosurus vulpecula* (33 records at two bridges), sugar gliders *Petaurus breviceps* (15 records at two bridges) and black rats *Rattus rattus* (19 records at two bridges). Two rope bridges across the highway (42–75 m long) were monitored at one site. Single bridges (each approximately 50 m long), crossing creeks underneath the highway at each of two sites, were monitored. At the fourth site, a rope bridge was suspended from a series of poles along a 70-m-long land bridge over the highway. Sites were up to 270 km apart.

Bridges, erected in 2004–2008, comprised rope mesh either laid flat or formed into tunnels. They were monitored by 1–3 camera traps/bridge for 42–503 nights/camera.

A replicated, site comparison study in 2007–2011 along a highway in Victoria, Australia (5) found that canopy rope bridges across highways, along with glider poles, were used by squirrel gliders *Petaurus norfolcensis*. Three of seven squirrel gliders crossed roads when canopy bridges were present. The proportion of squirrel gliders crossing roads where canopy bridges or glider poles were installed (29%) was higher than that which crossed roads when such structures were absent (0%). However more still (70%) crossed at a narrow, single-lane road with low traffic flows and no artificial crossing structures. Camera traps recorded 1,187 crossings at canopy bridges. It took 9–13 months for gliders to habituate to and use bridges. In July 2007, two rope bridges and three glider poles were installed at five sites along a 70-km-long section of a four-lane divided highway. Canopy rope bridges were 70 m long, 0.5 m wide and 6 m high. Camera traps monitored bridge (August 2007– May 2011; 787–873 nights/bridge) and pole use (December 2009–March 2011; 22–87 nights/pole crossing). In 2010–2011, 42 gliders were radio-tracked at sites with and without crossings and at a single-lane-road site (<10 m wide).

A study in 2008–2009 of a forested and urban area in Porto Alegre, Brazil (6) found that rope canopy bridges over roads were used by three mammal species. Rope canopy bridges were used by brown howler monkeys *Alouatta guariba clamitans* (4 of 6 bridges), porcupines *Sphiggurus villosus* (2 of 6 bridges) and white-eared opossums *Didelphis albiventris* (1 of 6 bridges). Six canopy bridges were installed in 2001– 2006 at sites close to a protected reserve where brown howler monkeys had been killed on roads or used power lines to cross them. Each bridge consisted of a horizontal 'ladder' made from rope and rubber hose (4 x 12 m parallel ropes with rubber hose 'steps' at 80 cm intervals and interlaced ropes forming a 'X' between each step). Camera traps and trained local observers monitored each of the six bridges for a total of 33–152 days during 6–15 months in 2008–2009.

A replicated study in 2012–2014 at five sites along a highway through eucalyptus forest in Victoria, Australia (7; an expansion of 5) found that canopy rope bridges were used by four species of arboreal marsupial to cross the road. Remote cameras detected 455 crossings of canopy bridges by squirrel gliders *Petaurus norfolcensis*, 229 by common brushtail possums *Trichosurus vulpecula*, 386 by common ringtail possums *Pseudocheirus peregrinus* and two by brush-tailed phascogales *Phascogale tapoatafa*. The study was conducted along two sections of the Hume Freeway, located 200 km apart. In 2007–2009, four 60–85-m-long canopy bridges, made of 15-mm-diameter rope woven into a flat net, 50 cm wide, were erected. They were 6 m above the road. A fifth bridge, 170 m long, was erected at ≥4 m high. Wildlife crossings were monitored between June 2012 and February 2013, using motion-triggered cameras.

A study in 2012–2016 in a forest site within a university campus in New South Wales, Australia (8) found that northern mountain brushtail possums *Trichosurus caninus* and common ringtail possums *Pseudocheirus peregrinus* used canopy bridges but koalas *Phascolarctos cinereus* and squirrel gliders *Petaurus norfolcensis* did not. Twenty-two passes of northern mountain brushtail possums and two of common ringtail possums were detected on rope bridges. Koalas were detected 75 times and squirrel gliders three times in two nearby trees but were not detected on rope bridges. The trial was conducted in a 30 × 100 m eucalyptus-dominated forest patch. Rope-bridges of four designs extended 8–11 m between different pairs of trees. One rope bridge had 8-cm gaps between rope strands, one was made of woven-mesh with 1-cm gaps between strands, one was a ladder wrapped around internal wires to produce a sausage shape and one consisted of a woven mesh bridge with rope-ladder sides. One or two camera traps were used to monitor each rope-bridge and single cameras were used on two nearby reference trees, for 2.8–3.1 years/tree, between December 2012 and February 2016.

A study in 2012–2013 at a forest site in the Lower Urubamba region, Peru (9) found that canopy bridges over a pipeline route were used by 25 arboreal mammal species with use increasing over 10 months, and crossing rates were higher over the bridges than at ground level. Twenty-five arboreal mammal species were recorded crossing over 13 canopy bridges (see original paper for details). Overall, use of the bridges increased over 10 months (total 40–55 crossings/100 nights). Crossing rates were higher over the bridges (total 45 crossings/100 nights) than below them at ground level (total 0.3 crossings/100 nights), although the difference was not tested for statistical significance. A gas pipeline route (10–25 m wide) was cleared through an area of native forest in June–August 2012. Thirteen canopy bridges (with branches from one or more trees connecting across the clearing) were preserved along a 5.2 km stretch of the route. Ten bridges remained functional by the end of the study in August 2013. Three failed due to exposure/tree damage. From September 2012, camera traps recorded crossing activity over the bridges (1–4 cameras/bridge) and at ground level below (2–3 cameras/ bridge) for 11–12 months.

A study in 2007–2015 at five points where a highway bisected woodland in Victoria, Australia (10) found that arboreal marsupials using rope bridges did not suffer high predation rates when doing so. Among 13,488 detections of arboreal marsupials (from rope bridges and glider pole crossings combined — separate figures not given in paper), there was one recorded predation attempt. This was an unsuccessful night-time predation attempt on a squirrel glider *Petaurus norfolcensis* using a rope bridge, by an unidentified bird. In July 2007, five crossing structures were installed along 70 km of highway. Two were rope mesh canopy bridges (70 m long, 5 m wide) and three were poles for gliders (one or two poles/crossing, 12–14 m tall). Crossings were monitored with motion and heat activated cameras, from July 2007 to February 2015. Cameras recorded 5–10 images, 3 s apart (2007–2011) or a 10–20 s video (2011–2015). Predation attempts were detectable when animals were ≤5 m from each end of a canopy bridge, and ≤1 m from the top of each glider pole.

(1) Goosem M., Weston N. & Bushnell S. (2005) *Effectiveness of rope bridge arboreal overpasses and faunal underpasses in providing connectivity for rainforest fauna.* Proceedings of the International Conference on Ecology and Transportation, Center for Transportation and the Environment, North Carolina State University, Raleigh NC, USA, 304–318.

(2) Weston N., Goosem M., Marsh H., Cohen M. & Wilson R. (2011) Using canopy bridges to link habitat for arboreal mammals: successful trials in the Wet Tropics of Queensland. *Australian Mammalogy*, 33, 93–105, https://doi.org/10.1071/am11003

(3) Taylor B.D. & Goldingay R.L. (2012) Restoring connectivity in landscapes fragmented by major roads: a case study using wooden poles as 'stepping stones' for gliding mammals. *Restoration Ecology*, 20, 671–678, https://doi.org/10.1111/j.1526-100x.2011.00847.x

(4) Goldingay R.L., Rohweder D. & Taylor B.D. (2013) Will arboreal mammals use rope-bridges across a highway in eastern Australia? *Australian Mammalogy*, 35, 30–38, https://doi.org/10.1071/am12006

(5) Soanes K., Lobo M.C., Vesk P.A., McCarthy M.A., Moore J.L. & van der Ree R. (2013) Movement re-established but not restored: Inferring the effectiveness of road-crossing mitigation for a gliding mammal by monitoring use. *Biological Conservation*, 159, 434–441, https://doi.org/10.1016/j.biocon.2012.10.016

(6) Teixeira F.Z., Printes R.C., Fagundes J.C.G., Alonso A.C. & Kindel A. (2013) Canopy bridges as road overpasses for wildlife in urban fragmented landscapes. *Biota Neotropica*, 13, 117–123, https://doi.org/10.1590/s1676-06032013000100013

(7) Soanes K., Vesk P.A. & van der Ree R. (2015) Monitoring the use of road-crossing structures by arboreal marsupials: insights gained from motion-triggered cameras and passive integrated transponder (PIT) tags. *Wildlife Research*, 42, 241–256, https://doi.org/10.1071/wr14067

(8) Goldingay R.L. & Taylor B.D. (2017) Targeted field testing of wildlife road-crossing structures: koalas and canopy rope-bridges. *Australian Mammalogy*, 39, 100–104, https://doi.org/10.1071/am16014

(9) Gregory T., Carrasco-Rueda F., Alonso A., Kolowski J. & Deichmann J.L. (2017) Natural canopy bridges effectively mitigate tropical forest fragmentation for arboreal mammals. *Scientific Reports*, 7, 3892, https://doi.org/10.1038/s41598-017-04112-x

(10) Soanes K., Mitchell B. & van der Ree R. (2017) Quantifying predation attempts on arboreal marsupials using wildlife crossing structures above a major road. *Australian Mammalogy*, 39, 254–257, https://doi.org/10.1071/am16044

5.10. Install one-way gates or other structures to allow wildlife to leave roadways

https://www.conservationevidence.com/actions/2558

- **Seven studies** evaluated the effects on mammals of installing one-way gates or other structures to allow wildlife to leave roadways. All seven studies were in the USA[1-7].

COMMUNITY RESPONSE (5 STUDIES)

- **Survival (5 studies):** Two before-and-after studies (one replicated), in the USA[2,3], found that barrier fencing with one-way gates reduced deer-vehicle collisions. One of two studies (one before-and-after and one replicated, controlled), in the USA[4,7], found that barrier fencing with escape gates along roads with one or more underpasses reduced moose-vehicle collisions[4], whilst the other found no reduction in total mammal road casualty rates[7]. A replicated, controlled, before-and-after study in USA[6] found that earth escape ramps reduced mammal road mortalities.

POPULATION RESPONSE (0 STUDIES)

BEHAVIOUR (4 STUDIES)

- **Use (4 studies):** One of two studies (one replicated) in the USA[1,5] found that one-way gates allowed mule deer to escape when trapped along highways with barrier fencing[1], whilst the other found that a small proportion used one-way gates[5]. A replicated, controlled, before-and-after study in the USA[6] found that earth escape ramps were used more often than were one-way escape gates to enable deer to escape highways with barrier fencing. A replicated, controlled study in the USA[7] found that barrier fencing with escape gates and underpasses facilitated road crossings by a range of mammals.

Background

Fencing alongside roads can prevent or reduce mammal access to roads and, thus, reduce vehicle collisions with mammals. However, mammals that do manage to access roads, either around fence ends or through defective sections of fence, can then become trapped on the road. One-way gates are intended to allow escape of such mammals from the road whilst not enabling additional animals to access the road. Other structures can serve a similar purpose, such as ramps up to fence-top height at one side.

See also: *Install barrier fencing along roads.*

A replicated study in 1970–1972 in Colorado, USA (1) found that one-way gates allowed mule deer *Odocoileus hemionus hemionus* to escape when trapped along highways with barrier fencing. A total of 558 passages were recorded through eight gates, with 96% in the one-way direction designed. Use of each gate ranged from seven to 335 passages. Track counts indicated that the gates enabled approximately 223 deer to escape the highway. There were also 3,293 tracks counted of deer approaching gates heading towards the highway but not passing through. During 31 trails, three types of one-way gate were tested (two at a time) along a fence between a field with a mule deer and one with its food. The location and direction of each gate was changed frequently. Eight gates, of the most effective design, were installed in 2.4-m-high barrier fencing along a 1.5-mile section of highway. Passages were monitored using track counts and mechanical counters. Gates along the highway were checked daily during migrations in 1970–1972.

A before-and-after study in the 1970s along two highways in California, USA (2) found that barrier fencing incorporating one-way gates reduced deer-vehicle collisions by 68–87%. Fewer deer *Odocoileus* spp. road mortalities were recorded after construction of the six fence sections (average 2/km/year) than before (average 11/km/year). Six different lengths (1.9–7.7 km) of 2.4-m fencing were installed along Interstate 70 and Colorado Highway 82. Five of the fences were only on one side of the road, the other was on both sides and connected to an underpass. Four of the fences had one-way gates to allow deer to escape from the highway. Deer carcasses found along the road were counted in each fenced area before and after installation. Cost-benefit analysis was also undertaken using pre-fence mortality (dead deer) and fence effectiveness and estimates of cost of vehicle repair, deer value, discount rate, cost of fence and cost of fence maintenance (see original article for results).

A replicated, before-and-after study in 1977–1979 along two highways in Minnesota, USA (3) found that barrier fencing with one-way gates decreased deer-vehicle collisions. Along two fenced road sections, 1.3 and 8 deer/year were killed compared to an estimated 20/year in the pre-fence period. One fence was installed in a ditch with 1 m of water, meaning 30% of gates could not be used to escape the highway. Overall, 69% of 51 passages through gates were in the correct direction, i.e. from the highway to outside the fenced corridor. Two sections of 2.4-m-high fence with one-way gates along new highways were monitored for 18 months. Fences were 4 and 5 km long with nine and 10 pairs of gates (30 m apart), respectively. Deer were monitored crossing through gates by using baler counters and track beds. Deer-vehicle collisions were monitored for one year before (along old adjacent highway) and 18 months after installation. Cost-benefit analysis was also carried out (see the original article for further details).

A before-and-after study in 1977–1990 in Alaska, USA (4) found that barrier fencing with one-way gates, along with an underpass and road lighting, reduced vehicle collisions with moose *Alces alces*. Effects of fencing, gates, lighting and the underpass could not be separated. There were fewer moose-vehicle collisions after installation of fencing with one-way gates, an underpass and lighting (0.7/year) than before (17/year). There was no significant difference in the distribution of moose in relation to the highway between after and before fence installation. A total of 17 moose were observed using one-way gates and tracks suggested gates were used frequently. However, this meant that moose were regularly getting onto the highway. The first gates installed stayed open if swung all the way open and gates got stuck open below 0°C, because of the lubricant used. In October 1987, road lighting was installed along 11.5 km of the highway. Fencing and 30 one-way gates were installed along 5.5 km of this section and an underpass was created. Moose-vehicle collisions were monitored before (1977–1987) and after (1987–1990) installation. One-way gates were monitored using track counts in snow.

A study in 1994–1995 along two highways through grassland and shrubland in Utah, USA (5) found that one-way gates were used by some mule deer *Odocoileus hemionus* to escape a highway, but most did not cross through them. From 243 instances in which deer approached gates from the highway, 40 deer (16%) used gates to leave the highway. None of 128 deer that approached from the side away from the highway passed through gates. In September 1994, five and four crossing points were installed along a two-and a four-lane highway respectively. Fencing, 2.3 m high, directed deer to crossing points. Warning signs alerted approaching motorists to crossing points. Four one-way gates were installed at each crossing to allow deer trapped along the road to escape. One-way gate specifications were not detailed in the paper. Earthen track beds at 12 randomly selected one-way gates were checked at least once each week from September 1994 to November 1995 (except January–March 1995).

A replicated, controlled, before-and-after study in 1997–1999 along two highways in Utah, USA (6) found that earth escape ramps reduced road mortalities and were used more often than one-way escape gates to enable deer to escape highways with 2.4-m-high barrier fencing. Road mortalities decreased more after ramp installations at two sites (after: 4.8 and 2.0 killed/km; before: 6.7 and 4.6 killed/km) than at a control site during this time (after: 4.0 killed/km; before: 5.2 killed/km). At one site, 188 successful ramp crossings were recorded. At the other, 192 were recorded. Combined values from both sites showed ramps were used 8–11 times more often than were one-way gates. Nine earth ramps (1.5-m drop-off) were installed along 2.4 km of highway in 1997 and seven along 2.4 km of another highway in 1998. Ten and eight one-way gates respectively were installed previously at these sites (installation date not stated). Animal movements across ramps and through gates were monitored from May–July until October in 1998 and 1999 using track plots. Road mortality and monthly spotlight counts of deer were carried out before and after construction of ramps along both sections, and along an 8-km control section (1-m fencing, no mitigation measures) in 1997–1999. Cost-benefit analysis was also carried out (see original article for results).

A replicated, controlled study in 2000–2007 along a highway in North Carolina, USA (7) found that barrier fencing with escape gates and underpasses facilitated road crossings by a range of mammals but did not reduce road casualties. A similar rate of mammal road casualties was recorded over one year on road sections with fencing, escape gates and underpasses (5.0/km) as on sections without (5.1/km). A four-lane highway was constructed with three underpasses. Barrier fencing, 3 m high, was installed ≥800 m along the highway from each underpass. Gates allowed trapped animals to escape the highway. Road deaths were recorded along 6 km of road with fencing and underpasses and 11 km without, twice/week, from July 2006–July 2007.

(1) Reed D.F., Pojar T.M. & Woodard T.N. (1974) Use of one-way gates by mule deer. *The Journal of Wildlife Management*, 38, 9–15, https://doi.org/10.2307/3800194

(2) Reed D.F., Beck T.D.I. & Woodward T.N. (1982) Methods of reducing deer–vehicle accidents: benefit–cost analysis. *Wildlife Society Bulletin*, 10, 349–354.

(3) Ludwig J. & Bremicker T. (1983) Evaluation of 2.4 m fences and one-way gates for reducing deer vehicle collisions in Minnesota. *Transportation Research Record*, 913, 19–22.

(4) McDonald M.G. (1991) Moose movement and mortality associated with the Glenn Highway expansion. *Alces*, 27, 208–219.

(5) Lehnert M.E. & Bissonette J.A. (1997) Effectiveness of highway crosswalk structures at reducing deer-vehicle collisions. *Wildlife Society Bulletin*, 25, 809–818.

(6) Bissonette J. & Hammer M. (2000) *Comparing the effectiveness of earthen escape ramps with one-way gates in Utah.* USGS Utah cooperative Fish and Wildlife Research Unit, Logan, Utah.

(7) McCollister M.F. & van Manen F.T. (2010) Effectiveness of wildlife underpasses and fencing to reduce wildlife-vehicle collisions. *The Journal of Wildlife Management*, 74, 1722–1731, https://doi.org/10.2193/2009-535

5.11. Install barrier fencing along roads

https://www.conservationevidence.com/actions/2567

- **Twelve studies** evaluated the effects on mammals of installing barrier fencing along roads. Eight studies were in the USA[1-6,9,10], one each was in Canada[7], Germany[8] and Brazil[11] and one spanned the USA, Canada and Sweden[12].

COMMUNITY RESPONSE (0 STUDIES)

POPULATION RESPONSE (9 STUDIES)

- **Survival (9 studies):** Three controlled studies, in the USA[6], Germany[8] and Brazil[11], found that roadside fencing or equivalent barrier systems reduced the numbers of mammals, including wildcats[8] and coypu[11], killed by vehicles on roads. Two before-and-after studies, in the USA[2,3], found that roadside fencing with one-way gates to allow escape from the road, reduced the number of collisions between vehicles and deer. A study in the USA[4] found that a 2.7-m-high fence did not reduce road-kills of white-tailed deer compared to a 2.2-m-high fence. A controlled, before-and-after study in the USA[5] found that barrier fencing with designated crossing points did not significantly reduce road deaths of mule deer. A replicated, controlled, before-and-after study in Canada[7] found that electric fences, (along with an underpass beneath one highway), reduced moose-vehicle collisions. A review of fencing studies from USA, Canada and Sweden[12], found that longer fencing along roadsides led to a greater reduction of collisions between large mammals and cars than did shorter fence sections.

BEHAVIOUR (5 STUDIES)

- **Behaviour change (5 studies):** A controlled, before-and-after study in the USA[1] found that 2.3-m-high fencing in good condition prevented most white-tailed deer accessing a highway. A replicated, controlled, before-and-after study in Canada[7] found that electric fences reduced moose access to highways. Three studies (two replicated), in the USA[4,9,10], found that higher fences (2.4–2.7 m) prevented more white-tailed deer from entering highways than did fences that were 2.2 m high[4], 1.2 m high with outriggers[9] or 1.2–1.8 m high[10].

Background

Wildlife barrier fencing aims to prevent animals from crossing roads. They are typically wire mesh fences 2–2.5 m high running parallel to the road. Although fencing may protect wildlife from traffic, it should not create an absolute barrier that prevents migration, isolates populations, fragments habitat, or causes injuries. Wildlife fencing is therefore usually combined with safe crossing opportunities such as wildlife underpasses and overpasses (see *Install overpasses over roads/railways, Install tunnels/culverts/underpass under railways, Install tunnels/culverts/underpass under roads*). Wildlife escapes, such as one-way gates, are often integrated with wildlife fencing to allow animals that do manage to cross the fence to escape from the fenced road (see: *Install one-way gates or other structures to allow wildlife to leave roadways*). Wildlife such as deer frequently try to pass through holes in fences and so fences must be well maintained (Ward 1982).

Studies included here are those that specifically assess fence effectiveness, sometimes in combination with other collision reduction actions, but not where effects of fencing cannot be separated from effects of road underpasses. For these interventions combined, see *Install barrier fencing and underpasses along roads*.

As well as the threat to wildlife from vehicles, fencing is often placed to reduce dangers and costs to motorists that can result from collisions with wildlife. Assessment of whether or not to install fences may be based on a cost-benefit analysis (e.g. Huijser 2009).

Ward A.L. (1982) Mule deer behavior in relation to fencing and underpasses on Interstate 80 in Wyoming. *Transportation Research Record*, 859, 8–13.

Huijser M.P., Duffield J.W., Clevenger A.P., Ament R.J. & McGowan P.T. (2009) Cost–benefit analyses of mitigation measures aimed at reducing collisions with large ungulates in the United States and Canada: a decision support tool. *Ecology and Society*, 14, article 15.

A controlled, before-and-after study in 1975 along a highway through mixed hardwood forest in Pennsylvania, USA (1) found that, provided it was in good repair, 2.3-m-high fencing prevented most white-tailed deer *Odocoileus virginianus* from crossing a highway. Significantly fewer deer crossed the fence once it had been repaired (0–6), compared to before (77–84) and once repairs were undone (23–153), and compared to control sections (on which repairs were not carried out) during the same periods (24–247; 111–141; 53–268 crossings respectively). The 2.3-m-high fences ran either side of a four-lane highway, with a top section angled 45° away from the highway. The study site comprised two 0.8-km control sections with a 1.6-km experimental section between. Fence defects included gaps under the fence and lowered or broken top wires. Tracks in snow and sand along the fence both sides of the highway were monitored before repairs, after repairs along the experimental section and after repairs were undone. This cycle was implemented once in both winter and spring 1975 and tracks were surveyed over five days during each period.

A before-and-after study in the 1970s along two highways in California, USA found that barrier fences, including one connected to an underpass, and others to one-way gates, reduced deer-vehicle collisions by 68–87%. Fewer deer *Odocoileus* spp. road mortalities were recorded after construction of the six fence sections (average 2/km/year) than before (average 11/km/year). Six different lengths (1.9–7.7 km) of 2.4-m fencing were installed along Interstate 70 and Colorado Highway 82. Five of the fences were only on one side of the road, the other was on both sides and connected to an underpass. Four of the fences had one-way gates to allow deer to escape from the highway. Deer carcasses found along the road were counted in each fenced area before and after installation. Cost-benefit analysis was also undertaken using pre-fence mortality (dead deer) and fence effectiveness and estimates of cost of vehicle repair, value of deer, discount rate, cost of fence and cost of fence maintenance (see the original article for results).

A replicated, before-and-after study in 1977–1979 along two highways in Minnesota, USA (3) found that barrier fencing with one-way gates decreased deer-vehicle collisions. Along two fenced road sections, 1.3 and 8 deer/year were killed compared to an estimated 20/year in the pre-fence period. One fence was installed in a ditch with 1 m of water, meaning 30% of gates could not be used to escape the highway. Overall, 69% of 51 passages through gates were in the correct direction, i.e. from the highway to outside the fenced corridor. Two sections of 2.4-m-high fence with one-way gates along new highways were monitored for 18 months. Fences were 4 and 5 km long with nine and 10 pairs of gates (30 m apart), respectively. Deer were monitored crossing through gates by using baler counters and track beds. Deer-vehicle collisions were monitored for one year before (along old adjacent highway) and 18 months after installation. Cost-benefit analysis was also carried out (see the original article for further details).

A study in 1981–1983 in forest in Pennsylvania, USA (4) found that a 2.7-m-high deer-proof fence reduced the number of white-tailed deer *Odocoileus virginianus* on the highway compared to a 2.2-m-high fence, but did not reduce road-kills. A total of 240 groups of deer were observed on the highway alongside 23 km of 2.7-m-high fence compared to 465 alongside 18 km of 2.2-m-high fence. Overall, 1,687 deer (82% of all sightings) were on highway verges. In 1981–1983, one hundred deer died on the highway (1.2 deer/km/year) and numbers did not differ between fence types. Deer were monitored along a 41-km section of a 4–6-lane highway, 23 km of which had a 2.7-m-high mesh fence and the remainder a 2.2-m-high fence with an overhang. Thirty-six spotlight surveys were undertaken along the highway from January 1981 to January 1983.

A controlled, before-and-after study in 1991–1995 along two highways in Utah, USA (5) found that barrier fencing with designated crossing points and warning signs did not reduce road deaths of mule deer *Odocoileus hemionus*. Deaths fell on both fenced and unfenced sections but the rate of fall was not significantly higher on fenced road sections (after: 36–46; before: 111–148) than on unfenced sections (after: 34–63; before: 75–123). The number of deer on road verges fell by 34–55% following fence installation. In September 1994, four and five crossing points were installed along a two- and a four-lane highway respectively.

Fencing, 2.3 m high, restricted access to roadsides and directed deer towards crossing points. At these points, deer could jump a 1-m-high fence into funnel shaped fencing (2.3 m high) with a narrow opening to the road. One-way gates allowed deer trapped along the road to escape. Three warning signs, spaced 152 m apart, and painted lines across the road at crossings, indicated to drivers that it was a crossing point. Road deaths (weekly) and behaviour were monitored along fenced and nearby unfenced roads before and after installation, from October 1991 to November 1995. Spotlight count surveys were undertaken twice/ month.

A controlled, before-and-after study in 1998–2002 along a highway in Florida, USA (6) found that a barrier wall-culvert system reduced mammal road-kills. After construction, 33 mammals of ≥12 species were recorded dead on the 2.8-km section of road with the barrier (2.8 km) compared to 50 mammals on a 400-m section without barriers. Of those killed along the barrier, 17 were rice rats *Oryzomys palustris*, which climbed adjacent vegetation to get over the barrier. In 2000–2001, a 1-m-high concrete wall with 15-cm overhanging lip was constructed along a 2.8-km section of a highway. Eight concrete culverts were spaced 200–500 m apart below the wall. Roadkills were monitored on three days/week before (August 1998–1999) and after (March 2001–March 2002) barrier wall construction.

A replicated, before-and-after study in 2003–2005 along two highways in Québec, Canada (7) found that electric fences, along with an underpass beneath one highway, reduced moose *Alces alces* access to highways and moose-vehicle collisions. There were fewer moose-vehicle collisions after fence construction (zero) than before (1–5/year) and moose tracks on the road decreased by 76–84%. Only 33% (of 53) of moose tracks on the road were from moose that had crossed a fence; most entered through vehicle access routes (31%) or at fence ends (7%). Fences prevented 78% (7/9) of radio-collared moose from crossing the highway. Electric fences (1.5 m high, cables 0.3 m apart) were installed along both sides of a 5-km section of Highway 175 in 2002 and a 10-km section of Highway 169 in 2004 (both two-lane). Moose were monitored along fenced and adjacent equal-length unfenced road sections using weekly track surveys in May–August of 2003–2005. GPS collars were fitted to 47 moose and locations recorded every 2–3 hours for 1–3 years.

An underpass was constructed along one highway (23 m long, 16 m wide, 7 m high) and a fence opening on the other (that triggered dynamic warning road signs).

A controlled study in 2001–2005 along a motorway through forest and agricultural land in Germany (8) found that installing roadside fencing designed to keep wildcats *Felis silvestris* off the road reduced road-related wildcat mortality. Wildcat mortality was lower where wildcat fencing was installed (0.07 deaths/km/year) than in areas with other types of fencing (0.41–0.44 deaths/km/year). This difference was not tested for statistical significance. In 2002, two-metre-high wildcat fencing, with 5 × 5 cm mesh, a 50-cm-wide metal sheet overhang and a board down to 30 cm below ground, was installed along 6.4 km of road. Fine-meshed fence (same specifications as the wildcat fence, but without the overhang) was installed along 4 km of road. Standard wildlife fencing was installed on 7 km of road. Wildcat mortality data collected by researchers was supplemented by reports from motorway authorities and members of the public.

A replicated, before-and-after study in 2009–2010 along a university campus road in Georgia, USA (9) found that a 2.4-m-high fence was more successful at preventing white-tailed deer *Odocoileus virginianus* accessing the road than was a 1.2-m-high fence with outriggers attached to the top. Fewer deer crossed the road in a section with 2.4-m-high fencing (<0.01 crossings/day) than in a section with 1.2-m-fence with 0.6-m outriggers (0.05 crossings/day). Before fence construction, deer made 0.3–1.0 crossings/day. In May–June 2009, a vertical wire fence (1.6 km long, 2.4-m-high) and an outrigger fence (1.6 km long, 1.2 m high with a 0.6-m-long outrigger at 45°, attached to the top and threaded with five wires) were erected. Between January 2009 and March 2010, movements of eight adult female deer were monitored using GPS collars. Four deer had home ranges that overlapping the 2.4-m-high fence and four overlapped the 1.2-m-high fence with outriggers.

A replicated, controlled study in 2008 in fields in Georgia, USA (10) found that white-tailed deer *Odocoileus virginianus* did not jump 2.4-m-high barrier fencing, at 1.8 m fewer jumped if fencing was opaque and 1.2-m-high fences with outriggers angled towards deer were jumped less than those angled away. Among deer that jumped the

1.2-m control fence, fewer jumped each subsequently taller fence (1.5 m: 92%; 1.8 m: 75%; 2.1 m: 42%; 2.4 m: 0%). In opaque fence trails, 90% jumped 1.2 and 1.5-m fences and 50% jumped the 1.8-m fence. With an outrigger, fewer jumped when this was angled towards deer (60%) than away (90%). Three treatment areas (0.1–0.2 ha) were bisected with a test fence. Designs were woven-wire fencing either alone (1.5, 1.8, 2.1 and 2.4 m high), covered with opaque fabric (fence 1.2, 1.5 and 1.8 m high), 1.2 m high with a 0.6-m 50% opaque plastic outrigger angled at 45°, or a 1.2-m-high control fence. Ten adult female deer were each tested with each design in each treatment area. After 48 hours of habituation and limited food, deer were enclosed on the opposite side of test fences from food. Deer were videoed throughout each 25-hour trial.

A controlled, before-and-after study in 1995–2002 along a highway through a wetland in Rio Grande do Sul, Brazil (11) found that roadside fencing and underpasses reduced the number of road-kills of coypu *Myocastor coypus*. Fewer coypu were killed by cars after fencing was installed (3.6 coypu/100 km/day) than before (8.3 coypu/100 km/day). The total number of animal road-kills (including all mammals, birds and reptiles) after fencing was installed (10.3 animals/100 km/day) was smaller than before fencing (15.3 animals/100 km/day) (this result was not tested for statistical significance). Road-kill rates fell in fenced sections but increased in the unfenced section (see paper for details). Two sections of a two-lane highway, totalling 10.2 km long, were fenced in 1998. The fence was 50–100-mm mesh, 1.10 m high. Between these sections was a 5.5-km-long unfenced section. Nineteen underpasses in total were also installed along these three road sections. Road-kills were counted from a car from July 1995 to June 2002. Monitoring was conducted at an average speed of 50 km/h, by 2–4 observers, along 15.7 km of highway. A total of 619 monitoring runs were made before fence installation (July 1995 to September 1998) and 571 afterwards (October 1998 to June 2002).

A 2016 review of fencing studies from USA, Canada and Sweden (12) found that longer fencing along roadsides led to a greater reduction of collisions between large mammals and cars than did shorter fence sections. Results were not tested for statistical significance. Fences reduced collisions between large mammals and cars more in road sections fenced along >5 km (average 84% reduction in relation to before fencing) than in sections fenced along <5 km (average 53% reduction). The review identified 21 fenced road sections (18 from the USA, two from Canada and one from Sweden). Fences were 0.6–33.8 km long and 2.1–2.5 m high. Large mammals targeted by surveys included white-tailed deer *Odocoileus virginianus*, moose *Alces alces*, roe deer *Capreolus capreolus*, mule deer *Odocoileus hemionus*, elk *Cervus canadensis* and bighorn sheep *Ovis canadensis*.

(1) Falk N.W., Graves H.B. & Bellis E.D. (1978) Highway right-of-way fences as deer deterrents. *The Journal of Wildlife Management*, 42, 646–650, https://doi.org/10.2307/3800834

(2) Reed D.F., Beck T.D.I. & Woodward T.N. (1982) Methods of reducing deer–vehicle accidents: benefit–cost analysis. *Wildlife Society Bulletin*, 10, 349–354.

(3) Ludwig J. & Bremicker T. (1983) Evaluation of 2.4 m fences and one-way gates for reducing deer vehicle collisions in Minnesota. *Transportation Research Record*, 913, 19–22.

(4) Feldhamer G.A., Gates J.E., Harman D.M., Loranger A.J. & Dixon K.R. (1986) Effects of Interstate highway fencing on white-tailed deer activity. *The Journal of Wildlife Management*, 50, 497–503, https://doi.org/10.2307/3801112

(5) Lehnert M.E. & Bissonette J.A. (1997) Effectiveness of highway crosswalk structures at reducing deer-vehicle collisions. *Wildlife Society Bulletin*, 25, 809–818, https://doi.org/10.2307/3808706

(6) Dodd C.K., Barichivich W.J. & Smith L.L. (2004) Effectiveness of a barrier wall and culverts in reducing wildlife mortality on a heavily travelled highway in Florida. *Biological Conservation*, 118, 619–631, https://doi.org/10.1016/j.biocon.2003.10.011

(7) Leblond M., Dussault C., Ouellet J.-P., Poulin M., Courtois R. & Fortin J. (2007) Electric fencing as a measure to reduce moose–vehicle collisions. *The Journal of Wildlife Management*, 71, 1695–1703, https://doi.org/10.2193/2006-375

(8) Klar N., Herrmann M. & Kramer-Schadt S. (2009) Effects and mitigation of road impacts on individual movement behavior of wildcats. *The Journal of Wildlife Management*, 73, 631–638, https://doi.org/10.2193/2007-574

(9) Gulsby W.D., Stull D.W., Gallagher G.R., Osborn D.A., Warren R.J., Miller K.V. & Tannenbaum L.V. (2011) Movements and home ranges of white-tailed deer in response to roadside fences. *Wildlife Society Bulletin*, 35, 282–290, https://doi.org/10.1002/wsb.38

(10) Stull D.W., Gulsby W.D., Martin J.A., D'Angelo G.J., Gallagher G.R., Osborn D.A., Warren R.J. & Miller K.V. (2011) Comparison of fencing designs for excluding deer from roadways. *Human-Wildlife Interactions*, 5, 47–57.

(11) Bager A. & Fontoura V. (2013) Evaluation of the effectiveness of a wildlife roadkill mitigation system in wetland habitat. *Ecological Engineering*, 53, 31–38, https://doi.org/10.1016/j.ecoleng.2013.01.006

(12) Huijser M.P., Fairbank E.R., Camel-Means W., Graham J., Watson V., Basting P. & Becker D. (2016) Effectiveness of short sections of wildlife fencing and crossing structures along highways in reducing wildlife–vehicle collisions and providing safe crossing opportunities for large mammals. *Biological Conservation*, 197, 61–68, https://doi.org/10.1016/j.biocon.2016.02.002

5.12. Install barrier fencing and underpasses along roads

https://www.conservationevidence.com/actions/2571

- **Fifty-five studies** evaluated the effects on mammals of installing barrier fencing and underpasses along roads. Twenty-seven were in the USA[1–8,15–19,21,25,30,35,39,41,43–45,47,51,52a,52b,53], nine were in Canada[9–11,13,22,23,28,46,54], seven were in Australia[14,20,29,36,48–50], two each were in Spain[24,32], Portugal[26,31], the UK[27,42] and Sweden[33,34], one each was in Denmark[12], Germany[37] and Croatia[38] and one was a review covering Australia, Europe and North America[40].

COMMUNITY RESPONSE (0 STUDIES)

POPULATION RESPONSE (15 STUDIES)

- **Survival (15 studies):** Eleven of 15 studies (including 12 before-and-after studies and two site comparisons), in the USA[1,5,8,16,21,35,39,44,45], Australia[29,36], Sweden[33,34] and Canada[13,28], found that installing underpasses and associated roadside barrier fencing reduced collisions between vehicles and mammals[1,5,13,28,29,33–36,44,45]. Three studies found that the roadkill rate was not reduced[8,16,39] and one study found that vehicle-mammal collisions continued to occur after installation[21].

BEHAVIOUR (52 STUDIES)

- **Use (52 studies):** Seventeen of 18 studies (including 10 before-and-after studies) in the USA[1–4,16–19,25,30,35,41,44,45,52b,53] Canada[28] and Sweden[33], which reported exclusively on ungulates, found that underpasses installed along with roadside barrier fencing were used by a range of ungulate species. These were mule deer[1,2,3,17,19,45,53], mountain goat[4], pronghorn[18], white-tailed deer[19,41,52b] elk[19,25], moose[28] and Florida Key deer[30,35,44]. The other study found that underpasses were not used by moose[33] whilst one of the studies that did report use by ungulates further reported that they were not used by white-tailed deer[16]. Further observations from these studies included that elk preferred more open, shorter underpasses to those that were enclosed or longer[25], underpass use was not

Background

Schemes designed to reduce collisions between vehicles and wild mammals may use multiple interventions. Two of the most common ones, installing barrier fencing and providing routes for mammals to travel underneath roads, are often employed within the same scheme. This may entail regular roadside fencing with entrances to underpasses set further back away from the road or fencing may be designed to adjoin the sides of underpass entrances. Sometimes, fencing may be installed to form a funnel leading towards underpass entrances.

This intervention includes studies where these two actions are in place at the same site. In most studies, all underpasses (where there are multiple crossings) are beneath stretches or roads that have barrier fencing. In a minority, just some of the underpasses monitored are along stretches with barrier fencing. Studies included use of either conventional fencing, electric fences or other barriers, such as walls. Most studies report solely on the use of crossings or trends in numbers of mammals killed on roads. There is an absence of studies reporting on wider population-level effects of the presence of these structures.

See *Install tunnels/culverts/underpass under roads* for studies where underpasses are either installed without use of barrier fencing or where it is not clear from the study that barrier fencing was installed. See also *Install barrier fencing along roads* for studies which, in some cases, included underpasses but where the specific effect of fencing was evaluated.

affected by traffic levels[41] and that mule deer used underpasses less than they used overpasses[53]. Thirty-four studies (including four before-and-after studies, seven replicated studies, three site comparisons and two reviews), in the USA[6–8,15,21,39,43,47,51,52a], Canada[9–11,22,23,46,54], Australia[14,20,29,36,48,49,50], Spain[24,32], Portugal[26,31], the UK[27,42], Denmark[12], Germany[37], Croatia[38] and across multiple continents[40], that either studied mammals other than ungulates or multiple species including ungulates, found that underpasses in areas with roadside fencing were used by mammals. Among these studies, one found that small culverts were used by mice and voles more than were larger underpasses[22], one found that bandicoots used underpasses less after they were lengthened[49] and one found that culverts were used by grizzly bears less often than were overpasses[54].

A before-and-after study in 1970–1973 along a highway in Colorado, USA (1; same experimental set-up as 2) found that an underpass, in areas with roadside fencing and one-way gates, reduced road mortalities and allowed most local mule deer *Odocoileus hemionus* to migrate safely under a highway. There were 14 deer-vehicle accidents/year within the fenced section compared to 36/year before installation of the underpass and fencing. On average, 345 mule deer (61% of the local population) used the culvert each season, with up to 17 crossings/day. Underpass use was not affected by artificial lighting. On average, 17% of deer used one-way gates to escape the highway and 17% went round the ends of fences or did not cross. In 1970, a concrete box underpass (3 × 3 × 30 m, with two skylights) was installed under a 3.2-km section of highway. The 2.4-m-high barrier fencing either side had eight one-way gates. Underpass-use was monitored by track counts and mechanical counters daily and a video camera at night during spring–summer and autumn migrations in 1970–1973. Artificial lighting was alternately turned on and off over 28 nights, in June and October 1973. Tracks at gates and deer movements along the fence were monitored each morning.

A study in 1974–1979 along a highway in Colorado, USA (2; same experimental set-up as 1) found that an underpass, in an area with roadside fencing, continued to be used by mule deer *Odocoileus hemionus*

4–9 years after installation Between 1.3 and 5.8 deer/morning (average 2.3) were observed exiting the underpass each year (total 298 deer). Deer behaviour suggested that 75% of animals exiting the underpass were reluctant, wary, or frightened. Eleven hesitated just inside the exit and 23 showed wariness or excitability after exiting the underpass. Behavioural responses of deer to the underpass were reported not to have changed substantially over 10 years (1970–1979) of spring-summer use. In 1970, a concrete box underpass (3 m high, 3 m wide, 30 m long) was installed under a 3.2-km section of highway. Entrances were separated from the road by 2.4-m-high barrier fencing. Deer were observed from 130 m away, at 05:00–07:00 h, on 9–30 days (average 16), during each spring/summer migration in 1974–1979. Behavioural responses were likened (but not compared numerically) with those from earlier monitoring that commenced in 1970.

A study in 1977–1979 along a highway through shrubland in Wyoming, USA (1) found that underpasses, in areas with roadside fencing, were used by mule deer *Odocoileus hemionus* to cross under the road. During four migration periods (two spring, two autumn–winter) immediately after underpasses were connected to a fence, >4,000 crossings through underpasses were made by deer (precise figure not stated). The study was conducted along a 7.8-mile stretch of highway constructed in late 1970. The highway was located on a migration route of 1,600–2,000 mule deer. Over four migratory periods, seven underpasses (length: 110–393 feet; width: 10–50 feet; height: 10–17 feet) were monitored for deer use. Underpasses were connected to 8-foot-high roadside fencing that guided animals towards entrances. From 1978, an attempt was made to attract deer to six of the seven underpasses by baiting with alfalfa hay, supplemented with apple pulp or by vegetable trimmings. Deer movements were monitored by track counts and surveillance cameras.

A before-and-after study in 1975–1981 in Montana, USA (4) found that two underpasses and roadside fencing increased highway crossing success by mountain goats *Oreamnos americanus*. After construction, 90% of highway crossing attempts were successful compared to 86% during and 74% before construction (unsuccessful attempts were when the crossing was temporarily thwarted). Crossing hesitations and run-backs decreased by 80% after underpass construction, delay time before crossing declined by about 30% and signs of fear (measured by an index) decreased. All crossings were successful when there was no disturbance, but success decreased to 85% when humans or traffic were present. A large underpass (3–8 m high, 23 m wide, 11 m long) was constructed where goats were observed crossing. In addition, a new road bridge included a ledge underneath for goats to cross (3 m high, 3 m wide, 11 m long). A sheer wall downhill and barrier fencing prevented goats crossing between underpasses. Old goat trails were removed and new trails to underpasses dug. Goat crossings were monitored before (1975), during (May–October 1980) and after underpass construction (October 1980–September 1981).

A before-and-after study in 1977–1990 along a highway in Alaska, USA (5) found that barrier fencing with one-way gates, along with an underpass and road lighting, reduced vehicle collisions with moose *Alces alces*. Effects of fencing and the underpass could not be separated from those of gates and lighting. There were fewer moose-vehicle collisions after installation of fencing with one-way gates, an underpass and lighting (0.7/year) than before (17/year). There was no significant difference in the distribution of moose in relation to the highway after and before fence installation. A total of 17 moose were observed using one-way gates and tracks suggested gates were used frequently. However, this meant that moose were regularly getting onto the highway. The first gates installed stayed open if swung all the way open and gates got stuck open below 0°C, because of the lubricant used. In October 1987, road lighting was installed along 11.5 km of the highway. Fencing and 30 one-way gates were installed along 5.5 km of this section and an underpass was created. Moose-vehicle collisions were monitored before (1977–1987) and after (1987–1990) installation. One-way gates were monitored using track counts in snow.

A study in 1994–1995 in Florida, USA (6) found that four underpasses beneath a highway, in areas with roadside fencing, were used by Florida panthers *Felis concolor coryi* and a range of other mammal species. Ten crossings were recorded through underpasses by panthers, as were 361 by white-tailed deer *Odocoileus virginianus*, 133 by bobcats *Lynx rufus*, 167 by raccoons *Procyon lotor* and two by black bears *Ursus americanus*. Panther records were thought to relate to two individuals. Four concrete bridge underpasses (21–26 m wide, 49 m long) were monitored along a 64-km stretch of a four-lane, divided highway. Barrier fencing, 3 m high, ran along the highway. Infrared game counters and cameras were used to monitor underpasses for 2, 10, 14 and 16 months in 1994–1995.

A replicated study in 1995 along two highways in Florida, USA (7) found that large underpasses and box culverts, in areas with roadside fencing, were used by a range of mammal species. Mammals recorded using large underpasses were white-tailed deer *Odocoileus virginianus* (5.1 crossings/month), panther *Felis concolor* (2.2), bobcat *Lynx rufus* (1.3) and raccoon *Procyon lotor* (1.4). Box culverts were additionally used by red foxes *Vulpes vulpes* and otters *Lontra canadensis*. Two box culverts (2.4 m high, 7 m wide, 15 m long) were monitored along a 6.4-km section of a highway. Two of nine large underpasses (21–25 m wide, 49 m long) with vegetation were monitored along a 15-km section of a different highway. Highways had barrier fencing 3.4 m high with a 1-m overhang. Underpasses were monitored from March or April 1995 (end date not stated) using an infra-red digital counter and camera and by counting tracks.

A before-and-after study in 1993–1995 of a highway in Florida, USA (8) found that an underpass beneath a highway, in an area with roadside fencing, was used by mammals but the road-kill rate was not reduced. Nine mammal species used the crossing. Most crossings were by rabbits *Sylvilagus palustris* (69 crossings), racoons *Procyon lotor* (61), armadillos *Dasypus novemcinctus* (36), opossums *Didelphis virginiana* (36), foxes *Vulpes vulpes* (29) and bobcats *Lynx rufus* (27). The number of mammals of squirrel size or larger killed on the fenced road section was not significantly different in the 11 months after fence installation (13 animals) relative to the 11 months before (10 animals). A wildlife crossing (14.3 m long, 7.3 m wide and 2.4 m tall) was constructed under the two-lane highway between summer and December 1994. A 3-m-high fence extended along both sides of the highway, 0.6 km in one direction and 1.1 km in the other. Underpass use was determined in December 1994 to December 1995 by footprint surveys and by using a motion-triggered camera. Road-kills were surveyed three times/week from November 1993 to December 1995.

A study in 1996–1997 along a highway through forest and grassland in Alberta, Canada (9; same experimental set-up as 11) found that underpasses, in areas with roadside fencing, were used by at least 10 species of medium-and large-sized mammals. Over 12 months at 11 underpasses, there were 1,338 detections of elk *Cervus canadensis*, 538 of deer *Odocoileus* spp., 373 of coyotes *Canis latrans*, 97 of black bears *Ursus americanus*, 77 of wolves *Canis lupus*, 29 of cougars *Puma concolor* and six of grizzly bears *Ursus arctos*. Most visits resulted in completed passages (96–100%, depending on species). Bighorn sheep *Ovis canadensis*, mountain goats *Oreamnos americanus* and moose *Alces alces* were also detected (frequency not reported). Elk, deer and coyotes used all 11 underpasses, black bears used nine, wolves used six, cougars used five and grizzly bears used three underpasses. The study was conducted along 27 km of a four-lane highway. Wildlife movements were monitored through seven cement open-span underpasses, under two bridges over creeks and through two metal culverts. Barrier fencing, 2.4 m high, ran alongside the highway. Underpasses, constructed in 1986–1991, were located in twinned highway sections. Animal tracks were monitored at each end of each crossing within a sand, silt and clay mix (2 × 4 m) every 3–4 days from November 1996 to October 1997.

A study in 1999 along a highway in Alberta, Canada (10) found that drainage culverts, in areas with roadside wildlife exclusion fencing, were used by small-and medium-sized mammals. Crossings at 24 culverts included snowshoe hare *Lepus americanus* (13 crossings at 8 culverts), red squirrel *Tamiasciurus hudsonicus* (6 crossings at 4 culverts), deer mouse *Peromyscus maniculatus* (161 crossings at 14 culverts), voles *Arvicolinae* spp. (5 crossings at 3 culverts) and shrews *Sorex* spp. (43 crossings at 16 culverts). Weasels *Mustela* sp., and martens *Martes americana* also used culverts. Culvert use positively correlated with traffic volume and road width (hare, squirrel, vole), road clearance (squirrel) and culvert length (hare, vole) and negatively correlated with distance to cover (vole), age (hare, squirrel) and openness (squirrel, vole). Shrews preferred larger, more open culverts. Vegetation cover effected use by hares, squirrels and voles. The Trans-Canada highway was expanded to four lanes, with 2.4-m-high wildlife exclusion fencing, in three sections, completed in 1986, 1988 and 1997. Twenty-four drainage culverts were monitored along a 55-km highway section, using multiple sooted track-plates (75 × 30 cm) in each culvert. Plates were checked weekly in January–March 1999. Structural and landscape variables were recorded at culverts.

A study in 1995–1998 along a highway in Alberta, Canada (11; same experimental set-up as 9) found that underpasses, in areas with roadside barrier fencing, were used by large herbivores and carnivores. A total of 8,959 elk *Cervus canadensis* appearances, 2,411 deer *Odocoileus* sp. appearances and two moose *Alces alces* appearances were recorded at 11 underpasses. There were also 193 appearances of black bears *Ursus americanus*, seven of grizzly bears *Ursus arctos*, 117 of cougars *Puma concolor* and 311 of wolves *Canis lupus*. On 98% of visits, the animal passed through. Features that positively influenced use of underpasses included increased length, noise level and distance to drainage. Increased width, openness, distance to forest and human activities negatively influenced their use. Nine cement open-span underpasses and two metal culverts (length: 26–96 m, width: 4–15 m, height: 2.5–4.0 m) were monitored along a 27-km stretch of the four-lane Trans-Canada Highway. Barrier fencing, 2.4 m high, ran alongside the highway. Tracks were monitored in sand or clay at each end of each crossing, every 3–4 days, from January 1995 to March 1996 and November 1996 to June 1998. Information about structure, landscape and human activity were recorded for each underpass.

A study in 1997 along a highway in Jutland, Denmark (12) found that an underpass, in an area with roadside barrier fencing, was used by four mammal species. These were red fox *Vulpes vulpes* (122 observations, 161 tracks), badger *Meles meles* (16 observations, 22 tracks), stone marten *Martes foina* (18 observations, 41 tracks) and roe deer *Capreolus capreolus* (20 observations, 41 tracks). The roe deer records were all accounted for by a single male, with other animals present in the area not using the underpass. Three brown hares *Lepus europaeus* were observed entering the underpass, but all turned around and did not pass through. The entrance of a tunnel underpass (13 m wide, 7.5 m high, 155 m long) was monitored using a video camera and two infra-red lamps for 30 days in April–May and in August–September 1997 (total 495 hours). Tracks in sand at either end of the stream through the underpass were recorded daily. There was 1.8-m-high fencing both sides of the highway, for 1 km in each direction from the underpass.

A before-and-after study in 1981–1999 in temperate mixed woodland forest and grassland in Alberta, Canada (13) found that underpasses and overpasses, along with roadside fencing, reduced road deaths of large mammals. Wildlife-vehicle collisions were significantly lower during the two years after fencing (5–28/year) compared to the two years before (18–93/year) for all three road sections, despite an increase in traffic flow. Ungulate casualties declined by 80%. Species included coyote *Canis latrans*, black bear *Ursus americanus*, wolf *Canis lupus*, bighorn sheep *Ovis canadensis*, moose *Alces alces*, deer *Odocoileus* spp. and elk *Cervus canadensis*. Most road deaths were within 1 km of the end of the fences. Deaths also occurred close to drainage structures. The Trans-Canada highway was expanded to four lanes and had 2.4-m-high wildlife exclusion fence installed in three phased sections, completed in 1984 (10 km), 1987 (16 km) and 1997 (18 km). Twenty-two wildlife underpasses and two overpasses were constructed along these sections. Wildlife-vehicle collisions were monitored from May 1981 to December 1999.

A study in 2002–2003 of a highway bisecting forest blocks in Victoria, Australia (14) found that an underpass, along with roadside fencing, was used by 13 native mammal species. These comprised 76% of mammal species recorded in the adjacent forest (bats not included). The underpass was used by koalas *Phascolarctos cinereus*, wombats *Lasiorhinus latifrons*, echidnas, macropods (e.g. kangaroos, wallabies), rodents and carnivorous marsupials (four of five species), and gliders and possums (four of seven species). In 1997, a 70-m wide underpass was built under a split dual-carriageway bridge. Some vegetation was retained and some planted within the underpass. Barrier fencing, 2 m high, ran the length of the highway (with koala escape poles). Intensive sampling was carried out for one week/month in July 2002–June 2003, within the underpass and at two forest sites, 100 m and 320 m from the underpass. Small mammal traps, hair tubes, nest boxes for arboreal mammals, spotlight counts, track surveys and scat surveys were used to monitor wildlife.

A replicated study in 2000–2003 along a highway in Pennsylvania, USA (15) found that a range of mammals used box culverts and bridge underpasses, some of which were in areas with roadside fencing. In the first phase, eight of nine culverts were used by mammals, with white-tailed deer *Odocoileus virginianus* (one culvert), raccoon *Procyon lotor* (seven), opossum *Didelphis marsupialis* (two), feral cat *Felis catus* (one), long-tailed weasel *Mustela frenata* (one), red fox *Vulpes fulva* (one), striped skunk *Mephitis mephitis* (one) and black bear *Ursus americanus* (one) recorded. In the second phase, white-tailed deer used nine of 20 larger culverts (with higher cross-section:length ratios). Black bears, opossums, raccoons and muskrats *Ondatra zibethicus* also used these culverts. Deer did not use culverts >90 m long, but use was not affected by substrate (concrete, natural or water). In September–November 2000, nine culverts were monitored using infrared-triggered cameras. Approximately half of the culverts had sediment on their floors. Twenty larger culverts that were considered suitable for deer (out of 70) were monitored using cameras, 10 in September–November 2002 and 10 in May–July 2003. Entrances to 13 of these were separated from roads by right-of-way fencing.

A before-and-after study in 2002–2003 along a highway in Arizona, USA (16; same experimental set-up as 25) found that two open-span bridge underpasses, in areas with roadside elk-proof fencing, were used by elk *Cervus canadensis* but not by white-tailed deer *Odocoileus virginianus* and vehicle-deer collisions did not decrease after installation. A total of 181 collisions were reported, with no difference in rates along the section before and after the two underpasses were constructed. GPS collars recorded 675 highway crossings by elk, only 6% of which were through underpasses. Overall, 62% of 1,435 elk, but only 0.4% of 257 white-tailed deer recorded on cameras at underpasses crossed through them. Two open-span bridge underpasses (<250 m apart) along the State Route 260 highway were monitored using video cameras and track counts (inside and 60 m from entrances). Cameras were also installed at the ends of the short sections of elk-proof fencing. Thirty elk were tracked using GPS collars (May 2002 to July 2003). Vehicle-deer collisions were recorded before and after underpass installation.

A study in 2001–2003 along two highways in Wyoming, USA (17) found that use of underpasses, in areas with roadside fencing, by mule deer *Odocoileus hemionus* decreased with a decrease in underpass width. Only one of the six underpasses was consistently used by mule deer, accounting for 91% of the 1,028 recorded crossings made through all underpasses. It had a high cross section:length ratio and was near a historic migration route. At an experimental underpass, the percentage of deer turning away from the underpass increased significantly as the cross section:length ratio decreased. Six (of 12) underpasses along a section of Interstate 80 were monitored. Four were box type and two were small gravel road underpasses. Use was assessed using infrared-triggered cameras and track surveys. One experimental underpass was installed in 2001. It was 18 m long. The width was experimentally manipulated from 3–6 m and height from 2–3 m. Video cameras recorded deer behaviour. Underpasses were monitored from autumn 2001 to spring 2003. Fences, 2.4 m high, ran alongside the highway.

A study in 2001–2002 along a highway in Wyoming, USA (18) found that an underpass, in an area with roadside deer-proof fencing, was used by pronghorn *Antilocapra americana*. A total of 70 pronghorns passed through the underpass over 11 occasions between December and April (group size 1–57). These animals did not hesitate before crossing. An additional 19 pronghorns approached the structure but did not cross. All but two crossings took place at dusk or pre-dawn and most were in the presence of mule deer *Odocoileus hemionus*. A 2.4-m-high deer-proof fence was constructed in 1989 alongside 11 km of United States Highway 30. In 2001, a wildlife underpass was constructed. Underpass use was monitored using motion sensors with infrared-triggered cameras at either end from October 2001 to May 2002.

A study in 2002–2003 along a highway in Montana, USA (19) found that seven bridge underpasses, in areas with roadside fencing, were used by white-tailed deer *Odocoileus virginianus*, mule deer *Odocoileus hemionus* and elk *Cervus canadensis*. White-tailed deer were photographed 791 times, mule deer 379 times and elk 100 times. Between 38 and 430 deer were recorded at each underpass, but none in culverts. Smaller numbers were recorded of striped skunk *Mephitis mephitis* (nine photographs), raccoon *Procyon lotor* (three), red fox *Vulpes vulpes* (one), coyote *Canis latrans* (three) and black bear *Ursus americanus* (one). There were no significant relationships between wildlife use and underpass structural features. Distribution of mammal road deaths was independent of underpass locations. Seven bridge underpasses and three culverts were monitored along an 80-km highway section from October 2002 to July 2003. Crossings connected with roadside fencing, though this was inadequately maintained and was permeable to deer. Heat-and motion-sensitive cameras were used at underpasses (for 101–700 camera days/ underpass). Details about location, structure, vegetation cover and human activities were recorded for each underpass. Road deaths were opportunistically recorded and combined with data collected by road maintenance crews (spanning 1998–2002).

A study in 2000–2001 in coastal lowlands in New South Wales, Australia (20) found that concrete wildlife culverts, in areas with roadside fencing, were used by small and medium-sized mammals. Mammal tracks made up 82% of all vertebrate tracks recorded. These were made by bandicoots Perameloidea (25% of all tracks), rats (25%), wallabies (13%), mice Muridae (10%), feral cat *Felis catus* (<2%) and red foxes *Vulpes vulpes* (<2%). Koala *Phascolarctos cinereus* tracks were recorded twice. In cage traps, house mouse *Mus musculus* (29 individuals) and swamp rat *Rattus lutreolus* (16 individuals) were the most common among six species (67 individuals) caught. Nine concrete culverts along a 2.5-km section of highway were monitored. They were 2.4 m wide, 1.2 m high and 18 m long. A 1.8-m-high fence ran along either side of the road. Tracks were recorded on sand in culverts from 22–30 September 2000 and 1–9 December 2000. Between 15 and 17 cage traps were set in and next to each culvert on four nights in September 2000 (560 trap-nights).

A study in 2001–2002 along a highway in Florida, USA (21) found that culverts, in areas with roadside barrier walls, were used by mammals but road casualties still occurred. Ten mammal species (and one species pair) were recorded using culverts. These included rice rat/hispid cotton rat *Oryzomys palustris/Sigmodon hispidus* (in five culverts), cotton mouse *Peromyscus gossypinus* (three culverts), round-tailed muskrat *Neofiber alleni* (three culverts) and southeastern short-tailed shrew *Blarina carolinensis* (two culverts). Other species used one culvert each. During the same period, ≥13 mammal species were recorded dead on the road. The most frequent casualties were rice rat (25), Virginia opossum *Didelphis virginianus* (15) and nine-banded armadillo *Dasypus novemcinctus* (10). Culverts reduced overall vertebrate road mortality, but separate mammal figures were not reported for before culverts were installed. Eight culverts (from 0.9 m diameter to 2.4 × 2.4 m cross-section, all 44 m long) were connected using prefabricated concrete barrier walls. Culverts were monitored from 14 March 2001 to 5 March 2002 using funnel traps, camera traps and sand track stations. Roadkills were monitored by walking the 3.2-km road over three consecutive days each week.

A study in 1999–2000 in Alberta, Canada (22) found that small culverts, in areas with roadside barrier fencing, were used by mice and voles more than were larger underpasses. More translocated animals returned to their capture location through 0.3-m-diameter culverts (deer mice *Peromyscus maniculatus*: 100% returned; red-backed voles *Clethrionomys gapperi*: 86%; meadow voles *Microtus pennsylvanicus*: 58%) than through 3-m-wide underpasses (69, 49, 10% respectively). More animals successfully returned through underpasses (and overpasses) with 100% vegetation cover at entrances (55–100% of animals returned) compared to those with 50% (20–76%) or no cover (0–66%). Animals crossed within 1–4 days. Nine vegetated soft-bottomed, unvegetated arch-shaped underpasses (64–73 m long) and nine metal drainage culverts with grass cover (63–72 m long) were studied. Crossings were linked to roadside fencing that limited movements of large animals. Territorial mice and voles were captured using Longworth live traps (166 caught), ear-tagged, coated with fluorescent powder, taken across the road, released at standardized distances from crossings (20, 40, 60 m) and followed as they returned. Vegetation cover 2 m inside and outside entrances was varied using spruce branches to 100%, 50% and no cover. Traps at original capture sites were monitored for four days after translocation. Monitoring was undertaken in July–October 1999 and 2000.

A study in 1997–2000 of a highway in Alberta, Canada (23) found that underpasses, in areas with roadside fencing, were used by large mammals. The 11 underpasses were visited by elk *Cervus canadensis* (1302 records), deer *Odocoileus* sp. (543), cougars *Puma concolor* (105), black bears *Ursus americanus* (103), wolves *Canis lupus* (43) and grizzly bears *Ursus arctos* (six). The majority of animals that visited underpasses crossed through the structures. Underpass height and width were both positively correlated with the number of animals using them. Two bridge underpasses (3 m high, 11 m wide), four concrete box underpasses (2.5 × 3.0 m) and five metal culverts (4 m high, 7 m wide) were monitored along an 18-km stretch of the four-lane Trans-Canada Highway. Barrier fencing, 2.4 m high, ran along the highway. Tracks were monitored at each end of each crossing, in a 2 × 4-m sand, silt and clay tracking station, every 3–4 days from November 1997 to August 2000. Information about each structure, the surrounding landscape, and human activity were recorded for each underpass.

A study in 2002 of a highway in Zamora, Spain (24; same experimental set-up as 32) found that underpasses and culverts, in areas with roadside barrier fencing, were used by mammals. Circular culverts were used by hedgehog *Erinaceus europaeus*, garden dormouse *Eliomys quercinus*, badger *Meles meles*, common genet *Genetta genetta* and red fox *Vulpes vulpes*. Adapted (enlarged) culverts were used by red squirrel *Sciurus vulgaris*, badger and red fox. Open-span underpasses were used by hedgehog, badger, red fox and red deer *Cervus elaphus*. Wildlife underpasses were used by hedgehog, badger, common genet and red fox. Crossings were also used by rodents and shrews, rabbit *Oryctolagus cuniculus*, Iberian hare *Lepus granatensis*, weasel *Mustela nivalis*, European wildcat *Felis silvestris* and wolf *Canis lupus* (see paper for details). Sixty-four underpasses/culverts (30–150 m long) under a 72-km section of motorway were monitored. These included 33 circular drainage culverts (2 m diameter), 10 wildlife-adapted box culverts (2–3 m wide, 2 m high), 14 open-span underpasses (rural tracks/paths, 4–9 m wide, 4–6 m high) and seven wildlife underpasses (20 m wide, 5–7 m high). The motorway was barrier-fenced. Animal tracks were monitored over 10 days in June–September 2002 using marble dust (1-m-wide cross). Camera traps verified species identifications in some underpasses.

A study in 2002–2005 along a highway through riparian meadows in Arizona, USA (25; same experimental set-up as 16) found that two open-span bridge underpasses, in areas with roadside ungulate-proof fencing, were used by Rocky Mountain elk *Cervus canadensis nelsoni*, with a more open, shorter underpass with natural sides being used most frequently. In total, 3,708 elk, in 1,266 groups, were recorded at the two underpasses (91% of all mammals recorded) with 2,612 elk in 905 groups passing through the underpasses. More elk groups passed through the shorter underpass (663 groups) than through the longer underpass (242 groups). Seven additional mammal species were recorded at the two underpasses (species not stated in paper). Two open-span bridge underpasses (<250 m apart), along the State Route 260 highway, were studied. Fencing, 2.4 m high, along 0.6 km of highway, funneled animals towards underpasses. Underpasses were monitored using four video cameras, in September 2002 to September 2005. The shorter underpass was 7 m high, 10 m wide and 53 m long, with open, natural sides. The longer underpass was 12 m high, 16 m wide and 111 m long, with concrete walls.

A replicated study in 2004 along two roads through agricultural land in Alentejo, Portugal (26) found that all 34 monitored culverts, some in areas with roadside fencing, were used by mammals. Crossings were made by small mammals (289 crossings, 34 culverts), hedgehogs *Erinaceus europaeus* (55 crossings, 15 culverts), hares and rabbits (71 crossings, 15 culverts), weasels *Mustela nivalis* (16 crossings, 9 culverts), stone martens *Martes foina* (93 crossings, 28 culverts), Eurasian badgers *Meles meles* (55 crossings, 10 culverts), otters *Lutra lutra* (2 crossings, 2 culverts), common genets *Genetta genetta* (65 crossings, 20 culverts), Egyptian mongooses *Herpestes ichneumon* (82 crossings, 21 culverts) and red foxes *Vulpes vulpes* (27 crossings, 12 culverts). A total of 34 culverts (<1.0 m wide, 8–25 m long) were monitored along two roads (17 culverts along each). Road sections studied were 16 and 30 km long. There was 1.5-m-high roadside fencing along the 30-km section. Tracks were monitored using marble dust (60–100 cm wide) which was placed inside each end of each culvert. Tracks were recorded on four days in each of spring, summer and autumn 2004 (total 408 culvert monitoring days).

A study in 2007 along a road, in Northumberland, UK (27) found that three underpasses, with entrances fenced off from the road, were used by several species of small and medium-sized mammals to make crossings. Tracks were identified of western hedgehog *Erinaceus europaeus*, brown rat *Rattus norvegicus*, badger *Meles meles* and American mink *Mustela vison*. The number of underpasses used and frequency of use was not detailed in the paper. Underpasses, 0.6–0.9 m wide, were constructed in 2003–2006 along a 46-km stretch of road and were fenced off from the road. Mammal use was monitored in August–October 2007. Clay-based drain seals (45 × 45-cm surface and 0.5 cm thick), used as footprint pads, were placed at entrances to three dry culverts and checked weekly for footprints.

A before-and-after study in 1990–2005 along a highway in Québec, Canada (28) found that an underpass was used by moose *Alces alces* and, along with electric fences, it reduced moose-vehicle collisions. Twenty-three sets of moose tracks were recorded in the underpass over three years. There were fewer moose-vehicle collisions after fence construction (zero) than before (1.4/year). An underpass (23 m long, 16 m wide, 7 m high) was established along both side of a river, under a bridge along the highway. Electric fences (1.5 m high, wires 0.3 m apart) were installed along both sides of a 5-km highway section, encompassing the underpass, in 2002. Data on moose-vehicle collisions before fence installation were collated by the Ministère des Transports du Québec, between 1990 and 2002. Details of monitoring collisions after installation are not given.

A before-and-after study in 2004–2007 along a highway through eucalypt woodland in Queensland, Australia (29) found that two underpasses, in areas with roadside barrier fencing, were used by mammals and the mammal road casualty rate fell after construction. There were three wild mammal road casualties over 29 months post-construction and six during four months pre-construction. This comparison was not tested for statistical significance. Tracks detected in underpasses were from rodents (370 tracks), house mice *Mus musculus* (115), Dasyurid sp. (most likely Common dunnart *Sminthopsis murina*) (17), northern brown bandicoots *Isoodon macrourus* (179), possums (16), red-necked wallabies *Macropus rufogriseus* (3), short-beaked echidnas *Tachyglossus aculeatus* (2) and from feral cats *Felis catus*, dogs *Canis lupus familiaris* and brown hares *Lepus europaeus*. Proportions of tracks representing full crossings varied by species with the highest figure for wild mammals being for possums (18–40% of records). In 2004, a 1.3-km section of highway was upgraded to four lanes and a variety of wildlife crossings constructed, linked by barrier fencing (2.5 m high). Use of two underpasses (2.4 m high, 2.5 m wide, 48 m long) with water flowing through and ledges attached to side walls, was monitored, starting six months after construction. Tracks were counted on sand within each entrance, twice weekly from August 2005–February 2006 and monthly from June 2006–June 2007. Road-kill was monitored twice weekly before (April–July 2004) and weekly after construction until June 2007.

A before-and-after study in 1996–2004 in Florida, USA (30, same experimental set-up as 35 and 44) found that two underpasses, along with roadside barrier fencing, reduced Florida Key deer *Odocoileus virginianus clavium* collisions with vehicles by 94%. There were 2 collisions/year over two years after fence construction compared to 12–20 collisions/year over five years before construction (total 79 collisions). Underpass use increased over time, with 22 photographs of deer/month over the first six months and 59/month over the following six months. Average annual deer ranges and core areas did not change after underpass construction. Only 45% (5/11) of radio-collared deer were located on both sides of the highway after construction compared to 100% (9/9) before. In 2002, two box underpasses (14 × 8 × 3 m) were constructed with 2.6-km-long barrier fencing (2.4 m high) and four deer guards (modified cattle guards) installed between them, along a two-lane highway. Deer mortalities on roads were recorded from 1996, by direct sightings, law enforcement reports and observations of vultures. Underpass use was monitored using infrared-triggered cameras from February 2003–January 2004. Deer were radio-tracked between January 1998 and December 2000 (44 deer) and between February 2003 and January 2004 (32 deer) and were located 6–7 times/week.

A replicated study in 2004 along two roads in southern Portugal (31) found that underpasses and culverts along roads bounded by livestock fencing were used by carnivore species to cross highways. Crossing rates of underpasses were similar to those of culverts for red fox *Vulpes vulpes* (underpasses: 0.25 crossings/day; culverts: 0.11), badger *Meles meles* (underpasses: 0.30; culverts: 0.15), genet *Genetta genetta* (underpasses: 0.15; culverts: 0.9) and Egyptian mongoose *Herpestes ichneumon* (underpasses: 0.29; culverts: 0.22). Stone marten *Martes foina* used underpasses more (0.22 crossings/day) than they used culverts (0.05 crossings/day). Fifty-seven passages under 252 km of two major roads were monitored. They comprised 1.2 circular culverts/km (1 and 1.5 m diameters), 0.3 box culverts/km (2 × 2 m to 5 × 5 m), and 0.5 underpasses/km (5 m high and 8 m wide). Crossing structures were 5–1,566 m apart. Livestock fencing, 1.5 m high, ran along both sides of both roads. A 1-m² plot of marble dust was placed at each end and in the middle of each passage. This was checked for tracks every five days, over 20 consecutive days of monitoring, in both spring and summer 2004.

A study in 2001 along a highway in Zamora province, Spain (32; same experimental set-up as 24) found that road underpasses and culverts, in areas with roadside barrier fencing, were used by mammals. Wildlife underpasses were the most used out of four structure types, by polecats *Mustela putorius* (detected on average on 0.2/10 days/underpass), roe deer *Capreolus capreolus* (0.4/10), red deer *Cervus elaphus* (0.4/10), wild boar *Sus scrofa* (0.6/10) and rabbits and hares (1.2/10). Open-span underpasses was the most used structure by small-spotted genets

Genetta genetta (0.3/10) and red foxes *Vulpes vulpes* (4.7/10). European badgers *Meles meles* (3.1/10) and rats (0.4/10) used wildlife-adapted box culverts more than other structure. Small mammals (1.6/10) were most frequently recorded in circular culverts. Thirty-three crossings were monitored. These comprised five wildlife underpasses (14–20 m wide, 5–8 m high, 30–96 m long), seven open-span underpasses (rural tracks/paths, 4–9 m wide, 4–6 m high, 32–72 m long), seven wildlife-adapted box culverts (2–4 m wide, 2–3 m high, 36–45 m long) and 14 circular drainage culverts (2 m diameter, 35–62 m long). The motorway had barrier fencing along its length. Animal tracks were recorded using marble dust (1-m-wide cross) over 10 days in March–June 2001.

A before-and-after study in 2002–2005 along a highway through mixed forest and farmland in southwestern Sweden (33; same experimental set-up as 34) found that following installation of an underpass, overpasses and barrier fencing, moose *Alces alces* road casualties declined but moose did not use the underpass. There were fewer moose-vehicle collisions after fence construction (zero/year) than before (2.7/year). During construction, 1.8 collisions/year were recorded. Moose were recorded crossing the highway 47 times before construction of crossing features, 76 during and 12 times after features were installed. All crossings after fencing prevented direct road access were via the two wildlife overpasses. Two 6-km sections of a highway were converted to a fenced four-lane highway in 2000–2004. The sections contained one wildlife underpass (35 m long, 4.7 m high, 13 m wide), two wildlife overpasses, three conventional road tunnels and two conventional bridges that could be crossed. Twenty-four moose were radio-collared. Locations were recorded every two hours before construction (February–September 2002), during construction (October 2002–May 2004) and after construction (June 2004–December 2005; 8,830 moose days).

A before-and-after study in 2000–2005 in forest and farmland in southwestern Sweden (34; same experimental set-up as 33) found that barrier fencing and three road crossings reduced moose *Alces alces* and roe deer *Capreolus capreolus* road-kills. Deaths were reduced 70% from averages of 2.7 moose killed/year and 5.3 roe deer killed/year over the 12 years pre-construction. In 2000–2004, a 12-km section of the European Highway 6 was converted from two to four lanes and 2.2-m-high exclusion fencing was installed along its length. Two overpasses and one underpass were also constructed. Moose and deer casualty rates were collated from casualties reported to police pre-construction (1990–2001) and post-construction (up to 2005).

A before-and-after study in 1996–2005 along a highway in Florida, USA (35; same experimental set-up as 30 and 44) found two underpasses with associated barrier fencing reduced vehicle collisions with Florida Key deer *Odocoileus virginianus clavium*. Fewer deer were killed on the fenced road section after underpass and fence installation (0–3/year) than before (11–20/year). There were more collisions on unfenced road sections after installation (40/year) than before (24/year), so collisions were not reduced overall. However, deer densities increased and the ratio of collisions to deer numbers suggested that risks of collisions decreased after construction. Deer use of two underpasses increased from the first year after construction (871 detections) to the second and third years (1,857 and 1,629 deer detections respectively). A 2.6-km-long system with two underpasses (dimensions not stated), 2.4-m-high fencing and four deer guards were constructed on US Highway 1. An infrared trail monitor and camera monitored deer passages at the centre of each underpass for three years post-construction (2003–2005). Deer-vehicle collisions were recorded (from 1996) from direct sightings, citizen and law enforcement reports and observations of vultures before (1996–2000) and after (2003–2005) fence and underpass construction.

A site comparison study in 2006 along a Highway in New South Wales, Australia (36) found that two underpasses were used by mammals and that presence of crossing-structures along with barrier fencing reduced road-kills. There were fewer road-kills over seven weeks along the section with crossing-structures (0.02/km of survey) than along a section without crossings (0.09/km of survey). The most frequently recorded road casualties were bandicoots (16 casualties) and kangaroos and wallabies (nine casualties). Bandicoots used the two underpasses more than they used the two overpasses (87 vs 28 tracks) as did rodents (82 vs 15). Kangaroos and wallabies used underpasses less than they used overpasses (36 vs 104 tracks). Use was similar between structure types for possums (14 vs 9). There were two concrete box culverts (3 × 3 m, 42–63 m long) and two wildlife bridges (9–37 m wide, with vegetation) with 5 km of exclusion fencing, along a 12-km section of dual-carriageway highway. Tracks were monitored on sand plots across each crossing. Road-kill surveys were conducted along the 12-km section and along a 51-km two-lane section without crossings or fencing. Track and road-kill surveys were conducted up to three times/ week over seven weeks in August–September 2006.

A study in 2001–2005 along a motorway through forest and agricultural land in Germany (37) found that most underpasses and overpasses, in areas with roadside fences, were used by wildcats *Felis silvestris* to cross roads. Wildcats used crossing structures on 18 of the 21 occasions on which they were recorded <50 m from the motorway. The three underpasses were each used by one cat from a total of eight wildcats that had underpasses located within their home ranges. One 40-m-wide underpass and two road underpasses (9–14 m wide), along with two open-span viaducts and two forest road overpasses, were monitored in 2002–2005. All underpasses were 29 m long. Underpasses were connected to fencing that was designed specifically to exclude wildcats from the road. Twelve wildcats were radio-collared between January 2001 and February 2005. Animals were tracked at night for 3–30 months each.

A study in 1999–2001 along a road through beech and fir forest in Gorski kotar, Croatia (38) found that an underpass below a section of road on a viaduct, and separated from the road by barrier fencing, was used by medium to large-sized mammals. Tracks were recorded of roe deer *Capreolus capreolus* (total 20 tracks), red deer *Cervus elaphus* (12) wild boar *Sus scrofa* (1), brown bear *Ursus arctos* (4), grey wolf *Canis lupus* (1) and Eurasian lynx *Lynx lynx* (1). However, the underpass had five times fewer mammal crossings/day than did three overpasses (100–835 m wide). A new highway was constructed in 1998–2004 with 44 wildlife crossings and 2.1-m barrier fencing along a 9-km section. An underpass (569 m wide, below a 25-m-high road viaduct) was monitored. Tracks (in snow, mud or sand) and other animal signs were counted 23 times in January 1999–January 2001.

A site comparison study in 2000–2007 along a highway in North Carolina, USA (39) found that underpasses and barrier fencing facilitated road crossings by a range of mammals but did not reduce road casualties. Camera traps showed crossings through the three underpasses by white-tailed deer *Odocoileus virginianus* (2,258 times), raccoon *Procyon lotor* (125), American black bear *Ursus americanus* (15), bobcat *Lynx rufus* (11), grey fox *Urocyon cinereoargenteus* (eight), Virginia opossum *Didelphis virginiana* (six), rabbits *Sylvilagus* spp. (two) and *Canis* spp. (two). Track counts indicated an additional 3,552 mammal crossings by 15 species, with 90% by white tailed deer. A similar number of mammals was killed over one year on road sections with underpasses and fencing (5.0/km) as on sections without (5.1/km). A four-lane highway was constructed with three underpasses. Barrier fencing, 3 m high, was installed ≥800 m along the highway from each underpass. Gates allowed trapped animals to escape the highway. Underpass use was monitored by 2–3 camera traps /underpass. Twice-weekly track surveys were conducted (on 2.5-m-wide plates across underpasses). Road deaths were recorded along 6 km of road with fencing and underpasses and 11 km without, twice/week, from July 2006–July 2007.

A review of 30 papers reporting on monitoring of 329 crossing structures in Australia, Europe and North America (40) found that mammals used most culverts and underpasses, among which some were in areas with roadside barrier fencing. Small mammals used pipes (demonstrated by 6/7 relevant studies), drainage culverts (5/5 studies), adapted culverts (5/5 studies), wildlife underpasses (3/4 studies) and bridge underpasses (2/3 studies). Arboreal mammals used pipes (1/1 studies), drainage culverts (4/4 studies), adapted culverts (4/4 studies) and bridge underpasses (1/1 studies). Medium-sized mammals used pipes (8/11 studies), drainage culverts (12/13 studies), adapted culverts (8/8 studies), wildlife underpasses (6/8 studies) and bridge underpasses (6/7 studies). Large mammals used pipes (6/9 studies), drainage culverts (11/12 studies), adapted culverts (11/11 studies), wildlife underpasses (24/24 studies) and bridge underpasses (14/15 studies). Larger mammals tended to use more open underpasses. Small and medium-sized mammals used underpasses with funnel-fencing or adjoining walls and those with vegetation cover close to entrances. Those with vegetation cover tended to be avoided by some ungulates. Thirty papers reporting monitoring of 329 crossing structures were reviewed. Fourteen papers investigated multiple structure types, resulting in a total of 52 studies of different structure types. Underpasses, from small drainage pipes to dry passage bridges, comprised 82% of crossings.

A study in 2003–2007 at six sites along a highway through forest and shrubland in Arizona, USA (41) found that underpasses, in areas with ungulate-proof fencing, were used by white-tailed deer *Odocoileus virginianus* and that underpass use was not affected by traffic levels. Crossing rates of white-tailed deer that approached underpasses did not differ significantly between traffic volume levels of 0 vehicles/minute (0.28 crossings/approach), 1–2 vehicles/minute (0.34 crossings/approach), 2–4 vehicles/minute (0.40 crossings/approach), 4–6 vehicles/minute (0.27 crossings/approach) and >6 vehicles/minute (0.28 crossings/approach). Deer passage rates and traffic flows were monitored at six wildlife underpasses beneath 27 km of an upgraded four-lane highway. Underpasses were 53–128 m long and 5–15 m high. Five underpasses had a fenced above-ground section (11–48 m long) between the two carriageways. Roadside fencing, 2.4 m high, was gradually installed with the full road section fenced by 2006. Four video cameras with infrared beams monitored traffic and deer at each underpass in 2003–2007. The number of deer approaching within 50 m of underpasses and the number crossing the highway through underpasses was counted.

A replicated study in 2010 at 38 sites along nine roads in England, UK (42) found that underpasses, in areas with roadside fencing, were used by badgers *Meles meles*, Eurasian otters *Lutra lutra*, red foxes *Vulpes vulpes*, European hedgehogs *Erinaceus europaeus* and brown rats *Rattus rattus* to cross roads. Of 38 underpasses monitored, 34 were used by badgers. Eurasian otters, red foxes, European hedgehogs and brown rats used underpasses, but the number of underpasses used or crossing frequencies are not reported. Badger footprints were recorded 7–8 times in 14 underpasses, 4–6 times in 11 underpasses and 1–3 times in 9 underpasses. Mammals were monitored in 38 underpasses, installed in 2003–2007, under single carriageway roads (16 underpasses), dual carriageways (20 underpasses), a motorway (one underpass) and a junction (one underpass). Underpasses were 20–120 m long, 0.3–1 m in diameter (most were 0.6 m diameter) and were made of concrete and corrugated iron. Roadside fence characteristics are not specified. Mammals were surveyed weekly, between August and October 2010, by monitoring footprints in a clay mat (45 × 45 cm) at the entrance of each underpass.

A replicated study in 2002–2008 along a highway in Arizona, USA (43) found that wildlife underpasses, in areas with roadside ungulate-proof fencing, were used by mammals. Six underpasses were approached 14,683 times by wild mammals, of 15 species. Of all animals recorded (which included also 450 records of domestic animals and one of a bird) 72% crossed through underpasses. Elk *Cervus canadensis* accounted for 70% of visits by wild mammals to underpasses, white-tailed deer *Odocoileus virginianus* for 13% and mule deer *Odocoileus hemionus* for 7%. Other crossings comprised coyote *Canis latrans* (1%), grey *fox Urocyon cinereoargenteus* (2%), raccoon *Procyon lotor* (2%) and other mammals (4%). Reconstruction of a 27-km stretch of State Route 260 was undertaken in 2000–2006 and included creation of 11 large wildlife underpasses, connected to ungulate-proof fencing. Six underpasses (34–41 m wide, 5–12 m high and 53–128 m long) were monitored for an average 4.7 (2.5–5.5) years using animal-triggered multi-camera video surveillance.

A before-and-after, site comparison study in 1996–2009 along a highway through woodland and developed areas in Florida, USA (44; same experimental set-up as 30 and 35) found that underpasses beneath the highway, along with roadside fencing, reduced vehicle collisions with Florida Key deer *Odocoileus virginianus clavium*. Fewer deer were killed on the road over seven years after underpass and fence installation (1.6/year) than in the five years before installation (15.6/year). Concurrently, along an unfenced section without underpasses, 43 deer/year were killed in the latter period and 24/year were killed in the earlier period. Underpass use increased from 185 passages during the first year after construction to 1,337 passages in the seventh year after construction. A highway was upgraded to increase vehicle capacity, with construction completed in 2002. Two box culvert underpasses (14 m long, 8 m wide, 3 m high) were installed under a 2.6-km-long fenced road section through undeveloped land. Deer-vehicle collisions were monitored along this section and along an adjacent 3.0-km-long unfenced section through a developed area, before culvert installation (1996–2000) and after (2003–2009). Culvert use was monitored using camera traps.

A before-and-after study in 1990–2011 of scrubland in Wyoming, USA (45) found that underpasses beneath a highway, in areas with roadside game-proof fencing, were extensively used by mule deer *Odocoileus hemionus* and collisions between deer and vehicles reduced. Over three years, 49,146 mule deer were recorded moving through seven underpasses. Passage rates through underpasses of deer approaching to ≤50 m increased over three years, from 54% to 92%. After underpass construction, there were 1.8 collisions/month between deer and vehicles compared to 9.8 collisions/month before. Underpasses were also used by elk *Cervus canadensis* (1,953 crossings), pronghorns *Antilocapra americana* (201), coyotes *Canus latrans* (13), bobcats *Lynx rufus* (77), badgers *Taxidea taxus* (9), moose *Alces alces* (13), raccoons *Procyon lotor* (3) and cougars *Puma concolor* (1). Seven concrete underpasses (approximately 6 m wide, 3 m high and 18 m long) and 21 km of fencing were installed in 2001–2008. Three camera traps/underpass were operated from 1 October (16 December in first year) to 31 May between 2008–2009 and 2010–2011. Vehicle-deer collision data were collated before (1 January 1990–1 October 2001) and after underpass construction (1 October 2008–1 May 2011).

A study in 2006–2008 of 18 wildlife crossings under a highway, along with roadside fencing, in a national park in Alberta, Canada (46) found that American black bears *Ursus americanus* and grizzly bears *Ursus arctos* used underpasses. Over three years, 218 crossings of American black bears and 153 of grizzly bears were detected. These were through 13 culverts (black bear: 44 crossings; grizzly bear: 36) and five open-span underpasses (black bear: 174 crossings; grizzly bear: 117). Bear crossings were monitored at 20 of 25 wildlife crossing structures in Bow Valley, Banff National Park, including 18 culverts and underpasses. Fencing (2.4 m high) was installed alongside the road. Bear tracks were counted in May–October 2006, April–October 2007 and April–October 2008 on track pads, comprising 1.5–2 m of sandy loam, spanning the width of the wildlife crossing. Track pads were checked every two days and the species, direction of travel and number of animals was recorded.

A study in 1997–2009 along a major road in California, USA (47) found that all 19 culverts under the road (most of which were in areas with roadside fencing) were used as road crossing points by coyotes *Canis latrans*, bobcats *Lynx rufus*, and mule deer *Odocoileus hemionus*. Coyotes used 18–19 of the 19 culverts studied, and bobcats used 13–19 culverts. Mule deer used 1–4 of the five underpasses considered suitable for them. Ranges represent the numbers of culverts used in each of two survey periods. Sixteen culverts were part of a road upgrade programme, conducted in 2005, that included installation of 3-m-high roadside fencing. From November 1997 to January 2000, remotely triggered cameras were placed in each culvert. Cameras were again placed in each culvert from August 2008 to September 2009. Between the two surveys, the road network was expanded and adjacent habitat was restored.

A review published in 2014 of eleven studies in Australia (48) found that underpasses, separated from roads by fencing, were used by red-necked wallabies *Macropus rufogriseus*, swamp wallabies *Wallabia bicolor*, red-legged pademelons *Thylogale stigmatica*, long-nosed potoroos *Potorous tridactylus* and Lumholtz's tree-kangaroos *Dendrolagus lumholtzi*. At all road underpasses, fencing was used to deter animals crossing roads rather than using underpasses. Underpasses in the study were 1.2–3.4 m high, 2.4–3.7 m wide, and 20–52 m long.

A before-and-after study in 2000–2008 along a highway through swamp and woodland in New South Wales, Australia (49) found that after being extended, underpasses beneath a newly constructed carriageway (in areas with roadside fencing), were used less by northern brown bandicoots *Isoodon macrourus* and long-nosed bandicoots *Perameles nasuta*. Bandicoot crossings through underpasses averaged 0.03/day after underpass extension, compared to 0.5/day during road widening and 1.1/day before widening. Construction of a single-carriageway by-pass finished in 1998. Six underpasses, 90–240 m apart, along 750 m of bypass, were studied. Underpasses were 2.4 m wide, 1.2 m high and 17–19 m long. In 2005–2006, an additional highway carriageway was constructed, with a 20–30-m-wide vegetated central strip. Four underpasses were extended, with an above-ground, enclosed section across the central strip, one underpass ran continuously under both carriageways and one linked with a creek bridge under the new carriageway. Crossings were 49–58 m long. Crossing entrances were separated from the road by 1.8-m-high fencing. Footprint sand pads were checked daily over 4–8 days to document tunnel passages. Underpasses were surveyed five times before widening (spring 2000 to autumn 2005), four times during widening (spring 2005 to spring 2006) and four times after widening (summer 2007 to autumn 2008). Not all underpasses were surveyed each time.

A study in 2012–2013 in six urban sites in Western Australia, Australia (50) found that underpasses, separated from roads by fencing, were used by mammals to cross the road. Southern brown bandicoots *Isoodon obesulus fusciventer* crossed 540 times, western grey kangaroos *Macropus fuliginosus* crossed 186 times and brushtail possums *Trichosurus vulpecula* crossed twice. Underpasses were also used by several invasive mammal species. Road crossings were monitored through 10 underpasses from May 2012 to May 2013, using camera traps. Underpasses were round (0.6–0.9 m diameter) or square culverts (0.6–1.2 m wide, 0.5–1.2 m high). They were 23–88 m long and separated from roads by 0.6–1.8-m-high fences. The time since construction ranged from two to 19 years.

A study in 2010–2012 of a desert region of California, USA (51) found that underpasses in areas with roadside fencing were used by a range of native mammals. There were 3,778 wildlife occurrences (mammals and birds) recorded over 4,279 monitoring days (where a monitoring day is one underpass monitored for one day). Rodents made up 32% of occurrences. Rabbits and hares, mainly desert cottontails *Sylvilagus audubonii*, made up 29%. Birds made up 27% of wildlife occurrences. Other mammals recorded included mule deer *Odocoileus hemionus*, mountain lion *Puma concolor*, bobcat *Lynx rufus*, coyote *Canis latrans* and ground squirrels (frequencies not reported). Seven underpasses, measuring 18–150 m wide, 3–9 m high and 12–112 m long, were studied. Roads were fenced, but gaps allowed animal passage and fences did not funnel animals towards underpasses. Wildlife movements were monitored from July 2010 to November 2012, using camera traps and track pads.

A replicated, site comparison study in 2013 along a highway in Montana, USA (52a) found that underpasses connected with long roadside fences were used by similar numbers of large mammals compared to those with no fences or very short fences. The rate of large mammal crossings through underpasses connected to 6.1–6.2-km-long roadside fences (0.44 mammals/underpass/day) and 1.4–2.7-km-long fences (0.77 mammals/underpass/day) was not significantly different to the rate crossing through underpasses with no fencing or with fences up to 0.4 km long (0.22 mammals/underpass/day). Mammals identified using underpasses were white-tailed deer *Odocoileus virginianus*, mule deer *Odocoileus hemionus*, American black bear *Ursus americanus*, mountain lion *Puma concolor*, grizzly bear *Ursus arctos* and elk *Cervus canadensis*. Twenty-three underpasses were monitored along US Hwy 93 North. Roads were fenced alongside underpasses for 0.0–6.2 km length with 2.4-m high fencing. Wildlife crossings were monitored using ≥1 camera trap/underpass in January–December 2013.

A study in 2012–2013 along a highway in Montana, USA (52b) found that underpasses in areas with roadside fencing were used by white-tailed deer *Odocoileus virginianus* for crossing the road more often than was the road surface. This result was not tested for statistical significance. There were 727 road crossings with 721 by white-tailed deer, three by American black bear *Ursus americanus* and three by either this species or grizzly bear *Ursus arctos*. Eighty-two percent of all crossings were through underpasses and 18% were above the road. Ten fenced underpasses were monitored along US Hwy 93 North. Underpasses were 2–5 m high and 4–40 m wide. Fences were 2.4 m high and 3–256 m long. The proportion of wildlife crossings did not change with fence length (data presented as regression results). Between June 2012 and October 2013, road crossings were monitored for two weeks/underpass using one camera trap at each fence end and at least one at an underpass entrance. Only highway crossings in which animals entered or exited underpasses or accessed or left the highway at a fence end (not returning within ≤3 minutes) were considered.

A study in 2010–2014 of two sites along a highway in Nevada, USA (53) found that underpasses, in areas with roadside fencing, were used by migratory mule deer *Odocoileus hemionus* to cross a road, but less so than were overpasses. Fewer mule deer crossed the road through three underpasses (44–629 deer crossings/underpass/season) than across two overpasses (234–4,007 deer crossings/overpass/season). Crossing structures, 1.5–2.0 km apart, at important crossings for migratory deer, were completed by August 2010 (August 2011 for one overpass). One site had two underpasses and one overpass. The other had one of each structure. Underpasses, 8 m wide, 28 m long and 6 m tall, were oval in cross-section. Concrete arch overpasses, were 31–49 m wide and 8–20 m long. All structures had soil bases. Fencing, 2.4 m high, deterred deer access to the highway between crossings and extended 0.8–1.6 km beyond crossings at each site. Crossings were monitored during eight mule deer migratory periods (autumn 2010 to spring 2014), using camera traps, over 10 weeks in each migration (15 September to 1 December and 1 March to 15 May). Cameras were positioned 12 m apart along crossing structures.

A study in 1996–2014 of a major highway in Alberta, Canada (54) found that culverts, in areas with roadside fencing, were used as crossing points by grizzly bears *Ursus arctos*, but less often than were overpasses, especially by family groups. Over 18 years, grizzly bears used culverts less often (122 crossings/structure) than they used overpasses (241 crossings/structure). Over eight years, bear family groups used culverts less often (0.0–0.3 family groups/year/structure) than they used overpasses (1.4 family groups/year/structure). In 1996–2006, 2-m-wide pads, were covered in sandy-loam soil to survey bear movements at 23 crossing structures. From 2008 to 2014, remote cameras were installed at all crossing structures. As more crossing structures were built in the area, they were added to the survey, up to a maximum of 19 culverts and 18 overpasses. Crossing structure entrances were separated from the road by fencing.

(1) Reed D.F., Woodard T.N. & Pojar T.M. (1975) Behavioral response of mule deer to a highway underpass. *The Journal of Wildlife Management*, 39, 361–367, https://doi.org/10.2307/3799915

(2) Reed D.F. (1981) Mule deer behavior at a highway underpass exit. *The Journal of Wildlife Management*, 45, 542–543.

(3) Ward A.L. (1982) Mule deer behavior in relation to fencing and underpasses on Interstate 80 in Wyoming. *Transportation Research Record*, 859. 8–13.

(4) Singer F.J. & Doherty J.L. (1985) Managing mountain goats at a highway crossing. *Wildlife Society Bulletin*, 13, 469–477.

(5) McDonald M.G. (1991) Moose movement and mortality associated with the Glenn Highway expansion. *Alces*, 27, 208–219.

(6) Foster M.L. & Humphrey S.R. (1995) Use of highway underpasses by Florida panthers and other wildlife. *Wildlife Society Bulletin*, 23, 95–100.

(7) Land D. & Lotz M. (1996) *Wildlife crossing designs and use by Florida panthers and other wildlife in southwest Florida*. Proceedings -Trends in addressing wildlife mortality: transportation related wildlife mortality seminar, Florida Department of Transportation, Tallahassee, Florida USA, 379–386.

(8) Roof J. & Wooding J. (1996) *Evaluation of the S.R. 46 wildlife crossing in Lake County, Florida*. Florida Game and Fresh Water Fish Commission, Wildlife Research Laboratory.

(9) Clevenger A. P. (1998) *Permeability of the Trans-Canada highway to wildlife in Banff National Park: importance of crossing structures and factors influencing their effectiveness.* Proceedings-International Conference on Wildlife Ecology and Transportation, Florida Department of Transportation, Tallahassee, Florida, USA, 109–119.

(10) Clevenger A.P. & Waltho N. (1999) *Dry drainage culvert use and design considerations for small-and medium–sized mammal movement across a major transportation corridor.* Proceedings -International Conference on Wildlife Ecology and Transportation, Florida Department of Transportation, Tallahassee, Florida, USA, 264–270.

(11) Clevenger A.P. & Waltho N. (2000) Factors influencing the effectiveness of wildlife underpasses in Banff National Park, Alberta, Canada. *Conservation Biology*, 14, 47–56, https://doi.org/10.1046/j.1523-1739.2000.00099-085.x

(12) Mathiasen R. & Madsen A.B. (2000) Infrared video-monitoring of mammals at a fauna underpass. International *Journal of Mammalian Biology*, 65, 59–61.

(13) Clevenger A.P., Chruszcz B. & Gunson K.E. (2001) Highway mitigation fencing reduces wildlife-vehicle collisions. *Wildlife Society Bulletin*, 29, 646–653, https://doi.org/10.2307/3784191

(14) Abson R.N. & Lawrence R.E. (2003) *Monitoring the use of the Slaty Creek wildlife underpass, Calder Freeway, Blackforest, Macedon, Victoria, Australia.* Proceedings -International Conference on Wildlife Ecology and Transportation, Center for Transportation and the Environment, North Carolina State University, Raleigh NC, USA, 303–308.

(15) Brudin C.O. (2003) *Wildlife use of existing culverts and bridges in north central Pennsylvania.* Proceedings -International Conference on Wildlife Ecology and Transportation, Center for Transportation and the Environment, North Carolina State University, Raleigh NC, USA, 334–352.

(16) Dodd N.L., Gagon J.W. & Schweinsburg R.E. (2003) *Evaluation of measures to minimize wildlife-vehicle collisions and maintain wildlife permeability across highways in Arizona, USA.* Proceedings -International Conference on Wildlife Ecology and Transportation, Center for Transportation and the Environment, North Carolina State University, Raleigh NC, USA, 353–354.

(17) Gordon K.M. & Anderson S.H. (2003) *Mule deer use of underpasses in western and southeastern Wyoming.* Proceedings -International Conference on Wildlife Ecology and Transportation, Center for Transportation and the Environment, North Carolina State University, Raleigh NC, USA, 309–318.

(18) Plumb R.E., Gordon K.M. & Anderson S.H. (2003) Pronghorn use of a wildlife underpass. *Wildlife Society Bulletin*, 31, 1244–1245, https://doi.org/10.2307/3784474

(19) Servheen C., Shoemaker R. & Lawrence L. (2003) *A sampling of wildlife use in relation to structure variables for bridges and culverts under I-90 between Alberton and St. Regis, Montana.* Proceedings -International Conference on

Wildlife Ecology and Transportation, Center for Transportation and the Environment, North Carolina State University, Raleigh NC, USA, 331–341.

(20) Taylor B.D. & Goldingay R.L. (2003) Cutting the carnage: wildlife usage of road culverts in north-eastern New South Wales. *Wildlife Research*, 30, 529–537, https://doi.org/10.1071/WR0106

(21) Dodd C.K., Barichivich W.J. & Smith L.L. (2004) Effectiveness of a barrier wall and culverts in reducing wildlife mortality on a heavily traveled highway in Florida. *Biological Conservation*, 118, 619–631, https://doi.org/10.1016/j.biocon.2003.10.011

(22) McDonald W. & St Clair C.C. (2004) Elements that promote highway crossing structure use by small mammals in Banff National Park. *Journal of Applied Ecology*, 41, 82–93, https://doi.org/10.1111/j.1365-2664.2004.00877.x

(23) Clevenger A.P. & Waltho N. (2005) Performance indices to identify attributes of highway crossing structures facilitating movement of large mammals. *Biodiversity and Conservation*, 121, 453–464, https://doi.org/10.1016/j.biocon.2004.04.025

(24) Mata C., Hervàs I., Herranz J., Suàrez F. & Malo J.E. (2005) Complementary use by vertebrates of crossing structures along a fenced Spanish motorway. *Biological Conservation*, 124, 397–405, https://doi.org/10.1016/j.biocon.2005.01.044

(25) Dodd N.I., Gagnon J.W., Manzo A.I. & Schweinsburg R.E. (2007) Video surveillance to assess highway underpass use by elk in Arizona. *The Journal of Wildlife Management*, 71, 637–645, https://doi.org/10.2193/2006-340

(26) Ascensão F. & Mira A. (2007) Factors affecting culvert use by vertebrates along two stretches of road in southern Portugal. *Ecological Research*, 22, 57–66, https://doi.org/10.1007/s11284-006-0004-1

(27) Baker A., Knowles M. & Latham D. (2007) Using clay drain seals to assess the use of dry culverts installed to allow mammals to pass under the A1 trunk road, Northumberland, England. *Conservation Evidence*, 4, 77–80

(28) Leblond M., Dussault C., Ouellet J.-P., Poulin M., Courtois R. & Fortin J. (2007) Electric fencing as a measure to reduce moose–vehicle collisions. *The Journal of Wildlife Management*, 71, 1695–1703, https://doi.org/10.2193/2006-375

(29) Bond A.R. & Jones N.J. (2008) Temporal trends in use of fauna-friendly underpasses and overpasses. *Wildlife Research*, 35, 103–112, https://doi.org/10.1071/WR07027

(30) Braden A.W., Lopez R.R., Roberts C.W., Silvy N.J., Owen C.B. & Frank P.A. (2008) Florida Key deer *Odocoileus virginianus clavium* underpass use and movements along a highway corridor. *Wildlife Biology*, 14, 155–163, https://doi.org/10.2981/0909-6396(2008)14[155:FKDOVC]2.0.CO;2

(31) Grilo C., Bissonette J.A., and Santos-Reis M. (2008) Response of carnivores to existing highway culverts and underpasses: implications for road planning and mitigation. *Biodiversity Conservation*, 17, 1685–1699.

(32) Mata C., Hervàs I., Herranz J., Suàrez F. & Malo J.E. (2008) Are motorway passages worth building? Vertebrate use of road-crossing structures on a Spanish motorway. *Journal of Environmental Management*, 88, 407–415, https://doi.org/10.1016/j.jenvman.2007.03.014

(33) Olsson M.P.O. & Widen P. (2008) Effects of highway fencing and wildlife crossings on moose *Alces alces* movements and space use in southwestern Sweden. *Wildlife Biology*, 14, 111–117, https://doi.org/10.2981/0909-6396(2008)14[111:EOHFAW]2.0.CO;2

(34) Olsson M.P.O., Widen P. & Larkin J.L. (2008) Effectiveness of a highway overpass to promote landscape connectivity and movement of moose and roe deer in Sweden. *Landscape and Urban Planning*, 85, 133–139, https://doi.org/10.1016/j.landurbplan.2007.10.006

(35) Parker I.D., Braden A.W., Lopez R.R., Silvy N.J., Davis D.S. & Owen C.B. (2008) Effects of US 1 Project on Florida Key deer mortality. *The Journal of Wildlife Management*, 72, 354–359.

(36) Hayes I. & Goldingay R.L. (2009) Use of fauna road-crossing structures in north-eastern New South Wales. *Australian Mammalogy*, 31, 89–95, https://doi.org/10.1071/AM09007

(37) Klar N., Herrmann M. & Kramer-Schadt S. (2009) Effects and mitigation of road impacts on individual movement behavior of wildcats. *The Journal of Wildlife Management*, 73, 631–638, https://doi.org/10.2193/2007-574

(38) Kusak J., Huber D., Gomerčić T., Schwaderer G. & Gužvica G. (2009) The permeability of highway in Gorski kotar (Croatia) for large mammals. *European Journal of Wildlife Research*, 55, 7–21, https://doi.org/10.1007/s10344-008-0208-5

(39) McCollister M.F. & van Manen F.T. (2010) Effectiveness of wildlife underpasses and fencing to reduce wildlife-vehicle collisions. *The Journal of Wildlife Management*, 74, 1722–1731, https://doi.org/10.2193/2009-535

(40) Taylor B.D. & Goldingay R.L. (2010) Roads and wildlife: impacts, mitigation and implications for wildlife management in Australia. *Wildlife Research*, 37, 320–331, https://doi.org/10.1071/wr09171

(41) Dodd N.L., & Gagnon J.W. (2011) Influence of underpasses and traffic on white-tailed deer highway permeability. *Wildlife Society Bulletin*, 35, 270–281, https://doi.org/10.1002/wsb.31

(42) Eldridge B. & Wynn J. (2011) Use of badger tunnels on Highway Agency schemes in England. *Conservation Evidence*, 8, 53–57.

(43) Gagnon J.W., Dodd N.L., Ogren K.S. & Schweinsburg R.E. (2011) Factors associated with use of wildlife underpasses and importance of long-term

monitoring. *The Journal of Wildlife Management*, 75, 1477–1487, https://doi. org/10.1002/jwmg.160

(44) Parker I.D., Lopez R.R., Silvy N.J., Davis D.S. & Owen C.B. (2011) Long-term effectiveness of US 1 crossing project in reducing Florida Key deer mortality. *Wildlife Society Bulletin*, 35, 296–302, https://doi.org/10.1002/wsb.45

(45) Sawyer H., Lebeau C. & Hart T. (2012) Mitigating roadway impacts to migratory mule deer—a case study with underpasses and continuous fencing. *Wildlife Society Bulletin*, 36, 492–498, https://doi.org/10.1002/ wsb.166

(46) Sawaya M.A., Clevenger A.P. & Kalinowski, S.T. (2013) Demographic connectivity for ursid populations at wildlife crossing structures in Banff National Park. *Conservation Biology*, 27, 721–730, https://doi.org/10.1111/ cobi.12075

(47) Alonso, R. S., Lyren, L. M., Boydston, E. E., Haas, C. D., & Crooks, K. R. (2014) Evaluation of road expansion and connectivity mitigation for wildlife in southern California. *The Southwestern Naturalist*, 59(2), 181–187, https:// doi.org/10.1894/f04-tal-51.1

(48) Bond A.R. & Jones D.N. (2014) Roads and macropods: interactions and implications. *Australian Mammalogy*, 36, 1–14, https://doi.org/10.1071/ am13005

(49) Taylor B.D. & Goldingay R.L. (2014) Use of highway underpasses by bandicoots over a 7-year period that encompassed road widening. *Australian Mammalogy*, 36, 178–183, https://doi.org/10.1071/am13034

(50) Chambers B. & Bencini R. (2015) Factors affecting the use of fauna underpasses by bandicoots and bobtail lizards. *Animal Conservation*, 18, 424–432, https://doi.org/10.1111/acv.12189

(51) Murphy-Mariscal M.L., Barrows C.W. & Allen M.F. (2015) Native wildlife use of highway underpasses in a desert environment. *The Southwestern Naturalist*, 60, 340–348, https://doi.org/10.1894/0038-4909-60.4.340

(52) Huijser M.P., Fairbank E.R., Camel-Means W., Graham J., Watson V., Basting P. & Becker D. (2016) Effectiveness of short sections of wildlife fencing and crossing structures along highways in reducing wildlife–vehicle collisions and providing safe crossing opportunities for large mammals. *Biological Conservation*, 197, 61–68, https://doi.org/10.1016/j.biocon.2016.02.002

(53) Simpson N.O., Stewart K.M., Schroeder C., Cox M., Huebner K. & Wasley, T. (2016) Overpasses and underpasses: Effectiveness of crossing structures for migratory ungulates. *The Journal of Wildlife Management*, 80, 1370–1378, https://doi.org/10.1002/jwmg.21132

(54) Ford A.T., Barrueto M. & Clevenger A.P. (2017) Road mitigation is a demographic filter for grizzly bears. *Wildlife Society Bulletin*, 41, 712–719, https://doi.org/10.1002/wsb.828

5.13. Install barrier fencing along railways

https://www.conservationevidence.com/actions/2590

- **One study** evaluated the effects on mammals of installing barrier fencing along railways. This study was in Norway[1].

COMMUNITY RESPONSE (0 STUDIES)

POPULATION RESPONSE (1 STUDY)

- **Survival (1 study):** A before-and-after study in Norway[1] found that fencing eliminated moose collisions with trains, except at the fence end.

BEHAVIOUR (0 STUDIES)

Background

Collisions with trains can cause substantial numbers of mammal deaths (e.g. Gundersen & Andreassen 1998). Barrier fencing alongside railways may reduce access to railway tracks by mammals and, thus, reduce the number of mammal-train collisions.

Gundersen H. & Andreassen H.P. (1998) The risk of moose *Alces alces* collision: A predictive logistic model for moose-train accidents. *Wildlife Biology*, 4, 103–110.

A before-and-after study in 1985–2003 in forest in southern Norway (1) found that 1 km of fencing eliminated moose *Alces alces* collisions with trains along that stretch. The exception was one killed at the fence end. Within the wider study area, there were 0.58 moose/km killed each winter during the study period. In 1995, a 1-km-long wire-mesh fence was erected alongside a railway line. Moose-train collisions along a 100-km stretch of the railway line were recorded from July 1985–April 2003.

(1) Andreassen H.P., Gundersen H. & Storaas T. (2005) The effect of scent-marking, forest clearing, and supplemental feeding on moose-train collisions. *The Journal of Wildlife Management*, 69, 1125–1132, https://doi.org/10.2193/0022-541x(2005)069[1125:teosfc]2.0.co;2

5.14. Install wildlife warning reflectors along roads

https://www.conservationevidence.com/actions/2591

- **Fifteen studies** evaluated the effects on mammals of installing wildlife warning reflectors along roads. Nine studies were in the USA[1–5,7,9,10,11], three were in Austalia[8,12,13], two were in Germany[14,15] and one was in Denmark[6].

COMMUNITY RESPONSE (0 STUDIES)

POPULATION RESPONSE (10 STUDIES)

- **Abundance (1 study):** A before-and-after study in Australia[8] found that, when warning reflectors were installed (along with speed restrictions, reflective wildlife signs, rumble strips, wildlife escape ramps and an educational pamphlet), a small population of eastern quoll re-established in the area.

- **Survival (10 studies):** Five of eight controlled or before-and-after studies in the USA[1,3,4,5,7,9,10] and Germany[15] found that wildlife warning reflectors did not reduce collisions between vehicles and deer[3,4,9,10,15]. Two studies found that vehicle-deer collisions were reduced by reflectors[1,7] and one found that collisions were reduced in rural areas but increased in suburban areas[5]. A before-and-after study in Australia[8] found that, when warning reflectors were installed (along with speed restrictions, reflective wildlife signs, rumble strips, wildlife escape ramps and an educational pamphlet), vehicle collisions with Tasmanian devils, but not eastern quolls, decreased. A review of two studies in Australia[13] found mixed responses of mammal road deaths to wildlife warning reflectors.

BEHAVIOUR (5 STUDIES)

- **Behaviour change (5 studies):** Three of four studies (including three controlled studies), in the USA[2,11], Denmark[6] and Germany[14], found that wildlife warning reflectors did not cause deer to behave in ways that made collisions with vehicles less likely (such as by avoiding crossing roads). The other study found that deer initially responded to wildlife reflectors with alarm and flight but then became habituated[6].

A replicated, controlled study in Australia[12] found that one of four reflector model/colour combinations increased fleeing behaviour of bush wallabies when lights approached. The other combinations had no effect and none of the combinations affected red kangaroos.

Background

Reflectors are installed on posts along the edge of the road, a certain distance apart and at the height of the average vehicle headlamp. At night, as vehicle lights approach, the reflectors glow brighter and create an 'optical fence' as light from headlights is reflected onto roadside habitat, which aims to deter wildlife from approaching the road until the vehicle has passed. Polished stainless-steel wildlife mirrors can also be installed to reflect the headlights from passing cars causing light to flicker sharp, pencil-like beams that aim to startle animals and stop them moving until the lights have passed.

A replicated, controlled study in 1981–1984 in a forest-grassland area in Washington, USA (1) found that wildlife reflectors reduced road deaths of deer *Odocoileus* sp. Fewer deer were killed when reflectors were uncovered (6 of the 58 killed overall) compared to when they were covered (52 of the 58 road-kills recorded). Four test sections were established along a highway (0.7–1.1 km long). Swareflex wildlife reflectors (17 × 5 cm; red) were mounted on 1-m posts, 20 m apart (10 m at bends) and 1 m from the edge of the highway. Reflectors in each section were alternately covered and uncovered at 1-week intervals during October–April from February 1981–April 1984. Intervals were extended to two weeks after December 1982. Alternate test sections were paired so that reflectors in each pair were covered while reflectors in adjacent sections were uncovered. Road-kills were recorded daily.

A controlled study in 1984 of captive deer in Michigan, USA (2) found that reflectors, angled to deflect car headlight illumination into adjacent habitat, did not affect crossing rates of white-tailed deer *Odocoileus virginianus*. There were no significant differences in crossing rates when the route was fitted with red reflectors (256 crossings), white reflectors (200 crossings) or no reflectors (264 crossings). Ten captive-born deer were housed in a 3.5-acre pen. Five posts were installed in a line at 66-foot intervals. A pair of car headlights was aimed alongside this line. Each night, one trial each was run using no reflectors, white reflectors and red reflectors. Reflectors were fastened 42 inches up posts. All treatment orders were replicated three times. Data were collected over 18 nights, between 20 August and 6 October 1984. Trials lasted 15 minutes. Water (to attract deer) was dispensed noisily, by remote control, at five and 10 minutes, first on one side of the post line, then the other. Water ran into containers with holes, which drained in 1.5 minutes. Crossings by deer were counted by observers in concealed positions.

A before-and-after study in 1977–1982 along a road through agricultural land in Illinois, USA (3) found that warning reflectors did not reduce deer-vehicle collisions. A similar number of white-tailed deer *Odocoileus virginianus* was killed overnight during a year with reflectors installed (six deer) as during the previous two years before reflectors were installed (5–6/year). The local deer population was reported to have decreased over this time. Behaviour of deer crossing the road or feeding at the roadside did not appear to be altered by reflectors. Eighty Swareflex wildlife warning reflectors were installed along each side of a 0.8-km section of a two-lane highway (speed limit 88 km/hour). Reflectors comprised two mirrors (5 × 17 cm) covered with red prism plates on posts 20 m apart, 3 m from the road edge. Collision data were provided by transportation personnel and direct observations.

A controlled study in 1986–1989 along a highway in Wyoming, USA (4) found that Swareflex reflectors did not reduce road deaths of mule deer *Odocoileus hemionus*. More deer were killed when reflectors were displayed (126) than when they were covered (64). During the same periods, there were 85 and 62 deer killed respectively in a control site without reflectors. After three years, only 215 (61%) of the reflectors were still in good condition. In October 1986, Swareflex reflectors were installed on both sides of a 3.2-km section of a highway (US 30). The 350 reflectors were on posts (height 61–91 cm), 20 m apart (10 m on bends) and 3 m from the road edge. Reflectors were covered and uncovered at 1-week intervals from October 1986 to February 1987 and then at 2-week intervals until May 1989. A control section (3.2 km) without reflectors was also monitored. Deer-vehicle collisions were monitored in October 1986–April 1987 (daily), November 1987–April 1988 and October 1988–May 1989 (each at 2–5-day intervals).

A replicated, before-and-after study in 1980–1994 along 16 highways in Minnesota, USA (5) found that reflectors reduced rural deer-vehicle collisions by 50–97%, but that collisions in suburban areas increased. Collisions were reduced by 90% along roads in the four coniferous forest areas (after installation: 2 collisions; before: 26), 79% along roads in the four 'farmland' areas (after installation: 9 collisions; before: 54) and 87% along roads in the four hardwood forest areas (after installation: 3 collisions; before: 25). However, collisions increased in four suburban areas (after installation: 4.4–7.3 collisions/year; before: 2.4–3.4). Swareflex brand red reflectors were installed along 16 highway sections through three different rural habitats and in a suburban area. Deer-vehicle collisions were monitored before (pre-1988) and after installation (1988–1994).

A study in 1996 in a forest in Zealand, Denmark (6) found that fallow deer *Dama dama* initially responded to wildlife reflectors with alarm and flight but became habituated to the light reflection. On the first night, using a low level of lighting, deer fled from the reflection in 99% of cases. On night five, using the same light level, only 16% fled and 74% did not react. On nights 6–7 with four light levels, 86–94% fled. However, on nights 16–17 only 30–37% fled and 38–48% showed no response. Following a one-night break, deer fled almost twice as much as they did the night before the break (35–90% vs 20–54%). Feeding deer were exposed to light reflections (WEGU reflector; two sloping mirrors within a cover) at predetermined time intervals and their behavioural responses were recorded. Data were collected over 17 nights (two with no lighting used) in April 1996. Only the lowest light level was used on the first five nights. Subsequently, four levels were used.

A replicated, randomized, controlled study in 1999–2005 along a highway in Indiana, USA (7) found that wildlife reflectors reduced deer-vehicle collisions by 19% overall, but there was no difference between different reflector colours, spacing or design. When reflector sites were combined and compared with sites without reflectors, there was a 19% reduction in deer-vehicle collisions with reflector use. However, there was no significant difference in numbers of collisions between different reflector combinations (colours, spacing, single/dual design, reflectors on central reservation or not) or between each reflector combination and sites without reflectors. The greatest decrease in collisions was associated with 30-m reflector spacing regardless of colour or design. In 1999, two replicates of 16 treatment combinations (randomized order) were installed along two 1.6-km-long road sections. Treatments were different reflector colour (red and blue/green), spacing (30 m and 45 m), design (single and dual reflectors) and whether or not the central reservation also had reflectors. There was a 1.6-km control section without reflectors at each end of each replicate. Numbers of deer-vehicle collisions were recorded in April–May and October–November in 1999–2005.

A before-and-after study in 1990–1998 in Tasmania, Australia (8) found that, following installation of wildlife warning reflectors, speed restrictions, reflective wildlife signs, rumble strips, wildlife escape ramps and publication of an educational pamphlet, an eastern quoll *Dasyurus viverrinus* population partially re-established and vehicle collisions with Tasmanian devils *Sarcophilus laniarius*, but not eastern quolls, decreased. Effects of the different actions were not investigated individually and results were not tested for statistical significance. Following local extinctions, 3–4 quolls re-colonised within six months of installation, increasing to ≥8 animals after two years. Road-kills for quolls were similar after implementation (1.5/year) compared to before (1.6/year), but decreased for Tasmanian devils (after: 1.5/year; before: 3.6). Following road widening in 1991, vehicle-wildlife collisions increased and quolls became locally extinct (from 19 animals). In 1996, reflective wildlife deterrents (Swareflex; 20 m intervals, 50 cm above ground) were installed, along with the other five interventions. Animals were surveyed using 60 cage traps for three nights during alternate months in October 1990–April 1993. Then, 10–20 traps were set for 20–100 trap nights in April, May and July 1995–1998. Spotlight counts were made once or twice in 1991, 1995, 1996 and 1998. Road-kills were recorded in 1990–1996.

A replicated, controlled study in 2000–2003 along 10 highways in Virginia, USA (9) found that warning reflectors did not reduce collisions between vehicles and deer *Odocoileus* sp. There was a similar rate of deer road casualties on sections with reflectors (4.6/mile/year) compared to sections without reflectors (4.8/mile/year). Deer warning reflectors (red) were installed on posts along 0.4–2.3-km sections of 10 highways (2–4 lane) from October 2000 to May 2002. Reflector sites and adjacent sites without reflectors were each monitored for 6–28 months. Deer road-kills data were collated by officials from the state Department of Transportation.

A replicated, controlled, before-and-after study in 1992–2000 along roads in Michigan, USA (10) found that wildlife warning reflectors did not reduce deer-vehicle collisions. The rate of collisions after reflectors were installed (8.5/year) was similar to that before reflectors were installed (8.2/year). This was also similar to the collision rate on another road section, at the same time, where reflectors were not installed (after: 13/year; before: 9.5/year). The total number of deer-vehicle collisions recorded was 279. In 1998, Swareflex wildlife warning reflectors were installed along three 3.2-km-long sections of road. Three additional 3.2-km-long road sections were controls with no reflectors. Collisions between 18:00 and 24:00 h, monitored by Michigan State Police, were compared before (1992–1997) and after (1998 and 2000) reflector installation.

A before-and-after study in 2004–2005 at a college campus in Georgia, USA (11) found that wildlife warning reflectors did not reduce white-tailed deer *Odocoileus virginianus* behaviours that were likely to cause collisions with vehicle. When red or blue-green reflectors were installed, there was a proportional increase in behaviours that were likely to cause deer–vehicle collisions. White or amber reflectors resulted in an increased rate both of responses that increase and that decrease collision likelihood. A total of 1,370 deer responses were recorded. A smaller proportion of animals stopped moving toward the road as a vehicle approached when reflectors were installed (red: 13%; white: 55%; blue-green: 14%; amber: 50%) compared to before reflectors were installed (64%). In two test areas (5 km apart), 15 posts were installed 15 m apart, staggered on opposite sides of the road. After two weeks, Strieter-Lite Wild Animal Highway Warning Reflectors were installed on posts (61–76 cm above road). Deer–vehicle interactions were observed using an infrared camera for four hours/night before (15 nights in November 2004–January 2005) and after installation of reflectors (January–May 2005). Two reflector colours were tested in each area for 15 nights each.

A replicated, controlled study in 2006 at two grassland sites in New South Wales, Australia (12) found that red Swareflex wildlife warning reflectors increased the proportion of bush wallabies *Macropus rufogriseus* fleeing approaching lights but red Strieter-Lite reflectors and white version of both types did not affect proportions of fleeing bush wallabies or red kangaroos *Macropus rufus*. A higher proportion of bush wallabies fled when lights shone at red Swareflex reflectors (8%) than when lights shone without reflectors (3%). There was no such response for red kangaroos (reflectors: 3%; no reflectors: 5%). There were no significant differences in fleeing response rates for bush wallabies when lights shone at red Strieter-Lite reflectors (with: 5%; without: 3%) or at white reflectors of either type (with: 5–6%; without: 3%). There were also no significant differences in fleeing response rates for red kangaroos when lights shone at red Strieter-Lite reflectors (with: 5%; without: 7%) or at white reflectors of either type (with: 3–5%; without: 5%). In two grassland enclosures, a 'road' strip was mown and had 55-W lights installed in pairs every 20 m. Sequentially activating these lights mimicked approaching cars. Wildlife warning reflectors (Swareflex and Strieter-Lite) were placed on either side of the road at 20-m intervals. Over three days, animals were exposed to one night with no lights, one night with lights and no reflectors and one night with lights and reflectors. This three-day sequence was repeated 15 times and fleeing behaviour was surveyed using infrared cameras.

A review of two studies in 2000–2010 in Australia (13) found that installing wildlife warning reflectors had mixed results regarding reducing road deaths of mammals. One study showed reflectors prompted increased vigilance and flight by red kangaroos *Macropus rufus*. Another study showed that reflectors did not reduce the number of Proserpine rock wallabies *Petrogale persephone* killed by collisions with vehicles.

A replicated, randomized, controlled study in 2002–2014 in two grassland sites and five roadside areas in Germany (14) found that wildlife warning reflectors along roads did not cause roe deer *Capreolus capreolus* to evade traffic more effectively. In two fenced grassland areas, there was no significant difference in successful evasion of traffic when wildlife reflectors were used and not used (data reported as model results). The same results were found in five roadside areas (data reported as model results). In two fenced grassland areas, reflectors and headlights (mimicking cars), headlights without reflectors and no reflectors or headlights were each in place for two periods of one week each. This was carried out four times between September 2012 and April 2014. The order of these combinations of reflectors and lights was varied randomly. Groups of three to six deer occupied each area. Their behaviour was monitored by infrared video cameras. At five sites, three thermal cameras were installed between June 2012 and June 2014 in trees close to roads at 3–4 m high. Between July 2012 and April 2014, wildlife warning reflectors were installed along both side of the roads. The behaviour of roe deer clearly visible in video recordings was documented.

A replicated, controlled study in 2014–2017 of 151 road sites in central Germany (15) found that four types of wildlife warning reflector did not reduce wildlife-vehicle collisions. The number of vehicle collisions was similar with and without four types of wildlife warning reflectors for three groups of mammals: deer (roe deer *Capreolus capreolus*, red deer *Cervus elaphus*, fallow deer *Dama dama*); wild boar *Sus scrofa*; and other mammals (badger *Meles meles*, red fox *Vulpes vulpes*, hare *Lepus europaeus*/rabbit *Oryctolagus cuniculus*, wildcat *Felis silvestris*, racoon *Procyon lotor*). Data are reported as statistical model results. Three types of wildlife warning reflectors were installed along 151 stretches of road (average 2 km long): dark-blue reflectors (51 sites); light-blue reflectors (50 sites) and multi-coloured reflectors (50 sites). In addition, one type of reflector (transparent/silver) with an acoustic warning (1.5 second sounds triggered by vehicle headlights) was installed along a 200 m stretch of road at 10 of the 101 sites with blue reflectors. Reflectors were installed on posts (55–100 cm high) spaced 25–50 m apart. Wildlife-vehicle collisions reported to the police (1,984 in total) were analysed for 12 months with the reflectors installed and 12 months without in 2014–2017.

(1) Schafer J.A. & Penland S.T. (1985) Effectiveness of Swareflex reflectors in reducing deer-vehicle accidents. *The Journal of Wildlife Management*, 49, 774–776.

(2) Zacks J.L. (1986) Do white-tailed deer avoid red? An evaluation of the premise underlying the design of swareflex wildlife reflectors. *Transportation Research Record*, 1075, 35–43.

(3) Waring G.H., Griffis J.L. & Vaughn M.E. (1991) White-tailed deer roadside behavior, wildlife warning reflectors and highway mortality. *Applied Animal Behaviour Science*, 29, 215–223.

(4) Reeve A.F. & Anderson S.H. (1993) Ineffectiveness of Swareflex reflectors at reducing deer vehicle collisions. *Wildlife Society Bulletin*, 21, 127–132.

(5) Pafko F. & Kovach B. (1996) *Experience with deer reflectors*. Proceedings of the Trends in addressing transportation related wildlife mortality: transportation related wildlife mortality seminar, FL-ER-58-96, Tallahassee, USA, 135–146.

(6) Ujvári M., Baagøe H.J. & Madsen A.B. (1998) Effectiveness of wildlife warning reflectors in reducing deer vehicle collisions: a behavioral study. *The Journal of Wildlife Management*, 62, 1094–1099.

(7) Gulen S., McCabe G. & Wolfe S.E. (2000) *Evaluation of wildlife reflectors in reducing vehicle-deer collisions on Indiana Interstate 80/90*. FHWA/IN/JTRP-2006/18 Unpublished Report. Indiana Department of Transportation

(8) Jones M.E. (2000) Road upgrade, road mortality and remedial measures: impacts on a population of eastern quolls and Tasmanian devils. *Wildlife Research*, 27, 289–296, https://doi.org/10.1071/wr98069

(9) Cottrell B.H. (2003) *Technical assistance report: evaluation of deer warning reflectors in Virginia*. VTRC 03-TAR6.

(10) Rogers E. (2004) An ecological landscape study of deer vehicle collisions in Kent County, Michigan. Report to Kent County Road Commission, Michigan, USA

(11) D'Angelo G.J., D'Angelo J.G., Gallagher G.R., Osborn D.A., Miller K.V. & Warren R.J. (2006) Evaluation of wildlife warning reflectors for altering white-tailed deer behavior along roadways. *Wildlife Society Bulletin*, 34, 1175–1183, https://doi.org/10.2193/0091-7648(2006)34[1175:eowwrf]2.0.co;2

(12) Ramp D. & Croft D.B. (2006) Do wildlife warning reflectors elicit aversion in captive macropods? *Wildlife Research*, 33, 583–590, https://doi.org/10.1071/wr05115

(13) Bond A.R. & Jones D.N. (2014) Roads and macropods: interactions and implications. *Australian Mammalogy*, 36, 1–14, https://doi.org/10.1071/am13005

(14) Brieger F., Hagen R., Kröschel M., Hartig F., Petersen I., Ortmann S. & Suchant, R. (2017) Do roe deer react to wildlife warning reflectors? A test combining a controlled experiment with field observations. *European Journal of Wildlife Research*, 63, 72, https://doi.org/10.1007/s10344-017-1130-5

(15) Benten A., Hothorn T., Vor T. & Ammer C. (2018) Wildlife warning reflectors do not mitigate wildlife–vehicle collisions on roads. *Accident Analysis & Prevention*, 120, 64–73, https://doi.org/10.1016/j.aap.2018.08.003

5.15. Install acoustic wildlife warnings along roads

https://www.conservationevidence.com/actions/2592

- **Two studies** evaluated the effects on mammals of installing acoustic wildlife warnings along roads. One study was in Demark[1] and one was in Australia[2].

COMMUNITY RESPONSE (0 STUDIES)

POPULATION RESPONSE (0 STUDIES)

BEHAVIOUR (2 STUDIES)

- **Behaviour change (2 studies)**: A before-and-after study in Denmark[1] found that sound from acoustic road markings did not alter fallow deer behaviour. A controlled study in Australia[2] found that Roo-Guard® sound emitters did not deter tammar wallabies from food and so were not considered suitable for keeping them off roads.

Background

Collisions with vehicles can be a major cause of mortality for wild mammals and, especially where larger mammal species are involved, a cause of injury, death and economic loss for motorists (Conover *et al.* 1995). A range of interventions may be employed to deter mammals for accessing roads. This can include use of acoustic warnings which can either be devices that emit sounds or modifications to the road surface that produce noise when vehicle tyres pass over them.

See also: *Fit vehicles with ultrasonic warning devices.*

Conover M.R., Pitt W.C., Kessler K.K., DuBow T.J. & Sanborn W.A. (1995) Review of human injuries, illnesses, and economic losses caused by wildlife in the United States. *Wildlife Society Bulletin*, 23, 407–414.

A before-and-after study in 1997 in a mixed hardwood forest in Zealand, Denmark (1) found that acoustic road markings did not alter the behaviour of fallow deer *Dama dama*. Behavioural responses varied among nights, but deer showed increasing indifference to sounds from road markings over 11 nights (i.e. deer appeared to become habituated). Behaviour differed before (flight: 2%, no reaction: 96–99%) and during playbacks, but deer reactions declined over 10 nights of playbacks (night 1: flight 13%; nights 8–10: flight 3–0%, no reaction 88–99%). An area of forest next to an unpaved road closed to vehicles was selected where a herd of 6–12 fallow deer were fed (maize). Recordings of a car passing two types of acoustic road markings which produced sounds when a vehicle's tyres passed over (low frequency longflex; higher spossflex), multiplied to 70 sequences (each 0.11–0.16 s) were made. Behavioural responses of deer to play-back sounds (58 decibels) at predetermined time intervals (exposure for: 5, 2, 7, 3, 1 and 2 minutes) were monitored over 11 nights in February–March 1997. Behaviour was also recorded every 15 minutes during the two nights before sound trials commenced.

A controlled study in 2005 in a grass enclosure in Western Australia, Australia (2) found that Roo-Guard® sound emitters did not deter tammar wallabies *Macropus eugenii* from food and so were not considered suitable for keeping them off roads. There was no significant difference between the use of the enclosure or food sources when the Roo-Guards were switched on or off. This was the case even when there was an alternative source of food available away from Roo-Guards. The device did not result in any obvious behavioural responses such as flight or distress. Nine tammars were kept in an enclosure (60 × 30 m), with a test area (60 × 20 m) divided into 12 squares. The remainder of the enclosure was covered in trees and bushes. Roo-Guard® Mk II high-frequency sound emitters were installed on the edge of the test area, 0.5 m off the ground. Animals were observed though a night-vision scope on three nights (18:00–21:00 h) with the Roo-Guard® turned on and three with it turned off, for each of four treatments: food 20 m from Roo-Guard®, or food 20 and 60 m from Roo-Guard®, and the same two treatments but with the sides with food and Roo-Guards swapped over.

(1) Ujvári M., Baagøe H.J. & Madsen A.B. (2004) Effectiveness of acoustic road markings in reducing deer-vehicle collisions: a behavioural study. *Wildlife Biology*, 10, 155–159, https://doi.org/10.2981/wlb.2004.011

(2) Muirhead S., Blache D., Wykes B. & Bencini R. (2006) Roo-Guard® sound emitters are not effective at deterring tammar wallabies (*Macropus eugenii*)

from a source of food. *Wildlife Research*, 33, 131–136, https://doi.org/10.1071/wr04032

5.16. Install wildlife crosswalks

https://www.conservationevidence.com/actions/2593

- **One study** evaluated the effects on mammals of installing wildlife crosswalks. This study was in the USA[1].

COMMUNITY RESPONSE (0 STUDIES)

POPULATION RESPONSE (1 STUDY)

- **Survival (1 study):** A replicated, before-and-after, site comparison study in the USA[1] found that designated crossing points with barrier fencing did not significantly reduce road deaths of mule deer.

BEHAVIOUR (0 STUDIES)

Background

Crosswalks are intended to guide wildlife across roads at specific crossing points along fenced stretches of highway and to provide drivers with warning signs indicating specific locations where animals are expected to cross. In this narrow crossing zone, animals walking on to the road are guided directly across the road by river cobbles and/or painted cattle guards.

A replicated, before-and-after, site comparison study in 1991–1995 along two highways in Utah, USA (1) found that designated crossing points with barrier fencing did not significantly reduce road deaths of mule deer *Odocoileus hemionus*. Deaths decreased on both fenced and unfenced sections but the rate of decline was not significantly higher on fenced road sections with crossings (after: 36–46 deer fatalities over 15 months; before: 111–148 over 36 months) than over the same period on unfenced sections (after: 34–63; before: 75–123). In September 1994, four and five crossing points were installed along a two-and a four-lane highway respectively. Fencing (2.3 m high) restricted access to roadside resources and directed deer to crossing points. At these points, deer could jump a 1-m-high fence into funnel shaped fencing (2.3 m high) with a narrow opening to the road. One-way gates allowed deer trapped along the road to escape. Three warning signs, 152 m apart before crossings, and painted lines across the road at crossings, indicated to drivers that it was a crossing point. Road deaths were monitored weekly along treatment and nearby control roads before and after crossing installation, from October 1991 to November 1995.

(1) Lehnert M.E. & Bissonette J.A. (1997) Effectiveness of highway crosswalk structures at reducing deer-vehicle collisions. *Wildlife Society Bulletin*, 25, 809–818.

5.17. Install wildlife exclusion grates/cattle grids

https://www.conservationevidence.com/actions/2594

- **Three studies** evaluated the effects on mammals of installing wildlife exclusion grates or cattle grids. All three studies were in the USA[1,2,3].

COMMUNITY RESPONSE (0 STUDIES)

POPULATION RESPONSE (0 STUDIES)

BEHAVIOUR (3 STUDIES)

- **Behaviour change (3 studies):** Two of three studies (including two replicated, before-and-after studies), in the USA[1,2,3], found that steel grates largely prevented crossings by deer[2,3] whilst two found that they did not prevent crossings by deer and elk[1] or black bears[3]. In one of the studies, only one of three designs prevented crossings[2].

Background

Wildlife exclusion grates or cattle grids are designed to discourage wildlife, particularly ungulates, from walking through a gap in a fence where an access road approaches a larger road with higher traffic volume and vehicle speeds for example. If effective, they could reduce animal mortality and also collision-related risks for motorists.

See also: *Agriculture & Aquaculture -Install metal grids at field entrances to prevent mammals entering to reduce human-wildlife conflict.*

A study in 1972–1973 of two fences in Colorado, USA (1) found that steel rail deer guards did not prevent crossings through vehicle openings by mule deer *Odocoileus hemionus hemionus* or elk *Cervus canadensis*. In test conditions, 16 of 18 mule deer released adjacent to 12, 18 or 24-foot-wide guards, crossed the guards, in an average time of 173 s. During natural encounters, 11 mule deer and one elk crossed a 24-ft-long guard and four mule deer crossed a 12-ft-long guard. There were at least 11 approaches by mule deer and three by elk in which animals did not then cross. Guards, at vehicle openings in 8-foot-high fences, comprised flat steel rails, 0.5 inches wide, 4 inches high and 120 inches long, set 4 inches apart. Rails were perpendicular to the traffic direction. Eighteen deer were released in situations where crossing guards provided the only exit. Deer and elk tracks, from natural encounters with two guards, were examined periodically, from 29 June 1972 to 19 April 1973.

A replicated, before-and-after study in 2001 in Florida, USA (2) found that one of three deer exclusion grates excluded Florida Key deer *Odocoileus virginianus clavium*. Only one deer crossed the grate that incorporated diagonal cross-members into the metal grid, compared to 305 that crossed when the grate was covered over with plywood. Fifty deer crossed the two grate designs without diagonal cross-members, compared to 199 that crossed when covered over. Males were more successful at crossing than females. In 2001, three types of grate were tested for deer-exclusion efficiency. All grates were 6.1 × 6.1 m, each with a different grate pattern: grid of 10 × 13 cm rectangles with diagonal cross member through each rectangle and 8 × 10 cm or 10 ×

8 cm rectangles without diagonal cross member. Food was provided within a fenced area accessible only by crossing the grate. Grates were covered (therefore, easily crossable) for 1–2 weeks and then uncovered for one week, three times (for two designs) or once (third design). Infrared cameras were used to monitor deer crossings.

A replicated, before-and-after study in 2003–2010 at two roadside areas in Montana, USA (3) found that wildlife exclusion grates reduced crossings of a major highway by deer *Odocoileus* spp., but not by black bears *Ursus americanus*. After installing wildlife exclusion grates, a lower proportion of deer approaching the road subsequently crossed it (6%) than did so before grates were installed (44%). The proportion of black bears crossing the road, out of those approaching it, was not significantly different after grates were installed (62%) compared to before they were installed (87%). Between October 2004 and November 2010, fencing was installed along the roadside. Single exclusion grates were fitted at each of two junctions with minor roads. Grates were 6.8 m wide and 6.6 m long. In June–October of 2003–2005, eight 100 × 2 m areas were coated with sand to record animal tracks. Using these data, the percentage of animals that crossed the road was calculated. Wildlife cameras were placed at both grates between July 2008 and July 2010. The number of times an animal was ≤2 m from grates and whether it subsequently crossed were recorded.

(1) Reed D.F., Pojar T.M. & Woodard T.N. (1974) Mule deer responses to deer guards. *Journal of Range Management*, 27, 111–113.

(2) Peterson M.N., Lopez R.P., Silvy N.J., Owen C.B., Frank P.A. & Braden A.W. (2003) Evaluation of deer-exclusion grates in urban areas. *Wildlife Society Bulletin*, 31, 1198–1204.

(3) Allen T.D., Huijser M.P. & Willey D.W. (2013) Effectiveness of wildlife guards at access roads. *Wildlife Society Bulletin*, 37, 402–408, https://doi.org/10.1002/wsb.253

5.18. Reduce legal speed limit

https://www.conservationevidence.com/actions/2596

- **One study** evaluated the effects on mammals of reducing the legal speed limit. This study was in Canada[1].

COMMUNITY RESPONSE (0 STUDIES)

POPULATION RESPONSE (1 STUDY)

- **Survival (1 study):** A controlled, before-and-after study in Canada[1] found that speed limit reductions and enforcement did not reduce vehicle collisions with bighorn sheep or elk.

BEHAVIOUR (0 STUDIES)

Background

High vehicle speed is generally considered to be a substantial contributing factor in wildlife-vehicle collisions. Speed limits can be reduced in areas where there are high numbers of collisions, either permanently or during seasonal migrations.

A controlled, before-and-after study in 1983–1998 along a highway in Alberta, Canada (1) found that speed limit reductions and enforcement did not reduce vehicle collisions with bighorn sheep *Ovis canadensis* or elk *Cervus canadensis*. Sheep collision rates were similar in the reduced speed zones after limits were reduced (10.4 collisions/year) compared to before (10.3/year). Concurrently, in control areas where the speed limit was not reduced, there were fewer collisions in this second period (2.5 collisions/year) than the first period (3.4/year). Elk collisions increased with the speed limit reduction (after: 9.6/year; before: 7.8/year) but increased by more in the control zone (after: 14.3/year; before: 7.8/year). The local elk population increased 178% during the study. In 1991, the speed limit along a rural two-lane highway was reduced from 90 km/h to 70 km/h on three road sections (2.5, 4.0 and 9.0 km long). Monitoring in 1995 indicated that <20% of vehicles obeyed the 70 km/h limit. On average, 5,475 speeding tickets were issued/year. Animal-vehicle collisions were monitored for eight years before and eight years after speed limits were reduced, on three 2–3-km-long road sections for sheep and one 30-km-long section for elk. Vehicle speeds were monitored along two road sections in 1995.

(1) Bertwistle J. (1999) *The effects of reduced speed zones on reducing bighorn sheep and elk collisions with vehicles on the Yellowhead Highway in Jasper National*

Park. Proceedings -Third International Conference on Wildlife Ecology and Transportation. Tallahassee, Florida, USA, 89–97.

5.19. Install traffic calming structures to reduce speeds

https://www.conservationevidence.com/actions/2598

- **One study** evaluated the effects on mammals of installing traffic calming structures to reduce speeds. This study was in Australia[1].

COMMUNITY RESPONSE (0 STUDIES)

POPULATION RESPONSE (1 STUDY)

- **Abundance (1 study):** A before-and-after study in Australia[1] found that following installation of barriers to create a single lane, rumble strips, reflective wildlife signs, reflective wildlife deterrents, wildlife escape ramps and production of an educational pamphlet, a small population of eastern quoll population re-established in the area.

- **Survival (1 study):** A before-and-after study in Australia[1] found that following installation of barriers to create a single lane, rumble strips, reflective wildlife signs, reflective wildlife deterrents, wildlife escape ramps and production of an educational pamphlet, vehicle collisions with Tasmanian devils, but not eastern quolls decreased.

BEHAVIOUR (0 STUDIES)

Background

Reducing the design speed of a road can be used to reduce vehicle speed rather than reducing the legal speed limit. Traffic calming methods include speed bumps, rumble strips, curb or pavement extensions (to reduce road width) and raised central medians/ islands. Such structures get the attention of drivers and encourage them to slow down, which may help to reduce wildlife-vehicle collisions.

A before-and-after study in 1990–1998 in Tasmania, Australia (1) found that following installation of barriers to create a single lane, rumble strips, reflective wildlife signs, reflective wildlife deterrents, wildlife escape ramps and publication of an educational pamphlet, an eastern quoll *Dasyurus viverrinus* population partially re-established and vehicle collisions with Tasmanian devils *Sarcophilus laniarius*, but not eastern quolls, decreased. Results were not tested for statistical significance. Following local extinction, 3–4 quolls re-colonised within six months of installation, increasing to ≥8 animals after two years. Road-kills were similar for quolls before and after implementation (1.6 vs 1.5/year), but decreased for Tasmanian devils (3.6 vs 1.5/year). Vehicle speeds declined by 20 km/h (17–35% reduction) at the site centre and by 3–7% at edges. Following road widening in 1991, vehicle-wildlife collisions increased and quolls became locally extinct (from 19 animals). In 1996, four 'slow points' (barriers, creating a single give-way lane, rumble strips and four other interventions) were created. Animals were surveyed using 60 cage traps for three nights in alternate months in October 1990–April 1993. Then, 10–20 traps were set for 20–100 trap nights in each April, May and July of 1995–1998. Spotlight counts were made once or twice in 1991, 1995, 1996 and 1998. Road-kills were recorded in 1990–1996. Vehicle speeds were recorded at four locations.

(1) Jones M.E. (2000) Road upgrade, road mortality and remedial measures: impacts on a population of eastern quolls and Tasmanian devils. *Wildlife Research*, 27, 289–296, https://doi.org/10.1071/wr98069

5.20. Modify vegetation along roads to reduce collisions with mammals by enhancing visibility for drivers

https://www.conservationevidence.com/actions/2599

- We found no studies that evaluated the effects of modifying vegetation along roads to reduce collisions with mammals by enhancing visibility for drivers.

'We found no studies' means that we have not yet found any studies that have directly evaluated this intervention during our systematic journal and report searches. Therefore, we have no evidence to indicate whether or not the intervention has any desirable or harmful effects.

Background

Collisions with vehicles can be a major cause of mortality for wild mammals and, especially where larger mammal species are involved, a cause of injury, death and economic loss for motorists (Conover *et al.* 1995). A range of interventions can be employed to in an attempt to reduce the animal-vehicle collision rate. One option may be to cut back vegetation along roadsides in areas with high collision rates. This could give motorists a clearer sight of animals at the roadside ahead and, hence, more chance to take avoiding action if they see an animal moving onto the road.

Conover M.R., Pitt W.C., Kessler K.K., DuBow T.J. & Sanborn W.A. (1995) Review of human injuries, illnesses, and economic losses caused by wildlife in the United States. *Wildlife Society Bulletin*, 23, 407–414.

5.21. Modify the roadside environment to reduce collisions by reducing attractiveness of road verges to mammals

https://www.conservationevidence.com/actions/2600

• **One study** evaluated the effects of modifying the roadside environment to reduce collisions by reducing attractiveness of road verges to mammals. This study was in Canada[1].

COMMUNITY RESPONSE (0 STUDIES)

POPULATION RESPONSE (0 STUDIES)

BEHAVIOUR (1 STUDY)

• **Behaviour change (1 study):** A replicated, before-and-after, site comparison study in Canada[1] found that draining roadside salt pools and filling them with rocks reduced the number and duration of moose visits.

Background

Collisions with vehicles can be a major cause of mortality for wild mammals and, especially where larger mammal species are involved, a cause of injury, death and economic loss for motorists (Conover *et al.* 1995). A range of interventions can be employed to in an attempt to reduce the animal-vehicle collision rate. One option may be to modify the roadside environment to make it less attractive to mammals. This could involve removing vegetation that provides mammals with feeding or shelter resources, planting vegetation that is unattractive to mammals or removing other roadside features that are known to attract mammals and create accident hotspots.

Conover M.R., Pitt W.C., Kessler K.K., DuBow T.J. & Sanborn W.A. (1995) Review of human injuries, illnesses, and economic losses caused by wildlife in the United States. *Wildlife Society Bulletin*, 23, 407–414.

A replicated, before-and-after, site comparison study in 2003–2005 in mixed coniferous and deciduous forest in Québec, Canada (1) found that draining roadside salt pools and filling them with rocks reduced the number and duration of visits by moose *Alces alces*. There was a lower overall visit rate to salt pools at night after some were drained and filled with rocks (0.2 visits/100 hours) than before (1.5 visits/100 hours). This decline was due to a fall in visits to drained pools with visit rates to undrained pools not changing significantly (see paper for details). Daytime visits did not decrease (after: 0.2/100 hours; before: 0.2–0.5). The average length of time spent at pools decreased (after: 0.02 hours/100 hours; before: 0.11–0.18). Before management, 57% (113/198) of recorded visits were of moose that drank the salty water. After management, no moose drank at drained pools. Moose were monitored at 12 roadside salt pools from mid-May to mid-August in 2003–2005. In autumn 2004, seven salt pools (those near most moose-vehicle collisions) were drained and filled with rocks (10–30 cm diameter) to deter moose. The other five were left untreated. Moose were monitored using movement and heat detectors that triggered a video camera or photo camera with infrared lights.

(1) Leblond M., Dussault C., Ouellet J.-P., Poulin M., Courtois R. & Fortin J. (2007) Management of roadside salt pools to reduce moose–vehicle collisions. *The Journal of Wildlife Management*, 71, 2304–2310, https://doi.org/10.2193/2006-459

5.22. Remove roadkill regularly to reduce kill rate of predators/scavengers

https://www.conservationevidence.com/actions/2601

- We found no studies that evaluated the effects of removing roadkill regularly to reduce the kill rate of predators/scavengers.

'We found no studies' means that we have not yet found any studies that have directly evaluated this intervention during our systematic journal and report searches. Therefore, we have no evidence to indicate whether or not the intervention has any desirable or harmful effects.

Background

Animals killed on roads provide a food source for scavengers and some predators. These scavengers and predators then become vulnerable to being killed in collisions with vehicles themselves. Removing carcasses of road-killed animals thus removes a source of attraction towards roads for these species.

5.23. Modify vegetation along railways to reduce collisions by reducing attractiveness to mammals

https://www.conservationevidence.com/actions/2603

- **Two studies** evaluated the effects of modifying vegetation along railways to reduce collisions by reducing attractiveness to wildlife. Both studies were in Norway[1,2].

COMMUNITY RESPONSE (0 STUDIES)

POPULATION RESPONSE (2 STUDIES)

- **Survival (2 studies):** Two site comparison studies in Norway[1,2] found that clearing vegetation from alongside railways reduced moose-train collisions.

BEHAVIOUR (0 STUDIES)

Background

Wild mammals may be at increased risk of collisions with trains if they spend time on or close to the railway. Vegetation alongside railways may provide a feeding resource that attracts animals while, at the same time, obscuring views of oncoming trains. Removing vegetation in areas with high recorded collision rates may reduce attractiveness of such areas to mammals and, thus, reduce the risk of collision with trains.

A before-and-after study, site comparison study in 1980–1988 along a railway through boreal forest in Nord-Trøndelag County, Norway (1) found that vegetation removal alongside the railway reduced moose *Alces alces* deaths. Fewer moose were killed by trains after vegetation clearance (22 moose) than before (87 moose). Numbers also fell along uncleared sections but to a lesser extent with 27 killed after vegetation was cleared in experimental sections compared to 47 before. Vegetation clearance was estimated to be cost effective if more than 0.28 moose/km/year were expected to be killed in absence of clearance. Moose deaths were recorded along a 61-km section of railway in April–November of 1980–1988. In 1984, two sections with the highest casualties (totalling 22 km), had all bushes and trees removed from 20 m either side of the railway and all those <4 m high removed from a further 10 m width. Additional vegetation was removed at bends and on areas of browse attractive to moose. In 1986, cleared areas were sprayed with herbicide (Roundup) to reduce vegetation re-growth.

A site comparison study in 1985–2003 along a railway through forest in Hedmark County, Norway (2) found that vegetation clearance alongside the railway reduced moose *Alces alces* collisions with trains.

Fewer moose were killed after clearance (1.3/km/year) than before (2.6/km/year). Providing feeding stations away from the railway during winter in addition to clearing vegetation alongside the railway did not significantly further reduce collisions (5% reduction) compared to clearing vegetation alone. Before clearance, there were 2.5 times more moose killed/km/year within treatment sections compared to comparison sections. Numbers killed/km in treatment sections were fairly constant but casualties tended to increase in comparison sections over the study period (see paper for details). Eight forest clearings (1–14 km long) were established from 1990–2002 along a 100-km-long railway section. Vegetation >30 cm high was cut each year from alongside the railway. Sections without treatments were monitored as comparison sites (49 km). Moose-train collisions were recorded from July 1985– April 2003.

(1) Jaren V., Andersen R., Ulleberg M., Pedersen P.H. & Wiseth B. (1991) Moose-train collisions: the effects of vegetation removal with a cost–benefit analysis. *Alces*, 27, 93–99.

(2) Andreassen H.P., Gundersen H. & Storaas T. (2005) The effect of scent-marking, forest clearing, and supplemental feeding on moose-train collisions. *The Journal of Wildlife Management*, 69, 1125–1132, https://doi.org/10.2193/0022-541x(2005)069[1125:teosfc]2.0.co;2

5.24. Retain/maintain road verges as small mammal habitat

https://www.conservationevidence.com/actions/2604

- We found no studies that evaluated the effects of retaining or maintaining road verges as small mammal habitat.

'We found no studies' means that we have not yet found any studies that have directly evaluated this intervention during our systematic journal and report searches. Therefore, we have no evidence to indicate whether or not the intervention has any desirable or harmful effects.

Background

Roads can damage or destroy grassland habitats that host a range of mammal species, especially rodents and other small mammals. Roadside verges provide habitat that can at least partly mitigate this loss for a range of small mammal species (e.g. Ascensão *et al.* 2012; Bellamy *et al.* 2000).

Ascensão, F., Clevenger, A., Grilo, C., Filipe, J., & Santos-Reis, M. (2012). Highway verges as habitat providers for small mammals in agrosilvopastoral environments. *Biodiversity and Conservation*, 21, 3681–3697, https://doi.org/10.1007/s10531-012-0390-3

Bellamy P.E., Shore R.F., Ardeshir D., Treweek J.R. & Sparks T.H. (2000) Road verges as habitat for small mammals in Britain. *Mammal Review*, 30, 131–139, https://doi.org/10.1046/j.1365-2907.2000.00061.x

5.25. Fit vehicles with ultrasonic warning devices

https://www.conservationevidence.com/actions/2606

- **Three studies** evaluated the effects on mammals of fitting vehicles with ultrasonic warning devices. Two studies were in the USA[1,3] and one was in Australia[2].

COMMUNITY RESPONSE (0 STUDIES)

POPULATION RESPONSE (1 STUDY)

- **Survival (1 study):** A replicated, controlled study in Australia found that Shu Roo warning whistles did not reduce animal-vehicle collisions for eastern grey kangaroos or red kangaroos[2]

BEHAVIOUR (3 STUDIES)

- **Behaviour change (3 studies):** Three controlled studies (two replicated), in the USA[1,3] and Australia[2], found that ultrasonic warning devices did not deter mule deer[1], eastern grey kangaroos[2], red kangaroos[2] or white-tailed deer[3] from roads.

Background

Collisions between mammals such as deer and vehicles can result in death or injury to animals and humans alike. For example, it has been estimated that over 1 million deer-vehicle collisions occur annually in the USA alone (Conover *et al.* 1995). Wildlife warning whistles are designed to produce high frequency, ultrasonic noises to alert or frighten animals away from oncoming vehicles. Whistles can be mounted on vehicles, with the sound being emitted once the vehicle reaches a certain speed. Alternatively, whistles can be mounted on poles or small trees along roads and be activated by headlights of approaching cars.

See also: *Install acoustic wildlife warning along roads.*

Conover M.R., Pitt W.C., Kessler K.K., DuBow T.J. & Sanborn W.A. (1995) Review of human injuries, illnesses, and economic losses caused by wildlife in the United States. *Wildlife Society Bulletin*, 23, 407–414.

A controlled study in 1990 in sagebrush in Utah, USA (1) found that vehicle mounted wildlife warning whistles had no effect on the behaviour of mule deer *Odocoileus hemionus*. The proportions of deer that responded to the vehicle were 31% with a whistle and 39% without. Six percent of deer ran away from the vehicle with a whistle and 12% did so from the vehicle without a whistle. Authors reported that they did not know if the whistles produced any sound, nor if deer heard them. Two brands of wildlife warning whistles (Game Tracker's and Sav-a-life, producing 16–20 kHz) were mounted on the front of a truck. These were tested during late afternoon and early evening along 9.7 km of dirt road in January–February 1990. For each of 150 groups of deer (average six deer), a pass at 65 km/hour was made without and then with the whistle. Deer responses (none, head lifted, changed orientation, ran away, ran towards) and distances from the road were recorded for each pass (distances did not differ significantly between first and second passes).

A replicated, controlled study in 1997–2001 along roads in New South Wales, Queensland, Victoria and Western Australia, Australia (2) found that Shu Roo warning whistles did not alter behaviour of eastern grey kangaroos *Macropus giganteus* or red kangaroos *Macropus rufus* and did not reduce kangaroo-vehicle collisions. There was no significant difference in the number of kangaroos hit by vehicles with or without whistles (22% with; 7% without). Vigilance responses did not differ significantly for either species when whistles were turned on (60–65%) or off (40–75%) and no animals fled in response. The Shu Roo was not purely ultrasonic (4–19 kHz) and was only detected at 50 m. The whistle was not detectable above the noise of the four vehicles tested. The Shu Roo (two speakers in a rectangular metal case) signal was tested in the lab and in the field at 20–400 m (static and mounted on four vehicle types). Responses of 31 captive kangaroos to the Shu Roo (turned on/ off), mounted on a vehicle at 20–50 m, was recorded on 15 occasions in July–September 1997. Fifteen companies, in which people travelled large distances (average 49,000 km) conducted surveys in four states in August 1999 to January 2001. Fifty-seven vehicles had a Shu Roo fitted and 40 vehicles did not.

A replicated, controlled study in 2006 at a college campus in Georgia, USA (3) found that high frequency sounds from moving vehicles did not reduce white-tailed deer *Odocoileus virginianus* behaviours that were likely to cause a deer–vehicle collision. At 0.28 kHz, there was a significant increase in the proportion of behaviours likely to cause a collision (13%) compared to a vehicle without treatment (5%). At four other frequencies, there was no significant difference in proportions of negative behavioural responses compared to the vehicle without treatment (1–28 kHz: 6–9%). The proportion of behaviours likely to decrease deer-vehicle collisions did not differ between different high frequencies and no high-frequency sound (0.28 kHz: 33%; 1 kHz: 37%; 8 kHz: 24%; 15 kHz: 33%; 28 kHz: 24%; no high-frequency sound: 35%;). Two road sections (≥ 5 km apart), 280 m and 220 m long, were studied. For each of 319 trials, a deer was observed before and during one of six randomly assigned treatments: 0.28, 1, 8, 15 or 28 kHz or no sound. The high-frequency sounds (within deer hearing range) were played at 70 decibels from front-mounted speakers on the vehicle (48 km/hr). Deer within 10 m of the road or ahead of the vehicle were monitored from an observation platform, from 06:00 to 09:00 h and 19:00 to 22:00 h, in April and June 2006.

(1) Romin L.A. & Dalton L.B. (1992) Lack of response by mule deer to wildlife warning whistles. *Wildlife Society Bulletin*, 20, 382–384.

(2) Bender H. (2001) *Deterrence of kangaroos from roadways using ultrasonic frequencies: efficacy of the Shu Roo*. University of Melbourne, Department of Zoology unpublished report.

(3) Valitzski S.A., D'Angelo G.J., Gallagher G.R., Osborn D.A., Miller K.V. & Warren R.J. (2009) Deer responses to sounds from a vehicle-mounted sound-production system. *The Journal of Wildlife Management*, 73, 1072–1076, https://doi.org/10.2193/2007-581

5.26. Install signage to warn motorists about wildlife presence

https://www.conservationevidence.com/actions/2608

- **Six studies** evaluated the effects on mammals of installing signage to warn motorists about wildlife presence. Four studies were in the USA[1,3,4,5] one was in Australia[2] and one was in Canada[6].

COMMUNITY RESPONSE (0 STUDIES)

POPULATION RESPONSE (6 STUDIES)

- **Abundance (1 study):** A before-and-after study in Australia[2] found that when wildlife signs were installed along with speed restrictions, rumble strips, reflective wildlife deterrents, wildlife escape ramps and an educational pamphlet, a small population of eastern quoll re-established in the area.

- **Survival (6 studies):** Three of five studies (including four controlled and three before-and-after studies), in the USA[1,3,4,5] and Canada[6], found that warning signs did not reduce collisions between vehicles and deer[1,3,5]. The other two studies found that warning signs did reduce collisions between vehicles and deer[4,6]. A before-and-after study in Australia[2] found that wildlife signs along with speed restrictions, rumble strips, reflective wildlife deterrents, wildlife escape ramps and an educational pamphlet, reduced collisions between vehicles and Tasmanian devils but not eastern quolls.

BEHAVIOUR (0 STUDIES)

OTHER (2 STUDIES)

• **Human behaviour change (2 studies):** Two controlled studies (one also replicated, before-and-after), in the USA[1,4], found that signs warning of animals on the road reduced vehicles speeds.

Background

Wildlife crossing signs alert drivers to the potential presence of wildlife on or near a road. They encourage drivers to be more alert and/or reduce the speed of their vehicle, with the goal of reducing animal-vehicle collisions. Motorists may become habituated to signs if they are present all year round, are too common or look similar to other signs. Solutions may be to use temporary seasonal signs, animated signs, flashing lights or flags to catch the attention of drivers. Animal detection warning systems have sensors that detect large animals on or near the road that are wired to flashing signs.

Studies that investigate the effect on vehicle speed of warning signs are not included here if they do not report relevant metrics on vehicle-mammal collision rates (e.g. Lehnert & Bissonette 1997; Al-Ghamdi & AlGadhi 2004) though information on changes in motorists' speed is reported here if the study also reports collision rates.

See also: *Reduce legal speed limit.*

Lehnert M.E. & Bissonette J.A. (1997) Effectiveness of highway crosswalk structures at reducing deer-vehicle collisions. *Wildlife Society Bulletin*, 25, 809–818.

Al-Ghamdi A.S. & AlGadhi S.A. (2004) Warning signs as countermeasures to camel–vehicle collisions in Saudi Arabia. *Accident Analysis & Prevention*, 36, 749–760, https://doi.org/10.1016/j.aap.2003.05.006

A controlled study in 1972–1973 in Colorado, USA (1) found that lighted, animated deer crossing signs reduced vehicle speeds but did not reduce deer-vehicle collisions. There was an average of one collision for each 57 deer-crossings when the signs were both on and off. Average vehicle speeds were lower with the signs on, but the reduction was by <5 km/h. Three deer carcasses at the highway edge (46, 98 and 107 m before signs) reduced speeds but the reduction did not differ between when signs were on (10 km/h reduction) or off (13 km/h reduction). Two deer crossing signs were installed along a 1.6-km-long highway section (with 97 km/h limit), where deer-vehicle collisions were frequent. Signs were reflective yellow diamonds (1.8 × 1.8 m) with four silhouettes of deer in neon tubing lighting across the sign. Signs were turned on and off for alternate weekly periods during January–March over four weeks in 1972 and 11 weeks in 1973. Numbers of deer crossing the highway were estimated by nightly spotlight counts. Collisions were recorded each night and morning. Vehicle speeds were measured at 0.2, 1.1 and 2.4 km behind the sign between 18:00 and 22:00 h.

A before-and-after study in 1990–1998 in Tasmania, Australia (2) found that following installation of reflective wildlife signs, speed restrictions, rumble strips, reflective wildlife deterrents, wildlife escape ramps and publication of an educational pamphlet, an eastern quoll *Dasyurus viverrinus* population partially re-established and vehicle collisions with Tasmanian devils *Sarcophilus laniarius*, but not eastern quolls, decreased. Results were not tested for statistical significance. Following local extinction, 3–4 quolls re-colonised within six months of installation, increasing to ≥8 animals after two years. Road-kills were similar for quolls before and after implementation (1.6 vs 1.5/ year), but decreased for Tasmanian devils (3.6 vs 1.5/year). Following road widening in 1991, vehicle-wildlife collisions increased and quolls became locally extinct (from 19 animals). In 1996, large, reflective signs displaying a wallaby, and the words 'Cradle Wildlife Zone' were installed, along with the other five interventions. Animals were surveyed using 60 cage traps for three nights in alternate months in October 1990– April 1993. Then, 10–20 traps were set for 20–100 trap nights each April, May and July in 1995–1998. Spotlight counts were made once or twice in 1991, 1995, 1996 and 1998. Road-kills were recorded in 1990–1996.

A replicated, controlled, before-and-after study in 1996–2000 along roads in three townships in Michigan, USA (3) found that deer warning signs (including some of a novel design) did not reduce deer-vehicle collisions. In one township, the overall collision rate after installing standard and novel warning signs (55/year) did not differ significantly from that before installation (69/year). At the same time, there was no change in average rates in three townships without warning signs (after: 41–62/year/township; before: 36–62/year/township). There was no significant difference in average collision rates 200 feet either side of signs on seven road stretches that just had the novel sign design (after installation: 9/year/stretch; before: 11/year/stretch). Vehicle speeds were not lower with signage than without along one road stretch and were <0.5 miles/hour lower along a second stretch. Two warning sign designs were installed around one township between October and January of 1998–2000. Eighteen novel signs, (leaping deer and car on an orange background and text stating 'High crash area') were installed on seven road stretches with high vehicle-deer collision rates. Fifty-two standard signs (leaping deer on orange background) were installed on other sections. Collisions, monitored by State Police, were compared in the township before (1996–1997) and after installation (1998 and 2000) and in three townships without signs. Vehicle speeds were monitored for 15–24-hour periods before (1,124 vehicles) and after installation (1,221 vehicles) on two road sections.

A replicated, controlled, before-and-after study in 1995–2002 along five highways in Utah, Nevada and Idaho, USA (4) found that temporary warning signs reduced vehicle speeds and collisions with mule deer *Odocoileus hemionus* during migrations. Fewer deer deaths occurred after signs were installed (3–12/migration) than before (7–35/ migration). Concurrently, deaths did not decline on a road section without signs (after: 3–13/migration; before: 3–11/migration). Once signs were installed, the proportion of vehicles speeding (8%) was lower than before they were installed (19%). There was no concurrent decline on a road section without signs (after: 19%; before: 25%). Signs affected speeds of heavy trucks more than of passenger vehicles. Sections of five highways, crossed by mule deer seasonal migrations, were studied. Each 6.5-km-long section was divided into two with each half randomly assigned as treatment or control. Treatment sections had temporary yellow and black warning signs (2 × 1 m) with reflective flags and solar-powered flashing amber lights installed at each end and smaller signs (1 m²) each mile. Deer-vehicle collisions were monitored daily during spring and autumn migrations, before (2–4 years) and after (1–4 years) signs were installed. Night-time vehicle speeds were monitored in 2000–2001.

A before-and-after study in 1989–2004 along 22 sections of highway in Kansas, USA (5) found that deer warning signs did not reduce vehicle collisions with white-tailed deer *Odocoileus virginianus*. The collision rate after signs were installed (0.83) did not differ from than in the 2–10 years before signs were installed (0.78; units not clear in report, but may refer to deer killed/km/year). However, the rate over just the three years after sign installation (0.71) was significantly lower than that in just the three years before installation (1.16). Numbers of collisions closely followed trends in deer populations, which increased to a peak in around 1999 and then decreased. Deer-vehicle collision data were obtained for 22 sections of highway (section lengths not stated) across seven counties for 2–10 years before and 2–5 years after deer warning signs were installed. Timing of sign installations was not known precisely but was assumed, in the report, to have been within six months of publication of Road Safety Reports, which were mostly published in 1999.

A replicated, randomized, controlled, before-and-after study in 2005–2008 at 26 urban sites around a city in Alberta, Canada (6) found that warning signs reduced the number of collisions between vehicles and white-tailed deer *Odocoileus virginianus*. At warning sign locations, there were fewer deer-vehicle collisions after sign installation (0.4 deer-vehicle collisions/location/year) than before (1.7 deer-vehicle collisions/location/year). Concurrently, at locations without warning signs, there was no significant difference in deer-vehicle collision rates after (1.0 deer-vehicle collisions/location/year) compared to before signs were installed (1.7 deer-vehicle collisions/location/year). Twenty-six road locations with high incidence of deer-vehicle collisions were selected. Pairs of reflective deer warning signs (90 × 90 cm, diamond shape) were mounted on 3-m-high posts, 1,600 m apart, facing opposite directions, at 13 locations (randomly selected) in June 2008. The other 13 locations had no signs installed. Deer carcasses (mostly white-tailed deer but possibly some mule deer *Odocoileus hemionus*) were monitored within an 800-m radius of each location from June to December in 2005–2007 (before sign installation) and in June–December 2008 (after sign installation).

(1) Pojar T.M., Prosencer R.A., Reed D.F. & Woodard T.N. (1975) Effectiveness of a lighted, animated deer crossing sign. *The Journal of Wildlife Management*, 39, 87–91.

(2) Jones M.E. (2000) Road upgrade, road mortality and remedial measures: impacts on a population of eastern quolls and Tasmanian devils. *Wildlife Research*, 27, 289–296, https://doi.org/10.1071/wr98069

(3) Rogers E. (2004) *An ecological landscape study of deer vehicle collisions in Kent County, Michigan*. Report to Kent County Road Commission, Michigan, USA.

(4) Sullivan T.L., Williams A.F., Messmer T.A., Hellinga L.A. & Kyrychenko S.Y. (2004) Effectiveness of temporary warning signs in reducing deer-vehicle collisions during mule deer migrations. *Wildlife Society Bulletin*, 32, 907–915, https://doi.org/10.2193/0091-7648(2004)032[0907:eotwsi]2.0.co;2

(5) Meyer E. (2006) *Assessing the effectiveness of deer warning signs*. Final report. KTRAN: KU-03-6.

(6) Found R. & Boyce M.S. (2011) Warning signs mitigate deer–vehicle collisions in an urban area. *Wildlife Society Bulletin*, 35, 291–295, https://doi.org/10.1002/wsb.12

5.27. Use road lighting to reduce vehicle collisions with mammals

https://www.conservationevidence.com/actions/2614

- **Two studies** evaluated the effects on mammals of using road lighting to reduce vehicle collisions with mammals. Both studies were in the USA[1,2].

COMMUNITY RESPONSE (0 STUDIES)

POPULATION RESPONSE (2 STUDIES)

- **Survival (2 studies):** One of two studies (one controlled and one before-and-after), in the USA[1,2], found that road lighting reduced vehicle collisions with moose[2]. The other study found that road lighting did not reduce vehicle collisions with mule deer[1].

BEHAVIOUR (0 STUDIES)

Background

The risk of wildlife-vehicle collisions was found to be six times higher at night and dawn than during the day (Lavsund & Sandegren 1991). Installing lighting along roads may increase visibility of animals to motorists and may, therefore, reduce the number of collisions. However, in areas where species are sensitive to human disturbance, they may avoid areas of roads with artificial lighting and, instead, cross elsewhere.

Lavsund S. & Sandegren F. (1991) Moose-vehicle relations in Sweden: a review. *Alces*, 27, 118–126.

A controlled study in 1974–1979 along a highway in Colorado, USA (1) found that highway lighting did not reduce vehicle collisions with mule deer *Odocoileus hemionus*. There was no significant difference between deer-vehicle collision rates with lights on (39 collisions from 2,611 crossings) or off (45 collisions from 2,480 crossings). Lighting did not alter the location of crossings, with accidents not occurring closer to the lights when they were off. Lighting did not alter vehicle speeds (lights on: 79 km/h; lights off: 80 km/h). Thirteen 37,000-lumen, 700-W, clear, mercury-vapour lamps (12 m high) were installed along 1.2 km of a four-lane highway (speed limit 88.5 km/h). Nine were spaced at 59–69-m intervals along 0.5 km of highway (full lighting) and two at each end were spaced at 119 and 302 m (transition lighting). Lights were alternately turned on and off for one-week periods in January–April of 1974–1979. Deer-vehicle collisions were recorded each morning and evening. Deer crossings were recorded during nightly spotlight surveys and using snow track counts. Deer behaviour was observed for two hours/night. Vehicle speeds were recorded during 35 nights in 1974.

A before-and-after study in 1977–1990 along a highway in Alaska, USA (2) found that road lighting reduced vehicle collisions with moose *Alces alces*. There were 65% fewer moose-vehicle collisions when lighting was installed compared to before its installation (actual numbers not stated). There were 95% fewer moose-vehicle collisions along the section with lighting, fencing with one-way gates and an underpass after they were installed (0.7/year) than before (17/year). Overall mortality along the entire stretch of road was lower after installation of lighting, barrier fencing and an underpass, with fewer collisions (12/year) than previously (38/year). In October 1987, road lighting was installed along 11.5 km of the highway. Fencing and 30 one-way gates were installed along 5.5 km of this section and an underpass was created. Moose-vehicle collisions were monitored before (1977–1987) and after (1987–1990) installation.

(1) Reed D.F. & Woodard T.N. (1981) Effectiveness of highway lighting in reducing deer-vehicle accidents. *The Journal of Wildlife Management*, 45, 721–726.

(2) McDonald M.G. (1991) Moose movement and mortality associated with the Glenn Highway expansion. *Alces*, 27, 208–219.

5.28. Use chemical repellents along roads or railways

https://www.conservationevidence.com/actions/2615

- **Five studies** evaluated the effects on mammals of using chemical repellents along roads or railways. Two studies were in Canada[2,3] and one each was in Germany[1], Norway[4] and Denmark[5].

COMMUNITY RESPONSE (0 STUDIES)

POPULATION RESPONSE (2 STUDIES)

- **Survival (2 studies):** Two studies (one before-and-after, one site comparison), in Germany and Norway[1,4], found that chemical-based repellents did not reduce collisions between ungulates and road vehicles[1] or trains[4].

BEHAVIOUR (4 STUDIES)

- **Behaviour change (4 studies):** Two of four studies (including three replicated, controlled studies), in Germany[1], Canada[2,3], and Denmark[5], found that chemical repellents, trialled for potential to deter animals from roads, did not deter ungulates[2,5]. The other two studies found mixed results with repellents temporarily deterring some ungulate species in one study[1] and one of three deterrents deterring caribou in the other[3].

Background

Large number of mammals, especially deer and other ungulate species, are killed in collisions with road vehicles (e.g. Conover *et al.* 1995) or trains. This could be reduced if the application of repellents could deter animals from accessing roads.

See also: *Agriculture & Aquaculture-Use repellents that smell bad ('area repellents') to deter crop or property damage by mammals to reduce human-wildlife conflict.*

Conover M.R., Pitt W.C., Kessler K.K., DuBow T.J. & Sanborn W.A. (1995) Review of human injuries, illnesses, and economic losses caused by wildlife in the United States. *Wildlife Society Bulletin*, 23, 407–414.

A before-and-after study in 1991–1996 at a research centre in Nordrhein-Westfalen, Germany (1) found that Duftzaun scent repellent temporarily deterred some but not all large mammal species and did not reduce vehicle collisions. Red deer *Cervus elaphus*, roe deer *Capreolus capreolus* and wild boar Sus scrofa were killed on the road. There was no significant difference between numbers killed on the road when repellent was used (18/year) compared with before (13/year) or after (9/year) use (data supplied by author). In enclosure trials, mufflon *Ovis orientalis* (seven animals) avoided scented posts for 15 minutes. Sika deer *Cervus nippon* (four) avoided posts for a few minutes and roe deer (four) approached posts cautiously. Red deer (one) and fallow deer *Dama dama* (four) were not deterred by repellent. Trials were held in six enclosures. Duftzaun (a mixture of 10 acids integrated into a ridged foam) was applied to tops of posts supporting 50% of daily feed and animals' behaviours were recorded. In November 1992, a Duftzaun 'scent fence' was installed along a 2.8-km-long highway section where deer crossed. Scent was re-injected after four weeks and then every three months. Vehicle-wildlife collisions were recorded for two years before installation (1991–1992), three years after installation (1993–1995) and one year post-trial (1996).

A replicated, controlled study in 1996–1998 in forest in Ontario, Canada (2) found that 18 scent repellents (trialled for potential to deter animals from roads) did not deter white-tailed deer *Odocoileus virginianris*, elk *Cervus canadensis nelsoni* or moose *Alces alces americana*. Animals used a similar proportion of trails with repellents applied (63–80%) and of trails without repellents (62–74%). Similarly, at mineral licks with repellents, there were fresh animal tracks on 59% of days, which was not significantly different to the 72% of days at mineral licks without repellents. Eighteen potential repellents were identified (from literature review) and tested on wild deer or deer, elk and moose. Repellents were mainly chemicals, including commercial repellents (Deer Away powder, Critter Ridder, mothballs) and those that simulated predators (e.g. wolf, coyote) or humans (soap, hair, clothing, sweat), but also included wolf and human silhouettes. Use of pairs of trails through snow (up to 240 pairs) with head-height repellents or without repellents, were monitored by counting tracks in winter 1997 or 1998. Repellents were also tested at a mineral lick. Use of this was monitored by track counts and an infra-red camera on days with and without repellents, in summer 1997.

A replicated, controlled study in 1998 in three captive facilities in Alberta, Canada (3) found that one of three repellents (trialled for potential to deter animals from roads) discouraged feeding by caribou *Rangifer tarandus*. Animals ate significantly less food treated with lithium chloride (day 1: 900 g consumed; days 2–5: 200–300 g/day) than untreated food (1,200 g/day). Caribou ate significantly less food treated with Deer Away Big Game Repellent® on day 1 (300 g consumed) but not days 2–5 (700–900 g/day) compared to untreated food (1,200 g/day). Wolfin® did not affect the amount eaten (days 1–5: 1,100 g/day; untreated: 1,100 g/day). Lithium chloride (a gastrointestinal toxicant), Deer Away Big Game Repellent® (olfactory and taste repellent) and Wolfin® (olfactory repellent stimulating wolf urine), which could each be added to salt-sand mixtures or placed along roads to discourage salt licking, were tested on 14 captive caribou at three sites. Big Game Repellent powder (12–15 g/kg pellets) and lithium chloride (150 mg/kg body mass) were put on pelleted food. Wolfin capsules (5 cm) were placed on 1-m-high posts, 2 m from pellets. Food was provided without repellent for two days before and after a five-day period with repellents, in February–May 1998.

A before-and-after, site comparison study in 1985–2003 along a railway through forest in Hedmark County, Norway (4) found that chemical scent-based repellent did not reduce moose *Alces alces* collisions with trains. In scent-marked areas, there was an average of 0.3 collisions/km/year when scent marks were applied compared to 1.8/km/year before. However, there was large variation in effectiveness between sections and the reduction was not statistically significant. Numbers killed/km/year in non-treated sections tended to rise over the study period (see paper for details). Along a 100-km-long stretch of railway, ten 500-m-long sections were sprayed with repellent during the winter of 1994–1995 and a further 10 in 1995–1996, during the first days when snow exceeded 20 cm depth. The repellent 'Duftzaun' (components from brown bear *Ursus arctos*, wolf *Canis lupus*, lynx *Lynx lynx* and humans) was sprayed on trees and bamboo canes at 5-m intervals. One treatment lasted 3–4 months. Sections without treatment (total 49 km) were also monitored. Moose-train collisions were recorded from July 1985–April 2003.

A replicated, controlled, before-and-after study in 2006 in a conifer plantation in Denmark (5) found that the repellents Mota FL and Wolf Urine (trialled for potential to deter animals from roads) did not reduce visits by deer. Roe deer *Capreolus capreolus* visited a similar number of Moto FL-treated plots after application (6–8 plots/day) and before (4–8 plots/day). Visit rates to untreated plots were similar after application in treatment plots (7–8 plots/day) compared to before (5–8 plots/day). The same pattern held for red deer *Cervus elaphus* treatment plots (after: 1–3 plots/day); before: 0–4 plots/day) and untreated plots (after: 2–4 plots/day; before: 0–3 plots/day). Roe deer visited a similar number of Wolf Urine-treated plots after application (7–9 plots/day) and before (7–9 plots/day). Visit rates to untreated plots were similar after application in treatment plots (6–9 plots/day) compared to before (6–9 plots/day). The same pattern held for red deer treatment plots (after: 1–4 plots/day; before: 1–3 plots/day) and untreated plots (after: 0–4 plots/day; before: 0–4 plots/day). Eighteen sand arenas (4 m diameter, ≥400 m apart) included nine for repellent treatments and nine controls. Arenas were baited with beet and maize every 3–4 days or as required, for two months. Deer tracks were monitored daily for seven days before repellent was sponged onto four scent posts at each treatment arena. Track monitoring continued for seven further days. Mota FL was assessed from 7–21 February 2006. Repellent posts were then cleaned with alcohol and Wolf Urine assessed from 8–22 March 2006.

(1) Lutz W. (1994) Ergebnisse der Anwendung eines sogenannten Duftzaunes zur Vermeidung von Wildverlusten durch den Straßenverkehr nach Gehege- und Freilandorientierungen. [Trial results of the use of a 'Duftzaun' (scent fence) to prevent game losses due to traffic a ccidents]. *Zeitschrift für Jagdwissenschaften*, 40, 91–108.

(2) Castiov F. (1999) *Testing potential repellents for mitigation of vehicle-induced mortality of wild ungulates in Ontario.* Thesis. School of Graduate Studies and Research, Laurentian University.

(3) Brown W.K., Hall W.K., Linton L.R., Huenefeld R.E. & Shipley L.A. (2000) Repellency of three compounds to caribou. *Wildlife Society Bulletin*, 28, 365–371.

(4) Andreassen H.P., Gundersen H. & Storaas T. (2005) The effect of scent-marking, forest clearing, and supplemental feeding on moose-train collisions. *The Journal of Wildlife Management*, 69, 1125–1132, https://doi.org/10.2193/0022-541x(2005)069[1125:teosfc]2.0.co;2

(5) Elmeros M., Winbladh J.K., Andersen P.N., Madsen A.B. & Christensen J.T. (2011) Effectiveness of odour repellents on red deer (*Cervus elaphus*) and roe deer (*Capreolus capreolus*): a field test. *European Journal of Wildlife Research*, 57, 1223–1226, https://doi.org/10.1007/s10344-011-0517-y

5.29. Use alternative de-icers on roads

https://www.conservationevidence.com/actions/2616

- We found no studies that evaluated the effects on mammals of using alternative de-icers on roads.

'We found no studies' means that we have not yet found any studies that have directly evaluated this intervention during our systematic journal and report searches. Therefore, we have no evidence to indicate whether or not the intervention has any desirable or harmful effects.

Background

Use of chloride salts as de-icers along roads in winter can attract wildlife and may therefore increase vehicle-wildlife collisions, particularly in areas without natural salt licks. The main de-icers used by highway agencies are chloride-based salts such as sodium chloride, calcium chloride or magnesium chloride, or acetate-based de-icers such as potassium, sodium or calcium magnesium acetate. Reducing the amount of salt used or using alternative de-icers without salt, particularly in areas with high vehicle-wildlife collision rates, may reduce the attractiveness of roadsides to wildlife.

A study in Canada found that filling roadside salt pools with rocks (thus rendering them unavailable as salt-lick sources) reduced the number and duration of visits by moose *Alces alces* (Leblond *et al.* 2007; see *Modify the roadside environment to reduce collisions by reducing attractiveness of road verges to mammals*).

Leblond M., Dussault C., Ouellet J.-P., Poulin M., Courtois R. & Fortin J. (2007) Management of roadside salt pools to reduce moose–vehicle collisions. *The Journal of Wildlife Management*, 71, 2304–2310, https://doi.org/10.2193/2006-459

5.30. Provide food/salt lick to divert mammals from roads or railways

https://www.conservationevidence.com/actions/2617

- **Three studies** evaluated the effects of providing food or salt licks to divert mammals from roads. One study was in the USA[1], one was in Norway[2] and one was a review of studies from across North America and Europe[3].

COMMUNITY RESPONSE (0 STUDIES)

POPULATION RESPONSE (2 STUDIES)

- **Survival (2 studies):** A replicated, controlled study in the USA[1] found that intercept feeding reduced mule deer road deaths along two of three highways in one of two years. A replicated, site comparison study in Norway[2] found that intercept feeding reduced moose collisions with trains.

BEHAVIOUR (1 STUDY)

- **Behaviour change (1 study):** A review of feeding wild ungulates in North America, and Europe[3] found that feeding diverted ungulates away from roads in one of three studies.

Background

'Intercept feeding' provides supplemental food sources in a particular location in an attempt to divert animals away from roads or railways. It is typically used as a technique aimed at ungulates, which can account for a large number of collisions between vehicles and wildlife (e.g. an estimated >1 million deer-vehicle collisions annually in the USA, Conover *et al.* 1995).

Conover M.R., Pitt W.C., Kessler K.K., DuBow T.J. & Sanborn W.A. (1995) Review of human injuries, illnesses, and economic losses caused by wildlife in the United States. *Wildlife Society Bulletin*, 23, 407–414.

A replicated, controlled study in 1985–1986 along three highways in Utah, USA (1) found that intercept feeding reduced mule deer *Odocoileus hemionus* road deaths along two of three highways in one of two years. In the first year, the numbers of mule deer killed on road sections with intercept feeding (8–19 deer killed) were not significantly different to the numbers killed on those without (14–31). The following year, roads kills were lower on two highway sections with intercept feeding (with feeding: 34–38 deer killed; without: 59–89), but higher with feeding on the third (feeding: 31; without: 13). Feeding stations were closer to this third highway (0.4 km) than to the others (0.8–1.2 km). Road-kill deer were recorded along three highways, within 21–24-km-long sections. Highways were divided into a treatment (feed) and control (no-feed) section of equal length (8.3 or 9.6 km), separated by a shorter buffer zone (4.2 or 4.8 km). Treatment and control sections were swapped in the second year. There were four feeding stations/treatment section. Alfalfa hay, deer pellets and apple mash were provided 1–3 times/3 days from January to mid-March of 1985 and 1986.

A replicated, site comparison study in 1985–2003 along a railway through forest in Hedmark County, Norway (2) found that intercept feeding stations reduced moose *Alces alces* collisions with trains. There was an estimated 40% collision reduction following feeding station establishment, equating to six fewer moose collisions/year. Providing intercept feeding stations and clearing vegetation >30cm high from alongside the railway did not significantly further reduce collisions (5% reduction) compared to implementing just one of these treatments. Before providing feeding stations, 2.5 times more moose were killed/km/year within treatment sections compared to comparison sections. Numbers killed/km in treatment sections were fairly constant but casualties increased in comparison sections over the study period. Moose feeding stations were established, in 1995, along a 100-km-long railway section. Feeding stations were in side-valleys, linked to three railway sections (4, 6 and 8 km long). Landowners provided food during the winter, using baled grasses and silage and/or herbs, from when snow accumulated until April–May. Sections without treatments were also monitored (total 49 km long). Moose-train collisions were recorded from July 1985–April 2003.

A review of evidence within studies looking at effects of feeding wild ungulates in North America, Fennoscandia and elsewhere in Europe (3) found that diversionary feeding diverted ungulates away from roads in one of three studies. No such effect was found in the other two studies. The review also assessed evidence for supplementary feeding affecting survival and morphological characteristics. In total, the review reported evidence from 101 studies that met predefined criteria from an initial list of 232 papers and reports. Three of these studies investigated the effectiveness of feeding for diverting ungulates away from roads.

(1) Wood P. & Wolfe M.L. (1988) Intercept feeding as a means of reducing deer-vehicle collisions. *Wildlife Society Bulletin*, 16, 376–380.

(2) Andreassen H.P., Gundersen H. & Storaas T. (2005) The effect of scent-marking, forest clearing, and supplemental feeding on moose-train collisions. *The Journal of Wildlife Management*, 69, 1125–1132, https://doi.org/10.2193/0022-541x(2005)069[1125:teosfc]2.0.co;2

(3) Milner J.M., van Beest F.M., Schmidt K.T., Brook R.K. & Storaas T. (2014) To feed or not to feed? Evidence of the intended and unintended effects of

feeding wild ungulates. *The Journal of Wildlife Management*, 78, 1322–1334, https://doi.org/10.1002/jwmg.798

5.31. Use reflective collars or paint on mammals to reduce collisions with road vehicles

https://www.conservationevidence.com/actions/2619

- We found no studies that evaluated the effects of using reflective collars or paint on mammals to reduce collisions with road vehicles.

'We found no studies' means that we have not yet found any studies that have directly evaluated this intervention during our systematic journal and report searches. Therefore, we have no evidence to indicate whether or not the intervention has any desirable or harmful effects.

Background

Fitting collars with reflective tape on animals to increase their visibility to drivers was considered in Canada for reintroduced wood bison *Bos bison* (Huijser *et al.* 2007). In Finland, a spray that reflects vehicle headlights was applied to the antlers of reindeer with the aim of making the animals more visible to motorists and so reducing collisions with vehicles (https://www.ibtimes.co.in/finnish-reindeer-given-glowing-antlers-to-prevent-accidents-539561).

Huijser M.P., McGowen P., Fuller J., Hardy A., Kociolek A., Clevenger A.P., et al. (2007) *Wildlife–vehicle collision reduction study*. Report to Congress. U.S. Department of Transportation, Federal Highway Administration, Washington D.C., USA.

5.32. Use wildlife decoy to reduce vehicle collisions with mammals

https://www.conservationevidence.com/actions/2620

- We found no studies that evaluated the effects of using wildlife decoys to reduce vehicle collisions with mammals.

'We found no studies' means that we have not yet found any studies that have directly evaluated this intervention during our systematic journal and report searches. Therefore, we have no evidence to indicate whether or not the intervention has any desirable or harmful effects.

Background

Animal silhouettes made of wood, Styrofoam or cardboard, or models or stuffed animals, placed along the edge of roads, may remind people to slow down in certain areas where animals are commonly hit. Reduced vehicles speeds may help to reduce vehicle-wildlife collisions. One small study found that a stuffed deer did reduce vehicle speeds (Reed & Woodard 1981) but did not assess whether or not this resulted in fewer collisions between vehicles and animals.

Reed D.F. & Woodard T.N. (1981) Effectiveness of highway lighting in reducing deer-vehicle accidents. *The Journal of Wildlife Management*, 45, 721–726.

5.33. Close roads in defined seasons

https://www.conservationevidence.com/actions/2626

- **One study** evaluated the effects on mammals of closing roads in defined seasons. This study was in the USA[1].

COMMUNITY RESPONSE (0 STUDIES)

POPULATION RESPONSE (0 STUDIES)

BEHAVIOUR (1 STUDY)

- **Use (1 study):** A site comparison study in the USA[1] found that closing roads to traffic during the hunting season increased use of those areas by mule deer.

Background

Some mammals may avoid areas around roads (e.g. Rost & Bailey 1979). Closing these roads to traffic, especially at times of the year when they most use the habitat that the road runs through, may increase their use of such areas and, hence, increase their access to natural resources such as food and shelter.

Rost G.R. & Bailey J.A. (1979) Distribution of mule deer and elk in relation to roads. *The Journal of Wildlife Management*, 43, 634–641.

A site comparison study in 2015 in a forest in Oregon, USA (1) found that closing roads to traffic during the hunting season increased use of those areas by mule deer *Odocoileus hemionus*. Mule deer positions were closer to closed roads (average 190 m) than to open roads (average 1,250 m). In March 2015, an unspecified number of mule deer were captured and fitted with GPS collars that recorded their location every 13 hours. Deer locations and distances to the nearest road were recorded in August–October 2015. During this period, an unspecified number of roads in the area were closed to vehicles, while others remained open. This period overlapped with the legal hunting season.

(1) Curtis A.M. & Du Toit J.T. (2017) Efficacy of travel management areas for reducing disturbance to mule deer during hunting seasons. *Wildlife Society Bulletin*, 41, 309–312, https://doi.org/10.1002/wsb.771

Utility & Service Lines

5.34. Install crossings over/under pipelines

https://www.conservationevidence.com/actions/2627

- **Three studies** evaluated the effects on mammals of installing crossings over/under pipelines. Two studies were in the USA[1,2] and one was in Canada[3].

COMMUNITY RESPONSE (0 STUDIES)

POPULATION RESPONSE (0 STUDIES)

BEHAVIOUR (3 STUDIES)

- **Use (3 studies):** A study in USA[1] found that buried pipeline sections were used more frequently than their availability as crossing points by caribou. A study in USA[2] found that pipeline sections elevated specifically to permit mammal crossings underneath were not used by moose or caribou more than were other elevated sections. A controlled study in Canada[3] found that a range of large mammal species used wildlife crossings over pipelines.

Background

Pipelines can extend hundreds of kms and may represent substantial barriers to mammal movements if they lie at or just above the surface of the ground. Crossing points can be either elevated sections of pipe with space for mammals to pass beneath, buried sections or sections with crossing ramps constructed over the pipe.

A study in 1981–1983 of three sites along a pipeline across tundra in Alaska, USA (1) found that buried pipeline sections were used more frequently than their availability as crossing points by caribou *Rangifer tarandus*. Buried pipeline sections accounted for 10 of 180 crossings (6%) at one site, 5 of 41 crossings (12%) at a second site and 65 of 732 crossings (9%) at a third site. These proportions were all higher than the proportion of pipeline that was buried at these sites (2%). Ramps (20–50 m wide) were installed across buried pipeline sections at three study sites. Sites covered 180–275 ha, each including 1.7–2.2 km of pipeline. Sections not buried were elevated 1.2–4.3 m above the ground. A crossing comprised one or more caribou crossing the pipeline, with >50% of group members successfully crossing. Crossings were documented by direct observations in late June to early August of 1981–1983.

A study in 1977–1978 of a pipeline across tundra in Alaska, USA (2) found that pipeline sections elevated specifically to permit crossings of animals underneath were not used by moose *Alces alces* or caribou *Rangifer tarandus* more than were other elevated sections. Of 81 crossing sections elevated to facilitate mammal crossings, 13 (16%) were used by moose, a similar rate to the 754 of 6,526 other elevated sections (12%) that were crossed. Caribou used four of 53 specifically elevated crossing sections (8%) available to them, a lower rate than the 10% of remaining elevated sections used as crossing points. Along a 145-km-long pipeline, 81 pipe sections were elevated specifically to permit large mammal passage underneath. These sections were ≥3 m high. Remaining sections, were of variable, but generally lower, height. All elevated pipe sections were 18.3 m long between supports. Animal passage was determined by footprint surveys after fresh snow. The pipe, separated into three sections, was surveyed on 11–15 occasions in October 1977–February 1978 and 1–5 occasions in March–April 1978.

A controlled study in 2006–2007 in boreal mixed-woodland in Alberta, Canada (3) found that mammals used wildlife crossings over oil pipelines. Camera-trapping showed that successful crossings were made by deer (white-tailed deer *Odocoileus virginianus* and mule deer *Odocoileus hemionus*) on 746 of 904 approaches (83%), by moose *Alces alces* on 157 of 178 approaches (88%) and by coyotes *Canis latrans* on 52 of 59 of approaches (88%). Crossings were also made by lynx *Lynx canadensis* and black bear *Ursus americanus* (twice each) and gray wolf *Canis lupus* (once). Snow-tracking showed that deer had a higher successful pipeline crossing rate at wildlife crossings (96% of approaches) than along pipeline sections without crossings (90%). Moose success rate at crossings (66%) was lower than on sections without crossings (77%). In March 2006, five crossing structures of soil and vegetation ($\geq$20 m long, $\geq$4 m wide, 2–3 m high) were installed along 5.5 km of pipeline. Use of these crossings, and of gaps under elevated sections along 1.6 km of pipeline, was monitored. Snow track surveys were carried out at three-week intervals in February–March 2006 and November 2006–April 2007. Camera traps were installed along each pipeline section with two at each crossing for one year (2006–2007).

(1) Curatolo J.A. & Murphy S.M. (1986) The effects of pipelines, roads, and traffic on the movements of caribou, *Rangifer tarandus. The Canadian Field-Naturalist*, 100, 218–224.

(2) Eide S.H., Miller S.D. & Chihuly M.A. (1986) Oil pipeline crossing sites utilized in winter by moose, *Alces alces*, and caribou, *Rangifer tarandus*, in Southcentral Alaska. *The Canadian Field-Naturalist*, 100, 197–207.

(3) Dunne B.M. & Quinn M.S. (2009) Effectiveness of above-ground pipeline mitigation for moose (*Alces alces*) and other large mammals. *Biological Conservation*, 142, 332–343, https://doi.org/10.1016/j.biocon.2008.10.029

Shipping Lanes

5.35. Install overpasses over waterways

https://www.conservationevidence.com/actions/2628

- **Two studies** evaluated the effects on mammals of installing overpasses over waterways. One study was in the USA[1] and one was in Spain[2].

COMMUNITY RESPONSE (0 STUDIES)

POPULATION RESPONSE (0 STUDIES)

BEHAVIOUR (2 STUDIES)

- **Use (2 studies):** Two studies (one replicated, one a site comparison) in the USA[1] and Spain[2], found that bridges and overpasses over waterways were used by desert mule deer, collared peccaries and coyotes[1] and by a range of large and medium-sized mammals[2].

Background

Waterways can separate populations of a species or provide barriers to movements. Artificial waterways (such as canals and aqueducts) can disrupt movements between previously connected habitat. This may result in genetic isolation of populations (e.g. Corlatti *et al.* 2009) or drownings, if animals attempt to cross waterways that have steep sides. Crossing points may be installed for use of animals in an attempt to maintain connectivity and free movement between sites or habitats.

See also: *Install barrier fencing along waterways* and *Provide mammals with escape routes from canals.*

Corlatti L., Hackländer K. & Frey-Roos F. (2009) Ability of wildlife overpasses to provide connectivity and prevent genetic isolation. *Conservation Biology*, 23, 548–556, https://doi.org/10.1111/j.1523-1739.2008.01162.x

A site comparison study in 1996–1997 along an aqueduct in Arizona, USA (1) found that overpasses over a waterway within a created wildlife corridor were used by desert mule deer *Odocoileus hemionus eremicus*, collared peccaries *Pecari tajacu* and coyotes *Canis latrans*. Mule deer and peccaries used all six wildlife overpasses inside the corridor. Bridges outside the corridor, not designed for wildlife, were also used. However, there were more mule deer tracks on wildlife overpasses inside the corridor (average 0.06–0.11 tracks/reading) than on bridges outside the corridor (0–0.01 tracks/reading). The same held for peccaries (wildlife overpasses: 0.15–0.21 tracks/reading; bridges: 0.06–0.17). There was no difference for coyotes (wildlife overpasses: (0.28–0.45 tracks/reading; bridges: 0.31–0.59). Aqueduct crossings were provided at five points within and one immediately adjacent to the corridor. Crossings were 9–173 m wide. Four crossings to the north were also monitored along 11 km of aqueduct. Crossings within the corridor contained natural soil and vegetation. Those outside were concrete overchutes or overpasses of water. Animal tracks were recorded on sand plots (2–22/crossing) on ≥7 consecutive days/month from August 1996 to July 1997 (total 117 checks/plot).

A replicated study in 1993–1998 along a canal in Guardo, northern Spain (2) found that all nine small bridges and six of 14 wider bridges designed for humans and livestock were used as crossing points by mammals. Crossings were made by roe deer *Capreolus capreolus* (four crossings), red deer *Cervus elaphus* (four), wild boar *Sus scrofa* (nine), wolf *Canis lupus* (three), fox (52) and by mustelids, mainly badgers *Meles meles* and stone martens *Martes foina* (14). Iberian hares *Lepus granatensis* and hedgehogs *Erinaceus europaeus* were also recorded. Small wildlife bridges were used more than were larger bridges by all mammals as a whole (see paper for details) and bridges near scrubland were used more (12 out of 13 used) than were those near cropland (one out of nine used). Despite crossings being available, 123 roe deer and 34 wild boars were found drowned over the five years. Fourteen concrete bridges (for humans and livestock; 5.0–7.5 m wide) and nine small wildlife bridges (2.5–3.6 m wide) along 24 km of a 5-m-wide concrete water canal were monitored. Tracks in sand and other animal signs were recorded on each bridge every three days from April to September 1998. Drowned mammals were monitored daily from April 1993 to October 1998.

(1) Popowski R.J. & Krausman P.R. (2002) Use of crossings over the Tucson aqueduct by selected mammals. *The Southwestern Naturalist*, 47, 363–371, https://doi.org/10.2307/3672494

(2) Peris S. & Morales J. (2004) Use of passages across a canal by wild mammals and related mortality. *European Journal of Wildlife Research*, 50, 67–72, https://doi.org/10.1007/s10344-004-0045-0

5.36. Install barrier fencing along waterways

https://www.conservationevidence.com/actions/2636

- We found no studies that evaluated the effects on mammals of installing barrier fencing along waterways.

'We found no studies' means that we have not yet found any studies that have directly evaluated this intervention during our systematic journal and report searches. Therefore, we have no evidence to indicate whether or not the intervention has any desirable or harmful effects.

Background

Mammals may be attracted to canals and other waterways for drinking. When such waterways have steep sides, mammals may fall in and be unable to escape. Waterways may also act as barriers to animal movements and mammals may attempt to cross them but be unable to exit the water. In such cases, mammals may be at risk of drowning (e.g. Peris & Morales 2004). At areas of high risk, barrier fencing could be installed in order to prevent mammals accessing waterways and so reduce the drowning risk.

Peris S. & Morales J. (2004) Use of passages across a canal by wild mammals and related mortality. *European Journal of Wildlife Research*, 50, 67–72, https://doi.org/10.1007/s10344-004-0045-0

5.37. Provide mammals with escape routes from canals

https://www.conservationevidence.com/actions/2638

- **Five studies** evaluated the effects on mammals of providing mammals with escape routes from canals. Two studies were in Germany[1,2] and one each was in the USA[3], the Netherlands[4] and Argentina[5].

COMMUNITY RESPONSE (0 STUDIES)

POPULATION RESPONSE (2 STUDIES)

- **Survival (2 studies):** One of two studies (one before-and-after), in Germany[2] and the USA[3], found that ramps and ladders reduced mule deer drownings[3] whilst the other study found that ramps and shallow-water inlets did not reduce mammal drownings[2].

BEHAVIOUR (3 STUDIES)

- **Use (3 studies):** Three studies (one replicated) in Germany[1], the Netherlands[4] and Argentina[5], found that ramps and other access or escape routes out of water were used by a range of medium-sized and large mammal species.

Background

Mammals may be attracted to canals and other artificial waterways for drinking. When such waterways have steep sides, mammals may fall in and be unable to escape. Such waterways may also act as barriers to animal movements and mammals may attempt to cross them but be unable to exit the water whilst some aquatic mammals may also enter deliberately but struggle to exit the water. In such cases, mammals may be at risk of drowning (e.g. Peris & Morales 2004). Escape routes may be installed to enable mammals that have fallen in or otherwise entered the water to escape back onto land. These may take the form or ramps, ladders, shallow inlets or other structures that mammals could use to climb out.

Peris S. & Morales J. (2004) Use of passages across a canal by wild mammals and related mortality. *European Journal of Wildlife Research*, 50, 67–72, https://doi.org/10.1007/s10344-004-0045-0

A study (year not stated) in a swimming pool and on a stretch of a canal in Lower Saxony, Germany (1) found that a platform was used by at least five mammal species to exit water and both metal ramps and vegetated islands by at least two species Roe deer *Capreolus capreolus*, red deer *Cervus elaphus*, wild boar *Sus scrofa*, red foxes *Vulpes vulpes* and badgers *Meles meles* used timber platforms to exit from waterways. Rabbits *Oryctolagus cuniculus* and hedgehogs *Erinaceus europaeus* used a ramp covered with meshed metal to exit from waterways. Red foxes and badgers used vegetated islands to leave water. Timber platforms were tested by releasing medium-sized (e.g. foxes) and large mammals (e.g. deer) into a swimming pool, and guiding them to a platform. A ramp covered with meshed metal was tested for small mammals (e.g. rabbits) and a 'vegetated island' (4.5 m × 2.5 m; 1.5 m above water level) was tested for deer, badgers and foxes. The vegetated island comprised timber beams 'planted' with leafy branches either fixed to the bank or anchored in the middle of a steep-banked stretch of canal.

A before-and-after, site comparison study in 1978–1982 of a steep-sided canal in Germany (2) found that installing shallow-water inlets and ramps did not reduce mammal drownings. There was no evidence of large mammals leaving the canal by inlets or of a reduction in the number drowned after inlet establishment (after: 15 individuals drowned in one year; before: 11 drowned in two years). There was no evidence of small mammals using ramps as exits. There was no significant difference in the density of drowned small mammals on canal sections with and without ramps where the length of canal surveyed without ramps was twice the length surveyed with ramps: hamster *Cricetus cricetus* (with: 50; without: 80), common vole *Microtus arvalis* (with: 14; without: 25), water vole *Arviola terrestris* (with: four; without: seven). Inlets were shallow shelving exit points (250–500 m apart) established in spring 1979. Sand at eight inlet entrances was checked daily in September 1979, and April–May of 1980 and 1981 for mammal footprints. The canal was searched every 2–3 days for drowned animals before and after inlet establishment (1978–1980). Ramps (≤50 m apart) were installed in May 1982. Sand at ramp exits was checked daily over 20 days in August for small mammal footprints. Live-trapping was conducted over 13 days.

A study in 1982–1985 in a canal between farmland and desert in Arizona, USA (1) found that ramps and ladders reduced mule deer *Odocoileus hemionus* drownings. Of at least 282 times that deer fell into the canal over a 40-month period, three deer drowned, 116 escaped via steps, 79 via ramps and eight via metal ladders. A further 50 escaped without using structures and 10 were pulled out alive. Exit points of 16 deer were not determined. Over two previous years, before escape routes were improved, 18 deer drowned on the same canal section. A 15-km-long canal section, 5.5–10 m wide was studied. There were six dams, five with existing escape stairs. In 1980–1981, three escape ramps (3 m wide, at 25° to the direction of water flow with a 25% slope) were added. There was also one 1.3-m-wide iron ladder and seven reinforcement-bar ladders (date of installation not stated). Wire cables (3 cm diameter) across the water surface, directed trapped deer toward each escape structure. Deer were monitored and reported by canal workers and by monitoring tracks at 1–3-day intervals in June 1982 to September 1985 (total 478 visits). Drownings in 1979–1980 were logged by canal staff.

A study in 2002–2005 in two wetland areas in the Netherlands (4) found that providing mammals with escape or access routes from and into canals resulted in their use by Eurasian otters *Lutra lutra*. In 2002–2005, twenty-four animals, comprising a mix of wild-caught and captive-bred individuals, were released at two sites. In one of the areas, modifications to canal banks were made to aid entry and exit by otters to and from the water. Use of exits from canals was monitored by direct observation, observation of tracks in the snow, and identification of otter faeces.

A replicated study in 2012–2015 of two irrigation canals in Jujuy, Argentina (5) found that at least three mammal species used escape ramps to exit from waterways. Two tapirs *Tapirus terrestris*, one collared peccary *Pecari tajacu* and one *Mazama americana* were recorded exiting water via ramps. Thirteen additional mammal species were detected on escape ramps though it is unclear if they used these to exit from water. Two irrigation canals were studied, one crossing a forest reserve and the other crossing sugar cane and citrus plantations. In 2012–2013, fifteen 3-m-wide escape routes with 20-cm-high steps were constructed. Escape routes were 0.15–1.8 km apart. Monitoring was conducted using camera traps set in October 2012, May 2013, March 2014 and December 2015. Camera traps were 2–3 m from escape routes and were set to take one photo every 5 minutes for approximately 40 days.

(1) Schneider V.E. & Waffel H.H. (1978) Vorschläge zu Schutzmaßnahmen für Wildtiere beim Ausbau von Schiffahrtskanälen und kanalisierten Binnenwasserstraßen [Suggested modes of protection for wild animals in connection with the construction of shipping canals and canalised inland waterways]. *Zeitschrift fur Jagdwissenschaft*, 24, 72–88.

(2) Wietfeld J. (1984) Die Wirksamkeit von Schutzmaßnahmen zur Verhinderung von Tierverlusten in verspundeten Gewässern [The effectiveness of protection measures to prevent animal losses in blocked waters]. *Zeitschrift fur Jagdwissenschaft*, 30, 176–184.

(3) Rautenstrauch K.R. & Krausman P.R. (1989) Preventing mule deer drownings in the Mohawk canal, Arizona. *Wildlife Society Bulletin*, 17, 280–286.

(4) Lammertsma D., Niewold F., Jansman H., Kuiters L., Koelewijn H.P., Haro M.P., van Adrichem M., Boerwinkel M-C. & Bovenschen J. (2006) Herintroductie van de otter: een succesverhaal [Reintroduction of the otter: a success story]? *De Levende Natuur*, 107, 42–46.

(5) Albanesi S.A., Jayat J.P. & Brown A.D. (2016) Mortalidad de mamíferos y medidas de mitigación en canales de riego del pedemonte de Yungas de la alta cuenca del río Bermejo, Argentina [Mortality of mammals and mitigation actions in irrigation canals of the Yungas piedmont of the High Bermejo River Basin, Argentina]. *Mastozoología Neotropical*, 23, 505–514.

6. Threat: Biological resource use

Background

Biological resource use (as defined in this synopsis) includes the killing of mammals for food or sporting purposes, as well as logging and wood harvesting and the impact that this has on wild mammals. While hunting has a direct effect on mammal survival, logging and wood harvesting indirectly threaten mammals through habitat destruction and fragmentation, disturbance and increased access for hunting.

Hunting & Collecting Terrestrial Animals

6.1. Prohibit or restrict hunting of a species

https://www.conservationevidence.com/actions/2597

- **Five studies** evaluated the effects of prohibiting or restricting hunting of a mammal species. One study each was in Norway[1], the USA[2], South Africa[3], Poland[4] and Zimbabwe[5].

COMMUNITY RESPONSE (0 STUDIES)

POPULATION RESPONSE (5 STUDIES)

- **Abundance (2 studies):** Two studies (including one before-and-after study), in the USA[2] and Poland[4], found that prohibiting hunting led to population increases of tule elk[2] and wolves[4].

- **Survival (3 studies):** A before-and-after study in Norway[1] found that restricting or prohibiting hunting did not alter the

   https://doi.org/10.11647/OBP.0234.06

number of brown bears killed. A study in Zimbabwe[5] reported that banning the hunting, possession and trade of Temminck's ground pangolins did not eliminate hunting of the species. A before-and-after study in South Africa[3] found that increasing legal protection of leopards, along with reducing human-leopard conflict by promoting improved animal husbandry, was associated with increased survival.

BEHAVIOUR (0 STUDIES)

Background

Hunting may in some cases lead to reductions or local extinctions of mammal species. This intervention covers prohibiting or restricting hunting specifically where hunting is the major threat to a population of a species. For legal protection aimed at other threats see *Habitat protection — Legally protect habitat for mammals.*

A before-and-after study in 1908–1918 in Sweden and one in 1967–1977 in Norway (1) found that the number of brown bears *Ursus arctos* reported killed did not change significantly after hunting was prohibited. The number of brown bears reported killed over five years after legal protection was introduced (Sweden: 6.8 bears/year; Norway: 1.2 bears/year) did not differ significantly to that over the five years before legal protection (Sweden: 7.2 bears/year; Norway: 1.6 bears/year). Numbers of bears killed were obtained from national harvesting records. Bears were protected on Crown land in 1913 in Sweden and fully protected in 1972 in Norway. Bears could still be killed to protect livestock and for self-defence.

A before-and-after study in 1971–1998 in California, USA (2) found that numbers of tule elk *Cervus canadensis nannodes* increased after hunting was prohibited. The tule elk population grew from approximately 500 individuals in 1971 when it received official protection against hunting, to 2,000 individuals in 1989 and >3,000 individuals in 1998. Tule elk became officially protected in 1971. The bill prohibited hunting until the population reached 2,000 individuals. No monitoring or habitat details are provided. Other management interventions (not detailed) were carried out by California Department of Fish and Game during the length of the study.

A before-and-after study in 2003–2007, in a mixed woodland and grassland area in KwaZulu-Natal, South Africa (3) found that increasing legal protection of leopards *Panthera pardus* along with reducing human-leopard conflict, by promoting improved animal husbandry, was associated with increased leopard survival. The annual mortality rate of leopards in the three years after increased protection and improved husbandry were introduced (12–17%) was lower than during the two previous years (33–47%). Conditions to be met before a permit was issued to kill leopards that predated livestock were tightened in January 2005. New regulations required that there had to be at least three verifiable predation incidents within two months and further livestock protection steps were required. Additionally, selling permits to sports hunters was banned. Workshops in January–July 2005 promoted best practice in protecting livestock from predation (including corralling vulnerable animals, guarding herds, regularly changing grazing paddocks and disposing of carcasses). Twenty-six leopards were monitored by radio-tracking before actions were introduced (2003–2004) and 28 after they were introduced (2005–2007).

A study in 2001–2013 in a forest within an agricultural landscape across western Poland (4) found that after hunting was prohibited, wolves *Canis lupus* increased in number. Fourteen years after hunting was banned, the wolf population (139 wolves) was higher than three years after the ban was introduced (7–9 wolves). After five years, the first cases of wolf reproduction in the study area were confirmed. Of the 28 wolf deaths recorded, 17 were caused by traffic and seven animals were killed illegally. Wolf field signs (tracks, droppings, scratch marks), camera-trapping and howling simulation surveys were used by trained personnel to locate territories. Mortality reports were collated and verified where possible. Surveys prioritised areas with wolf reports and areas identified as being the most suitable habitat.

A study in 2010–2015 in Zimbabwe (5) reported that banning the hunting, possession and trade of Temminck's ground pangolins *Smutsia temminckii* did not eliminate hunting of the species, but enforcement led to a higher number of confiscations. After a nationwide ban on hunting, possession and trade in 1975, a total of 65 Temminck's ground pangolin seizures were made in 2010–2015. The number of pangolins confiscated increased over this period from 0–1/six-month period in 2010–2011 up to 4–13/six-month period in 2014–2015. Of 53 live pangolins seized, 32 were released back into the wild. In 1975, the Temminck's ground pangolin was given full protection on Zimbabwe's Specially Protected Species list. During the study period, all pangolins were listed in Appendix II of CITES. Pangolin seizure data for the period between October 2010 and July 2015 were compiled from information from Zimbabwean wildlife management authorities and courts, from the media and from an NGO.

(1) Swenson J.E., Wabakken P., Sandegren F., Bjärvall A., Franzén R. & Söderberg, A. (1995) The near extinction and recovery of brown bears in Scandinavia in relation to the bear management policies of Norway and Sweden. *Wildlife Biology*, 1, 11–25.

(2) Adess N. (1998) *Tule elk; return of a species*. National Park Service Point Reyes National Seashore, California, USA.

(3) Balme G.A., Slotow R. & Hunter L.T.B. (2009) Impact of conservation interventions on the dynamics and persistence of a persecuted leopard (*Panthera pardus*) population. *Biological Conservation*, 142, 2681–2690, https://doi.org/10.1016/j.biocon.2009.06.020

(4) Nowak S. & Mysłajek R.W. (2016) Wolf recovery and population dynamics in Western Poland, 2001–2012. *Mammal Research*, 61, 83–98, https://doi.org/10.1007/s13364-016-0263-3

(5) Shepherd C.R., Connelly E., Hywood L. & Cassey P. (2017) Taking a stand against illegal wildlife trade: the Zimbabwean approach to pangolin conservation. *Oryx*, 51, 280–285, https://doi.org/10.1017/s0030605316000119

6.2. Ban private ownership of hunted mammals

https://www.conservationevidence.com/actions/2602

- **One study** evaluated the effects of banning private ownership of hunted mammals. This study was in Sweden[1].

COMMUNITY RESPONSE (0 STUDIES)

POPULATION RESPONSE (1 STUDY)

- **Survival (1 study):** A before-and-after study in Sweden found that fewer brown bears were reported killed after the banning of private ownership of hunted bears.

BEHAVIOUR (0 STUDIES)

Background

The hunting of some animals may be driven by demand from collectors who purchase animal skins or furs. Banning such private ownership may reduce incentives for hunting.

A before-and-after study in 1922–1932 in Sweden (1) found that after the banning of private ownership of hunted bears, fewer brown bears *Ursus arctos* were reported killed. Fewer brown bears were reported killed during the five years after the private ownership of hunted bears was banned (average 0.8 bears/county/year) than during the five years before the ban (8.2 bears/county/year). All killed brown bears became state property in 1927. Numbers of bears killed in 1922–1932 were obtained from national harvesting records.

(1) Swenson J.E., Wabakken P., Sandegren F., Bjärvall A., Franzén R. & Söderberg A. (1995) The near extinction and recovery of brown bears in Scandinavia in relation to the bear management policies of Norway and Sweden. *Wildlife Biology*, 1, 11–25.

6.3. Site management for target mammal species carried out by field sport practitioners

https://www.conservationevidence.com/actions/2605

- **One study** evaluated the effects of site management for a target mammal species being carried out by field sport practitioners. This study was in Ireland[1].

COMMUNITY RESPONSE (0 STUDIES)

POPULATION RESPONSE (1 STUDY)

- **Abundance (1 study):** A replicated, site comparison study in the Republic of Ireland[1] found that sites managed for the sport of coursing Irish hares held more of this species than did the wider countryside.

BEHAVIOUR (0 STUDIES)

> **Background**
>
> Hunters and field sport participants may manage sites specifically to maintaining populations of their target mammal species. Management could include predator control and management of habitat features.

A replicated, site comparison study in 2003–2007 on 17 improved farmland sites in County Donegal, Republic of Ireland (1) found that sites managed for the sport of coursing Irish hares *Lepus timidus hibernicus* held more of this species than did the wider countryside. Accounting for differences in habitat, hare densities on coursing sites (96 hares/km²) were higher than on wider countryside sites (31 hares/km²). Eight sites managed for hare coursing were compared with nine sites containing suitable hare habitat in the wider countryside. Management for hare coursing included predator control, poaching deterrence, retaining fine scale habitat features, such as rush patches, and administering veterinary attention while holding hares captive prior to coursing events. Hares flushed by lines of 20–30 beaters were counted, in September–December of 2003–2007.

(1) Reid N., Magee C. & Montgomery W.I. (2010) Integrating field sports, hare population management and conservation. *Acta Theriologica*, 55, 61–71, https://doi.org/10.4098/j.at.0001-7051.030.2009

6.4. Set hunting quotas based on target species population trends

https://www.conservationevidence.com/actions/2607

- **Three studies** evaluated the effects of setting hunting quotas for mammals based on target species population trends. One study each was in Canada[1], Spain[2] and Norway[3].

COMMUNITY RESPONSE (0 STUDIES)

POPULATION RESPONSE (3 STUDIES)

- **Abundance (2 studies):** Two studies, in Spain[2] and Norway[3], found that restricting hunting and basing quotas on population targets enabled population increases for Pyrenean chamois[2] and Eurasian lynx[3].

- **Survival (1 study):** A before-and-after study in Canada[1] found that setting harvest quotas based on population trends, and lengthening the hunting season, did not decrease the number of cougars killed by hunters.

BEHAVIOUR (0 STUDIES)

Background

Management of wildlife species that are regarded as game animals may involve setting hunting quotas that are designed to enable the population to reach or remain at a particular level. Whilst many hunting systems use quotas, the studies included here are those based on mammal species with particular local conservation concerns rather than where quotas are based purely on maximising the harvest.

A before-and-after study in 1990–1991 in boreal forest in Alberta, Canada (1) found that setting harvest quotas based on the population trends of the target species, and increasing the length of the hunting season, did not decrease the number of cougars *Puma concolor* killed by hunters. After setting harvest quotas, the number of cougars killed was higher (54 animals) than before setting of harvest quotas (33 animals). In 1981–1989 radio collars were attached to 44 cougars and data collected used to estimate the population size. The area was divided into 11 Cougar Management Areas and quotas were set at 10% of the estimated population for each area. A further quota of 50% of the total harvest quota was set for female cougars. When either quota was reached, the hunting season within a specific area was closed.

A study in 1995–2007 in mixed forest, cliffs and meadows across three mountain massifs in Navarre and Aragon, Spain (2) found that, following imposition of hunting restrictions, populations of Pyrenean chamois *Rupicapra pyrenaica pyrenaica* increased. Results were not tested for statistical significance. The population at one massif rose from at least 33 in 1995 to at least 136 (an average growth rate of 15%/year) in 2007 and, at another massif, from at least 144 in 1996 to at least 455 (11%/year) in 2007. A third massif was occupied by eight chamois from at least 2002, with 11 there in 2007. The first two massifs cross regional jurisdictions. Hunting did not occur in one region, but was allowed in the other up to 1993, when it was temporarily banned. Limited hunting resumed in this region in 2006, based on 5% annual harvest. Hunting was not carried out in the third massif. Chamois were surveyed from dawn until midday in June and November each year, in 1995–2007.

A study in 1996–2008 in primarily forested areas in Norway (3) found that adaptive management, including basing hunting quotas on population trends, enabled Eurasian lynx *Lynx lynx* populations to recover after a population decline. Three years after modification of hunting quotas, the population of Lynx was higher (453 animals) than prior to modifications (259 animals). Before modifications of quotas, lynx populations had dropped from 411–486 to 259 over an eight-year period. Lynx harvests were uncapped up to 1992. From 1994, responsibility for setting hunting quotas was devolved to 18 counties and then transferred to eight regional units in 2005. The number of lynx family groups was estimated by collating records of lynx tracks along with records of young animals found dead or killed by vehicles or hunters. These data were extrapolated to form overall population estimates for 1996–2008.

(1) Ross I.P., Jalkotzy M.G. & Gunson J.R. (1996) The quota system of cougar harvest management in Alberta. *Wildlife Society Bulletin*, 24, 490–494.

(2) Herrero J., Garin I., Prada C. & García-Serrano A. (2010) Inter-agency coordination fosters the recovery of the Pyrenean chamois *Rupicapra pyrenaica pyrenaica* at its western limit. *Oryx*, 44, 529–532, https://doi.org/10.1017/s0030605310000761

(3) Linnell J.D.C., Broseth H., Odden J. & Nilsen E.B. (2010) Sustainably harvesting a large carnivore? Development of Eurasian lynx populations in Norway during 160 years of shifting policy. *Environmental Management*, 45, 1142–1154, https://doi.org/10.1007/s00267-010-9455-9

6.5. Prohibit or restrict hunting of particular sex/ breeding status/age animals

https://www.conservationevidence.com/actions/2609

- **Two studies** evaluated the effects of prohibiting or restricting hunting of particular sex, breeding status or age animals. Both studies were in the USA[1,2].

COMMUNITY RESPONSE (0 STUDIES)

POPULATION RESPONSE (2 STUDIES)

- **Reproduction (2 studies):** Two replicated, before-and-after studies, in the USA[1,2], found that limiting hunting of male deer did not increase the numbers of young deer/adult female.

- **Population structure (1 study):** A replicated, before-and-after study in the USA[1] found that limiting hunting of older male elk resulted in an increased ratio of male:female elk.

BEHAVIOUR (0 STUDIES)

Background

Within some hunted populations of mammals, certain age or sex classes are favoured targets for hunters. This can result in altered population structures which can be detrimental to breeding success for example (e.g. Torres-Porras *et al.* 2014). Management of game mammals may, therefore, involve imposing specific hunting restrictions, so as to reduce or prohibit harvests of particular sex or age classes.

Torres-Porras J., Carranza J., Pérez-González J., Mateos C. & Alarcos S. (2014) The tragedy of the commons: unsustainable population structure of Iberian red deer in hunting estates. *European Journal of Wildlife Research*, 60, 351–357, https://doi.org/10.1007/s10344-013-0793-9

A replicated, before-and-after study in 1984–2000 in three forest and shrubland sites in Washington, USA (1) found that limiting hunting of adult male elk *Cervus canadensis* resulted in an increase in the numbers of males relative to females, but no change in numbers of calves relative to females. After hunting restrictions commenced, there were more male relative to female elk (6.7–12.9 males/100 females) than before hunting restrictions commenced (2.7–5.7 males/100 females). The abundance of calves relative to female elk did not change (after: 21–37 calves/100 females; before: 30–37 calves/100 females). The strategy of open-entry yearling hunting and limited hunting of elk ≥ 2.5 years old with branched antlers was introduced at one site in 1989 and at two sites in 1994. These sites were monitored in 1984–2000 and 1991–2000 respectively and covered 2,300–4,500 km². Elk were counted from helicopters, and categorised by age and sex, in late February or early March each year.

A replicated, before-and-after study in 1983–1998 of four deer management areas in a largely forested region of Colorado, USA (2) found that restricting the harvest of male mule deer *Odocoileus hemionus* did not increase the number of fawns/adult female deer. After introduction of hunting restrictions, the fawn:adult female deer ratio declined by 7.5 fawns:100 adult females (absolute numbers not presented). During this time, harvests of male deer fell from an average of 788/management area/year to 209/management area/year and the ratio of male:female deer increased by 4.5:100 female deer. Harvests of male deer were unlimited up to 1990. Commencing in 1991 (one area), 1992 (two areas) and 1995 (one area), restrictions were imposed on harvests of male deer, resulting in a fall in average harvests from 788/year pre-restriction to 209/year post-restriction. Aerial deer surveys were carried out in December–January.

(1) Bender L.C., Fowler P.E., Bernatowicz J.A., Musser J.L. & Stream L.E. (2002) Effects of open-entry spike-bull, limited-entry branched-bull harvesting on elk composition in Washington. *Wildlife Society Bulletin*, 30, 1078–1084.

(2) Bishop C.J., White G.C., Freddy D.J. & Watkins B.E. (2005) Effect of limited antlered harvest on mule deer sex and age ratios. *Wildlife Society Bulletin*, 33, 662–668, https://doi.org/10.2193/0091-7648(2005)33[662:eolaho]2.0.co;2

6.6. Incentivise species protection through licensed trophy hunting

https://www.conservationevidence.com/actions/2610

- **One study** evaluated the effects on mammals of incentivising species protection through licensed trophy hunting. This study was in Nepal[1].

COMMUNITY RESPONSE (0 STUDIES)

POPULATION RESPONSE (1 STUDY)

- **Abundance (1 study):** A study in Nepal[1] found that after trophy hunting started, bharal abundance increased, though the sex ratio of this species, and of Himalayan tahr, became skewed towards females.

BEHAVIOUR (0 STUDIES)

Background

Trophy hunting is the hunting of wild animals for recreation. Usually, this involves large or otherwise distinguished animals, such as large carnivores, or species with large antlers. The animal, or part of it, is kept by the hunter, often for display. Trophy hunting may provide financial support to local communities or conservation initiatives, through locally levied fees (Di Minin *et al.* 2016). This may increase the perceived value of maintaining populations of such species in the long term and may, hence, incentivise greater habitat and species protection in such areas.

Di Minin E., Leader-Williams N. & Bradshaw C.J.A. (2016) Banning trophy hunting will exacerbate biodiversity loss. *Trends in Ecology & Evolution*, 31, 99–102, https://doi.org/10.1016/j.tree.2015.12.006

A study in 1990–2011 in forest and grassland in a hunting reserve in Nepal (1) found that following commencement of trophy hunting, populations of bharal *Pseudois nayaur* increased, though the sex ratio of this species, and of Himalayan tahr *Hemitragus jemlahicus*, became skewed towards females. Twenty-one years after the establishment of trophy hunting, the estimated bharal population was higher (>1,500 animals) than three years after it was established (approximately 400 animals). The proportion of males to females was lower after 21 years (82:100) than three years after (129:100). A similar pattern was seen for the thar population (21 years after: 62:100; three years after: 214:100). The hunting reserve, covering 1,325 km², was established in 1987. Trophy hunters, especially from outside Nepal, pay for the right to hunt male bharal and tahr. Females are not hunted. Data were collated from a range of sources, primarily derived from vantage point counts.

(1) Aryal A., Dhakal M., Panthi S., Yadav B.P., Shrestha U.B., Bencini R. Raubenheimer D. & Ji W. (2015) Is trophy hunting of bharal (blue sheep) and Himalayan tahr contributing to their conservation in Nepal? *Hystrix, the Italian Journal of Mammalogy*, 26, 85–88.

6.7. Use selective trapping methods in hunting activities

https://www.conservationevidence.com/actions/2611

- We found no studies that evaluated the effects on non-target mammals of using selective trapping methods in hunting activities.

'We found no studies' means that we have not yet found any studies that have directly evaluated this intervention during our systematic journal and report searches. Therefore, we have no evidence to indicate whether or not the intervention has any desirable or harmful effects.

Background

Hunting using traps (such as snares, leg-hold traps or cage traps) can result in capture of rare, threatened or protected non-target species mammal (e.g. Andreasen *et al.* 2018). Measures to reduce such 'bycatch' might include setting a weight-sensitive release catch, placing traps in particular areas (or avoiding other areas) or only using specific baits.

Andreasen A.M., Stewart K.M., Sedinger J.S., Lackey C.W. & Beckmann J.P. (2018) Survival of cougars caught in non-target foothold traps and snares. *The Journal of Wildlife Management*, 82, 906–917, https://doi.org/10.1002/jwmg.21445

6.8. Use wildlife refuges to reduce hunting impacts

https://www.conservationevidence.com/actions/2612

- **Two studies** evaluated the effects on mammal species of using wildlife refuges to reduce hunting impacts. One study was in Canada[1] and one was in Mexico[2].

COMMUNITY RESPONSE (0 STUDIES)

POPULATION RESPONSE (2 STUDIES)

- **Abundance (2 studies):** One of two replicated site comparison studies in Canada[1] and Mexico[2] found more moose in areas with limited hunting than in more heavily hunted areas[1]. The other study found mixed results with only one of five species being more numerous in a non-hunted refuge[2].

BEHAVIOUR (0 STUDIES)

Background

To help protect or sustain populations of hunted species, refuges may be designated that have limited or not hunting. This intervention covers studies that assess the impact of such refuges where they lie adjacent to hunted areas.

See also: *Habitat Protection — Legally protect habitat for mammals.*

A replicated, site comparison study in 1984 of 24 forest blocks in Quebec, Canada (1) found more moose *Alces alces* in game reserves with limited hunting than in more heavily hunted areas. Games reserves held 0.28 moose/km² compared to 0.06/km² in adjacent hunted areas and 0.14/km² in hunted areas ≥50 km away. Dispersal from game reserves was reported to sustain moose harvests in adjacent areas. Moose density was estimated by surveying 24 plots of 60 km² each. Twelve plots were in areas that overlapped between game reserves with limited hunting (108 hunter-days/100 km²/year) and more heavily hunted adjacent areas (518 hunter-days/100 km²/year). Twelve plots were in hunting areas ≥50 km from a reserve (with 315 hunter-days/100 km²/year). Twelve transect lines/plot were surveyed from fixed-wing aircraft in January 1984.

A replicated, site comparison study in 2001 of four forest areas in Campeche, Mexico (2) found that one of five ungulate species was more numerous in a non-hunted refuge area compared to in hunted areas and two were more numerous in hunted areas. There were more white-lipped peccaries *Tayassu pecari* in non-hunted (0.24 tracks/km) than hunted (0.08 tracks/km) areas. White-tailed deer *Odocoileus virginianus* were more numerous in hunted areas (non-hunted: 0.24; hunted: 0.88 tracks/km) as was Central American tapir *Tapirus bairdii* (non-hunted: 0.03; hunted: 0.42 tracks/km). No differences between areas were found for brocket deer *Mazama* sp. (non-hunted: 6.4; hunted: 6.7 tracks/km) or collared peccary *Pecari tajacu* (non-hunted: 0.9; hunted: 1.0 tracks/km). Transects were established on land not hunted on since the 1980s, and on three adjacent hunted sites with similar habitat. Transects were ≥3 km from villages and had start points ≥2 km apart. Twenty-eight transects (total 57 km) were walked in the non-hunted area and 18–24 transects (35–70 km/site), were walked in hunted areas. Transects were walked in February–July 2001. Ungulate tracks within 1 m of transects were counted and recorded to species.

(1) Crête M. & Jolicoeur H. (1985) Comparing two systems of moose management for harvest. *Wildlife Society Bulletin*, 13, 464–469.

(2) Reyna-Hurtado R. & Tanner G.W. (2007) Ungulate relative abundance in hunted and non-hunted sites in Calakmul Forest (Southern Mexico). *Biodiversity and Conservation*, 16, 743–756, https://doi.org/10.1007/s10531-005-6198-7

6.9. Provide/increase anti-poaching patrols

https://www.conservationevidence.com/actions/2618

- **Seven studies** evaluated the effects of providing or increasing anti-poaching patrols on mammals. Two studies were in Thailand[1,4] and one each was in Brazil[2], Iran[3], Lao People's Democratic Republic[5], South Africa[6] and Tajikistan[7].

COMMUNITY RESPONSE (0 STUDIES)

POPULATION RESPONSE (7 STUDIES)

- **Abundance (6 studies):** Two studies, in Thailand[1] and Iran[3], found more deer and small mammals[1] and more urial sheep and Persian leopards[3] close to ranger stations (from which anti-poaching patrols were carried out) than further from them. One of three before-and-after studies, in Brazil[2], Thailand[4] and Lao People's Democratic Republic[5], found that ranger patrols increased mammal abundance[2]. The other two studies found that patrols did not increase tiger abundance[4,5]. A site comparison study in Tajikistan[7] found more snow leopard, argali, and ibex where anti-poaching patrols were conducted.

- **Survival (1 study):** A study in South Africa[6] found that anti-poaching patrols did not deter African rhinoceros poaching.

BEHAVIOUR (0 STUDIES)

Background

Poaching is the illegal killing or taking of mammals or other wildlife species. It can lead to population declines or push species towards local extinctions (e.g. Wittemyer *et al.* 2014). In absence of enforcement, anti-poaching legislation may be insufficient to prevent declines (e.g. López-Bao *et al.* 2015). Patrols may be instigated to deter or to apprehend poachers.

López-Bao J.V., Blanco J.C., Rodríguez A., Godinho R., Sazatornil V., Alvares F., García E.J., Llaneza L., Rico M., Cortés Y., Palacios V. & Chapron G. (2015) Toothless wildlife protection laws. *Biodiversity and Conservation*, 24, 2105–2108, https://doi.org/10.1007/s10531-015-0914-8

Wittemyer G., Northrup J.M., Blanc J., Douglas-Hamilton I., Omondi P. & Burnham K.P. (2014) Illegal killing for ivory drives global decline in African elephants. *Proceedings of the National Academy of Sciences*, 111, 13117–13121, https://doi.org/10.1073/pnas.1403984111

A study in 2003–2007 in forest in a national park in central Thailand (1) found that, close to ranger stations, deer and small mammals were more abundant than further away. Sambar deer *Rusa unicolor*, red muntjac *Muntiacus muntjak* and a range of small prey species were more likely to be found close to ranger stations than further away (modelled result — data not presented). Poachers were also more likely to be found within 5 km of ranger stations than further away within the national park. Authors suggest that this may be due to roads making ranger stations more accessible and possibly complicity of ranger staff. The national park was 2,168 km² in area. Camera traps were operated in 217 locations over 6,260 total trap nights from October 2003 to March 2007, to survey animals and poacher presence. Cameras were placed across 22 park management zones.

A before-and-after study in 1997–2008 in a protected area dominated by secondary Atlantic forest in Brazil (2) found that implementing ranger patrols increased mammal abundance and reduced hunting pressure. After the introduction of patrols by rangers, mammal abundance was higher (8.7 encounters/10 km walked) than before ranger patrols (5.1 encounters/10 km walked) and hunting pressure was lower (after: six encounters; before: 24 encounters). In May 1997–August 2004 and October 2007–November 2008, forest trails were censused for medium-sized and large mammals. A single observer walked at approximately 1 km/hour along trails 3–5 km long, pausing every 50 m to listen for animal sounds, and using binoculars and a headlamp at night to detect animals. Day censuses began within 1 hour of sunrise and night censuses within 1 hour of sunset. In total, 233 km of transects were walked.

A study in 2011–2013 in a steppe site in a national park in Iran (3) found that presence of ranger stations, which were bases for anti-poaching patrols, was associated with increased numbers of urial sheep *Ovis vignei* and Persian leopards *Panthera pardus saxicolor*. The density of urial sheep decreased with increasing distance from ranger station. This distance was also the best predictor of sheep flock sizes, which were larger closer to ranger stations. Leopards were also more likely to be found closer to ranger stations, though leopard abundance was best explained by urial sheep density. Results were presented as model coefficients. Urial sheep numbers and distribution were determined by distance sampling, along 186 km of line transects, surveyed from 22 January–19 February 2013, 15 August–8 September 2013 and 21–24 February 2014. Leopards were surveyed using 29 camera traps in January–March 2011.

A before-and-after study in 2005–2012 in a tropical dry forest reserve in the Western Forest Complex, Thailand (4) found that as anti-poaching patrols intensified, poaching incidents decreased, but the estimated tiger *Panthera tigris* abundance did not change significantly over seven years. The estimated tiger abundance was similar seven years after poaching patrols started to increase (56 tigers) compared to the year before poaching patrols started to increase (51 tigers). In the final two years of the study, when patrols were at their highest levels, there were 22 poaching incidents detected/1,000 km patrolled, compared to 24–30 incidents/1,000 km patrolled over the previous five years. The study was conducted in a 2,780-km² reserve, adjacent to approximately 30 villages. In 2006–2012, there was an increase each year in the number of patrol days/year (from 1,031 in 2006 to 3,316 in 2012) and distance patrolled/year (5,979 km in 2006 to 12,907 km in 2012). Tigers were surveyed annually between 2005 and 2012, using camera traps across 524–1,094 km² (137–2,000 locations/year, 910–3,869 camera-trap days/year). Paired camera traps were positioned along anticipated tiger travel routes.

A before-and-after study in 2007–2012 in a mainly grassland and forest protected area in Lao People's Democratic Republic (5) found that increasing patrol intensity did not lead to higher tiger *Panthera tigris* abundance. Patrol effort was positively correlated with funding, but not with tiger abundance trends. The number of large tiger tracks (pads >7 cm wide) at the end of the six-year study period (3/1,000 km patrolled) was lower than that over the first three years (8/1,000 km patrolled). The proportion of collected carnivore scats that were from tigers decreased to 3.6% at the end of the study from 15.4–15.6% in the first two years. Patrol effort in a 5,950 km² protected area increased from 1.7 days/part-time team in 2005–2007 to a peak of 22.7 days/full-time team in 2008–2009, then dropped by 4.2% in 2009–2012. Track data and scats were collected by foot patrols and other fieldworkers. Scats were identified to species by DNA analysis.

A study in 2011–2013 in a protected area in South Africa (6) found that where anti-poaching patrols were more common, poaching of African rhinoceros was also more common, but that there was no relationship between the amount of time rangers spent in a location and the likelihood of a poaching event. In areas that rangers visited more frequently, poaching of rhinoceros was more likely to occur. However, in areas where rangers spent more time patrolling, poaching was no more likely to occur. Data were reported as model results. Authors suggest that a range of factors, such as practicalities of access, may result in both more ranger visits and more poaching. Between September 2011 and September 2013, ranger locations were recorded at three-minute intervals in 0.25-km^2 grid cells across the protected area. The location of rhinoceros poaching events, identified from monitoring by park authorities, was overlaid on to the same grid. The average frequency and duration of visits by rangers was calculated for each area where rhinoceros poaching occurred.

A site comparison study in 2012–2013 in two tundra sites in Tajikistan (7) found that, in an area where anti-poaching patrols were carried out, densities of snow leopard *Panthera uncia*, argali *Ovis ammon polii*, and ibex *Capra sibirica* were higher than in an area where no patrols were carried out. The area where anti-poaching patrols were carried out had a higher snow leopard density (0.7 individuals/100 km^2) than where no patrols were carried out (0.5 individuals/100 km^2). The same was true for argali (patrols: 11.0; no patrols: 0.1 individuals/100 km^2) and ibex (patrols: 4.3; no patrols: 2.0 individuals/100 km^2). One site was patrolled by 3–5 rangers year-round. The other site was not patrolled. In June and September 2012, thirty-seven camera traps were deployed at the patrolled site and 34 in the unpatrolled site. Photographs were used to identify individual snow leopards. In September–October 2013, at both sites, 20 randomly selected locations were surveyed for 90 minutes and the abundance of all ungulate species was recorded.

(1) Jenks K.E., Howard J. & Leimgruber P. (2012) Do ranger stations deter poaching activity in national parks in Thailand? *Biotropica*, 44, 826–833, https://doi.org/10.1111/j.1744-7429.2012.00869.x

(2) Flesher K.M. & Laufer J. (2013) Protecting wildlife in a heavily hunted biodiversity hotspot: a case study from the Atlantic Forest of Bahia, Brazil. *Tropical Conservation Science*, 6, 181–200, https://doi.org/10.1177/194008291300600202

(3) Ghoddousi A., Hamidi A.K., Soofi M., Khorozyan I., Kiabi B.H. & Waltert M. (2015) Effects of ranger stations on predator and prey distribution and abundance in an Iranian steppe landscape. *Animal Conservation*, 19, 273–280, https://doi.org/10.1111/acv.12240

(4) Duangchantrasiri S., Umponjan M., Simcharoen S., Pattanavibool A., Chaiwattana S., Maneerat S., Kumar N.S., Jathanna D., Srivathsa A. & Karanth K.U. (2016) Dynamics of a low-density tiger population in Southeast Asia in the context of improved law enforcement. *Conservation Biology*, 30, 639–648, https://doi.org/10.1111/cobi.12655

(5) Johnson A., Goodrich J., Hansel T., Rasphone A., Saypanya S., Vongkhamheng C., Venevongphet & Strindberg S. (2016) To protect or neglect? Design, monitoring, and evaluation of a law enforcement strategy to recover small populations of wild tigers and their prey. *Biological Conservation*, 202, 99–109, https://doi.org/10.1016/j.biocon.2016.08.018

(6) Barichievy C., Munro L., Clinning G., Whittington-Jones B. & Masterson G. (2017) Do armed field-rangers deter rhino poachers? An empirical analysis. *Biological Conservation*, 209, 554–560, https://doi.org/10.1016/j.biocon.2017.03.017

(7) Kachel S.M., McCarthy K.P., McCarthy T.M. & Oshurmamadov, N. (2017) Investigating the potential impact of trophy hunting of wild ungulates on snow leopard *Panthera uncia* conservation in Tajikistan. *Oryx*, 5, 597–604, https://doi.org/10.1017/s0030605316000193

6.10. Make introduction of non-native mammals for sporting purposes illegal

https://www.conservationevidence.com/actions/2621

- We found no studies that evaluated the effects on native mammals of making introduction of non-native mammals for sporting purposes illegal.

'We found no studies' means that we have not yet found any studies that have directly evaluated this intervention during our systematic journal and report searches. Therefore, we have no evidence to indicate whether or not the intervention has any desirable or harmful effects.

Background

Mammals introduced for sporting purposes may negatively affect native species. This may be through such processes as predation of native mammals (Saunders *et al.* 2010), through competition for resources or through hybridising with native species (Reid & Montgomery 2007). Banning importation of non-native mammals for sporting purposes could reduce or prevent further such threats.

Reid N. & Montgomery W.I. (2007) Is naturalisation of the brown hare in Ireland a threat to the endemic Irish hare? *Biology and Environment: Proceedings of the Royal Irish Academy*, 107B, 129–138, https://doi.org/10.3318/bioe.2007.107.3.129

Saunders G.R., Gentle M.N. & Dickman C.R. (2010) The impacts and management of foxes *Vulpes vulpes* in Australia. *Mammal Review*, 40, 181–211https://doi.org/10.1111/j.1365-2907.2010.00159.x

6.11. Commercially breed for the mammal production trade

https://www.conservationevidence.com/actions/2622

- We found no studies that evaluated the effects of commercially breeding mammals for trade on wild populations of those species.

'We found no studies' means that we have not yet found any studies that have directly evaluated this intervention during our systematic journal and report searches. Therefore, we have no evidence to indicate whether or not the intervention has any desirable or harmful effects.

Background

Some mammal species have economic value for products derived from them, such as fur. Captive breeding of these species, on a commercial scale, could reduce incentives for hunting or trapping wild individuals. This could, in turn, relieve pressures on populations of rare or threatened species.

6.12. Promote sustainable alternative livelihoods

https://www.conservationevidence.com/actions/2623

- We found no studies that evaluated the effects of promoting sustainable alternative livelihoods on mammals.

'We found no studies' means that we have not yet found any studies that have directly evaluated this intervention during our systematic journal and report searches. Therefore, we have no evidence to indicate whether or not the intervention has any desirable or harmful effects.

Background

Conserving biodiversity and eliminating poverty are linked global challenges. The poor, particularly the rural poor, depend on nature for many elements of their livelihoods, including food, fuel, shelter and medicines. By promoting sustainable alternative livelihoods, and/or livelihood diversification, the aim is to provide or encourage other sources of income that reduce pressure on natural resources, such as mammals, to sustainable levels. There is a wide diversity of potential alternative sources of income, which depend on the situation, but include activities such as the development of other small-scale productions systems, eco-tourism or craft work for example. Working alongside people who will ultimately benefit from conservation can build social capital, improve accountability, reduce poverty and result in more effective biodiversity conservation.

6.13. Promote mammal-related ecotourism

https://www.conservationevidence.com/actions/2624

- We found no studies that evaluated the effects on mammals of promoting mammal-related ecotourism.

'We found no studies' means that we have not yet found any studies that have directly evaluated this intervention during our systematic journal and report searches. Therefore, we have no evidence to indicate whether or not the intervention has any desirable or harmful effects.

Background

Watching mammals as a recreational activity and has grown considerably in popularity over recent years (Dinets & Hall 2018) with nature-based tourism in general increasing in most countries (Balmford *et al.* 2009). This may result in conservation benefits, such as increased revenue to local conservation projects and assistance with collection of data. Negative impacts can include disturbance pressures at popular sites or increased development to support tourism-related activities. Assessing the net benefits of mammal-related ecotourism may be hampered by lack of population data (Buckley *et al.* 2016).

Balmford A., Beresford J., Green J., Naidoo R., Walpole M. & Manica A. (2009) A global perspective on trends in nature-based tourism. *PLoS Biology*, 7, e1000144, https://doi.org/10.1371/journal.pbio.1000144

Buckley R.C., Morrison C. & Castley J.G. (2016) Net effects of ecotourism on threatened species survival. *PLoS ONE*, 11, e0147988, https://doi.org/10.1371/journal.pone.0147988

Dinets V. & Hall J. (2018) Mammalwatching: A new source of support for science and conservation. *International Journal of Biodiversity and Conservation*, 10, 154–160, https://doi.org/10.5897/ijbc2017.1162

6.14. Ban exports of hunting trophies

https://www.conservationevidence.com/actions/2625

- **One study** evaluated the effects of banning exports of hunting trophies on wild mammals. This study was in Cameroon[1].

COMMUNITY RESPONSE (0 STUDIES)

POPULATION RESPONSE (1 STUDY)

- **Abundance (1 study):** A before-and-after study in Cameroon[1] found similar hippopotamus abundances before and after a ban on exporting hippopotamus hunting trophies.

BEHAVIOUR (0 STUDIES)

Background

Trophy hunting is the hunting of wild animals for recreation. Usually, this involves large or otherwise distinguished animals, such as large carnivores, or species with large antlers. The animal, or part of it, is kept by the hunter, often for display. Some trophy hunting provides financial support to local communities or conservation, through locally levied fees (Minin *et al*. 2016). However, permitting exports of hunting trophies (often from developing countries to developed countries) may provide incentives for hunting at unsustainable levels (Lindsey *et al*. 2016) or may provide a route for importing illegally hunted trophies. Bans on trophy hunting exports are designed to remove this incentive and, hence, reduce incentives for the hunting of relevant species.

Lindsey P.A., Balme G.A., Funston P.F., Henschel P.H & Hunter L.T.B. (2016) Life after Cecil: channelling global outrage into funding for conservation in Africa. *Conservation Letters*, 9, 296–301, https://doi.org/10.1111/conl.12224

Minin E.D., Leader-Williams N. & Bradshaw C.J.A. (2016) Banning trophy hunting will exacerbate biodiversity loss. *Trends in Ecology & Evolution*, 31, 99–102, https://doi.org/10.1016/j.tree.2015.12.006

A before-and-after study in 2000–2014 along a river within and around Faro National Park, Cameroon (1) found similar numbers of hippopotamuses *Hippopotamus amphibious* before and after a ban on exporting of hippopotamus hunting trophies. Results were not tested for statistical significance. Two years after a ban on exporting hippopotamus hunting trophies, 685 hippopotamuses were counted, compared with 647 hippopotamuses counted 12 years before the ban and 525 counted four years before the ban. CITES (Convention on International Trade in Endangered Species of Wild Fauna and Flora) suspended exports of hippopotamus trophies from Cameroon in 2012. In March 2014, hippopotamuses were counted over three days in the dry season, along 97 km of the Faro River. Animals were counted between 07:30 and 17:30 h, by two teams of 2–3 observers. Observers walked through the riverbed at a speed of 1–4 km/hour. Similar counting methods were used in 2000 and 2008 (twelve and four years before the ban respectively) but precise details are not given.

(1)   Scholte P., Nguimkeng F. & Iyah E. (2017) Good news from north-central Africa: largest population of Vulnerable common hippopotamus *Hippopotamus amphibius* is stable. *Oryx*, 51, 218–221, https://doi.org/10.1017/s0030605315001258

Logging & Wood Harvesting

6.15. Use selective harvesting instead of clearcutting

https://www.conservationevidence.com/actions/2637

- **Eight studies** evaluated the effects on mammals of using selective harvesting instead of clearcutting. Four studies were in Canada[1,3,6,8], three were in the USA[2,4,5] and one was a review of studies in North America[7].

COMMUNITY RESPONSE (1 STUDY)

- **Richness/diversity (1 study):** A replicated, site comparison study in Canada[8] found that harvesting trees selectively did not result in higher small mammal species richness compared to clearcutting.

POPULATION RESPONSE (7 STUDIES)

- **Abundance (7 studies):** One of six replicated, controlled or replicated, site comparison studies in the USA[4,5] and Canada[1,3,6,8] found more small mammals in selectively harvested forest stands than in fully harvested, regenerating stands[4]. Three studies found that selective harvesting did not increase small mammal abundance relative to clearcutting[1,5,8]. The other two studies found mixed results with one of four small mammal species being more numerous in selectively harvested stands[3] or in selectively harvested stands only in some years[6]. A systematic review in North American forests[7] found that partially harvested forests had more red-backed voles but not deer mice than did clearcut forests.

BEHAVIOUR (1 STUDY)

- **Use (1 study):** A site comparison study in the USA[2] found that partially harvested forest was not used by snowshoe hares more than was largely clearcut forest.

Background

Clearcutting of large areas of forest can have substantial impacts on associated fauna. Selective logging is the removal of selected trees within a forest based on criteria such as diameter, height or species. Remaining trees are left in the stand, as opposed to clearcutting where all trees are felled. This intervention is similar to several others that involve harvesting some, but not all, trees. In this case, tree removal was largely based on forestry specifications, rather than designed spatially to retain undisturbed patches. This intervention covers a wide range of tree removal intensities. In some cases, management is for shelterwood, a specific forestry practice that involves gradually removing mature trees to allow growing space for younger trees that initially germinate in partial shade.

See also *Fell trees in groups, leaving surrounding forest unharvested, Retain undisturbed patches during thinning operations, Use thinning of forest instead of clearcutting* and *Use patch retention harvesting instead of clearcutting.*

A replicated, site comparison study in 1980 of a forest in Nova Scotia, Canada (1) found that selectively harvested plots, cut as shelterwood, did not host more small mammals than did clearcut plots. In shelterwood plots, average capture rates (10–31 small mammals/100 trap nights) did not differ significantly from those in clearcuts (12–27 small mammals/100 trap nights). The forest had regrown following fire 80 years previously. Three plots (average 3.6 ha) were clearcut 3–5-years previously and two plots (average 1.9 ha) were shelterwood cut, entailing removing a proportion of harvestable timber. Shelterwood plots had an average tree stem basal area of 9.4 m²/ha (compared to 25.9 m²/ha in adjacent unharvested forest). Small mammals were surveyed using snap traps for four consecutive nights and days, one or twice in each plot in July–August 1980.

A site comparison study in 1974–1977 of three mixed forest blocks in Maine, USA (2) did not find more snowshoe hares *Lepus americanus* in partially harvested forest than in largely clearcut forest. In a partially harvested forest, a lower proportion of transect sections (7.9%) contained hare tracks compared to in a largely clearcut forest (17.6%). However, patches of unharvested trees were included within the clearcut forest sampled, and tracks were most numerous in or close to these. Hare tracks were surveyed, 1–2 days after snowfall, over the winters of 1974–1975, 1975–1976 and 1976–1977. Tracks were counted on 15-m sections along 50 km of permanent lines through clearcut and partially harvested forest. Partial harvesting occurred in 1974–1977 and the clearcut forest was harvested in 1960–1975.

A replicated, controlled, before-and-after study in 1994–1998 of a coniferous forest in British Colombia, Canada (3) found that, when forest was harvested by single tree selection, one of four small mammal species was more abundant relative to clearcutting. Populations of all species did not differ between plots assigned for different treatments in the year before harvesting. After harvesting, there were more southern red-backed voles *Clethrionomys gapperi* in single tree selection plots (20.8–44.0/ha) than in clearcuts (0.1–10.8/ha). Long-tailed vole *Microtus longicaudus* was less abundant in single tree selection than clearcut plots (0.0–3.4 vs 2.6–16.2/ha) as was northwestern chipmunk *Tamias amoenus* (0.8–1.4 vs 1.9–6.0/ha). Deer Mouse *Peromyscus maniculatus* numbers were similar between treatments (single tree selection: 0.4–4.0/ha; clearcuts: 0.8–5.0/ha). Forest stands were c.30 ha. There were three replicates each of single tree selection (removing 33% of timber volume) and 10-ha clearcuts, harvested in winter 1994–1995. Small mammals were live-trapped in 1994–1998, over two consecutive nights, at 3-week intervals, from June or July to August or September.

A replicated, controlled study in 1997–1998 of a forest in Maine, USA (4) found more small mammals in selectively harvested forest stands than fully harvested, regenerating stands. Annual average catches were higher in partially harvested than fully harvested stands for the three most abundant species; red-backed vole *Clethrionomys gapperi* (partially harvested: 12.4–22.1; fully harvested: 2.5–5.0 voles/grid), deer mouse *Peromyscus maniculatus* (partially harvested: 4.9–12.5; fully harvested: 0–2.5 mice/grid) and short-tailed shrew *Blarina brevicauda* (partially harvested: 4.3–5.0; fully harvested: 0–3.0 shrews/grid). These comparisons were not tested for statistical significance. Seven stands were selectively harvested between 1992 and 1995, with 52–59% of basal tree area removed and 13 m²/ha basal area remaining. Two forest stands were clearcut between 1974 and 1984 and treated with the herbicide, glyphosate, 3–8 years post-harvest. Small mammals were surveyed in live trap grids, between 22 June and 28 July 1997 and between 21 June and 31 July 1998.

A replicated, controlled, before-and-after study in 1991–1997 of two second-growth forests in Arkansas and Oklahoma, USA (5) found that selectively harvesting isolated trees did not increase small mammal abundance relative clearcutting. Before harvesting, average small mammal abundances did not differ significantly between stands planned for different treatments (single tree selection: 2.7 small mammals/100 trap nights; clearcut: 0.9). Similarly, after harvesting, small mammal numbers did not differ significantly between single tree selection stands (6.4/100 trap nights) and clearcut stands (10.7). In each of four blocks of second-growth forest (59–69 years old at start of study), one stand was managed by single tree selection and one was clearcut, harvested in summer 1993. Tree basal area after harvesting was 15–16 m²/ha in single tree selection plots (compared to 24–32 m²/ha in unharvested forest). Stand extent was 13–28 ha. Small mammals were surveyed using an average of 67 Sherman live traps/stand, pre-harvest in 1991 and 1992, and post-harvest in 1995, 1997 and 1999. Traps were operated for seven consecutive nights during winter (December–January).

A replicated, controlled study in 1994–1997 of Douglas-fir *Pseudotsuga menziesii* forest in British Colombia, Canada (6) found that selective harvesting of trees increased one of four small mammal species abundance in the third and fourth, but not first and second, year after harvesting relative to clearcutting. There were more southern red-backed voles *Myodes gapperi* in the third and four year in all selectively logged treatments (6–17/plot) than in clearcut stands (0–1/plot), but similar numbers between treatments in the first two years (selective cut: 33–42/plot; clearcut: 13–34/plot). There were no differences between treatments for deer mouse *Peromyscus maniculatus* (selective cut: 1–15/

plot; clearcut: 6–21/plot) or northwestern chipmunk *Tamias amoenus* (selective cut: 0–6/plot; clearcut: 0–6/plot). There were more meadow voles *Microtus pennsylvanicus* in clearcut stands (selective cut: 0–2/plot; clearcut: 3–14/plot). Forest stands, 20–25 ha in extent, were partially harvested in winter 1993–1994. Two each had 20% of timber volume removed by individual-tree selection, 35% removed by individual-tree selection on 50% of the area and 50% volume removed by individual-tree selection. These were compared with two 1.6-ha clearcut areas. Small mammals were live-trapped, at 2–4-week intervals, in May–October of 1994, 1995, and 1996 and in April–May 1997.

A systematic review in 2008 of 56 studies of small mammal responses to partial harvesting, clearcutting or wildfire in North American forests (7) found that partially harvested forests had more red-backed voles *Myodes gapperi*, but not deer mice *Peromyscus maniculatus* than did clearcut forests. Absolute abundances are not presented but vole numbers in partially harvested stands, 1–9 years after harvesting, were significantly higher than in clearcut stands. Deer mouse abundances did not differ significantly between partially harvested and clearcut stands. Meta-analyses were carried out on studies identified following a defined literature search procedure.

A replicated, site comparison study in 2006–2007 in a mixed temperate forest in Quebec, Canada (8) found that harvesting trees selectively did not result in higher small mammal species richness or abundance compared to clearcutting. Small mammal species richness did not vary along a gradient of retained conifer basal area that resulted from different felling densities (result presented as statistical model coefficient). The combined abundances of red-backed voles *Myodes gapperi*, masked shrews *Sorex cinereus* and deer mice *Peromyscus maniculatus* (which comprised 92% of individuals caught) did not vary with conifer basal area (result presented as statistical model coefficient). Four tree blocks were harvested in 2004–2005. Three or four harvesting treatments (each 20-ha extent) were applied in each block. Selective harvesting resulted in retention of 17–23%, 57–69% or 60–73% of standing timber. Clearcut areas had <10% of timber remaining. Small mammals were live-trapped, between 3 July and 25 August in 2006 and 2007.

(1) Swan D., Freedman B. & Dilworth T. (1984) Effects of various hardwood forest management practices on small mammals in central Nova Scotia. *The Canadian Field-Naturalist*, 98, 362–364.

(2) Monthey R.W. (1986) Responses of snowshoe hares, *Lepus americanus*, to timber harvesting in northern Maine. *The Canadian Field-Naturalist*, 100, 568–570.

(3) Klenner W. & Sullivan T.P. (2003) Partial and clearcut harvesting of high-elevation spruce–fir forests: implications for small mammal communities. *Canadian Journal of Forest Research*, 33, 2283–2296, https://doi.org/10.1139/x03-142

(4) Fuller A.K., Harrison D.J. & Lachowski H.J. (2004) Stand scale effects of partial harvesting and clearcutting on small mammals and forest structure. *Forest Ecology and Management*, 191, 373–386, https://doi.org/10.1016/j.foreco.2004.01.014

(5) Perry R.W. & Thill R.E. (2005) Small-mammal responses to pine regeneration treatments in the Ouachita Mountains of Arkansas and Oklahoma, USA. *Forest Ecology and Management*, 219, 81–94, https://doi.org/10.1016/j.foreco.2005.09.001

(6) Klenner W. & Sullivan T.P. (2009) Partial and clearcut harvesting of dry Douglas-fir forests: Implications for small mammal communities. *Forest Ecology and Management*, 257, 1078–1086, https://doi.org/10.1016/j.foreco.2008.11.012

(7) Zwolak R. (2009) A meta-analysis of the effects of wildfire, clearcutting, and partial harvest on the abundance of North American small mammals. *Forest Ecology and Management*, 258, 539–545, https://doi.org/10.1016/j.foreco.2009.05.033

(8) Le Blanc M-L., Fortin D., Darveau M. & Ruel J-C. (2010) Short term response of small mammals and forest birds to silvicultural practices differing in tree retention in irregular boreal forests. *Ecoscience*, 17, 334–342, https://doi.org/10.2980/17-3-3340

6.16. Use patch retention harvesting instead of clearcutting

https://www.conservationevidence.com/actions/2639

- **Three studies** evaluated the effects on mammals of using patch retention harvesting instead of clearcutting. Two studies were in Canada[1,3] and one was in Australia[2].

COMMUNITY RESPONSE (0 STUDIES)

POPULATION RESPONSE (3 STUDIES)

- **Abundance (3 studies):** Two replicated, controlled, before-and-after studies and a replicated, site comparison study in Canada[1,3] and Australia[2] found that retaining patches of unharvested trees instead of clearcutting whole forest stands increased or maintained numbers of some but not all small mammals. Higher abundances where tree patches were retained were found for southern red-backed voles[1,3], bush rat[2] and for female agile antechinus[2]. No benefit of retaining forest patches was found on abundances of deer mouse[1], meadow vole[1] and male agile antechinus[2].

BEHAVIOUR (0 STUDIES)

Background

Removing trees, through clearcutting or clearfelling, can have substantial, usually negative, effects on forest mammals, through alteration of habitat and removal of food and shelter. Patch retention is the act of leaving groups of trees during harvesting, which may act as refugia to support forest fauna and enable its re-colonisation of the remainder of the forest as it regrows.

A replicated, controlled, before-and-after study in 1993–1996 of a boreal forest area in Alberta, Canada (1) found that retaining patches of unharvested trees enhanced numbers of red-backed voles *Clethrionomys gapperi*, but not of deer mice *Peromyscus maniculatus* or meadow voles *Microtus pennsylvanicus*, relative to those in fully harvested areas. Following harvesting, yearly peak red-backed vole population estimates were higher with retained tree patches (101–172 voles/plot) than without (53–91 voles/plot). Deer mice had similar abundance between treatments (patches: 107–148 mice/plot; no patches: 71–115 mice/plot). Meadow vole numbers were higher in fully harvested plots (patches: 0–24 voles/grid; no patches: 36–118). In a 6 × 6-km study area, four plots were managed during winter 1993–1994. In two plots, trees were felled, but leaving undisturbed 40-m diameter patches, comprising 10% of total tree basal area. In two other plots, trees were felled entirely. Small mammals were surveyed using 60 or 120 Longworth live traps/6 ha block. Traps were set for three nights and two days, at fortnightly or longer intervals, from May or June to August or September, in 1993–1996.

A replicated, controlled, before-and-after study in 2002–2009 of forest across three districts in Victoria, Australia (2) found that retaining forest islands when clearfelling reduced subsequent abundance declines after brash burning for some small mammal relative to in clearfelled areas. Average bush rat *Rattus fuscipes* abundance declined less following burning in island retention patches (before: 2.1; after: 1.6/grid) than in clearfelled patches (before: 1.2; after: 0.4/grid). Female agile antechinus *Antechinus agilis* abundance declined less following burning in island retention patches (before: 2.2; after: 1.5/grid) than in clearfelled patches (before: 1.0; after: 0.1/grid). However, male agile antechinus abundance declines were similar following burning in island retention patches (before: 1.1; after: 0.4/grid) and clearfelled patches (before: 0.5; after: 0.2/grid). Forest patches (coupes) of ≥15 ha were established in six blocks. In each block, one patch was entirely clearfelled, one was clearfelled, but retaining a 1.5-ha forest island and one was clearfelled, but retaining three 0.5-ha islands. Post-felling, blocks were prescribed burned to clear brash. Small mammals were surveyed using four live-trap grids in each patch. Three grids/patch were in retained forest islands. Surveys took place before felling, after felling and after burning. Treatments were staggered, so surveys spanned 2002 to 2009.

A replicated, site comparison study in 2015–2016 of a coniferous forest site in British Columbia, Canada (3) found that retaining patches of trees when harvesting sustained higher southern red-backed voles *Myodes gapperi* populations compared to clearfelling. Nineteen to 20 years post-harvest, there were more red-backed voles in patch retention plots (5.7/ha) than in clearfelled plots (3.3/ha). Harvesting, in 1996, comprised three replicate plots each of tree patch retention (10 m²/ha basal area, retained as a group — group sizes not stated) and clearfelling. Plot sizes ranged from 3.6–12.8 ha. Forest overstorey was mostly lodgepole pine *Pinus contorta* and Douglas fir *Pseudotsuga menziesii*, of average ages of 82–228 years. Following harvesting, sites were planted with lodgepole pine, Douglas fir and interior spruce *Picea glauca* × *engelmannii* seedlings in 1997. Small mammals were sampled at four-week intervals in May–October of 2015 and 2016. One live-trapping grid (49 traps across 1 ha) was located in each plot. Traps were set for two nights and one full day on each occasion.

(1) Moses R.A. & Boutin S. (2001) The influence of clearcut logging and residual leave material on small mammal populations in aspen-dominated boreal mixedwoods. *Canadian Journal of Forest Research*, 31, 483–495, https://doi. org/10.1139/x00-186

(2) Lindenmayer D.B., Knight E., McBurney L., Michael D. & Banks S.C. (2010) Small mammals and retention islands: An experimental study of animal response to alternative logging practices. *Forest Ecology and Management*, 260, 2070–2078, https://doi.org/10.1016/j.foreco.2010.08.047

(3) Sullivan T.P. & Sullivan D.S. (2017) Green-tree retention and recovery of an old-forest specialist, the southern red-backed vole (*Myodes gapperi*), 20 years after harvest. *Wildlife Research*, 44, 669–680, https://doi.org/10.1071/wr17065

6.17. Retain undisturbed patches during thinning operations

https://www.conservationevidence.com/actions/2640

- **Two studies** evaluated the effects on mammals of retaining undisturbed patches during thinning operations. Both studies were in the USA[1,2].

COMMUNITY RESPONSE (0 STUDIES)

POPULATION RESPONSE (0 STUDIES)

BEHAVIOUR (2 STUDIES)

- **Use (2 studies):** Two randomized, replicated, controlled studies (one also before-and-after) in the USA[1,2] found that snowshoe hares[1] and tassel-eared squirrels[2] used retained undisturbed forest patches more than thinned areas.

Background

Thinning is a forestry practice that involves the selective removal of trees to reduce tree density and improve the growth rate, health and timber quality of remaining trees. Thinning has been done historically to maximize timber production but may have ecological benefits, such as opening up the canopy and allowing more light in, which may benefit some species. However, some species may benefit from the shelter available within retained undisturbed forest patches.

A randomized, replicated, controlled, before-and-after study in 2001–2003 of a coniferous forest in Montana, USA (1) found that snowshoe hares *Lepus americanus* used retained undisturbed patches more than they used thinned forest. More hare tracks were counted in undisturbed patches than in thinned areas when patches comprised 8% (undisturbed: 106; thinned: 25 tracks/km) and 35% (undisturbed: 107; thinned: 15 tracks/km) of the stand. The same was found for faecal pellet counts in 8% (undisturbed: 1.0; thinned: 0.2 pellets/tray) and 35% (undisturbed: 1.4; thinned: 0.1 pellets/tray) retention patches. After treatments were applied, hares increased use of undisturbed (before treatment: 29; after: 144 tracks/km) and mature (before treatment: 64–80; after: 88–181 tracks/km) stands, suggesting movements into these areas. Five conifer stands (10.5–14.0 ha), regenerating naturally after felling in 1985, were selected. Treatments were applied in June 2002 and comprised: thinning with five 0.2-ha unthinned patches (8%) retained (two stands), thinning with five 0.8-ha unthinned patches (35%) retained (two stands) and one undisturbed stand. Conifer density was 5,350–7,050/ha before and 656–750/ha after thinning. Two adjacent mature stands represented pre-harvest conditions. Hare-track density was assessed from December–March in 2001–2002 (prior to thinning) and 2002–2003 (after thinning). Faecal pellets were surveyed each winter within 50 trays in each stand, into which pellets accumulated during April snowmelt.

A randomized, replicated, controlled study in 2005–2007 of a ponderosa pine *Pinus ponderosa* forest in Northern Arizona, USA (2) found that tassel-eared squirrels *Sciurus aberti* made greater use of undisturbed than thinned forest. In winter 57% and during the rest of the year 51% of squirrel home range areas fell within undisturbed forest compared to 39% availability by extent in the study area. Squirrels also showed a preference for dense canopies. In winter, canopies with 51–75% cover accounted for 53% of squirrel use compared to 44% of resource availability. Thinning was carried out from 1998–2000. Seventeen-hectare blocks within a 10-km² area were randomly assigned to no thinning and to low, medium and high-intensity thinning. A combination of these managements was applied to four additional blocks of approximately 40 ha each. Squirrel locations were monitored by radio-tracking from December 2005 to July 2007.

(1) Ausband D.E. & Baty G.R. (2005) Effects of precommercial thinning on snowshoe hare habitat use during winter in low-elevation montane forests. *Canadian Journal of Forest Research*, 35, 206–210, https://doi.org/10.1139/x04-152

(2) Loberger C.D., Theimer T.C., Rosenstock S.S. & Wightman C.S. (2011) Use of restoration-treated ponderosa pine forest by tassel-eared squirrels. *Journal of Mammalogy*, 92, 1021–1027, https://doi.org/10.1644/10-mamm-a-321.1

6.18. Clear or open patches in forests

https://www.conservationevidence.com/actions/2641

- **Four studies** evaluated the effects on mammals of clearing or opening patches in forests. Two studies were in the USA[2,3], one was in Bolivia[1] and one was in Canada[4].

COMMUNITY RESPONSE (0 STUDIES)

POPULATION RESPONSE (4 STUDIES)

- **Abundance (4 studies):** Two of four replicated studies (including three controlled studies and a site comparison study), in Bolivia[1], the USA[2,3] and Canada[4], found that creating gaps or open patches within forests did not increase small mammal abundance[1,2] relative to uncut forest. One study found that it did increase small mammal abundance[4] and one found increased abundance for one of four small mammal species[3].

BEHAVIOUR (0 STUDIES)

Background

Gaps in forests can be natural features that add diversity to the habitat. They can be created by natural events, such as mature trees falling, and maintained by grazing animals. In absence of natural gaps (such as in a younger forest) artificially creating gaps may mimic the same conditions. This intervention considers some cases where gaps are created primarily as a conservation action and others where gaps are created as part of timber harvesting.

A replicated, site comparison study in 1998 of tropical forest in Bolivia (1) found that creating forests gaps, by selective felling, did not increase small mammal abundance relative to that in undisturbed forest. The number of small mammals trapped did not differ between large gaps (7.0/plot), small gaps (6.8/plot) and undisturbed forest (5.2/plot). Similarly, total species richness did not differ between large gaps (four species), small gaps (five species) and undisturbed forest (five species). Trees were harvested selectively, creating gaps, in June-October 1997. Within each of six blocks, one small gap (average 247 m^2), one large gap (average 811 m^2) and one undisturbed area (400 m^2) were studied. Treatments in a block were separated by <100 m. Small mammals were monitored using eight Sherman live traps and a larger cage trap, set in each gap or undisturbed forest area, for six days each in April, July, and November 1998.

A replicated, controlled study in 1995–1997 of three stands in a coniferous forest in Washington, USA (2) found that creating gaps in forests did not increase abundances of most small mammal species. Species responses to treatments were not tested for statistical significance. Five to six years after gap creation, there were no clear treatment preferences among the most frequently recorded species, Trowbridge's shrew *Sorex trowbridgii* (large gaps: 0.5–3.5/100 trap nights; forest: 0.0–3.8), Keen's mouse *Peromyscus keeni* (large gaps: 3.1–5.4/100 trap nights; forest: 1.9–5.9) and southern red-backed vole *Clethrionomys gapperi* (large gaps: 0.5–1.9/100 trap nights; forest: 0.4–1.9). Seven years after gap creation, there was a similar lack of clear treatment preferences among the shrew species, montane shrew *Sorex monticolus* (medium gaps: 0.0–4.2/100 trap nights; large gaps: 0.3–0.6; forest: 0.6–1.2), Trowbridge's shrew (medium gaps: 1.8–7.7/100 trap nights; large gaps: 1.2–5.7; forest: 2.1–4.8) and vagrant shrew *Sorex vagrans* (medium gaps: 0.0/100 trap nights; large gaps: 0.0–0.6; forest: 0.0–0.3). Gaps were created in 1990 in three Douglas-fir *Pseudotsuga menziesii* dominated stands, c.90, 140 and 500 years old. Gap diameters were 1 (large) and 0.6 and 0.4 (medium) times the average surrounding tree height. There were two replicates of each size/stand. Differing combinations of treatments and stands was sampled for small mammals in summer and autumn 1995–1997 using live traps, killing traps and pitfall traps.

A replicated, controlled, before-and-after study in 1994–1998 of a coniferous forest in British Colombia, Canada (3) found a greater abundance of one small mammal species when forest was harvested in small patches, relative to clearcutting, but not of three other species. Populations of all species did not differ between treatment plots in the pre-treatment year. After harvesting, there were more southern red-backed voles *Clethrionomys gapperi* in patch harvesting plots (0.1-ha patches: 18.7–49.7/ha; 1-ha patches: 18.0–38.1/ha) than in clearcuts (0.1–10.8/ha). Long-tailed voles *Microtus longicaudus* were less abundant in patch harvesting plots than clearcut plots (0.1-ha patches: 0.4–4.5/ha; 1-ha patches: 0.2–2.6/ha; clearcuts: 2.6–16.2/ha). Abundances were similar between treatments for northwestern chipmunk *Tamias amoenus* (0.1-ha patches: 2.9–3.4/ha; 1-ha patches: 2.2–2.4/ha; clearcuts: 3.7–6.0/ha) and deer mouse *Peromyscus maniculatus* (0.1-ha patches: 0.4–5.1/ha; 1-ha patches: 2.0–4.5/ha; clearcuts: 0.8–5.0/ha). Forest stands were c.30 ha. There were three replicate stands each harvested in winter 1994/95, with 0.1-ha patches, 1-ha patches and 10-ha clearcuts. Each involved removing 30% volume of timber. Small mammals were live-trapped in 1994–1998, over two consecutive nights, at 3-week intervals, from June or July to August or September.

A replicated, controlled, before-and-after study in 1991–1997 of two second-growth forests in Arkansas and Oklahoma, USA (4) found that felling small groups of trees increased small mammal abundance relative to unharvested stands, but not to clearcut stands. Before harvesting, average small mammal abundances were similar between stands planned for different treatments (unharvested: 2.5 small mammals/100 trap nights; small group felling: 2.2; clearcut: 0.9). After harvesting, more small mammals were caught in small group felling stands (6.7/100 trap nights) than in unharvested stands (1.7) but a similar number was caught in clearcut stands (10.7). In each of four blocks of second-growth forest (59–69 years old at start of study), one stand was managed by felling trees to create 3–10 openings of 0.04–1.9 ha, covering 6–14% of stand area, one was clearcut and one was unharvested. Harvesting was conducted in summer 1993. Stands covered 13–28 ha. Small mammals were surveyed using an average of 66.5 Sherman live traps/stand, pre-harvest in 1991 and 1992, and post-harvest in 1995, 1997 and 1999. Traps were operated for seven consecutive nights during winter (December–January).

(1) Fredericksen N.J., Fredericksen T.S., Flores B. & Rumiz D. (1999) Wildlife use of different-sized logging gaps in a tropical dry forest. *Tropical Ecology*, 40, 167–175.

(2) Gitzen R.A. & West S.D. (2002) Small mammal response to experimental canopy gaps in the southern Washington Cascades. *Forest Ecology and Management*, 168, 187–199, https://doi.org/10.1016/s0378-1127(01)00745-9

(3) Klenner W. & Sullivan T.P. (2003) Partial and clearcut harvesting of high-elevation spruce–fir forests: implications for small mammal communities. *Canadian Journal of Forest Research*, 33, 2283–2296, https://doi.org/10.1139/x03-142

(4) Perry R.W. & Thill R.E. (2005) Small-mammal responses to pine regeneration treatments in the Ouachita Mountains of Arkansas and Oklahoma, USA. *Forest Ecology and Management*, 219, 81–94, https://doi.org/10.1016/j.foreco.2005.09.001

6.19. Retain dead trees after uprooting

https://www.conservationevidence.com/actions/2642

- **One study** evaluated the effects on mammals of retaining dead trees after uprooting. This study was in the USA[1].

COMMUNITY RESPONSE (0 STUDIES)

POPULATION RESPONSE (0 STUDIES)

BEHAVIOUR (1 STUDY)

- **Use (1 study):** A replicated, controlled study in the USA[1] found that areas where trees were uprooted but left on site were used more by desert cottontails than were cleared areas.

Background

Management or restoration of some habitats involves removing trees. This may occur, for example, in sites where fire suppression has caused woodland to become denser than it has been historically. Retaining uprooted trees can increase structural diversity at ground level. This may in turn increase cover available to some mammal species.

A replicated, controlled study in 1965–1968 of pinyon-juniper forest at a site in New Mexico, USA (1) found that where trees were uprooted but left on site, more desert cottontail *Sylvilagus audubonii* faecal pellets were counted than in fully cleared areas. Results were not tested for statistical significance. Where uprooted trees were left, there were 3.2 cottontail pellets/ft^2 compared to 1.0 pellets/ft^2 where trees were uprooted and burned. In each of two blocks, there was one plot with all trees uprooted and left on site and one with all trees uprooted, piled up and burned. Plots covered 300–500 acres each. Treatments were carried out in 1965. Cottontail pellets were counted on randomly selected sample points on belts of 1/400 acre within the middle of each plot, in 1968.

(1) Kundaeli J.N. & Reynolds H.G. (1972) Desert cottontail use of natural and modified pinyon-juniper woodland. *Journal of Range Management*, 25, 116–118.

6.20. Use thinning of forest instead of clearcutting

https://www.conservationevidence.com/actions/2643

- **One study** evaluated the effects on mammals of using thinning of forest instead of clearcutting. This study was in the USA[1].

COMMUNITY RESPONSE (0 STUDIES)

POPULATION RESPONSE (0 STUDIES)

BEHAVIOUR (1 STUDY)

- **Use (1 study):** A replicated, controlled study in the USA[1] found that thinned forest areas were used more by desert cottontails than were fully cleared or uncleared areas.

Background

Harvesting of timber within forests can be carried out by clearcutting sites or by various methods of harvesting a proportion of trees. By thinning, rather than felling a whole forest, larger areas would need to be managed in order to achieve the same timber harvest though some degree of forest cover can be retained over that area. Thinning forest may benefit some species that prefer an open forest structure whilst not having detrimental effects on forest mammals that clearcutting would be likely to have.

See also *Thin trees within forest* for where thinning is an intervention in woodland that would otherwise be left without removing trees.

A replicated, controlled study in 1965–1968 of pinyon-juniper forest at a site in New Mexico, USA (1) found that in areas where trees were thinned, more desert cottontail *Sylvilagus audubonii* faecal pellets were counted than in fully cleared areas or uncleared areas. Results were not tested for statistical significance. In thinned plots, there were 2.7 cottontail pellets/ft^2 compared to 1.0 pellets/ft^2 where trees were cleared (uprooted and burned) and 2.4 pellets/ft^2 where trees were left unmanaged. In each of two blocks, there was one plot with trees thinned to 100 trees/acre, one with all trees uprooted, piled up and burned and one with trees left unmanaged. Plots covered 300–500 acres each. Treatments were carried out in 1965. Cottontail pellets were counted at randomly selected sample points in treatment plots in 1968.

(1) Kundaeli J.N. & Reynolds H.G. (1972) Desert cottontail use of natural and modified pinyon-juniper woodland. *Journal of Range Management*, 25, 116–118.

6.21. Remove competing vegetation to allow tree establishment in clearcut areas

https://www.conservationevidence.com/actions/2644

- **Three studies** evaluated the effects on mammals of removing competing vegetation to allow tree establishment in clearcut areas. Two studies were in Canada[2,3] and one was in the USA[1].

COMMUNITY RESPONSE (0 STUDIES)

POPULATION RESPONSE (0 STUDIES)

BEHAVIOUR (3 STUDIES)

- **Use (3 studies):** One of three studies (including two controlled studies and one site comparison study), in the USA[1] and Canada[2,3], found that, where competing vegetation was removed to allow tree establishment in clearcut areas, American martens used the areas more[3]. One study found mixed results for moose[1] and one found no increase in site use by snowshoe hares[2].

Background

Following felling of trees, for timber harvesting, a range of actions may be employed to accelerate forest regrowth. Tree establishment (either through natural regeneration or planting) may be inhibited by rapid growth of herbaceous or scrubby vegetation. This vegetation may be controlled or removed by use of herbicides or by using tools, such as brushsaws, to physically remove such vegetation. Using such techniques to allow or encourage forest regrowth in clearcut areas may speed up the time until such habitat becomes suitable for forest-dwelling mammals.

A randomized, controlled, before-and-after study in 1991–1993 in a coniferous forest in Maine, USA (1) found that moose *Alces alces* did not use herbidice-treated forest clearcuts more than untreated clearcuts 1–2 years after treatment but foraging and sleeping signs were more numerous on treated than untreated clearcuts 7–11 years after treatment. Moose track quantity was similar between plots in the year before herbicide application (treatment plots: 0.07 track groups/ha; untreated: 0.08). One to two years after treatment, there were no significant differences in total number of track groups (treated: 1.6–3.0/km; untreated: 2.6–5.1), pellet groups (treated: 0.1–0.2/km; untreated: 0.2–0.4) or moose beds (treated: 0.03–0.05/km; untreated: 0.13–0.26), but there were fewer foraging tracks in treated plots (treated: 0.4 track groups/km; untreated: 1.0 tracks/km). After 7–11 years, there were more foraging tracks in treated (2.1–4.3/km) than untreated (1.1–1.8) plots and more moose beds (treated: 0.35–0.55/km; untreated: 0.12–0.31). There were no differences between treatments for total track groups (treated: 5.3–7.7/km; untreated: 3.4–4.2) or pellet groups (treated: 0.8–0.9/km; untreated: 0.4–0.5). Six of 12 clearcuts (18–89 ha), harvested 4.5–8.5 years previously, were herbicide-treated in August 1991. Six of 11 different clearcuts (21–73 ha) were glyphosate-treated 7–10 years before sampling. Treated plots in this second group averaged 19 years post-felling and, untreated plots, 16 years. Across all 23 plots, groups of moose foraging tracks and all tracks, moose beds and faecal pellet clumps were counted 5–7 times/year in January–March of 1992 and 1993, along 2-m-wide transects, 3–7 days after snowfall.

A replicated, controlled study in 1991–1996 of a coniferous forest in Québec, Canada (2) found that, up to nine years after clearcutting, snowshoe hares *Lepus americanus* were not more numerous in replanted areas where competing vegetation had been removed than in naturally regenerating clearcuts. Data were not fully reported, nor were results of statistical analyses. However, hares seldom used removal plots. Only 5% of vegetation removal plots contained hare faecal pellets during any one survey and no preference for removal plots over those regenerating naturally was identified. Twenty-five sites (6–9 ha) were studied. Ten were clearcut in 1987, replanted in spring 1990, and competing vegetation removed in August 1992. In five sites vegetation was removed using brushsaws, and five using herbicide solution. Fifteen naturally regenerated sites, clearcut between 1987 and 1989, were controls. Hare faecal pellets were counted and cleared in 1 × 5-m plots, in June and September, 1991–1996.

A replicated, site comparison study in 2001–2002 of boreal forest stands in Ontario, Canada (3) found that stands subject to herbicide treatment and tree planting after logging were used more by American martens *Martes americana* than were naturally regenerating stands. The effects of herbicide and planting were not separated in the study. Radio-tracked martens made greater use of herbicide-treated and planted stands than they did of naturally regenerating stands (data not presented). However, the live-capture rate of martens in herbicide-treated and planted stands (5.6 martens/100 trap nights) was not significantly different to that in regenerating stands (1.9 martens/100 trap nights). Stands were all 35–45 years old and located in a 600-km^2 forestry area. Forest stands were either herbicide-treated and planted following logging or were left to regenerate naturally after logging. Martens were live-trapped in 2003–2007, and monitored subsequently by radio-tracking.

(1) Eschholz W.E., Servello F.A., Griffith B., Raymond K.S. & Krohn W.B. (1996) Winter use of glyphosate-treated clearcuts by moose in Maine. *The Journal of Wildlife Management*, 60, 764–769.

(2) de Bellefeuille S., Bélanger L., Huot J. & Cimon A. (2001) Clearcutting and regeneration practices in Quebec boreal balsam fir forest: effects on snowshoe hare. *Canadian Journal of Forest Research*, 31, 41–51, https://doi.org/10.1139/x00-140

(3) Thompson I.D., Baker J.A., Jastrebski C., Dacosta J., Fryxell J. & Corbett D. (2008) Effects of post-harvest silviculture on use of boreal forest stands by amphibians and marten in Ontario. *Forestry Chronicle*, 84, 741–747, https://doi.org/10.5558/tfc84741-5

6.22. Retain understorey vegetation within plantations

https://www.conservationevidence.com/actions/2645

- **One study** evaluated the effects on mammals of retaining understorey vegetation within plantations. This study was in Chile[1].

COMMUNITY RESPONSE (1 STUDY)

- **Richness/diversity (1 study):** A replicated, controlled, before-and-after study in Chile[1] found that areas with retained understorey vegetation had more species of medium-sized mammal, compared to areas cleared of understorey vegetation.

POPULATION RESPONSE (0 STUDIES)

BEHAVIOUR (1 STUDY)

• **Use (1 study):** A replicated, controlled, before-and-after study in Chile[1] found that areas with retained understorey vegetation had more visits from medium-sized mammals, compared to areas cleared of understorey vegetation.

Background

Understorey vegetation may compete for resources with planted trees, especially when trees are young, and is, therefore, sometimes removed as part of commercial forest management. However, retaining understorey vegetation has the potential to support native mammals (e.g. Carrilho *et al.* 2017) and may form part of a suite of actions that could attract premium payments for timber products marketed as being biodiversity-friendly.

Carrilho, M., Teixeira, D., Santos-Reis, M., & Rosalino, L. M. (2017). Small mammal abundance in Mediterranean Eucalyptus plantations: how shrub cover can really make a difference. *Forest Ecology and Management, 391*, 256–263, https://doi.org/10.1016/j.foreco.2017.01.032

A replicated, controlled, before-and-after study in 2009–2012 of a Monterey pine *Pinus radiata* plantation in central Chile (1) found that retaining understorey vegetation resulted in there being a greater number and higher visit rate of medium-sized mammal species, compared to areas cleared of understorey vegetation. Before clearance, the same four species were recorded both in plots designated to be uncleared and cleared; *Leopardus guigna*, culpeo *Pseudalopex culpaeus*, Molina's hog-nosed skunk *Conepatus chinga* and southern pudu *Pudu puda*. After understorey clearance, all four species remained in uncleared plots but just southern pudu occurred in cleared plots. There were also fewer visits to cleared plots after understorey removal (visit rates presented as response ratios). Thirteen plots (≥300 m apart) were monitored using camera traps for four to five nights, monthly, from October 2009 to July 2012. In February 2011, understorey vegetation was removed from 1,600 m² around cameras in five plots. Regrowth was controlled in February 2012.

(1) Simonetti J.A., Grez A.A. & Estades C.F. (2013) Providing habitat for native mammals through understory enhancement in forestry plantations. *Conservation Biology*, 27, 1117–1121, https://doi.org/10.1111/cobi.12129

6.23. Leave standing deadwood/snags in forests

https://www.conservationevidence.com/actions/2646

- **One study** evaluated the effects on mammals of leaving standing deadwood or snags in forests. This study was in the USA[1].

COMMUNITY RESPONSE (0 STUDIES)

POPULATION RESPONSE (1 STUDY)

- **Abundance (1 study):** A replicated, controlled study in the USA[1] found that increasing the quantity of standing deadwood in forests increased the abundance of one of three shrew species, compared to removing deadwood.

BEHAVIOUR (0 STUDIES)

Background

Snags or standing dead trees and other dead wood can provide habitat or resources for some species within forest. Retaining or increasing provision of these features may benefit some forest mammal species.

A replicated, controlled study in 2007–2008 of three stands of loblolly pine *Pinus taeda* in South Carolina, USA (1) found that increasing the amount of forest standing deadwood increased the abundance of one of three shrew species compared to removing dead wood but not compared to in unmanipulated plots. More southeastern shrews *Sorex longirostris* were caught in plots with increased standing deadwood quantities (0.046 shrews/m of drift fence) than in plots cleared of fallen debris (0.013). Neither treatment differed significantly from the quantity in unmanipulated plots (0.026). There were no significant differences between treatments for southern short-tailed shrew *Blarina carolinensis* (standing deadwood: 0.069 shrews/m of drift fence; debris cleared: 0.051; unmanipulated: 0.058) or North American least shrew *Cryptotis parva* (standing deadwood: 0.004 shrews/m of drift fence; debris cleared: 0.014; unmanipulated: 0.015). Three plots, each 9.3 ha, were located in each of three loblolly pine stands, planted in 1950–1953. In each stand, standing deadwood quantities were increased tenfold in one plot in 2001, by ringbarking and injecting herbicide into trees, in another plot woody debris ≥10 cm across and ≥60-cm long was removed annually from 1996 and one plot was unmanipulated. Shrews were sampled across plots for 14 days, on seven occasions, from January 2007 to August 2008. Shrews were caught in 19-l plastic buckets, connected by drift fencing.

(1) Davis J.C., Castleberry S.B. & Kilgo J.C. (2010) Influence of coarse woody debris on the soricid community in southeastern Coastal Plain pine stands. *Journal of Mammalogy*, 91, 993–999, https://doi.org/10.1644/09-mamm-a-170.1

6.24. Leave coarse woody debris in forests

https://www.conservationevidence.com/actions/2647

- **Three studies** evaluated the effects on mammals of leaving coarse woody debris in forests. One study was in Canada[1], one was in the USA[2] and one was in Malaysia[3].

COMMUNITY RESPONSE (1 STUDY)

- **Richness/diversity (1 study):** A replicated, site comparison study, in Malaysia[3] found more small mammal species groups in felled forest areas with woody debris than without.

POPULATION RESPONSE (3 STUDIES)

- **Abundance** (**3 studies**): One out of three replicated studies (two controlled, one site comparison, one before-and-after) in Canada[1], the USA[2] and Malaysia[3] found that retaining or adding coarse woody debris did not increase numbers or frequency of records of small mammals[1,3]. The other study found that two of three shrew species were more numerous in areas with increased volumes of coarse woody debris than areas without coarse woody debris[2].

BEHAVIOUR (0 STUDIES)

Background

Coarse woody debris consists of fallen dead trees and cut branches that are left after tree harvesting. Coarse woody debris increases the structural diversity at the forest floor. Sometimes, debris may be removed as part of forestry operations, such as for use as biofuel. However, retained coarse woody debris may provide resources on the forest floor that benefit woodland species.

This intervention covers studies where coarse woody debris is left evenly distributed. See also *Gather coarse woody debris into piles after felling*.

A replicated, controlled, before-and-after study in 1993–1996 of a boreal forest area in Alberta, Canada (1) found that retaining woody debris following harvesting did not enhance numbers of three small mammal species, relative to those in cleared areas. This was the case for estimated annual peak populations of red-backed vole *Clethrionomys gapperi* (debris: 53–91 voles/plot; no debris: 91–99 voles/plot), deer mouse *Peromyscus maniculatus* (debris: 71–115 mice/plot; no debris: 79–151 mice/plot) and meadow vole *Microtus pennsylvanicus* (debris: 36–118 voles/plot; no debris: 7–146 voles/plot). In a 6 × 6-km study area, trees across four plots were clearfelled during winter 1993–1994. In two plots, woody brash was spread by bulldozer to form a strip, approximately 50 m wide and 0.5 m deep, generally along block centres. Woody debris was removed entirely from the other two plots. Small mammals were surveyed using 60 or 120 Longworth live traps/6 ha block. Traps were operated for three nights and two days, at fortnightly or longer intervals, from May or June to August or September in 1993–1996.

A replicated, controlled study in 2007–2008 of three stands of loblolly pine *Pinus taeda* in South Carolina, USA (2) found that increasing coarse wood debris quantity increased the abundance of two of three shrew species compared to removing debris, but not compared to leaving debris as it fell. More southeastern shrews *Sorex longirostris* were caught in plots with increased coarse woody debris quantities (0.057 shrews/m of drift fence) than in plots cleared of fallen debris (0.013). Numbers in neither treatment differed significantly from those in unmanipulated plots (0.026). The same pattern was seen for southern short-tailed shrew *Blarina carolinensis* (increased debris: 0.105 shrews/m of drift fence; debris cleared: 0.051; unmanipulated: 0.058). However, there were no differences between treatments for North American least shrew *Cryptotis parva* (increased debris: 0.012 shrews/m of drift fence; debris cleared: 0.014; unmanipulated: 0.015). Three plots, each 9.3 ha, were located in each of three loblolly pine stands planted in 1950–1953. In each stand, woody debris quantities were increased fivefold in one plot in 2001 by felling trees, decreased in one plot by annually removing woody debris ≥10 cm across and ≥60 cm long from 1996 and left as it fell in one plot. Shrews were sampled across plots for 14 days, during seven seasons, from January 2007 to August 2008. Shrews were caught in 19-l plastic buckets connected by drift fencing.

A replicated, site comparison study in 2013 of a tropical forest in Malaysia (3) found more small mammal species groups, but not individual small mammals, where woody debris was left after selective logging than in areas lacking woody debris. On average, six small mammal species groups were recorded at sites with debris compared to four at sites without. No significant difference was detected for average numbers of small mammal recorded at sites with debris (43) compared to sites without (39). Sites were compared with respect to tree density, canopy openness, understorey vegetation cover, distance to road and slope and no differences in these measures were detected between sites with and without debris. Trees were selectively logged, within a 200-ha area, in 2010–2011. Single camera traps were set, around two years later, for 10 days each at 17 locations with logging woody debris and 17 without. Camera locations were ≥50 m from logging roads and were baited.

(1) Moses R.A. & Boutin S. (2001) The influence of clearcut logging and residual leave material on small mammal populations in aspen-dominated boreal mixed-woods. *Canadian Journal of Forest Research*, 31, 483–495, https://doi. org/10.1139/x00-186

(2) Davis J.C., Castleberry S.B. & Kilgo J.C. (2010) Influence of coarse woody debris on the soricid community in southeastern Coastal Plain pine stands. *Journal of Mammalogy*, 91, 993–999, https://doi.org/10.1644/09-mamm-a-170.1

(3) Yamada T., Yoshida S., Hosaka T. & Okuda T. (2016) Logging residues conserve small mammalian diversity in a Malaysian production forest. *Biological Conservation*, 194, 100–104, https://doi.org/10.1016/j.biocon.2015.12.004

6.25. Gather coarse woody debris into piles after felling

https://www.conservationevidence.com/actions/2653

- **Two studies** evaluated the effects on mammals of gathering coarse woody debris into piles after felling. Both studies were in Canada[1,2].

COMMUNITY RESPONSE (1 STUDY)

- **Richness/diversity (1 study)**: A randomized, replicated, controlled study in Canada[2] found higher mammal species richness where coarse woody debris was gathered into piles.

POPULATION RESPONSE (2 STUDIES)

- **Abundance (2 studies)**: One of two randomized, replicated, controlled studies in Canada[1,2] found higher counts of San Bernardino long-tailed voles where coarse woody debris was gathered into piles[1]. The other study found higher small mammal abundance at one of three plots where debris was gathered into piles[2].

BEHAVIOUR (0 STUDIES)

Background

Coarse woody debris consists of fallen dead trees and cut branches that are left during tree harvesting. Gathering coarse woody debris into piles, either at a single point or as a line of debris across the forest floor, can increase structural diversity on a forest scale relative to evenly spreading the material.

A randomized, replicated, controlled study in 2006–2009, of a lodgepole pine *Pinus contorta*-dominated forest in British Colombia, Canada (1) found that gathering coarse woody debris from tree harvest waste into piles resulted in higher counts of San Bernardino long-tailed voles *Microtus longicaudus* than where debris was uniformly dispersed. There were more voles in plots where woody debris was gathered into piles at single points (9 voles/ha) or piles comprising rows of debris (7 voles/ha) than in plots where it was dispersed evenly (1 vole/ha). Within plots where woody debris was gathered in piles, more were caught within the piles (11–16 voles/ha) than on open ground (3 voles/ha). Plots were largely clearfelled in October 2006. Course woody debris was gathered into piles or uniformly dispersed. There were three replicate plots of each treatment, 0.2–3.0 km apart. Voles were sampled over two nights, at 4-week intervals, in May–October of 2007, 2008, and 2009, using Longworth live traps in a grid of 49 points across 1 ha in each plot.

A randomized, replicated, controlled study in 2005–2010 of three forest sites in British Colombia, Canada (2) found that plots with piles of coarse woody debris had greater small mammal abundance than did plots where woody debris was evenly spread at one of the three sites and that species richness was higher with debris in piles across all sites or in one of three sites, depending on survey method used. More small mammals were trapped in plots with course woody debris in single piles (38/plot) or arranged in lines (37/plot) than with evenly dispersed woody debris (21/plot) at one site. There were no differences at the two other sites (piles: 18–27; dispersed: 14–23/plot). Species richness of trapped mammals followed a similar pattern at the site with an abundance difference, with more species in plots with woody debris piles (4.3–4.6/plot) than with dispersed woody debris (3.7/plot). There was no difference at the other two sites (piles: 3.3–3.9; dispersed: 3.1–3.6). However, snow-tracking surveys recorded more mammal species in plots with course woody debris piles (2.7–3.4/plot) than with dispersed woody debris (1.7/plot). Trees (dominated by lodgepole pine *Pinus contorta*) were harvested at three sites in 2005–2007. Each site had three randomly assigned replicates of course woody debris gathered into single piles (2–3 piles/ha, 1–3 m high), debris gathered into rows (1–3 m high) and evenly dispersed debris. Plots within a site averaged 0.6–0.8 km apart. Small mammals were live-trapped for three nights and two days, at 4–8-week intervals, in May–October of 2007–2009. Mammal tracks were surveyed, generally three days after snowfall, twice each winter, from 2007–2008 to 2009–2010.

(1) Sullivan T.P. & Sullivan D.D. (2012) Woody debris, voles, and trees: Influence of habitat structures (piles and windrows) on long-tailed vole populations and feeding damage. *Forest Ecology and Management*, 189–198, https://doi.org/10.1016/j.foreco.2011.09.001

(2) Sullivan T.P., Sullivan D.S., Lindgren P.M.F. & Ransome D.B. (2012) If we build habitat, will they come? Woody debris structures and conservation of forest mammals. *Journal of Mammalogy*, 93, 1456–1468, https://doi.org/10.1644/11-mamm-a-250.1

6.26. Retain riparian buffer strips during timber harvest

https://www.conservationevidence.com/actions/2652

- We found no studies that evaluated the effects on mammals of retaining riparian buffer strips during timber harvesting.

'We found no studies' means that we have not yet found any studies that have directly evaluated this intervention during our systematic journal and report searches. Therefore, we have no evidence to indicate whether or not the intervention has any desirable or harmful effects.

Background

Retained riparian forest buffer strips can help to shield waterways from potentially negative impacts of tree harvesting, such as sedimentation. Such retained habitat may also enable persistence of forest mammals following clearfelling (Hannon *et al.* 2002).

Hannon S.J., Paszkowski C.A., Boutin S., DeGroot J., Macdonald S.E., Wheatley M. & Eaton B.R. (2002) Abundance and species composition of amphibians, small mammals, and songbirds in riparian forest buffer strips of varying widths in the boreal mixed-wood of Alberta. *Canadian Journal of Forest Research*, 32, 1784–1800, https://doi.org/10.1139/x02-092

6.27. Retain wildlife corridors in logged areas

https://www.conservationevidence.com/actions/2651

- **Two studies** evaluated the effects on mammals of retaining wildlife corridors in logged areas. One study was in Australia[1] and one was in Canada[2].

COMMUNITY RESPONSE (0 STUDIES)

POPULATION RESPONSE (0 STUDIES)

BEHAVIOUR (2 STUDIES)

- **Use (2 studies):** A replicated study in Australia[1] found that corridors of trees, retained after harvesting, supported seven species of arboreal marsupial. A replicated, controlled study in Canada[2] found that lines of woody debris through clearcut areas that were connected to adjacent forest were not used more by red-backed voles than were isolated lines of woody debris.

Background

Corridors are areas of habitat that are contiguous or isolated (i.e. linkages or stepping stones) that enable species to disperse and migrate through the landscape. In a managed forest environment, corridors may enable recolonization of isolated forest blocks. This intervention includes corridors of natural unharvested vegetation and of cover provided by arrangement of felling debris.

A replicated study (year not stated) of forest at 49 sites in Victoria, Australia (1) found that linear corridors of unharvested trees retained after tree harvesting operations supported seven species of arboreal marsupial. From 402 tree hollows surveyed, 69 arboreal marsupials were recorded, at 54 trees. Greater glider *Petauroides volans* and mountain brushtail possum *Trichosurus caninus* were the most frequently recorded species, accounting for 78% of observations. Sites were chosen where forest had regrown for around 50 years, following wildfires in 1939, and then been felled years <4 years before mammal observations, but leaving a linear strip. Strips were 125–762 m long and had average widths of 30–264 m. Forty-three strips comprised *Eucalyptus regnans* stands and six were of *Eucalyptus delegatensis*. Strips had 1–29 trees with hollows. Marsupial occupation of tree hollows was determined by direct observations.

A replicated, controlled study in 2010–2012 of forest at three sites in British Colombia, Canada (2) found that following tree harvesting, rows of woody debris connected to adjacent forest were not used more by red-backed voles *Myodes gapperi* than were isolated rows of woody debris. The average number of voles/trapping session in rows of woody debris attached to forest (9.0) did not differ from the number in those that were isolated (9.3). However, both had more voles than did unharvested forest (4.4). Seventeen plots were spread across three sites of 42–47 ha extent. Eight plots contained rows of woody debris attached to forest edge, six had isolated woody debris rows in clearcut areas and three were unharvested mature or old-growth forest. Plots averaged 0.23–0.40 km apart. Rows of woody debris averaged 136–344 m long, 1–3 m high and 6–9 m diameter or width. Felling and establishment of rows of woody debris occurred in autumn 2009. Voles were sampled using Longworth live traps, at 4-week intervals (two sites) or 4–8-week intervals (one site), from May to October 2010–2012. Traps were set for one day and two nights each time.

(1) Lindenmayer D.B., Cunningham R.B., & Donnelly C.F. (1993) The conservation of arboreal marsupials in the montane ash forests of the central highlands of Victoria, south-east Australia, iv. the presence and abundance of arboreal marsupials in retained linear habitats (wildlife corridors) within logged forest. *Biological Conservation*, 66, 207–221.

(2) Sullivan T.P. & Sullivan D.S. (2014) Responses of red-backed voles (*Myodes gapperi*) to windrows of woody debris along forest–clearcut edges. *Wildlife Research*, 41, 212–221, https://doi.org/10.1071/wr14050

6.28. Thin trees within forest

https://www.conservationevidence.com/actions/2650

- **Twelve studies** evaluated the effects on mammals of thinning trees within forests. Six studies were in Canada[2,4,8–11] and six were in the USA[1,3,5,6,7,12].

COMMUNITY RESPONSE (2 STUDIES)

- **Species richness (2 studies):** A replicated, site comparison study in the USA[1] found that in thinned tree forest stands, there was similar mammal species richness compared to in unthinned stands. A replicated, controlled study in Canada[8] found that thinning of regenerating lodgepole pine did not result in greater small mammal species richness 12–14 years later.

POPULATION RESPONSE (8 STUDIES)

- **Abundance (8 studies):** Two of eight replicated or replicated and controlled, site comparison studies, in the USA[1,3,5,6,7,12] and Canada[4,8], found that thinning trees within forests lead to higher numbers of small mammals[1,5,7]. Two studies showed increases for some, but not all, small mammal species[3,6] with a further study showing an increase for one of two squirrel species in response to at least some forest thinning treatments[4].

The other two studies showed no increases in abundances of small mammals[8] or northern flying squirrels[12] between 12 and 14 years after thinning.

BEHAVIOUR (4 STUDIES)

- **Use (4 studies):** Three of four controlled and comparison studies (three also replicated, one randomized) in Canada[2,9,10,11] found that thinning trees within forests did not lead to greater use of areas by mule deer[9,10,11], moose[9,10,11] or snowshoe hares[10,11]. The other study found that a thinned area was used more by white-tailed deer than was unthinned forest[2].

Background

Thinning is a forestry operation that involves removing some trees in order to allow remaining trees to grow faster, or straighter or otherwise to produce better quality timber. It may especially be applied in young forest, a few years after onset of regeneration or planting. Thinning increases light that reaches the forest floor, potentially adding to habitat diversity, and may enable remaining trees to produce higher quality forage for herbivores.

The evidence summarised for this intervention includes one case where trees were selectively thinned to increase overwinter browse availability for deer and one where combinations of thinning and felling of groups of trees were combined. See also *Use thinning of forest instead of clearcutting.*

A replicated, site comparison study in 1990–1991 of aspen *Populus tremuloides* forest at four sites in Minnesota, USA (1) found that in thinned tree stands, there was a greater abundance of small mammals, but a similar species richness compared to in unthinned stands. The average yearly site abundance of small mammals was greater in thinned stands (12–29 individuals/grid) than in unthinned stands (9–19 individuals/grid). Species richness did not differ between stand treatments (thinned: 2.8–5.3 species/grid; unthinned: 3.0–5.7 species/grid). Aspen stands at four sites had been growing for 9–11 years at time of thinning. Two had been thinned one year prior to sampling, one seven years previously and one 11 years previously. Unthinned stands were also surveyed at each site. Stands were 6–74 ha in extent. Small mammals were surveyed using snap traps, over two nights and one day, in July–September 1990 and 1991. Stands had 2–7 grids, of 64 traps each.

A site comparison study in 1996 of forest in Quebec, Canada (2) found that, following tree thinning through a partial forest cut aimed at increasing browse availability, white-tailed deer *Odocoileus virginianus* made proportionally greater use of the cut area than of the forest as a whole. Deer use of the cut area (estimated at 15,170 deer-days/km²) was higher than in the forest as a whole (estimated 2,808 deer-days/km²). However, deer did not move home ranges and only animals whose ranges overlapped the cut area used it. A partial forest cut, across 43 ha, was made in January–February 1996. This thinned the forest by removing approximately 40% of deciduous tree stems (with conifers and understorey trees retained). Deer use of the cut area was determined by counting pellet groups, on 27 and 28 April 1996, in eighty-four 2 × 40-m plots. This was compared with estimated pellet density in the whole forest area (total 25 km²) that was based on pellet production from an estimate of the overall deer population. Habitat selections of 30 individual deer were monitored by radiotracking, in January–April 1996.

A replicated, controlled, before-and-after study in 1994–1996 of four coniferous forest sites and a replicated, site comparison study in 1995–1996 of eight coniferous forest sites, all in Oregon, USA (3) found that thinning trees increased abundances of some small mammal species. Out of 12 species, abundances of three, deer mouse *Peromyscus maniculatus*, creeping vole *Microtus oregoni* and white-footed vole *Arborimus albipes*, increased in thinned plots during the two years post-thinning relative to in unthinned plots. Pacific jumping mouse *Zapus trinotatus* increased in thinned plots relative to in unthinned plots between the first and second years post-thinning. Seven species had similar abundances in each treatment. Western red-backed vole *Clethrionomys californicus* was less common in thinned than in unthinned plots. Capture rates did not significantly differ between plots before thinning. See paper for data. Of nine species, five, Pacific shrew *Sorex pacificus*, Trowbridge's shrew *Sorex trowbridgii*, vagrant shrew *Sorex vagrans*, creeping vole and Pacific jumping mouse, were more abundant in plots thinned 7–24 years previously than in unthinned plots. See paper for data. Four sites, each with three 35–45-year-old Douglas-fir stands (26–40 ha/stand) were studied. Two stands/site were thinned in 1994–1995 (to averages of 193–267 trees/ha) and one was unthinned (average 500 trees/ha). Also, at eight pairs of stands, 52–100 years old and <1 mile apart, one stand (10–28 ha) had been thinned 7–24 years before surveying and one (20–73 ha) was unthinned. Small mammals were surveyed within the controlled study using pitfall traps for six weeks/year in 1994 (before thinning) and in 1995 and 1996 (after thinning). In the site comparison study, pitfall traps were operated for 40 consecutive days in each 1995 and 1996.

A replicated, controlled study in 2000–2002 of three coniferous forest sites in British Columbia, Canada (4) found thinning of lodgepole pine *Pinus contorta* stands resulted in higher numbers of northern flying squirrels *Glaucomys sabrinus* when resultant tree density was high, whilst thinning did not affect abundances of red squirrels *Tamiasciurus hudsonicus*. Average northern flying squirrel abundance was highest in thinned stands where remaining trees were at high density (4.6 squirrels/stand), intermediate in medium density stands (3.3/stand) and lowest in low density (1.3/stand) and unthinned (1.8/stand) stands. Red squirrel abundance did not differ between treatments (high density: 10.8/stand; medium density: 9.7/stand; low density: 13.5/stand; unthinned: 11.3/stand). In each of three sites, four forest stands, regenerating following felling and/or wildfire in 1960–1972, were studied. In 1988–1989, one stand each in each site was thinned to approximately 500 (low), 1,000 (medium), and 2,000 (high) stems/ha and one was unthinned (with 4,700–6,000 stems/ha in 1988). Squirrels were surveyed using Tomahawk live traps, at 4-week intervals, from May–October 2000 and 2001 and at 8-week intervals in 2002. One trapping grid (9 ha, 50 traps) was located in each stand.

A randomized, replicated, controlled, before-and-after study in 1994–2001 in a pine and oak forest area in Missouri, USA (5) found that thinning and partial harvesting of trees led to a higher abundance of *Peromyscus* mice spp. Two to five years after harvesting, the annual average number of mice caught in uneven-aged harvesting compartments, where single trees and small groups were felled (8.5–27.0 mice) and even-aged harvesting compartments, involving limited clearcutting and thinning (11.4–31.5 mice) were higher than in uncut compartments (5.9–10.0 mice). Catch data from two *Peromyscus* spp. were combined. Mice were live-trapped, in two blocks of 144 traps each, in nine compartments (312–514 ha), over six nights each year in April or May of 1994–1995 and 1998–2001. Compartments were grouped in three replicate blocks. Uneven-aged harvesting (three compartments) involved cutting single trees and small groups. Even-aged harvesting (three compartments) involved clearcutting and thinning 10–15 % of trees. Three compartments were uncut. Harvesting was carried out in 1996. Biomass removal was similar between harvesting treatments.

A replicated, site comparison study in 2000–2001 of coniferous forest across seven townships in Maine, USA (6) found that thinned regrowing forest stands had more red-backed voles *Clethrionomys gapperi* and masked shrews *Sorex cinereus*, but not deer mouse *Peromyscus maniculatus* or short-tailed shrew *Blarina brevicauda* than did unthinned stands. More red-backed voles were caught in thinned (9.1/survey) than in unthinned (3.8/survey) stands. The same pattern held for masked shrew (6.8 vs 1.2). No significant abundance differences were detected for deer mouse (3.6 vs 4.4) or short-tailed shrew (6.0 vs 4.4). Twenty-four stands were felled in 1967–1983, herbicide-treated in 1977–1988 and thinned in 1984–1999. Thirteen stands were felled in 1974–1982 and herbicide-treated in 1982–1988 but not thinned. Small mammals were surveyed at 64 live-trapping stations/stand for six consecutive 24-h periods during June–August 2000 and again in 2001.

A replicated, controlled, before-and-after study in 1991–1997 of two second-growth forests in Arkansas and Oklahoma, USA (7) found that thinning trees increased small mammal abundance relative to unthinned stands, but not to clearcut stands. Before management, average small mammal abundances were similar between stands planned for different treatments (thinning: 2.4 small mammals/100 trap nights; no thinning: 2.5; clearcut: 0.9). After management, more small mammals were caught in thinned stands (9.3/100 trap nights) than in unthinned stands (1.7) but a similar number was caught in clearcut stands (10.7). In each of four blocks of second-growth forest (59–69 years old at start of study), one stand was thinned, retaining 49–99 of the largest trees/ha, one was not thinned and one was clearcut. Tree removal was conducted in summer 1993. Stand extent was 13–28 ha. Small mammals were surveyed using an average of 67 Sherman live traps/stand, pre-management in 1991 and 1992, and post-management in 1995, 1997 and 1999. Traps were operated for seven consecutive nights during winter (December–January).

A replicated, controlled study in 2000–2002 of three coniferous forests in British Columbia, Canada (8) found that thinning of regenerating lodgepole pine *Pinus contorta* stands did not result in higher small mammal abundance or species richness 12–14 years later. Small mammal abundance varied between years but not between treatments (low remaining tree density: 13–26 individuals/stand; medium density: 11–23 individuals/stand; high density: 15–27 individuals/stand; unthinned: 10–26 individuals/stand). Similarly, species richness did not differ between treatments (low tree density: 2.3–4.3 species/stand; medium density: 3.7–3.9 species/stand; high density: 3.0–3.4 species/ stand; unthinned: 2.5–3.7 species/stand). In each of three sites, four forest stands, regenerating following felling and/or wildfire in 1960– 1972, were studied. In 1988–1989, one stand each in each site was thinned to approximately 500 (low), 1,000 (medium), and 2,000 (high) stems/ ha and one was unthinned (with 4,700–6,000 stems/ha in 1988). Small mammals were live-trapped, over two nights and one day, at 4-week intervals, from May–October of 2000, 2001, and 2002. One trapping grid (1 ha, 49 trap stations) was located in each stand.

A replicated, site comparison study in 1999–2003 of two pine forest sites in British Columbia, Canada (9) found that thinning lodgepole pine *Pinus contorta* stands did not lead to greater use by mule deer *Odocoileus hemionus* or moose *Alces alces*. The average number of mule deer faecal pellet groups did not differ between thinned and unthinned stands in summer (thinned stands: 219/ha; unthinned stands: 73/ha) or winter (thinned: 378/ha; unthinned: 190/ha). Similarly, there was no significant difference between stands in the quantity of moose faecal pellet groups in summer (thinned: 7/ha; unthinnged: 7/ha) or winter (thinned: 16/ha; unthinned: 30 pellet groups/ha). Across the two sites, three forest stands in total were thinned in 1993 (to 1,000 stems/ha) and three were left unthinned. Stands had been clearcut in 1978–1982 and lodgepole pine had regenerated naturally. Faecal pellet groups were counted over a two-week period, five times in May and four times in October, in 55–145 plots/stands (plots were circles of 1.26 m radius), in 1999–2003.

A randomized, replicated, controlled study in 2000–2004 of five second-growth lodgepole pine *Pinus contorta* forests in British Colombia, Canada (10) found that in thinned stands, the abundances of snowshoe hare *Lepus americanus*, mule deer *Odocoileus hemionus* and moose *Alces alces* were not greater than in unthinned stands. Faecal pellet counts for snowshoe hares were not significantly different between low-density thinned plots (70,000 pellets/ha), medium-density thinned plots (60,000 pellets/ha), high-density thinned plots (38,000 pellets/ha) or unthinned plots (13,000 pellets/ha). Similarly, despite large count variations, no significant differences between treatments were detected for mule deer (low: 259 pellet clumps/ha; medium: 79; high: 33; unthinned: 13) or moose (low: 365 pellet clumps/ha; medium: 133; high: 188; unthinned: 93). In each of three areas, four stands (17–27 years old) were studied. One stand each was thinned to low (approximately 500 stems/ha), medium (1,000 stems/ha) and high (2,000 stems/ha) tree density in 1988–1989. One was unthinned (4,700–6,000 stems/ha at time of thinning). Treatments were assigned randomly within study areas. Mammal faecal pellets and clumps were surveyed in one hundred 5-m² plots in each stand. Plots were cleared of pellets in early October 2000. Pellets were counted in spring 2004.

A replicated, controlled study in 2003–2008 of four lodgepole pine *Pinus contorta* forests in British Colombia, Canada (11) found that thinning did not increase forest stand use by snowshoe hares *Lepus americanus*, mule deer *Odocoileus hemionus* or moose *Alces alces*, relative to unthinned stands, 15–20 years after thinning. Hare faecal pellet density did not differ significantly between low (26,000 pellets/ha), medium (25,000 pellets/ha) or high (49,000 pellets/ha) density thinning or unthinned forest (106,000 pellets/ha). Similarly, there were no significant differences between treatments for mule deer (low: 495 pellet-groups/ha; medium: 500; high: 447; unthinned: 195) or moose (low: 190 pellet-groups/ha; medium: 88; high: 131; unthinned: 71). Naturally regenerated young lodgepole pine stands were studied at four sites. Stands were thinned, in 1988–1993, to target densities of 500 (low), 1,000 (medium) and 2,000 (high) stems/ha. Unthinned stands had >3,000 stems/ha. Mammal faecal pellets and pellet-groups were surveyed in 5-m² plots (55–145 plots/stand). Plots were cleared of pellets in autumn 2003. New pellets and pellet-groups were counted in spring 2008.

A replicated, controlled study in 2007–2008 of a Douglas-fir *Pseudotsuga menziesii* forest in Oregon, USA (12) found that, 11–13 years after thinning, northern flying squirrels *Glaucomys sabrinus* were not more numerous in thinned than in unthinned stands. Flying squirrel density was lower in thinned (0.4 squirrels/ha) than unthinned (2.0/ha) stands. Among thinned stands, there were more flying squirrels in those that were lightly thinned with gaps (0.5/ha) than in heavily thinned stands (0.2/ha). The numbers in lightly thinned stands without gaps (0.4/ha) did not differ significantly from that in lightly thinned stands with gaps. Treatments were applied to 16 stands (15–53 ha), in four blocks (2.5–21 km apart), of 55–65-year-old forest, in 1994–1997. In each block, treatments were heavy thinning (to 125–137 trees/ha), light thinning (250–275 trees/ha), light thinning with gaps (as light thinning but also with 20% of the stand harvested leaving 0.2-ha gaps) and unthinned. Flying squirrels were surveyed using 100 traps/stand for four nights and three days, between late September and late November, in 2007 and 2008.

(1) Christian D.P., Reuvers-House M., Hanowski J.M., Niemi G.J., Blake J.G. & Berguson W.E. (1996) Effects of mechanical strip thinning of aspen on small mammals and breeding birds in northern Minnesota, U.S.A. *Canadian Journal of Forest Research*, 26, 1284–1294.

(2) St-Louis A., Ouellet J.-P, Crête M. Maltais J. & Huot J. (2000) Effects of partial cutting in winter on white-tailed deer. *Canadian Journal of Forest Research*, 30, 655–661.

(3) Suzuki N. & Hayes J.P. (2003) Effects of thinning on small mammals in Oregon coastal forests. *The Journal of Wildlife Management*, 67, 352–371, https://doi.org/10.2307/3802777

(4) Ransome D.B., Lindgren P.M.F., Sullivan D.S. & Sullivan T.P. (2004) Long-term responses of ecosystem components to stand thinning in young lodgepole pine forest. I. Population dynamics of northern flying squirrels and red squirrels. *Forest Ecology and Management*, 202, 355–367, https://doi.org/10.1016/j.foreco.2004.08.002

(5) Fantz D.K. & Renken R.B. (2005) Short-term landscape-scale effects of forest management on *Peromyscus* spp. mice within Missouri Ozark forests. *Wildlife Society Bulletin*, 33, 293–301, https://doi.org/10.2193/0091-7648(2005)33[293:sleofm]2.0.co;2

(6) Homyack J.A., Harrison D.J. & Krohn WB. (2005) Long-term effects of precommercial thinning on small mammals in northern Maine. *Forest Ecology and Management*, 205, 43–57, https://doi.org/10.1016/j.foreco.2004.10.005

(7) Perry R.W. & Thill R.E. (2005) Small-mammal responses to pine regeneration treatments in the Ouachita Mountains of Arkansas and Oklahoma, USA. *Forest Ecology and Management*, 219, 81–94, https://doi.org/10.1016/j.foreco.2004.10.005

(8) Sullivan T.P., Sullivan D.S., Lindgren P.M.F. & Ransome D.B. (2005) Long-term responses of ecosystem components to stand thinning in young lodgepole pine forest II. Diversity and population dynamics of forest floor small mammals. *Forest Ecology and Management*, 205, 1–14.

(9) Sullivan T.P., Sullivan D.S., Lindgren P.M.F. & Ransome D.B. (2006) Influence of repeated fertilization on forest ecosystems: relative habitat use by mule deer and moose. *Canadian Journal of Forest Research*, 36, 1395–1406, https://doi.org/10.1139/x06-033

(10) Sullivan T.P., Sullivan D.S., Lindgren P.M.F. & Ransome D.B. (2007) Long-term responses of ecosystem components to stand thinning in young lodgepole pine forest: IV. Relative habitat use by mammalian herbivores. *Forest Ecology and Management*, 240, 32–41.

(11) Sullivan T.P., Sullivan D.S., Lindgren P.M.F. & Ransome D.B. (2010) Long-term responses of mammalian herbivores to stand thinning and fertilization in young lodgepole pine (*Pinus contorta* var. latifolia) forest. *Canadian Journal of Forest Research*, 40, 2302–2312, https://doi.org/10.1139/x10-173

(12) Manning T., Hagar J.C. & McComb B.C. (2012) Thinning of young Douglas-fir forests decreases density of northern flying squirrels in the Oregon Cascades. *Forest Ecology and Management*, 264, 115–124, https://doi.org/10.1016/j.foreco.2011.09.043

6.29. Apply fertilizer to trees

https://www.conservationevidence.com/actions/2649

- **Three studies** evaluated the effects on mammals of applying fertilizer to trees. All three studies were in Canada[1,2,3].

COMMUNITY RESPONSE (0 STUDIES)

POPULATION RESPONSE (0 STUDIES)

BEHAVIOUR (3 STUDIES)

- **Use (3 studies):** One of three replicated studies (including one controlled study and two site comparison studies), in

Canada[1,2,3], found that thinned forest stands to which fertilizer was applied were used more by snowshoe hares in winter but not in summer over the short-term[2]. The other studies found that forest stands to which fertilizer was applied were not more used by snowshoe hares in the longer term[3] or by mule deer or moose[1,3].

Background

Chemical fertilizers (nitrogen, phosphorus and potassium) are frequently applied to newly planted or regenerating trees. They increase soil fertility and may, therefore, enhance tree growth and nutritional content of foliage available to browsing herbivores. This could increase use of such areas by herbivores, leading to enhanced survival or abundance of these species.

A replicated, site comparison study in 1999–2003, in two pine forest sites in British Columbia, Canada (1, same experimental set-up as 2 and 3) found that applying fertilizer to thinned stands of lodgepole pines *Pinus contorta* did not increase their use by mule deer *Odocoileus hemionus* or moose *Alces alces*. Mule deer use of stands did not differ significantly between fertilized and unfertilized stands in summer (fertilized: 185–700 faecal pellet groups/ha; unfertilized: 5–276) or winter (fertilized: 392–472 faecal pellet groups/ha; unfertilized: 111–261). Similarly, for moose, there was no significant difference in stand use in summer (fertilized: 13–87 faecal pellet groups/ha; unfertilized: 3–31) or winter (fertilized: 29–90 faecal pellet groups/ha; unfertilized: 21–66). Across the two sites, six forest stands in total were felled in 1978–1982 and lodgepole pine then regenerated naturally. The stands were thinned in 1993 (to 1,000 stems/ha). Three stands were then fertilized six times in 1994–2003. Faecal pellet groups were counted over two-week periods, five times in May and four times in October, in 1999–2003, in 55–145 plots/stands (plots were circles of 1.3 m radius).

A replicated, controlled study, in 1999–2003, of three lodgepole pine *Pinus contorta* forests in British Columbia, Canada (2, same experimental set-up as 1 and 3) found that adding fertilizer to thinned forest stands increased their use by snowshoe hares *Lepus americanus* in winter but not in summer. In winter, the average density of hare faecal pellets across fertilized stands (7,000–62,000/ha) was higher than that across unfertilized stands (1,400–28,000/ha). In summer, there was no significant difference in the density of hare faecal pellets between fertilized stands (800–21,000/ha) and unfertilized stands (600–11,000/ ha). Within each of the three sites, blocks of commercially grown lodgepole pines were thinned to 2,000, 1,000, 500 and 250 stems/ha in 1993. Half of each stand was fertilized five times in 1994–2003. Hare faecal pellets on 5-m² permanent plots were counted in summer (May–September) and winter (October–April) 1999–2003.

A replicated, site comparison study in 2003–2008 of two lodgepole pine *Pinus contorta* forests in British Colombia, Canada (3; same experimental set-up as 1 and 2) found that repeated fertilization of thinned forest stands did not increase their use by snowshoe hares *Lepus americanus*, mule deer *Odocoileus hemionus* or moose *Alces alces*. Hare faecal pellet density and mule deer and moose pellet-group density did not differ between fertilized and unfertilized stands (data not presented). Naturally regenerated young lodgepole pine stands were studied at two sites. At each site, two stands were thinned, in 1993, to each of 2,000, 1,000, 500 and 250 stems/ha. Treatment stands were fertilized five times, in 1994–2003, using fertilizer blends which included 100–200 kg nitrogen/ha. Control stands were not fertilized. Mammal faecal pellets and pellet-groups were surveyed in 5-m² plots (55–145 plots/stand). Plots were cleared of pellets in autumn 2003. New pellets and pellet-groups were counted in spring 2008.

(1) Sullivan T.P., Sullivan D.S., Lindgren P.M.F. & Ransome DB (2006) Influence of repeated fertilization on forest ecosystems: relative habitat use by mule deer and moose. *Canadian Journal of Forest Research*, 36, 1395–1406, https:// doi.org/10.1139/x06-033

(2) Sullivan T.P., Sullivan D.S., Lindgren P.M.F. & Ransome D.B. (2006) Influence of repeated fertilization on forest ecosystems: relative habitat use by snowshoe hares (*Lepus americanus*). *Canadian Journal of Forest Research*, 36, 2080–2089, https://doi.org/10.1139/x06-093

(3) Sullivan T.P., Sullivan D.S., Lindgren P.M.F. & Ransome D.B. (2010) Long-term responses of mammalian herbivores to stand thinning and fertilization in young lodgepole pine (*Pinus contorta* var. *latifolia*) forest. *Canadian Journal of Forest Research*, 40, 2302–2312, https://doi.org/10.1139/x10-173

6.30. Fell trees in groups, leaving surrounding forest unharvested

https://www.conservationevidence.com/actions/2648

- **Three studies** evaluated the effects on mammals of felling trees in groups, leaving surrounding forest unharvested. Two studies were in Canada[1,2] and one was in the UK[3].

COMMUNITY RESPONSE (0 STUDIES)

POPULATION RESPONSE (3 STUDIES)

- **Abundance (2 studies):** One of two replicated studies (including one controlled study and one site comparison study), in Canada[1,2], found that felling groups of trees within otherwise undisturbed stands increased the abundance of one of four small mammal species relative to clearcutting. The other study found that none of four small mammal species monitored showed abundance increases.

- **Survival (1 study):** A study in the UK[3] found that when trees were felled in large groups with surrounding forest unaffected, there was less damage to artificial hazel dormouse nests than when trees were felled in small groups or thinned throughout.

BEHAVIOUR (0 STUDIES)

Background

When timber harvesting or woodland management operations take place, trees may be clearfelled across a large area, thinned throughout the woodland or cut in patches, leaving surrounding forest unharvested. Felling in groups will produce a lower timber harvest than clearfelling but will leave more forest unaffected, which may help to sustain populations of some species. It will also affect less of the woodland area overall than does thinning of trees or selecting individual trees scattered throughout the forest to fell.

A replicated, controlled study in 1994–1997 of Douglas-fir *Pseudotsuga menziesii* forest in British Colombia, Canada (1) found that felling groups of trees within otherwise undisturbed stands increased southern red-backed vole *Myodes gapperi* abundance in some years relative to clearcutting but did not increase abundances of three other small mammal species. There were more southern red-backed voles in the third and fourth year after felling in group cut stands (7–14/stand) than in clearcuts (0.3–0.7/stand) but similar numbers between treatments in the first two years (group cut: 27–51/stand; clearcut: 13–34/stand). There were no differences between treatments for deer mouse *Peromyscus maniculatus* (group cut: 2–13/stand; clearcut: 6–21) or northwestern chipmunk *Tamias amoenus* (group cut: 1–8/stand; clearcut: 0.3–6/stand). There were fewer meadow voles *Microtus pennsylvanicus* in 20% group cut stands (1–3/stand) than in 50% group cut stands (0.8–4/stand) or clearcut stands (3–14/stand). Forest stands (20–25 ha) were partially harvested in winter 1993/94. Two each had 20% volume removed by cutting patches of 0.1–1.6 ha and 50% volume removed by cutting patches of 0.1–1.6 ha. Abundances across these stands were compared with that in two clearcuts of 1.6 ha. Small mammals were sampled by live-trapping at 2–4-week intervals, from May–October in 1994, 1995, and 1996 and from April–May 1997.

A replicated, site comparison study in 2006 in four forest sites in British Columbia, Canada (2) found that harvesting trees in 1 ha blocks did not result in higher small mammal abundance compared to clearcutting large areas. The average number of red-backed voles *Myodes gapperi* caught in 1-ha cuts (19.0 individuals) was not significantly different to that caught in clearcuts (8.4 individuals). Numbers caught also did not differ significantly between felling types for dusky shrew *Sorex monticolus* (1-ha cuts: 34.0 individuals; clearcuts: 44.3 individuals), deer mouse *Peromyscus maniculatus* (1-ha cuts: 9.6 individuals; clearcuts: 11.6 individuals) or common shrew *Sorex cinereus* (1-ha cuts: 7.3; clearcuts: 7.0). A 1-ha area was harvested in each of four sites. These were compared with two large (>30 ha) clearcut areas. Trees were harvested in 1992–1993. Small mammals were live-trapped every three weeks in June–October 2006 (five sessions). Traps were operated for two nights and, if daytime temperatures were ≤25°C, the intervening day.

A study in 2003 of a forest in Worcestershire, UK (3) found that, when trees were felled in large groups with surrounding forest unaffected, there was less damage to artificial hazel dormouse *Muscardinus avellanarius* nests than when trees were felled in small groups or thinned throughout. A lower proportion of artificial nests was damaged during large group felling (31%) than small group felling (62–66%) or thinning (73%). Non-native Corsican pines *Pinus nigra* were cleared from one third of the area of each of four plots (3 ha each) in a forest undergoing restoration to ancient woodland vegetation. Plot treatments, executed in late autumn/winter 2003, were clearance of small groups (12–14 trees) using chainsaws, clearance of small groups using a mechanised harvester, thinning throughout using a harvester and large group fells (c.0.4 ha each) using a harvester. Artificial dormouse nests comprised spheres of florists' 'oasis' (7–10 cm diameter) on the ground mimicking natural nests.

(1) Klenner W. & Sullivan T.P. (2009) Partial and clearcut harvesting of dry Douglas-fir forests: Implications for small mammal communities. *Forest Ecology and Management*, 257, 1078–1086, https://doi.org/10.1016/j.foreco.2008.11.012

(2) Ransome D.B., Lindgren P.M.F., Waterhouse M.J., Armleder H.M. & Sullivan T.P. (2009) Small-mammal response to group-selection silvicultural systems in Engelmann spruce — subalpine fir forests 14 years postharvest. *Canadian Journal of Forest Research*, 39, 1698–1708, https://doi.org/10.1139/x09-095

(3) Trout R.C., Brooks S.E., Rudlin P. & Neil J. (2012) The effects of restoring a conifer plantation on an Ancient Woodland Site (PAWS) in the UK on the habitat and local population of the hazel dormouse (*Muscardinus avellanarius*). *European Journal of Wildlife Research*, 58, 635–643, https://doi.org/10.1007/s10344-012-0611-9

6.31. Coppice trees

https://www.conservationevidence.com/actions/2635

- We found no studies that evaluated the effects of coppicing trees on mammals.

'We found no studies' means that we have not yet found any studies that have directly evaluated this intervention during our systematic journal and

report searches. Therefore, we have no evidence to indicate whether or not the intervention has any desirable or harmful effects.

Background

Coppicing is a management practice typical of Eurasian northern temperate zone deciduous woodlands and wood pastures, in which stems of tree species, such as hazel *Corylus avellana* and sweet chestnut *Castanea sativa*, are cut near ground level once every few years, often in defined coppice compartments. These then regrow from the cut 'stool' giving a sustainable yield of woody material harvested on a rotational basis. Coppicing maintains a mosaic of woodland areas with differing amounts of daylight reaching the forest floor and, therefore, promotes a variety of ground vegetation conditions. This may benefit mammals that require either open canopy woodland or a mix of open and more closed woodland in close proximity. Coppicing has declined over the last century and some former coppice woodlands are no longer actively managed.

6.32. Allow forest to regenerate naturally following logging

https://www.conservationevidence.com/actions/2634

- **One study** evaluated the effects on mammals of allowing forest to regenerate naturally following logging. This study was in Canada[1].

COMMUNITY RESPONSE (0 STUDIES)

POPULATION RESPONSE (1 STUDY)

- **Abundance (1 study):** A replicated, site comparison study in Canada[1] found that, natural forest regeneration increased moose numbers relative to more intensive management in the short-to medium-term but not in the longer term.

BEHAVIOUR (0 STUDIES)

Background

After logging of forests, cut areas may be left to regenerate naturally or may be subject to management aimed at accelerating tree planting. Allowing natural regeneration may facilitate formation of more natural vegetation which could improve habitat and resource availability for mammals.

A replicated, site comparison study, in 2008–2009, on three large adjacent coniferous forest sites in Ontario, Canada (1) found that, following clearcutting, large-scale natural forest regeneration increased moose *Alces alces* numbers relative to more intensive silvicultural practices (mechanical ground preparation, replanting and herbicide application) 10 years after felling but not 30 years after felling. The number of moose faecal pellet clumps was positively correlated with the extent of naturally regenerating forest that was felled 10 years previously in areas of 10, 20 and 40 km^2 around the stand, but not with the extent subject to more intensive silviculture, nor with the extent felled 30 years previously and subject to either management practice (data not presented). Ten forest stands were felled 10 years previously (five regenerating naturally and five subject to intensive silviculture) and ten were felled 30 years previously (five regenerating naturally and five subject to intensive silviculture). Moose faecal pellet clumps were counted within five circles of 5.65 m radius in each stand between July and early September of 2008 or 2009.

(1) Baon J.J., McLaren B.E. & Malcolm J.R. (2011) Influence of post-harvest silviculture on understory vegetation: Implications for forage in a multiungulate system. *Forest Ecology and Management*, 262, 1704–1712, https://doi.org/10.1016/j.foreco.2011.07.022

6.33. Harvest timber outside mammal reproduction period

https://www.conservationevidence.com/actions/2633

- We found no studies that evaluated the effects of harvesting timber outside the mammal reproduction period.

'We found no studies' means that we have not yet found any studies that have directly evaluated this intervention during our systematic journal and report searches. Therefore, we have no evidence to indicate whether or not the intervention has any desirable or harmful effects.

Background

Tree-felling poses risks to woodland-dwelling mammals. For species with young in a nest or den, tree felling could cause death of these young through injury or abandonment. Planning timber harvesting for times outside the period when young are at their most vulnerable may reduce such direct casualties of felling operations.

6.34. Control firewood collection in remnant native forest and woodland

https://www.conservationevidence.com/actions/2632

- We found no studies that evaluated the effects on mammals of controlling firewood collection in remnant native forest and woodland.

'We found no studies' means that we have not yet found any studies that have directly evaluated this intervention during our systematic journal and report searches. Therefore, we have no evidence to indicate whether or not the intervention has any desirable or harmful effects.

Background

Firewood is an important fuel for heating and cooking in some homes and communities. However, wood that may be collected as firewood, such as from fallen trees, may provide an important element of the habitat for some forest floor species. This is most likely to be the case in forests that have been least affected by management. Thus, collection of firewood may be controlled in remnant native forests and woodland to benefit woodland biodiversity, including mammals.

6.35. Plant trees following clearfelling

https://www.conservationevidence.com/actions/2631

- **One study** evaluated the effects on mammals of planting trees following clearfelling. This study was in Canada[1].

COMMUNITY RESPONSE (0 STUDIES)

POPULATION RESPONSE (0 STUDIES)

BEHAVIOUR (1 STUDY)

- **Use (1 study):** A replicated, site comparison study in Canada[1] found that forest stands subject to tree planting and herbicide treatment after logging were used more by American martens compared to naturally regenerating stands.

Background

Following felling of tees, for timber harvesting, a range of actions may be employed to accelerate forest regrowth. These include treating with herbicide (to supress herbaceous vegetation) and planting of trees.

A replicated, site comparison study in 2001–2002 of boreal forest stands in Ontario, Canada (1) found that forest stands subject to tree planting and herbicide treatment after logging were used more by American martens *Martes americana* than were naturally regenerating stands. The effects of planting and herbicide use were not separated in the study. Radio-tracked martens made greater use of planted and herbicide-treated stands than they did of naturally regenerating stands (data not presented). However, the live-capture rate of martens in planted and herbicide-treated stands (5.6 martens/100 trap nights) was not significantly different to that in regenerating stands (1.9 martens/100 trap nights). Stands were 35–45 years old and located in a 600-km² forestry area. Forest stands were either regenerating naturally following logging or planted following logging and treated with herbicide. Martens were live-trapped in 2003–2007, and monitored subsequently by radio-tracking.

(1) Thompson I.D., Baker J.A., Jastrebski C., Dacosta J., Fryxell J. & Corbett D. (2008) Effects of post-harvest silviculture on use of boreal forest stands by amphibians and marten in Ontario. *Forestry Chronicle*, 84, 741–747, https://doi.org/10.5558/tfc84741-5

6.36. Use tree tubes/small fences/cages to protect individual trees

https://www.conservationevidence.com/actions/2630

- We found no studies that evaluated the effects of using tree tubes, small fences or cages to protect individual trees from mammals.

'We found no studies' means that we have not yet found any studies that have directly evaluated this intervention during our systematic journal and report searches. Therefore, we have no evidence to indicate whether or not the intervention has any desirable or harmful effects.

Background

A range of mammals, including rodents and ungulates, can cause substantial damage to trees, especially young trees, through browsing activities on foliage and by stripping bark from trees. As well as damage to natural habitats, this can cause financial losses to the forestry industry (Huitu *et al.* 2009). In an attempt to reduce such conflict, trees may be protected from attack using a range of barriers to prevent mammals from accessing them. If successful, this could reduce incentives for carrying out lethal control on these mammals.

Huitu O., Kiljunen N., Korpimäki E., Koskela E., Mappes T., Pietiäinen H., Pöysä H. & Henttonen H. (2009) Density-dependent vole damage in silviculture and associated economic losses at a nationwide scale. *Forest Ecology and Management*, 258, 1219–1224, https://doi.org/10.1016/j.foreco.2009.06.013

6.37. Provide supplementary feed to reduce tree damage

https://www.conservationevidence.com/actions/2629

- **One study** evaluated the effects of providing supplementary feed on the magnitude of tree damage caused by mammals. This study was in USA[1].

COMMUNITY RESPONSE (0 STUDIES)

POPULATION RESPONSE (0 STUDIES)

BEHAVIOUR (0 STUDIES)

OTHER (1 STUDY)

- **Human-wildlife conflict (1 study):** A replicated, randomized, paired sites, controlled, before-and-after study in USA[1] found that supplementary feeding reduced tree damage by black bears.

Background

Supplementary feeding may be offered to reduce the incentive for animals to damage trees when they are in search of food. If the intervention is successful in reducing tree damage, it may reduce incentives for carrying out lethal control of such mammal species.

A replicated, randomized, paired sites, controlled, before-and-after study in 1999–2002 in 14 coniferous forest sites in Washington, USA (1) found that supplementary feeding reduced tree damage caused by black bears *Ursus americanus*. The number of trees damaged by bears in sites where supplementary feeding was used was lower (3–10 trees/year) than in sites where no supplementary feeding was used (15–26 trees/year). When supplementary feeding was stopped at one site, the number of trees damaged by bears increased from 6 to 40/year. In March 1999, in fourteen 16–20-ha sites, bear-damaged trees were marked with paint. Sites with similar amounts of damage were paired. In April 1999, one site/pair was randomly chosen to have two plastic drums containing food pellets placed in it, while the other site had no supplementary food provided. Plastic drums were refilled weekly in April–July with 100 kg of pellets. In the first year, at sites where supplementary feed was provided, beaver *Castor canadensis* carcasses were hung from trees to attract bears. In July 2000, supplementary feeding was stopped at two of the seven sites (results not presented for the second site due to the feeding station not being maintained prior to this). Sites were surveyed for bear damage to trees in July of 1999–2002.

(1) Ziegltrum G. I. (2004) Efficacy of black bear supplemental feeding to reduce conifer damage in western Washington. *The Journal of Wildlife Management*, 68, 470–474, https://doi.org/10.2193/0022-541x(2004)068[0470:eobbsf]2.0 .co;2

7. Threat: Human intrusions and disturbance

Background

In addition to large-scale disturbances from activities such as agriculture, building developments, energy production and biological resource use, disturbance of mammal populations can come from smaller scale human intrusions. This chapter also includes some interventions aimed at reducing human-wildlife conflict where wild terrestrial mammals and humans come into contact. Such interventions, if effective, may reduce motivations or justifications for carrying out lethal control of mammals.

7.1. Use signs or access restrictions to reduce disturbance to mammals

https://www.conservationevidence.com/actions/2325

- **One study** evaluated the effects of using signs or access restrictions to reduce disturbance to mammals. This study was in the USA[1].

COMMUNITY RESPONSE (0 STUDIES)

POPULATION RESPONSE (0 STUDIES)

BEHAVIOUR (1 STUDY)

- **Use (1 study):** A replicated, paired sites, site comparison study in the USA[1] found that removing or closing roads increased use of those areas by black bears.

    https://doi.org/10.11647/OBP.0234.07

Background

Access to areas by people can cause disturbance to some mammals. This may cause them to alter behaviour, including through reducing their use of such areas. To limit this disturbance, access may be restricted, including through using signage or physical barriers.

See also: *Exclude or limit number of visitors to reserves or protected areas.*

A replicated, paired sites, site comparison study in 2006–2009 of a forest in Idaho, USA (1) found that removing or closing roads increased use of those areas by black bears *Ursus americanus*. More bears were detected on former roads that had been removed (4.6 detections/100 camera-trap days) than on paired open roads (0.3). Similarly, there were more on closed than on paired open roads when roads were closed by a barrier (closed: 1.5; open: 0.6 detections/100 camera-trap days) or by a gate (closed: 0.5; open: 0 detections/100 camera-trap days). Eighteen closed roads were paired with open roads. Closed roads included seven removed by reprofiling in the previous 10 years, five closed by barriers and six that were gated. Closed roads were sampled by camera-trapping 1.6 km along from their intersection with the paired open road. Open roads were sampled <100 m along from this intersection. One camera trap was used at each site between 1 April and 30 June and again between 30 August and 3 November, annually in 2006–2009.

(1) Switalski T.A. & Nelson C.R. (2011) Efficacy of road removal for restoring wildlife habitat: Black bear in the Northern Rocky Mountains, USA. *Biological Conservation*, 144, 2666–2673, https://doi.org/10.1016/j.biocon.2011.07.026

7.2. Set minimum distances for approaching mammals

https://www.conservationevidence.com/actions/2327

- We found no studies that evaluated the effects on mammals of setting a minimum permitted distance to which they can be approached.

'We found no studies' means that we have not yet found any studies that have directly evaluated this intervention during our systematic journal and report searches. Therefore, we have no evidence to indicate whether or not the intervention has any desirable or harmful effects.

Background

At some sites, such as at national parks where safaris are a popular means of visitors observing animals, large numbers of people or vehicles closely approaching mammals may cause them disturbance or cause changes in their behaviour. This may restrict areas that these animals use or affect hunting or feeding opportunities. Limits, including through voluntary guidelines, exist in some areas on the minimum distance to which people or vessels may approach sea mammals (e.g. Inman *et al.* 2016). Similar regulation or guidelines may also lessen such potential impacts for mammals.

Inman A., Brooker E., Dolman S., McCann R., Wilson A.M.W. (2016) The use of marine wildlife-watching codes and their role in managing activities within marine protected areas in Scotland. *Ocean & Coastal Management*, 132, 132–142, https://doi.org/10.1016/j.ocecoaman.2016.08.005

7.3. Set maximum number of people/vehicles approaching mammals

https://www.conservationevidence.com/actions/2328

- We found no studies that evaluated the effects on mammals of setting a maximum to the number of people or vehicles permitted to approach mammals.

'We found no studies' means that we have not yet found any studies that have directly evaluated this intervention during our systematic journal and report searches. Therefore, we have no evidence to indicate whether or not the intervention has any desirable or harmful effects.

Background

At some sites, such as at national parks where safaris are a popular means of visitors observing animals, large numbers of people or vehicles approaching mammals may cause them disturbance or cause changes in their behaviour. This may restrict areas that these animals use or affect hunting or feeding opportunities. Setting limits on the numbers of people or vehicles permitted to be in close proximity to such animals may lessen such potential impacts.

7.4. Exclude or limit number of visitors to reserves or protected areas

https://www.conservationevidence.com/actions/2330

- **Five studies** evaluated the effects on mammals of excluding or limiting the number of visitors to reserves or protected areas. Three studies were in the USA[1,2,3], one was in Ecuador[4] and one was in Thailand[5].

COMMUNITY RESPONSE (0 STUDIES)

POPULATION RESPONSE (2 STUDIES)

- **Abundance (1 study):** A site comparison study in Ecuador[4] found that a road with restricted access had a higher population of medium-sized and large mammals compared to a road with unrestricted access.

- **Survival (1 study):** A before-and-after study in the USA[1] found that temporarily restricting visitor access resulted in fewer bears being killed to protect humans.

BEHAVIOUR (3 STUDIES)

- **Use (3 studies):** Three studies (one a before-and-after study), in the USA[2,3] and Thailand[5], found that restricting human access to protected areas resulted in increased use of these areas by grizzly bears[2,3] and leopards[5].

Background

Protected areas are often popular with visitors as they may contain extensive areas suited for outdoor recreation. However, high visitor numbers can damage features that reserves and protected areas are established to protect. Some mammals are shy and are thus deterred by visitors whilst others, such as bears, may come into conflict with human visitors. A policy of excluding, restricting or otherwise limiting human visitors may be put in place to reduce the potential impact of such access on wildlife, including mammals.

A before-and-after study in 1968–1972 in Montana, USA (1) found that temporarily restricting visitor access, along with translocation, awareness raising and enforcement of regulations, resulted in fewer bears being killed to protect humans. After restricting visitor access, the rate of bear killings (1/year) was lower than in the preceding 13 years, when there were no visitor restrictions (1.5/year). Following implementation of visitor restrictions, three bears were also translocated away for visitor safety reasons. In 1968–1972 visitor restrictions, such as temporary trail closures or campsite closures, were imposed following verified reports of human-bear encounters. Numbers of bears killed following restrictions was compared to that prior to implementation of restrictions. The programmme also included awareness raising and policing of adherence to local regulations.

A study in 1984–1988 in a meadow and forest area in Wyoming, USA (2) found that restricting human access resulted in greater use of areas by grizzly bears *Ursus arctos*. Bears were found further from cover during closed and restricted periods (average 293–304 m) than during open periods (average 228 m). Bears were recorded close to campsites more frequently when the campsites were not in use than when they were in use, when sightings were reduced by 67%. Within a 4,850-ha study area, containing 14–23 grizzly bears, meadows and open areas were scanned regularly from a vantage point, for bear and human activity, from May–June through to July–September of 1984–1988. At different periods during this time, the area was classed as open (allowing day-use and overnight camping), restricted (allowing day-use only, but no overnight camping) or closed (no recreational use).

A study in 2006–2009 in temperate forest in a national park in Wyoming, USA (3) found that restricting human access allowed increased use by grizzly bears *Ursus arctos*. When human access was restricted more bears used human recreation areas (9.4–10.8% of satellite collar locations) than when human access was unrestricted (4.4–9.1% of satellite collar locations). During restricted periods, human use was lower (5 recreational users/day) than during unrestricted periods (147 recreational users/day). Human and bear activity was monitored across 81,176 ha, in April–September of 2006–2009. Human recreational areas were areas that humans used more than random areas and covered 7.7% of the study area. Peak human activity times were 08.00–18.59 hrs, during which ≥10% of groups were active. Recreational access was prohibited other than on a small number of backcountry campsites and trails during low tourist season (15 April–30 June) but was unrestricted in peak season (1 July–30 September). Fourteen bears were monitored using satellite collars and 385 recreational groups, totalling 1,341 people, carried GPS loggers while using the area.

A site comparison study in 2005–2006 of forest at three sites in Ecuador (4) found that a road with restricted access had a higher population of medium-sized and large mammals relative to a road with unrestricted access. Differences between sites were not tested for statistical significance. Primates, ungulates and large rodents were more abundant along the restricted access road (98 animals/km²) than they were along the unrestricted access road (48 animals/km²). However, there were more still at an undisturbed site (233 animals/km²). A 142-km-long oil exploration road was constructed in 1992. Road access for outsiders was restricted (details not provided), though the area was occupied by indigenous Waorani people, who settled and hunted along the road. At a different site, an oil exploration road, constructed in 1972, attracted colonists, leading to 4% annual forest loss in its vicinity. A third, undisturbed site was studied. Sites were ≤89 km apart. Mammals >1 kg were surveyed using distance sampling techniques along six 2-km transects at each site, in the morning and evening on eight occasions from April 2005 to July 2006.

A before-and-after study in 2003–2004 of a forest national park in Thailand (5) found that closing the park to visitors resulted in leopards using larger areas of the park. At least six leopards *Panthera pardus* were recorded and the density did not differ between when the park was closed or open to visitors. However, leopards occurred in more locations during the closed period (22 camera-trap locations) than in the open period (13 camera-trap locations). Additionally, there was a 45% higher daily detection rate during the closed than during the open period. Human presence was lower during the closed period (nine photos) than the open period (68 photos). Following flooding in October 2003, the park was closed to visitors. Camera traps were placed for three weeks at each of 72 locations, which were approximately 2 km apart, between November 2003 and January 2004. Previously, the same monitoring strategy had been implemented during March–May 2003, when the park was open to visitors.

(1) Martinka C. J. (1974) Preserving the natural status of grizzlies in Glacier National Park. *Wildlife Society Bulletin*, 2, 13–17.

(2) Gunther K.A. (1990) Visitor impact on grizzly bear activity in Pelican Valley, Yellowstone National Park. *Bears: Their Biology and Management, Eighth International Conference on Bear Research and Management, Victoria, British Columbia, Canada*, 8, 73–78.

(3) Coleman T.H., Schwartz C.C., Gunther K.A. & Creel S. (2013) Grizzly bear and human interaction in Yellowstone National Park: an evaluation of bear management areas. *The Journal of Wildlife Management*, 77, 1311–1320, https://doi.org/10.1002/jwmg.602

(4) Suárez E., Zapata-Ríos G., Utreras V., Strindberg S. & Vargas J. (2013) Controlling access to oil roads protects forest cover, but not wildlife communities: a case study from the rainforest of Yasuní Biosphere Reserve (Ecuador). *Animal Conservation*, 16, 265–274, https://doi.org/10.1111/j.1469-1795.2012.00592.x

(5) Ngoprasert D., Lynam A.J. & Gale G.A. (2017) Effects of temporary closure of a national park on leopard movement and behaviour in tropical Asia. *Mammalian Biology*, 82, 65–73, https://doi.org/10.1016/j.mambio.2016.11.004

7.5. Provide paths to limit extent of disturbance to mammals

https://www.conservationevidence.com/actions/2337

- We found no studies that evaluated the effects on mammals of providing paths to limit the extent of disturbance to mammals.

'We found no studies' means that we have not yet found any studies that have directly evaluated this intervention during our systematic journal and report searches. Therefore, we have no evidence to indicate whether or not the intervention has any desirable or harmful effects.

Background

In open habitats that are popular with human visitors for recreation, providing paths for people to use may reduce the overall area of ground on which mammals are vulnerable to human disturbance.

7.6. Use voluntary agreements with locals to reduce disturbance

https://www.conservationevidence.com/actions/2339

- We found no studies that evaluated the effects on mammals of using voluntary agreements with locals to reduce disturbance.

'We found no studies' means that we have not yet found any studies that have directly evaluated this intervention during our systematic journal and report searches. Therefore, we have no evidence to indicate whether or not the intervention has any desirable or harmful effects.

Background

Human access can be a major cause of disturbance to wild mammals. In some cases, disturbance can be reduced by restricting access using regulations or laws. In other instances, local communities may have long-standing access rights or traditions and voluntary agreements drawn up in consultation with such stakeholders may be attempted.

See also: *Agriculture and aquaculture — Relocate local pastoralist communities to reduce human-wildlife conflict.*

7.7. Habituate mammals to visitors

https://www.conservationevidence.com/actions/2340

- **One study** evaluated the effects of habituating mammals to visitors. This study was in the USA[1].

COMMUNITY RESPONSE (0 STUDIES)

POPULATION RESPONSE (0 STUDIES)

BEHAVIOUR (0 STUDIES)

OTHER (1 STUDY)

- **Human-wildlife conflict (1 study)**: A study in the USA[1] found that brown bears that were highly habituated to humans showed less aggression towards human visitors than did non-habituated bears.

Background

Some mammals can show aggressive behaviour towards humans. This can be a problem especially where the species is one in high demand from humans for opportunities to watch them, and one that is capable of causing serious injury or death to humans if it does attack. This is most likely to involve large charismatic carnivores. Where animals are predictable in their movements, there may be opportunities for habituating them to humans, thus reducing the risk to visitors. This may also, then, reduce instances in which there are pressures on wildlife managers to carry out lethal control of animals that show aggressive behaviours.

A study in 1973–1993 in a riverine and grassland site in Alaska, USA (1) found that brown bears *Ursus arctos* that were highly habituated to humans showed less aggression towards human visitors than did non-habituated bears. Results were not tested for statistical significance. No intense charges were made at people by highly habituated bears compared to eight by bears that were not highly habituated (four by 'wary' and four by 'partially habituated' bears). No human injuries from bears were recorded. All charges, other aggressive displays and bear visits to the campsite were averted by actions such as loud noises or, occasionally, use of non-lethal rubber shot. The programme operated in a 999-km^2 protected area in which bear hunting was prohibited. Bears were habituated by being in proximity to people in non-threatening interactions (see paper for details; numbers of bears not provided). Human visitors away from the campground were restricted to 10/ day, usually from early June to late August. Visitors were in groups, escorted by park staff and were instructed in exhibiting non-threatening behaviour, such as avoiding loud noises or sudden movements.

(1) Aumiller L.D. & Matt C.A. (1994) Management of McNeil River State Game Sanctuary for viewing of brown bears. *Bears: Their Biology and Management,* 9, 51–61.

7.8. Translocate mammals that have habituated to humans (e.g. bears)

https://www.conservationevidence.com/actions/2341

- **Two studies** evaluated the effects of translocating mammals that have habituated to humans. One study was in the USA[1] and one was in the USA and Canada[2].

COMMUNITY RESPONSE (0 STUDIES)

POPULATION RESPONSE (0 STUDIES)

BEHAVIOUR (0 STUDIES)

OTHER (2 STUDIES)

- **Human-wildlife conflict (2 studies):** A study in the USA[1] found that almost half of the translocated 'nuisance' black bears returned to their capture locations. A review of studies in the USA and Canada[2] found that black bears translocated away from sites of conflict with humans were less likely to return to their capture site if translocated as younger bears, over greater distances, or across geographic barriers.

Background

Some animals, such as bears, may exhibit 'nuisance behaviour' that may bring them into conflict with humans. For example, animals may attempt to raid foodstuffs at campgrounds and such individuals may then be perceived as representing a threat to humans. Animals may be translocated away from sites where issues arise, as an alternative to lethal control. Such translocations are deemed to be successful if the animal survives and resumes natural behaviour at the release site, does not return to the capture site and does not exhibit 'nuisance behaviour' elsewhere.

See also: *Residential and Commercial Development — Translocate problem mammals away from residential areas (e.g. habituated bears) to reduce human-wildlife conflict* for situations where habituated animals are removed from established settlements rather than recreation areas. Also see *Use non-lethal methods to deter carnivores from attacking humans*.

A study in 1967–1974 in forest and grassland in a national park straddling Tennessee and North Carolina, USA (1) found that after initial translocation, almost half of the 'nuisance' black bears *Ursus americanus* returned to their capture locations. Of 76 translocated bears, 36 were subsequently caught or seen within ≤8 km of their original capture location at least once (all except two of these were ≤2 km from their capture location). In a 2,072-km² national park with high recreational use, bears were translocated if they exhibited nuisance behaviour (such as accessing human food). Seventy-six bears (66 male, 10 female) were moved a total of 155 times (1–13 times/bear). Bears were released 6–65 km from capture sites. Translocated bears were ear-tagged and data were collated in 1967–1974, from sightings or recaptures.

A review of 19 studies in forested areas in 16 states and provinces in the USA and Canada (2) found that black bears *Ursus americanus* translocated away from sites of conflict with humans were less likely to return to their capture site if translocated as younger bears, over greater distances, or across geographic barriers. Of 15 sub-adult male bears translocated 32–85 km (pooled from two studies), one returned to its capture site, compared to 106 returns out of 145 bears >2 years old translocated 8–120 km (pooled from 12 studies). In data pooled from 12 studies, fewer bears (34 of 79 bears -43%) that were translocated 64–271 km returned to capture locations than bears translocated <64 km (81 of 100 bears — 81%). In one study of bears translocated ≤80 km, fewer returned when released at locations separated from capture sites by mountains or numerous ridges (5 of 27 bears — 19%) than when released across more uniform terrain (104 of 143 bears — 73%). Translocation and movement data were summarized from 19 studies (16 published in 1961–1984 and three unpublished) of bears translocated due to nuisance behaviour. Bears were considered to have returned home if found within 8–20 km of their capture site (this varied by study).

(1) Beeman L.E. & Pelton M.R. (1976) Homing of black bears in the Great Smoky Mountains National Park. *Bears: Their Biology and Management*, 3, 87–95.

(2) Rogers L.L. (1986) Effects of translocation distance on frequency of return by adult black bears. *Wildlife Society Bulletin*, 14, 76–80.

7.9. Treat mammals to reduce conflict caused by disease transmission to humans

https://www.conservationevidence.com/actions/2342

• **One study** evaluated the effects of treating mammals to reduce conflict caused by disease transmission to humans. This study was in Germany[1].

COMMUNITY RESPONSE (0 STUDIES)

POPULATION RESPONSE (0 STUDIES)

BEHAVIOUR (0 STUDIES)

OTHER (1 STUDY)

• **Human-wildlife conflict (1 study):** A controlled, before-and-after study in Germany[1] found that following a worming programme, proportions of red foxes infested with small fox tapeworm fell.

Background

Outbreaks of diseases that can be spread between animals and humans (zoonotic diseases) may result in calls for lethal control of the relevant carrier species. Motivations for lethal control may be reduced if the prevalence of diseases or parasites can be reduced by carrying out treatments in wild populations. This intervention specifically considers ways of reducing the risk of disease transmission to humans rather than ways of reducing the direct impact of disease on wild mammal populations.

A controlled, before-and-after study from 2005–2007 in rural and urban areas in Starnberg, Germany (1) found that following a worming programme, proportions of red foxes *Vulpes vulpes* infested with small fox tapeworm *Echinococcus multicularis* decreased. From four to 15 months after worming, a lower proportion of foxes (0.8%) was infested with tapeworms than was infested in untreated areas (33%). Before worming, the proportion infested was similar in areas to be treated (35%) and not treated (43%). From December 2005–March 2007, fox baits (Droncit®) laced with 50 mg of praziquantel were distributed by air in agricultural and recreational areas and by hand in towns and villages. Baits were distributed once every four weeks, over a 213-km² area, at a density of 50 baits/km². Additional bait was left around 100 den sites in January–February and June–August. No bait was distributed in a 238-km² control area. Tapeworm infestation levels were diagnosed in dissected foxes killed by hunters (133 before baiting and 123 after baiting). Small fox tapeworm causes alveolar echinococcosis in humans.

(1) König A., Romig T., Janko C., Hildenbrand R., Holzhofer E., Kotulski Y., Ludt C., Merli M., Eggenhofer S., Thoma D. & Vilsmeier J. (2008) Integrated-baiting concept against *Echinococcus multilocularis* in foxes is successful in southern Bavaria, Germany. *European Journal of Wildlife Research*, 54, 439–447, https://doi.org/10.1007/s10344-007-0168-1

7.10. Use conditioned taste aversion to reduce human-wildlife conflict in non-residential sites

https://www.conservationevidence.com/actions/2384

- **Two studies** evaluated the effects on mammals of using conditioned taste aversion to reduce human-wildlife conflict in non-residential sites. Both studies were in the USA[1,2].

COMMUNITY RESPONSE (0 STUDIES)

POPULATION RESPONSE (0 STUDIES)

BEHAVIOUR (0 STUDIES)

OTHER (2 STUDIES)

- **Human-wildlife conflict (2 studies):** Two studies, in the USA[1,2], found that lacing foodstuffs with substances that induce illness led to these foods being avoided by coyotes[1] and black bears[2].

Background

Some animals, such as bears, may exhibit 'nuisance behaviour' that may bring them into conflict with humans. This may especially be caused by animals attempting to raid foodstuffs, such as at campgrounds, picnic areas or other places that people gather. As well as causing damage to property and spreading rubbish, such animals may then be perceived as representing a threat to humans. As an alternative to lethal control, attempts may be made to make these animals associate human food sources with pain or discomfort by lacing foodstuffs with substances that cause gastrointestinal upset. If successful, such animals may subsequently avoid seeking out human sources of food.

Studies considered under this intervention are those concerning human-wildlife conflict away from permanent settlements. For related interventions, see also the Chapter, *Residential & commercial development*.

A study in 1977–1978 at a campsite in California, USA (1) found that using conditioned taste aversion reduced the number of coyotes *Canis latrans* that begged for food. Three months after adding lithium chloride (which induces gastrointestinal discomfort) to bait there had been no reported begging problems at the campsite, compared to >12 coyotes begging for food in the month prior to use of lithium chloride baits. Bait was consumed by coyotes 31 times over a 14-day period. From December 1977 to January 1978, meat bait was mixed with lithium chloride at a rate of 10 g/396 g of meat. Bait was left on paper plates at the campsite or thrown to individual coyotes. Animal calls were used to attract coyotes. During baiting, campsite visitors were asked not to feed coyotes. Methods for surveying coyotes were unclear in the original paper.

A study in 1992–1994 in a predominantly forested area in Minnesota, USA (2) found that inducing conditioned taste aversion through lacing military-issue meals with thiabendazole led to black bears *Ursus americanus* subsequently avoiding these foods. Consumption of laced meals induced illness in bears in <90 minutes. Thereafter, over 2–122 days post-treatment, bears did not consume military-issue meals offered during 32 of 41 trials and partially consumed such meals during nine trials. Only once did partial consumption comprise >50% of the meal. Other foodstuffs were, at least partially, consumed in 78% of trials. One year later, two of the bears did not consume military-issue meals in any of seven trials. However, one more year later, in a single trial, one of the bears fully consumed a military-issue meal. In May 1992, two adult female bears and three yearlings that were resident on a military reservation were each given a military-issue meal laced with thiabendazole (72–165 mg/kg bear). Bears were habituated to humans and could be studied closely without disturbance. Meals were ready to eat and consisted of a range of foods, each in sealed pouches and all in a sealed brown plastic bag. Subsequent trials involved military-issue meals and other foodstuffs (raw bacon, jelly, or peanut butter and jelly on bread).

(1) Cornell D. & Cornely J.E. (1979) Aversive conditioning of campground coyotes in Joshua Tree National Monument. *Wildlife Society Bulletin*, 7, 129–131.

(2) Ternent M. & Garshelis D. (1999) Taste-aversion conditioning to reduce nuisance activity by black bears in a Minnesota military reservation. *Wildlife Society Bulletin*, 27, 720–728.

7.11 Use non-lethal methods to deter carnivores from attacking humans

https://www.conservationevidence.com/actions/2385

- **Eight studies** evaluated the effects of using non-lethal methods to deter carnivores from attacking humans. Three studies were in the USA[1,2,3], two were in Australia[6,8], one was in the USA and Canada[4], one was in Austria[5] and one was in Bangladesh[7].

COMMUNITY RESPONSE (0 STUDIES)

POPULATION RESPONSE (1 STUDY)

- **Survival (1 study):** A study in Bangladesh[7] found that when domestic dogs accompanied people to give advance warning of tiger presence, fewer tigers were killed by people.

BEHAVIOUR (0 STUDIES)

OTHER (8 STUDIES)

- **Human-wildlife conflict (8 studies):** Two studies, in the USA[1,4] and Canada[4], found that pepper spray caused all[1] or most[4] American black bears[1,4] and grizzly bears[4] to flee or cease aggressive behaviour. One of these studies also showed that tear gas repelled half of American black bears[1]. Two studies in the USA[3] and Austria[5] found that grizzly/brown bears were repelled by rubber bullets[3] or by a range of deterrents including rubber bullets, chasing, shouting and throwing items[5]. A study in the USA[2] found that hikers wearing bear bells were less likely to be approached or charged by grizzly bears than were hikers without bells. A replicated, controlled study in Australia[6] found that ultrasonic sound deterrent units did not affect feeding location choices of dingoes. A study in Bangladesh[7] found that domestic dogs accompanying people gave advance warning of tiger presence, enabling people to take precautionary actions. A study in Australia[8] found that a motorised water pistol caused most dingoes to change direction or speed or move ≥5 m away, but sounding a horn did not.

Background

Actual or perceived danger to humans from carnivores can prompt lethal control of such animals. If measures can be introduced to reduce these threats, or threatening behaviour, this could reduce human-wildlife conflict and motivations for carrying out lethal control.

For related studies, see *Habituate mammals to visitors* and *Use conditioned taste aversion to reduce human-wildlife conflict in non-residential sites*. Additionally, several other interventions aim to reduce behaviour by wild mammals deemed to be a nuisance (such as raiding garbage containers) and, by reducing the extent to which carnivores and humans come into conflict, may also reduce the chances of attacks on humans. See, for example, *Residential & commercial development -Scare or otherwise deter mammals from human-occupied areas to reduce human-wildlife conflict* and *Residential & commercial development -Prevent mammals accessing potential wildlife food sources or denning sites to reduce nuisance behaviour and human-wildlife conflict*.

A study (year not stated) at campgrounds and garbage dumps in Minnesota and Michigan, USA (1) found that pepper spray repelled all American black bears *Ursus americanus* and tear gas repelled half of bears. Four out of five bears sprayed once in the eyes with pepper spray fled 7–20 m away and did not return. The fifth bear, a male, only fled after being sprayed four times (although on two occasions, the spray did not reach the bear's eye). Four bears exposed to tear gas left the site. However, two returned within a few minutes. No animals exhibited signs of aggression. The study was conducted in sites (number not stated) where black bears were reported to be taking food from people. Five black bears were sprayed in the eyes with pepper spray from distances of 1.5–3.0 m and four were sprayed with tear gas.

A study in 1980–1981 in forest in a national park in Wyoming, USA (2) found that hikers wearing bear bells were less likely to be approached or charged by grizzly bears *Ursus arctos*. Of initially motionless bears spotted ≤150 m from hikers, a higher proportion (67%) moved away from hikers with bells than from hikers without bells (26%). No bears charged at hikers with bells, whereas 14% of bears spotted by hikers without bells charged at the hikers. Hikers reported 97 observations of bears within 150 m. In 24% of encounters, hikers wore bells. Human-bear encounters in a 154-km² study area were surveyed from 3 June–15 September 1980 and 14 June–22 September 1981. Bell-wearing rates were assessed during timed counts of hikers on trails, at 15-day intervals. Hikers were questioned about bear encounters.

A study in 1986–1989 at seven sites in two national parks dominated by temperate forest in Wyoming, USA (3) found that using rubber bullets to scare problem grizzly bears *Ursus arctos* caused all bears to flee from study sites, at least for a short period. Five bears were shot at using rubber bullets, 41 times in total, with 27 hits recorded and bears fled each time. Bears were generally deterred from returning to the study area for 2–4 weeks. However, two bears continued to exhibit nuisance behaviour and repeatedly exploited sources of human food. Rubber bullets were fired at bears that had been seeking human food or foraging close to habitation. The behaviour of each bear was noted before and after the firing of bullets, as well as whether the bear fled from an area with a radius of approximately 100 m.

A study in 1984–1994 across the USA (primarily Alaska and Montana) and Canada (primarily British Columbia and Alberta) (4) found that after being sprayed with pepper spray, most brown bears *Ursus arctos* and American black bears *Ursus americanus* changed their behaviour. Fifteen out of 16 (94%) brown bears and all four (100%) black bears involved in close-range aggressive encounters with people changed the behaviour after being sprayed. However, in six cases (38%), brown bears continued to act aggressively and in three cases (19%) bears attacked the person spraying. Black bears did not leave the area after being sprayed. Sixty-six records of bear-human interactions involving pepper-spray use were collected from agencies throughout Canada and the USA and from individuals that used spray to deter bears. Results reported here are those involving close-range encounters with aggressive bears. Sprays used were thought to likely contain 10% capsicum extract.

A study in 1995–2000 of seven animals across a mixed, but mostly forested, landscape in central Austria (5) found that shooting rubber bullets, chasing, shouting and throwing items to reduce brown bears' *Ursus arctos* habituation to humans was partially successful. After 16 aversive conditioning treatments on seven bears, they returned to the site of treatment within <1 day to >6 months. The time to their next observed habituated behaviour (being ≤50 m from an observer and behaving in an indifferent or curious manner) was one week to three years. Aversive treatments, some in combination, included five capture events, 11 discharges of rubber bullets, four uses of cracker shells and two of fireworks and warning shots. Bears were monitored through reported sightings and footprint tracking. Three bears were also tracked using radio-collars and ear transmitters, but these became detached from two bears.

A replicated, controlled study (year not stated) on captive animals in Queensland, Australia (6) found that ultrasonic sound deterrent units, tested as potential deterrents for dingoes *Canis lupus dingo*, did not affect feeding location choices. Dingoes first selected bait in front of one ultrasonic unit (unit 1 of two) on 21% of occasions when it was turned on. This did not differ significantly from the 29% of occasions that unit 1 was selected first when it was turned off and unit 2 was turned on. Four captive dingoes were housed in pens, opening onto a communal area. Two ultrasonic units (Weitech Yard and Garden Protector) were positioned back to back, with 5 g of tuna in front of each. One unit (selected randomly) was turned on. Dingoes, individually in random order, were released into the communal area, and bait selection order noted. Sixty such trials were conducted.

A study in 2005–2007 in a mangrove area in Bangladesh (7) found that domestic dogs *Canis lupus familiaris* accompanying people gave advance warning of tiger *Panthera tigris* presence, enabling people to take precautionary actions. Of the responses by dogs to apparent tiger presence 62% were verified as accurate. One tiger was killed by people during 2006 (within the study period), compared to 12 in the preceding four years (most of which was before the study period). Four humans were killed by tigers during 2006, compared to 75 over the preceding four years. Forty domestic dogs were each taken into the forest 18 times between August 2005 and January 2007. Each dog, tethered to a person, accompanied a group of 5–7 people (plant-product harvesters, honey gatherers or fishermen). Dogs responded to most wild animals with excitement, quick movements and vocalisations, though apparent responses to tigers were fear and low noise and moving close to the owner without barking. The presence of tigers or other wild animals was verified immediately by observation, or the next day by locating pugmarks or scats.

A study in 2015 at a beach in Queensland, Australia (8) found that a motorised water pistol caused dingoes *Canis dingo* to display aversive responses (change direction or speed or move ≥5 m away) in most cases but sounding a horn did not. The water pistol produced more aversive responses (32 from 43 trials involving seven animals) than did blowing a whistle, a treatment assumed not to deter dingoes (one aversive response from 23 trials involving nine dingoes). The air horn produced no aversive responses in 13 trials involving six animals. Trials were conducted along a beach, in daylight, during 1–15 December 2015. With dingoes ≤5 m from an observer, a whistle was blown on the first trial, involving nine animals. For subsequent trials for these animals, the whistle was followed by sounding an air horn or firing a mechanical water pistol. Some trials for individual dingoes were repeated after short gaps (2–11 trials during 1–55 minutes).

(1) Rogers L.L. (1984) Reactions of free-ranging black bears to capsaicin spray repellent. *Wildlife Society Bulletin*, 12, 59–61.

(2) Jope K.J. (1985) Implications of grizzly bear habituation to hikers. *Wildlife Society Bulletin*, 13, 32–37.

(3) Gillin C.M., Hammond F.M. & Peterson, C.M. (1994) Evaluation of an aversive conditioning technique used on female grizzly bears in the Yellowstone Ecosystem. *Bears: Their Biology and Management*, 9, 503–512.

(4) Herrero S. & Higgins A. (1998) Field use of capsicum spray as a bear deterrent. *Ursus*, 10, 533–537.

(5) Rauer G., Kaczensky P. & Knauer F. (2003) Experiences with aversive conditioning of habituated brown bears in Austria and other European countries. *Ursus*, 14, 215–224.

(6) Edgar J.P., Appleby R.G. & Jones D.N. (2007) Efficacy of an ultrasonic device as a deterrent to dingoes (*Canis lupus dingo*): a preliminary investigation. *Journal of Ethology*, 25, 209–213, https://doi.org/10.1007/s10164-006-0004-1

(7) Khan M.M.H. (2009) Can domestic dogs save humans from tigers *Panthera tigris*? *Oryx*, 43, 44–47, https://doi.org/10.1017/s0030605308002068

(8) Appleby R., Smith B., Mackie J., Bernede L. & Jones D. (2017) Preliminary observations of dingo responses to assumed aversive stimuli. *Pacific Conservation Biology*, 23, 295–301, https://doi.org/10.1071/pc17005

8. Threat: Natural system modifications

Background

This chapter includes interventions to address threats that convert or degrade habitat as part of the management of natural or semi-natural systems, often to improve human welfare. This includes suppressing or increasing the intensity of fires and changing the natural flow of water.

8.1. Use prescribed burning

https://www.conservationevidence.com/actions/2388

- **Thirty-seven studies** evaluated the effects on mammals of using prescribed burning. Twenty-five studies were in the USA[1,3,4,6–10,12–16,18,20–24,26,27,29–31,34], three each were in Canada[2,5,25] and South Africa[17,19,36], two each were in Spain[11,37] and Tanzania[28,35] and one each was in France[32] and Australia[33].

COMMUNITY RESPONSE (2 STUDIES)

- **Richness/diversity (2 studies):** A replicated, randomized, controlled study in the USA[14] found similar small-mammal species richness after prescribed burning compared to in unburned forest. A replicated, site-comparison study in Australia[33] found that prescribed burns early in the dry season resulted in higher small-mammal species richness relative to wildfires later in the season.

    https://doi.org/10.11647/OBP.0234.08

POPULATION RESPONSE (16 STUDIES)

- **Abundance (11 studies):** Five of 10 replicated studies (of which eight were controlled and two were site comparisons), in the USA[4,10,14,20,22,24,27], Spain[11,37] and Australia[33], found that prescribed burning did not increase abundances of small mammals[4,14,20,22,24]. Three studies found mixed effects, on cottontail rabbits[10] and small mammals[27,37] and two found that burning increased numbers of European rabbits[11] and small mammals[33]. A systematic review in the USA[31] found that two mammal species showed positive responses (abundance or reproduction) to prescribed burning, while three showed no response.

- **Reproductive success (1 study):** A before-and-after, site comparison study in South Africa[19] found that 92% of Cape mountain zebra foals were produced in the three years post-fire compared to 8% in the three years pre-fire.

- **Condition (1 study):** A replicated, controlled study, in the USA[9], found that prescribed burning did not reduce bot fly infestation rates among rodents and cottontail rabbits.

- **Occupancy/range (3 studies):** Two of three studies (including two site comparisons and one controlled study) in the USA[21,34] and Canada[25], found that prescribed burning resulted in larger areas being occupied by black-tailed prairie dog colonies[21] and smaller individual home ranges of Mexican fox squirrels[34]. The third study found that prescribed burning did not increase occupancy rates of beaver lodges[25].

BEHAVIOUR (22 STUDIES)

- **Use (21 studies):** Ten of 21 studies (including eight controlled studies and eight site comparisons with a further four being before-and-after studies), in the USA[1,3,6,7,8,12,13,15,16,18,23,26,29,30], Canada[2,5], South Africa[17,19,36], Tanzania[28] and France[32], found that prescribed burning increased use of areas (measured either as time spent in areas or consumption of food resources) by bighorn sheep[1,5,16], mule deer[2], pronghorn antelope[6], elk[8,23], plains bison[15], Cape mountain zebras[19]and mouflon[32]. Six studies found mixed effects, with responses differing among different ages or sexes of white-tailed deer[12], bison[13] and elk[30], differing among different large herbivore species[17] or varying over time for elk[26], while swift foxes denned more but did not hunt more in burned areas[29]. The other five studies showed that prescribed burning did not increase use or herbivory by elk[3], black-tailed deer[7], white-tailed deer[18] or mixed species groups of mammalian herbivores[28,36].

- **Behaviour change (1 study):** A site comparison study in Tanzania[35] found that vigilance of Thomson's gazelles did not differ between those on burned and unburned areas.

Background

Fire is an integral part of the management and natural dynamics of some ecosystems. Some habitats are naturally fire-prone while in others, habitats are shaped by long-term traditional management (Bowman 1998). Some habitats are now managed through prescribed burning, partly to reduce the risk of uncontrolled wildfires. In other areas, burns are being introduced, following long periods of fire suppression, sometimes accompanied by mechanical clearance of woody material. Whilst burning can have a dramatic effect on the landscape, reducing cover and short-term food resources, feeding on new plant growth within burned areas can also increase an animal's nutritional intake, with concentrations of proteins in particular being elevated (Hobbs & Spowart 1984).

The studies featured generally compare prescribed burning with no management (which in one case means allowing wildfires) but, in some cases, comparisons are with mechanical clearance.

See also: *Burn at specific time of year.*

Hobbs N.T. & Spowart R.A. (1984) Effects of prescribed fire on nutrition of mountain sheep and mule deer during winter and spring. *The Journal of Wildlife Management*, 48, 551–560.

Bowman D.M.J.S. (1998) Tansley Review No. 101. The impact of Aboriginal landscape burning on the Australian biota. *New Phytologist*, 140, 385–410.

A site comparison in 1975–1978 of shrubland and grassland at a site in Idaho, USA (1) found that bighorn sheep *Ovis canadensis* consumed bluebunch wheatgrass *Agropyron spicatum* growing in burned areas more than they consumed on unburned areas. In the first summer after burning, a higher proportion of bluebunch wheatgrass stems was grazed on burned areas (73%) than on unburned areas (8%). The same pattern was observed, though with reducing magnitude, two years after burning (66 vs 25%), three years after burning (30 vs 10%) and four years after burning (36 vs 22%). Within an 86-km² study area, seven areas (0.05–0.45 ha, total area 1.51 ha) had controlled burns carried out in September 1974. One hundred randomly selected bluebunch wheatgrass stems from burned and unburned areas were inspected each year to calculate the proportion that was grazed.

A replicated, controlled study in 1975–1977 on grassland in British Columbia, Canada (2) found that in burned areas, bluebunch wheatgrass *Agropyron spicatum* was consumed more by foraging mule deer *Odocoileus hemionus* than it was in unburned areas. Deer took more bites/observation of bluebunch wheatgrass in burned plots (average 22 bites) than in unburned plots (average two bites). Plots were studied at two sites in sagebrush and two in Douglas fir *Pseudotsuga menziesii* forest. At each site, plots (1.25 × 5 m) were established in a block. In each block, in October 1975, three plots were burned and three were not burned. In April 1976, three deer were fenced onto the block and their selection between plots was assessed through direct observations at intervals through the day. The same three deer were used on all blocks and observed twice/block for one day each time. In April 1977, four deer were observed, on two blocks combined, over four days.

A randomized, replicated, controlled study in 1971–1974 of a grassland in Washington, USA (3) found that burning grass did not increase overwinter use by Rocky Mountain elk *Cervus canadensis nelsoni*. Overwinter use by elk totalled 47–80 elk days/ha on burned areas and 42–79 elk days/ha on unburned areas. Within each of six plots, one 3.1-ha subplot was randomly assigned for burning and one was not burned. Burning was carried out once, in late-autumn 1971. Elk pellets were counted in spring to assess use of plots in the winters of 1971–1972, 1972–1973 and 1973–1974.

A replicated, controlled study in 1981–1983 of a pinyon-juniper woodland in New Mexico, USA (4) found that felled forest areas that were burned did not have more small mammals than did felled unburned areas 13–18 years after treatment. A similar number of small mammals was caught in stands that were bulldozed and burned (408) as in stands that were bulldozed without burning (433). Fewer were caught in undisturbed stands (246). Treatment plots, c.120 ha each, were established in each of two woodland blocks, one in 1965, one in 1970. In each block, one plot was bulldozed (trees pushed over and left), one was bulldozed with trees pushed and piled, then burned and one was undisturbed. Small mammals were trapped in the second and third week of September, each year, in 1981–1983. Each plot was sampled for four days each year.

A replicated, site-comparison study in 1980 of forest in Alberta, Canada (5) found that previously burned areas were used more by Rocky Mountain bighorn sheep *Ovis canadensis canadensis* than were unburned areas. In all five comparisons, at different distances below the treeline, more sheep pellets were found in burned areas (14–424 pellet groups/ha) than in unburned areas (0–108 pellet groups/ha). Three fire-modified sites (burned in 1919–1970) and three unburned sites (average forest age of 81–256 years old) were studied. At each site, three transects ran downslope from the treeline, to the valley bottom. Relative use by sheep of each area was assessed by counting pellet-groups in randomly located plots along these transects in 1980.

A site comparison study in 1985–1986 of prairie in Alberta, USA (6) found that pronghorn antelope *Antilocapra americana* made greater use of burned areas, relative to their availability, than of unburned areas in five of the 12 months surveyed. The number of pronghorn groups on burned areas was greater than expected in September, October, November, January and April. During these months, 5–22 pronghorn groups were found on burned areas, from totals of 38–97 pronghorn groups overall. If no preference was shown for or against burned ground, 5% of groups would be expected on it. Pronghorns especially favoured burns containing pricklypear cactus *Opuntia polyacantha*. Areas were burned in July–August 1985. Pronghorn were surveyed monthly, from July 1985 to June 1986. Groups <1 km away were mapped along a 138-km route, travelled by vehicle at <50 km/hour.

A site comparison study in 1983–1985 of a shrubland site in California, USA (7) found that prescribed burning did not increase use of such areas by black-tailed deer *Odocoileus hemionus*, relative to unburned areas. There was no significant difference in density of faecal pellet groups between burned and unburned plots over the two years following burning (data not reported). In an area of chaparral shrubland, approximately 20% (7 km²) was burned in November 1983. Twenty-three circular plots, each 100 m², were surveyed for faecal clumps. Eleven plots were in burned areas and 12 were in unburned areas. Faecal pellet clumps were cleared and counted at end of each wet (November–April) and dry (May–October) season from November 1983 to October 1985.

A replicated, randomized, controlled study in 1983–1987 of a rough fescue *Festuca scabrella*-dominated grassland in Montana, USA (8) found that burning increased grazing of rough fescue by elk *Cervus canadensis nelsoni* in the first, but not subsequent, winters following burning. The proportion of rough fescue plants grazed was greater in burned plots (39%) than in unburned plots (15%) over the winter following autumn burning. Over the subsequent three winters, the proportion grazed did not differ between burned plots (including spring burns; 98–100%) and unburned plots (87–97%). Additionally, a higher proportion of rough fescue biomass was utilized over the first two winters following burning (82–86%) than was utilized in unburned plots (24–69%). Six plots were burned on 17 October 1983 and six on 15 April 1984. Three plots were not burned. Plots were 2 ha in extent. Treatments were assigned randomly. Elk utilization of rough fescue was assessed by inspecting the closest plant to 50 points along each of two transects/plot and determining the proportion that was grazed. Additionally, six caged and six non-caged samples on each treatment were clipped, in April 1985 and 1986, to determine elk utilization by biomass.

A replicated, controlled study in 1986–1988 in a wooded area in Oklahoma, USA (9) found that prescribed burning did not reduce bot fly *Cuterebra* infestation rates among rodents and cottontail rabbits *Sylvilagus floridanus*. The percentage of animals infested with *Cuterebra* larvae did not differ significantly between areas that were burned and sprayed with herbicide (14–20% of animals) and areas sprayed but not burned (18–20%). Eight plots (32.4 ha each) were burned annually in

April, from 1985, and eight were not burned. Four burned and four unburned plots were sprayed with the herbicide, tebuthiuron (at 2.2 kg/ha), in March 1983. Remaining plots were treated with the herbicide, triclopyr (at 2.2 kg/ha), in June 1983. Rodents were surveyed using snap traps, in July–September and December–March 1986–1988. Cottontail rabbits were collected by shooting in January and July 1987–1988. Animal carcasses were opened up and examined for *Cuterebra* presence.

A replicated, controlled study in 1986–1988 of a forest and grassland site in Oklahoma, USA (10) found that burning and spraying pastures with herbicide had mixed effects on cottontail rabbit *Sylvilagus floridanus* abundance when compared with spraying with herbicide alone. In seven of 10 comparisons, there was no significant difference between the number of cottontails found in pastures that were burned compared to those not burned. For three of 10 comparisons, there were more cottontails on burned areas (0.1–1.40 cottontails/ha) than on unburned areas (0–0.4). Eight 32-ha pastures were treated with the herbicides tebuthiuron or triclopyr (at 2.2 kg/ha in March 1983 or June 1983). Four of these pastures were burned in April 1985, 1986 and 1987. Rabbit density was estimated by walking transects, three times each in July and February, from July 1986 to February 1988.

A replicated, controlled, paired sites study in 1991–1992 of scrubland in a national park in southern Spain (11) found that burned plots hosted higher densities of European rabbits *Oryctolagus cuniculus* than did unburned plots. More rabbit pellets were counted in burned plots both in wet scrubland (burned: 11.6 pellets/plot/year; unburned: 9.8) and in dry scrubland (burned: 6.8; unburned: 1.6). Four plots each in wet and dry scrubland were burned in summer 1989. Each was paired with an unburned plot 1 km away, in the same habitat. Plots measured 300 × 200 m. Rabbit pellets were counted monthly in 1991 and 1992 at random sample locations in each plot.

A randomized, replicated, controlled study in 1988–1989, in a mixed forest and prairie site in Oklahoma, USA (12) found that burning areas of forest had mixed effects on use by white-tailed deer *Odocoileus virginianus*, depending on season and sex. Female deer preferred burned plots in spring and summer, but unburned plots in winter. Male deer preferred burned plots in summer and autumn. There was no habitat selection for other sex/season combinations. Data presented as proportions of radio-tracking locations. See paper for details. Four blocks, each containing five 32-ha plots, were studied. In each block, two plots were sprayed with herbicide and burned, two were sprayed with herbicide but not burned and one was not sprayed or burned. Burning was done each April in 1985–1987. Herbicide was applied in 1983. Ten female and seven male deer were radio-tracked in 1988–1989, and the use of burned and unburned areas relative to their size was assessed.

A study in 1993–1995 of a prairie site in Oklahoma, USA (13) found that burned areas were selected for grazing by mixed-age and mixed-sex bison *Bison bison* groups, but were avoided by mature bull groups. Burned areas were selected in a higher proportion than their availability by mixed groups during 23% of observations and avoided during 13%. Unburned areas were selected in 0% of observations and avoided in 63%. Burned areas were selected by bull groups in 4% of observations, and avoided in 46%. Unburned areas were selected in 29% of observations, and avoided in 14%. Three hundred bison were reintroduced into a 1,973-ha study area in October 1993. The area was expanded by 292 ha in 1995. Controlled burns were carried out five times between September 1993 and December 1995. Bison usage of burned and unburned areas was surveyed 4–12 times/month in 1994 and 1995. Herds were generally mature (>5 year-old) bulls and mixed groups of cows, calves and younger bulls.

A replicated, randomized, controlled study in 1992–1994 of pine forest in a mountainous area of Arkansas, USA (14) found similar small mammal numbers and species richness after prescribed burning compared to in unburned forest. Small mammal capture rates in burned stands (animals caught on 2.3–7.1% of trap nights) did not significantly differ to those in unburned stands (3.9–7.4%). Average species richness did not differ between burned (2.7–4.3 species/plot) and unburned plots (1.7–4.7/plot). In nine plots (14–45 ha), mid-storey trees were removed and, the following spring, plots were burnt. In three plots, mid-storey trees were not removed and plots were not burned. Management was carried out to benefit red-cockaded woodpeckers *Picoides borealis*. Small mammals were surveyed using 80 live-trap stations/plot from 27 December to 4 January in 1992–1993 and 1993–1994.

A replicated, site comparison study in 1990–1996 of one prairie site in each of Nebraska and Oklahoma, USA (15) found that plains bison *Bison*

bison bison preferentially selected burned areas in most years. At one
site, bison made more use of burned areas, relative to their availability,
in five of seven years. There was no consistent pattern in the second
site but, in the largest single range (65% of the site), bison selected
burned areas in two of three years. In both cases, results were presented
as deviation from modelled predictions. At one site, monitored from
1990–1996, approximately 13.5% of the site was burned each year. The
second site, monitored from 1993–1996, had approximately 20% burned
each year. Locations of mixed bison groups (females, bulls <4 years old
and occasionally adult bulls) were determined at least monthly during
respective monitoring periods at each site.

A before-and-after, site comparison study in 1986–1991 of a mixed
grassland, shrubland and woodland site in Utah, USA (16) found that
prescribed burning of sagebrush-grass shrublands and pinyon-juniper
woodland increased use of these areas by Rocky Mountain bighorn
sheep *Ovis canadensis*. Use of burned areas by sheep increased by 148%
and use of unmanaged areas decreased by 45%. Following burning,
more sheep used the area (82 sheep groups; average of 14 sheep/group)
than before burning (117 sheep groups; average of nine sheep/group).
On a 353-ha study area, 18% was burned and 49% was unmanaged.
Additionally, 32% was clearcut (results not presented here). Sheep-use
patterns were assessed pre-treatment, from June 1986 to September
1988, by observing 25–30 radio-collared sheep daily. Post-treatment use
was assessed in June–September 1991, by counting sheep 62 times from
an 11-km transect.

A before-and-after study in 2000 of a shrubland ranch in South
Africa (17) found that prescribed burning of an area increased its use
by roan antelope *Hippotragus equinus* and tsessebe *Damaliscus lunatus*,
but not by Lichtenstein's hartebeest *Alcelaphus lichtensteini* or sable
antelope *Hippotragus niger*. Roan were seen more frequently on burned
areas (113 sightings) than on unburned areas (81 sightings) relative to
their availability (31% of the study area was burned). Tsessebe showed
a similar pattern (burned: 77 sightings; unburned: 54 sightings) as did
zebra *Equus burchelli* (burned: 96 sightings; unburned: 24) There was
no consistent selection for burned areas by hartebeest (burned: 27;
unburned: 24) or sable antelope (burned: 12; unburned: 27). See paper
for further details of timings of use of burned areas. Rare herbivores

were farmed on a 2,700-ha game ranch. A 280-ha area was burned in October 2001 and a 565-ha area was burned in November 2001. Animal positions were surveyed from roads in early morning and late afternoon from October to December 2000.

A randomized, paired-sites, before-and-after study in 2001–2002 of a shrubland site in Texas, USA (18) found that burning plots already subject to mechanical vegetation clearance did not increase plot utilization by white-tailed deer *Odocoileus virginianus* relative to carrying out a second mechanical clearance. There was no significant difference in deer track counts between plots before (burning: 36; mechanical clearance: 37 track crossings/km) or after (burning: 43; mechanical clearance: 47 track crossings/km) treatments were applied. Ten plots (3–9 ha), established in a 6,154-ha study area, were paired by size, soil and vegetation. In March–April 1999, all plots were cleared of brush using a mechanical aerator pulled by a tractor. In September 2000, one plot from each pair was burned and the other was mechanically cleared a second time. Treatment assignment within pairs was random. Deer utilization was assessed by counting tracks along prepared track lanes, over three days, before and after treatments were applied.

A before-and-after, site comparison study in 1982–1997 in a shrubland in the Western Cape, South Africa (19) found that Cape mountain zebra *Equus zebra zebra* used burned areas more than unburned areas, and 92% of foals were produced in the three years post-fire compared to 8% in the three years pre-fire. Mountain zebras with access to burned areas used those areas 83% of the time (data not provided). By comparison, whilst the total areas burned were not stated, 23% of fires in the south-east section and 89% of fires in the north burned ≤25% the area. Of the foals produced within three years of a fire, 24 were produced in the three years post-fire compared to two pre-fire. Mountain zebras were monitored in two of three sections of the 9,428-ha nature reserve, the north (2,263 ha) and south-east (3,583 ha), where zebras mostly occurred. One of nine fires recorded since establishment of the reserve in 1974 was a prescribed fire (year not stated); others were natural fires (average interval between fires was seven years). Use of burned and unburned areas was monitored between the fires of 1992 and 1996. The number of foals produced was monitored three years before and after the fires of 1982, 1992, 1996–1997.

A replicated, randomized, controlled, before-and-after study in 2001–2003 in North Carolina, USA (20) found that prescribed burning did not alter the abundance of eight small mammal species. After burning, the numbers of captures of eight small mammal species did not differ significantly between burned (0–28 animals/plot) and unburned plots (0–17 animals/plot). Similarly, before burning, numbers did not differ between plots assigned for burning (0–24 animals/plot) and unburned plots (0–19 animals/plot). See paper for full breakdown of species abundances. Three blocks were established, containing plots of >14 ha. In each block, one plot was burned in March 2003 and one plot was not burned. Small mammals were live-trapped over 10 consecutive days and nights in July and August of 2001–2003.

A replicated, controlled, before-and-after study in 2002–2003 in a national park in North Dakota, USA (21) found that burning and clearing woody vegetation led to greater areas occupied by black-tailed prairie dog *Cynomys ludovicianus* colonies and more prairie dog burrows. The study does not distinguish between the effects of burning and mechanical vegetation clearance. After one year, prairie dog colonies occupied a greater area in plots where vegetation was cleared and burned (18–70% of available habitat) than in plots that were not cleared or burned (0–5%). Cleared and burned plots also had more new burrows (191–458) than did plots that were not cleared or burned (41–116). In each of three prairie dog colonies, a 2-ha plot just beyond the colony boundary underwent prescribed burning in May 2002 and mechanical brush removal in June 2002. Similar 2-ha plots that were not burned or cleared were used for comparison. Colony boundaries were mapped in May–September 2002 and May–August 2003. New burrows were mapped monthly during these periods.

A replicated, randomized, controlled, before-and-after study in 2001–2004 of a coniferous woodland in California, USA (22) found that prescribed fire did not increase the abundance of small mammals. Deer mouse *Peromyscus maniculatus* abundance was not significantly higher on burned than on unburned plots (results presented as modelled effect size). Similarly, lodgepole chipmunk *Neotamias speciosus* abundance and total small mammal biomass were not significantly higher in burned than in unburned plots. Nine plots, 15–20 ha in area, were studied. Three were burned between 28 September and 28 October 2001 and three were burned on 20 or 27 June 2002. Three plots were not burned. Treatments were allocated randomly to plots. Small mammals were sampled by live-trapping over eight consecutive nights and days each year. Sampling occurred in June–August 2001 (pre-treatment) and in June–September of 2002 and 2003 and June–August 2004.

A controlled study in 1984–1999 in a sagebrush shrubland in Montana, USA (23) found that prescribed burning increased use of the area by elk *Cervus canadensis*. Elk used areas that had been burned more frequently (163–628 elk-use days) than they used areas that had not been burned (32–298 elk-use days). Burned areas had higher grass and forb cover and lower sagebrush cover than unburned areas. In October 1984, a 40-ha area of sagebrush shrubland was burned and, in April 1988, a 30-ha area was burned. Five permanent 404-m² plots (20.1 × 20.1 m) were established in each burned area and another five placed within the unburned portion, one in 1988 and four more in 1993. In June 1988–1993 and 1999, elk use of plots was estimated by counting the number of pellets within 1 m of six transects laid in each plot. Vegetation cover was estimated within five 25 × 51 cm randomly placed quadrats each plot. No livestock were present in the study area.

A replicated, randomized, controlled study in 2003–2004 in a mixed forest site in North Carolina, USA (24) found that prescribed fire did not alter abundances of four shrew species. In both sampling years, numbers of northern short-tailed shrews *Blarina brevicauda* caught did not differ between plots that were burned (2–6 animals/plot) and plots that were not burned (3–10 animals/plot). The same pattern was seen for smoky shrews *Sorex fumeus* (1–2 animals/plot vs 1–2 animals/plot), American pygmy shrews *Sorex hoyi* (2–4 animals/plot vs 0–2 animals/plot), and southeastern shrew *Sorex longirostris* (1–4 animals/plot vs 1–5 animals/plot). In each of three blocks, established in 2001, one plot was burned in March 2003 and one plot was not burned. Plots were >14 ha. Shrews were surveyed using pitfall traps and drift fencing over 123 nights in 2003 and 125 nights in 2004.

A before-and-after, site comparison study in 1989–2001 within a mixed forest national park in Alberta, Canada (25) found that prescribed burning did not increase occupancy of lodges by beavers *Castor canadensis*. For lodges subject to prescribed burning once, the occupancy rate in the year after burning (25%) was lower than in the year before burning (41%). Some lodges were burned more than once and the odds of occupancy decreased by 58% for each additional burn. In a 194-km^2 national park, occupancy of 734 beaver lodges, located between 1989 and 2001, was monitored by aerial or ground surveys, every 1–3 years. There were 121 prescribed fires (1–1,059 ha in extent) from 1979–2001. All but six (in October) were lit between April and June. Around 49% of the park was not burned in the study period.

A site comparison study in 1989–1999 in a sagebrush shrubland in Montana, USA (26) found that prescribed burning was associated with a short-term, but not long-term, increase in elk *Cervus canadensis* usage. In the first year after burning, elk use of burned plots increased (from 116 to 210 elk-use days) and declined on unburned plots (from 189 to 120 elk-use days). After 10 years, elk use declined and was similar on both burned plots (72 elk-use days) and unburned plots (56 elk-use days). A 50-ha prescribed burn was made in April 1989, while 200 ha of the site was not burned. Five plots (404 m^2 each) were established each in burned and unburned areas. Unburned plots, damaged by wildfire in 1991, were replaced in 1993 by three plots on remaining unburned land. Elk used the site from November–May. Elk pellet groups were counted in June 1989–1991, 1993, and 1999 along transects across each plot.

A replicated, randomized, controlled, before-and-after study in 2001–2003 of a forest in California, USA (27) found that prescribed burning increased abundance of deer mouse *Peromyscus maniculatus*, but not California ground squirrels *Spermophilus beecheyi*, long-eared chipmunks *Tamias quadrimaculatus* or brush mice *Peromyscus boylii*. Deer mouse abundance increased with fire (after: 2.0; before: 0.5/100 trap nights) and declined at the same time in unburned plots (after: 1.3; before: 1.9/100 trap nights). Changes in capture rates from before to after treatments did not differ between burned and unburned plots for California ground squirrel, long-eared chipmunk or brush mouse (see paper for data). Forests stands were 14–29 ha each. Four stands were burned in October–November 2002 and four stands were not burned. Small mammals were live-trapped over nine consecutive days and nights in July–August 2001 (pre-burn) and 2003 (post-burn).

A replicated, controlled study in 2003–2004 of savanna grassland in a national park in Tanzania (28) found that prescribed burning did not result in a higher level of herbivory by mammals. The amount consumed by herbivores varied by season but the overall average amount in burned plots (223 g/m^2) was not significantly different to that in unburned plots (176 g/m^2). Six study areas (each ≥10 ha, 1–40 km apart) were selected. Each consisted of one patch burned in May–July 2003 and one unburned patch. Herbivore consumption was measured monthly, from September 2003 to July 2004, as biomass differences between caged and uncaged areas in study plots.

A before-and-after study in 2003–2005 of grassland in Colorado, USA (29) found that after a prescribed fire, swift foxes *Vulpes velox* denned more in the burned area but hunting use of the area did not significantly increase. Two foxes with core home ranges in the burn area denned inside the burn area more after the burn (100% of denning locations) than before (60–75% of locations). For four foxes with home ranges overlapping the burn area, the proportion of times they were located hunting in the burn area was not significantly higher after burning (45% of locations inside burn area) than before (32%). In January 2003– December 2004, ten foxes were radio-collared. Location was recorded ≥three times/week in 2003–2005. In March 2005, an area of 260 ha was burned by prescribed fire. Sufficient locations were obtained from four foxes to determine pre-and post-burn home range use.

A replicated, controlled study in 2005–2006 in a coniferous forest site in Oregon, USA (30) found that thinning, followed by prescribed burning was associated with mixed effects on use by North American elk *Cervus canadensis*, depending on season, stand age and sex. Thinning and burning were carried out on the same plots, so their influences could not be separated. Female elk used plots burned two and three years previously, proportionally to their availability, preferentially selected 4-year-old burns, and avoided 5-year-old burns. Male elk spent less time in all burned plots relative to their availability (data presented as selection ratios). In 2001–2003, twenty-six forest stands (average 26 ha) were thinned between May and October, followed by prescribed fire during September or October of either the same or the following year. Twenty-seven similar stands (average 55 ha) were not thinned or burned. Radio-collars were fitted on 18 female and five male elk in spring 2005, and 30 female and nine male elk in spring 2006. Locations were recorded automatically, within 1 hour of sunset or sunrise.

A systematic review in 2008 of management aimed at restoring natural processes in conifer forests in southwestern USA (31) found that, in forests where a prescribed burn of low to moderate severity followed thinning, two mammal species showed positive responses (abundance or reproduction) compared to in unmanaged forests, while three showed no response. Responses of tassel-eared squirrel *Sciurus aberti* and deer mouse *Peromyscus maniculatus* to burning after thinning were positive. No significant responses were detected for golden-mantled ground squirrel *Spermophilus lateralis*, gray-collared chipmunk *Tamias cinereicollis* or Mexican woodrat *Neotoma mexicana*. The specific effects of thinning versus burning were not separated, though a different part of the same study found no response by tassel-eared squirrel or deer mouse to thinning (without burning) by removal of small-to intermediate-diameter trees. The review used evidence from 22 studies and considered responses of species recorded in ≥5 studies. Responses of species to five ways of managing ponderosa pine *Pinus ponderosa* forests to recreate natural conditions and forest dynamics, and reduce wildfire risk, were assessed against responses to unmanaged controls. A controlled study in 2004–2008 of heather moorland at a site in southern France (32) found that burning heather (*Calluna vulgaris* and *Erica tetralix*) resulted in greater use of the moorland by mouflon *Ovis gmelini musimon* × *Ovis* sp. The average density of feeding mouflon (modelled to account for temperature-driven variations) was higher on burned plots (36/ha) than on unburned plots (5/ha). Before burning, each 360 × 80-m plot, had not been modified for >40 years. Two plots were burned in spring 2004 and two were left unburned. Mouflon use of plots was determined by counting feeding animals in each plot, at 20-minute intervals, for two hours up to sunset. In total, 668 such counts were made in 2004–2008.

A replicated, site comparison study in 2004–2010 of grassland in Western Australia, Australia (33) found that prescribed burns early in the dry season resulted in higher abundance and species richness of small mammals relative to extensive mid-to late-dry season wildfires. More mammals were found in plots with prescribed burning (5.7/ plot) than in areas subject to wildfire (3.5/plot). The same was true for species richness (prescribed burning: 1.4/plot; wildfire: 1.1/plot). Fire history was determined from satellite imagery from 1999–2010.

Prescribed burning was initiated in 2004. Areas burned less frequently than average were regarded as being managed by prescribed burning earlier in the dry season. Areas burned more frequently than average were regarded as being wildfire areas, burned later in the dry season. Forty small mammal traps/0.25-ha plot were operated for 120 trap-nights/year. The number of plots surveyed is not stated.

A site comparison study in 2002–2003 in a shrubland site in Arizona, USA (34) found that prescribed burning resulted in smaller individual home ranges and shorter daily movements for Mexican fox squirrels *Sciurus nayaritensis chiricahuae* than did fire suppression. The average home range in prescribed burning areas (2.9 ha) was smaller than in fire-suppression areas (6.6 ha). Average daily movements were lower in prescribed burning areas (212 m) than in fire-suppression areas (336 m). In a 5,000-ha protected area, prescribed burning was initiated in 1976. In 1980–2001, there were 33 fires, over 260-ha total extent. Forty-three squirrels were live-trapped. Adults were radio-collared and data were analysed from 11 male and nine females, with ≥30 location fixes per season, from May 2002 to September 2003. Daily movements were measured by locating animals three times from 05:00 h to 11:00 h.

A site comparison study in 2007 of savanna grassland in a park in Tanzania found that vigilance (a measure of perceived predation risk) of Thomson's gazelles *Gazella thomsonii* did not differ between those on burned and unburned areas. There was no difference between burned and unburned areas in group vigilance, individual vigilance or reaction time in the presence of a model cheetah (data not presented). Gazelles were observed in July–August 2007 on 10 burned areas (burned after mid-April with 2 cm average new grass growth) and nine unburned grassland areas. Vigilance was defined as an animal raising its head above shoulder height. Group vigilance was the average proportion of individuals vigilant in a group at 5-minute intervals over 1 hour. Individual vigilance was recorded for randomly selected females, over 2 minutes. Reaction to a model cheetah was timed following model placement from a vehicle 60 m away from the group. A before-and-after study in 2009–2010 on a shrubland reserve in South Africa found that burning reduced the number of locations in which herbivores were present. In each of two main habitats, the proportion of locations at which impala *Aepyceros melampus*, kudu *Tragelaphus strepsiceros*, and zebra *Equus burchelli* were found was lower after burning than before. In one of two habitats, wildebeest *Connochaetes taurinus* and giraffes were present at a higher proportion of sites after burning than before burning (see paper for full details). Two habitat types were studied, based on underlying quartzite and sandy soils. Mammal presence was quantified by determining presence or absence of faecal pellets for each species in plots along transects through each habitat. Pellets were counted in April–May 2009, burns were carried out in June–November 2009 and plots were resampled in June 2010.

A replicated, site comparison study in 2006–2007 of scrubland at a site in Spain (37) found more of some small mammal species at edges of old burned plots but not in plot centres or in younger plots, relative to unburned plots. In two of four comparisons, there were more Algerian mice *Mus spretus* in burned plots (64–109 captures/1,000 trap nights) than in unburned plots (32 captures/1,000 trap nights). For two of four comparisons, there was no significant difference (burned: 8–22 captures/1,000 trap nights; unburned 32 captures/1,000 trap nights). In three of four comparisons, there was no difference in the abundance of wood mice *Apodemus sylvaticus* between burned (2–7 captures/1,000 trap nights) and unburned areas (2 captures/1,000 trap night). In one of four comparisons there were more wood mice in burned areas (burned: 14 captures/1,000 trap nights; unburned 2 captures/1,000 trap night). There were no significant differences in the abundance of greater white-toothed shrew *Crocidura russula* or garden dormouse *Eliomys quercinus* between burned and unburned plots. Three plots were burned in winter 2003 (three years before sampling), three plots were burned in winter 2006 (one year before sampling) and three were not burned. Plots covered 1 ha and were ≥1 km apart. Small mammals were surveyed by live-trapping in unburned plots and in centres and edges of burned plots, once each in summer, autumn, winter and spring from summer 2006 to spring 2007). Traps were operated for seven consecutive nights (and closed in the day).

(1) Peek J.M., Riggs R.A. & Lauer J.L. (1979) Evaluation of fall burning on bighorn sheep winter range. *Journal of Range Management*, 32, 430–432.

(2) Willms W., Bailey A.W. & McLean A. (1980) Effect of burning or clipping *Agropyron spicatum* in the autumn on the spring foraging behaviour of mule deer and cattle. *Journal of Applied Ecology*, 17, 69–84.

(3) Skovlin J.M., Edgerton P.J. & McConnell B.R. (1983) Elk use of winter range as affected by cattle grazing, fertilizing, and burning in Southeastern Washington. *Journal of Range Management*, 36, 184–189.

(4) Severson K.E. (1986) Small mammals in modified pinyon-juniper woodlands, New Mexico. *Journal of Range Management*, 39, 31–34.

(5) Bentz J.A. & Woodard P.M. (1988) Vegetation characteristics and bighorn sheep use on burned and unburned areas in Alberta. *Wildlife Society Bulletin*, 16, 186–193.

(6) Courtney R.F. (1989) Pronghorn use of recently burned mixed prairie in Alberta. *The Journal of Wildlife Management*, 53, 302–305.

(7) Klinger R.C., Kutilek M.J. & Shellhammer H.S. (1989) Population responses of black-tailed deer to prescribed burning. *The Journal of Wildlife Management*, 53, 863–871.

(8) Jourdonnais C.S. & Bedunah D.J. (1990) Prescribed fire and cattle grazing on an elk winter range in Montana. *Wildlife Society Bulletin*, 18, 232–240.

(9) Boggs J.F., Lochmiller R.L., McMurry S.T., Leslie D.M. & Engle D.M. (1991) *Cuterebra* infestations in small-mammal communities as influenced by herbicides and fire. *Journal of Mammalogy*, 72, 322–327.

(10) Lochmiller R.L., Boggs J.F., Mcmurry S.T., Leslie Jr, D.M. & Engle D.M. (1991) Response of cottontail rabbit populations to herbicide and fire applications on cross timbers rangeland. *Journal of Range Management*, 44, 150–155.

(11) Moreno S. & Villafuerte R. (1995) Traditional management of scrubland for the conservation of rabbits *Oryctolagus cuniculus* and their predators in Doñana National Park, Spain. *Biological Conservation*, 73, 81–85.

(12) Leslie Jr. D.M., Soper R.B., Lochmiller R.L. & Engle D.M. (1996) Habitat use by white-tailed deer on cross timbers rangeland following brush management. *Journal of Range Management*, 49, 401–406.

(13) Coppedge B.R. & Shaw J.H. (1998) Bison grazing patterns on seasonally burned tallgrass prairie. *Journal of Range Management*, 51, 258–264.

(14) Masters R.E., Lochmillern R.L., McMurry S.T. & Bukenhofer G.A. (1998) Small mammal response to pine-grassland restoration for red-cockaded woodpeckers. *Wildlife Society Bulletin*, 26, 148–158.

(15) Biondini M.E., Steuter A.A. & Hamilton R.G. (1999) Bison use of fire-managed remnant prairies. *Journal of Range Management*, 52, 454–461.

(16) Smith T.S., Hardin P.J. & Flinders J.T. (1999) Response of bighorn sheep to clearcut logging and prescribed burning. *Wildlife Society Bulletin*, 27, 840–845.

(17) Gureja N. & Owen-Smith N. (2002) Comparative use of burnt grassland by rare antelope species in a lowveld game ranch, South Africa. *South African Journal of Wildlife Research*, 32, 31–38.

(18) Rogers J.O., Fulbright T.E. & Ruthven D.C. III (2004) Vegetation and deer response to mechanical shrub clearing and burning. *Journal of Range Management*, 57, 41–48, https://doi.org/10.2307/4003953

(19) Watson L.H., Odendaal H.E., Barry T.J. & Pietersen J. (2005) Population viability of Cape mountain zebra in Gamka Mountain Nature Reserve, South Africa: the influence of habitat and fire. *Biological Conservation*, 122, 173–180, https://doi.org/10.1016/j.biocon.2004.06.014

(20) Greenberg C.H., Otis D.L., Waldrop T.A. (2006) Response of white-footed mice (*Peromyscus leucopus*) to fire and fire surrogate fuel reduction treatments in a southern Appalachian hardwood forest. *Forest Ecology and Management*, 234, 355–362, https://doi.org/10.1016/j.foreco.2006.07.022

(21) Milne-Laux S. & Sweitzer R.A. (2006) Experimentally induced colony expansion by black-tailed prairie dogs (*Cynomys ludovicianus*) and implications for conservation. *Journal of Mammalogy*, 87, 296–303, https://doi.org/10.1644/05-mamm-a-056r2.1

(22) Monroe M.E. & Converse S.J. (2006) The effects of early season and late season prescribed fires on small mammals in a Sierra Nevada mixed conifer forest. *Forest Ecology and Management*, 236, 229–240, https://doi.org/10.1016/j.foreco.2006.09.008

(23) Van Dyke F. & Darragh J.A. (2006) Short-and long-term changes in elk use and forage production in sagebrush communities following prescribed burning. *Biodiversity and Conservation*, 15, 4375–4398, https://doi.org/10.1007/s10531-005-4383-3

(24) Greenberg C.H., Miller S. & Waldrop T.A. (2007) Short-term response of shrews to prescribed fire and mechanical fuel reduction in a Southern Appalachian upland hardwood forest. *Forest Ecology and Management*, 243, 231–236, https://doi.org/10.1016/j.foreco.2007.03.003

(25) Hood G.A., Bayley S.E. & Olson W. (2007) Effects of prescribed fire on habitat of beaver (*Castor canadensis*) in Elk Island National Park, Canada. *Forest Ecology and Management*, 239, 200–209, https://doi.org/10.1016/j.foreco.2006.12.005

(26) Van Dyke F. & Darragh J.A. (2007) Response of elk to changes in plant production and nutrition following prescribed burning. *The Journal of Wildlife Management*, 71, 23–29, https://doi.org/10.2193/2005-464

(27) Amacher A.J., Barrett R.H., Moghaddas J.J. & Stephens S.L. (2008) Preliminary effects of fire and mechanical fuel treatments on the abundance of small mammals in the mixed-conifer forest of the Sierra Nevada. *Forest Ecology and Management*, 255, 3193–3202, https://doi.org/10.1016/j.foreco.2007.10.059

(28) Hassan S.N., Rusch G.M., Hytteborn H., Skarpe C. & Kikula I. (2008) Effects of fire on sward structure and grazing in western

Serengeti, Tanzania. *African Journal of Ecology*, 46, 174–185, https://doi.org/10.1111/j.1365-2028.2007.00831.x

(29) Thompson C.M., Augustine D.J. & Mayers D.M. (2008) Swift fox response to prescribed fire in shortgrass steppe. *Western North American Naturalist*, 68, 251–256, https://doi.org/10.3398/1527-0904(2008)68[251:sfrtpf]2.0.co;2

(30) Long R.S., Rachlow J.L. & Kie J.G. (2009) Sex-specific responses of North American elk to habitat manipulation. *Journal of Mammalogy*, 90, 423–432, https://doi.org/10.1644/08-mamm-a-181.1

(31) Kalies E.K., Chambers C.L. & Covington W.W. (2010) Wildlife responses to thinning and burning treatments in southwestern conifer forests: A meta-analysis. *Forest Ecology and Management*, 259, 333–342, https://doi.org/10.1016/j.foreco.2009.10.024

(32) Cazau M., Garel M. & Maillard D. (2011) Responses of heather moorland and Mediterranean mouflon foraging to prescribed-burning and cutting. *The Journal of Wildlife Management*, 75, 967–972, https://doi.org/10.1002/jwmg.117

(33) Legge S., Murphy S., Kingswood R., Maher B. & Swan D. (2011) EcoFire: restoring the biodiversity values of the Kimberley region by managing fire. *Ecological Management & Restoration*, 12, 84–92, https://doi.org/10.1111/j.1442-8903.2011.00595.x

(34) Pasch B. & Koprowski J.L. (2011) Impacts of fire suppression on space use by Mexican fox squirrels. *Journal of Mammalogy*, 92, 227–234, https://doi.org/10.1644/10-mamm-a-133.1

(35) Eby S. & Ritchie M.E. (2013) The impacts of burning on Thomson's gazelles', *Gazella thomsonii*, vigilance in Serengeti National Park, Tanzania. *African Journal of Ecology*, 51, 337–342, https://doi.org/10.1111/aje.12044

(36) Isaacs L., Somers M.J. & Dalerum F. (2013) Effects of prescribed burning and mechanical bush clearing on ungulate space use in an African savannah. *Restoration Ecology*, 21, 260–266, https://doi.org/10.1111/j.1526-100x.2012.00877.x

(37) Moreno S. & Rouco C. (2013) Responses of a small-mammal community to habitat management through controlled burning in a protected Mediterranean area. *Acta Oecologica*, 49, 1–4, https://doi.org/10.1016/j.actao.2013.02.001

8.2. Burn at specific time of year

https://www.conservationevidence.com/actions/2416

- **Two studies** evaluated the effects on mammals of burning at a specific time of year. One study was in Australia[1], and one was in the USA[2].

COMMUNITY RESPONSE (0 STUDIES)

POPULATION RESPONSE (1 STUDY)

- **Abundance (1 study):** A replicated, randomized, controlled, before-and-after study in the USA[2] found that carrying out prescribed burns in autumn did not increase small mammal abundances or biomass relative to burning in summer.

- **Survival (1 study):** A randomized, replicated, controlled study in Australia[1] found that in forest burned early in the dry season, northern brown bandicoot survival rate declined less than in forests burned late in the dry season.

BEHAVIOUR (0 STUDIES)

Background

Fire is an integral part of the management and natural dynamics of some ecosystems. Some habitats are naturally fire-prone, while in others, habitats are shaped by long-term traditional management (Bowman 1998). Some habitats are now managed through prescribed burning, partly to reduce the risk of uncontrolled wildfires. The timing of such burns may impact the mammal fauna with changes to different burning dates potentially being beneficial (or, at least, less damaging) to some species.

See also: *Use prescribed burning.*

Bowman D.M.J.S. (1998) Tansley Review No. 101. The impact of Aboriginal landscape burning on the Australian biota. *New Phytologist*, 140, 385–410.

A randomized, replicated, controlled study in 1989–1995 of a forest site in Northern Territory, Australia (1) found that in forest burned early in the dry season, northern brown bandicoot *Isoodon macrourus* survival rate declined less than in forests burned late in the dry season. In early burn sites, the bimonthly survival rate fell during the study from 0.76 to 0.59 compared to a larger reduction in sites burned later in the year, from 0.78 to 0.19. Four compartments each extended across 15–20 km². Two were burned early in the dry season

(May–June) and two were burned late in the dry season (September–October, mimicking wildfire). Treatments were assigned randomly to compartments and were applied annually in 1990–1994. Bandicoots were surveyed by live-trapping in each compartment, over two nights, bimonthly, from July 1989 to May 1995.

A replicated, randomized, controlled, before-and-after study in 2001–2004 of a coniferous woodland in California, USA (2) found that carrying out prescribed burns in autumn did not increase small mammal abundances or biomass relative to burning in summer. Timing of burning did not significantly affect abundances of deer mice *Peromyscus maniculatus* or lodgepole chipmunks *Neotamias speciosus* or overall small-mammal biomass (results presented as model outputs). Nine plots, 15–20 ha in area, were studied. Three were burned between 28 September and 28 October 2001 and three were burned on 20 or 27 June 2002. Three plots were not burned. Treatments were allocated randomly to plots. Small mammals were sampled by live-trapping over eight consecutive nights and days each year. Sampling occurred in June–August 2001 (pre-treatment) and in June–September of 2002 and 2003 and June–August 2004.

(1) Pardon L.G., Brook B.W., Griffiths A.D. & Braithwaite R.W. (2003) Determinants of survival for the northern brown bandicoot under a landscape-scale fire experiment. *Journal of Animal Ecology*, 72, 106–115, https://doi.org/10.1046/j.1365-2656.2003.00686.x

(2) Monroe M.E. & Converse S.J. (2006) The effects of early season and late season prescribed fires on small mammals in a Sierra Nevada mixed conifer forest. *Forest Ecology and Management*, 236, 229–240, https://doi.org/10.1016/j.foreco.2006.09.008

8.3. Provide shelter structures after fire

https://www.conservationevidence.com/actions/2418

- We found no studies that evaluated the effects on mammals of providing shelter structures after fire.

'We found no studies' means that we have not yet found any studies that have directly evaluated this intervention during our systematic journal and report searches. Therefore, we have no evidence to indicate whether or not the intervention has any desirable or harmful effects.

Background

Fire is an integral part of the dynamics of some ecosystems. It can clear out woody material, creating ideal conditions for new growth or herbaceous plants and small trees that are utilized by mammalian grazers and browsers. However, fire can also be disruptive to species, by removing cover. It may make them more vulnerable to effects of extreme weather and to predation and can cause them to seek out remaining vegetated areas that provide some degree of shelter (e.g. Pereoglou *et al.* 2011). For rare or otherwise valued species, shelters, such as low boards with space underneath, might be distributed across the burn area to help mitigate these effects.

Pereoglou F., Macgregor C., Banks S.C., Ford F., Wood J. & Lindenmayer D.B. (2011) Refuge site selection by the eastern chestnut mouse in recently burnt heath. *Wildlife Research*, 2011, 38, 290–298, https://doi.org/10.1071/wr11007

8.4. Thin trees to reduce wildfire risk

https://www.conservationevidence.com/actions/2477

- **Three studies** evaluated the effects on mammals of thinning trees to reduce wildfire risk. All three studies were in the USA[1,2,3].

COMMUNITY RESPONSE (0 STUDIES)

POPULATION RESPONSE (2 STUDIES)

- **Abundance (2 studies):** A replicated, controlled, before-and-after study in the USA[1] found that reducing tree density increased abundances of two of four small mammal species. A systematic review in the USA[3] found that, in thinned forests, two mammal species were recorded in higher densities compared to in unmanaged forests, while three species showed no effect.

BEHAVIOUR (1 STUDY)

- **(1 study):** A replicated, controlled study in the USA[2] found that thinning followed by prescribed burning did not increase use of forest areas by North American elk in most season, stand age and sex comparisons.

Background

Through fire suppression, some forest areas have become denser than was the case under natural fire regimes. To reduce fuel loads and associated wildfire risk, trees may be thinned. By creating a more open woodland structure, this may encourage growth of herbaceous plants, shrubs and trees at lower levels, thus potentially providing increased resources for mammalian herbivores.

See also *Biological resource use — Thin trees within forest*, in which thinning is usually done for extraction of merchantable timber though reducing fuel loads may sometimes be a secondary motivation.

A replicated, controlled, before-and-after study in 1998–2003 of ponderosa pine *Pinus ponderosa* forest in Arizona, USA (1) found that reducing tree density increased abundances of two of four small mammal species. Deer mouse *Peromyscus maniculatus* and gray-collared chipmunk *Tamias cinereicollis* captures were both positively associated with decreasing tree density in treatment plots, but golden-mantled ground squirrel *Spermophilus lateralis* and Mexican woodrat *Neotoma exicana* captures showed no such relationship. Results were presented as statistical model outputs. Three blocks, each with four 14-ha plots, were studied. Treatments comprised removal of all trees except those dating from pre-European settlement and, within 18 m of those trees, retention of 1.5, 2 or 3 trees with dbh ≥41 cm (or twice this many trees with smaller dbh, if larger trees not available). Thinning was conducted in 1999. Most woody debris was then piled up and burned, followed by prescribed burning of the whole plot in April–July 2000. The fourth plot in each block was unmanaged. Small mammals were live-trapped in August–October in 1998–1999 and 2001–2003.

A replicated, controlled study in 2005–2006 in a coniferous forest in Oregon, USA (2) found that thinning followed by prescribed burning did not increase use of areas by North American elk *Cervus canadensis*, in most season, stand age and sex comparisons. Thinning and burning were carried out on the same plots, so their influences could not be separated. In spring, female elk used plots burned two and three years previously, proportionally to their availability, preferentially selected 4-year-old burns, and avoided 5-year-old burns. They showed no preference for thinned and burned plots in summer. Male elk did not show preference for any thinned and burned plots, relative to their availability, in spring or summer. Stands not thinned and burned were avoided by females and selected by males in spring. In summer they were selected by females and males showed no preference. Data all presented as selection ratios. In 2001–2004, twenty-six forest stands (average 26 ha) were thinned between May and October, followed by prescribed burning during September or October of either the same or the following year. Twenty-seven similar stands (average size 55 ha) were not thinned or burned. Eighteen female and five male elk were radio-collared in spring 2005 with 30 female and nine male elk radio-collared in spring 2006. Locations were recorded automatically, within 1 hour of sunset or sunrise.

A systematic review in 2008 of management aimed at restoring natural processes in conifer forests in southwestern USA (3) found that, in forests thinned by removing small-to medium-diameter trees, two mammal species were recorded in higher densities compared to in unmanaged forests, while three species showed no effect. Higher densities associated with thinning were seen in gray-collared chipmunk *Tamias cinereicollis* and Mexican woodrat *Neotoma mexicana*. No significant responses to thinning were detected for tassel-eared squirrel *Sciurus aberti*, deer mouse *Peromyscus maniculatus* or golden-mantled ground squirrel *Spermophilus lateralis*. The review used evidence from 22 studies and considered responses of species recorded in ≥5 studies. Densities of species in ponderosa pine *Pinus ponderosa* forests managed in five ways, to recreate natural conditions and forest dynamics and reduce wildfire risk, were compared with densities in unmanaged forest.

(1) Converse S.J., Block W.M. & White G.C. (2006) Small mammal population and habitat responses to forest thinning and prescribed fire. *Forest Ecology and Management*, 228, 263–273, https://doi.org/10.1016/j.foreco.2006.03.006

(2) Long R.S., Rachlow J.L. & Kie J.G. (2009) Sex-specific responses of North American elk to habitat manipulation. *Journal of Mammalogy*, 90, 423–432, https://doi.org/10.1644/08-mamm-a-181.1

(3) Kalies E.K., Chambers C.L. & Covington W.W. (2010) Wildlife responses to thinning and burning treatments in southwestern conifer forests: A meta-analysis. *Forest Ecology and Management*, 259, 333–342, https://doi.org/10.1016/j.foreco.2009.10.024

8.5. Remove burnt trees and branches after wildfire

https://www.conservationevidence.com/actions/2478

- **One study** evaluated the effects on mammals of removing burnt trees and branches after wildfire. This study was in Spain[1].

COMMUNITY RESPONSE (0 STUDIES)

POPULATION RESPONSE (1 STUDY)

- **Abundance (1 study):** A replicated, randomized, controlled study in Spain[1] found that removing burned trees and branches after wildfire did not increase European wild rabbit numbers compared to removing burned trees but leaving branches in place.

BEHAVIOUR (0 STUDIES)

Background

After wildfires, a frequent management option is to remove burnt trees but to leave branches on the ground for economic reasons and to prevent soil erosion. However, some mammals are thought to benefit from areas with a low density of woody material at ground level (e.g. Beja *et al.* 2007) so removing branches might benefit these species.

Beja P., Pais M. & Palma L. (2007) Rabbit *Oryctolagus cuniculus* habitats in Mediterranean scrubland: the role of scrub structure and composition. *Wildlife Biology*, 13, 28–37, https://doi.org/10.2981/0909-6396(2007)13[28: rochim]2.0.co;2

A replicated, randomized, controlled study in 2006–2008 of a pine-dominated forest in Catalonia, Spain (1) found that removing burned trees and branches after wildfire did not alter European wild rabbit *Oryctolagus cuniculus* numbers compared to removing burned trees but leaving branches in place. There was no significant difference between rabbit pellet numbers in plots with trees and branches removed (1,400–5,100 pellets/plot) and those with trees removed but branches left in place (3,100–7,700 pellets/plot). High-intensity wildfire in summer 2003 burned 4,600 ha of forest. Plots (100 × 100 m) were established, 200–6,615 m apart. All plots had burnt trees trunks removed in 2004. In 20 plots, branches were left on the ground. In 10 plots, branches were initially left on the ground, but most were then removed in spring 2006, though some were piled up and left in the plots. Rabbit relative abundance was assessed in June of 2006, 2007 and 2008 by counting latrines in 500 × 2 m transects.

(1) Rollan A. & Real J. (2011) Effect of wildfires and post-fire forest treatments on rabbit abundance. *European Journal of Wildlife Research*, 57, 201–209, https://doi.org/10.1007/s10344-010-0412-y

8.6. Remove mid-storey vegetation in forest

https://www.conservationevidence.com/actions/2480

- **One study** evaluated the effects on mammals of removing mid-storey vegetation in forest. This study was in the USA[1].

COMMUNITY RESPONSE (1 STUDY)

- **Richness/diversity (1 study):** A randomized, replicated, controlled study in the USA[1] found that after removing mid-storey vegetation, mammal species richness increased.

POPULATION RESPONSE (1 STUDY)

- **Abundance (1 study)**: A randomized, replicated, controlled study in the USA[1] found that after removing mid-storey vegetation, mammal abundance increased.

BEHAVIOUR (0 STUDIES)

> **Background**
>
> Through fire suppression, some forest areas have developed denser mid-storey vegetation (trees at intermediate height between the ground layer and the forest canopy) than was formerly the case. To reduce wildfire risk and for habitat restoration purposes, mid-storey vegetation may be removed either mechanically or through prescribed burning. This intervention considered specifically manual or mechanical removal of mid-storey vegetation and how this may affect forest mammals.
>
> See also: *Use prescribed burning*.

A randomized, replicated, controlled study in 1992–1994 of pine-grassland in a mountainous area of Arkansas, USA (1) found that after removing mid-storey vegetation, mammal abundance and species richness increased. Small-mammal-trapping success was higher in mid-storey-removal plots (caught in 3.8–7.4% of traps) than in unmanaged plots (0.9–2.2% of traps). Average species richness was higher in mid-storey removal plots (1.7–4.7 species) than in unmanaged plots (1.3–2.7 species). Forest mid-storey was mechanically removed in 14–45-ha plots. Management timing is unclear, but the practice was initiated in the study area in 1990, primarily to benefit red-cockaded woodpeckers *Picoides borealis*. Small mammals were live-trapped at 80 stations/plot from 27 December to 4 January. Surveys were conducted in three plots in 1992–1993 and three different plots in 1993–1994. At the same time, sampling was conducted in three plots with retained mid-storey vegetation.

(1) Masters R.E., Lochmillern R.L., McMurry S.T. & Bukenhofer G.A. (1998) Small mammal response to pine-grassland restoration for red-cockaded woodpeckers. *Wildlife Society Bulletin*, 26, 148–158.

8.7. Remove understorey vegetation in forest

https://www.conservationevidence.com/actions/2482

- **Three studies** evaluated the effects on mammals of removing understorey vegetation in forest. All three studies were in the USA[1,2,3].

COMMUNITY RESPONSE (0 STUDIES)

POPULATION RESPONSE (3 STUDIES)

- **Abundance (3 studies):** Three replicated, randomized, controlled studies (two also before-and-after), in the USA[1,2,3], found that compared to prescribed burning, mechanically removing understorey vegetation growth in forests did not increase abundances of white-footed mice[1], shrews[2] or four rodent species[3].

BEHAVIOUR (0 STUDIES)

Background

Through fire suppression, some forest areas have developed denser understorey vegetation than was the case under natural fire regimes. To reduce fuel loads and associated wildfire risk, understorey vegetation may be removed. Prescribed burning is one option for doing this, but the rapid habitat change that this causes, together with potential loss of food resources and shelter, could negatively impact forest-floor mammals. This intervention, therefore, considers specifically manual or mechanical removal of understorey vegetation as an alternative to prescribed burning, and how this affects forest mammals.

See also: *Use prescribed burning.*

A replicated, randomized, controlled, before-and-after study in 2001–2003 in North Carolina, USA (1) found that mechanically removing understorey vegetation in forest, to reduce fuel load and associated wildfire risk, did not increase white-footed mouse *Peromyscus leucopus* abundance compared to using prescribed fire. Mouse abundance increased across all treatments during the study, but the rate of increase in understorey removal plots (from 14 to 30 mice/plot) was not significantly different to that in prescribed burning plots (from 9 to 36 mice/plot). Plots (each >14 ha) were established in three blocks. In each block, understorey growth was mechanically felled in one plot in winter 2001–2002 and prescribed burning was carried out in a different plot in March 2003. Mice were live-trapped over 10 consecutive days and nights in July and August of 2001 (before management) and 2003 (after management).

A replicated, randomized, controlled study in 2003–2004 in North Carolina, USA (2) found that mechanically removing understorey vegetation in forest, to reduce fuel load and associated wildfire risk, did not increase shrew abundance compared to using prescribed fire. The number of shrews caught did not differ significantly between understorey removal plots and prescribed burning plots in the first year (understorey removal: 22 shrews/plot; burning: 15) or the second year (understorey removal: 7 shrews/plot; burning: 6) after treatments were applied. Plots (each >14 ha) were established in three blocks. Within each block, understorey growth was mechanically felled in one plot in winter 2001–2002 and prescribed burning was carried out in a different plot in March 2003. Shrews were surveyed using pitfall traps and drift fencing over 123 nights in 2003 and 125 nights in 2004.

A replicated, randomized, controlled, before-and-after study in 2001–2003 of a forest in California, USA (3) found that mechanically removing understorey vegetation in forest, to reduce fuel load and associated wildfire risk, did not increase abundances of California ground squirrels *Spermophilus beecheyi*, long-eared chipmunks *Tamias quadrimaculatus*, brush mice *Peromyscus boylii* or deer mice *Peromyscus maniculatus*, compared to using prescribed burning. Changes in capture rates between before and after treatments did not differ significantly between understorey removal plots and fire plots for California ground squirrel (understorey removal: 2.6 to 11.0; fire: 4.2 to 7.6/100 trap nights), long-eared chipmunk (understorey removal: 0.7 to 2.4; fire: 0.7 to 1.7/100 trap nights) or brush mouse (understorey removal: 0.6 to 1.4; fire: 0.1 to 1.4/100 trap nights). Deer mouse abundance declined with understorey removal (from 2.0 to 1.2/100 trap nights) compared to an increase with fire (from 0.5 to 2.0/100 trap nights). Forests stands were 14–29 ha each. In four stands, 90% of understorey trees were removed in 2001–2002. Four different stands were burned in October–November 2002. Small mammals were live-trapped over nine consecutive days and nights in July–August of 2001 (pre-treatment) and 2003 (post-treatment).

(1) Greenberg C.H., Otis D.L., Waldrop T.A. (2006) Response of white-footed mice (*Peromyscus leucopus*) to fire and fire surrogate fuel reduction treatments in a southern Appalachian hardwood forest. *Forest Ecology and Management*, 234, 355–362, https://doi.org/10.1016/j.foreco.2006.07.022

(2) Greenberg C.H., Miller S. & Waldrop T.A. (2007) Short-term response of shrews to prescribed fire and mechanical fuel reduction in a Southern Appalachian upland hardwood forest. *Forest Ecology and Management*, 243, 231–236, https://doi.org/10.1016/j.foreco.2007.03.003

(3) Amacher A.J., Barrett R.H., Moghaddas J.J. & Stephens S.L. (2008) Preliminary effects of fire and mechanical fuel treatments on the abundance of small mammals in the mixed-conifer forest of the Sierra Nevada. *Forest Ecology and Management*, 255, 3193–3202, https://doi.org/10.1016/j.foreco.2007.10.059

8.8. Remove trees and shrubs to recreate open areas of land

https://www.conservationevidence.com/actions/2483

- **Two studies** evaluated the effects on mammals of removing trees and shrubs to recreate open areas of land. Both studies were in the USA[1,2].

COMMUNITY RESPONSE (0 STUDIES)

POPULATION RESPONSE (1 STUDY)

- **Abundance (1 study):** A controlled study in the USA[1] found that where Ashe juniper trees were removed, there were higher abundances of three rodent species.

BEHAVIOUR (1 STUDY)

- **Use (1 study):** A before-and-after, site comparison study in the USA[2] found that removing trees increased use of areas by Rocky Mountain bighorn sheep.

Background

Through fire suppression, some forest areas have spread onto previously open ground or have developed denser understorey vegetation than was the case under natural fire regimes. To reduce fuel loads and restore more open habitats for mammalian herbivores, trees and shrubs may be removed. Specifically, this intervention includes studies where the intention is to recreate open areas on land onto which forest and scrub has spread.

For interventions that remove just limited vegetation layers within forests, or reduce tree density but leave forest cover, see *Remove mid-storey vegetation in forest*, *Remove understorey vegetation in forest* and *Thin trees to reduce wildfire risk*. For interventions looking to benefit mammals through management of longer-established forest, especially where these are carried out through timber harvesting operations, see *Biological Resource Use*.

A controlled study in 1995–1997 at a former savanna in Texas, USA (1) found that where Ashe juniper *Juniperus ashei* trees were removed, there were higher abundances of three rodent species. Results were not tested for statistical significance. There were more white-ankled mice *Peromyscus pectoralis* in areas where Ashe juniper were cut (96 mice caught) than in areas where no trees were cut (10 caught). The same pattern was true for white-footed mouse *Peromyscus leucopus* (cut: 22 mice caught; uncut: 1 mouse) and for hispid cotton rat *Sigmodon hispidus* (cut: 4 rats caught; uncut: 0 rats). In 1995–1996, Ashe juniper in three areas was cut with a chainsaw. In two further areas, no trees were cut. In all areas, native oak trees *Quercus* spp. were left uncut. In October 1995–May 1996, once a month, 20 traps baited with oats were laid along a 100-m-long transect in one cut area and similarly in two areas that had not been cut. In October 1996 to March 1997, three to four times each month, three cut areas and two uncut areas were monitored in the same way. Traps were set in the morning and checked at dawn. Animals caught were ear-tagged to enable identification of recaptures.

A before-and-after, site comparison study in 1986–1991 of a mixed grassland, shrubland and woodland site in Utah, USA (2) found that removing ponderosa pine *Pinus ponderosa* and mountain mahogany *Cercocarpus* spp. trees increased use of these areas by Rocky Mountain bighorn sheep *Ovis canadensis*. In areas where trees were removed, sheep activity increased by 165%, but in areas where no trees were cut, sheep activity declined by 45%. Across a 353-ha study area, 32% was clearcut, 49% was unmanaged and 18% was burned (results of burning treatment not present here). Sheep use patterns were assessed, before cutting or burning, from June 1986 to September 1988, by observing 25–30 radio-collared sheep daily. After burning and cutting, use was assessed in June–September 1991, by counting sheep, 62 times, from an 11-km transect.

(1) Schnepf K.A., Heselmeyer J.A. & Ribble D.O. (1998) Effects of cutting Ashe juniper woodlands on small mammal populations in the Texas Hill Country. *Natural Areas Journal*, 18, 333–337.

(2) Smith T.S., Hardin P.J. & Flinders J.T. (1999) Response of bighorn sheep to clearcut logging and prescribed burning. *Wildlife Society Bulletin*, 27, 840–845.

8.9. Provide artificial waterholes in dry season

https://www.conservationevidence.com/actions/2484

- **Three studies** evaluated the effects on mammals of providing artificial waterholes in the dry season. One study was in South Africa[1], one was in Tanzania[2] and one was in Jordan[3].

COMMUNITY RESPONSE (1 STUDY)

- **Richness/diversity (1 study):** A site comparison study in Tanzania[2] found that artificial waterholes were used by a similar number of large mammal species as was a natural waterhole.

POPULATION RESPONSE (0 STUDIES)

BEHAVIOUR (2 STUDIES)

- **Use (2 studies):** A study in South Africa[1] found that areas around artificial waterholes were used more by eight out of 13 mammalian herbivore species than was the wider landscape. A study in Jordan[3] found that artificial waterholes were used by striped hyenas.

Background

In response to reduced availability of natural water sources for mammals, artificial water holes may be constructed. These can help to enhance survival during drought periods. However, there are also concerns about negative effects on mammals from artificial waterholes, such as there being increased numbers of some common water-dependent species at expense of rarer herbivores (Smuts 1978), or that artificial waterholes maintain high populations that are then vulnerable to starvation (Walker *at al.* 1987).

Smuts G.L. (1978) Interrelations between predators, prey and their environment. *Bioscience*, 28, 316–320.

Walker B.H., Emslie R.H., Owen-Smith R.N., Scholes R.J. (1987) To cull or not to cull: lessons from a southern African drought. *Journal of Applied Ecology*, 24, 381–401.

A study in 1987–1993 in a mostly dry savanna protected area in the eastern Lowveld region, South Africa (1) found that, during the dry season, areas around artificial waterholes were used by higher numbers of animals of eight out of 13 mammalian herbivore species than was the wider landscape. Higher abundances near waterholes than across the wider landscape were recorded for eland *Taurotragus oryx*, Burchell's zebra *Equus burchelli*, buffalo *Syncerus caffer*, blue wildebeest *Connochaetes taurinus*, sable *Hippotragus niger*, white rhinoceros *Ceratotherium simum*, tsessebe *Damaliscus lunatus*, and roan *Hippotragus equinus* (data expressed as model results). However, the abundance of waterbuck *Kobus elipsiprimnus*, kudu *Tragelaphus strepsiceros*, giraffe *Giraffa camelopardalis*, impala *Aepyceros melampus* and elephant *Loxondonta africana* was lower near waterholes than across the wider landscape (data expressed as model results). In the 1930–1980s, more than 300 boreholes were drilled, 50 earth dams were constructed and seasonal and perennial rivers were dammed across Kruger National Park (>20,000 km^2). Mammals were counted during daytime by four observers, from a fixed-wing aircraft, during the dry season (May–August), in 1987–1993. Counts were made within 800-m wide transects, from 65–70 m high, flying at 95–100 knots.

A site comparison study in 2006 in a national park comprising woodland and savanna in Tanzania (2) found that artificial waterholes were used by a similar number of large mammal species as was a natural waterhole. Results were not tested for statistical significance. The number of species recorded at artificial waterholes (4–5 species) was similar to the number at the natural waterhole (three). Average numbers of impala *Aepyceros melampus* were considerably higher at one artificial waterhole (64 impalas) than at the natural waterhole (9). Giraffe *Giraffa camelopardalis* numbers were also higher at one artificial waterhole (26 giraffes) than at the natural waterhole (8). Two artificial waterholes and one natural waterhole were monitored. Large mammal numbers were estimated, in November 2006, by counting footprints and droppings in three 100-m^2 quadrats at each waterhole and by direct observation, for one day, from a vehicle.

A study in 2010–2012 in desert in a national park in Jordan (3) found that artificial waterholes were used by striped hyenas *Hyaena hyaena*. In the first year of monitoring, an estimated nine hyenas visited the two artificial waterholes with 10 hyenas visiting in the second year. Within a 320-km^2 national park, one artificial waterhole was created in 2003 and one in 2010. They were approximately 1 m in diameter and located 460 m apart. Hyenas were monitored using one camera trap at each water hole through August and September of 2010 and 2012. The park also contained approximately 60 permanent and semi-permanent natural waterholes and springs.

(1) Smit I.P.J., Grant C.C. & Devereux B.J. (2007) Do artificial waterholes influence the way herbivores use the landscape? Herbivore distribution patterns around rivers and artificial surface water sources in a large African savanna park. *Biological Conservation*, 136, 85–99, https://doi.org/10.1016/j.biocon.2006.11.009

(2) Epaphras A.M., Gereta E., Lejora I.A., Ole Meing'ataki G.E., Ng'umbi G., Kiwango Y., Mwangomo E., Semanini F., Vitalis L., Balozi J. & Mtahiko M.G.G. (2008) Wildlife water utilization and importance of artificial waterholes during dry season at Ruaha National Park, Tanzania. *Wetlands Ecology and Management*, 16, 183–188, https://doi.org/10.1007/s11273-007-9065-3

(3) Attum O., Rosenbarger D., Al awaji M., Kramer A & Eida E. (2017) Population size and artificial waterhole use by striped hyenas in the Dana Biosphere Reserve, Jordan. *Mammalia*, 81, 415–419, https://doi.org/10.1515/mammalia-2015-0155

8.10. Use fencing to protect water sources for use by wild mammals

https://www.conservationevidence.com/actions/2493

• We found no studies that evaluated the effects of using fencing to protect water sources for use by wild mammals.

'We found no studies' means that we have not yet found any studies that have directly evaluated this intervention during our systematic journal and report searches. Therefore, we have no evidence to indicate whether or not the intervention has any desirable or harmful effects.

Background

Water, from natural or artificial sources, can be an important resource, shaping the distribution of wild mammals in arid areas. Fencing may be installed to protect these water sources from domestic or wild animals whilst still permitting entry of smaller mammals (e.g. Gaudioso Lacasa *et al.* 2010).

Gaudioso Lacasa V., Sánchez García-Abad C., Prieto Martín R., Bartolomé Rodríguez D.J., Pérez Garrido J.A. & Alonso de La Varga M.E. (2010) Small game water troughs in a Spanish agrarian pseudo steppe: visits and water site choice by wild fauna. *European Journal of Wildlife Research*, 56, 591–599, https://doi.org/10.1007/s10344-009-0352-6

8.11. Provide supplementary food after fire

https://www.conservationevidence.com/actions/2494

• **One study** evaluated the effects on mammals of providing supplementary food after fire. This study was in the USA[1].

COMMUNITY RESPONSE (0 STUDIES)

POPULATION RESPONSE (1 STUDY)

• **Survival (1 study):** A replicated, randomized, controlled study in the USA[1] found that supplementary feeding did not increase survival of hispid cotton rats following prescribed fire.

BEHAVIOUR (0 STUDIES)

Background

This intervention specifically covers cases where supplementary food is provided in an attempt to offset threats associated with fire. Natural or prescribed fires, whilst being integral parts of some ecosystems, can temporarily reduce or remove available food. Supplementary food may be provided for rare or otherwise valued mammal species.

A replicated, randomized, controlled study in 2005–2009 of woodland in Georgia, USA (1) found that supplementary feeding did not increase survival rates of hispid cotton rats *Sigmodon hispidus* following prescribed fire. Survival rates over a 13-week post-fire period during which supplementary food was offered (0.02–0.04) were similar to those with no supplementary food offered (0.02–0.04). Eight plots (40 ha each) were studied. Four plots (exclosures) were surrounded by electric fencing to deter predator entry. All plots were burned in February of 2005, 2007, and 2009. From June 2007 to August 2009, two exclosures and two non-fenced plots received supplementary feed of rabbit chow. No food was provided at the other four plots. Pairs of grids were live-trapped four times/year from January 2005 to June 2007 and eight times/year from July 2007 to June 2009.

(1) Morris G., Hostetler J.A., Conner L.M. & Oli M.K. (2011) Effects of prescribed fire, supplemental feeding, and mammalian predator exclusion on hispid cotton rat populations. *Oecologia*, 167, 1005–1016, https://doi.org/10.1007/s00442-011-2053-6

9. Threat: Invasive alien and other problematic species

Background

Invasive and other problematic species of animals, plants and diseases have caused significant declines in many mammal species worldwide. Invasive species may prey on mammals, provide competition for resources, alter habitats or infect mammals with new diseases. This chapter describes the evidence from interventions designed to reduce the threat from invasive and other problematic species and disease.

See also: *Species management — Release translocated/captive-bred mammals to islands without invasive predators.*

For interventions that involve reducing predation by domestic cats *Felis catus* and dogs *Canis lupus familiaris* see the chapter *Threat: Residential and commercial development -Keep cats indoors or in outside runs to reduce predation of wild mammals, Use collar-mounted devices to reduce predation by domestic animals, Keep dogs indoors or in outside enclosures to reduce threats to wild mammals and Keep domestic cats and dogs well-fed to reduce predation of wild mammals.*

9.1. Use fencing to exclude grazers or other problematic species

https://www.conservationevidence.com/actions/2495

- **Three studies** evaluated the effects on mammals of using fencing to exclude grazers or other problematic species. One study was in each of the USA[1], Australia[2] and Spain[3].

 https://doi.org/10.11647/OBP.0234.09

COMMUNITY RESPONSE (1 STUDY)

- **Richness/diversity (1 study):** A controlled, before-and-after study in Australia[2] found that after fencing to exclude introduced herbivores, native mammal species richness increased.

POPULATION RESPONSE (3 STUDIES)

- **Abundance (3 studies):** Two controlled studies (including one replicated, paired sites study) in Spain[3] and Australia[2] found that using fences to exclude large[3] or introduced[2] herbivores increased the abundance of Algerian mice[3] and native mammals[2]. A replicated, paired sites study in the USA[1] found that in areas fenced to exclude livestock grazing and off-road vehicles, abundance of black-tailed hares was lower compared to in unfenced areas.

BEHAVIOUR (0 STUDIES)

Background

In areas that are occupied by non-native grazers or where domestic animals range freely over large areas, fencing may be used to prevent grazing in some areas. This may benefit some native mammals, such as herbivores that may otherwise be outcompeted for food resources.

A replicated, paired sites study in 1994–1995 in the Western Mojave Desert, California, USA (1) found that within an area fenced to exclude livestock grazing and off-road vehicles, abundance of black-tailed hares *Lepus californicus* was lower compared to unfenced areas. Fewer black-tailed hares were found in fenced plots (0–1.5 animals/transect; 1.5 droppings/1,250 cm^2) than in unfenced plots (1–4 animals/transect; 3–4 droppings/1,250 cm^2). In the Desert Tortoise Research Natural Area, off-road vehicles were prohibited from 1973, sheep grazing from 1978, and a 1 m high wire fence protecting the area was constructed by 1980. Two sites were selected near the north eastern and southern boundary. At each site, two 2.25-ha plots were established, one ≥400m inside the fenced area and one outside the fence (used by off-road vehicles until 1980 and grazed by sheep until 1994). Plots were matched for environmental variables. In each plot, hare numbers were estimated along four 1.2-km transects in May and July 1994, and at the north eastern site by counting pellets in 120 quadrats (40 × 50-cm) in April 1994 and 1995.

A controlled, before-and-after study in 2004–2007 in a woodland savannah in north-west Australia (2) found that after fencing to exclude introduced herbivores, the overall abundance and species richness of small-and medium-sized native mammals increased. After three years, the average number of mammals and mammal species/ plot was higher in sites from which introduced herbivores were excluded (abundance: 6.1–16.7 animals; species richness: 2.5–3.2 species) than in non-removal sites (abundance: 0.1–3.3 animals; species richness: 0.1–1.4 species). Overall abundance varied with habitat type and abundance increased with years since destocking for four of seven species (see original paper for details). In 2004–2005, a 40,300-ha area of Mornington Wildlife Sanctuary was fenced and cleared of large herbivores. Before 2004, the area had >2,000 cattle *Bos taurus* and >200 horses *Equus ferus caballus* and donkeys *Equus africanus asinus*. In 2007, less than 200 cattle remained. Native mammals were surveyed in twenty 0.25-ha plots in 2004 and in 42–43 plots annually in 2005–2007 (total 49 separate plots, most surveyed 3–4 times). By 2006 and 2007, sixteen plots still contained herbivores, and herbivores had been removed from the other plots (1–3 years previously). Each plot was surveyed using 20 box traps, four medium-sized cage traps and eight pitfall traps, for three consecutive nights each year. Fur was clipped to exclude recaptures.

A replicated, controlled, paired sites study in 2010–2012 in Holm oak *Quercus ilex* woodland in Cabañeros National Park, Central Spain (3) found that excluding large herbivores using fences increased the abundance of Algerian mice *Mus spretus*. The abundance of Algerian mice and the percentage of trees occupied by mice were higher inside exclosures (103 individuals caught; 60% of trees occupied) than outside (55 individuals caught; 30% of trees occupied). However, mice had higher levels of physiological stress indicators (faecal corticosterone metabolites) inside (33,041 ng/g dry faeces) than outside exclosures (29,225 ng/g). One 3 ha section of a 150 ha exclosure established in 1995 and a 4.7 ha exclosure established in 2008 were paired with grazed areas of equal size. Exclosures were fenced (2 m high) with a 32 x 16 cm mesh width that allowed movement of rodent predators but not of large herbivores. Mice were sampled during two consecutive nights in November 2010 and 2011 and February 2011 and 2012 using two Sherman traps placed under all 170 trees in the study sites. Fresh faecal samples from 92 different captured individuals were used to monitor faecal corticosterone metabolites.

(1) Brooks M. (1999) Effects of protective fencing on birds, lizards, and black-tailed hares in the western Mojave Desert. *Environmental Management*, 23, 387–400.

(2) Legge S., Kennedy M.S., Lloyd R.A.Y., Murphy S.A. & Fisher A. (2011) Rapid recovery of mammal fauna in the central Kimberley, northern Australia, following the removal of introduced herbivores. *Austral Ecology*, 36, 791–799, https://doi.org/10.1111/j.1442-9993.2010.02218.x

(3) Navarro-Castilla, Á., Diaz, M., & Barja, I. (2017). Does ungulate disturbance mediate behavioural and physiological stress responses in Algerian mice (*Mus spretus*)? A wild exclosure experiment. *Hystrix*, 28, 283–291.

9.2. Use fencing to exclude predators or other problematic species

https://www.conservationevidence.com/actions/2497

- **Ten** studies evaluated the effects on mammals of using fencing to exclude predators or other problematic species. Four studies were in Australia[2,3,8,10], four were in the USA[4,6,7,9] and two were in Spain[1,5].

COMMUNITY RESPONSE (1 STUDY)

- **Richness/diversity (1 study):** A site comparison study in Australia[3] found that fencing which excluded feral cats, foxes and rabbits increased small mammal species richness.

POPULATION RESPONSE (10 STUDIES)

- **Abundance (4 studies):** Two of three of studies (including two replicated, controlled studies), in Spain[1], Australia[3] and the USA[4], found that abundances of European rabbits[1] and small mammals[3] were higher within areas fenced to exclude predators or other problematic species, compared to in unfenced areas. The third study found that hispid cotton rat abundance was not higher with predator fencing[4]. A replicated, controlled study in Spain[5] found that translocated European rabbit abundance was higher in fenced areas that excluded both terrestrial carnivores and raptors than in areas only accessible to raptors.

- **Reproductive success (1 study):** A replicated, controlled study in USA[9] found that predator exclosures increased the number of white-tailed deer fawns relative to the number of adult females.

- **Survival (7 studies):** Four of six studies (including four replicated, controlled studies) in Spain[1], Australia[2,8,10] and the USA[4,6], found that fencing to exclude predators did not increase survival of translocated European rabbits[1], hispid cotton rats[2], southern flying squirrels[6] or western barred bandicoots[10]. The other two studies found that persistence of populations of eastern barred bandicoots[2] and long-haired rats[8] was greater inside than outside fences. A controlled, before-and-after study in the USA[7] found that electric fencing reduced coyote incursions into sites frequented by black-footed ferrets.

BEHAVIOUR (0 STUDIES)

Background

Predators can drive declines or local extinctions of vulnerable mammal species. Non-native predators may be a particular problem for native mammals that lack sufficient predator avoidance behaviours (e.g. Jones *et al.* 2004). Native predators can also threaten populations of mammals that persist in low numbers. Predator control may be impractical to sustain on a sufficient scale or may attract opposition on animal welfare grounds. Fencing, including electric fencing, may be a viable or more effective alternative in some situations.

See also *Species Management — Release translocated mammals into fenced areas* and *Release captive-bred mammals into fenced areas.*

Jones M.E., Smith G.C. & Jones S.M. (2004) Is anti-predator behaviour in Tasmanian eastern quolls (*Dasyurus viverrinus*) effective against introduced predators? *Animal Conservation*, 7, 155–160, https://doi.org/10.1017/s136794300400126x

A replicated, controlled study in 2002–2003 in four grassland and shrubland sites in south-west Spain (1) found that the survival of translocated European rabbits *Oryctolagus cuniculus* was similar between a plot fenced to exclude predators and an unfenced plot, but that abundance was higher in fenced plots. Three months after translocation, rabbit survival in fenced plots (40%) was not significantly different to survival in unfenced plots (57%). However, four months after translocation, the relative abundance of rabbits was higher in fenced than in unfenced plots (data presented as log pellet abundance/plot). Four translocation plots (>1 km apart), each 4 ha with 18 artificial warrens surrounded by low fencing, were established in the south of Sierra Norte of Seville Natural Park. Two plots were fenced (1 m below and 2.5 m above ground, with an electric wire on top) and two unfenced. A total of 724 wild rabbits were released in similar numbers into each plot distributed evenly between warrens. Rabbit survival was based on 45 radio-collared rabbits (19 in one fenced and 26 in one unfenced plot) located 5–7 times/week for 15 weeks. Abundance was estimated four months after translocation by counting pellets in ten 18-cm-diameter circles/warren.

A review of translocation studies in 1989–2005 in eight grassland and forest sites in Victoria, Australia (2) found that translocated eastern barred bandicoot *Perameles gunnii* populations released inside predator barrier-fencing persisted more successfully than did those translocated into unfenced areas. All three populations translocated into fenced areas persisted at the end of the study (1–26 years post-release). Only one out of five populations translocated to unfenced areas was known to persist at the end of the study (6–13 years post-release). Two populations were presumed extinct and the status was unclear, but with few recent records, at two other sites. Between 22 and 174 bandicoots were translocated into three fences sites (100–585 ha) and between 50 and 103 into five unfenced sites (85–500 ha) in 1989–2005. Translocated animals were both captive-bred and wild-born. Five sites had community involvement with the control of invasive red foxes *Vulpes vulpes*. Released bandicoots were provided with supplementary food for up to 10 days, in at least two sites. In most sites, bandicoots were monitored by trapping, but frequency and methods are not described.

A site comparison study in 1993–2007 on a shrubland site in South Australia (3) found that using fencing to exclude feral mammals (cats *Felis catus*, foxes *Vulpes vulpes* and rabbits *Oryctolagus cuniculus*) increased the abundance and species richness of small mammals. Small mammal abundance in the absence of feral mammals (10.3 individuals/sample) was higher than where feral mammals were present (3.6 individuals/ sample). Species richness followed a similar pattern (feral mammals absent: 3.0 species/sample; feral mammals present: 1.7 species/sample). An area of approximately 5 × 5 km was fenced to exclude feral mammals and cattle in 1999. An adjacent area, approximately 9× 9 km, was fenced in 1986 to exclude cattle, but not feral mammals. Small mammals were sampled using pitfall traps for a 10-day period in either December or January. Three points in the feral mammal and cattle exclosure were sampled in 2007. Five points in the cattle-only exclosure were sampled in 1993–1996 and again in 2007.

A replicated, randomized, controlled study in 2005–2009 in eight woodland sites in Georgia, USA (4) found that excluding predators did not increase survival, transition to reproductive states or abundance of hispid cotton rats *Sigmodon hispidus*. In non-fire periods, estimated 13-week survival in exclosures (0.16–0.39) were similar to that outside exclosures (0.16–0.38). The same pattern applied in fire periods (exclosures: 0.02–0.04; outside exclosures: 0.02–0.04). Rates of transition to reproductive states varied considerably with season and fire status but were not affected by predator exclusion (exclosure: 0.06–0.59; outside exclosure: 0.06–0.59). Averaged across all plots, predator exclusion did not change abundance (data not presented). Eight plots (40 ha each) were studied. Four were exclosures, with electric fencing to deter predator entry, and four were unfenced. All plots were burned in February 2005, 2007, and 2009. Pairs of grids were live-trapped four times/year from January 2005 to June 2007 and eight times/year from July 2007 to June 2009.

A replicated, controlled study in 2010 at a site in Sierra Morena, Spain (5) found that the abundance of translocated European rabbits *Oryctolagus cuniculus* was higher in areas fenced to exclude both terrestrial carnivores and raptors (top-closed) than in areas only accessible to raptors (top-open) during the six weeks after release. The weekly abundance of rabbits in top-closed plots (1.2–4.8 pellet abundance index) was higher than in top-open plots (0.7–3.2 pellet abundance index). The highest difference in rabbit abundance between top-closed and top-open plots was attained in the first 2 weeks. Five 0.5-ha plots, close together, were fenced (0.5 m below and 2 m above the ground with two electric wires and a floppy overhang) to exclude terrestrial carnivores. Each had five artificial warrens. Two plots had top net (top-closed) and three had no top net (top-open). Twenty-five adult wild rabbits (20 female) were released in each exclosure in February 2010. Rabbit abundance was estimated through pellet counts in 20 fixed 0.5-m^2 circular sampling sites each week for six weeks after translocation.

A replicated, controlled study in 2005–2009 in four woodland sites in Georgia, USA (6) found that using fencing to exclude predators did not increase survival of southern flying squirrels *Glaucomys volans*. Monthly survival rates for squirrels was similar in areas that were fenced to exclude predators and areas that were not fenced (data reported as model results). Four plots were fenced with a 1.2-m tall, electrified, fence while four plots were not fenced. Plots were 36–49 ha. One-hundred and forty-four traps baited with oats and bird feed were placed on the ground in each plot and 24 traps were placed in trees. Between January 2005 and June 2007, trapping was carried out four times a year and, in July 2007–September 2009, trapping was carried out eight times a year. Trapping was conducted over four consecutive nights. Animals caught were marked with ear tags.

A controlled, before-and-after study in 2010 at a grassland in Montana, USA (7) found that electric fencing reduced coyote *Canis latrans* incursions into black-tailed prairie dog *Cynomys ludovicianus* colonies that supported breeding black-footed ferrets *Mustela nigripes*. There was a lower rate of coyote incursions with the fence in place (four incursions during 84 search nights — 7% of coyote sightings during this period) than before it was installed (eight from 24 search nights — 42% of sightings) and after it was removed (20 from 34 search nights — 47% of sightings). Black-footed ferrets were reintroduced to the site in 1994. Two electric (electronet) fences, totalling 7.7 km and enclosing 108 ha, were erected on 27 July 2010 and removed on

2 October 2010. Fencing comprised nine horizontal poly-conductors, 10 cm apart, alternating between grounded and charged. Conductive polytape (2 cm wide) was strung above this at 107 cm high. Coyote sightings were noted inside fenced areas and in two unfenced areas during spotlight ferret surveys from 28 June to 26 July (pre-exclosure), 27 July to 2 October (exclosure) and 3 October to 24 October (post-exclosure). Coyotes found inside exclosures were expelled through temporarily lowered fence sections.

A replicated, paired sites, controlled study in 2011–2013 in two tropical savanna sites in the Northern Territory, Australia (8) found that fencing to exclude cats *Felis silvestris catus* prevented the local extirpation of released long-haired rats *Rattus villosissimus*. After 18 months, rats persisted in enclosures not accessible to cats (3.1–8.7 rats/enclosure) but were absent in compartments accessible to cats (0.0 rats/enclosure). Two 12.5-ha enclosures were established 13 km apart in Wongalara Wildlife Sanctuary. One half of each enclosure was surrounded by a 0.9-m-high fence that allowed access to cats and dingoes *Canis dingo* and the other half by a 2-m electrified 'floppy-top' fence that excluded cats and dingoes. Enclosures had a 40-cm barrier that prevented rats from moving in or out. Fifteen to 23 long-haired rats were introduced to each of the four compartments in October 2011 or April 2012. Rat abundance was monitored until June 2013 by live-trapping at two-month intervals (from 2 or 6 months after release) using 36 box traps in each compartment, deployed over 2–4 consecutive nights.

A replicated, controlled study in 2011–2012 of a forest in Georgia, USA (9) found that predator exclosures increased the fawn:adult female ratio of white-tailed deer *Odocoileus virginianus*. The average annual fawn:adult female ratio recorded was greater inside exclosures (0.19) than outside (0.09) exclosures. Authors reported that figures were relative rather than absolute ratios, as some fawns may have been too small to travel with their mothers at the time of sampling. Four 40-ha plots were fenced to exclude predators. The fence was 1.2 m tall and was electrified. Predators inside exclosures were live-trapped and released outside. Deer ≥12 weeks old were able to jump the fence. Four similar plots were established, but without a predator exclusion fence. Fawn and adult female ratios were determined using two camera traps in each plot, for two weeks in August 2011 and two weeks in August 2012.

A study in 1995–2010 on a shrubland-dominated peninsula in Western Australia, Australia (10) found that a translocated population of western barred bandicoots *Perameles bougainville* did not persist despite fencing to exclude invasive red foxes *Vulpes vulpes* and cats *Felis catus*. Nine years after being translocated into a fenced area, bandicoot numbers increased to an estimated 467 but over the next three years, the population fell to zero. Fourteen bandicoots were initially translocated in 1995–1996 from an offshore island to a 17-ha enclosure within a 1,200-ha section of a mainland peninsula, fenced to exclude foxes and feral cats. The peninsular fence was built in 1989 and despite being rebuilt and repaired several times, it was never an effective barrier to foxes and cats. Throughout the study period, foxes and cats were controlled inside the fenced area by baiting (using 1080 poison) and cats were also trapped and shot. Starting in May 1997 and over 10 years, 82 bandicoots were released from the enclosure to the fenced peninsula. Bandicoots were monitored along a 40 km track network, with cage traps set at 100-m intervals over two nights each three months from August 1995-October 2002 and then twice/year until September 2010 (25,000 trap-nights).

(1) Rouco C., Ferreras P., Castro F. & Villafuerte R. (2008) The effect of exclusion of terrestrial predators on short-term survival of translocated European wild rabbits. *Wildlife Research*, 35, 625–632, https://doi.org/10.1071/wr07151

(2) Winnard A.L. & Coulson G. (2008) Sixteen years of Eastern Barred Bandicoot *Perameles gunnii* reintroductions in Victoria: a review. *Pacific Conservation Biology*, 14, 34–53, https://doi.org/10.1071/pc080034

(3) Read J.L. & Cunningham R. (2010) Relative impacts of cattle grazing and feral animals on an Australian arid zone reptile and small mammal assemblage. *Austral Ecology*, 35, 314–324, https://doi.org/10.1111/j.1442-9993.2009.02040.x

(4) Morris G., Hostetler J.A., Conner L.M. & Oli M.K. (2011) Effects of prescribed fire, supplemental feeding, and mammalian predator exclusion on hispid cotton rat populations. *Oecologia*, 167, 1005–1016, https://doi.org/10.1007/s00442-011-2053-6

(5) Guerrero-Casado J., Ruiz-Aizpurua L. & Tortosa F. S. (2013) The short-term effect of total predation exclusion on wild rabbit abundance in restocking plots. *Acta Theriologica*, 58, 415–418, https://doi.org/10.1007/s13364-013-0140-2

(6) Karmacharya B., Hostetler J.A., Conner L.M., Morris G. & Oli M.K. (2013) The influence of mammalian predator exclusion, food supplementation, and prescribed fire on survival of *Glaucomys volans*. *Journal of Mammalogy*, 94, 672–682, https://doi.org/10.1644/12-mamm-a-071.1

(7) Matchett M.R., Breck S.W. & Callon J. (2013) Efficacy of electronet fencing for excluding coyotes: a case study for enhancing production of black-footed ferrets. *Wildlife Society Bulletin*, 37, 893–900, https://doi.org/10.1002/wsb.348

(8) Frank A.S., Johnson C.N., Potts J.M., Fisher A., Lawes M.J., Woinarski J.C., Tuft K., Radford I.J., Gordon I.J., Collis M.A. & Legge S. (2014) Experimental evidence that feral cats cause local extirpation of small mammals in Australia's tropical savannas. *Journal of Applied Ecology*, 51, 1486–1493, https://doi.org/10.1111/1365-2664.12323

(9) Conner L.M., Cherry M.J., Rutledge B.T., Killmaster C.H., Morris G. & Smith L.L. (2016) Predator exclusion as a management option for increasing white-tailed deer recruitment. *The Journal of Wildlife Management*, 80, 162–170, https://doi.org/10.1002/jwmg.999

(10) Short J. (2016) Predation by feral cats key to the failure of a long-term reintroduction of the western barred bandicoot (*Perameles bougainville*). *Wildlife Research*, 43, 38–50, https://doi.org/10.1071/wr15070

Invasive Non-Native/Alien Species/Diseases

9.3. Remove/control non-native amphibians (e.g. cane toads)

https://www.conservationevidence.com/actions/2498

- We found no studies that evaluated the effects on mammals of removing or controlling non-native amphibians.

'We found no studies' means that we have not yet found any studies that have directly evaluated this intervention during our systematic journal and report searches. Therefore, we have no evidence to indicate whether or not the intervention has any desirable or harmful effects.

Background

Whilst there are relatively few documented examples of non-native amphibians having direct detrimental impacts on native mammals, the spread of cane toads *Bufo marinus* in Australia is reported to have accelerated declines in northern quoll *Dasyurus hallucatus* which are poisoned in predation attempts on the toads (Woinarski *et al.* 2011). A range of methods for controlling cane toads, including biological control, have been proposed (e.g. Shanmuganathan *et al.* 2010; Ward-Fear *et al.* 2010).

Shanmuganathan T., Pallister J., Doody S., McCallum H., Robinson T., Sheppard A., Hardy C., Halliday D., Venables D., Voysey R., Strive T., Hinds L. & Hyatt A. (2010) Biological control of the cane toad in Australia: a review. *Animal Conservation*, 13(S1), 16–23, https://doi.org/10.1111/j.1469-1795.2009.00319.x

Ward-Fear G., Brown G.P. & Shine R. (2010) Using a native predator (the meat ant, *Iridomyrmex reburrus*) to reduce the abundance of an invasive species (the cane toad, *Bufo marinus*) in tropical Australia. *Journal of Applied Ecology*, 47, 273–280, https://doi.org/10.1111/j.1365-2664.2010.01773.x

Woinarski J.C.Z., Legge S., Fitzsimons J.A., Traill B.J., Burbidge A.A., Fisher A., Firth R.S.C., Gordon I.J., Griffiths A.D., Johnson C.N., McKenzie N.L., Palmer C., Radford I., Rankmore B., Ritchie E.G., Ward S. & Ziembicki M. (2011) The disappearing mammal fauna of northern Australia: context, cause, and response. *Conservation Letters*, 4, 192–201, https://doi.org/10.1111/j.1755-263x.2011.00164.x

9.4. Remove/control non-native invertebrates

https://www.conservationevidence.com/actions/2501

- **One study** evaluated the effects on mammals of removing or controlling non-native invertebrates. This study was in the USA[1].

COMMUNITY RESPONSE (0 STUDIES)

POPULATION RESPONSE (1 STUDY)

- **Abundance (1 study):** A replicated, controlled, before-and-after study the USA[1] found that after the control of red imported fire ants, capture rates of northern pygmy mice increased.

BEHAVIOUR (0 STUDIES)

Background

Non-native invertebrates can affect mammals in a number of ways. Alterations to habitats and predation on other species could reduce feeding resources available to mammals and, in some cases, direct predation on mammals can occur (Masser & Grant 1986). Such effects can lead to mammals avoiding areas occupied by non-native invertebrates (Killion & Grant 1993). Control of such species may be carried out in an attempt to reverse these impacts.

Masser M.P. & Grant W.E. (1986) Fire ant-induced trap mortality of small mammals in east-central Texas. *The Southwestern Naturalist*, 31, 540–542.

Killion M.J. & Grant W.E. (1993) Scale effects in assessing the impact of imported fire ants on small mammals. *The Southwestern Naturalist*, 38, 393–396.

A replicated, controlled, before-and-after study in 1989–1990 in coastal grassland and shrubland in Texas, USA (1) found that after the control of red imported fire ants *Solenopsis invicta*, capture rates of northern pygmy mice *Baiomys taylori* increased. Northern pygmy mouse capture rates increased more where red fire ants were controlled (from 6–9/plot during first three months (over winter) of ant control to 19–25/plot nine months later) than in uncontrolled areas (8–9/plot during first three months of ant control to 11–15/plot nine months later). Captures were similar between plots in the summer before treatments began (19–27 mice/plot). In June 1989, two 110 × 130-m plots were established at the Welder Wildlife Foundation refuge. Each plot was divided into a treatment area and an untreated area. In treatment areas, an aerosol insecticide (active ingredient 0.7% pyrethrin) was injected directly into ant mounds while a bait insecticide (active ingredient 0.88% amidinohydrazone) was deployed monthly, from November 1989 to October 1990. Between June 1989 and October 1990, mice were sampled for four days/month using 108 baited Sherman live traps/plot. Animals were marked at first capture, and only included in analysis when caught for a second time.

(1) Killion M.J., Grant W.E. & Vinson S.B. (1995) Response of *Baiomys taylori* to changes in density of imported fire ants. *Journal of Mammalogy*, 76, 141–147.

9.5. Remove/control non-native mammals

https://www.conservationevidence.com/actions/2504

- **Twenty-five studies** evaluated the effects on non-controlled mammals of removing or controlling non-native mammals. Twenty-one studies were in Australia[1–5,6a–f,7,10a,10b,12–18], and one was in each of France[8], the UK[9], Equador[11] and the USA[19].

COMMUNITY RESPONSE (0 STUDIES)

POPULATION RESPONSE (24 STUDIES)

- **Abundance (21 studies):** Ten of 18 controlled, before-and-after or site comparison studies, in Australia[1–4,6a–f,7,10a,10b,12,13,14,16,17], found that after controlling red foxes, abundances, densities or trapping frequencies increased for rock-wallaby spp.[1,6a,14,16] eastern grey kangaroo[3], woylie[6b], brush-tail possum[6b,6c,6d,6f,12] tammar wallaby[6b,6c,6d,6f], chuditch[12] and quenda[12]. Seven studies found mixed results with increases in some species but not others[6e,7,10a,10b,13], increases followed by declines[17] or increases only where cats as well as foxes were controlled[4]. The other study found no increase in bush rat numbers with fox control[2]. One of three replicated, before-and-after studies (including two controlled studies), in Australia[5], France[8] and Ecuador[11], found that control of invasive rodents increased numbers of lesser white-toothed shrews and greater white-toothed shrews[8]. One study found that Santiago rice rat abundance declined less with rodent control[11] and one found mixed results, with increased numbers of short-tailed mice at one out of four study sites[5].

- **Survival (1 study):** A replicated, controlled study in Australia[3] found that controlling red foxes increased survival of juvenile eastern grey kangaroos.

- **Occupancy/range (3 studies):** Three studies (two before-and-after, one controlled), in the UK[9] and Australia[15,18], found that after controlling non-native American mink[9], red foxes[15] and European rabbits[18], there were increases in ranges or proportions of sites occupied by water vole[9], common brushtail possum, long-nosed potoroo and southern brown bandicoot[15] and four native small mammal species[18].

BEHAVIOUR (1 STUDY)

- **Behaviour change (1 study):** A before-and-after study in the USA[19] found that following removal of feral cats, vertebrate prey increased as a proportion of the diet of island foxes.

Background

Non-native species are a threat to native fauna worldwide. Among mammals, non-native carnivores, typically transported by early European settlers, are especially a threat through predation of native species, including mammals, in locations with few native carnivore species. Non-native herbivores can have dramatic habitat impacts, and thus alter suitability of locations or food availability for native species. Control of non-native species can be expensive and benefits may be difficult to maintain, except in island situations where total elimination might be achievable. Nonetheless, actions aimed at reducing populations of non-native mammals may be carried out on an ongoing basis for the benefit of native species, including mammals.

See also: *Remove/control non-native mammals within a fenced area* and *Remove or control predators.*

A replicated, controlled, before-and-after study in 1979–1990 in four granite outcrop sites in Western Australia, Australia (1) found that after red fox *Vulpes vulpes* control, numbers of rock wallabies *Petrogale lateralis* increased. Results were not tested for statistical significance. In the two sites where fox control was carried out, there were more rock-wallabies after eight years of fox control (50–116 wallabies) than prior to fox control (10–29 wallabies). Over the same period, in the two sites where there was no fox control, wallaby populations declined (after: 0–13; before 7–32). Foxes were initially controlled by shooting and, later, by baiting with fowl eggs dosed with 4.5 mg of 1080 poison. Baiting occurred during the dry seasons of 1980–1983. In 1986–1990, baits were laid along tracks every four to five weeks. Rock-wallabie numbers were estimated by the frequency of recaptures in 1979, 1986 and 1990.

A controlled, before-and-after study in 1993–1995 in four mountain forest sites in the Australian Capital Territory, Australia (2) found that after baiting with poison to control invasive red foxes *Vulpes vulpes*, bush rat *Rattus fuscipes* numbers did not increase. Bush rat numbers at the end of the study were higher in sites with fox control (11–14 animals) compared to without (6–8 animals). However, in sites with control, bush rat numbers were similar 22 months after fox control began (11–14 animals) compared to immediately beforehand (11–12 animals; results not statistically tested). Four 10–28 km² sites were studied in Namadgi National Park. Fox control started in two sites in July 1993 using 1080 poison bait, and in two sites there was no fox control. Red fox numbers in baited sites were reduced from 2.8–3.4/km to <0.5/km in six months and to almost zero over the following 12 months, while fox density remained stable and approximately five times higher in unbaited sites. Bush rats were monitored on two plots in unbaited sites (>2 km apart) and in one plot in baited sites. In total, two trap lines (25 m apart) of 15 Elliott live traps were set at 10–14 m intervals for three consecutive nights, every two months from June 1993 to March 1995 (6,480 trap nights). Foxes were surveyed using spotlights along transects.

A replicated, controlled study in 1993–1995 in four open grassy sites in the Australian Capital Territory, Australia (3) found that controlling invasive red foxes *Vulpes vulpes* increased eastern grey kangaroo *Macropus giganteus* population growth rates and juvenile survival. Kangaroo population growth rates were higher in fox control sites than in uncontrolled sites (data reported as statistical model outputs). In sites with fox control the proportion of females with pouch young was similar at the end of pouch emergence (0.87–0.88 females with young) compared to at the beginning (0.78–0.80 females with young), whereas in sites without fox control, the proportion of females with young declined by 50% by the end of the pouch emergence phase (0.55–0.61 females with young) compared to the beginning (0.94–0.97 females with young). Foxes were removed from two sites within Namadgi National Park using 35 g FOXOFF baits (containing 0.3 mg of 1080 poison). Baiting commenced in July 1993 and reduced fox numbers from 2.8–3.4/km to <0.5/km within six months and to almost zero over the following 12 months. Fox numbers in two unbaited sites remained relatively constant (0.8–2/km). Kangaroos were counted in four sites (two with fox control and two without) 1 hour before dusk from a slow moving car (<5 km/h) along 1.5–2 km transects (400–700 m wide). Surveys were conducted in August, October and December 1993 and then monthly until March 1995. Transects were surveyed twice each survey period.

A before-and-after, controlled study in 1990–1994 in three sites in Western Australia, Australia (4) found that where both cats *Felis catus* and foxes *Vulpes vulpes* were controlled, captures of small mammals increased but where only foxes were controlled, they decreased. Combined fox and cat control doubled small mammal abundances (after: 93; before: 42 individuals captured), but counts fell by 80% where only foxes were controlled (after: 7; before: 55 individuals captured). Small mammal abundances remained similar where no predators were controlled. See original paper for full results. In 1991, a mainland peninsula was divided in three areas in which 1) both cats and foxes were controlled by using an electrified fence, poison baiting (dried meat or cat food with 4.5 mg 1080 poison or via secondary poisoning by poisoning rabbits *Oryctolagus cuniculus*), and trapping or shooting (12 km^2), 2) foxes were controlled by baiting (120–200 km^2) but cats were not targeted or 3) no control occurred. Predators were surveyed over 3–4 nights in vehicles using spotlights (transect length: 7.5–20 km). Small mammals were monitored with six pitfall-trap grids in each area. Each grid had eight pitfall traps, 30–50 m apart. Sampling was conducted over three consecutive days in March–April and June–July in 1990–1994 in predator control areas and 1992–1994 in the area without predator control.

A replicated, controlled, before-and-after study in 1999 at six shrub and grassland sites on an island in Western Australia, Australia (5) found that baiting to control invasive house mice *Mus domesticus* increased the density of short-tailed mice *Leggadina lakedownensis* in one out of four comparisons. Twenty-two days after baiting, the minimum abundance of short-tailed mice was higher in one site with bait deployed every 10 m than before baiting (12.7 vs 7.0 mice). Short-tailed mouse numbers were low in all other sites (baited and unbaited) and were similar after baiting compared to before (see original paper for details). House mice numbers declined on all baited sites (pre-baiting: 5.8–6.2 mice/ha; post baiting: 2.5–2.7 mice/ha). Six grids were established in individual sites at least 1 km apart in May 1999. Two sites were baited with 'Talon' (15-g wax blocks containing 0.005% brodifacoum) at 10 m intervals (117 bait stations/grid), two were baited at 20 m intervals (45 bait stations/grid) and two were unbaited. Bait was replenished every two days for seven days and then again on the fourteenth day. Each site had 25 trap stations arranged in a 5 x 5 pattern, each with one pitfall trap and associated 5 m drift-fencing and one Elliott trap. Sites were monitored for two nights before baiting and up to 22 nights after baiting.

A before-and-after, site comparison study in 1979–1990 on two islands in Western Australia, Australia (6a) found that following control of red foxes *Vulpes vulpes* using poisoned baits, numbers of Rothschild's rock wallaby *Petrogale rothschildi* increased. Results were not tested for statistical significance. After six years of fox control, wallaby numbers were higher (8.8 sightings/hour) than before control (0.3 sightings/hour). During the same period, numbers remained stable on a nearby fox free island (before: 18.7; after: 19.2 sightings/hour). Foxes were controlled by baiting on Dolphin island (3,203 ha), Dampier Archipelago. Meat baits or intact fowl eggs, laced with 1080-poison, were deployed manually in limited areas in October 1980 and May 1981 and then deployed aerially on a larger scale, three times from September 1984 to October 1989. Foxes were also controlled on neighbouring islands and the nearby mainland to prevent immigration (see original paper for details). In 1979–1980 and in 1990, spotlight counts of rock-wallabies were carried out on both Dolphin Island and the nearby fox-free Enderby Island (3,290 ha). Surveys were conducted on foot using a long range 100-W spotlight (1979–1980: 10; 1990: 4 hours of surveying). No fox abundance data are provided.

A before-and-after study in 1979–1998 in a forest reserve in Western Australia, Australia (6b) found that after baiting with poison to control red foxes *Vulpes vulpes*, numbers of woylies *Bettongia penicillata,* brush-tail possums *Trichosurus vulpecula* and tammar wallabies *Macropus eugenii* increased. Results were not tested for statistical significance. After eight years of fox control, numbers were higher than before control for woylies (after: 1.3; before: 0.0 sightings/hour, after: 0.2–0.3; before: 0.0 individuals/trap night), brush-tail possums (after: 7.7; before: 0.4 sightings/hour) and tammar wallabies (after: 9.4; before: 0.4 sightings/ hour). Numbers of tammar wallabies continued to increase up to 14 years after the start of fox control (40 sightings/hour). Foxes were controlled by baiting from 1984 in Tutanning Nature Reserve (2,200 ha). Baits (1080-poison meat baits or intact fowl eggs) were deployed monthly. Mammals were surveyed in 1979–1998 by repeated spotlight counts along 50 circuits near to the boundary of the reserve (circuit length is not provided). Woylies were also monitored using cage traps at

100 m intervals on 1 km-long transects (380 trap nights in 1979; 322 trap nights in 1984; 320 trap nights in 1989; 266 trap nights in 1992). Spotlight searches were conducted using long range 100-W lights.

A before-and-after study in 1987–1998 in a forest reserve in Western Australia, Australia (6c) found that after baiting with poison to control red foxes *Vulpes vulpes*, numbers of brush-tail possums *Trichosurus vulpecula* and tammar wallabies *Macropus eugenii* increased. Results were not tested for statistical significance. Three years after the start of fox control, numbers of tammar wallaby (105.2 sightings/hour) and brush-tail possums (10.5 sightings/hour) increased compared to prior to fox control (wallabies: 4.8 sightings/hour; brush-tail possums: 0 sightings/hour). Numbers of tammar wallabies (61.7 sightings/hour) and brush-tail possums (6.3 sightings/hour) remained higher nine years after fox control started. Foxes were controlled using poison baits (1080-poison meat baits or intact fowl eggs) from 1989 in a separate annex of Tutanning Nature Reserve (114 ha). Mammals were surveyed in 1987, 1992 and 1998 by repeated spotlight counts using long range 100 W lights.

A before-and-after, site comparison study in 1985–1996 in a forest reserve in Western Australia, Australia (6d) found that after baiting with poison to control red foxes *Vulpes vulpes*, numbers of brush-tail possums *Trichosurus vulpecula* and tammar wallabies *Macropus eugenii* increased and translocated woylies *Bettongia penicillata* were still present. Results were not tested for statistical significance. Numbers of brush-tail possums and tammar wallabies were higher in an area where foxes had been baited for seven years than in an area baited for three years (brush-tail possums: 9.1 vs 0.3; tammar wallabies: 1.8 vs 0.0). Four years after translocation, woylies, which were absent prior to fox control, were found to number eight individuals on the east side and 59 on the west side. Foxes were controlled by baiting from 1985 in the east area of the Boyagin Nature Reserve (4,780 ha) and from 1989 in the west. Baits (1080-poison meat baits or intact fowl eggs) were deployed monthly. Mammals were surveyed in 1989–1992 by repeated spotlight counts using long range 100-W lights and cage traps at 100 m intervals on 1 km-long transects in 1992 and 1996 (150 trap nights/area). In total 40 woylies were translocated in 1992 (20 released in the east and 20 in the west area).

A before-and-after study in 1970–1992 in a forest reserve in Western Australia, Australia (6e) found that after baiting with poison to control red foxes *Vulpes vulpes*, numbers of woylies *Bettongia penicillata* and brush-tail possums *Trichosurus vulpecula* increased, but tammar wallabies *Macropus eugenii* numbers did not. Results were not tested for statistical significance. Three years after the start of widespread fox control, overall numbers of individuals were higher than before control for woylies (after: 27.7; before: 1.2 sightings/hour) and brush-tail possums (after: 22.3; before: 2.8 sightings/hour) but tammar wallaby sightings remained infrequent (0 sightings/hour). Ten years after baiting began in a restricted area where fox control was tested before widespread control commenced, numbers of individuals were higher than before control for woylies (after: 23; before: 0.4 sightings/hour), brush-tail possums (after: 9.9; before 2.0 sightings/hour) and tammar wallabies (after: 1.23; before: 0.5 sightings/hour). Foxes were controlled by baiting in a restricted area from 1982, and across the whole reserve from 1989 in a 12,000 ha forest fragment in Dryandra Woodlands. Baits (1080-poison meat baits or intact fowl eggs) were deployed monthly. Mammals were surveyed before fox control in 1970–1971 (75 hours), once the restricted area baiting trial had commenced in 1987 (5 hours) and 1989 (8 hours), and after baiting had been extended to the whole reserve in 1990 (4.5 hours) and 1992 (5.7 hours). Repeated spotlight surveys were conducted along 49 routes using long range 100-W lights (route length is not provided). Woylies were also trapped in cages (see original paper for details).

A site comparison study in 1991–1998 in a national park in Western Australia, Australia (6f) found that after baiting with poison to control red fox *Vulpes vulpes*, numbers of brush-tail possums *Trichosurus vulpecula* and tammar wallabies *Macropus eugenii* increased. Results were not tested for statistical significance. Four years after the start of fox control, brush-tail possum and wallaby numbers were higher in areas where foxes were controlled than in areas where they were not (possums: 19.3 vs 1.1 sightings/hour; wallabies: 5.47 vs 0.0 sightings/hour). Trapping success rates for brush-tail possums were higher in baited compared to unbaited areas and increased every year in fox control areas (see original paper for details). Foxes were controlled in half of the 329,000-ha Fitzgerald River National Park. The other half of the park was left unbaited. Baits (dried meat with 4.5 mg of 1080 poison) were distributed aerially twice a year in 1991–1995 at a density of six baits/km². Supplementary bait was also distributed in some areas by vehicle in 1995–1996. Mammals were surveyed by repeated spotlight surveys using long range 100-W lights (unbaited area: 9.4 hours in 1994–1995; baited area: 17.1 hours in 1993–1996) and trapping (possums only) in 1994–1998 (4 km long trap lines with 40 traps set at 100 m intervals).

A replicated, site comparison study (year not stated) in eight swamp shrubland sites in Western Australia, Australia (7) found that controlling non-native red foxes *Vulpes vulpes* had mixed effects on quokka *Setonix brachyurus* populations. Results were not tested for statistical significance. In 10 of 15 comparisons, sites where foxes were controlled had higher quokka densities than did areas where foxes were not controlled (0.1–4.3 vs 0 quokkas/ha). In five of 15 comparisons, there were fewer or equal numbers of quokkas in fox-control and uncontrolled sites (0–0.07 vs 0–1.1 quokkas/ha). Starting in an unspecified year, once a month, at five sites, meat laced with 1080 poison was laid at 100-m intervals. At three sites, no bait was laid. Five baits/km² were also dropped from aircraft in the area surrounding baited sites. In each site two wire cage traps were placed every 50–100 m along a stream. One trap, measuring 0.90 × 0.45 × 0.45 m, was baited with apples. The other trap, measuring 0.59 × 0.205 × 0.205 m, was baited with peanut butter, rolled oats, honey, and pilchards. Quokkas were caught and released over an eight-day period at each site and were fitted with transponder microchips to allow individual identification.

A replicated, before-and-after study in 1994–2004 on five temperate oceanic islands in northern France (8) found that after the eradication of Norway rats *Rattus norvegicus*, the abundance of lesser white-toothed shrews *Crocidura suaveolens* increased on four islands and greater white-toothed shrews *Crocidura russula* increased on one island. No statistical analyses were performed. Ten years after rat eradication, the abundance of lesser white-toothed shrews on four islands was greater than that before rat eradication (after: 0.09–0.14 shrews/trap night; before: 0.00–0.01). One and two years after rat eradication on a further island, the abundance of greater white-toothed shrews was greater than that before rat eradication (after: 0.31 shrews/trap night; before: 0.02). In total, Norway rats were eradicated from seven islands (0.2–21 ha) in 1994–2002 by trapping and baiting with anticoagulant rodenticide (Bromadiolone©) or using strychnine poisoning (one island in 1951). Monitoring results from five islands are reported here. Small mammal sampling was conducted with 7–269 trap stations at 6–30 m intervals in 1994–2004. Each station had two live traps and was checked daily for 3–7 days.

A before-and-after study in 1997–2005 along a river in Norfolk, UK (9) found that after controlling invasive American mink *Mustela vison*, the proportion of sites occupied by water voles *Arvicola terrestris* increased. Results were not tested for statistical significance. After two years of mink control, a higher proportion of sites were occupied by water voles (27 of 59 sites, 46%) than before control (21 of 62 sites, 35%). No mink signs were found at any survey sites in 2005. Over 280 mink were trapped and euthanised along the River Wensum and its tributaries using traps on banks (1.3–1.6 mink/traps over 3 years, 262 individual mink) and rafts (1.8–2.2 mink/raft over 2 years, 18 individual mink). Between 200 and 220 bank traps (in 2004–2006) and 5–10 raft traps (in 2004–2005) were deployed. Raft traps were arranged in clusters of two to four with clusters at 1–5 km intervals. Water voles were surveyed in 1997 (62 sites), 2003 (60 sites) and 2005 (59 sites) by searching for water vole signs (e.g. latrines, burrows) along 500 m sections of waterway.

A before-and-after study in 1995–2002 in heath and forest habitats in New South Wales, Australia (10a) found that after controlling invasive red foxes *Vulpes vulpes*, one of seven mammal species increased. After four years of fox control, more common ringtail possums *Pseudocheirus peregrinus* were detected than before control (after: 1.8; before: 0.7 individuals/100 m). However, numbers remained similar between fox control and pre-control periods for long-nosed bandicoots *Perameles nasuta* (1.5 vs 0/transect), bush rats *Rattus fuscipes* (1.5 vs 0/transect), brown antechinus *Antechinus stuartii* (3.8–7.6 vs 3.2–3.6/transect), sugar gliders *Petaurus breviceps* (0.1–0.3 vs 0.1–0.2/100 m), black rats *Rattus rattus* (0.9–3.9 vs 2.6–5.8/transect) and common brushtail possum *Trichosurus vulpecula* (0.1–0.3 vs 0.0–0.1/100 m). Control, initiated in 1996, was performed over two weeks, in March and August, using FOXOFF® baits containing 3 mg of 1080 poison. Baits were placed 300–900 m apart. Terrestrial mammals were surveyed two years prior to fox control starting (1995–1996) and up to six years afterwards (in

1999, 2000, 2002). Trapping was over four nights between January and March, along five transects, using 20–25 Elliott live traps/transect and 3–4 possum traps/transect, set 20 m apart. Arboreal mammals were surveyed one year prior to fox control starting (1995) and up to 6 years afterwards (in 1996, 1999, 2000, 2002), along five 500-m-long spotlight transects, 1–2 hours after dark.

A site comparison study in 1999–2003 in New South Wales, Australia (10b) found that controlling invasive red foxes *Vulpes vulpes* increased abundances of four out of five small mammal species. After four years of fox control, numbers of brown antechinus *Antechinus stuartii*, bush rat *Rattus fuscipes*, black rat *Rattus rattus* and long-nosed bandicoot *Perameles nasuta*, but not of common brushtail possum *Trichosurus vulpecula*, were higher than in a site where foxes were not controlled (antechinus: 35 vs 17; bush rat: 29 vs 1; black rat: 1 vs 0; bandicoot: 3 vs 0; possum: 0 vs 4; results not tested for statistical significance). At Booderee National Park, fox control was conducted twice a year between 1999 and 2003 in March and August, using 3 mg 1080 FOXOFF® poison baits, 300–1,000 m apart. No control occurred at Jervis Bay National Park. In both parks, mammals were surveyed over five days in May 2003, along eight 120 m transects, using six Elliott live traps, three possum cage traps and three wire bandicoot traps, spaced 10 m apart. Transects were located at least 500 m apart.

A randomized, replicated, controlled, before-and-after study in 2002–2003 in arid shrubland on an island in Ecuador (11) found that control of invasive black rats *Rattus rattus* reduced the rate of seasonal declines in the abundance of Santiago rice rats *Nesoryzomys swarthi*. Rice rat abundance declined in all sites regardless of black rat control (with control: from 11 to 8–9; without control: from 18–19 to 11–12 rats), but the rate of decline was slower in sites where black rats were controlled (data presented as statistical model outputs). The rate of immigrating female rice rats was higher where black rats were controlled (data presented as statistical model outputs). Black rat numbers decreased more in sites with black rat control (from 18 to 1 rat) compared to sites without black rat control (from 14 to 3 rats). Three sites were selected in Santiago Island, Galapagos. In each site, two trapping grids were set up (98 traps set in pairs at 30 m intervals), in one grid all black rats caught were euthanised and in the other black rats were released after capture. Six trapping sessions were carried out between December 2002 and September 2003 in which each site was trapped for five nights. Additional trapping was conducted 8–10 days after the normal trapping to remove 'immigrant' black rats. Supplementary food (5 kg of rolled oats, 750 ml of vegetable oil and 600 g of peanut butter) was distributed in each site every six days.

A before-and-after study in 1980–2005 across an area of former bauxite mines in jarrah forest of Western Australia, Australia (12) found that controlling non-native red foxes *Vulpes vulpes* on restored mine areas resulted in increased abundance of chuditch *Dasyurus geoffroii*, quenda *Isoodon obesulus* and brushtail possum *Trichosurus vulpecula*. Results were not tested for statistical significance. Chuditch were caught in 0.2% of traps immediately after fox removal compared to none before, and in 1.4% of traps six years later. Quenda were caught in 2.7% of traps immediately after fox removal compared to none before, but they were also absent six years after fox removal. Brushtail possum were caught in 2.3% of traps six years after fox removal, compared to up to 0.5% before. Control of foxes, using poisoned baits, was carried out from 1994 and fox sightings decreased from 15 that year to none in 1999 and 2000. Mined areas were revegetated using various techniques. Mammals were monitored using wire cage traps, large and medium aluminium box traps and pit traps in 1980, 1993, 1997 and 2005.

A replicated, paired sites, controlled, before-and-after study in 1997–2003 in six forest sites in Australia (13) found that controlling invasive red foxes *Vulpes vulpes* increased overall native mammal abundance and abundances of three out of five species. The average number of trapped mammals was higher in fox-control (11.0) than in non-control sites (5.2). Average numbers of individuals trapped/session were higher in fox-control than in non-control sites for long-nosed potoroos *Potorous tridactylus* (5.1 vs 2.3), southern brown bandicoots *Isoodon obesulus* (2.3 vs 1.2) and common brushtail possums *Trichosurus vulpecula* (3.1 vs 1.0), but not for ringtail possums *Pseudocheirus peregrinus* or long-nosed bandicoots *Perameles nasuta* (numbers not given). Increases in abundance over time were found for long-nosed potoroos and ringtail possums, but not for southern brown bandicoots, common brushtail possums or long-nosed bandicoot (results from statistical models). In 1999–2003, foxes were controlled in three out of six forest sites (7,000–16,500 ha) and no control was conducted in the remaining three sites. From February 1999, baits (Foxoff Econbaits, containing 3 mg of 1080 poison) were buried at 15 cm depth every four weeks, at 1-km intervals. At each site, native mammals were surveyed over seven nights, along an 18-km transect, using 60 baited traps, set at 300-m intervals. Trapping was conducted twice before fox-control started (1997–1998) and 12 times after control started (July 1999–May 2003).

A replicated, before-and-after study in 1979–2007 at four sites in Western Australia, Australia (14) found that controlling non-native red foxes *Vulpes vulpes* resulted in an increase in the number of rock wallabies *Petrogale* spp. At all four sites, 10–24 years after fox control began, rock wallaby populations were higher (33–300 animals), than before fox control began (1–32 animals). Starting in 1982, baits containing 1080 poison were laid monthly around four wildlife reserves. At each site, where there were signs of rock wallabies, 30 live traps were baited with apples over a three-day period. Traps were set each evening and checked at dawn, in December–April and February–March of 1979–2007. All rock wallabies caught were tagged, weighed, and released near their capture site.

A replicated, controlled study in 2005–2013 in six forest areas in Australia (15) found that after using poison bait to control invasive red foxes *Vulpes vulpes*, occupancy rates of common brushtail possum *Trichosurus vulpecula*, long-nosed potoroo *Potorous tridactylus* and southern brown bandicoot *Isoodon obesulus* increased. The number of sites occupied by common brushtail possum (51), long-nosed potoroo (20) and southern brown bandicoot (25) was higher in areas where foxes were controlled than in other areas (common brushtail possum: 44; long-nosed potoroo: 7; southern brown bandicoot: 13). Six areas with no previous fox control where selected. From October 2005–November 2013, foxes were baited in three areas (4,703–9,750 ha) using FoxOff® (containing 3 mg of 1080 poison). Every 1 km, one bait was buried at a depth of 10 cm and replaced fortnightly. Three other areas (4,659–8,520 ha) were left unbaited. In each of the six areas, mammals were monitored annually at 40 sampling sites using hair tubes. Tubes were set for four days in spring 2005 and 2008–2013 and winter 2006 and 2007, and species were identified from hairs.

A replicated, controlled, before-and-after study in 1980–2012 in four mixed eucalyptus woodland and shrubland in southern Australia (16) found that after control of invasive red foxes *Vulpes vulpes*, population growth rates of yellow-footed rock wallabies *Petrogale xanthopus* increased. In the two populations exposed to fox control, rock-wallaby population growth rates were higher after fox control commenced than before (data presented as statistical model outputs). Over the same time periods, rock-wallaby population growth rates were similar in colonies where foxes were not controlled (data presented as statistical model outputs). In New South Wales, the number of rock-wallabies counted increased two years after fox control began (at start of fox control: 7; after: 16 animals), while in the site without fox control numbers remained similar. Two sites in New South Wales and two in South Australia were studied. In each state, foxes were controlled in one site and not controlled in the other site. Baiting strategy differed by location (see original paper for details). Bait stations (219 in New South Wales and 100 in South Australia) were baited using Foxoff Econobaits® or fresh or dried meat laced with 1080 poison. Baits were deployed from June 1995 in New South Wales and from June 2004 in South Australia. Wallabies were surveyed annually, over three mornings in the winter months, from a helicopter. Surveys were conducted in 1980–2001 (New South Wales) and 2000–2012 (South Australia).

A replicated, before-and-after study in 1970–2009 in two forest sites in Western Australia, Australia (17) found that controlling invasive red foxes *Vulpes vulpes* initially increased the abundance of woylies *Bettongia penicillata*, but woylie numbers returned to pre-control levels after about 25 years. Results were not tested for statistical significance. After 25 years of fox control, the trapping success of woylies (caught in 3–8% of traps from 2002–2006) was only marginally higher than pre-control levels (2–3% from 1970–1975). However, trapping success had increased up to 28–65% during the first 20 years after the start of fox control. Between April 2006 and October 2009, more woylies were killed by cats *Felis catus* (65%) than by foxes (21%). Foxes were controlled from the mid-1970s at two reserves (2–6,800 ha) by baiting (either dry meat with 3 mg of 1080 poison or Pro-baits) with 5 baits/km² every four weeks. No details about long-term woylie trapping are provided. Between April 2006 and October 2009, 146 woylies were radio-collared, of which 89 died. Cause of death was determined by DNA analysis and predation characteristics.

A before-and-after study in 1970–2014 in an arid region in South Australia, Australia (18) found that control of invasive European rabbits *Oryctolagus cuniculus*, using rabbit hemorrhagic disease virus, increased the area occupied by four native small mammal species. The extent of occurrence and area of occupancy (both expressed in thousands of km²) was greater after outbreaks of rabbit hemorrhagic disease than before for spinifex hopping mouse *Notomys alexis* (extent: 276–356 vs 180; area: 7–8 vs 3), dusky hopping mouse *Notomys fuscus* (extent: 105–130 vs 23; area: 6–11 vs 2), plains mouse *Pseudomys australis* (extent: 217–252 vs 63; area: 4–6 vs 2) and crest-tailed mulgara *Dasycercus cristicauda* (extent: 98–133 vs 1; area: 12–13 vs 1). After the first virus outbreak, rabbit abundance decreased by 85% (raw data not provided) in one site and from 139 to 22 rabbits/km² in the other site. Cat *Felis catus* and fox *Vulpes vulpes* numbers followed rabbit population trends. Occurrence records over a 615,000 km² region were compiled from published sources and divided into periods covering before the outbreak (1970–1995) and after first and second outbreaks (1996–2009 and 2010–2014). Area of occupancy was calculated from occupied 10 × 10 km grid squares. Extent of occurrence was calculated from minimum convex polygons around species records. Rabbit abundance was monitored in two long-term study sites using spotlight transects.

A before-and-after study in 2006–2012 of scrubland on an island in California, USA (19) found that following removal of feral cats *Felis catus*, vertebrate prey increased as a proportion of the diet of island foxes *Urocyon littoralis*. The frequency of deer mice *Peromyscus maniculatus* in fox scats was higher after cat removal (40%) than before (11%). The same pattern held for birds (after: 12% of scats; before: 6% of scats). Lizard frequency in fox scats was not significantly higher after cat removal (10%) than before (5%) and there were not significant changes in frequencies of arthropods, snails or fruit. Authors indicated that increased deer mouse and bird frequency suggests that foxes and cats had been competing for prey. However, fox abundance was more linked to precipitation levels, and declined over the study period. On a 5,896-ha island, feral cats were eradicated in 2009–2010. Fox scats collected before cat removal (1,180 scats, autumn 2006–summer 2009) and after removal (508 scats, autumn 2010–summer 2012) were analysed for food remains.

(1) Kinnear J.E., Onus M.L. & Sumner N.R. (1998) Fox control and rock-wallaby population dynamics—II. An update. *Wildlife Research*, 25, 81–88.

(2) Banks P.B. (1999) Predation by introduced foxes on native bush rats in Australia: do foxes take the doomed surplus? *Journal of Applied Ecology*, 36, 1063–1071.

(3) Banks P.B., Newsome A.E. & Dickman C.R. (2000) Predation by red foxes limits recruitment in populations of eastern grey kangaroos. *Austral Ecology*, 25, 283–291, https://doi.org/10.1046/j.1442-9993.2000.01039.x

(4) Risbey D.A., Calver M.C., Short J., Bradley J.S. & Wright I.W. (2000) The impact of cats and foxes on the small vertebrate fauna of Heirisson Prong, Western Australia. II. A field experiment. *Wildlife Research*, 27, 223–235, https://doi.org/10.1071/wr98092

(5) Moro D. (2001) Evaluation and cost-benefits of controlling house mice (*Mus domesticus*) on islands: an example from Thevenard Island, Western Australia. *Biological Conservation*, 99, 355–364, https://doi.org/10.1016/s0006-3207(00)00231-7

(6) Kinnear J.E., Sumner N.R. & Onus M.L. (2002) The red fox in Australia—an exotic predator turned biocontrol agent. *Biological Conservation*, 108, 335–359, https://doi.org/10.1016/s0006-3207(02)00116-7

(7) Hayward M.W., Paul J., Dillon M.J. & Fox B.J. (2003) Local population structure of a naturally occurring metapopulation of the quokka (*Setonix brachyurus* Macropodidae: Marsupialia). *Biological Conservation*, 110, 343–355, https://doi.org/10.1016/s0006-3207(02)00240-9

(8) Pascal M., Siorat F., Lorvelec O., Yésou P. & Simberloff D. (2005) A pleasing consequence of Norway rat eradication: two shrew species recover. *Diversity and Distributions*, 11, 193–198, https://doi.org/10.1111/j.1366-9516.2005.00137.x

(9) Thompson H. (2006) The use of floating rafts to detect and trap American mink *Mustela vison* for the conservation of water voles *Arvicola terrestris* along the River Wensum in Norfolk, England. *Conservation Evidence*, 3, 114–116

(10) Dexter N., Meek P., Moore S., Hudson M. & Richardson H. (2007) Population responses of small and medium sized mammals to fox control at Jervis Bay, Southeastern Australia. *Pacific Conservation Biology*, 13, 283–292, https://doi.org/10.1071/pc070283

(11) Harris D.B. & Macdonald D.W. (2007) Interference competition between introduced black rats and endemic Galapagos rice rats. *Ecology*, 88, 2330–2344, https://doi.org/10.1890/06-1701.1

(12) Nichols O.G. & Grant C.D. (2007) Vertebrate fauna recolonization of restored bauxite mines–key findings from almost 30 years of monitoring and research. *Restoration Ecology*, 15, S116–S126, https://doi.org/10.1111/j.1526-100x.2007.00299.x

(13) Dexter N. & Murray A. (2009) The impact of fox control on the relative abundance of forest mammals in East Gippsland, Victoria. *Wildlife Research*, 36, 252–261, https://doi.org/10.1071/wr08135

(14) Kinnear J.E., Krebs C.J., Pentland C., Orell P., Holme C. & Karvinen R. (2010) Predator-baiting experiments for the conservation of rock-wallabies in Western Australia: a 25-year review with recent advances. *Wildlife Research*, 37, 57–67, https://doi.org/10.1071/wr09046

(15) Robley A., Gormley A.M., Forsyth D.M. & Triggs B. (2014) Long-term and large-scale control of the introduced red fox increases native mammal occupancy in Australian forests. *Biological Conservation*, 180, 262–269, https://doi.org/10.1016/j.biocon.2014.10.017

(16) Sharp A., Norton M., Havelberg C., Cliff W. & Marks A. (2014) Population recovery of the yellow-footed rock-wallaby following fox control in New South Wales and South Australia. *Wildlife Research*, 41, 560–570, https://doi.org/10.1071/wr14151

(17) Marlow N.J., Thomas N.D., Williams A.A., Macmahon B., Lawson J., Hitchen Y., Angus J. & Berry O. (2015) Cats (*Felis catus*) are more abundant and are the dominant predator of woylies (*Bettongia penicillata*) after sustained fox (*Vulpes vulpes*) control. *Australian Journal of Zoology*, 63, 18–27, https://doi.org/10.1071/zo14024

(18) Pedler R.D., Brandle R., Read J.L., Southgate R., Bird P. & Moseby K.E. (2016) Rabbit biocontrol and landscape-scale recovery of threatened desert mammals. *Conservation Biology*, 30, 774–782, https://doi.org/10.1111/cobi.12684

(19) Cypher B.L., Kelly E.C., Ferrara F.J., Drost C.A., Westall T.L. & Hudgens B.R. (2017) Diet patterns of island foxes on San Nicolas Island relative to feral cat removal. *Pacific Conservation Biology*, 23, 180–188, https://doi.org/10.1071/pc16037

9.6. Remove/control non-native mammals within a fenced area

https://www.conservationevidence.com/actions/2528

- **One study** evaluated the effects on native mammals of removing or controlling non-native mammals within a fenced area. This study was in Australia[1].

COMMUNITY RESPONSE (1 STUDY)

- **Richness/diversity (1 study):** A site comparison study in Australia[1] found that in a fenced area where invasive cats, red foxes and European rabbits were removed, native mammal species richness was higher than outside the fenced area.

POPULATION RESPONSE (1 STUDY)

- **Abundance (1 study):** A site comparison study in Australia[1] found that in a fenced area where invasive cats, red foxes and European rabbits were removed, native mammals overall and two out of four small mammal species were more abundant than outside the fenced area.

BEHAVIOUR (0 STUDIES)

Background

Control of non-native mammals may be carried out to reverse detrimental impacts of such species on native plants and animals. Total elimination of non-native mammals may be difficult or impossible to carry out on a large scale, with control programmes often being confined to small islands, where elimination may be achievable. However, away from islands, a similar benefit might be realised if non-native mammals can be removed from within an area that is fenced to prevent their recolonization.

A site comparison study in 1997–2005 in a dune and shrubland site in South Australia, Australia (1) found that in a fenced area where invasive cats *Felis catus*, red foxes *Vulpes vulpes* and European rabbits *Oryctolagus cuniculus* were removed, native mammal species richness and abundance, and abundance of two out of four small mammal species were greater than outside the fenced area. Two to six years after the removal of cats, foxes and rabbits began, native mammal species richness and overall abundance was higher inside than outside the fenced removal area (data presented on log scales). Also, more spinifex hopping mice *Notomys alexis* and Bolam's mice *Pseudomys bolami* were caught in removal areas (spinifex: 13–51; Bolam's: 5–38) than in non-removal areas (spinifex: 3–4; Bolam's: 1–2). Numbers caught did not significantly differ in removal vs non-removal areas for fat-tailed dunnart *Sminthopsis crassicaudata* (0.3 vs 0.8) and stripe-faced dunnart *Sminthopsis macroura* (0.3–2.8 vs 1.1). Between 1997 and 2005, a 78-km² exclosure was established in five stages, inside which rabbits, cats and foxes were removed from 1999. Locally extinct mammals were reintroduced into the first area (14-km²) in 1999–2001. Twelve locations inside the exclosure and 12 outside (60–7,000-km apart) were sampled over four nights annually, in 1998–2005, using a line of six pitfall traps and 15 Elliott live traps.

(1) Moseby K.E., Hill B.M. & Read J.L. (2009) Arid Recovery–A comparison of reptile and small mammal populations inside and outside a large rabbit, cat and fox-proof exclosure in arid South Australia. *Austral Ecology*, 34, 156–169, https://doi.org/10.1111/j.1442-9993.2008.01916.x

9.7. Remove/control non-native plants

https://www.conservationevidence.com/actions/2529

- **Two studies** evaluated the effects on mammals of removing or controlling non-native invasive plants. Both studies were in the USA[1,2].

COMMUNITY RESPONSE (1 STUDY)

- **Richness/diversity (1 study):** A replicated study in the USA[2] found that control of introduced saltcedar did not change small mammal species richness.

POPULATION RESPONSE (1 STUDY)

- **Abundance (1 study):** A site comparison study in the USA[1] found that partial removal of velvet mesquite did not increase abundances of six mammal species.

BEHAVIOUR (0 STUDIES)

Background

Invasive plants can out compete established plant species and alter habitat structure. This may alter resource availability for mammals. Some mammal species may benefit but, for others, invasive plants may reduce available food or shelter or change the nature of the environment such that they are at increased risk of predation. Removal or control of non-native invasive plants may be carried out in an attempt to reverse these effects.

A site comparison study in 1976–1978 in three desert sites in Arizona, USA (1) found that partial removal of velvet mesquite *Prosopis juliflora* var. *velutina* did not increase abundances of six mammal species, and complete removal reduced the abundance of two species. The abundance of black-tailed jackrabbits *Lepus californicus* was higher in the undisturbed (0.37/km) and partially cleared mesquite sites (0.36/km) than in the cleared, mesquite-free, site (0.06/km). The same pattern held for antelope jackrabbit *Lepus alleni* (0.37 and 0.56 vs 0.09/km). However, abundances were similar in the undisturbed, partially and fully cleared sites for desert mule deer *Odocoileus hemionus crooki* (0.30, 0.24 and 0.16/km), javelina *Dicotyles tajacu* (0.24, 0.15 and 0.00/km), coyote *Canis latrans* (0.05, 0.06 and 0.01/km) and desert cottontail *Sylvilagus audubonii* (0.04, 0.02 and 0.03/km). Mesquite was cleared from one 300 ha site in 1955 using diesel oil, and partially removed from a second 300 ha site by clearing seven 2.8–30.4 ha patches by chaining in July 1976. At the third 300 ha site, mesquite was left undisturbed. Mammals were counted monthly along four 1,200-m transects between September 1976 and June 1978.

A replicated study in 2001–2012 in three sites in Nevada, USA (2) found that control of introduced saltcedar *Tamarix ramosissima*, did not change small mammal species richness. Ten years after saltcedar control commenced, small mammal species richness (3–6 species) was similar to that when control started (3–7 species). Small mammals were trapped annually in May or June for three consecutive nights between 2001 and 2011–2012 at three sites along waterways. An additional trapping period of three nights was conducted in July or August 2001–2004 at one site, and 2001–2006 at two sites. Each night at each site, 2–4 parallel rows of 25 Sherman® live traps, baited with wild birdseed mix, were set with 10 m between traps and 25–100 m between rows. Saltcedar was controlled by leaf beetles *Diorhabda* spp. released at the sites in 2001–2002.

(1) Germano D.J., Hungerford R. & Martin S.C. (1983) Responses of selected wildlife species to the removal of mesquite from desert grassland. *Journal of Range Management*, 36, 309–311.

(2) Longland W.S. (2014) Biological control of saltcedar (*Tamarix* spp.) by saltcedar leaf beetles (*Diorhabda* spp.): effects on small mammals. *Western North American Naturalist*, 74, 378–385, https://doi.org/10.3398/064.074.0403

9.8. Control non-native/problematic plants to restore habitat

https://www.conservationevidence.com/actions/2530

- We found no studies that evaluated the effects on mammals of controlling invasive or problematic plants to restore habitat.

'We found no studies' means that we have not yet found any studies that have directly evaluated this intervention during our systematic journal and report searches. Therefore, we have no evidence to indicate whether or not the intervention has any desirable or harmful effects.

Background

Invasive plant species can drive large scale changes to habitats. These changes can make habitats less suitable for use by mammal species. Control of invasive or problematic plants might be undertaken to recreate suitable conditions for fauna, including target mammals (e.g. Dumalisile Somers 2017).

Dumalisile L. & Somers M.J. (2017) The effects of an invasive alien plant (*Chromolaena odorata*) on large African mammals. *Nature Conservation Research*, 2, 102–108, https://doi.org/10.24189/ncr.2017.048

9.9. Reintroduce top predators to suppress and reduce the impacts of smaller non-native predator and prey species

https://www.conservationevidence.com/actions/2531

- We found no studies that evaluated the effects on mammals of reintroducing top predators to suppress and reduce the impacts of smaller non-native predator and prey species.

'We found no studies' means that we have not yet found any studies that have directly evaluated this intervention during our systematic journal and report searches. Therefore, we have no evidence to indicate whether or not the intervention has any desirable or harmful effects.

Background

Small and medium-sized non-native predators can have severe detrimental impacts on native fauna, including mammals (e.g. Doherty *et al.* 2017). Some evidence suggests that their numbers can be reduced, to the benefit of native fauna, if top predator conservation is promoted, such as through reintroductions (e.g. Nimmo *et al.* 2015).

Nimmo D.G., Watson S.J., Forsyth D.M. & Bradshaw C.J.A. (2015) Dingoes can help conserve wildlife and our methods can tell. *Journal of Applied Ecology*, 52, 281–285.

Doherty T.S., Dickman C.R., Johnson C.N., Legge S.M., Ritchie E.G. & Woinarski J.C.Z. (2017) Impacts and management of feral cats Felis catus in Australia. *Mammal Review*, 47, 83–97.

9.10. Control non-native prey species to reduce populations and impacts of non-native predators

https://www.conservationevidence.com/actions/2532

- We found no studies that evaluated the effects on mammals of controlling non-native prey species to reduce populations and impacts of non-native predators.

'We found no studies' means that we have not yet found any studies that have directly evaluated this intervention during our systematic journal and report searches. Therefore, we have no evidence to indicate whether or not the intervention has any desirable or harmful effects.

Background

The impact of non-native predators on native mammals can be more severe than that of native predators (Salo *et al.* 2007). Non-native predators may also feed on non-native prey and, in some situations, reducing non-native prey availability may lead to reductions in numbers of their non-native predators (Murphy *et al.* 1998; Mutze *et al.* 2017). This has potential to reduce the impact of non-native predators on native mammalian prey species.

Murphy E.C., Clapperton B.K., Bradfield P.M.F. & Speed H.J. (1998) Effects of rat-poisoning operations on abundance and diet of mustelids in New Zealand podocarp forests. *New Zealand Journal of Zoology*, 25, 315–328.

Salo P., Korpimäki E., Banks P.B., Nordström M. & Dickman C.R. (2007) Alien predators are more dangerous than native predators to prey populations. *Proceedings of the Royal Society B*, 274, 1237–1243, https://doi.org/10.1098/rspb.2006.0444

Mutze G. (2017) Continental-scale analysis of feral cat diet in Australia, prey-switching and the risk: benefit of rabbit control. *Journal of Biogeography*, 44, 1679–1681, https://doi.org/10.1111/jbi.12859

9.11. Provide artificial refuges for prey to evade/escape non-native predators

https://www.conservationevidence.com/actions/2533

- We found no studies that evaluated the effects on mammals of providing artificial refuges for prey to evade/escape non-native predators.

'We found no studies' means that we have not yet found any studies that have directly evaluated this intervention during our systematic journal and report searches. Therefore, we have no evidence to indicate whether or not the intervention has any desirable or harmful effects.

Background

This intervention considers use of small scale refuges rather than larger predator-free areas protected by fences. Artificial refuges, such as small shelters in otherwise open landscapes, could provide cover for native mammals to escape predation. Refuges are more often employed for reptile conservation, though at least one study found that they were insufficient to mitigate effects of non-native predators (Lettink *et al.* 2010). For mammals, refuges might entail small shelters, boxes or artificial burrows.

See also: *Habitat restoration and creation — Provide artificial refuges/breeding sites.*

Lettink M., Norbury G., Cree A., Seddon P.J., Duncan R.P., Schwarz C.J. (2010) Removal of introduced predators, but not artificial refuge supplementation, increases skink survival in coastal duneland. *Biological Conservation*, 143, 72–77, https://doi.org/10.1016/j.biocon.2009.09.004

9.12. Remove/control non-native species that could interbreed with native species

https://www.conservationevidence.com/actions/2534

- We found no studies that evaluated the effects on mammals of removing or controlling non-native species that could interbreed with native species.

'We found no studies' means that we have not yet found any studies that have directly evaluated this intervention during our systematic journal and report searches. Therefore, we have no evidence to indicate whether or not the intervention has any desirable or harmful effects.

Background

Hybridisation of non-native mammals with closely related native species can threaten local populations (e.g. Biedrzycka *et al.* 2020; Nussberger *et al.* 2014). Attempts may be made to reduce the risk through carrying out lethal control of the non-native species. The strategy can be difficult to execute, due to difficulties in separating hybrids from parent species.

Biedrzycka A., Solarz W. & Okarma H. (2012) Hybridization between native and introduced species of deer in Eastern Europe. *Journal of Mammalogy*, 93, 1331–1341, https://doi.org/10.1644/11-mamm-a-022.1

Nussberger B., Wandeler P., Weber D. & Keller L.F. (2014) Monitoring introgression in European wildcats in the Swiss Jura. *Conservation Genetics*, 15, 1219–1230, https://doi.org/10.1007/s10592-014-0613-0

9.13. Modify traps used in the control/eradication of non-native species to avoid injury of non-target mammal

https://www.conservationevidence.com/actions/2535

- **One study** evaluated the effects of modifying traps used in the control or eradication of non-native species to avoid injury of non-target mammals. This study was in the USA[1].

COMMUNITY RESPONSE (0 STUDIES)

POPULATION RESPONSE (1 STUDY)

- **Condition (1 study):** A before-and-after study in the USA[1] found that modifying traps used for catching non-native mammals reduced moderate but not severe injuries among incidentally captured San Nicolas Island foxes.

BEHAVIOUR (0 STUDIES)

Background

A range of live-trapping techniques is used in control activities aimed at non-native species. As traps may capture species additional to the targeted non-native species, using live traps enables release of those non-target captures. However, restrained mammals are at risk of suffering injuries prior to being released. This intervention considers cases where modifications might be made to live traps with the intention of reducing such incidental injuries.

A before-and-after study in 2006–2010 on an offshore island in California, USA (1) found that modifying traps used to control non-native cats *Felis catus* reduced moderate but not severe injuries among San Nicolas Island foxes *Urocyon littoralis dickeyi*. These results were not tested for statistical significance. A lower proportion of San Nicolas Island foxes that were caught in modified traps (4%) suffered moderate injuries than when unmodified traps were used (25%). However, the rates of severe and very severe injuries in San Nicolas Island foxes were similar (around 5%) between the periods when modified and unmodified traps were used. The study was conducted on a 5,896-ha island. During 20 days in 2006, sixty-four San Nicolas Island foxes were caught with leg-hold traps deployed to catch non-native cats. Between June 2009 and January 2010, using modified leg-hold traps, 1,011 Nicolas Island foxes were caught. Trap modifications included a shorter anchor cable and chain, lighter spring, and additional swivels to allow unrestricted rotation of the trapped animal. Traps were checked remotely 24 hours a day to reduce the time foxes spent in the traps.

(1) Jolley W.J., Campbell K.J., Holmes N.D., Garcelon D.K., Hanson C.C., Will D., Keitt B.S., Smith G. & Little A.E. (2012) Reducing the impacts of leg hold trapping on critically endangered foxes by modified traps and conditioned trap aversion on San Nicolas Island, California, USA. *Conservation Evidence*, 9, 43–49.

9.14. Use conditioned taste aversion to prevent non-target species from entering traps

https://www.conservationevidence.com/actions/2536

- **One study** evaluated the effects on mammals of using conditioned taste aversion to prevent non-target species from entering traps. This study was in the USA[1].

COMMUNITY RESPONSE (0 STUDIES)

POPULATION RESPONSE (0 STUDIES)

BEHAVIOUR (1 STUDY)

- **Behaviour change (1 study):** A replicated, controlled study in the USA[1] found that using bait laced with lithium chloride reduced the rate of entry of San Clemente Island foxes into traps set for feral cats.

Background

Animals may be trapped for a variety of reasons. In cases, such as where trapping is aimed at non-native species, a large number of traps might be set across the landscape. If there is a risk of catching non-target species, these will typically be live traps, from which individuals of non-target species can be released. However, trapping of animals usually entails at least some risk of injury to the animal as well as further risks, such as keeping parents away from their young. Furthermore, a trap holding a non-target animal is generally not then available for capturing the target animal until next visited by an operator. Conditioned taste aversion may be attempted, to try to make non-target mammals that are at risk of capture avoid traps because they associate them with an unpleasant taste or sensation.

A replicated, controlled study in 1992–1993 on an island in California, USA (1) found that lacing bait with lithium chloride reduced the rate of entry of San Clemente Island foxes *Urocyon littoralis clementae* into traps for feral cats *Felis catus*. In the first year, fewer foxes were recaptured using lithium chloride bait in traps (at 200 mg dose/kg of fox -9% recaught) than using unlaced bait (52% recaught). In the second year, fewer foxes were recaptured in traps using lithium chloride bait (3% recaught) than using unlaced bait (30% recaught). In sites where lithium chloride bait was used for 41 days and then switched to non-laced baits, recapture rates remained low for around 10 days after the switch, and then increased. Baits were placed in cage traps on a 146-km² island. In 1992, two areas received lithium chloride baits (which induce gastrointestinal discomfort) and unlaced baits were used in three areas. In 1993, two areas received lithium chloride baits which were then switched to unlaced baits after 41 days and seven areas received unlaced baits throughout. Eight to 20 traps were used/area. Baits comprised 50 g of mixed cat food, tuna and raw hamburger, placed in traps from February through to July–August in 1992–1993.

(1) Phillips R.B. & Winchell C.S. (2011) Reducing nontarget recaptures of an endangered predator using conditioned aversion and reward removal. *Journal of Applied Ecology*, 48, 1501–1507, https://doi.org/10.1111/j.1365-2664.2011.02044.x

9.15. Use reward removal to prevent non-target species from entering traps

https://www.conservationevidence.com/actions/2537

- **One study** evaluated the effects on mammals of using reward removal to prevent non-target species from entering traps. This study was in the USA[1].

COMMUNITY RESPONSE (0 STUDIES)

POPULATION RESPONSE (0 STUDIES)

BEHAVIOUR (1 STUDY)

- **Behaviour change (1 study):** A replicated, controlled study in the USA[1] found that when reward removal was practiced, the rate of San Clemente Island fox entry into traps set for feral cats was reduced.

Background

Animals may be trapped for a variety of reasons. In some cases, such as where trapping is aimed at non-native species, a large number of traps might be set across the landscape. If there is a risk of catching non-target species, these will typically be live traps, from which individuals of non-target species can be released. However, trapping of animals usually entails at least some risk of injury to the animal as well as further risks, such as keeping parents away from their young. Furthermore, a trap holding a non-target animal is generally not then available for capturing the target animal until next visited by an operator. Reward removal may be attempted, whereby strong-smelling bait is left in a form or situation where it is unavailable to animals, to consume. The intention is that non-target species will learn not to pursue that smell.

A replicated, controlled study in 1992 and 1994 on an island in California, USA (1) found that providing inaccessible bait inside a perforated can conditioned San Clemente Island foxes *Urocyon littoralis clementae* to avoid feral cat *Felis catus* traps. In the first year, fewer foxes were recaptured in traps with perforated can baits (8% recaught) than with accessible baits (52%). In the second year, fewer foxes were recaptured in traps using perforated can baits (1% recaptured) than those using accessible baits (27%). When bait treatments were switched between areas, recapture rates increased in those then receiving accessible bait and fell in those with perforated cans. Cat capture efficiency remained high throughout trials. Baits were placed in 8–20 cage traps/area on a 146-km² island. In 1992, perforated can baits were used in two areas and accessible baits were used in three areas. In 1994, two areas received perforated can baits and accessible baits were used in three areas. Treatments were swapped over in these five areas after 41 days. Inaccessible baits were perforated cat food canisters (1992) or perforated plastic canisters containing cat food, tuna, raw hamburger and a fish oil scent (1994). Accessible baits were cat food, tuna and raw hamburger. Baits were used in traps from February through to June–July in 1992 and 1994.

(1) Phillips R.B. & Winchell C.S. (2011) Reducing nontarget recaptures of an endangered predator using conditioned aversion and reward removal. *Journal of Applied Ecology*, 48, 1501–1507, https://doi.org/10.1111/j.1365-2664.2011.02044.x

Problematic Native Species/Diseases

9.16. Remove or control predators

https://www.conservationevidence.com/actions/2613

- **Ten studies** evaluated the effects on non-controlled mammals of removing or controlling predators. Seven studies were in North America[2,5–10], one was in Finland[1], one in Portugal[3] and one in Mexico[4].

COMMUNITY RESPONSE (0 STUDIES)

POPULATION RESPONSE (10 STUDIES)

- **Abundance (6 studies):** Three of six studies (including three controlled, one before-and-after and one replicated, paired sites study), in Finland[1] Portugal[3], Mexico[4] and the USA[2,5,6], found that removing predators increased abundances of pronghorns[5], moose[6] and European rabbits and Iberian hares[3]. One of these studies also found that mule deer abundance did not increase[5]. The other three studies found that removing predators did not increase mountain hare[1], caribou[2] or desert bighorn sheep abundance.

- **Reproductive success (2 studies):** Two replicated, before-and-after studies (one also controlled), in the USA[5,8], found that predator removal was associated with increased breeding productivity of white-tailed deer[8] and less of a productivity decline in pronghorns[5]. However, one of these studies also found that there was no change in breeding productivity of mule deer[3].

- **Survival (5 studies):** Two of five before-and-after studies (including two controlled studies and one replicated study), in the USA[2,6,7], Canada[10] and the USA and Canada combined[9], found that controlling predators did not increase survival of caribou calves[2], or of calf or adult female caribou[9]. Two studies found that moose calf survival[6] and woodland caribou calf survival[10] increased with predator control. The other study found mixed results with increases in white-tailed deer calf survival in some but not all years with predator control[7].

BEHAVIOUR (0 STUDIES)

Background

Predators can limit population sizes of prey species. Changes in habitat or land management can lead to increases in predator populations which might negatively affect prey. Removing or controlling predators, especially native predators, for the benefit of their wild prey species can be a controversial management strategy. In many situations, it is more likely to occur for game management than directly for species conservation. Nonetheless, there is potential for such management to lead to increases in the abundance, survival or reproduction success of prey species.

A replicated, paired sites, controlled study in 1993–1998 of boreal forest in three areas in Finland (1) found that removing predators did not increase numbers of mountain hares *Lepus timidus*. In two of three areas, mountain hare numbers increased in both predator removal and predator protection sites, with the rate of increase being higher in the predator protection site than the removal site in one of those areas. In the third area, hare numbers declined each year in predator removal sites but increased in two of five years in protection sites. Data are presented as track count indices. In each of three areas, a predator removal and predator protection site were established, ≥5 km apart. Sites each covered 48–116 km². Predator removal, carried out by hunters during normal hunting seasons, commenced in August 1993, targeting red fox *Vulpes vulpes*, pine marten *Martes martes*, stoat *Mustela ermine* and raccoon dog *Nyctereutes procyonoides*. Hares were monitored by snow track counts, annually from 15 January to 15 March, in 1993–1998.

A controlled, before-and-after study in 1990–2000 in alpine tundra and subalpine shrubland in Alaska, USA (2) found that wolf *Canis lupus* culling did not increase calf survival or population size of caribou *Rangifer tarandus*. Between 1992–1993 (before the wolf cull) and 1994–1995 (after the cull), the increase in calf:cow ratio within the cull area (before: 7.4:100; after: 21.5:100) was no greater than in a similar sized herd in an area without wolf culling (before: 11.2:100; after: 19.5:100). However, the change was greater than in a smaller sized herd in an area without wolf culling, where the calf:cow ratio declined (before: 15.8:100; after: 11.5:100). The long-term (1993–2000) change in caribou numbers in the population where wolves were controlled (before: 3,661; after: 3,227) was comparable to the population change in one of the areas without culling (before: 1,970; after: 1,730), but not to the other (before: 500; after: 675), although no statistical tests were carried out. Autumn calf:cow ratios were monitored annually between 1990 and 2000 from a helicopter, guided by radio-collared females. See original paper for methods for estimating population size. In 1993–1994, 60–62% of wolves were controlled by trapping, snaring and shooting. Smaller numbers (20–40%) were culled in subsequent years by local hunters.

A replicated, paired sites study in 2000–2001 of 24 games estates and hunting areas in Alentejo, Portugal (3) found that controlling predators resulted in greater numbers of European rabbits *Oryctolagus cuniculus* and Iberian hares *Lepus granatensis*. Game estates that controlled predators had a greater number of European rabbits (5.9 rabbits/10 km) and Iberian hares (1.7 hares/10 km) than paired hunting areas without predator control (0.5 rabbits/10 km; 0.3 hares/10 km). Twelve game estates that controlled predators (with box traps, shooting, snares) for >3 years were paired with 12 hunting areas without predator control. Paired sites (average 12 km²) were mostly grazed woodlands and farmland. Species controlled were red foxes *Vulpes vulpes* (11 estates), Egyptian mongooses *Herpestes ichneumon* (six estates), feral cats *Felis catus* and dogs *Canis familiaris* (two estates), common genets *Genetta genetta* (one estate), stone martens *Martes foina* (one estate) and azure-winged magpies *Cyanopica cyanus* (one estate). Each site within a pair was sampled once on consecutive days in May–June 2000 or 2001. Rabbits and hares and/or their signs (faeces, footprints) were counted along walked transects (average 12 km long).

A replicated study in 1951–2007 in nine desert sites in Arizona and New Mexico, USA, and the Gulf of California, Mexico (4) found that controlling mountain lions *Puma concolor* did not increase the population size of desert bighorn sheep *Ovis canadensis*. No bighorn sheep populations at sites where mountain lions were controlled increased in size (data not presented). Data were obtained from historical records for 10 sites with long-term survey and hunting information. Data included counts of bighorn sheep from both surveys and hunter harvests, and annual mountain lion harvests. No information on the number of mountain lions controlled is provided.

A replicated, paired sites, controlled, before-and-after study in 2007–2008 in 12 rangeland sites in Wyoming and Utah, USA (5) found that after coyotes *Canis latrans* were removed, pronghorn *Antilocapra americana* abundance was higher and productivity declined less in removal than non-removal sites, but for mule deer *Odocoileus hemionus* abundance and productivity did not differ. After eight months of coyote control, the abundance of pronghorn was higher and decline in productivity smaller in removal (abundance: 4.4 pronghorn/km²; change in productivity: -6.5 fawns/100 adult females) than in non-removal sites (abundance: 2.5 pronghorn/km²; change in productivity: -22 fawns/100 adult females). However, mule deer abundance and productivity did not differ between removal (abundance: 3.5 mule deer/km²; productivity: 56 fawns/100 adult females) and non-removal sites (abundance: 4.9 mule deer/km^2; productivity: 62 fawns/100 adult females). Six pairs of sites in similar habitat were selected. Site areas totalled 10,517 km^2. Between late July 2007 and March 2008, an average of 195 coyotes/1,000 km^2 were removed from one site in each pair by trapping and shooting. Pronghorn and mule deer were counted by driving 17–27 km-long transects at 25 km/ hr weekly during July and August and fortnightly in September, in 2007 and 2008.

A before-and-after study in 2001–2007 in a mosaic of shrub, forest and taiga in Alaska, USA (6) found that control of American black bear *Ursus americanus*, brown bear *Ursus arctos* and wolf *Canis lupus* increased moose *Alces alces* abundance and calf survival. Moose abundance and calf survival were higher after predator control (abundance: 0.56 moose/km^2; calf/adult ratio: 51–63 calves/100 adult females) than before control (abundance: 0.38 moose/km^2; calf/adult ratio: 34 calves/100 adult females). In May 2003 and 2004, 109 black and nine brown bears were translocated at least 240 km from a 1,368-km^2 area, reducing the populations by approximately 96% and 50% respectively. In 200–2008, wolf numbers were reduced by 11–33 animals/year across a wider 8,314-km^2 area by aircraft-assisted shooting, conventional hunting and trapping (density in 2001: 5.1 wolves/1,000 km^2; density in 2006: 1.3 wolves/1,000 km^2). Aircraft surveys (3.1 min/km^2) were used to monitor moose numbers and calf/adult ratios annually, in autumn, at 87 sites within the study area, each of 15.7 km^2.

A replicated, before-and-after study in 2006–2012 in three forest sites in South Carolina, USA (7) found that control of coyotes *Canis latrans* increased fawn survival in white-tailed deer *Odocoileus virginianus* in two out of three years. The annual survival rate of deer calves was higher one year (0.51) and three years (0.43) after the start of coyote control than before control (0.23), but did not differ two years (0.20) after the start of coyote control. The percentage of fawn mortalities that resulted from predation by coyotes was similar after (73%) compared to before control (80%). Between mid-January and early April 2010–2012, four hundred and seventy-four coyotes were removed from three 32-km^2 sites (1.6 coyotes /km^2/year) by trapping. The survival of 216 fawns (91 before and 125 after coyote control) was monitored using motion-sensitive radio-collars. Calves were monitored every eight hours if younger than four weeks, 1–3 times/day up to 12 weeks of age, weekly up to 16 weeks and 1–4 times/month up to 12 months.

A replicated, before-and-after study in 2010–2013 in two forest sites in Georgia, USA (8) found that controlling coyotes *Canis latrans* increased the number of young white-tailed deer *Odocoileus virginianus* relative to adult females in one of two sites. In one of two sites the number of young white-tailed deer was higher after coyote control (1.01 fawns/ adult female) compared to before control (0.63 fawns/adult female). However, in one site there was no significant difference (after control: 0.85 fawns/adult female; before control: 0.84 fawns/adult female). Coyote abundance was lower after control (4–16 animals/site) than before control (16–21 animals/site). In March–June 2011, professional trappers controlled coyotes in both sites. In January and February of 2010–2013, infrared cameras were arranged in a grid pattern, over a 2,000-ha area, at a density of 1 camera/65 ha at each site. Cameras were baited with corn and took a photograph every 15 minutes for 10 days. The number of pictures of young deer relative to pictures of adult females was calculated.

A before-and-after study in 1994–2002 in a large forest and shrubland area in Alaska, USA and Yukon, Canada (9) found that trapping and removing or sterilizing wolves *Canis lupus* did not reduce caribou *Rangifer tarandus* mortality. The annual mortality of caribou calves (≤1 year old) did not differ after wolf removal or sterilization commenced (50–67%) compared to before (39–65%). Adult female (≥1 year old) annual mortality was also similar after wolf removal or sterilization commenced (9–10%) compared to before (9%). In a 50,000-km² study area, 52–78 newborn caribou calves/year were radio-collared in May 1994–2002. Caribou were monitored during ≥3 flights/year. In 15 wolf packs, the dominant pair was sterilized in November 1997 and remaining wolves in those packs were translocated, mainly in April 1998. Eight additional packs were similarly treated over the following two winters. Caribou mortality was measured over four years before and five years after wolf control commenced.

A controlled, before-and-after study in 2008–2013 in four boreal forest, peatland and heath sites in Newfoundland, Canada (10) found that controlling coyotes *Canis latrans* increased caribou *Rangifer tarandus* calf survival. Caribou calf survival was higher when coyotes were controlled (70-day survival: 41%; 182-day survival: 32%) compared to before coyote control was carried out (70-day survival: 9%; 182-day survival: 7%). Survival rates across these two periods at sites without coyote control were stable (70-day survival: 52–58%; 182-day survival: 47%). At one site (covering 480 km²), lethal neck snares were set in March or April of 2012 and 2013 and were removed one week before caribou calving commenced in May. Forty coyotes were removed over these two years. Coyotes were not controlled at three other caribou calving sites. Caribou calves were radio-collared in late May to early June of 2008–2009 (193 calves) and 2012–2013 (103 calves), when 1–5-days old, and were monitored by radio-tracking through to November.

(1) Kauhala K., Helle P., Helle E. & Korhonen J. (1999) Impact of predator removal on predator and mountain hare populations in Finland. *Annales Zoologici Fennici*, 36, 139–148.

(2) Valkenburg P., McNay M.E. & Dale B.W. (2004) Calf mortality and population growth in the Delta caribou herd after wolf control. *Wildlife Society Bulletin*, 32, 746–756, https://doi.org/10.2193/0091-7648(2004)032[0746:cmapgi]2.0.co;2

(3) Beja P., Gordinho L., Reino L., Loureiro F., Santos-Reis M., & Borralho R. (2009) Predator abundance in relation to small game management in southern Portugal: conservation implications. *European Journal of Wildlife Research*, 55, 227–238, https://doi.org/10.1007/s10344-008-0236-1

(4) Wakeling B.F., Lee R., Brown D., Thompson R., Tluczek M. & Weisenberger M. (2009) The restoration of desert bighorn sheep in the Southwest, 1951–2007: factors influencing success. *Desert Bighorn Council Transactions*, 50, 1–17.

(5) Brown D.E. & Conover M.R. (2011) Effects of large-scale removal of coyotes on pronghorn and mule deer productivity and abundance. *The Journal of Wildlife Management*, 75, 876–882, https://doi.org/10.1002/jwmg.126

(6) Keech M.A., Lindberg M.S., Boertje R.D., Valkenburg P., Taras B.D., Boudreau T.A. & Beckmen K. B. (2011) Effects of predator treatments, individual traits, and environment on moose survival in Alaska. *The Journal of Wildlife Management*, 75, 1361–1380, https://doi.org/10.1002/jwmg.188

(7) Kilgo J.C., Vukovich M., Ray H.S., Shaw C.E. & Ruth C. (2014) Coyote removal, understory cover, and survival of white-tailed deer neonates. *The Journal of Wildlife Management*, 78, 1261–1271, https://doi.org/10.1002/jwmg.764

(8) Gulsby W.D., Killmaster C.H., Bowers J.W., Kelly J.D., Sacks B.N., Statham M.J. & Miller K.V. (2015) White-tailed deer fawn recruitment before and after

experimental coyote removals in central Georgia. *Wildlife Society Bulletin*, 39, 248–255, https://doi.org/10.1002/wsb.534

(9) Boertje R.D., Gardner C.L., Ellis M.M., Bentzen T.W. & Gross J.A. (2017) Demography of an increasing caribou herd with restricted wolf control. *The Journal of Wildlife Management*, 81, 429–448, https://doi.org/10.1002/jwmg.21209

(10) Lewis K.P., Gullage S.E., Fifield D.A., Jennings D.H. & Mahoney S.P. (2017) Manipulations of black bear and coyote affect caribou calf survival. *The Journal of Wildlife Management*, 81, 122–132, https://doi.org/10.1002/jwmg.21174

9.17. Sterilize predators

https://www.conservationevidence.com/actions/2573

- **One study** evaluated the effects on potential prey mammals of sterilizing predators. This study was in the USA and Canada[1].

COMMUNITY RESPONSE (0 STUDIES)

POPULATION RESPONSE (1 STUDY)

- **Survival (1 study)**: A before-and-after study in the USA and Canada[1] found that sterilising some wolves (combined with trapping and removing others) did not increase caribou survival.

BEHAVIOUR (0 STUDIES)

Background

Predators can limit population sizes of prey species. Changes in habitat or land management can lead to increases in predator populations which might negatively affect prey. Removing or controlling predators, especially native predators, for the benefit of their wild prey species can be a controversial management strategy. Nonetheless, there is potential for such management to lead to increases in the abundance, survival or reproduction success of prey species. Sterilization of predators may be proposed as an alternative strategy that may be regarded as being more acceptable than removal or lethal control.

A before-and-after study in 1994–2002 in a large forest and shrubland area in Alaska, USA and Yukon, Canada (1) found that sterilising some wolves *Canis lupus* (and trapping and removing others) did not reduce caribou *Rangifer tarandus* mortality. The annual mortality of caribou calves (≤1 year old) did not differ after wolf sterilization and removal commenced (50–67%) compared to before (39–65%). Adult female (≥1 year old) annual mortality was also similar after wolf sterilization and removal commenced (9–10%) compared to before (9%). In a 50,000-km² study area, 52–78 newborn caribou calves/year were radio-collared in May 1994–2002. In fifteen wolf packs, the dominant pair was sterilized in November 1997 and remaining wolves in those packs were translocated, mainly in April 1998. Eight additional packs were similarly treated over the following two winters. Caribou mortality was measured over four years before and five after wolf control commenced during ≥3 aerial surveys/year.

(1) Boertje R.D., Gardner C.L., Ellis M.M., Bentzen T.W. & Gross J.A. (2017) Demography of an increasing caribou herd with restricted wolf control. *The Journal of Wildlife Management*, 81, 429–448, https://doi.org/10.1002/jwmg.21209

9.18. Remove or control competitors

https://www.conservationevidence.com/actions/2575

- **Two studies** evaluated the effects on non-controlled mammals of removing or controlling competitors. One study was across Norway and Sweden[1] and one was in Norway[2].

COMMUNITY RESPONSE (0 STUDIES)

POPULATION RESPONSE (1 STUDY)

- **Reproductive success (1 study):** A replicated, controlled study in Norway and Sweden[1] found that red fox control, along with supplementary feeding, was associated with an increase in arctic fox litters.

BEHAVIOUR (1 STUDY)

- **Use (1 study):** A controlled study in Norway[2] found that where red foxes had been controlled arctic foxes were more likely to colonize.

> **Background**
>
> The range occupied by a species may be limited by the presence of competitors. In many cases, removing native competitors may be a controversial management strategy. However, abundance increases or range expansion of a competitor species, due to habitat or land management changes, may motivate removal or control of this species if its presence negatively impacts on another species that is deemed to be a higher conservation priority.

A replicated, controlled study in 1999–2011 at 10 tundra sites in Norway and Sweden (1) found that the number of arctic fox *Vulpes lagopus* litters increased after control of red foxes *Vulpes vulpes*, along with supplementary winter feeding at arctic fox dens. Where red foxes were intensively controlled, the number of active artic fox dens in winter increased more than at sites where no control or a low level of control was undertaken (data reported as statistical model results). The same response was found in the number of arctic fox litters produced, and with more litters produced when food was provided at den sites (data reported as statistical model results). Three sites were intensive control sites, with an average of 19–92 red foxes culled, and supplementary feeding provided for an average of 11–13.5 arctic fox dens at two of those sites. Three sites had low levels of control, with 1.5–7 red foxes culled and 1–3 dens fed at each of those sites. Four sites had no fox control and only one den was fed at one site. Red foxes were controlled during winter from 1999. The number of arctic fox litters was counted in known arctic fox dens during July and August 1999–2011.

A controlled study in 2005–2010 in 25 tundra sites in Finnmark, Norway (2) found that the probability of colonization by arctic fox *Vulpes lagopus* was higher in sites where red foxes *Vulpes vulpes* had been controlled. Arctic foxes colonized some sites where red foxes were controlled but their probability of colonizing sites without fox control was zero (reported as statistical model results). Between 2005 and 2010, intensive culling removed 885 red foxes from the Varanger peninsula. Foxes were monitored annually, over a 2-month period in late winter, using automatic digital cameras in front of a frozen block of reindeer remains. Fifteen camera sites were located across the area where red foxes were controlled and 10 areas without control (Nordkynn peninsula and Ifjordfjellet). Each camera took photographs of the carcass and its close surroundings every 10 min.

(1) Angerbjörn A., Eide N.E., Dalén L., Elmhagen B., Hellström P., Ims R.A., Killengreen S., Landa A., Meijer T., Mela M. & Niemimaa J. (2013) Carnivore conservation in practice: replicated management actions on a large spatial scale. *Journal of Applied Ecology*, 50, 59–67, https://doi.org/10.1111/1365-2664.12033

(2) Hamel S., Killengreen S.T., Henden J.A., Yoccoz N.G. & Ims R.A. (2013) Disentangling the importance of interspecific competition, food availability, and habitat in species occupancy: recolonization of the endangered Fennoscandian arctic fox. *Biological Conservation*, 160, 114–120, https://doi.org/10.1016/j.biocon.2013.01.011

9.19. Provide diversionary feeding for predators

https://www.conservationevidence.com/actions/2578

- **One study** evaluated the effects on potential prey mammals of providing diversionary feeding for predators. This study was in Canada[1].

COMMUNITY RESPONSE (0 STUDIES)

POPULATION RESPONSE (1 STUDY)

- **Survival (1 study):** A controlled, before-and-after study in Canada[1] found that diversionary feeding of predators appeared to increase woodland caribou calf survival.

BEHAVIOUR (0 STUDIES)

Background

Predators can limit population sizes of prey species. Changes in habitat or land management can lead to increases in predator populations, which might negatively affect prey. Removing or controlling predators, especially native predators, for the benefit of their wild prey species can be a controversial management strategy. Nonetheless, there is potential for reduced predator activities to lead to increases in the abundance, survival or reproductive success of prey species. Supplementary feeding of predators may be proposed as an alternative strategy that may be regarded as being more acceptable than removal or lethal control.

A controlled, before-and-after study in 2008–2011 in four boreal forest, peatland and heath sites in Newfoundland, Canada (1) found that diversionary feeding of predators appeared to increase woodland caribou *Rangifer tarandus* calf survival. However, the significance of the intervention was not explicitly tested. Caribou calf survival during diversionary feeding (70-day survival: 23%; 182-day survival: 14%) appeared to be higher than before diversionary feeding commenced (70-day survival: 9%; 182-day survival: 7%) though there was high variability in these data. Survival rates across these two periods at sites without diversionary feeding were stable (70-day survival: 56–59%; 182-day survival: 41–47%). Supplementary food was mostly taken by American black bears *Ursus americanus* which, along with coyotes *Canis latrans*, were the most frequent confirmed predators of caribou calves. At one site, 500-kg bags of bakery waste were distributed in a grid of 4.5 × 4.3-km quadrats, covering most of the caribou calving area. Food was provided from before 25 May until mid-July in 2010 and 2011 and was replenished weekly as required. In 2011, food was supplemented with beaver *Castor canadensis* carcasses. Three other caribou calving sites received no supplementary food. Across all sites, 313 caribou calves were radio-collared in late May to early June of 2008–2011, when 1–5 days old, and were monitored by radio-tracking through to November.

(1) Lewis K.P., Gullage S.E., Fifield D.A., Jennings D.H. & Mahoney S.P. (2017) Manipulations of black bear and coyote affect caribou calf survival. *The Journal of Wildlife Management*, 81, 122–132, https://doi.org/10.1002/jwmg.21174

9.20. Sterilise non-native domestic or feral species (e.g. cats and dogs)

https://www.conservationevidence.com/actions/2579

- We found no studies that evaluated the effects on mammals of sterilising non-native domestic or feral species (e.g. cats and dogs).

'We found no studies' means that we have not yet found any studies that have directly evaluated this intervention during our systematic journal and report searches. Therefore, we have no evidence to indicate whether or not the intervention has any desirable or harmful effects.

Background

Domestic animals may present a range of problems for wild mammals. These can include predation (e.g. Woods *et al.* 2013), disease transmission and hybridization between closely related species (Nussberger *et al.* 2014). Culling (especially feral animals) may be an option for reducing these threats but can be controversial on animal rights or animal welfare grounds. Sterilizing such animals is an alternative strategy that may reduce impacts of non-native species in the longer term and may also be possible to achieve on a large scale among domestic animals, by liaising with their owners.

Woods M., Mcdonald R. & Harris S. (2003) Predation of wildlife by domestic cats *Felis catus* in Great Britain. *Mammal Review*, 33, 174–188, https://doi.org/10.1046/j.1365-2907.2003.00017.x

Nussberger B., Wandeler P., Weber D. & Keller L.F. (2014) Monitoring introgression in European wildcats in the Swiss Jura. *Conservation Genetics*, 15, 1219–1230, https://doi.org/10.1007/s10592-014-0613-0

9.21. Train mammals to avoid problematic species

https://www.conservationevidence.com/actions/2580

- **Two studies** evaluated the effects of training mammals to avoid problematic species. Both studies were in Australia[1a,1b].

COMMUNITY RESPONSE (0 STUDIES)

POPULATION RESPONSE (1 STUDY)

- **Survival (1 study):** A controlled study in Australia[1b] found that training greater bilbies to avoid introduced predators did not increase their post-release survival.

BEHAVIOUR (2 STUDIES)

- **Behaviour change (2 studies):** One of two controlled studies in Australia found that greater bilbies trained to avoid introduced predators showed more predator avoidance behaviour[1a], the second study found no difference in behaviour between trained and untrained bilbies[1b].

Background

Mammals raised in areas free of non-native predators may be poorly adapted for use in translocations into areas where they have a greater chance of encountering such predators. This intervention includes cases where attempts are made to expose them to non-native predator cues with the intent that they will be able to avoid these after release. This intervention covers specifically training attempts on wild-born mammals. For captive-born mammals, see: *Species management — Train captive-bred mammals to avoid predators.*

A controlled study in 2005 in a desert reserve in South Australia, Australia (1a) found that greater bilbies *Macrotis lagotis* which had been trained to avoid invasive mammalian predators showed more predator avoidance behaviour than bilbies which had not received such training. Seven bilbies which had been trained to avoid predators changed burrow more frequently (5.7 times in 11 nights) than seven bilbies without such training (1.4 times). Trained bilbies also moved further between successive burrows (trained: 1,387 m; untrained: 158 m) and selected burrows with more entrance holes (trained: 3.6 entrances; untrained: 2.2 entrances) than untrained individuals. Additionally, all seven trained bilbies changed burrow the night after cat *Felis catus* scent was sprayed at their burrow entrance, but none of the untrained bilbies changed burrow. In May–June 2005, 14 bilbies were caught in a predator-free area of the Arid Recovery Reserve. Upon capture, seven individuals were exposed to a mock attack by a cat carcass and to cat urine and faecal matter and seven were not. Bilbies were then released at the capture site. All bilbies were equipped with microchips and radio-transmitters. Bilbies were radio-tracked daily to locate their diurnal burrow. Three days after capture, bilbies were located in their diurnal burrows and cat scent was sprayed at the entrance within four hours of sunset.

A controlled study in 2007–2009 in a desert reserve in South Australia, Australia (1b) found that post-release survival and predator avoidance behaviour of greater bilbies *Macrotis lagotis* with and without training to avoid invasive mammalian predators did not differ. Nine of 10 bilbies trained to avoid predators and eight of 10 without such training survived over six months after release. The trained bilby that died was either predated or scavenged by a wedge-tailed eagle *Aquila audax*. One bilby without training was killed by a cat *Felis catus* and one died of natural causes. Four months after release, the number of bilbies which changed burrow the night after cat scent was sprayed at their burrow entrance was the same for trained and untrained individuals (3 of 5 bilbies in each group). The population became extinct 19 months after release. In August 2007, twenty bilbies were caught in a predator-free area of the Arid Recovery Reserve and released, within three hours, into a 200-km² unfenced area with invasive cats and foxes *Vulpes vulpes*. Upon capture, 10 individuals were exposed to a mock attack by a cat carcass and to cat urine and faecal matter and 10 were not. All bilbies were equipped with radio-transmitters. Daily attempts were made to locate bilbies during the first month and weekly mortality checks were made for at least the following six months. Four months after release, bilbies were located in their diurnal burrows and cat scent was sprayed at the entrance within four hours of sunset.

(1) Moseby K.E., Cameron A. & Crisp H.A. (2012) Can predator avoidance training improve reintroduction outcomes for the greater bilby in arid

Australia? *Animal Behaviour*, 83, 1011–1021, https://doi.org/10.1016/j.anbehav.2012.01.023

9.22. Treat disease in wild mammals

https://www.conservationevidence.com/actions/2581

- **Three studies** evaluated the effects on mammals of treating disease in the wild. Two studies were in the USA[2,3] and one was in Germany[1].

COMMUNITY RESPONSE (0 STUDIES)

POPULATION RESPONSE (2 STUDIES)

- **Condition (2 studies):** A replicated study in Germany[1] found that medical treatment of mouflons against foot rot disease healed most infected animals. A before-and-after study in the USA[2] found that management which included vaccination of Yellowstone bison did not reduce prevalence of brucellosis.

BEHAVIOUR (1 STUDY)

- **Uptake (1 study):** A study in the USA[3] found that a molasses-based bait was readily consumed by white-tailed deer, including when it contained a dose of a disease vaccination.

Background

Treatment of diseases in wild mammals can be problematic. It can be difficult to diagnose causes of illness and the administration of medicines directly to target individuals can be challenging. Except in cases of highly threatened species, treatment of disease in wild mammals is usually only carried out when there are potential economic costs of not treating, such as a risk of transmission to domestic animals or reductions in numbers or health of animals that have sporting value. This intervention includes cases where animals are confined for treatment (and one study on captive animals that trials a delivery mechanism for treatments that might be administered to wild mammals) but in all cases, the aim is to improve the health of wild populations.

See also: *Use vaccination programme.*

A replicated study in 1994–2005 in three forest sites in Hessen and Rheinland-Pfalz, Germany (1) found that medical treatment of mouflons *Ovis gmelini musimon* against foot rot disease healed most infected animals. No statistical analyses were performed. All 152 infected individuals fully treated for foot rot disease recovered with no signs of reinfection. No data are provided for 13 individuals that only received partial treatment. Two hundred and fifty mouflons were caught using a fenced kraal or net trap and kept in a corral for six weeks. All were injected with penicilline–streptomycine (1–3 ml of Tardomyocel III comp®), had an anti-parasitic treatment (0.2 mg/kg of Ivomec®) and, in cases of bad general condition (e.g. fever) a supplementary treatment was administered (see paper for details). A total of 165 animals with foot rot were treated by trimming the wounded hooves and covering them in antiseptic fluid (Kodan®-Tincture). Some were treated with an additional antibiotic injection (5.0 ml Procain Penicillin G® solution). If needed, a second treatment was conducted after two or three days. Four to six weeks after treatment, a final trimming of the hooves was undertaken before the animals were released.

A before-and-after study in 2001–2010 on grasslands in and around a national park in Wyoming, USA (2) found that intensive management, including vaccination, of Yellowstone bison *Bison bison bison* did not reduce prevalence of brucellosis *Brucella abortus*. The proportion of adult female bison testing positive for brucellosis increased or remained constant during the period at approximately 60%. However, transmission of brucellosis from bison to domestic cattle was almost eliminated. Bison were intensively managed, which included separating them from cattle on winter pastures, herding them into the park in spring, and periodic culls where these aims could not be achieved. A proportion of bison was tested for brucellosis and animals that tested positive were slaughtered. Bison, especially adult females, were vaccinated either when captured or by remote vaccine delivery. During 2001–2010, 1,643 bison that tested positive for brucellosis were slaughtered and 18 were released. A total of 1,517 bison that tested negative or were untested were also slaughtered. The overall population ranged from 2,432 to 5,015 during this period.

A study in 2012 on captive animals in Iowa, USA (3) found that white-tailed deer *Odocoileus virginianus* readily consumed a molasses-based bait, including when it contained a dose of a disease vaccination.

In 48 of 50 trials, all baits were consumed within three hours. However, on >62% of occasions, all baits in one serving were consumed by a single deer. All baits containing *Mycobacterium bovis* bacillus Calmette–Guerin (BCG) vaccine were consumed. Baits, containing flour, cane molasses, sugar, water, shortening, sodium bicarbonate and sodium chloride, were baked into 8-g pellets. Seven pellets were fed to deer in addition to their usual feed, in each of five pens (three each containing three deer, one with four deer and one with 50 deer) daily for 10 days. Consumption was observed using camera traps. Additionally, five baits containing 0.2 ml BCG were offered to three deer during January 2012, in addition to their usual feed.

(1) Volmer K., Hecht W., Weiß R. & Grauheding D. (2008) Treatment of foot rot in free-ranging mouflon (*Ovis gmelini musimon*) populations—does it make sense? *European Journal of Wildlife Research*, 54, 657–665, https://doi.org/10.1007/s10344-008-0192-9

(2) White P.J., Wallen R.L., Geremia C., Treanor J.T. & Blanton D.W. (2011) Management of Yellowstone bison and brucellosis transmission risk — Implications for conservation and restoration. *Biological Conservation*, 144, 1322–1334, https://doi.org/10.1016/j.biocon.2011.01.003

(3) Palmer M.V., Stafne M.R., Waters W.R., Thacker T.C. & Phillips G.E. (2014) Testing a molasses-based bait for oral vaccination of white-tailed deer (*Odocoileus virginianus*) against *Mycobacterium bovis*. *European Journal of Wildlife Research*, 60, 265–270, https://doi.org/10.1007/s10344-013-0777-9

9.23. Use vaccination programme

https://www.conservationevidence.com/actions/2582

- **Seven studies** evaluated the effects on mammals of using vaccination programmes. Three studies were in the UK[5a,5b,6] and one study was in each of Belgium[1], Spain[2], Poland[3] and Ethiopia[4].

COMMUNITY RESPONSE (0 STUDIES)

POPULATION RESPONSE (7 STUDIES)

- **Abundance (1 study):** A before-and-after study in Poland[3] found that following an anti-rabies vaccination programme, red fox numbers increased.

- **Condition (6 studies):** Five studies (including three replicated, three controlled and two before-and-after studies) in Belgium[1], Spain[2] and the UK[5a,5b,6] found that following vaccination, rabies was less frequent in red foxes[1], numbers of Eurasian badgers[5a,5b,6] infected with tuberculosis was reduced and European rabbits[2] developed immunity to myxomatosis and rabbit haemorrhagic disease. One of the studies[5a] also found that vaccination reduced the speed and extent of infection in infected Eurasian badgers. A study in Ethiopia[4] found that following vaccination of Ethiopian wolves, a rabies outbreak halted.

BEHAVIOUR (0 STUDIES)

Background

Vaccinating wild mammals can be challenging, due to difficulties in administering vaccines in appropriate doses to target animals. Only in particular cases, such as when animals may be affected by a zoonotic disease, that could spread to humans or domestic livestock, or when particularly endangered mammal populations are threatened, is vaccination likely to be attempted.

A study in 1989–1991 in a rural region of Luxembourg, southern Belgium (1) found that vaccinating red foxes *Vulpes vulpes* against rabies reduced the occurrence of rabies. After one vaccination attempt, six out of nine (67%) rabid and 11 of 14 (79%) healthy foxes tested had consumed the bait. After the second attempt, 25 of 31 (81%) adult foxes and 27 of 55 (49%) juvenile foxes tested had consumed bait, and all 86 were healthy. After the third vaccination phase, 64 of 79 (81%) foxes had consumed bait and only one tested positive for rabies (authors note that it was found at the edge of the vaccination area, and had not taken bait). Additionally, the number of cases of rabies reported in livestock every six months fell from 7–61 before the second vaccination attempt (January 1985–June 1990) to zero in the year afterwards (reporting of rabies in livestock is mandatory in Belgium). In November 1989, April 1990 and October 1990, a total of 25,000 field vaccine-baits containing

VVTGgRAB and a tetracycline biomarker were dropped by helicopter across a 2,200 km² area at a density of 15/km (excluding urban areas). After each vaccination period (January–March 1990, April–October 1990, November 1990–April 1991) a total of 188 foxes which were found dead or shot by hunters were tested for both rabies and the presence of tetracycline (which would indicate that they had consumed the bait).

A replicated, before-and-after study in 1999–2002 in Cadiz province, Spain (2) found that most vaccinated European wild rabbits *Oryctolagus cuniculus* developed immunity to myxomatosis and rabbit haemorrhagic disease. Of 32 rabbits which initially had no immunity to myxomatosis, 26 (81%) had developed immunity 2–4 weeks after vaccination. Of 81 rabbits which initially had no immunity to rabbit haemorrhagic disease, 68 (84%) had developed immunity 2–4 weeks after vaccination. The development of immunity did not differ between males and females, nor did it vary with time spent in captivity. Between November 1999 and March 2002, six groups of 14–46 wild-caught rabbits (some of which already had natural immunity to one or both diseases) were vaccinated against myxomatosis and rabbit haemorrhagic disease with commercial vaccines, and held in captivity for two, three or four weeks. Blood samples were taken from each rabbit both before vaccination, and two days prior to release, to test for immunity to each disease.

A before-and-after study in 1980–2005 in a rural area near Rogów, Central Poland (3) found that following an anti-rabies vaccination programme, red fox *Vulpes vulpes* numbers increased. The density of fox tracks was higher after the start of the vaccination programme than before (11.0 vs 5.9 snow tracks/km/day). The same pattern held for fox density as recorded by surveys from vehicles (2.6 vs 1.2 foxes/km²) and for active dens (15.0 vs 9.3 dens with young/year). However, there were fewer cubs/den after vaccination (3.4) than before (3.8). Anti-rabies vaccinations started in 1995–1996. Between 1980 and 2005, fox densities were estimated annually within an 89-km² area. Estimates were from counts of tracks in snow (average annual transect length was 90 km before and 55 km after the vaccination programme), individuals seen from vehicles in forest habitats, and location of dens and number of cubs within the dens.

A study in 2003–2004 in alpine habitat in a national park in Ethiopia (4) found that vaccinating Ethiopian wolves *Canis simensis* successfully halted a rabies outbreak. Of 69 wolves vaccinated in the 'intervention zone' (beyond the boundaries of the outbreak) between one to four months after rabies was confirmed, all 19 animals sampled one month later had protective levels of rabies antibodies. Six months after initial vaccinations, two wolves that received a booster vaccination at 30 days still had protective levels of antibodies while one wolf that did not receive a booster had levels below those regarded as providing protection. Of five wolves sampled 12 months after initial vaccinations, one that received a booster still had protective levels of rabies antibodies while four that received only initial vaccinations did not have protective levels. The last confirmed rabies death was two months after the start of the vaccination programme. Rabies was first confirmed on 28 October 2003 from wolf mortalities since mid-August. Sixty-nine wolves were vaccinated in the intervention zone, between November 2003 and February 2004. A further eight were vaccinated during follow-up recapture (March–November 2004). Mortality in the affected sub-population was 76%.

A replicated, controlled study in 2006–2009 on 15 wild-caught, captive Eurasian badgers *Meles meles* in England, UK (5a) found that vaccinating badgers against tuberculosis reduced the likelihood of tuberculosis infection, and reduced both the speed and the extent of infection in infected animals. Three out of nine badgers vaccinated with Bacillus Calmette-Guérin (BCG) became infected with tuberculosis, compared to six out of six badgers which had not been vaccinated. The time taken for infection to develop was longer in vaccinated badgers (two, eight and 12 weeks), than in non-vaccinated badgers (2–4 weeks). Vaccinated badgers had fewer lesions (median score: 4) than non-vaccinated badgers (median score: 9–12.5). Fifteen tuberculosis-free wild badgers were caught and housed in groups of up to four. Nine badgers were injected with 1 ml of Bacillus Calmette-Guérin (BCG) Danish strain 1331 vaccine and six were not vaccinated. After 17 weeks, all 15 badgers were infected with tuberculosis. Every 2–3 weeks badgers were anaesthetized and examined for tuberculosis infection and, 29 weeks after vaccination, the badgers were killed and examined for tuberculosis infection. (Years of study assumed from information provided, as not specified).

A replicated, randomized, controlled study in 2006–2009 in an area of mixed woodland and farmland in Gloucestershire, UK (5b, same experimental set-up as 6) found that vaccinating Eurasian badgers *Meles meles* against tuberculosis reduced the number of animals infected. Vaccination with Bacillus Calmette-Guérin (BCG) reduced the number of badgers with tuberculosis in vaccinated groups (15/179 infected, 8%) compared to non-vaccinated groups (18/83 infected, 22%). In 2009, badgers were caught in cage traps, set for two consecutive nights, twice a year, at every active sett in a 55 km² study area. Badgers were tested for tuberculosis using three tests. Social groups were randomly allocated to 'vaccinated' or 'not vaccinated' treatments. Every badger caught in a vaccination group was injected with 1 ml of Bacillus Calmette-Guérin (BCG) Danish strain 1331 vaccine once per year. A total of 179 badgers from 38 social groups were vaccinated, while 83 badgers from 26 social groups were unvaccinated.

A randomized, controlled, before-and-after study in 2006–2009 in an area of mixed woodland and farmland in Gloucestershire, UK (6, same experimental set-up as 5b) found that vaccinating Eurasian badgers *Meles meles* against tuberculosis reduced the number of animals infected. Three years after vaccination with Bacillus Calmette-Guérin (BCG) began, the number of badgers infected with tuberculosis (119 of 342 tested, 35%) was lower than before vaccination began (156 of 294 tested, 53%). Vaccination reduced the likelihood of individual badgers testing positive for tuberculosis by 54%. Unvaccinated badgers from vaccinated social groups were less likely to have tuberculosis (adults: 35%, cubs: 21% infected) than badgers from unvaccinated social groups (adults: 52%, cubs: 33% infected). Additionally, unvaccinated cubs were 79% less likely to become infected with tuberculosis when at least one third of the adults in their social group were vaccinated. However the probability of an unvaccinated adult having tuberculosis did not change when more group members were vaccinated. From June 2006–October 2009, badgers were caught in baited steel mesh traps, set for two consecutive nights, twice a year at every active sett in a 55 km² study area. Badgers were tested for tuberculosis using three tests. Social groups were randomly allocated to 'vaccinated' or 'not vaccinated' treatments. Badgers in vaccination groups were injected with 1 ml of Bacillus Calmette-Guérin (BCG) Danish strain 1331 vaccine once/year.

(1) Brochier B., Kieny M.P., Costy F., Coppens P., Bauduin B., Lecocq J.P., Languet B., Chappuis G., Desmettre P., Afiademanyo K., Libois R. & Pastoret P.-P. (1991) Large-scale eradication of rabies using recombinant vaccinia-rabies vaccine. *Nature*, 354, 520–522.

(2) Cabezas S., Calvete C. & Moreno S. (2006) Vaccination success and body condition in the European wild rabbit: applications for conservation strategies. *Journal of Wildlife Management*, 70, 1125–1131, https://doi. org/10.2193/0022-541x(2006)70[1125:vsabci]2.0.co;2

(3) Goszczyński J., Misiorowska M. & Juszko S. (2008) Changes in the density and spatial distribution of red fox dens and cub numbers in central Poland following rabies vaccination. *Acta Theriologica*, 53, 121–127, https://doi. org/10.1007/bf03194245

(4) Knobel D.L., Fooks A.R., Brookes S.M., Randall D.A., Williams S.D., Argaw K., Shiferaw F., Tallents L.A. & Laurenson M.K. (2008) Trapping and vaccination of endangered Ethiopian wolves to control an outbreak of rabies. *Journal of Applied Ecology*, 45, 109–116, https://doi. org/10.1111/j.1365-2664.2007.01387.x

(5) Chambers M.A., Rogers F., Delahay R.J., Lesellier S., Ashford R., Dalley D., Gowtage S., Davé D., Palmer S., Brewer J., Crawshaw T., Clifton-Hadley R., Carter S., Cheeseman C., Hanks C., Murray A., Palphramand K., Pietravalle S., Smith G.C., Tomlinson A., Walker N.J., Wilson G.J., Corner L.A.L., Rushton S.P., Shirley M.D.F., Gettinby G., McDonald R.A. & Hewinson R.G. (2011) Bacillus Calmette-Guérin vaccination reduces the severity and progression of tuberculosis in badgers. *Proceedings of the Royal Society of Biology*, 278, 1913–1920.

(6) Carter S.P., Chambers M.A., Rushton S.P., Shirley M.D.F. Schuchert P., Pietravalle S., Murray A., Rogers F., Gettinby G., Smith G.C., Delahay R.J., Hewinson R.G. & McDonald R.A. (2012) BCG vaccination reduces risk of tuberculosis infection in vaccinated badgers and unvaccinated badger cubs. *PLoS One*, 7, e49833, https://doi.org/10.1371/journal.pone.0049833

9.24. Eliminate highly virulent diseases early in an epidemic by culling all individuals (healthy and infected) in a defined area

https://www.conservationevidence.com/actions/2585

- We found no studies that evaluated the effects on mammals of eliminating highly virulent diseases early in an epidemic by culling all individuals (healthy and infected) in a defined area.

'We found no studies' means that we have not yet found any studies that have directly evaluated this intervention during our systematic journal and report searches. Therefore, we have no evidence to indicate whether or not the intervention has any desirable or harmful effects.

Background

Culling is a well-established approach for the management of some diseases in domestic animals, and although it has been used in an attempt to eliminate disease or reduce rates of transmission in a range of wild mammal species (Carter et al. 2009), the culling of diseased wild mammals for conservation is rarely attempted, probably due to ethical and ecological considerations (Woodroffe 1999). Nonetheless, prompt culling of all animals in an area might have potential to control or eliminate disease outbreaks and reduce longer-term negative impacts of disease on populations (McCallum 2008).

Carter S.P., Roy, S.S., Ji, W.H., Cowan, D.P., Smith, G.C., Delahay, R.J., Rossi, S. and Woodroffe, R. (2008) Options for the control of disease 2: Targeting hosts. Pages 121–146 in: R.J. Delahay, G.C. Smith & M.R. Hutchings (eds) *Management of disease in wild mammals*. Springer, UK, https://doi. org/10.1007/978-4-431-77134-0_7

Woodroffe R. (1999) Managing disease threats to wild mammals. *Animal Conservation*, 2, 185–193, https://doi.org/10.1111/j.1469-1795.1999.tb00064.x

McCallum H. (2008) Tasmanian devil facial tumour disease: lessons for conservation biology. *Trends in Ecology and Evolution*, 23, 631–637, https:// doi.org/10.1016/j.tree.2008.07.001

9.25. Cull disease-infected animals

https://www.conservationevidence.com/actions/2586

- **One study** evaluated the effects on mammals of culling disease-infected animals. This study was in Tasmania[1].

COMMUNITY RESPONSE (0 STUDIES)

POPULATION RESPONSE (1 STUDY)

- **Condition (1 study):** A before-and-after, site comparison study in Tasmania[1] found that culling disease-infected Tasmanian devils resulted in fewer animals with large tumours associated with late stages of the disease.

BEHAVIOUR (0 STUDIES)

Background

When mammal populations are threatened by disease, one potential action is to remove contact between diseased and disease-free animals. However, it is rarely attempted, possibly due to ethical and ecological concerns (Woodroffe 1999).

Woodroffe R. (1999) Managing disease threats to wild mammals. *Animal Conservation*, 2, 185–193, https://doi.org/10.1111/j.1469-1795.1999.tb00064.x

A before-and-after and site comparison study in 2004–2007 on two peninsulas in Tasmania (1) found that culling disease-infected Tasmanian devils *Sarcophilus harrisii* resulted in fewer animals with large tumours associated with late stages of the disease. One year after intensive culling commenced, the proportion of trapped Tasmanian devils with large tumours (22%) was lower than during the first month of intensive culling (67%; numbers not reported). Tasmanian devil density remained constant during this time (1.6 devils/km²) compared to a similar site without culling where density declined (from 0.9 to 0.6 devils/km²), although statistical tests were not carried out. Tasmanian devils infected with Devil Facial Tumour Disease were culled during an 18-month pilot study commencing in June 2004, and an intensive 12-month trapping program commencing in January 2006. Tasmanian devils were trapped within a 160-km² area on the peninsula during 4–5 x 10-day trips/year. Infected individuals or those with signs of the disease were euthanized. Numbers with large tumours (>4 cm) were counted in February 2006 and January 2007. Tasmanian devil density was recorded in the study area and at a similar 160-km² peninsula on the same coast (methods not reported).

(1) Jones M.E., Jarman P.J., Lees C.M., Hesterman H., Hamede R.K., Mooney N.J., Mann D., Pukk C.E., Bergfield J. & McCallum H. (2007) Conservation

Management of Tasmanian Devils in the Context of an Emerging, Extinction-threatening Disease: Devil Facial Tumor Disease. *EcoHealth*, 4, 326–337, https://doi.org/10.1007/s10393-007-0120-6

9.26. Use drugs to treat parasites

https://www.conservationevidence.com/actions/2587

- **Seven studies** evaluated the effects on mammals of using drugs to treat parasites. Three studies were in the USA[2,3,4], two were in Spain[5a,5b], one was in Germany[1] and one was in Croatia[6].

COMMUNITY RESPONSE (0 STUDIES)

POPULATION RESPONSE (7 STUDIES)

- **Survival (1 study):** A randomized, replicated, controlled study the USA[4] found that medical treatment of Rocky Mountain bighorn sheep against lungworm did not increase lamb survival.

- **Condition (6 studies):** Three of four before-and-after studies (one controlled), in Germany[1], the USA[2,3] and Croatia[6], found that after administering drugs to mammals, parasite burdens were reduced in roe deer[1] and in wild boar piglets[6] and numbers of white-tailed deer[3] infected were reduced. A third study found that levels of lungworm larvae in bighorn sheep faeces were reduced one month after drug treatment but not after three to seven months[2]. One of these studies also found that the drug treatment resulted in increased body weight in roe deer fawns[1]. A replicated, controlled, before-and-after study in Spain[5a] found that higher doses of ivermectin treated sarcoptic mange in Spanish ibex faster than lower doses, and treatment was more effective in animals with less severe infections. A replicated, before-and-after study in Spain[5b] found that after injecting Spanish ibex with ivermectin to treat sarcoptic mange a mange-free herd was established.

BEHAVIOUR (0 STUDIES)

Background

High levels of parasites in wild mammals may reduce fitness and lead to higher levels of mortality (e.g. Cooper *et al.* 2012). Drugs are readily available to reduce infestation levels of a wide range of parasites, though they are more frequently used to treat domestic animals. Attempts to treat wild mammals are most likely to be made where there is specific economic value to the wild mammal, such as among species that are valued for sporting purposes. In such cases, drug treatments may be administered through adding to baits or supplementary food left for animals.

Cooper N., Kamilar J.M. & Nunn C.L. (2012) Host longevity and parasite species richness in mammals. *PLoS ONE*, 7, e42190, https://doi.org/10.1371/journal. pone.0042190

A before-and-after study in 1979–1986 in a forest area in Middle Rhine, Germany (1) found that supplementing food with a drug to reduce parasitic worms reduced parasite burdens and increased body weights in roe deer *Capreolus capreolus*. After seven years of treatment, nematode burdens were reduced by 95% in fawns and 99% in adult deer, compared to levels before treatments began. Average weights of fawns killed for venison increased during this time to 9.4 kg, from 4.9 kg prior to treatment with the drug. Following discovery of high nematode burdens and associated mortality in 1979, winter fodder of deer (bran, mill leftovers and maize silage) was supplemented with anthelmintic powder (Fenbendazole, containing 4% Panacur) for seven years in a dose of 5 mg/kg body weight. Parasite burdens were assessed from faecal samples and from 90 carcasses collected before and 57 after treatments.

A replicated, controlled, before-and-after study in 1987–1988 in a state park in South Dakota, USA (2) found that following medical treatment, lungworm larvae levels in bighorn sheep *Ovis canadensis* faeces reduced over the following month, but not 3–7 months after treatment. In the month following treatment, average concentrations of lungworm larvae in faeces of bighorn sheep treated with one dose (50–250 larvae/g faeces) or two doses of ivermectin (50–300 larvae/g faeces) were lower than in untreated sheep (500–1,400 larvae/g faeces). However, by 3–7 months after treatments, average concentrations of lungworm larvae did not differ significantly between treated (600–1,300 larvae/g faeces) and untreated sheep (300–600 larvae/g faeces). One group of free-ranging female sheep received alfalfa treated with the anti-parasitic drug ivermectin in February 1987 and 1988 (four and six individuals, respectively) and another group received it in both February and March 1987 and January and February 1988 (seven and 14 sheep respectively). Five (1987) and nine (1988) sheep were untreated. Each treatment was administrated over two successive days at a rate of 2 ml ivermectin/sheep, and sheep were pre-baited with untreated alfalfa two weeks prior to each treatment. Parasite counts were made through analysing sheep faeces collected weekly from January to March and June to August in 1987–1988.

A controlled, before-and-after study in 1987–1989 in a grassland wildlife refuge in Texas, USA (3) found that feeding white-tailed deer *Odocoileus virginianus* medicated corn reduced trematode *Fascioloides magna* parasite infection by 63%. Four weeks after treatment with triclabendazole, fewer white-tailed deer were infected with live parasites (2/23) than in baited control (15/24) and unbaited control areas (24/30). Before treatment, the number of infected deer was similar (area to be treated: 8/9; baited control: 4/8; unbaited control: 5/8). In winter 1987–1989, at each of 10 sites across a 391-ha treatment pasture and 10 sites across 421-ha of baited control pasture, untreated corn was distributed for 3–4 weeks, before corn containing triclabendazole (500 ml triclabendazole/23 kg corn) was used in the treatment pasture for a further week. The estimated dose was 11 mg/kg body weight/deer/ day for seven days. Corn was placed at dusk, and deer were counted at each bait site between 2100–2300 hr. At a third, 439-ha unbaited control pasture, no corn was distributed. In January 1987, before baiting began, 13 fawns and 12 adult deer were shot across the three areas. In 1987– 1989, four weeks after baiting finished, 6–15 adult deer were shot on each pasture. The liver of each deer was examined for parasites.

A randomized, replicated, controlled study in 1991–1995 in two mountain ranges in Colorado, USA (4) found that medical treatment of Rocky Mountain bighorn sheep *Ovis canadensis canadensis* against lungworm did not increase lamb survival. Average annual recruitment did not differ between herds treated for lungworm (0.5–0.7 lambs/ adult female) and untreated herds (0.6–0.7 lambs/adult female). Adult bighorn females of four herds were captured in February–March 1991– 1995 and were marked and radio-collared. Between 1991 and 1995 the herds were either fed for 8–10 weeks each winter with 2 kg/individual/ day of alfalfa hay and 1 kg/individual/day of apple pulp, fed with alfalfa hay and apple pulp with two treatments of a drug to reduce parasitic worms (Fenbendazole, 3 g/adult female) added to the apple pulp late in the feeding period, given Fenbendazole-treated salt blocks (1.65 g Fenbendazole/kg) from December to April, or not given food or Fenbendazole-treated salt blocks. Treatments were rotated annually under a predetermined, randomly selected scheme. Lamb survival for 11–18 marked adult females/herd was assessed every two weeks between May and October.

A replicated, controlled, before-and-after study in 1988 in a mountainous National Park in southern Spain (5a) found that injecting Spanish ibex *Capra pyrenaica hispanica* with higher doses of ivermectin treated sarcoptic mange *Sarcoptes scabiei* faster than lower doses, and treatment was more effective in animals with less severe infections. All nine ibex with limited mange recovered after being treated with ivermectin. Six animals injected with 0.4 mg/kg body weight had no scabs or mites 21 days after treatment, and three animals injected with 0.2 mg/kg body weight had no scabs or mites four and five weeks after treatment, respectively. However, only three of six ibex with severe infection recovered following treatment, and two died. The sixth animal was still carrying mites two months after treatment. From September–December 1988, wild Spanish ibex were caught, sedated, and treated with Foxim anti-mange treatment (500 mg/l of water). Fifteen adult (>2-years old) female ibex with sarcoptic mange were divided into five treatment groups: 1) ibex with limited mange, given a single dose of ivermectin (0.4 mg/kg body weight) by syringe injection; 2) ibex with limited mange given a single dose of ivermectin (0.4 mg/kg body weight) by rifle dart injection; 3) ibex with limited mange given a single dose of ivermectin (0.2 mg/kg body weight) by syringe; 4) ibex with severe mange given two doses of ivermectin (0.2 mg/kg body weight) by syringe, two weeks apart; 5) ibex with severe mange given two doses of ivermectin (0.4 mg/kg body weight) by syringe, two weeks apart. Infection was classified into four levels of severity, and treatment tested on the worst two: limited ('consolidation': affected skin limited to a few body parts) and severe ('chronic': severe skin disease covering much of the body). Ibex were examined for two months to monitor recovery.

A replicated, before-and-after study in 1989 in a mountainous National Park in southern Spain (5b) found that after injecting Spanish ibex *Capra pyrenaica hispanica* with ivermectin to treat sarcoptic mange *Sarcoptes scabiei*, a mange-free herd was established. All 32 Spanish ibex treated with ivermectin showed no signs of mange six weeks after treatment began. After joining 65 mange-free ibex (at least 12 of which were treated in an earlier program, and 17 of which were mange-free on capture), the total population of 97 ibex showed no signs of mange for at least a year. From February–March 1989, sixty-three Spanish ibex were caught, sedated and examined for sarcoptic mange. The 14 ibex with chronic mange were injected with ivermectin (0.4 mg/kg body weight) and released at the capture site. The 49 remaining ibex, including healthy animals, were injected with ivermectin (0.4 mg/kg body weight) and a foxim spray (500 mg/l), and examined for mites. The 17 animals without mites were placed in 'quarantine' pens, and 32 with mites were kept in 'treatment' pens and injected with ivermectin (0.2 mg/kg body weight) two-and four-weeks later before joining the 'quarantine' pens. After two weeks in quarantine, ibex showing no symptoms of mange were given a final dose of ivermectin and released into a 400-ha enclosure in Nava de San Pedro Park which already contained 48 ibex.

A replicated, before-and-after study in three sites in Slavonia, Croatia (6) found that using drugs to treat parasites reduced the number of parasite eggs in the dung of wild boar *Sus scrofa* piglets. These results were not tested for statistical significance. After 14 days, parasite eggs were found in 0–10% of piglet faecal samples compared to 70–100% before treatment. The anti-parasitic drug ivermectin (0.6% formulation) was mixed with piglet feed at a concentration of 9 parts per million. An unspecified number of piglets in three sites were offered the feed for seven days using semi-automated piglet feeders, which were refilled twice each week. Faecal samples from the piglets were examined before the treatment and after seven and 14 days.

(1) Düwel D. (1987) Repeated treatment of roe deer (*Capreolus capreolus*) with Panacur in winter for control of nematode infection. *Zeitschrift für Jagdwissenschaft*, 33, 242–248

(2) Easterly T.G., Jenkins K.J. & McCabe T.R. (1992) Efficacy of orally administered ivermectin on lungworm infection in free-ranging bighorn sheep. *Wildlife Society Bulletin*, 20, 34–39.

(3) Qureshi T., Drawe D.L., Davis D.S. & Craig T.M. (1994) Use of bait containing triclabendazole to treat *Fascioloides magna* infections in free-ranging white-tailed deer. *Journal of Wildlife Diseases*, 30, 346–350.

(4) Miller M.W., Vayhinger J.E., Bowden D.C., Roush S.P., Verry T.E., Torres A.N. & Jurgens V.D. (2000) Drug treatment for lungworm in bighorn sheep: reevaluation of a 20-year-old management prescription. *The Journal of Wildlife Management*, 64, 505–512, https://doi.org/10.2307/3803248

(5) León-Vizcaíno L., Cubero M.J., González-Capitel E., Simón M.A., Pérez L., de Ybáñez M.R.R., Ortíz J.M., Candela M.G. & Alonso F. (2001) Experimental ivermectin treatment of sarcoptic mange and establishment of a mange-free population of Spanish ibex. *Journal of Wildlife Diseases*, 37, 775–785, https://doi.org/10.7589/0090-3558-37.4.775

(6) Rajkovi-Janje R., Manojlovi L. & Gojmerac T. (2004) In-feed 0.6% ivermectin formulation for treatment of wild boar in the Moslavina hunting ground in Croatia. *European Journal of Wildlife Research*, 50, 41–43, https://doi.org/10.1007/s10344-003-0033-9

9.27. Establish populations isolated from disease

https://www.conservationevidence.com/actions/2588

- **One study** evaluated the effects on mammals of establishing populations isolated from disease. The study was in sub-Saharan Africa[1].

COMMUNITY RESPONSE (0 STUDIES)

POPULATION RESPONSE (1 STUDY)

- **Condition (1 study):** A site comparison study throughout sub-Saharan Africa[1] found that fencing reduced prevalence of canine distemper but not of rabies, coronavirus or canine parvovirus in African wild dogs.

BEHAVIOUR (0 STUDIES)

Background

When mammal populations are threatened by disease, a short-to medium-term management option may be to establish wild-living or captive populations that are isolated from sources of the disease, such as on islands or in large fenced enclosures (e.g. Jones *et al.* 2007). These could aid persistence of the species and provide stock for reintroductions, should the disease be eliminated or sufficiently controlled in the originally affected areas.

Jones M.E., Jarman P.J., Lees C.M., Hesterman H., Hamede R.K., Mooney N.J., Mann D., Pukk C,.E., Bergfeld J. & McCallum H. (2007) Conservation management of Tasmanian devils in the context of an emerging, extinction-threatening disease: devil facial tumor disease. *EcoHealth*, 4, 326–337, https://doi.org/10.1007/s10393-007-0120-6

A site comparison study in 1988–2010 of 16 sites throughout sub-Saharan Africa (1) found that fencing reduced prevalence of canine distemper but not of rabies, coronavirus or canine parvovirus in African wild dogs *Lycaon pictus*. Prevalence of canine distemper was lower in fenced protected sites (0.04 seroprevalence) than in unfenced protected sites (0.28) or unfenced and unprotected sites (0.20). However, the prevalence of rabies, coronavirus or parvovirus did not change significantly between fenced protected sites (rabies: 0.02; coronavirus: 0.03; parvovirus: 0.22 seroprevalence), unfenced protected sites (rabies: 0.06; coronavirus: 0.11; parvovirus: 0.19) and unfenced and unprotected sites (rabies: 0.12; coronavirus: 0.18; parvovirus: 0.21). Blood samples were collected from 268 African wild dogs in 1988–2009 across 16 sites representing five unconnected wild dog populations: South Africa (2 unconnected populations; 7 protected-fenced sites, 3 unprotected-unfenced), Zimbabwe, Botswana (1 population; 2 protected-unfenced site, 2 unprotected-unfenced), Tanzania (1 protected-unfenced site) and Kenya (1 unprotected-unfenced site). Protected-fenced sites had game fencing likely to exclude domestic dogs. Seroprevalence (proportion of animals with detectable antibodies against a disease) was determined from blood samples.

(1) Prager K.C., Mazet J.A.K., Munson L., Cleaveland S., Donnelly C.A., Dubovi E.J., Szykman Gunther M., Lines R., Mills G., Davies-Mostert H.T., Weldon McNutt J., Rasmussen G., Terio K., Woodroffe R. (2012) The effect of protected areas on pathogen exposure in endangered African wild dog (*Lycaon pictus*) populations. *Biological Conservation*, 150, 15–22, https://doi.org/10.1016/j.biocon.2012.03.005

9.28. Control ticks/fleas/lice in wild mammal populations

https://www.conservationevidence.com/actions/2589

- **Two studies** evaluated the effects of controlling ticks, fleas or lice in wild mammal populations. Both studies were in the USA[1,2].

COMMUNITY RESPONSE (0 STUDIES)

POPULATION RESPONSE (2 STUDIES)

- **Condition (2 studies):** A replicated, paired sites, controlled study in the USA[1] found that a grain-bait insecticide product did not consistently reduce flea burdens on Utah prairie dogs. A controlled study the USA[2] found that treating wolves with ivermectin cleared them of infestations of biting dog lice.

BEHAVIOUR (0 STUDIES)

Background

Although the effects of parasites, such as ticks, fleas and lice, on their hosts are often undetectable, there can be serious adverse health effects of high parasite burdens, including reduced reproductive output and increased mortality (Wall 2007). Furthermore, in some cases, parasites can be carriers of disease that can have severe adverse effects on populations (e.g. Biggins & Kosoy 2001). Treatments, developed primarily for domestic animals, may be administered to wild mammals to reduce parasite burdens. The administering of such treatments, though, can be challenging.

Biggins D.E. & Kosoy M.Y. (2001) Influences of introduced plague on North American mammals: implications from ecology of plague in Asia. *Journal of Mammalogy*, 82, 906–916, https://doi.org/10.1644/1545-1542(2001)082 <0906:ioipon>2.0.co;2

Wall R. (2007) Ectoparasites: future challenges in a changing world. *Veterinary Parasitology*, 148, 62–74, https://doi.org/10.1016/j.vetpar.2007.05.011

A replicated, paired sites, controlled study in 2009–2010 on six grasslands in Utah, USA (1) found that following treatment with a grain-bait insecticide product, there was no consistent reduction in flea burdens on Utah prairie dogs *Cynomys parvidens*. After one summer, fewer fleas were recorded on prairie dogs in treated than untreated colonies at two sites, there was no difference at one site and more fleas were recorded in treated than untreated colonies at one site. After the second summer (with treatments applied twice) there were fewer fleas on prairie dogs in treated than untreated colonies at one site and no difference at two sites. See paper for full data. At six sites with prairie dog colonies, treatment and control plots were established, covering 2–190 ha, depending on animal density. Four sites were monitored in 2009 and three in 2010. In 2009, 56 g of imidacloprid-treated oat grain bait (Kaput®) was scattered within 2.4 m of each burrow in treatment colonies, once in May–June. Imidacloprid is an insecticide that can reduce burdens of fleas and, thus, reduce the risk of transmission of plague. In 2010, the treatment was applied twice, five days apart, in April–May. Prairie dogs were trapped monthly, using 100 live traps for five days in both treatment and control areas at each site, in June–October, and combed to count fleas.

A controlled study in 2002–2010 in a forested area of Alaska, USA (2) found that treating wolves *Canis lupus* with ivermectin cleared them of infestations of biting dog lice *Trichodectes canis*. All of 12–19 wolf packs treated with ivermectin, were lice-free in the winter following treatment. In spring, 15–50% of packs were infested over the three years of treatments, 5% were infested the following spring, with 0% spring infestation in the last two years of monitoring. Three untreated packs remained infested throughout four years of monitoring. In a 13,000-km² study area, lice infestation in two packs was confirmed by inspecting animal hides harvested by trappers in 2002–2005. Moose or lynx meat, injected with ivermectin, was distributed aerially at den and rendezvous sites of 12–19 wolf packs at 10–20 day intervals in 2005–2007. Infestation status and responses to treatments were determined by live-trapping wolves, direct observations and by inspection of hides obtained from trappers during 2005–2010.

(1) Jachowski D.S., Brown N.L., Wehtje M., Tripp D.W., Millspaugh J.J. & Gompper M.E. (2012) Mitigating plague risk in Utah prairie dogs: Evaluation of a systemic flea-control product. *Wildlife Society Bulletin*, 36, 167–175, https://doi.org/10.1002/wsb.107

(2) Gardner C.L., Beckmen K.B., Pamperin N.J. & Del Vecchio P. (2013) Experimental treatment of dog lice infestation in interior Alaska wolf packs. *Wildlife Management*, 77, 626–632, https://doi.org/10.1002/jwmg.495

10. Threat: Pollution

Background

Pollution, of many diverse types, has direct and indirect impacts on mammals. Water-borne pollutants can devastate otherwise productive wetland and coastal habitats. Many pesticides linked to mammal deaths are still in widespread use and especially those targeting rodents may pass up through the food chain to predatory mammals. Oil spills remain a threat to some mammals of aquatic habitats, while solid waste is an increasing problem. Little is known of the long-term effects of many pollutants, including those that persist and accumulate in the environment. Organic farming, with reduced or zero input of pesticides, herbicides or artificial fertilizers, is included in this chapter.

10.1. Reduce pesticide or fertilizer use

https://www.conservationevidence.com/actions/2539

- **Three studies** evaluated the effects on mammals of reducing pesticide, herbicide or fertilizer use. Two studies were in the UK[1], one was in Italy[2] and one was in Argentina[3].

COMMUNITY RESPONSE (1 STUDY)

- **Richness/diversity (1 study):** A replicated, site comparison study in Argentina[3] found that farming without pesticides or fertilizers did not increase small mammal species richness in field margins.

POPULATION RESPONSE (2 STUDIES)

     https://doi.org/10.11647/OBP.0234.10

- **Abundance (2 studies):** One of two site comparison studies, in the UK[1] and Italy[2], found that reducing pesticide or fertilizer use, by farming organically, increased wood mouse abundance[1]. The other study found that it did not increase European hare abundance[2].

BEHAVIOUR (1 STUDY)

- **Use (1 study):** A replicated, site comparison study in Argentina[3] found that farming without pesticides or fertilizers did not increase small mammal use of field margins.

Background

Pesticides (including insecticides, herbicides and fungicides) and fertilizers, used especially in agriculture, but also in horticulture, amenity grassland, gardens and other situations, may have a negative effect on wildlife. Through reducing plant and insect diversity, or through direct toxicity, they may also natively impact mammals. Organic farming, an agricultural system that excludes the use of synthetic fertilizers and pesticides and relies on techniques such as crop rotation, compost and biological pest control, is included within this intervention.

A site comparison study in 1994–1996 on arable land in Gloucestershire, UK (1) found that reducing pesticide, herbicide or fertilizer use by farming organically was associated with higher numbers of wood mice *Apodemus sylvaticus*. More wood mice were caught on an organic farm (monthly averages of 19–24 individuals) than on a conventional farm (8–17 individuals). This result was not tested for statistical significance, though there were significantly more juvenile mice on the organic farm compared to the conventional farm and female mice on the organic farm were significantly heavier in two out of three years (data not presented). On one organic farm and one conventional farm, wood mice were surveyed using 56 Longworth live traps in each of two fields, at each farm, each year, in 1994–1996.

A replicated, site comparison study in 2011 on 26 mainly arable farms in Tuscany, Italy (2) found that reducing pesticide, herbicide or fertilizer use, by farming organically, did not increase abundances of European hares *Lepus europaeus*. The density of hares on organic farms (14 hares/ km²) was lower than on conventional farms (24 hares/km²). Higher hare density appeared, instead, to be more strongly positively related to increased habitat diversity, including crop diversity. Half of the 26 study farms, average size 6.1 km², were organic and half were non-organic farms. Organic farms complied with European Union organic farming requirements. Hare density was estimated using spotlight counts from a car, two or three times at each farm, in early March 2011.

A replicated, site comparison study in 2011–2013 of three arable farms in Córdoba, Argentina (3) found that farming without herbicides, fertilizers, or fungicides did not increase small mammal use of field margins or small mammal species richness in margins. Average annual small mammal capture rates on margins not treated with pesticides or fertilizers (2.5–2.9 individuals/20 traps) did not significantly differ from those on conventionally farmed margins (2.4–3.2 individuals/20 traps). Average annual small mammal species richness without pesticides and fertilizers (1.1–1.2 species/20 traps) did not differ from that with conventional farming (1.1–1.2 species/20 traps). Organic fields were managed without herbicides, fertilizers or fungicides for 10–19 years. A range of these chemicals was used on conventionally farmed fields. Small mammals were live-trapped, using lines of 20 traps in 1.5–2.5-m-wide vegetated field margin strips on three farms. Trapping was carried out over four consecutive nights, once each in spring, summer and autumn, from November 2011 to June 2013. There were 106–116 trap lines/ sampling period (proportion in each margin management type not stated).

(1) Macdonald D.W., Tattersall F.H., Service K.M., Firbank L.G. & Feber R.E. (2007) Mammals, agri-environment schemes and set-aside — what are the putative benefits? *Mammal Review*, 37, 259–277, https://doi. org/10.1111/j.1365-2907.2007.00100.x

(2) Santilli F. & Galardi L. (2016) Effect of habitat structure and type of farming on European hare (*Lepus europaeus*) abundance. *Hystrix, the Italian Journal of Mammalogy*, 27(2).

(3) Coda J., Gomez D., Steinmann A.R. & Priotto J. (2015) Small mammals in farmlands of Argentina: Responses to organic and conventional farming. *Agriculture, Ecosystems & Environment*, 211, 17–23, https://doi.org/10.1016/j. agee.2015.05.007

10.2. Leave headlands in fields unsprayed

https://www.conservationevidence.com/actions/2540

- **Two studies** evaluated the effects on mammals of leaving headlands in fields unsprayed. One study was in the UK[1] and one was in the Netherlands[2].

COMMUNITY RESPONSE (0 STUDIES)

POPULATION RESPONSE (0 STUDIES)

BEHAVIOUR (2 STUDIES)

- **Use (2 studies):** Two replicated studies (one also controlled) in the UK[1] and the Netherlands[2], found that crop edge headlands that were not sprayed with pesticides were used more by mice than were sprayed crop edges[1,2].

Background

Conservation headland management may involve restricting fertiliser, herbicide and insecticide spraying along a strip through a sown arable crop. Typically, as under agri-environment schemes practiced in Europe, this may be a 6-m-wide strip with selected herbicide applications permitted to control certain weeds or invasive species.

A replicated study in 1986–1987 in an arable field, in Oxfordshire, UK (1) found that not spraying herbicide on headlands of crop at the field edge was associated with higher use of those areas by wood mice *Apodemus sylvaticus*. The proportion of location fixes obtained for mice in unsprayed or sprayed plots indicated greater selection of unsprayed plots relative to their availability within home ranges (data presented as preference indices). Plots extended 10 m into a winter wheat field and were 20 m long. Plots were either sprayed or not sprayed with a range of agricultural herbicides. Application of other chemicals (insecticides, fungicides, growth regulators and fertilizers) were the same across all plots. Wood mouse movements were monitored by radio-tracking 15 mice, between June and August in each of 1986 and 1987.

A replicated, controlled study in 1990–1993 of six arable farms in the Netherlands (2) found that unsprayed crop edge headlands were used more by field mice *Apodemus* spp. than were crop edges sprayed with herbicides and insecticides. Results were not tested for statistical significance. More field mice were caught in unsprayed crop edges (38 mice caught) than in sprayed edges (27 mice caught). Strips 3–6 m wide, 100–450 m long, along the edges of crops, were left unsprayed by herbicides and insecticides and were compared to sprayed crop edges in the same field. Small mammals were surveyed using pitfall traps during 13 weeks in 1990 and 12 weeks in 1991 (all in May–July). The number of strips on which small mammals were surveyed is unclear.

(1) Tew T.E., Macdonald D.W. & Rands M.R.W. (1992) Herbicide application affects microhabitat use by arable wood mice (*Apodemus sylvaticus*). *Journal of Applied Ecology*, 29, 532–539.

(2) de Snoo G.R. (1999) Unsprayed field margins: effects on environment, biodiversity and agricultural practice. *Landscape and Urban Planning*, 46, 151–160.

10.3. Establish riparian buffers

https://www.conservationevidence.com/actions/2541

- We found no studies that evaluated the effects on mammals of establishing riparian buffers.

'We found no studies' means that we have not yet found any studies that have directly evaluated this intervention during our systematic journal and report searches. Therefore, we have no evidence to indicate whether or not the intervention has any desirable or harmful effects.

Background

Uncultivated strips of vegetation at the edge of waterways are often used to help reduce pollution entering the water within agricultural and forestry systems. These buffer strips may, therefore, help to enhance environmental quality for aquatic and semi-aquatic mammal species.

See also: *Biological resource use — Retain riparian buffer strips during timber harvest.*

10.4. Translocate mammals away from site contaminated by oil spill

https://www.conservationevidence.com/actions/2542

- **One study** evaluated the effects of translocating mammals away from a site contaminated by oil spill. This study was in the USA[1].

COMMUNITY RESPONSE (0 STUDIES)

POPULATION RESPONSE (1 STUDY)

- **Survival (1 study):** A study in the USA[1] found that after being translocated in a trial of responses to a hypothetical pollution incident, most sea-otters survived for the duration of monitoring.

BEHAVIOUR (1 STUDY)

- **Behaviour change (1 study):** A study in the USA[1] found that after being translocated in a trial of responses to a hypothetical pollution incident, most sea-otters did not return to their capture location.

Background

Where there is a large pollution event that has potential to affect wild mammals, one intervention option may be to translocate these mammals to another site. In such event, the translocation would be an emergency action, carried out with minimal planning. It would only be likely to be considered where the survival chances of mammals would be very low otherwise.

A study in 1988–1989 in coastal waters of California, USA (1) found that after being translocated in a trial of responses to a hypothetical pollution incident, most sea-otters *Enhydra lutris* survived for the duration of monitoring and did not return to their capture location. Seventeen of 19 translocated sea otters survived for at least 16–87 days after release. Two died at the release site, after 21 and 28 days after release. Five of

19 translocated sea otters were recorded back at their capture location during the monitoring period. Twelve were last recorded at a site 27 km from the release site. Nineteen sea otters were caught between May 1988 and May 1989 and were released 291 km further north. Nine were released immediately on arrival and 10 were held for 48 hours in floating pens before release. Sea otters were radio-tracked from the ground or air for 16–87 days after release.

(1) Ralls K., Doroff A. & Mercure A. (1992) Movements of sea otters relocated along the California coast. *Marine Mammal Science*, 8, 178–184.

11. Threat: Climate change and severe weather

Background

Climate change, extreme weather and geological events can be very large-scale threats. Most interventions used in response to them, therefore, are general conservation interventions, such as providing artificial den sites, discussed in *Habitat restoration and creation*, and translocations and captive breeding, discussed in *Species Management*.

11.1. Retain/provide migration corridors

https://www.conservationevidence.com/actions/2551

- We found no studies that evaluated the effects on mammals of retaining or providing migration corridors.

'We found no studies' means that we have not yet found any studies that have directly evaluated this intervention during our systematic journal and report searches. Therefore, we have no evidence to indicate whether or not the intervention has any desirable or harmful effects.

https://doi.org/10.11647/OBP.0234.11

11.2. Protect habitat along elevational gradients

https://www.conservationevidence.com/actions/2552

- We found no studies that evaluated the effects on mammals of protecting habitat along elevational gradients.

'We found no studies' means that we have not yet found any studies that have directly evaluated this intervention during our systematic journal and report searches. Therefore, we have no evidence to indicate whether or not the intervention has any desirable or harmful effects.

Chen I.C., Hill J.K., Ohlemüller R., Roy D.B & Thomas C.D. (2011) Rapid range shifts of species associated with high levels of climate warming. *Science*, 333, 1024–1026, https://doi.org/10.1126/science.1206432

Myers, P., Lundrigan, B. L., Hoffman, S. M., Haraminac, A. P., & Seto, S. H. (2009). Climate-induced changes in the small mammal communities of the Northern Great Lakes Region. *Global Change Biology*, 15, 1434-1454, https://doi.org/10.1111/j.1365-2486.2009.01846.x

Rowe, R. J., Finarelli, J. A., & Rickart, E. A. (2010). Range dynamics of small mammals along an elevational gradient over an 80-year interval. *Global Change Biology*, 16, 2930-2943, https://doi.org/10.1111/j.1365-2486.2009.02150.x

11.3. Translocate animals from source populations subject to similar climatic conditions

https://www.conservationevidence.com/actions/2553

- **One study** evaluated the effects of translocating mammals from source populations subject to similar climatic conditions. This study was in the USA[1].

COMMUNITY RESPONSE (0 STUDIES)

POPULATION RESPONSE (1 STUDY)

- **Reproductive success (1 study):** A study in the USA[1] found that bighorn sheep translocated from populations subject to a similar climate to the recipient site reared more offspring than did those translocated from milder climatic areas.

BEHAVIOUR (0 STUDIES)

Background

As human-induced climate change leads to increasing temperatures, species shift their distributions to higher latitudes and elevations (Hickling *et al.* 2006). However, some species cannot disperse quickly enough, or may not be able to cross human or man-made barriers (Thomas 2011). This results in some animals being present in areas that represent poor quality habitat, resulting in increased mortality rates that may risk local or even global extinction. One solution that has been suggested for this problem is the translocation of animals to areas where climatic conditions are similar to those formerly found in their natural ranges (Thomas 2011).

Hickling R., Roy D.B., Hill J.K., Fox R. & Thomas C.D. (2006) The distributions of a wide range of taxonomic groups are expanding polewards. *Global Change Biology*, 12, 450–455, https://doi.org/10.1111/j.1365-2486.2006.01116.x

Thomas C.D. (2011) Translocation of species, climate change, and the end of trying to recreate past ecological communities. *Trends in Ecology & Evolution*, 26, 216–221, https://doi.org/10.1016/j.tree.2011.02.006

A study in 2006–2011 of scrubland across a large area in North Dakota, USA (1) found that bighorn sheep *Ovis canadensis* translocated from populations subject to a similar climate to the recipient site reared more offspring, compared to those translocated from areas with a milder climate. Sheep from an area with a climate similar to the recipient site had a higher average annual recruitment (0.6 juveniles/adult female) than did sheep originating from a milder climate area (0.2 juveniles/ adult female). Thirty-nine bighorn sheep originating from Montana, where climate was similar to the recipient site, were release in North Dakota in 2006–2007. Their annual recruitment was compared with that of sheep released between 1956 and 2004, which originated from stock from British Columbia, Canada. Recruitment was assessed by direct observations of radio-tracked sheep, annually, in late summer and the following March of 2006–2011.

(1) Wiedmann B.P. & Sargeant G.A. (2014) Ecotypic variation in recruitment of reintroduced bighorn sheep: implications for translocation. *The Journal of Wildlife Management*, 78, 394–401, https://doi.org/10.1002/jwmg.669

11.4. Provide dams/water holes during drought

https://www.conservationevidence.com/actions/2554

- We found no studies that evaluated the effects on mammals of providing dams or water holes during drought.

'We found no studies' means that we have not yet found any studies that have directly evaluated this intervention during our systematic journal and report searches. Therefore, we have no evidence to indicate whether or not the intervention has any desirable or harmful effects.

Background

Climate change may increase the frequency of droughts. Populations of some mammal species that are reliant on availability of water may be buffered against effects of drought by artificial provision of water. This could be through digging holes down to the water table or building dams, to store water for use in times of drought.

For cases where provision of water as an intervention is a response to water shortage caused by other human-induced activities, rather than directly via climate change, see *Natural system modifications — Provide artificial waterholes in dry season*.

11.5. Apply water to vegetation to increase food availability during drought

https://www.conservationevidence.com/actions/2555

- **One study** evaluated the effects on mammals of applying water to vegetation to increase food availability during drought. This study was in the USA[1].

COMMUNITY RESPONSE (0 STUDIES)

POPULATION RESPONSE (0 STUDIES)

BEHAVIOUR (1 STUDY)

- **Use (1 study)**: A controlled, before-and-after study in the USA[1] found that watering scrub during drought increased its use by adult Sonoran pronghorns for feeding.

Background

Drought can cause plants to die as a result of a lack of water. Dieback of vegetation may in turn negatively affect mammal populations by reducing the availability of food. Applying water during a drought may help to reduce some of these negative consequences.

A controlled, before-and-after study in 2005 in a desert enclosure in Arizona, USA (1) found that watering scrub during drought increased its use for feeding by adult Sonoran pronghorns *Antilocapra american sonoriensis*. In winter (January–March), before plots were watered, pronghorns selected plots to be watered and unwatered in proportion to their availability. After watering commenced, pronghorns fed more in watered plots than their availability in spring (April–June), summer (July–September) and autumn (October–December). Use of watered plots was highest in autumn, when 48% of observations were in these plots, which covered 5% cover of the enclosure. Seven adult pronghorns were held in a 130-ha enclosure. Eight desert scrub plots, c.8,000 m² each, were watered at least once every two weeks from April–December 2005, by applying c.13 cm of water. Autumn rainfall during the study period was low (4 mm, compared to average of 16 mm). Pronghorn feeding area selection was determined by watching from a partially concealed viewpoint, from 23 January to 2 December 2005. Observations were recorded at 2-minute intervals, four to five days/week during either first light to noon or noon to last light, giving 38,900 individual observations.

(1) Wilson R.R., Krausman P.R. & Morgart J.R. (2010) Forage enhancement plots as a management tool for Sonoran pronghorn recovery. *The Journal of Wildlife Management*, 74, 236–239, https://doi.org/10.2193/2009-191

11.6. Remove flood water

https://www.conservationevidence.com/actions/2557

- We found no studies that evaluated the effects on mammals of removing flood water.

'We found no studies' means that we have not yet found any studies that have directly evaluated this intervention during our systematic journal and report searches. Therefore, we have no evidence to indicate whether or not the intervention has any desirable or harmful effects.

Background

Climate change increases the risk of extreme weather events, including flooding. Flood waters may cover habitat normally used by mammal species. For example, more than half of China's mammal species were found to be exposed to risks from flooding (Ameca y Juárez & Jiang 2016). In addition to direct casualties from effects of water (such as drowning) flood water may alter the habitat, for example through changes to vegetation. Furthermore, mammal mortality may be higher when flood water persists for longer (Wuczyński & Jakubiec 2013). Enabling rapid removal of flood water, such as through creating drainage routes, may lessen such impacts.

Ameca y Juáreza E.I. & Jianga Z. (2016) Flood exposure for vertebrates in China's terrestrial priority areas for biodiversity conservation: Identifying internal refugia. *Biological Conservation*, 199, 137–145, https://doi.org/10.1016/j.biocon.2016.04.021

Wuczyński A. & Jakubiec Z. (2013) Mortality of game mammals caused by an extreme flooding event in south-western Poland. *Natural Hazards*, 69, 85–97, https://doi.org/10.1007/s11069-013-0687-x

12. Habitat protection

Background

Habitat destruction is the largest single threat to biodiversity and habitat fragmentation and degradation often reduces the quality of remaining habitat. Habitat protection is therefore one of the most frequently used conservation interventions, particularly in the tropics and in other areas with large patches of surviving natural vegetation.

Habitat protection can be through the designation of legally protected areas, using national or local legislation. It can also be through the designation of community conservation areas or similar schemes, which do not provide formal protection but may increase the profile of a site and make its destruction less likely. Alternatively, protection can be of entire habitat types, for example through the European Union's Habitats Directive. On a smaller scale, habitat protection may involve ensuring areas of important habitat are retained during detrimental activities.

12.1. Legally protect habitat for mammals

https://www.conservationevidence.com/actions/2559

- **Seven studies** evaluated the effects of legally protecting habitat for mammals. One study each was in Zambia[1], the USA[2], Tanzania[3], Brazil[4], Nepal[6] and India[7] and one was a systematic review of sites with a wide geographic spread[5].

COMMUNITY RESPONSE (0 STUDIES)

 https://doi.org/10.11647/OBP.0234.12

POPULATION RESPONSE (7 STUDIES)

- **Abundance (7 studies):** A systematic review of protected areas across the globe[5] found that 24 of 31 studies reported an increase in mammal populations in protected areas relative to unprotected areas. Three studies (including two site comparison studies), in Zambia[1], the USA[2] and Nepal[6], found that populations of red lechwe[1], black bears[2] and one-horned rhinoceros[6] grew following site protection or were higher than in adjacent non-protected sites. One of three site comparison studies, in Tanzania[3], Brazil[4] and India[7], found that populations of more mammal species increased inside protected areas than in adjacent unprotected areas[3]. One study found that populations of only three of 11 species were higher on protected than on unprotected land[7] whilst the third study found that 13 of 16 species were less abundant in a protected area than in a nearby unprotected area[4].

BEHAVIOUR (0 STUDIES)

Background

Legally protecting habitat may reduce its conversion and degradation by humans. This may in turn increase the abundance and diversity of mammals that make use of that habitat.

Assessing the effectiveness of protected areas is particularly difficult. For example, protected and unprotected areas might start off with different quality habitats (protection being granted to the best quality habitat). Protected areas are also more likely to be in remote areas, so less accessible to threats such as harvesting (Joppa & Pfaff 2009). Finally, effectiveness is best monitored over long timescales, but this increases the chance that other factors influence the ecosystem. The most reliable studies would compare protected and unprotected areas over time, and possibly correct for some of the biases.

See also: *Biological resource use — Use wildlife refuges to reduce hunting impacts.*

Joppa L.N. & Pfaff A. (2009) High and far: biases in the location of protected areas. *PLoS ONE*, 4, e8273, https://doi.org/10.1371/journal.pone.0008273

A review of the Kafue National Park in Zambia (1) found that following establishment of a national park, the population of red lechwe *Kobus leche leche* increased. In 1950, when the national park was established, there were approximately 100 red lechwe. By 1985, the population was estimated at 3,400 animals. Methods used by studies to estimate the population in 1950 were not given but, in 1985, a study used aerial surveys to determine abundance.

A site comparison study in 1981–1990 in a mixed forest area in North Carolina, USA (2) found that there were more black bears *Ursus americanus* in a bear sanctuary than on adjacent non-sanctuary land. Bears were detected at a higher rate in the bear sanctuary (0.01–0.04 bear visits/station/day) than outside the sanctuary (0–0.01 bear visits/station/day). In 1981, a total of 136 bait stations (68 in the sanctuary and 68 on adjacent non-sanctuary land) were established. The two parts of the study area were approximately equal in size and, combined, covered >400 km². In 1981–1990, at each station, two open cans of sardines were nailed to a tree. After five days, bait stations were revisited and any signs of bear visits noted. It was unclear how often the bait stations were baited each year.

A replicated, paired, site comparison study in 1990–2001 in seven savanna areas in Tanzania (3) found that populations of more mammal species increased inside protected national parks than in adjacent unprotected areas, but that population declines were also more frequent in protected than unprotected areas. In all seven comparisons, populations of more mammal species increased in national parks (0–20%) than in unprotected areas (0–5%). However, in six of seven comparisons, populations of more mammal species also declined in national parks (5–62%) than in unprotected areas (0–21%). In one of seven comparisons, the opposite was found (national parks: 0%, unprotected areas: 22%). Between May 1990 and May 2001, large mammals in seven zones, each spanning a national park and surrounding area, were surveyed from aeroplanes. Planes followed transects and two observers recorded numbers of animals seen between parallel rods attached to the aircraft. Population densities were calculated and assigned to cells covering the area surveyed. Population estimates over 10 years in each cell were used to determine changes in both protected and unprotected areas.

A site comparison study in 2005–2007 in two sites mostly composed of secondary forest in Pará, Brazil (4) found that 13 of 16 species were less abundant in a protected area than in a nearby unprotected area. Results were not tested for statistical significance. Populations of 13 of 16 species were lower in the protected area (0–4.5 photos/100 camera-trap nights) than in a nearby unprotected area (0.1–5.0 photos/100 camera-trap nights). Three of the 16 species were more abundant in the protected area (0.2–4.5 photos/100 camera-trap nights) than in the unprotected area (0.2–4.1 photos/100 camera-trap nights). Vegetation in the protected area was largely secondary rainforest and, in the unprotected area, 65% was secondary forest and 35% was pasture. Five camera-trap surveys were carried out between July 2005 and November 2007 at 10–22 locations in a protected area and 10–22 locations in a nearby unprotected area. Cameras were placed 50–70 cm above ground level at each location. Each camera took one photograph every 5 minutes. Relative abundance of species was estimated by dividing the number of photos of a species by the number of trap-nights.

A systematic review in 2013 of the effectiveness of protected areas across the globe, but especially in Latin America (5) found that 24 of 31 studies reported an increase in mammal populations in protected areas relative to unprotected areas. Seven of 31 studies reported a decline or no change in mammal populations in protected areas relative to unprotected areas. Twelve studies used a before-and-after methodology and 19 studies were site comparisons.

A before-and-after study in 1950–2011 in an area dominated by forest and grassland in western Nepal (6) found that greater one-horned rhinoceros *Rhinoceros unicornis* numbers more than tripled over 38 years after the establishment of a national park. Rhinoceros numbers declined >80% (from 800 in 1950 to 147 in 1972) during the 23 years before the establishment of the national park. However, during the 38 after the establishment of the national park, numbers increase by >70% (from 147 in 1972 to 534 in 2011). The study area became the Chitwan National Park in 1973. Since 1975, rhinoceroses were protected by the Nepal Army and, in 2007, a nationwide anti-poaching programme was launched. In

1986–2003, eighty-three rhinoceroses were translocated from Chitwan National Park to other reserves. Monitoring details are not provided.

A site comparison study in 2011–2013 in two agricultural and forest areas in north-eastern India (7) found that the number of species and abundance of seven of 11 large mammal species did not differ between a protected wildlife sanctuary area and community managed land. The number of species was similar in the protected (17 species) and the community managed areas (16 species). Seven of 11 large mammal species had similar abundances in the protected area and on community managed land (data reported as model results). Three species were more abundant in the protected area and one was more abundant on the community managed land. In October–November 2011 and August– September 2012, eleven sites were established in the wildlife sanctuary and 14 sites in the community managed land. At each site, a 500 × 5-m U-shaped transect, divided into 20-m segments, was surveyed by two observers for signs of mammal presence. In April–June 2013, twenty-two infrared cameras were deployed in the wildlife sanctuary and 18 were deployed in the community managed areas. Cameras were attached to trees, 25 cm above ground. They operated 24 hours/day and were baited with rotting bananas and smoked dried fish.

(1) Howard G.W. & Chabwela H.N. (1987) The red lechwe of Busanga Plain, Zambia—a conservation success. *Oryx*, 21, 233–235.

(2) Powell R.A., Zimmerman J.W., Seaman D.E. & Gilliam J.F. (1996) Demographic analyses of a hunted black bear population with access to a refuge. *Conservation Biology*, 10, 224–234.

(3) Stoner C., Caro T.I.M., Mduma S., Mlingwa C., Sabuni G. & Borner M. (2007) Assessment of effectiveness of protection strategies in Tanzania based on a decade of survey data for large herbivores. *Conservation Biology*, 21, 635–646, https://doi.org/10.1111/j.1523-1739.2007.00705.x

(4) Negroes N., Revilla E., Fonseca C., Soares A.M., Jácomo A.T. & Silveira L. (2011) Private forest reserves can aid in preserving the community of medium and large-sized vertebrates in the Amazon arc of deforestation. *Biodiversity and Conservation*, 20, 505–518, https://doi.org/10.1007/s10531-010-9961-3

(5) Geldmann J., Barnes M., Coad L., Craigie I.D., Hockings M. & Burgess N.D. (2013) Effectiveness of terrestrial protected areas in reducing habitat loss and population declines. *Biological Conservation*, 161, 230–238, https://doi.org/10.1016/j.biocon.2013.02.018

(6) Thapa K., Nepal S., Thapa G., Bhatta S.R. & Wikramanayake E. (2013) Past, present and future conservation of the greater one-horned rhinoceros *Rhinoceros unicornis* in Nepal. *Oryx*, 47, 345–351, https://doi.org/10.1017/s0030605311001670

(7) Velho N., Srinivasan U., Singh P. & Laurance W.F. (2016) Large mammal use of protected and community-managed lands in a biodiversity hotspot. *Animal Conservation* 19, 199–208, https://doi.org/10.1111/acv.12234

12.2. Encourage habitat protection of privately-owned land

https://www.conservationevidence.com/actions/2560

- We found no studies that evaluated the effects on mammals of encouraging habitat protection of privately-owned land.

'We found no studies' means that we have not yet found any studies that have directly evaluated this intervention during our systematic journal and report searches. Therefore, we have no evidence to indicate whether or not the intervention has any desirable or harmful effects.

Background

Most land is privately-owned by individuals or businesses. Whilst most of this land is not managed for wildlife conservation, some areas are operated as private nature reserves (e.g. Lanhholz 1996), or as part of larger protected areas, including corridors and buffer zones (e.g. Environmental Law Institute 2003, Figgis 2004). On other land, a wide range of individual actions may be taken to promote or conserve wildlife. The effectiveness of these individual actions is covered under those specific interventions. This intervention more generally considers the effectiveness of promoting habitat conservation among private landowners.

Environmental Law Institute (2003) *Legal tools and incentives for private lands conservation in Latin America: building models for success.* Environmental Law Institute, Washington, USA Figgis, P. (2004) *Conservation on private lands: the Australian Experience.* International Union for Conservation of Nature, Gland, Switzerland.

Langholz J. (1996) Economics, objectives, and success of private nature reserves in Sub-Saharan Africa and Latin America. *Conservation Biology*, 10, 271–280.

12.3. Build fences around protected areas

https://www.conservationevidence.com/actions/2561

- **Two studies** evaluated the effects on mammals of building fences around protected areas. One study was in Kenya[1] and one was in Mozambique[2].

COMMUNITY RESPONSE (1 STUDY)

- **Richness/diversity (1 study):** A before-and-after study in Kenya[1] found that after a fence was built around a protected area, mammal species richness initially increased in both study sites, but subsequently declined at one of the sites.

POPULATION RESPONSE (2 STUDIES)

- **Abundance (2 studies):** A paired sites study in Mozambique[2] found that inside a fenced sanctuary there were more mammal scats than outside the sanctuary. A before-and-after study in Kenya[1] found that after a fence was built around a protected area, mammal abundance initially increased in both study sites, but it subsequently declined at one of the sites.

BEHAVIOUR (0 STUDIES)

Background

Fences may be constructed around protected areas to keep out poachers or predators, including invasive species (e.g. Hayward & Kerley 2009). They may also prevent other potentially damaging incursions, such as by off-road vehicles that may damage habitat, or casual entry by people on foot who may disturb mammals. Where protected areas are surrounded by land in which there are greater threats to wild mammals, such as persecution of carnivores, fences may reduce losses of such species by preventing them encountering these threats. Possible disadvantages of fences include inhibiting species' dispersal, potentially leading to reductions in genetic diversity.

Hayward M.W. & Kerley G.I.H. (2009) Fencing for conservation: Restriction of evolutionary potential or a riposte to threatening processes? *Biological Conservation*, 142, 1–13, https://doi.org/10.1016/j.biocon.2008.09.022

A before-and-after study in 1963–2011 at two montane forest and alpine grassland sites within a conservation area in central Kenya (1) found that after installing fencing around the protected area, mammal abundance and species richness increased initially but, at one site, abundance and richness subsequently declined. At both sites, following fence installation around the protected area, a declining trend in mammal abundance and species richness changed to an increasing trend (data reported as model results). However, at one of these sites, eight years after the fence was installed, abundance and species richness had again declined significantly, though there was no significant decline at the other site (data reported as model results). Nightly censuses of wildlife at watering holes and salt licks were carried out between approximately 15:00 h 08:00 h, at two lodges in Aberdare Conservation Area, in 1963–2011. In 1991, fencing was built around the 38 km perimeter of the park closest to the study sites and, by 2009, the entire conservation area was fenced.

A paired sites study in 2014 in a savanna reserve in Sofala, Mozambique (2) found that inside a fenced sanctuary there were more mammal scats than outside the sanctuary. More mammal scats were collected inside the fenced sanctuary (268 scats) than outside of it (207 scats). Scats were produced by 24 species, including nine antelope species, at least three carnivores, two primates, blue wildebeest *Connochaetes taurinus*, zebra *Equus quagga*, porcupine *Hystrix africaeaustralis*, scrub hare *Lepus saxatilis*, warthog *Phacochoerus africanus*, bushpig *Potamochoerus larvatus* and African buffalo *Syncerus caffer*. In June–August 2014, mammal scats were collected along ten 5 km × 5-m transects in Gorongosa National Park. Five transects, >1 km apart, were located inside a 62-km² fenced wildlife sanctuary and five were located outside of it. The fence was constructed between August 2006 and September 2014. Scats were detected by two observers and the identity of species that produced the scat was determined by direct observation or based on the experience of the local rangers or field guides.

(1) Massey A.L., King A.A. & Foufopoulos J. (2014) Fencing protected areas: A long-term assessment of the effects of reserve establishment and fencing on African mammalian diversity. *Biological Conservation*, 176, 162–171, https://doi.org/10.1016/j.biocon.2014.05.023

(2) Correia M., Timóteo S., Rodríguez-Echeverría S., Mazars-Simon A. & Heleno R. (2017) Refaunation and the reinstatement of the seed-dispersal function in Gorongosa National Park. *Conservation Biology*, 31, 76–85, https://doi.org/10.1111/cobi.12782

12.4. Retain buffer zones around core habitat

https://www.conservationevidence.com/actions/2562

- We found no studies that evaluated the effects on mammals of retaining buffer zones around core habitat.

'We found no studies' means that we have not yet found any studies that have directly evaluated this intervention during our systematic journal and report searches. Therefore, we have no evidence to indicate whether or not the intervention has any desirable or harmful effects.

Background

Protected areas are usually subject to the influence of activities in surrounding areas. Buffer zones around core habitat in protected areas are usually areas of land which do not receive full protection and are not subject to the same management intensity of core areas, but on which there may be some degree of limit to activities such as hunting, agriculture and development. In some cases, buffer zones themselves can provide additional habitat for mammals (Paolino *et al.* 2016) though this can also expose them to a higher level of human-related threats (van der Meer *et al.* 2013).

van der Meer E., Fritz H., Blinston P. & Rasmussen G.S.A. (2013) Ecological trap in the buffer zone of a protected area: effects of indirect anthropogenic mortality on the African wild dog *Lycaon pictus*. *Oryx*, 48, 285–293, https://doi.org/10.1017/s0030605312001366

Paolino R.M., Versiani N.F., Pasqualotto N., Rodrigues T.F., Krepschi V.G. & Chiarello A.G. (2016) Buffer zone use by mammals in a Cerrado protected area. *Biota Neotropica*, 16, e20140117, https://doi.org/10.1590/1676-0611-bn-2014-0117

12.5. Increase size of protected area

https://www.conservationevidence.com/actions/2563

- **One study** evaluated the effects on mammals of increasing the size of a protected area. This study was in South Africa[1].

COMMUNITY RESPONSE (0 STUDIES)

POPULATION RESPONSE (0 STUDIES)

BEHAVIOUR (1 STUDY)

- **Behaviour change (1 study):** A before-and-after study in South Africa[1] found that expanding a fenced reserve resulted in the home range of a reintroduced group of lions becoming larger but the core range becoming smaller.

Background

Large protected areas may be better able to support viable populations of mammals than are smaller areas. However, protected area effectiveness may also be linked to sites being surrounded by similar habitat, having strong public support, effective law enforcement, low human population densities and sufficient financial resources (Struhsaker *et al.* 2005). Where these are not in place, factors such as activities of surrounding human populations may have a greater impact on species survival (Parks & Harcourt 2002).

Parks S.A, & Harcourt A.H. (2002) Reserve size, local human density, and mammalian extinctions in U.S. protected areas. *Conservation Biology*, 16, 800–808, https://doi.org/10.1046/j.1523-1739.2002.00288.x

Struhsaker T.T., Struhsaker P.J. & Siex K.S. (2005) Conserving Africa's rain forests: problems in protected areas and possible solutions. *Biological Conservation*, 123, 45–54, https://doi.org/10.1016/j.biocon.2004.10.007

A before-and-after study in 2000–2001 at a primarily savanna site in South Africa (1) found that expanding a fenced reserve resulted in the home range of a reintroduced group of lions *Panthera leo* becoming larger but the core range becoming smaller. Following fence removal, the home range was larger (74 km²) than prior to fence removal (38 km²). The opposite was true for the core range (after fence removal: 2 km²; before fence removal: 11 km²). In December 1994, a pride of five lions was reintroduced to the fenced Greater Makalali Conservancy, where lions had previously become extinct. Two male lions were subsequently removed and replaced by two new males in 1999. In October 2000, the fenced area was enlarged from 11,089 ha to 13,600 ha, by removing a fence between the conservancy and a neighbouring game reserve. Lions were monitored through visual observations for six months before and six months after fence removal. The home range was defined as the smallest area containing 95% of the distribution used and the core range was the smallest area containing 50% of distribution used.

(1) Druce D., Genis H., Braak J., Greatwood S., Delsink A., Kettles R., Hunter L. & Slotow R. (2004) Population demography and spatial ecology of a reintroduced lion population in the Greater Makalali Conservancy, South Africa. *Koedoe*, 47 103–118, https://doi.org/10.4102/koedoe.v47i1.64

12.6. Increase resources for managing protected areas

https://www.conservationevidence.com/actions/2564

- **One study** evaluated the effects on mammals of increasing resources for managing protected areas. This study was in Tanzania[1].

COMMUNITY RESPONSE (1 STUDY)

- **Species richness (1 study):** A site comparison study in Tanzania[1] found that mammal species richness was higher in a well-resourced national park, than in a less well-resourced forest reserve.

POPULATION RESPONSE (1 STUDY)

- **Abundance (1 study):** A site comparison study Tanzania[1] found that there were greater occupancy rates or relative abundances of most mammal species in a well-resourced national park than in a less well-resourced forest reserve.

BEHAVIOUR (0 STUDIES)

Background

Enforcement of regulations, such as those regarding hunting, can be a challenge for protected areas. This intervention covers increases in those resources, such as funding sufficient staff.

A site comparison study in 2013–2014 in two forested protected areas in the Udzungwa Mountains, Tanzania (1) found that in a well-resourced protected national park, there was greater mammal species richness and occupancy rates or relative abundances for most mammal species compared to those in a forest reserve managed with fewer resources. Estimated mammal species richness was higher in the national park (29 species) than in the forest reserve (18 species). Modelled occupancy rates (a measure of the proportion of sites used by species) were higher in the national park compared to the forest reserve for three species and were lower for one species. For species occurring at both sites, but in insufficient numbers to perform occupancy modelling, relative abundances were higher in the national park compared to the forest reserve for five species and were lower for one species. One site was a 177-km² forest within a well-resourced national park where poaching was considered to be rare. The other was a 200-km² forest reserve, managed with fewer resources and where poaching for bushmeat occurred. Each area was surveyed using camera traps, over 917 camera-trap days in the national park and 850 camera-trap days in the forest reserve, between July 2013 and February 2014.

(1) Hegerl C., Burgess N.D., Nielsen M.R., Martin E., Ciolli M. & Rovero F. (2017) Using camera trap data to assess the impact of bushmeat hunting on forest mammals in Tanzania. *Oryx*, 51, 87–97, https://doi.org/10.1017/s0030605315000836

13. Habitat restoration and creation

Background

Habitat destruction is one of the largest threats to mammal species and populations and habitat protection remains one of the most important and frequently used conservation interventions. However, in many parts of the world, restoring damaged habitats, improving habitats through altering management regimes or creating areas of new habitat may also be possible.

Habitat restoration or creation is often required by law as a response to activities that destroy large areas of natural habitats. Restoration activities may include planting vegetation, removing invasive species or creating breeding or shelter habitats, for example.

Studies describing the effects of interventions that involve restoration through processes such as fire and water management are discussed in the chapter *Threat: Natural system modifications*, and those that involve the control of invasive species in the chapter *Threat: Invasive and other problematic species and diseases*.

13.1. Remove topsoil that has had fertilizer added to mimic low nutrient soil

https://www.conservationevidence.com/actions/2544

- We found no studies that evaluated the effects on mammals of removing topsoil that has had fertilizer added to mimic low nutrient soil.

 https://doi.org/10.11647/OBP.0234.13

'We found no studies' means that we have not yet found any studies that have directly evaluated this intervention during our systematic journal and report searches. Therefore, we have no evidence to indicate whether or not the intervention has any desirable or harmful effects.

Background

Removing topsoil may help to reduce fertility of soils as well as removing seeds that are found in topsoil. Both of these outcomes may help the establishment of native plant species, which may in turn influence the abundance of mammal species.

13.2. Manage vegetation using livestock grazing

https://www.conservationevidence.com/actions/2545

- **Six studies** evaluated the effects on mammals of managing vegetation using livestock grazing. Four studies were in the USA[1-4], one was in Norway[5] and one was in Mexico[6].

COMMUNITY RESPONSE (0 STUDIES)

POPULATION RESPONSE (1 STUDY)

- **Abundance (1 study):** A replicated, controlled, before-and-after study in the USA[4] found that introduction of livestock grazing increased the abundance of Stephens' kangaroo rat after two years.

BEHAVIOUR (5 STUDIES)

- **Use (4 studies):** One of four studies (three replicated controlled studies and a before-and-after study), in the USA[1,2,3] and Norway[5], found that sheep-grazed pasture was used by feeding reindeer more than was ungrazed pasture[5]. One found mixed effects on Rocky Mountain elk use of grazed plots[1] and another found no response of Rocky Mountain elk to spring cattle grazing[2]. The forth study found cattle grazing to increase the proportion of rough fescue biomass utilized by elk in the first, but not second winter after grazing[3].

- **Behaviour change (1 study):** A replicated, paired sites study in Mexico[6] found that in pastures grazed by cattle, Tehuantepec jackrabbits spent more time feeding than they did in pastures not grazed by cattle.

Background

Using grazing to manage vegetation can limit succession that would otherwise lead to an increase in woody plant species. This may help to increase the abundance of mammal species that depend on early-succession habitats.

A before-and-after study in 1948–1974 in a predominantly grassland wildlife management area in Oregon, USA (1) found that when cattle grazing was reintroduced, there was a mixed effect on Rocky Mountain elk *Cervus canadensis* abundance. Four years after cattle were first reintroduced, elk numbers (325) were similar to those before cattle reintroduction (120–500), although disturbance by snowmobiles during this period may have reduced abundance. After nine years, elk numbers (1,191) were higher than before reintroduction (120–500). In 1960 the site was designated as a wildlife management area. Cattle grazed ceased in 1960 but was reintroduced in 1965 at a rate of 340 animal unit months (AUMs — a grazing measure based on forage requirement). Cattle grazing was increased to 700 AUMs in 1967 and 900 AUMS in 1969–1974. Cattle grazing was managed to optimise forage conditions and prevent accumulation of residual unpalatable vegetation. Elk were counted from horseback, along fixed routes, five times each winter, in 1948–1974.

A randomized, replicated, controlled study in 1971–1974 of a grassland in Washington, USA (2) found that spring grazing by cattle did not increase pasture use by Rocky Mountain elk *Cervus canadensis nelsoni* the following winter. There were no significant differences in the numbers of elk using cattle-grazed and ungrazed plots in the first winter (grazed: 60; ungrazed: 68 elk days/ha) or third winter (grazed: 38; ungrazed: 51 elk days/ha) after cattle grazing commenced. In the second winter, fewer elk used grazed plots (71 elk days/ha) than used ungrazed plots (98 elk days/ha). Three plots (9.3 ha each) were randomly assigned to be grazed by cattle and three were ungrazed. Grazing was at a rate of one mature cow or equivalent/2.4 ha, from mid-April to early-June in 1971–1973. Elk pellets were counted each spring to assess elk use of plots in winters of 1971–1972, 1972–1973, and 1973–1974.

A replicated, controlled study in 1983–1987 of a rough fescue *Festuca scabrella*-dominated grassland in Montana, USA (3) found that cattle grazing increased the proportion of rough fescue biomass utilized by elk *Cervus canadensis nelsoni* in the first, but not second winter after grazing. Over the first winter, a higher proportion of rough fescue was utilized by elk in cattle-grazed plots (58%) than in non-cattle-grazed plots (24%). There was no difference between plots the following winter (cattle grazed: 78%; ungrazed: 69%). Additionally, the proportion of rough fescue plants grazed by elk over the four years from outset of the experiment did not differ between plots grazed (26–98%) or ungrazed (15–97%) by cattle. Cattle-grazing entailed 104 cow/calf pairs on a 104-ha pasture, from 18 October 1983 to 22 December 1983. There were three ungrazed control plots, 2 ha each in extent. Six caged and six non-caged samples on each treatment were clipped in April 1985 and 1986 to determine elk utilization by biomass. Additionally, utilization of rough fescue was assessed by determining the proportion of plants grazed by elk by inspecting the closest plant to 50 points along each of two transects per plot.

A replicated, controlled, before-and-after study in 1998–2000 in five grassland sites in California, USA (4) found that using livestock grazing to manage vegetation had mixed effects on the abundance of Stephens' kangaroo rat *Dipodomys stephensi*. One year after grazing started, there was no difference in the density of Stephens' kangaroo rat (9 animals/ ha) compared to before grazing started (9 animals/ha). However, after two years, their density had increased to 22 animals/ha. Areas that were grazed had a lower density of kangaroo rats both before grazing started and after one year when compared to ungrazed areas (9 animals/ha vs 28 animals/ha), but after two years there was no longer a significant difference (22 animals/ha vs 28 animals/ha). In 1998 and 1999, two sites were grazed by sheep for between four hours and three days, and two sites were not grazed in either year. An unspecified number of Sherman live traps were placed in each site. In 1996–2000, at unspecified times of year, trapping was conducted over three consecutive nights. Traps were opened in the evening and checked at midnight and at dawn and animals caught were individually marked.

A replicated, controlled study in 2003–2005 of pasture at a site in northern Norway (5) found that sheep-grazed pasture was used by feeding reindeer *Rangifer tarandus* more than was ungrazed pasture. Reindeer spent more time feeding in low-intensity sheep grazed plots (30% of all feeding observations) and high-intensity sheep grazed plots (28%) than in ungrazed plots (17%). Sixteen plots were established in each of two 0.3-ha fields. Each field contained four plots of each high-intensity sheep grazing, low-intensity sheep grazing and ungrazed pasture. Low-and high-intensity sheep grazing comprised two (ewe and yearling) and four (ewe and three lambs) sheep respectively, for 10 days at the beginning of July in 2003 and 2004, contained within temporary internal fencing. Four 2-year-old male reindeer were grazed on each field for two weeks in autumn 2003, spring and autumn 2004 and spring 2005. Reindeer feeding patch choice was determined by timed observations.

A replicated, paired sites study in 2014 in 10 pastures in Oaxaca, Mexico (6) found that in pastures grazed by cattle, Tehuantepec jackrabbits *Lepus flavigularis* spent more time feeding than they did in pastures not grazed by cattle. When in pastures with cattle, Tehuantepec jackrabbits spent more time feeding (75%) than when in pastures without cattle (66%). The study was conducted in five pastures with cattle (average of 16 cows/pasture) and five pastures without. Pastures averaged 11 ha extent and were located next to each other. Cattle moved freely within each pasture. In March 2014, twenty-two adult jackrabbits were captured, radio-tagged and released at the capture site. Animals were followed for ≤10 days in March and September 2014. Additionally, jackrabbit behaviour was recorded from five fixed observation sites throughout the study area. The behaviour (eating, resting and socializing) of jackrabbits was recorded between 6:00–10:00 h and 17:00–20:00 h in pastures with or without cattle.

(1) Anderson E.W. & Scherzinger R.J. (1975) Improving quality of winter forage for elk by cattle grazing. *Journal of Range Management*, 28, 120–125.

(2) Skovlin J.M., Edgerton P.J. & McConnell B.R. (1983) Elk use of winter range as affected by cattle grazing, fertilizing, and burning in Southeastern Washington. *Journal of Range Management*, 36, 184–189.

(3) Jourdonnais C.S. & Bedunah D.J. (1990) Prescribed fire and cattle grazing on an elk winter range in Montana. *Wildlife Society Bulletin*, 18, 232–240.

(4) Kelt D.A., Konno E.S. & Wilson J.A. (2005) Habitat management for the endangered Stephens' kangaroo rat: the effect of mowing and grazing. *The Journal of Wildlife Management*, 69, 424–429, https://doi.org/10.2193/0022-541x(2005)069<0424:hmftes>2.0.co;2

(5) Colman J.E., Mysterud A., Jørgensen N.H. & Moe S.R. (2009) Active land use improves reindeer pastures: evidence from a patch choice experiment. *Journal of Zoology*, 279, 358–363, https://doi.org/10.1111/j.1469-7998.2009.00626.x

(6) Luna-Casanova A., Rioja-Paradela T., Scott-Morales L. & Carrillo-Reyes A. (2016) Endangered jackrabbit *Lepus flavigularis* prefers to establish its feeding and resting sites on pasture with cattle presence. *Therya*, 7, 277–284, https://doi.org/10.12933/therya-16-393

13.3. Manage vegetation using grazing by wild herbivores

https://www.conservationevidence.com/actions/2548

- **Two studies** evaluated the effects on mammals of managing vegetation using grazing by wild herbivores. One study was in the USA[1] and one was in South Africa[2].

COMMUNITY RESPONSE (0 STUDIES)

POPULATION RESPONSE (2 STUDIES)

- **Abundance (2 studies):** A site comparison study in the USA[1] found that areas with higher numbers of wild herbivore grazers hosted more small mammals than did areas grazed by fewer wild herbivores. A study in South Africa[2] found that grazing by Cape mountain zebras did not lead to a higher population of bontebok.

BEHAVIOUR (0 STUDIES)

Background

Using grazing to manage vegetation can limit succession that would otherwise lead to an increase in woody plant species. This may help to increase the abundance of mammal species that depend on early-succession habitats. As well as managing vegetation using domestic herbivores, in some cases wild herbivore numbers can be manipulated with similar aims.

A site comparison study in 1998–1999 at a forest site in Tennessee, USA (1) found that in areas grazed by high numbers of wild herbivores, of three species, there were more small mammals than in areas grazed by fewer wild herbivores with just one species present. More small mammals were caught in areas with high wild herbivore abundance (145 small mammals) than in areas with low wild herbivore abundance (96 small mammals). Numbers caught in areas with high and low herbivore abundance were: white-footed mouse *Peromyscus leucopus* (130 vs 69), northern short-tailed shrew *Blarina brevicauda* (8 vs 22), woodland vole *Microtus pinetorum* (2 vs 5), golden mouse *Ochrotomys nuttalli* (4 vs 0), southern flying squirrel *Glaucomys volans* (1 vs 0) (species-level results were not statistically tested). Small mammals were surveyed at six plots inside a 324-ha enclosure, where elk *Cervus canadensis* and bison *Bison bison* were released in 1994, and six plots outside the enclosure, where no elk or bison occurred. White-tailed deer *Odocoileus virginianus* occurred both inside and outside the enclosure. Herbivore density was 46/km² inside the enclosure and 6–10/km² outside the enclosure. Small mammals were sampled 13 times at each plot, from June 1998 to May 1999, using 15 Sherman live traps, along a 100-m transect, for three nights each time.

A study in 1987–2009 in a shrubland protected area in Western Cape, South Africa (2) found that following the introduction of Cape mountain zebras *Equus zebra zebra* to manage vegetation and facilitate improved grazing for bontebok *Damaliscus pygargus pygargus*, numbers of bontebok did not increase. Twenty-two years after Cape mountain zebras were introduced, bontebok numbers were approximately one-third lower (187) than at the time of zebra introduction (298). Authors suggest that zebras and bonteboks may compete for similar resources. In 1987–1990, twelve Cape mountain zebras were translocated into a 3,435-ha national park. Between 1987–1990 and 2009, zebra numbers increased from 12 to 48 individuals. Population monitoring details for bonteboks and zebras are not provided.

(1) Weickert C.C., Whittaker J.C. & Feldhamer G.A. (2001) Effects of enclosed large ungulates on small mammals at Land Between The Lakes, Kentucky. *The Canadian Field-Naturalist*, 115, 247–250.

(2) Watson L.H., Kraaij T. & Novellie P. (2011) Management of rare ungulates in a small park: habitat use of bontebok and Cape mountain zebra in Bontebok National Park assessed by counts of dung groups. *South African Journal of Wildlife Research*, 41, 158–166, https://doi.org/10.3957/056.041.0202

13.4. Replant vegetation

https://www.conservationevidence.com/actions/2549

- We found no studies that evaluated the effects on mammals of replanting vegetation.

'We found no studies' means that we have not yet found any studies that have directly evaluated this intervention during our systematic journal and report searches. Therefore, we have no evidence to indicate whether or not the intervention has any desirable or harmful effects.

Background

Planting vegetation can help to relatively rapidly re-establish habitats after human disturbance. As a result, this replanting may help to increase mammal species richness and abundance.

13.5. Remove vegetation by hand/machine

https://www.conservationevidence.com/actions/2550

- **Twenty studies** evaluated the effects on mammals of removing vegetation by hand or machine. Eleven studies were in the USA[1,3–6,8,9,10,16,18,19], and one each was in Canada[2], South

Africa[15], Israel[7], Norway[11], Portugal[12], France[13], Spain[14], the Netherlands[17] and Thailand[20].

COMMUNITY RESPONSE (1 STUDY)

- **Richness/diversity (1 study):** A site comparison study in the USA[3] found that mechanically clearing trees within woodland reduced small mammal diversity.

POPULATION RESPONSE (12 STUDIES)

- **Abundance (11 studies):** Eight of 11 site comparison or controlled studies (nine of which were replicated), in the USA[1,3,4,5,9,10,19], Israel[7], Portugal[12], Spain[14] and the Netherlands[17], found that clearing woody vegetation[3,5,10,12,14,19] or herbaceous and grassland vegetation[4,9] benefitted target mammals. Population or density increases were recorded for small mammals[3,5], European rabbits[12,14] and Stephens' kangaroo rat[9] while black-tailed prairie dog[10] and California ground squirrel[19] colonies were larger or denser and Utah prairie dog colonies established better than in uncleared areas[4]. Two studies found mixed results of clearing woody vegetation, with hazel dormouse abundance declining, then increasing[17] and small mammal abundance increasing, then declining in both cleared and uncleared plots alike[1]. One study found no effect of scrub clearance from sand dunes on habitat specialist small mammals[7].

- **Survival (1 study):** A replicated, site comparison study in the USA[16] found that mechanical disturbance of woody vegetation within forest (combined with reseeding, follow-up herbicide application and further seeding) increased overwinter survival of mule deer fawns.

BEHAVIOUR (8 STUDIES)

- **Use (8 studies):** Four of seven studies (of which six were site comparisons or controlled), in the USA[6,8,18], Canada[2], Norway[11], France[13] and Thailand[20], found that areas cleared of woody vegetation[13,20] or herbaceous and grassland vegetation[2,11] were utilized more by mule deer[2], reindeer[11], mouflon[13] and gaur[20]. One study found that clearing woody vegetation promoted increased use by white-tailed deer in some but not all plots[6], one found that it did not increase use by mule deer[18] and one found that carrying out a second clearance on previously cleared plots did not increase use by white-tailed deer[8]. A before-and-after study in South Africa[15] found that clearing woody vegetation from shrubland increased wildebeest and zebra abundance following subsequent burning but not when carried out without burning whilst other mammals did not show consistent responses.

Background

Regular disturbance may maintain vegetation in a desirable, semi-natural state — particularly in early-successional habitats. Removal of vegetation may help to maintain habitats in an early-successional state, which may benefit mammal species that depend on such habitats.

This intervention includes removal of annual vegetation (e.g. herbs and grasses removed by mowing) as well as scrubby vegetation and trees. Tree clearance studies included here are those where woodland had colonised previously open areas and was cleared for conservation purposes, without being part of commercial forest management. For studies of partial clearance in long-established or commercially managed forest, see *Biological Resource Use — Clear or open patches in forests*.

A replicated, controlled, before-and-after study in 1966–1970, of pinyon-juniper woodland and grassland at six sites in Utah, USA (1) found that, after clearance of pinyon-juniper and seeding with grassland species, small mammal abundances in both cleared and uncleared plots followed similar patterns. Comparisons between treatments were not tested for statistical significance. Two years after clearance and seeding, more deer mice *Peromyscus maniculatus* were caught in cleared plots (107–118 from 180 trap nights) and in uncleared plots (89 from 180 trap nights) than were caught before clearance and seeding (19 from 270 trap nights). However, after three to four years, abundance in cleared plots declined (16–37 mice from 180 trap nights) and abundances in uncleared plots also declined (27–30 from 180 trap nights). Trees were cleared by dragging a heavy chain or were bulldozed. Aerial seeding followed. Felled wood was gathered into lines and left in place or burned, or was dispersed during a second pass of the chain. In 1966–1970, small mammals were sampled using snap-traps over a range of dates in August–November.

A replicated, controlled study in 1975–1977 on grassland in British Columbia, Canada (2) found that in mown areas, bluebunch wheatgrass *Agropyron spicatum* was consumed more by foraging mule deer *Odocoileus hemionus* than in unmown areas. Deer took a higher average number of bites/observation of bluebunch wheatgrass in mown plots (12 bites) than in unmown plots (two bites). Plots were studied at two sites in sagebrush and two in Douglas fir *Pseudotsuga menziesii* forest. At each site, plots (1.25 × 5 m) were established in a block. In each block, in October 1975, three plots were clipped using a lawnmower and electric-powered sickle and three were uncut. In April 1976, three deer were fenced onto the block and their selection between plots was assessed through direct observations at intervals through the day. The same three deer were used on all blocks and observed twice/block for one day each time. In April 1977, four deer were observed, on two blocks combined, over four days.

A site comparison study in 1977 of five areas within a pinyon-juniper woodland in Colorado, USA (3) found that mechanically clearing trees increased small mammal abundance but reduced diversity. More small mammals were caught in area cleared areas (175–295 individuals) than in the uncleared area (102 individuals). However, diversity was lower in cleared areas than in the uncleared area (results reported as Shannon-Weaver diversity index). Small mammals were sampled in four study areas ($\leq$28 km apart). One area was mature pinyon-juniper woodland whilst other areas comprised woodland that had been cleared by chaining (a heavy anchor chain was dragged between two bulldozers) 1, 8, and 15 years previously. Small mammals were live-trapped on three grids in each area (32 trap stations/grid). Trapping was conducted concurrently on all areas, during two trapping sessions of eight days each, in mid-July and mid-August 1977.

A controlled study in 1978–1981 of grassland at four sites in a national park in Utah, USA (4) found that mechanical disturbance of vegetation promoted establishment of translocated Utah prairie dogs *Cynomys parvidens*. In the first year of translocation, more prairie dogs (8–16) were counted on sites where vegetation was disturbed than on sites where vegetation was not disturbed (0.3). The same pattern held over the second year (disturbed: 9–14; undisturbed: 0 prairie dogs) and third year (disturbed: 15–16; undisturbed: 0 prairie dogs) after translocation. In August 1978, vegetation in one site was disturbed using a rotobeater. In another site, four railroad rails were dragged twice over the site. Vegetation was not disturbed at a third site. Sites were 5 ha each. On each site, 200 artificial burrows were created. In early-summer 1979, a total of 200 prairie-dogs were translocated and released across four sites (these three sites and a fourth site, not detailed here). Counts were conducted through summer and autumn of 1979 and in summer 1980–1981.

A replicated, controlled study in 1981–1983 of a pinyon-juniper woodland in New Mexico, USA (5) found that 13–18 years after treatment, felled or thinned stands had more small mammals than did undisturbed stands. The number of animals caught in stands that were thinned (432) or bulldozed (433) did not differ from each other but both were greater than the number in undisturbed stands (246). Species composition differed, with more grassland species in bulldozed stands (bulldozed: 95–175; thinned: 35; undisturbed: 46) and more woodland mice in thinned stands (thinned: 58; bulldozed: 6–11; undisturbed: 26). Plots, approximately 120 ha each, were established in each of two woodland blocks, one in 1965, one in 1970. In each block, one plot was thinned (trees ≥6.1 m apart), one was bulldozed (trees pushed over and left) and one was undisturbed. Small mammals were trapped in the second and third week of September, each year, in 1981–1983. Each plot was sampled for four days each year.

A randomized, controlled, before-and-after study in 1981–1983 of forest and grassland on a ranch in Texas, USA (6) found that after partial clearing of woody vegetation, there was a mixed response in white-tailed deer *Odocoileus virginianus* use of these areas. Changes in use of partially cleared areas were not tested for statistical significance. In two of four plots that were partially cleared, average deer numbers increased (after: 22–24 deer/100 ha; before: 3–13 deer/100 ha). In the other two plots that were partially cleared, average deer number declined (after: 11–15 deer/100 ha; before: 13–15 deer/100 ha). In the plot that was not cleared, deer numbers declined (after: 20 deer/100 ha; before: 27 deer/100 ha). On a 20,000 ha ranch, five plots (120 ha each, ≥4 km apart) were studied. Two tractors dragged a heavy-duty chain in a U-shape to partly clear four plots of woody vegetation in May–June 1981. Plots had 30, 50, 70, and 80% of woody vegetation cleared. Uprooted woody material was removed by burning in July 1981. A fifth plot remained uncleared. Treatments were assigned randomly to plots. Deer were counted from helicopter transects, every three months, from March 1981 to March 1983.

A replicated, controlled study in 1995–1996 of a coastal sand dune in Israel (7) found that removing scrub did not increase abundances of habitat specialist sand-living small mammals. The total number of Anderson's gerbils *Gerbillus allenbyi* in cleared plots (124) did not significantly differ from that in uncleared plots (107). The same applied for Tristram's jird *Meriones tristrami*, (cleared: 3; uncleared: 8). However, scrub clearance reduced numbers of invasive house mice *Mus musculus* (cleared: 6; uncleared: 109). All aboveground woody vegetation was removed from two 50 × 50-m plots, in September 1995. Plots were >200 m apart. Uncleared plots were located 50–200 m from each cleared plot. Small mammals were surveyed using 36 Sherman live traps in each plot, over four nights, each month, from December 1995 to September 1996.

A before-and-after study in 2001–2002 of a shrubland site in Texas, USA (8) found that carrying out a second mechanical vegetation clearance of plots already subject to an earlier mechanical clearance did not increase their utilization by white-tailed deer *Odocoileus virginianus*. There was no significant difference in deer track counts between plots before (37 track crossings/km) or after (47 track crossings/km) the second mechanical clearance. Plots (3–9 ha), were established in a 6,154-ha study area. In March–April 1999, five plots were cleared of woody vegetation using a mechanical aerator pulled by a tractor. Plots were mechanically cleared again in September 2000. Deer utilization was assessed by counting tracks along prepared track lanes, over three days on four occasions. Surveys were conducted once before clearance, in late-May to July 2000, and three times after clearance, in December 2000 to January 2001, May 2001 and June–July 2001.

A replicated, controlled, before-and-after study in 1996–2000 of a grassland area in California, USA (9) found that after vegetation mowing commenced, Stephens' kangaroo rat *Dipodomys stephensi* abundance increased. More animals were estimated to be in mown plots two years after mowing began (mown: 21; before mowing 18) and in plots that were both mown and grazed (mown: 15; before mowing: 8). Plots that were neither grazed nor mown contained more animals than mown or mown and grazed plots, although the number after management of other plots commenced did not differ from that before management (28 vs 28 kangaroo rats). Seven plots (80 × 80 m) were surveyed. Two were mown in 1998 and 1999, three were mown in 1998 and grazed by sheep in 1999 and two were not grazed or mowed. Mowing cut vegetation as short as the mower allowed. Cut vegetation was left on site. Grazing removed all available forage. Kangaroo rats were surveyed using grids of Sherman live traps, over three consecutive nights, bimonthly, from November 1996 to October 2000.

A replicated, controlled, before-and-after study in 2002–2003 in a national park in Dakota, USA (10) found greater areas occupied by black-tailed prairie dog *Cynomys ludovicianus* colonies and more prairie dog burrows, in plots that were burned and mechanically cleared of woody vegetation than in plots that were not cleared or burned. The study does not distinguish between the effects of mechanical vegetation clearance and burning. At the end of the second summer after vegetation clearance, prairie dog colonies had expanded more (into 18–70% of available habitat) in burned and cleared plots compared to unmanaged plots (0–5%). In burned and cleared plots, there were more new burrows (191–458) after two summers than in unmanaged plots (41–116). At each of three prairie dog colonies, a 2-ha treatment plot, just beyond the colony boundary, underwent prescribed burning in May 2002 and mechanical removal of woody vegetation in June 2002. Similarly, selected 2-ha plots were left unmanaged. Colonies boundaries were mapped in May–September 2002 and May–August 2003. New burrows were mapped monthly during these periods.

A replicated, controlled study in 2003–2005 of pasture at a site in northern Norway (11) found that mown pasture was selected by feeding reindeer *Rangifer tarandus* more than was unmown pasture. Reindeer spent more time feeding in mown plots (25% of all feeding observations) than in unmown plots (17%). Sixteen plots were established in each of two 0.3-ha fields. Each field contained four replicate plots of high-intensity sheep grazing, low-intensity sheep grazing, mowing and unmanaged. Sheep grazing treatments are not reported on in the paper. Mown plots were cut in July, to 5 cm height, with cuttings removed. Four 2-year-old male reindeer grazed in each field for two weeks in autumn 2003, spring and autumn 2004 and spring 2005. Reindeer feeding patch choice was determined during timed observations.

A replicated, controlled, before-and-after study in 2000–2002 on scrubland in a nature reserve in southwest Portugal (12) found that clearing scrub (through establishing firebreaks) increased densities of European rabbits *Oryctolagus cuniculus*. In areas where firebreaks were established average annual rabbit pellet densities ($1.1–3.6/m^2$) were higher than prior to establishment of firebreaks ($0.5–1.5/m^2$). Pellet densities were also higher than in areas where no firebreaks were established (firebreaks: $1.1–3.6/m^2$; no firebreaks: $0.4–2.2/m^2$). Four 300-ha sites, ≥3 km apart, were studied. In February 2001, areas of grassland were restored by cutting 5-m-wide firebreak strips through scrub. The other two sites remained unmanaged. Rabbit pellets were counted, monthly, at fixed points along transects, from May 2001 to October 2002.

A controlled study in 2004–2008 of heather moorland at a site in southern France (13) found that cutting heather (*Calluna vulgaris* and *Erica tetralix*) resulted in greater use of it by mouflon *Ovis gmelini musimon* × *Ovis* sp. Average density of feeding mouflon was higher on cut plots (27/ha) than on uncut plots (5/ha). Prior to the study, each 360 × 80-m plot had not been modified for >40 years. Two plots were cut in spring 2004, to an average height of 5 cm, and two were left uncut. Mouflon use of plots was determined by counting feeding animals in each plot, at 20-minute intervals, for two hours up to sunset. In total, 668 such counts were made in 2004–2008.

A replicated, site comparison study in 2008–2012 in grassland and scrubland along a mountain chain in Andalusia, Spain (14) found that removing scrubland vegetation to create pasture increased abundances of translocated European rabbits *Oryctolagus cuniculus* in areas of high scrub coverage but not of medium-or low-scrub coverage. In high scrub cover areas, there were more rabbits around plots where scrub was cleared (5.9 latrines/km) than where scrub was not cleared (2.6 latrines/km). There was no significant difference in rabbit abundance in areas of medium cover scrub (scrub clearance: 7.1 pellets/km; no scrub clearance: 5.0 pellets/km) or low scrub cover (scrub clearance: 1.6 pellets/km; no scrub clearance: 2.1 pellets/km). In autumn and winter of 2008–2009, between 75 and 90 rabbits/ha were released into fenced plots (0.5–7.7 ha). Wooden branches and artificial warrens were added within a 500-m radius outside plots and, at some, scrubland was cleared to create pasture (number of plots/treatment and pasture sizes not reported). At the end of each breeding season in 2009–2011, small gates allowed rabbits to disperse through fences into adjacent areas. Rabbit abundance was estimated by latrine counts in four 500-m-long transects around each plot, in summer 2012.

A before-and-after study in 2009–2010 on savannah in South Africa (15) found that in areas cleared of woody vegetation, wildebeest *Connochaetes taurinus* and zebra *Equus burchelli* abundance was higher than in uncleared areas after areas were burned, but not before burning, whilst other mammals did not show consistent responses. Wildebeest faecal pellet prevalence was higher in cleared than in uncleared plots after burning (cleared: in 4–7% of plots; uncleared: 1%) but not before (cleared: 0%; uncleared: 2%). Similarly, zebra pellet prevalence was higher in cleared than in uncleared plots after burning (cleared: in 18–30% of plots; uncleared: 7%) but not before (cleared: 16–19%; uncleared: 20%). Impala *Aepyceros melampus*, kudu *Tragelaphus strepsiceros* and giraffe *Giraffa camelopardalis* did not show consistent differences between responses in cleared versus uncleared land. Herbivore abundance was determined by establishing presence or absence of faecal pellets for each species in plots along transects through areas on sandy soils subject to mechanical clearance of woody vegetation by barko crawler, bosvreter and chainsaw (date of clearance not stated) and uncleared areas. Pellets were counted in April–May 2009, prescribed burns were carried out in June–November 2009 and plots were resampled in June 2010.

A replicated, site comparison study in 2005–2008 of a pine-juniper forest in Colorado, USA (16) found that mechanical disturbance of vegetation (combined with reseeding, follow-up herbicide application and further seeding — referred to as advanced management) increased overwinter survival of mule deer *Odocoileus hemionus* fawns. Management actions were not carried out individually, so their relative effects cannot be determined. Average overwinter survival was highest under advanced management (77%), intermediate under mechanical disturbance and seeding without follow-up actions (69%) and lowest with no habitat management (67%). Mechanical management, commencing in 1998–2004, involved removing and mulching trees to create open areas. These were seeded with grasses and flowering plants. Follow-up actions in advanced management plots, two to four years later, involved controlling weeds with herbicide and further seeding with deer browse species. Fawns were radio-collared on eight study plots; two advanced management plots, four mechanical management plots and two unmanaged plots. Survival was assessed by monitoring fawns from capture (1 December to 1 January) until 15 June, in winters of 2004–2005 through to 2007–2008, three to six years after mechanical treatments.

A replicated, before-and-after, site comparison study in 2009–2013 at six forest sites in the Netherlands (17) found that after clearance of most mature trees, hazel dormouse *Muscardinus avellanarius* nest abundance declined briefly but then increased relative to areas where no trees were cleared. Dormouse nest numbers in cleared plots fell in the year after clearing to 32% of pre-clearance levels. Two to four years after clearance, nest numbers were higher, at 374–803% of pre-clearance levels. Data were presented as standardised indices. In uncleared plots, there was a declining trend throughout with, at the end of the study, nest numbers 21% of the count made at the start of the study. Dormouse nests were counted along transects in September and November each year in 2009–2013. In 10 arbitrarily chosen 'managed' segments along transects (average 92 m long), 75–100% of mature trees were cut in winter 2009–2010. Ten unmanaged transect sections (average 181 m long) were monitored as controls.

A replicated, site comparison study in 2006–2009 of pine and juniper forests interspersed with meadows on a plateau in Colorado, USA (18) found that mule deer *Odocoileus hemionus* densities did not differ between plots where trees were cleared and those where trees were not cleared. Average deer density was 6–37 deer/km^2 on plots where trees were cleared and 5–85 deer/km^2 on plots where no trees were cleared.

Tree clearance was carried out on four plots, two to eight years prior to deer surveys. This comprised uprooting trees with a bulldozer, followed by mechanical roller chopping to break vegetation into smaller pieces, or hydro-axing, whereby individual trees were mulched to ground level. In two plots, no trees were cleared. Deer numbers were estimated by resighting marked individuals, in late winter each year in 2006–2009, from aerial surveys. Surveys were conducted over 15–94 km²/plot.

A replicated, controlled, paired sites study in 2011–2014 of two areas of grassland and scrubland in southern California, USA (19) found that in mown areas, California ground squirrel *Otospermophilus beecheyi* burrow densities were higher compared to in unmown areas. Three years after management commenced, there were more squirrel burrows in mown (11–122/subplot) compared to in unmown (12–54/subplot) areas. Each of six plots comprised a circle covering 0.8 ha, divided into three equal wedge-shaped subplots. One subplot in each plot was mown in May, for two years, at 7.5–15 cm height, with cut material removed and one was unmown. (Management details for the third subplot are not relevant to this intervention). Management commenced in 2011 (two plots) and 2012 (four plots). Squirrels were translocated into plots at a target rate of 30–50/plot. Squirrel abundance was determined by counting squirrel burrows.

A site comparison study in 2010–2012 in two secondary forest plots in Nakhon Ratchasima Province, Thailand (20) found that clearing vegetation using chainsaws increased the density of gaur *Bos gaurus* using these areas. Average gaur density was higher in a plot where pioneer trees were felled (8.6 individuals/km²/day) than in a plot where the vegetation was left unmanaged (4.0 individuals/km²/day). The study was conducted within an 8-km² area, reforested since 1994. In May–September 2010, a total of 407 pioneer *Macaranga siamensis* trees were felled with chainsaws to open up 28% of a 5.7-ha plot. Trees were not felled in a nearby 4.7-ha plot. The ground within the felled and unfelled plots was cleared, using a tractor, in June and December 2011. Gaur dung piles were counted monthly, between February 2011 and March 2012, with the exception of June and December 2011. Dung piles were counted by 9–10 volunteers along 50-m-long transects (number not stated) with counts used to estimate guar usage of plots.

(1) Baker M.F. & Frischknecht N.C. (1973) Small mammals increase on recently cleared and seeded juniper rangeland. *Journal of Range Management*, 26, 101–103.

(2) Willms W., Bailey A.W. & McLean A. (1980) Effect of burning or clipping *Agropyron spicatum* in the autumn on the spring foraging behaviour of mule deer and cattle. *Journal of Applied Ecology*, 17, 69–84.

(3) O'Meara T.E., Haufler J.B., Stelter L.H. & Nagy J.G. (1981) Nongame wildlife responses to chaining of pinyon-juniper woodlands. *The Journal of Wildlife Management*, 45, 381–389.

(4) Player R.L. & Urness P.J. (1982) Habitat manipulation for reestablishment of Utah prairie dogs In Capitol Reef National Park. *Great Basin Naturalist*, 42, 517–523.

(5) Severson K.E. (1986) Small mammals in modified pinyon-juniper woodlands, New Mexico. *Journal of Range Management*, 39, 31–34.

(6) Rollins D., Bryant F.C., Waid D.D. & Bradley L.C. (1988) Deer response to brush management in central Texas. *Wildlife Society Bulletin*, 16, 277–284.

(7) Kutiel P., Peled Y. & Geffen E. (2000) The effect of removing shrub cover on annual plants and small mammals in a coastal sand dune ecosystem. *Biological Conservation*, 94, 235–242, https://doi.org/10.1016/s0006-3207(99)00172-x

(8) Rogers J.O., Fulbright T.E. & Ruthven D.C. III (2004) Vegetation and deer response to mechanical shrub clearing and burning. *Journal of Range Management*, 57, 41–48, https://doi.org/10.2307/4003953

(9) Kelt D.A., Konno E.S. & Wilson J.A. (2005) Habitat management for the endangered Stephens' kangaroo rat: the effect of mowing and grazing. *The Journal of Wildlife Management*, 69, 424–429, https://doi.org/10.2193/0022-541x(2005)069<0424:hmftes>2.0.co;2

(10) Milne-Laux S. & Sweitzer R.A. (2006) Experimentally induced colony expansion by black-tailed prairie dogs (*Cynomys ludovicianus*) and implications for conservation. *Journal of Mammalogy*, 87, 296–303, https://doi.org/10.1644/05-mamm-a-056r2.1

(11) Colman J.E., Mysterud A., Jørgensen N.H. & Moe S.R. (2009) Active land use improves reindeer pastures: evidence from a patch choice experiment. *Journal of Zoology*, 279, 358–363, https://doi.org/10.1111/j.1469-7998.2009.00626.x

(12) Ferreira C. & Alves P.C. (2009) Influence of habitat management on the abundance and diet of wild rabbit (*Oryctolagus cuniculus algirus*) populations in Mediterranean ecosystems. *European Journal of Wildlife Research*, 55, 478–496, https://doi.org/10.1007/s10344-009-0257-4

(13) Cazau M., Garel M. & Maillard D. (2011) Responses of heather moorland and Mediterranean mouflon foraging to prescribed-burning and cutting. *The Journal of Wildlife Management*, 75, 967–972, https://doi.org/10.1002/jwmg.117

(14) Guerrero-Casado J., Carpio A.J., Ruiz-Aizpurua L. & Tortosa F.S. (2013) Restocking a keystone species in a biodiversity hotspot: Recovering the European rabbit on a landscape scale. *Journal for Nature Conservation*, 21, 444–448, https://doi.org/10.1016/j.jnc.2013.07.006

(15) Isaacs L., Somers M.J. & Dalerum F. (2013) Effects of prescribed burning and mechanical bush clearing on ungulate space use in an African savannah. *Restoration Ecology*, 21, 260–266, https://doi.org/10.1111/j.1526-100x.2012.00877.x

(16) Bergman E.J., Bishop C.J., Freddy D.J., White G.C. & Doherty P.F. (2014) Habitat management influences overwinter survival of mule deer fawns in Colorado. *The Journal of Wildlife Management*, 78, 448–455, https://doi.org/10.1002/jwmg.683

(17) Ramakers J.J.C., Dorenbosch M. & Foppen R.P.B. (2014) Surviving on the edge: a conservation-oriented habitat analysis and forest edge manipulation for the hazel dormouse in the Netherlands. *European Journal of Wildlife Research*, 60, 927–931, https://doi.org/10.1007/s10344-014-0849-5

(18) Bergman E.J., Doherty P.F., White G.C. & Freddy D.J. (2015) Habitat and herbivore density: response of mule deer to habitat management. *The Journal of Wildlife Management*, 79, 60–68, https://doi.org/10.1002/jwmg.801

(19) McCullough Hennessy S., Deutschman D.H., Shier D.M., Nordstrom L.A., Lenihan C., Montagne J.-P., Wisinski C.L. & Swaisgood R.R. (2016) Experimental habitat restoration for conserved species using ecosystem engineers and vegetation management. *Animal Conservation*, 19, 506–514, https://doi.org/10.1111/acv.12266

(20) Prayong N. & Srikosamatara S. (2017) Cutting trees in a secondary forest to increase gaur *Bos gaurus* numbers in Khao Phaeng Ma Reforestation area, Nakhon Ratchasima Province, Thailand. *Conservation Evidence*, 14, 5–9.

13.6. Remove vegetation using herbicides

https://www.conservationevidence.com/actions/2565

- **Six studies** evaluated the effects on mammals of removing vegetation using herbicides. All six studies were in the USA[1-6].

COMMUNITY RESPONSE (0 STUDIES)

POPULATION RESPONSE (4 STUDIES)

- **Abundance (2 studies):** Two controlled studies (one replicated) in the USA[1,6] found that applying herbicide did not increase numbers of translocated Utah prairie dogs[1] or alter mule deer densities in areas of tree clearance[6].

- **Survival (1 study):** A replicated, site comparison study in the USA[5] found that applying herbicide, along with mechanical disturbance and seeding, increased overwinter survival of mule deer fawns.

- **Condition (1 study):** A replicated, controlled study in the USA[2] found that applying herbicide did not reduce bot fly infestation rates of rodents and cottontail rabbits.

BEHAVIOUR (2 STUDIES)

- **Use (2 studies):** Two replicated, controlled studies in the USA[3,4] found that applying herbicide increased forest use by female, but not male, white-tailed deer[4] and increased pasture use by cottontail rabbits in some, but not all, sampling seasons[3].

Background

Removal of vegetation may help to maintain habitats in an early-successional state. Herbicides may also be used to control some colonising plants species in favour of others that are more attractive as food plants. This may benefit mammal species that depend on such habitats.

A controlled study in 1979–1981 at two grassland sites in a national park in Utah, USA (1) found that herbicide application did not increase establishment of translocated Utah prairie dogs *Cynomys parvidens*. In the first year of translocation, the average number of prairie dogs counted on the site sprayed with herbicide (1.7) was not significantly different to that on the unsprayed site (0.3). In the second and third year, no prairie dogs were counted on either site. One site was treated with the herbicide, 2,4-D, at a rate of 2.2 kg active ingredient/ha (date of treatment not given) and one site was not sprayed. Sites were 5 ha each. On each site, 200 artificial burrows were created. In early-summer 1979, two hundred prairie dogs were translocated and released across four sites (the sprayed and unsprayed sites and two further sites not detailed in this summary). Counts were conducted through summer and fall of 1979 and in summer 1980–1981.

A replicated, controlled study in 1986–1988 of a woodland in Oklahoma, USA (2, same experimental set-up as 3 and 4) found that applying herbicide did not reduce bot fly *Cuterebra* infestation rates of rodents and cottontail rabbits *Sylvilagus floridanus*. Prevalence of bot fly did not differ between plots treated with herbicide (present on 64 of 342 animals examined, 19%), or untreated plots (25 of 133 animals examined, 19%). Eight 32.4-ha plots were treated with the herbicides, tebuthiuron or triclopyr (at 2.2 kg/ha), in March or June 1983 and four plots were not sprayed with herbicide. Rodents were collected using snap traps in July–September and December–March during 1986–1988. Cottontail rabbits were collected by shooting in January and July of 1987–1988. Animals were examined for bot fly burden.

A replicated, controlled study in 1986–1988 of forest and grassland at a site in Oklahoma, USA (3, same experimental set-up as 2 and 4) found that herbicide-treated pastures hosted more cottontail rabbits *Sylvilagus floridanus* than did untreated pastures during some, but not all, sampling seasons. In three of 10 comparisons, cottontails were more abundant in herbicide-treated pastures than in untreated pastures (0.8–1.1 vs 0.1–0.2 rabbits/ha), in two cases they were less abundant on treated than untreated pastures (0.0 vs 1.9 rabbits/ha) and for the other five comparisons no difference was detected. Four 32.4-ha pastures were treated with the herbicides tebuthiuron or triclopyr at a rate of 2.2 kg/ha in March or June 1983 and two were untreated control pastures. Rabbit density was estimated by walking transects three times each July and February, from July 1986 to February 1988.

A randomized, replicated, controlled study in 1988–1989 of an upland hardwood forest with tallgrass prairie in Oklahoma, USA (4 same experimental set-up as 2 and 3) found that applying herbicide increased forest use by female, but not male, white-tailed deer *Odocoileus virginianus*. Female deer preferentially selected herbicide-treated plots over untreated plots in spring, summer and autumn, but there was no difference in winter. Males showed no preference between treated or untreated plots (see original paper for full results). Four blocks, each consisting of five 32-ha plots, were studied. In each block, the herbicides, tebuthiuron and triclopyr, were sprayed in 1983 in one plot each, as well as in two plots that were also burned each April, in 1985–1987. One plot was not burned or sprayed with herbicide. Two additional pastures that were burned but not sprayed along with adjacent areas that were not burned or sprayed were also monitored. Ten female and seven male deer were radio-tracked, in 1988–1989.

A replicated, site comparison study in 2005–2008 of a pine-juniper forest in Colorado, USA (5) found that herbicide application (combined with seeding and preceded by mechanical disturbance and initial seeding — referred to as advanced management) increased overwinter survival of mule deer *Odocoileus hemionus* fawns. Management actions were not carried out individually, so their relative effects cannot be determined. Average overwinter survival was highest under advanced management (77%), intermediate under mechanical disturbance and seeding without follow-up actions (69%) and lowest with no habitat management (67%). Mechanical management, commencing in 1998–2004, involved removing and mulching trees to create open areas. These were seeded with grasses and forbs. In advanced management plots, follow-up actions, two to four years later, involved controlling weeds with herbicide and further seeding with deer browse species. Fawns were radio-collared on eight study plots; two advanced management plots, four mechanical management plots and two unmanaged plots. Survival was assessed by monitoring fawns from capture (1 December to 1 January) until 15 June, in winters of 2004–2005 through to 2007–2008, three to six years after mechanical treatments.

A replicated, site comparison study in 2006–2009 in six pine and juniper forest sites in Colorado, USA (6) found that treatment with herbicide, alongside clearance of trees and sowing seed, did not alter mule deer *Odocoileus hemionus* densities compared to clearance of trees alone. The effects of herbicide and reseeding could not be separated in this study. In areas that were sprayed with herbicide, cleared, and sown with seeds, deer density was not higher (5–31 deer/km^2) than in plots that were cleared but not treated with herbicide or sown with seed (6–37 deer/km^2). Six sites were cleared of trees, two to eight years before deer surveys, using a bulldozer and by chopping vegetation into smaller pieces, or mulching individual trees to ground level by hydro-axing. On two of these sites, unpalatable grasses were controlled with herbicides and seeds of plant species eaten by mule deer were sown. The four remaining sites were not further managed after tree clearance. Deer numbers were estimated by sighting marked individuals during aerial surveys, in late winter each year of 2006–2009. Areas surveyed were 15–84 km²/site.

(1) Player R.L. & Urness P.J. (1982) Habitat manipulation for reestablishment of Utah prairie dogs In Capitol Reef National Park. *Great Basin Naturalist*, 42, 517–523.

(2) Boggs J.F., Lochmiller R.L., McMurry S.T., Leslie D.M., & Engle D.M. (1991) *Cuterebra* infestations in small-mammal communities as influenced by herbicides and fire. *Journal of Mammalogy*, 72, 322–327.

(3) Lochmiller R.L., Boggs J.F., Mcmurry S.T., Leslie Jr, D.M. & Engle D.M. (1991) Response of cottontail rabbit populations to herbicide and fire applications on cross timbers rangeland. *Journal of Range Management*, 44, 150–155.

(4) Leslie Jr. D.M., Soper R.B., Lochmiller R.L. & Engle D.M. (1996) Habitat use by white-tailed deer on cross timbers rangeland following brush management. *Journal of Range Management*, 49, 401–406.

(5) Bergman E.J., Bishop C.J., Freddy D.J., White G.C. & Doherty P.F. (2014) Habitat management influences overwinter survival of mule deer fawns in Colorado. *The Journal of Wildlife Management*, 78, 448–455, https://doi.org/10.1002/jwmg.683

(6) Bergman E.J., Doherty P.F., White G.C. & Freddy D.J. (2015) Habitat and herbivore density: response of mule deer to habitat management. *The Journal of Wildlife Management*, 79, 60–68, https://doi.org/10.1002/jwmg.801

13.7. Restore or create grassland

https://www.conservationevidence.com/actions/2566

- **Three studies** evaluated the effects on mammals of restoring or creating grassland. One study each was in Portugal[1], the USA[2] and Hungary[3].

COMMUNITY RESPONSE (1 STUDY)

- **Richness/diversity (1 study):** A replicated, site comparison study in Hungary[3] found that grassland restored on former cropland hosted a similar small mammal species richness compared to native grassland.

POPULATION RESPONSE (3 STUDIES)

- **Abundance (2 studies):** A controlled, before-and-after study in Portugal[1] found that sowing pasture grasses into areas cleared of scrub did not increase European rabbit densities. A replicated, site comparison study in Hungary[3] found that grassland restored on former cropland hosted a similar abundance of small mammals compared to native grassland.

- **Survival (1 study):** A replicated, site comparison study in the USA[2] found that seeding with grassland species as part of a suite of actions including mechanical disturbance and herbicide application increased overwinter survival of mule deer fawns.

BEHAVIOUR (0 STUDIES)

Background

Many grasslands have been lost to agricultural intensification through conversion to cropland or through agricultural abandonment, whereby colonization by woodland or scrub may occur. Agri-environment schemes in Europe and North America support preservation or restoration of grasslands for agricultural, conservation and carbon storage reasons. Restoration of these grasslands may benefit some mammal species that are associated with them.

See also: *Restore or create savannas.*

A controlled, before-and-after study in 2000–2002 on a scrubland in southwest Portugal (1) found that sowing pasture grasses into areas cleared of scrub did not increase densities of European rabbits *Oryctolagus cuniculus*. Rabbit pellet density after sowing of seeds (1.6–3.6 pellets/m^2) did not differ significantly from that before sowing (1.5 pellets/m^2). Trends in rabbit density were similar on an area not sown with seed (after: 1.1–1.3 pellets/m^2; before: 0.5 pellets/m^2). Two 300-ha study areas were located at least 3 km apart. In February 2001, scrub was cleared in 5-m-wide strips at both sites. Cleared strips at one site were then sown with two pasture grasses, rye *Secale cereale* and slender oat *Avena barbat,* and with subterranean clover *Trifolium subterraneum*. At the second site, no seeds were sown. Rabbit pellets were counted monthly, at fixed points along transects, from May 2001 to October 2002.

A replicated, site comparison study in 2005–2008 of a pine-juniper forest in Colorado, USA (2) found that seeding with grassland species as part of a suite of actions including mechanical disturbance and herbicide application (referred to as advanced management) increased overwinter survival of mule deer *Odocoileus hemionus* fawns. Average overwinter survival was highest under advanced management (77%), intermediate under mechanical disturbance and reseeding but without follow-up actions (69%) and lowest with no habitat management (67%). Mechanical management, commencing in 1998–2004, involved removing and mulching trees to create open areas. These were reseeded with grasses and other flowering plants. Follow-up actions in advanced management plots, two to four years later, involved controlling weeds with herbicide and further seeding with deer browse species. Management actions were not carried out individually, so their relative effects cannot be determined. Fawns were radio-collared on eight study plots; two advanced management plots, four mechanical management plots and two unmanaged plots. Survival was assessed by monitoring fawns from capture (1 December to 1 January) until 15 June, in winters of 2004–2005 to 2007–2008, three to six years after mechanical treatments.

A replicated, site comparison study in 2011–2012 in a marsh and grassland site in Hungary (3) found that grassland restored on former cropland hosted a similar species richness and abundance of small mammals compared to native grassland. The average species richness in restored grassland plots (0–5.9/survey) did not differ significantly from native grassland (0–6.0/survey). Likewise, the average total small mammal catch did not differ between restored grassland (0–40/survey) and native grassland (0–48/survey). However, among restored plots, June-mown restorations had more individuals (1–40/survey) than did August-mown (0–17/survey) or sheep-grazed (0–9/survey) restorations. Restoration was carried out in 2005–2008 on former cropland. Within a 4,073-ha site, eight restored grassland plots and two natural grassland plots were studied. Plots covered 16–300 ha. Small mammals were surveyed using 36 Sherman live traps/site, over five nights and days, in spring and autumn of 2011 and 2012.

(1) Ferreira C. & Alves P.C. (2009) Influence of habitat management on the abundance and diet of wild rabbit (*Oryctolagus cuniculus algirus*) populations in Mediterranean ecosystems. *European Journal of Wildlife Research*, 55, 478–496, https://doi.org/10.1007/s10344-009-0257-4

(2) Bergman E.J., Bishop C.J., Freddy D.J., White G.C. & Doherty P.F. (2014) Habitat management influences overwinter survival of mule deer fawns in Colorado. *The Journal of Wildlife Management*, 78, 448–455, https://doi.org/10.1002/jwmg.683

(3) Mérő T.O., Bocz R., Polyák L., Horváth G. & Lengyel S. (2015) Local habitat management and landscape-scale restoration influence small-mammal communities in grasslands. *Animal Conservation*, 18, 442–450, https://doi.org/10.1111/acv.12191

13.8. Restore or create savannas

https://www.conservationevidence.com/actions/2568

- **Two studies** evaluated the effects on mammals of restoring or creating savannas. One study was in Senegal[1] and one was in the USA[2].

COMMUNITY RESPONSE (1 STUDY)

- **Richness/diversity (1 study):** A replicated, randomized, paired sites, controlled study in the USA[2] found that restoring savannas by removing trees increased small mammal diversity.

POPULATION RESPONSE (2 STUDIES)

- **Abundance (2 studies):** A study in Senegal[1] found that in a population of dorcas gazelle translocated into a fenced enclosure where vegetation had been restored, births outnumbered deaths. A replicated, randomized, paired sites, controlled study in the USA[2] found that restoring savannas by removing trees did not, in most cases, change small mammal abundance.

BEHAVIOUR (0 STUDIES)

Background

Through under-grazing or burning suppression, savannah vegetation can revert to denser scrubland. Restoring savanna may benefit mammals typically associated with the habitat.

See also: *Restore or create grassland.*

A study in 2009–2013 in a savanna site in Katané, Senegal (1) found that in a population of dorcas gazelle *Gazella dorcas neglecta* translocated into a fenced enclosure where vegetation had been restored, births outnumbered deaths. It is not clear whether these effects were a direct result of vegetation restoration or translocation into a fenced area. Over four years after release, more births (31) than deaths (4) of dorcas gazelles were recorded. Twenty-three (nine male and 14 female) dorcas gazelles were translocated between two reserves in northern Senegal in March 2009. Vegetation was restored prior to the translocation but no details regarding the restoration are provided. Gazelles were released into a 440-ha fenced enclosure that was enlarged to 640 ha in 2010. The translocated dorcas gazelles shared the enclosure with scimitar-horned oryx *Oryx dammah*, mhorr gazelle *Nanger dama mhorr* and red-fronted gazelle *Eudorcas rufifrons*. The enclosure fence was not impermeable to small-to-medium sized animals, including predators. Dorcas gazelles were ear-tagged and monitored from June 2009 to March 2013.

A replicated, randomized, paired sites, controlled study in 2008–2013 in five areas in a former oak savanna in Michigan, USA (2) found that restoring savannas by removing trees resulted in no change in small mammal abundance in 18 of 21 comparisons, but that small mammal diversity increased. After five years, in 18 of 21 comparisons small mammal abundance did not differ between areas where trees were removed (0.0–4.2 animals/area) and areas where trees were retained (0.0–0.6 animals/area). However, in three of 21 comparisons there were more small mammals (trees removed: 1.8–4.6 animals/area; trees retained: 0.0–1.8 animals/area). Small mammal diversity increased where trees were removed, but it declined where trees were retained (data reported as model results). In June–July 2008, five 3.2-ha blocks, each comprising four 0.8-ha plots, were designated. In each block, trees were removed from three plots and retained in one plot. In July 2010 the entire area was burnt in a prescribed burn. Once a year, in October 2008–July 2013, nine live traps baited with sunflower seeds were placed in each plot. Traps were set at 17:00–20:00 and checked at 6:00–11:00. Captured animals were individually marked to enable identification of re-captures.

(1) Abáigar T., Cano M., Djigo C.A., Gomis J., Sarr T., Youm B., Fernández-
    Bellon H. & Ensenyat C. (2016) Social organization and demography

of reintroduced Dorcas gazelle (*Gazella dorcas neglecta*) in North Ferlo Fauna Reserve, Senegal. *Mammalia*, 80, 593–600, https://doi.org/10.1515/mammalia-2015-0017

(2) Larsen A.L., Jacquot J.J., Keenlance P.W. & Keough H.L. (2016) Effects of an ongoing oak savanna restoration on small mammals in Lower Michigan. *Forest Ecology and Management*, 367, 120–127, https://doi.org/10.1016/j.foreco.2016.02.016

13.9. Restore or create shrubland

https://www.conservationevidence.com/actions/2569

- **Three studies** evaluated the effects on mammals of restoring or creating shrubland. Two studies were in the USA[1,3] and one was in Mexico[2].

COMMUNITY RESPONSE (2 STUDIES)

- **Richness/diversity (2 studies):** Two site comparison studies, in the USA[1] and Mexico[2], found that following desert scrub[1] or shrubland[2] restoration, mammal species richness was similar to that in undisturbed areas.

POPULATION RESPONSE (1 STUDY)

- **Abundance (1 study):** A site comparison study in the USA[1] found that restored desert scrub hosted similar small mammal abundance compared to undisturbed desert scrub.

BEHAVIOUR (1 STUDY)

- **Use (1 study):** A replicated, site comparison study in the USA[3] found that restoring shrubland following tree clearance did not increase usage of areas by mule deer compared to tree clearance alone.

Background

Loss of shrubland may be due to a range of factors, including too many grazing animals inhibiting regeneration of shrubs, too few grazing animals or fire suppression leading to reversion to woodland, or invasion by non-native species. Shrubland restoration or creation may benefit mammals associated with the habitat.

A site comparison study in 1995 in a desert site in California, USA (1) found that restored desert scrub hosted similar small mammal species richness and abundance compared to undisturbed desert scrub. Five small mammal species were recorded in restored desert scrub, similar to the seven recorded in undisturbed desert scrub. Additionally, the average number of individuals caught of each species did not differ significantly between restored and undisturbed desert scrub (San Diego pocket mouse *Chaetodipus fallax*: 2.9 vs 3.5 individuals/night; spiny pocket mouse *Chaetodipus spinatus*: 2.9 vs 1.4; Merriam's kangaroo rat *Dipodomys merriami*: 0.0 vs 0.1; desert woodrat *Neotoma lepida*: 7.4 vs 8.0; cactus mouse *Peromyscus eremicus*: 5.8 vs 3.4; deer mouse *Peromyscus maniculatus*: 4.5 vs 2.8; California ground squirrel *Spermophilus beecheyi*: 0.0 vs 0.1). Small mammals were caught in a 20-acre desert scrub site restored after construction of a dam, and in surrounding undisturbed desert scrub. During eight nights in March–May 1995, small mammals were captured with 180 Sherman live traps, divided equally between restored and undisturbed desert scrub. Traps were set in different locations each trap-night. Desert scrub was restored by topsoil replacement, direct seeding of shrubs and planting of shrub seedlings.

A site comparison study in 2009–2010 of scrubland at three sites in Mexico City, Mexico (2) found that where native shrubland vegetation was restored on degraded areas, mammal species richness was similar to that in a natural area, but more species were non-native. No statistical analyses were performed. In restored areas mammal species richness was similar (8–10 species) to that in an undisturbed shrubland (7 species). However, the restored areas had more non-native species (4 species) than did the undisturbed area (1 species). In 2005–2006, in two sites, non-native plants were removed and native shrubland vegetation was established. A nearby undisturbed shrubland was used for comparison. Small mammals were surveyed using 16 Sherman live traps on each site, over two consecutive nights, every three months, from February 2009 to May 2010. Medium-sized mammals were surveyed on day and night visits, every two weeks, from May 2009 to May 2010. Mammal latrine samples were identified to species.

A replicated, site comparison study in 2006–2009 of pine and juniper forests interspersed with grassland in Colorado, USA (3) found that restoring shrubland by sowing seeds and applying herbicide following tree clearance, did not increase densities of mule deer *Odocoileus hemionus* using these plots compared to plots that were cleared of trees alone. The effects of seeding and herbicide could not be separated in this study. Deer densities in cleared plots that were seeded and sprayed with herbicide (5–31 deer/km²) were not significantly different from those in plots that were just cleared (6–37 deer/km²). Six plots were cleared of trees, 2–8 years before deer surveys commenced, using a bulldozer and by chopping vegetation, or mulching trees to ground level, by hydro-axing. On two plots, at the same time as deer surveys, unpalatable grasses were controlled with herbicides and seeds, mainly of shrub species eaten by mule deer, were sown. The four remaining plots were not further managed after tree clearance. Deer numbers were estimated by sighting marked individuals during aerial surveys, in late winter each year, in 2006–2009 (not all plots were surveyed each year). Areas surveyed were 15–84 km²/plot.

(1) Patten M.A. (1997) Reestablishment of a rodent community in restored desert scrub. Restoration Ecology, 5, 156–161.

(2) San-José M., Garmendia A. & Cano-Santana Z. (2013) Vertebrate fauna evaluation after habitat restoration in a reserve within Mexico City. *Ecological Restoration*, 31, 249–252, https://doi.org/10.3368/er.31.3.249

(3) Bergman E.J., Doherty P.F., White G.C. & Freddy D.J. (2015) Habitat and herbivore density: response of mule deer to habitat management. *The Journal of Wildlife Management*, 79, 60–68, https://doi.org/10.1002/jwmg.801

13.10. Restore or create forest

https://www.conservationevidence.com/actions/2570

- **Five studies** evaluated the effects on mammals of restoring or creating forest. Two studies were in the USA[1,2] and one each were in Colombia[3], Italy[4] and Australia[5].

COMMUNITY RESPONSE (2 STUDIES)

- **Richness/diversity (2 studies):** Two site comparison studies (one replicated) in the USA[1] and Colombia[3] found that mammal species richness in restored forest was similar to that in established forest.

POPULATION RESPONSE (2 STUDIES)

- **Abundance (2 studies):** One of two replicated studies (one a site comparison) in Australia[5] and Italy[4] found that replanted or regrowing forest supported a higher abundance of hazel dormice than did coppiced forest[4]. The other study found only low numbers of common brushtail possums or common ringtail possums by 7–30 years after planting[5].

BEHAVIOUR (1 STUDY)

- **Usage (1 study):** A replicated, site comparison study in the USA[2] found that restored riparian forest areas were visited more by carnivores than were remnant forests when restored areas were newly established, but not subsequently, whilst restored areas were not visited more frequently by black-tailed deer.

Background

Restoring or creating forest and woodland may provide important habitat for forest-dependant mammal species, particularly in disturbed or fragmented landscapes. Trees grow slowly and therefore the effects of forest restoration may not be evident for decades or even longer after restoration begins. Care must therefore be taken when interpreting the results of these studies.

A replicated, site comparison study in 1999–2001 of riparian forest at a site in California, USA (1) found that mammal species richness in restored riparian forest was similar to that in natural riparian forest. Mammal species richness in restored sites did not differ from that in natural sites during any season of sampling (data not reported). There was also no significant difference in species richness of small mammals (rodents and shrews) between restored (2–3 species) and natural (3–5 species) sites. Restoration, which included planting of woody riparian species, commenced between 1996 and 1998. Small mammals were surveyed between December 1999 and February 2001, using 16 Sherman live traps/ha. Other mammals were caught in larger live traps (cross section 7.6 × 8.9 cm) between November 1999 and April 2001.

A replicated, site comparison study in 2010–2012 of 16 riparian forest sites in California, USA (2) found that restored riparian forest areas were visited more by carnivores than were remnant forests when restored areas were newly established, but not subsequently, whilst restored areas were not visited more frequently by black-tailed deer *Odocoileus hemionus columbianus*. More mammalian carnivore species were detected in young restored forests (3.4/plot) than in remnant forests (1.8/plot) but neither figure differed from that in old restored forests (2.1/plot). Coyotes *Canis latrans* made more visits to young restored forests than to remnant forests (data not presented). No differences were detected between visit rates to the three forest stages for raccoon *Procyon lotor*, bobcat *Felis rufus* or black-tailed deer. Five young restored forests (restored in 2003–2007), six old restored forests (restored in 1991–2001) and five natural forest remnants were sampled. Camera traps were operated over two consecutive years in December–March and May–July, starting in December 2010 and finishing in July 2012.

A site comparison study in 2013–2014 in a forest in Caldas department, Colombia (3) found that mammal species richness was similar in an area reforested with flooded gum *Eucalyptus grandis* compared to native forest, though there were differences in occurrence rates of individual species between forest types. Mammal species richness did not differ significantly between the reforested (9 species) and native forest (11 species) areas. Nine-banded armadillos *Dasypus novemcinctus* were recorded less frequently in the reforested site (10 records) than in native forest (30 records) as were South American coatis *Nasua nasua* (23 vs 48 records). Western mountain coatis *Nasuella olivacea* was recorded more frequently (43 records) in the reforested site than in native forest (10 records). There were no differences in the number of records of red-tailed squirrel *Sciurus granatensis* or dwarf red brocket *Mazama rufina* between forest types (data not reported). A 93-ha area, reforested in the 1960s, was compared with a 146-ha native forest block. Mammals were surveyed using four camera traps each in the two forest blocks, from September 2013 to February 2014.

A replicated study in 2010–2012 of 10 deciduous woodland sites in a protected area in central Italy (4) found that forest regrowing on previously cultivated and/or grazed land had a greater abundance of hazel dormice *Muscardinus avellanarius*, and they had greater survival rates, than in coppiced forest. Peak abundance was higher in regrowing forest plots (17 dormice/plot) than in recent coppice (0–1/plot) and old coppice (1–7/plot). Monthly survival probability in regrowing forest (0.75) was higher than in old coppice (0.43). Too few dormice were recorded in young coppice to calculate survival. Forest type did not affect average litter size (regrowing forest: 4.5 young/litter; old coppice: 4.8 young/litter; no litters found in new coppice). Hazel dormice were surveyed within a grid of 36 tree-mounted wooden nest boxes/plot. Two recently coppiced plots (1–5 years since coppicing), three old coppice plots (20–30 years since coppicing) and two regrowing plots (formerly cultivated and/or grazed areas, unmanaged for 20 years) were sampled.

A replicated, site comparison study in 2002–2011 of 137 forest sites in New South Wales, Australia (5) found that replanted forest supported few common brushtail possums *Trichosurus vulpecula* or common ringtail possums *Pseudocheirus peregrinus* by 7–30 years after planting. The probability of a replanted site holding brushtail possums when surveyed 7–30 years after planting (0.02) was lower than that in old growth forest (0.44). For ringtail possums, the probability of occupancy in replanted forest 7–30 years after planting (0.07) was also lower than that in old growth forest (0.75). Greater tree cover in the surrounding area did not increase the probability of subsequent colonisation for either species (result presented as model coefficient). Sixty-five replanted forests and 72 old growth forests were surveyed. Most replanted forests were 7–30 years old and comprised local and exotic Australian plant species. Old growth forests were ≥200 years old. Marsupials were surveyed by spotlight, whilst walking at an average 3 km/h, 1–5 hours after dusk. At each site a 200-m transect was surveyed for 20 min. Sites were surveyed in winter 2002, 2003, 2008, 2009 and 2011.

(1) Queheillalt D.M. & Morrison M.L. (2006) Vertebrate use of a restored riparian site: a case study on the central coast of California. *The Journal of Wildlife Management*, 70, 859–866, https://doi.org/10.2193/0022-541x(2006)70[859:vuoarr]2.0.co;2

(2) Derugin V.V., Silveira J.G., Golet G.H. & LeBuhn G. (2016) Response of medium-and large-sized terrestrial fauna to corridor restoration along the middle Sacramento River. *Restoration Ecology*, 24, 128–136, https://doi.org/10.1111/rec.12286

(3) Ramírez-Mejía A.F. & Sánchez F. (2016) Activity patterns and habitat use of mammals in an Andean forest and a Eucalyptus reforestation in Colombia. *Hystrix*, 27, 11319.

(4) Sozio G., Iannarilli F., Melcorea I., Boschetti M., Fipaldini D., Luciani M., Roviani D., Schiavano A. & Mortelliti A. (2016) Forest management affects individual and population parameters of the hazel dormouse *Muscardinus avellanarius*. *Mammalian Biology*, 81, 96–103, https://doi.org/10.1016/j. mambio.2014.12.006

(5) Lindenmayer D.B., Mortelliti A., Ikin K., Pierson J., Crane M., Michael D. & Okada S. (2017) The vacant planting: limited influence of habitat restoration on patch colonization patterns by arboreal marsupials in south-eastern Australia. *Animal Conservation*, 20, 294–304, https://doi.org/10.1111/acv.12316

13.11. Restore or create wetlands

https://www.conservationevidence.com/actions/2572

- **Four studies** evaluated the effects on mammals of restoring or creating wetlands. Three studies were in the USA[1,2,3] and one was in the UK[4].

COMMUNITY RESPONSE (2 STUDIES)

- **Community composition (1 study):** A site comparison study in the USA[2] found that the composition of mammal species present differed between a created and a natural wetland.

- **Richness/diversity (2 studies):** Two site comparison studies (one replicated) in the USA[2,3], found that mammal species richness did not differ between created and natural wetlands[2,3].

POPULATION RESPONSE (2 STUDIES)

- **Abundance (1 study):** A before-and-after study in the USA[1] found that following marshland restoration, muskrat abundance increased.

- **Survival (1 study):** A replicated, controlled, before-and-after study in the UK[4], found that water voles persisted better in wetlands that were partially restored using mechanical or manual methods than they did in wetlands undergoing complete mechanical restoration.

BEHAVIOUR (0 STUDIES)

Background

Wetland habitats are often drained or degraded during the development of agriculture or expansion of urban areas or other land uses. Restoration of these wetland habitats can help to increase local species richness and abundance of mammal species that depend on wetlands.

A before-and-after study from the 1960s to 1981 of a marshland alongside Lake Erie, Ohio, USA (1) found that marshland restoration was associated with increased numbers of muskrat *Ondatra zibethicus*. Population trends were not tested statistically. Four to five years after marsh restoration started, the average number of muskrat pelts collected in the annual harvest (3,657–5,583) was higher than four years prior to restoration (376). The number of pelts was similar to that 10 years prior to restoration, before the marshland was degraded by high water levels (3,681 pelts). Muskrat pelt prices did not significantly affect harvest size. Marsh was restored by reconstructing dikes to facilitate water level control. Muskrat harvest figures were obtained from trappers, who traditionally trapped the same areas each year. The harvest was not directly regulated.

A site comparison study in 1994–1995 of two forested wetlands in Maryland, USA (2) found that a created forested wetland had the same mammal species richness as a nearby natural site, but different species composition. No statistical analyses were performed. Four mammal species were recorded both on the created site and the natural site. Meadow vole *Microtus pennsylvanicus* was more abundant at the created site (0.17–0.58 individuals/trap/day) than at the natural site (0 individuals/trap/day). The same pattern was seen for House mouse *Mus musculus*, and domestic cat *Felis catus* (no data reported). White-footed mouse *Peromyscus leucopus* was less abundant at the created site (0–0.17 individuals/trap/night) than at the natural site (0.14–0.67 individuals/trap/night). Pine vole *Pitymys pinetorum*, gray squirrel *Sciurus carolinensis* and opossum *Didelphis virginiana* were found only in the natural site. Forest wetland (5.5 ha) was created on a former firing range. The site was graded in December 1993 and planted with native vegetation in spring and summer 1994. Mammals were live-trapped from November 1994 to March 1995 on the created site and adjacent natural forest wetland, using Sherman traps and larger box traps. Tracks were monitored in sand pits in summer 1995.

A replicated, site comparison study in 1999–2000 of 17 wetlands in South Dakota, USA (3) found that mammal species richness was similar in created, restored and enhanced wetlands compared to in natural wetlands. There was no significant difference in the average number of species found in created (2.7 species), restored (2.4 species) and enhanced wetlands (1.9 species) and in natural wetlands (1.4 species). Four created, four restored, four enhanced and five natural wetlands were sampled. Wetland creation involved either impounding a small stream or excavating a basin. Restoration included plugging drainage ditches or breaking sub-surface drainage tiles. Enhancement included manipulating water levels to increase wetland size or changing vegetation structure. Wetland creation, restoration and enhancement was carried out within the previous 10 years. Monitoring was undertaken in spring and autumn in 1999–2000. Sampling at each site included live-trapping (four transects, each with five traps spaced 5 m apart), complemented with pitfall traps and sightings.

A replicated, controlled, before-and-after study in 2008–2010 on a wetland near Peterborough, UK (4), found that partial pond restoration using mechanical or manual methods led to greater persistence of water voles *Arvicola amphibius* than did complete mechanical restoration. No statistical analyses were performed. After management, the number of pond visits (out of 12: four visits to each of three ponds) revealing water vole presence at partial manual restoration ponds (nine) and partial mechanical restoration ponds (nine) was greater than at full mechanical restoration ponds (two) and similar to that at unmanaged ponds (10). Before management, water voles were present at all ponds set to undergo restoration and at two of three unmanaged ponds. Pond restoration took place between October 2008 and January 2009, on a 126-ha site. Four ponds were restored by complete mechanical excavation of edge and bottom vegetation, four by mechanical clearance of 15 m of pond edge, four by manual clearance of 15 m of pond edge and four were unmanaged. Ponds were in three replicate clusters. Monitoring entailed searches for water vole feeding signs or latrines in autumn 2008 (pre-restoration) and in June, September and October 2009 and March 2010 (post-restoration).

(1) Kroll R.W. & Meeks R.L. (1985) Muskrat population recovery following habitat re-establishment near Southwestern Lake Erie. *Wildlife Society Bulletin*, 13, 483–486.

(2) Perry M.C., Sibrel C.B. & Gough G.A. (1996) Wetlands mitigation: partnership between an electric power company and a federal wildlife refuge. *Environmental Management*, 20, 933–939.

(3) Juni S. & Berry C.R. (2001) A biodiversity assessment of compensatory mitigation wetlands in eastern South Dakota. *Proceedings of the South Dakota Academy of Science*, 80, 185–200

(4) Furnborough P., Kirby P., Lambert S., Pankhurst T., Parker P. & Piec D. (2011) *The effectiveness and cost efficiency of different pond restoration techniques for bearded stonewort and other aquatic taxa*. Report on the Second Life for Ponds project at Hampton Nature Reserve in Peterborough, Cambridgeshire. The Froglife Trust, Peterborough, UK.

13.12. Manage wetland water levels for mammal species

https://www.conservationevidence.com/actions/2574

- **One study** evaluated the effects of managing wetland water levels for mammal species. This study was in the USA[1].

COMMUNITY RESPONSE (0 STUDIES)

POPULATION RESPONSE (1 STUDY)

- **Abundance (1 study):** A replicated, site comparison study in the USA[1] found that managing wetland water levels to be higher in winter increased the abundance of muskrat houses.

BEHAVIOUR (0 STUDIES)

Background

Some wetland mammal species may benefit from specific management of water levels. Water levels may affect factors such as predation rates, food availability and access to shelter. Management of wetland levels will affect a range of wetland species, so decisions regarding such management should be taken with regard to this full assemblage where possible.

A replicated, site comparison study in 2000–2006 at three wetland sites on the St Lawrence River, USA (1) found that managing wetland water levels to be higher in winter increased the abundance of muskrat *Ondatra zibethicus* houses. This result was not analysed for statistical significance. At wetlands where water levels were managed to be higher in winter, muskrat house density was higher (3.0 houses/ha) than in wetlands where water levels were not managed (0.7 houses/ha). At two wetland sites, in 2000–2004 and 2004–2006, water control structures were installed to increase water levels during winter. At a third site, no such structure was installed. Where water levels were not managed, they were lower during winter. Muskrat houses were counted at all sites in winters of 2001–2006, using unspecified methodologies.

(1) Toner J., Farrell J.M. & Mead J.V. (2010) Muskrat abundance responses to water level regulation within freshwater coastal wetlands. *Wetlands*, 30, 211–219, https://doi.org/10.1007/s13157-010-0034-x

13.13. Create or maintain corridors between habitat patches

https://www.conservationevidence.com/actions/2576

- **Four studies** evaluated the effects on mammals of creating or maintaining corridors between habitat patches. One study was in each of Canada[1], the USA[2], Norway[3] and the Czech Republic[4].

COMMUNITY RESPONSE (0 STUDIES)

POPULATION RESPONSE (0 STUDIES)

BEHAVIOUR (4 STUDIES)

- **Use (4 studies):** Four studies (three replicated) in Canada[1], the USA[2], Norway[3] and the Czech Republic[4] found that corridors between habitat patches were used by small mammals[1,2,3,4]. Additionally, North American deermice moved further through corridors with increased corridor width and connectivity[2] and root voles moved further in corridors of intermediate width[3].

Background

Corridors are areas of natural habitat that are contiguous or isolated (i.e. linkages or stepping stones; Rouget *et al.* 2006). They may enable animals to disperse and migrate between intact habitat patches, which may increase their chances of survival. They may be particularly important in landscapes where there is relatively little remaining natural habitat.

Rouget M., Cowling R.M., Lombard A.T., Knight A.T. & Kerley G.I.H. (2006) Designing large-scale conservation corridors for pattern and process. *Conservation Biology*, 20, 549–561.

A replicated, site comparison study in 1989 of woodland blocks and connecting woodland and grassland corridors at a site in Ontario, Canada (1) found that wooded corridors were used by both resident and transient eastern chipmunks *Tamias striatus*. In total there were 530 captures of 119 chipmunks (68 males, 51 females). Chipmunks were resident (caught in >1 trapping session) in all four woods and were trapped in 14 of the 18 corridors. They were trapped in all 13 corridors that were characterised by mature trees. Just one was caught among the five grass-dominated corridors that largely lacked trees or shrubs. Chipmunks were live-trapped in four woods and 18 corridors across 220 ha of farmland (mostly pasture and crops). Corridors were field margins alongside fences with vegetation ranging from long grass, through shrubs to mature woodland trees. Four trapping sessions were conducted in May–September 1989. Each session comprised four consecutive days trapping in woods and, the following week, four consecutive days trapping in corridors.

A randomized, replicated study in 1992 of woodland corridors in a national park in Wyoming, USA (2) found that increased corridor continuity and greater corridor width increased movements of North American deermice *Peromyscus maniculatus*. Travel along corridors by deermice was greater in continuous corridors than those with gaps and was greater in wide than narrow corridors. However, vegetation characteristics (tree density, ground cover and fallen log density) were more important in determining deermouse movements (results presented as statistical model). Twelve corridors were studied, these being linear stands of aspen *Populus tremuloides*, surrounded by sagebrush *Artemesia* sp. Three corridors were wide (20–27 m) with a 10-m gap partway along, three were wide and continuous, three were narrow (10–16 m) with a 10-m gap and three were narrow and continuous. Deermice were monitored by live-trapping over 10 days, in May–July 1992, at each side of gaps and equivalent spacing in continuous corridors.

A replicated study in 1992 of a grassland in southeast Norway (3) found that root voles *Microtus oeconomus* used habitat corridors, but moved further in intermediate-width than in narrow or wide corridors. In intermediate (1-m-wide) corridors, voles moved an average of 205 m along the corridor in 12 hours. In narrow (0.4-m-wide) corridors, average movement was 35 m and, in wide (3 m-wide) corridors, was 75 m. Two 5 × 5-m habitat patches were connected by a 310 m-long corridor. Patches and corridor comprised dense, homogeneous meadow vegetation. Adult male voles were released, one in each habitat patch, at 08:00 h and the trial was terminated at 18:00 h. Fieldwork spanned August–October 1992, starting with the wide corridor. Corridor width was then reduced by mowing and herbicide use. Vole movements were monitored by radio tracking and footprint plates.

A site comparison study in 1992–1996 in an agricultural landscape in Moravia, Czech Republic (4) found that corridors created between habitat patches were used by eight small mammal species. Eight small mammal species were recorded in the corridor, five of which were also present in a nearby native woodland. In 1991, native trees and shrubs were planted in agricultural fields to create a 10-m-wide corridor. To survey small mammal populations in the corridor, 100 snap-traps were placed at 3-m intervals, and 50 snap-traps were placed in a nearby forest. Each trap was baited with a wick soaked in fat and left for three nights. Traps were set twice each year, in spring and autumn, in 1992–1996, apart from in 1994, when sampling was also carried out in summer.

(1) Bennett A.F., Henein K. & Merriam G. (1994) Corridor use and the elements of corridor quality: chipmunks and fencerows in a farmland mosaic. *Biological Conservation*, 68, 155–165.

(2) Ruefenacht B. & Knight R.L. (1995) Influences of corridor continuity and width on survival and movement of deermice. *Biological Conservation*, 71, 269–274.

(3) Andreassen H.P., Halle S. & Ims R.A. (1996) Optimal width of movement corridors for root voles: not too narrow and not too wide. *Journal of Applied Ecology*, 33, 63–70.

(4) Bryja J. & Zukal J. (2000) Small mammal communities in newly planted biocorridors and their surroundings in southern Moravia (Czech Republic). *Folia Zoologica-Praha*, 49, 191–197.

13.14. Apply fertilizer to vegetation to increase food availability

https://www.conservationevidence.com/actions/2577

- **Two studies** evaluated the effects on mammals of applying fertilizer to vegetation to increase food availability. One study was in Canada[1] and one was in the USA[2].

COMMUNITY RESPONSE (0 STUDIES)

POPULATION RESPONSE (0 STUDIES)

BEHAVIOUR (2 STUDIES)

- **Use (2 studies):** Two replicated, controlled studies, in Canada[1] and the USA[2], found that applying fertilizer increased the use of vegetation by pronghorns[1] and Rocky Mountain elk[2].

Background

Adding fertilizer to a habitat often increases the growth of plants. As a result this could potentially increase the amount of food available to herbivorous mammals.

A replicated, controlled study in 1977 on a sagebrush grassland site in Alberta, Canada (1) found that fertilizing sagebrush increased its usage by pronghorns *Antilocapra americana*. There were 21% more pronghorn faecal pellets on fertilized sagebrush than on unfertilized sagebrush (counts not presented). The proportion of sagebrush leaders browsed by proghorns in fertilized plots (34%) was higher than in unfertilized plots (18%). Twenty-two pronghorns were retained in a 256-ha enclosure from April 1975 to November 1977. Twelve plots, each 6 × 15 m, were fertilized, with 84–252 kg N/ha and 39–118 kg P/ha, on 29 April 1975. For each plot, two unfertilized control plots were established. In November 1977, pronghorn use of plots was assessed by faecal pellet counts and by assessing the proportion of sagebrush leaders that was browsed.

A randomized, replicated, controlled study in 1971–1974 of a grassland in Washington, USA (2) found that applying fertilizer increased overwintering numbers of Rocky Mountain elk *Cervus canadensis nelsoni* the following winter, but not in subsequent winters. After one year, elk use was higher in fertilized areas (82 elk days/ha) than in unfertilized areas (55 elk days/ha). There was no difference in use by elk in the second (fertilized: 79; unfertilized: 90 elk days/ha) or third winters (fertilized: 45; unfertilized: 42 elk days/ha) following fertilizer application. Within each of six plots, one subplot was randomly assigned for fertilizer application and one was unfertilized. Subplots measured 3 ha. Fertilizer was applied once, in autumn 1971, at 56 kg N/ha. Elk pellets were counted in spring, to assess use of plots in the winters of 1971–1972, 1972–1973 and 1973–1974.

(1) Barrett M.W. (1979) Evaluation of fertilizer on pronghorn winter range in Alberta. *Journal of Range Management*, 32, 55–59.

(2) Skovlin J.M., Edgerton P.J. & McConnell B.R. (1983) Elk use of winter range as affected by cattle grazing, fertilizing, and burning in Southeastern Washington. *Journal of Range Management*, 36, 184–189.

13.15. Provide artificial refuges/breeding sites

https://www.conservationevidence.com/actions/2583

- **Eight studies** evaluated the effects on mammals of providing artificial refuges/breeding sites. Two studies were in each of the USA[3,8], Spain[4,5] and Portugal[6,7] and one was in each of Argentina[1] and Australia[2].

COMMUNITY RESPONSE (0 STUDIES)

POPULATION RESPONSE (4 STUDIES)

- **Abundance (3 studies):** Two studies (one controlled), in Spain[4] and Portugal[7], found that artificial warrens increased European rabbit abundance. A replicated, randomized, controlled, before-and-after study in Argentina[1] found that artificial refuges did not increase abundances of small vesper mice or Azara's grass mice.

- **Survival (1 study):** A study in USA[3] found that artificial escape dens increased swift fox survival rates.

BEHAVIOUR (4 STUDIES)

- **Use (4 studies):** Four studies (two replicated), in Australia[2], Spain[5], Portugal[6] and the USA[8], found that artificial refuges, warrens or nest structures were used by fat-tailed dunnarts[2], European rabbits[5,6], and Key Largo woodrats and Key Largo cotton mice[8].

Background

Natural dens can reduce the vulnerability of animals to attack. Providing artificial dens and refuges may mimic natural dens, thereby reducing mortality as a result of predation. Refuges and dens may also provide protection from extreme weather conditions.

This intervention specifically covers situations where refuges or breeding sites are provided for existing wild mammal populations. For provision of refuges for translocated mammals, see *Species Management — Release translocated/captive-bred mammals into area with artificial refuges/breeding sites*. See also *Provide artificial dens or nest boxes on trees* for the specific intervention of providing boxes attached to trees.

A replicated, randomized, controlled, before-and-after study in 1995 in a sunflower field in Buenos Aires Province, Argentina (1) found that providing artificial refuges did not increase abundances of small vesper mice *Calomys laucha* or Azara's grass mice *Akodon azarae*. The number of small vesper mice one to two months after refuges were placed did not differ significantly between plots with (4) and without refuges (5–8), and had not differed before refuges were placed (refuge plots: 14; no refuges: 18). Similarly, the number of Azara's grass mice did not differ between plots with (9–30) and without refuges (5–20) one to two months after refuges were placed, and had not differed before they were placed (refuge plots: 37; no refuges: 34). In July 1995, 60 artificial shelters (12 cm long, 10 cm diameter tins with one entrance hole, provided with cottonwool and wrapped in paper and nylon bags) were half-buried at each of three randomly selected plots. Three other plots received no shelters. Mice were live-trapped for three consecutive nights in all six plots, one week before shelters were provided (late-July) and twice after (mid-August and early-September) using Sherman traps baited with peanut butter, laid 10 m apart in grids of 15 × 4 traps.

A study in 2000–2001 in a grassland and woodland reserve in Victoria, Australia (2) found that artificial log refuges were used by fat-tailed dunnarts *Sminthopsis crassicaudata*. Fat-tailed dunnarts were found beneath both recently placed (20 of 408 refuges) and old refuges (9 of 271 refuges) in grassland. However, introduced house mice *Mus musculus* were more often found beneath recently placed (10 of 408 refuges) than old refuges (1 of 271 refuges) in grassland. Fat-tailed dunnarts preferred Eucalyptus (34 of 447 refuges) to cypress-pine (9 of 684 refuges) posts, and preferred wider, more decayed posts with more holes (see paper for details). In May 2000, between 12 and 20 old white cypress-pine *Callitris glaucophylla* and Eucalyptus *Eucalyptus sp.* fence posts were placed in each of 91 quadrats (total 1,131 new refuges) throughout a 3,780-ha national park in grassland and woodland. Mammals were surveyed monthly, beneath both new refuges and beneath 271 old fence posts which had lain in the same grassland sites for more than 15 years. Surveys were conducted from June 2000 to January 2001 and between 08:00 h and 20:00 h.

A study in 2002–2004 in a grassland site in Texas, USA (3) found that artificial escape dens increased swift fox *Vulpes velox* survival rates. Average annual survival in plots with artificial escape dens (81%) was higher than in areas without such dens (52%). Six of 11 confirmed mortalities were due to predation by coyotes *Canis latrans*, three were of unknown causes, one died of natural causes and one was predated by a raptor. All mortalities were outside artificial den plots. Thirty-six artificial escape dens were installed 322 m apart in each of three 2.6-km² plots within a 100-km² study area. Two plots had established swift fox populations while the third did not. Each den was a covered, 4-m long, 20-cm diameter corrugated-plastic pipe with open ends. Fifty-five foxes were radio-collared and tracked, 2–4 times/week, for up to two years, between January 2002 and August 2004. Survival was estimated from 41 adult foxes (28 in artificial burrow plots, 13 in the study area but outside artificial burrow plots).

A controlled study in 2005–2007 in an open forest and scrubland site in Córdoba province, Spain (4) found that a plot with artificial warrens, water provision and fencing to excluding ungulate herbivores had more European rabbits *Oryctolagus cuniculus* than did a plot without these interventions. The three interventions were all carried out in the same plot, so their relative effects could not be determined. Average rabbit pellet counts were higher in the plot where the interventions were deployed (first year: 0.33 pellets/m²/day; second year: 1.08 pellets/m²/day) than in the plot without these interventions (first year: 0.02 pellets/m²/day; second year: 0.03 pellets/m²/day). A 2-ha plot was fenced to exclude ungulates in March 2005. Rabbits and predators could pass through the fence. Five artificial warrens were installed and water was provided at one place. No interventions were deployed in a second, otherwise similar, plot. Rabbit density was determined by monthly counts of pellets, from March 2005 to March 2007, in 0.5-m² circles, every 100 m, along a 1-km transect in each plot.

A replicated, site comparison study in 2007 of pasture and scrubland on 14 estates in central Spain (5) found higher usage of artificial warrens where rabbit *Oryctolagus cuniculus* abundance was highest and that occupancy of tube warrens was higher than of stone warrens or pallet warrens. In grid squares where artificial warrens were used by rabbits, more rabbit latrines were found (13.5 latrines/km) than in squares where artificial warrens were not used (3.2 latrines/km). Authors report that it is unclear if artificial warrens boosted populations or if warren usage reflected pre-existing population levels. Occupancy of tube warrens (67% occupied) was greater than of stone or pallet warrens (54% occupied). Tube warrens (120 installed) comprised a labyrinth of concrete tubes 1 m underground. Stone warrens (207) were c.5 m diameter, with stones arranged to leave galleries and holes. Pallet warrens (198) were at least four wooden pallets, covered with soil. Rabbit latrines were surveyed along fixed routes within 98 squares in a 500 × 500 m grid, spread across 14 estates, in February–March 2007.

A replicated study in 2007–2009 in six agroforestry sites in Alentejo and Algarve, Portugal (6) found that European rabbits *Oryctolagus cuniculus* used most available artificial shelters. European rabbits used 65 out of 100 artificial shelters. Rabbit numbers were higher in areas where a higher percentage of artificial shelters were used (data presented as correlation). Between 2007 and 2009, a total of 100 artificial shelters were constructed across six agroforestry estates dominated by cork oak *Quercus suber*. Artificial shelters were clustered in groups of 6–8. Each shelter had six entrance points but no more details about shelters were provided. Shelters were surveyed once every three months during the first year after construction and once every six months thereafter. Shelters were considered in use if pellets were detected near their entrances. Rabbit relative abundance was assessed by the density of pellets within a 300-m radius around the shelter.

A study in 2007–2009 of a mixed woodland, scrub and agricultural area in southern Portugal (7) found that installing artificial warrens, along with other habitat management, increased presence and abundance of European rabbits *Oryctolagus cuniculus*. Rabbit presence and abundance were each higher within 100 m of artificial warrens than at greater distances (data reported as statistical model results). Rabbit numbers increased steadily through the study and artificial warrens achieved a 64% occupancy rate by 2009. A range of habitat management actions for rabbits was carried out from 2006 to 2009. These comprised managing scrubland, creating pastures and building 28 artificial warrens (constructed from wood pallets and vegetation remains, covered with soil). Rabbit presence and relative abundance were determined through latrine counts in 45 plots, located around two areas of rabbit activity. Counts were carried out in most months from July 2007 to June 2009.

A study in 2004–2013 in a forest reserve in Florida, USA (8) found that Key Largo woodrats *Neotoma floridana smalli* and Key Largo cotton mice *Peromyscus gossypinus allapaticola* used artificial nest structures. Out of 284 artificial nests, Key Largo woodrats were detected at 65 (23%) and Key Largo cotton mice at 175 (62%). Between 2004 and

2013, over 760 artificial nest structures for woodrats and cotton mice were built in the Crocodile Lake National Wildlife Refuge. Artificial nest structures ranged from boulders and rubble piles to recycled jet-ski structures, cinder blocks with PVC pipes, tin, and natural materials, and 1–2 m segments of plastic culvert pipes cut in half longitudinally and covered in natural materials. In April–May 2013, two hundred and eighty-four artificial nests were monitored using camera traps. One camera trap was set 0.5–3.0 m away from each nest. Cameras recorded for 5–6 nights/nest.

(1) Hodara K., Busch M. & Kravetz F.O. (2000) Effects of shelter addition on *Akodon azarae* and *Calomys laucha* (Rodentia, Muridae) in agroecosystems of central Argentina during winter. *Mammalia*, 64, 295–306, https://doi.org/10.1515/mamm.2000.64.3.295

(2) Michael D.R., Lunt I.D. & Robinson W.A. (2004) Enhancing fauna habitat in grazed native grasslands and woodlands: use of artificially placed log refuges by fauna. *Wildlife Research*, 31, 65–71, https://doi.org/10.1071/wr02106

(3) McGee B.K., Ballard W.B., Nicholson K.L., Cypher B.L., Lemons P.R. & Kamler J.F. (2006) Effects of artificial escape dens on swift fox populations in Northwest Texas. *Wildlife Society Bulletin*, 34, 821–827, https://doi.org/10.2193/0091-7648(2006)34[821:eoaedo]2.0.co;2

(4) Catalán I., Rodríguez-Hidalgo P. & Tortosa F.S. (2008) Is habitat management an effective tool for wild rabbit (*Oryctolagus cuniculus*) population reinforcement? *European Journal of Wildlife Research*, 54, 449–453, https://doi.org/10.1007/s10344-007-0169-0

(5) Fernández-Olalla M., Martínez-Jauregui M., Guil F. & San Miguel-Ayanz A. (2010) Provision of artificial warrens as a means to enhance native wild rabbit populations: what type of warren and where should they be sited? *European Journal of Wildlife Research*, 56, 829–837, https://doi.org/10.1007/s10344-010-0377-x

(6) Loureiro F., Martins A.R., Santos E., Lecoq M., Emauz A., Pedroso N.M. & Hotham P. (2011) O papel do programa lince (lpn/ffi) na recuperação do habitat e presas do lince-ibérico no sul de portugal. *Galemys*, 23:17–25.

(7) Godinho S., Mestre F., Ferreira J.P., Machado R. & Santos P. (2013) Effectiveness of habitat management in the recovery of low-density populations of wild rabbit. *European Journal of Wildlife Research*, 59, 847–858, https://doi.org/10.1007/s10344-013-0738-3

(8) Cove M.V., Simons T.R., Gardner B., Maurer A.S. & O'Connell A.F. (2017) Evaluating nest supplementation as a recovery strategy for the endangered rodents of the Florida Keys. *Restoration Ecology*, 25, 253–260, https://doi.org/10.1111/rec.12418

13.16. Provide artificial dens or nest boxes on trees

https://www.conservationevidence.com/actions/2584

- **Thirty studies** evaluated the effects on mammals of providing artificial dens or nest boxes on trees. Fourteen studies were in Australia[8,9,12,13,15,16,18,19,21,22,24,27,29,30], nine were in the USA[1–7,14,25], three were in the UK[10,11,28], one was in each of Canada[17], Lithuania[20], South Africa[23] and Japan[26].

COMMUNITY RESPONSE (0 STUDIES)

POPULATION RESPONSE (6 STUDIES)

- **Abundance (5 studies):** Three of five controlled studies (three also replicated) in the USA[2,14], the UK[10], Canada[17] and Lithuania[20], found that provision of artificial dens or nest boxes increased abundances of gray squirrels[2] and common dormice[10,20]. The other two studies found that northern flying squirrel[14,17] and Douglas squirrel[17] abundances did not increase.

- **Condition (1 study):** A replicated, randomized, paired sites, controlled, before-and-after study in Canada[17] found that nest boxes provision did not increase body masses of northern flying squirrel or Douglas squirrel.

BEHAVIOUR (27 STUDIES)

- **Use (27 studies):** Twenty-seven studies, in Australia[8,9,12,13,15,16,18,19,21,22,24,27,29,30], the USA[1,3–7,14,25], the UK[11,28], Canada[17], South Africa[23] and Japan[26] found that artificial dens or nest boxes were used by a range of mammal species for roosting and breeding.

Background

Some mammals use cavities in trees for denning, roosting or breeding. Woodland management for timber extraction may disproportionately remove trees that are sufficiently mature to have developed such cavities. Nest boxes, usually made of wood and attached to tree trunks, may provide an environment that mimics natural tree cavities and is adopted by such mammals. This intervention includes creation of artificial cavities within the tree, by excavating a quantity of wood and replacing a front plate with a constricted opening. This intervention specifically includes artificial dens or nest boxes in or on trees. For provision of structures in other situations, see *Provide artificial refuges/breeding sites*.

A study in 1940–1947 in a forest site in Michigan, USA (1) found that artificial dens were used by raccoons *Procyon lotor*. Over the four years that 15 dens were monitored, 2–13 of them showed signs of being occupied by racoons. Fifteen dens were made of wood and measured 36 × 36 × 31 cm, with entrances measuring 10 × 15 cm. Dens were attached to trees in July 1940, at 7.5–12 m high. They were inspected for signs of racoon use in August, October, and November 1940, May 1941, June 1946, and June 1947.

A replicated, controlled, before-and-after study in 1963–1965 of a forest in Maryland, USA (2) found that areas with artificial dens had more gray squirrels *Sciurus carolinensis* than did areas without dens. No statistical analyses were performed. There were more gray squirrels after dens were installed (1.0–1.8 squirrels/acre) than before installation (0.6–0.9 squirrels/acre). Numbers were stable through this period in plots where dens were not installed (0.8–0.9 squirrels/acre over two years in one plot and 1.0–1.2 squirrels/acre over three years in another). Squirrels were surveyed by live-trapping in five woodland plots (9.5–26 acres extent) in January–February. Three plots were sampled in 1963 and all five in 1964 and 1965. Artificial dens (one den/1.25 acres) were attached to trees in one plot after surveys in 1963 and in two plots after surveys in 1964. Dens comprised half a car tyre, folded and fastened into a kidney-shaped box, with an entrance at the top.

A study in 1974–1977 in a forest plantation site in Utah, USA (3) found that nest boxes were used by Abert's squirrel *Sciurus aberti* and red squirrel *Tamiasciurus hudsonicus*. After three years all 12 nest boxes installed were used by Abert's squirrels. Additionally, a red squirrel was detected in one next box, one year after installation. In May 1974, twelve nest boxes (30 × 30 × 40 cm) were placed in a forest area. Boxes were secured 7.6–14 m high, to ponderosa pine *Pinus ponderosa*, and were checked periodically for signs of use until October 1977.

A replicated study in 1973–1975 of two stands of young hardwood trees in Ohio and Illinois, USA (4) found that nest boxes were used by gray squirrels *Sciurus carolinensis* at one site and by flying squirrels *Glaucomys volans* at both sites. At a 21–23-year-old forest stand, gray squirrels did not make active use of any of 10 boxes but flying squirrels occupied 7–10 boxes over six inspections. At a 32–36-year-old forest stand, gray squirrels occupied 7–18 boxes across five inspections and flying squirrels occupied 2–6 boxes. Ten boxes were installed in autumn 1973 in the 21–23-year-old stand, which covered 1.9 ha. They were inspected six times from April 1974 to November 1975. Twenty boxes were installed in April 1973 in the 32–36-year-old stand, which covered 4 ha. They were inspected five times from August 1973 to March 1975.

A study in 1977–1979 in three riverine forest sites in Louisiana and Mississippi, USA (5) found that nest boxes were used by Virginia opossums *Didelphis virginiana*, southern flying squirrels *Glaucomys volans*, fox squirrels *Sciurus niger*, gray squirrels *Sciurus carolinensis*, golden mice *Ochrotomys nuttalli* and eastern woodrats *Neotoma floridana*. Virginia opossums, southern flying squirrels and fox squirrels were more frequently detected in nest boxes than in natural cavities (opossums: 1.2% vs 0.2 of inspections; flying squirrels: 2.1% vs 0.2; fox squirrels: 0.7% vs <0.1%). Gray squirrels were detected with more similar frequencies in nest boxes (1.6 % of inspections) and natural cavities (1.1%). These comparisons were not subjected to statistical tests. Golden mice and eastern woodrats used next boxes rarely (<0.05% of box inspections). Boxes were erected in hardwood and hardwood/pine forests and were of three sizes: large (60 x 30 x 30 cm, 13 cm diameter entrance), medium (45 x 20 x 20 cm, 7.5 cm diameter entrance) and small (30 x 15 x 15 cm, 5 x 7 rectangle entrance). Fifty boxes were installed at two sites and 90 at the other. All boxes had 5–10 cm of pine shavings in the bottom.

Boxes and natural cavities were inspected every month from April 1977 to February 1979.

A study in 1977–1980 in a range of agricultural, woodland and suburban areas across two counties in Tennessee, USA (6) found that nest boxes were used by gray squirrels *Sciurus carolinensis*, southern flying squirrels *Glaucomys volans* and occasionally opossums *Didelphis virginianus*. Over three years, gray squirrels were detected in 4–34% of boxes in agricultural sites, 0–19% in woodland and 12–49% in suburban areas. Southern flying squirrels were detected in 0–6% of boxes in agricultural sites, 0–26% in woodland and 0–9% in suburban areas. Opossums were detected only in 2% of boxes in suburban sites during the winter of one year. In 1977, one hundred and fifty wooden nest boxes were erected. Fifty were installed across an unstated number of agricultural sites (at a density of 1 box/1.4 ha), fifty were installed across three woodland sites (1 box/2.0 ha) and fifty were installed across three suburban areas around one city (1 box/2.5 ha). Boxes were 48 cm high, had a 7.6-cm diameter entrance hole and were nailed 4.6–6.1 m high on trees. They were inspected during March-June (spring) and December-February (winter) from 1978 and 1980.

A study in 1979 in a forest in Maryland, USA (7) found that artificial den cavities were used by southern flying squirrels *Glaucomys volans* and white-footed mice *Peromyscus leucopus*. Within 12 months, 84% of artificial cavities had been used by rodents or birds (data provided for both groups combined). Southern flying squirrels nested in the 40 artificial cavities six times and white-footed mice once. In July–August 1979, forty artificial cavities were created in a forest dominated by chestnut oak *Quercus prinus*. Cavities were created in 37 oaks, two pitch pines *Pinus rigida* and one white ash *Fraxinus americana*. Trees averaged 28 cm diameter at breast height. Cavities were 1.5 m above ground, were 15 × 13 cm across and 15 cm deep. The slab of wood initially removed from the tree surface was reattached across the front of the cavity with a 3.8-cm-diameter entrance hole.

A replicated study in 1977–1980 in two forest sites in Victoria, Australia (8) found that nest boxes were used by brown antechinus *Antechinus stuartii*, bobucks *Trichosurus caninus*, feathertail gliders *Acrobates pygmaeus*, sugar gliders *Petaurus breviceps* and greater gliders *Petauroides volans*. Out of the total of 240 nest boxes across the two sites, brown antechinus used 13 (5%), bobucks used seven (3%), feathertail gliders used 20 (8%), sugar gliders used 16 (7%) and greater gliders used one (<1%). Preference for diameter of entrance hole and height of box was significant for brown antechinus (tended to use 5 cm hole; avoided 8 m height) and sugar glider (tended to use 5 cm hole; selected 8 m height), but no other mammal species. In July 1977, 120 nest boxes were installed in each of two 4-ha forest sites dominated by eucalyptus. Sites were located 6.5 km apart. Boxes were made of 13-mm wide wood, were 22 × 31 cm across and 45 cm high. Entrance hole sizes were 5, 8, 12 or 15-cm in diameter and boxes were attached at heights of 1.5, 4 or 8 m on tree trunks. Nest boxes were installed 20 m apart. Each contained a 50-mm layer of wood shavings. They were inspected fortnightly, for six months after installation and then approximately monthly until January 1980.

A replicated study in 1982–1984 in woodland at four sites in Western Australia, Australia (9) found that nest boxes were used by mardos *Antechinus flavipes*. Within a 16-year-old regenerating block, all 36 boxes were used at least once, with 2–34 boxes being used across the 18 inspections. Single visits also revealed use of 7/34 boxes in virgin forest and 5/34 in streamside trees, but 0/34 were used in a 50-year-old regenerating block. Thirty-six nest boxes (internal volumes of 0.003–0.017 m³) were erected in each of four areas in June 1982. The 16-year-old block was 47-ha of regenerating karri forest. This was clearfelled in 1966 and prescribed burned in 1967. Boxes were fixed 3–5.5 m up trees. Further sites were virgin forest, retained streamside trees within a four-year-old regenerating block and a 50-year-old regenerating block. Boxes at these sites were set at 4.5–6.5 m height. Boxes were checked in the 16-year-old block monthly, from September 1982 to August 1983, then six further times to May 1984. Boxes at other sites were checked once, in May 1983.

A controlled study in 1986 in a woodland in Somerset, UK (10) found that nest boxes increased dormouse *Muscardinus avellanarius* abundance after 2–3 months. In woodland plots with nest boxes, more dormice were caught (8–11 dormice/plot) than in plots without nest boxes (3–6 dormice/plot). Within a 4-ha woodland, nest boxes were installed in two plots (0.8 and 1.2 ha), and two similar plots did not have nest boxes installed. Boxes, had internal dimensions of 115 ×130 ×

120 mm and a 35-mm entrance hole. They were installed in May 1986, with the hole facing the tree, at a density of c.30 boxes/ha. Relative dormouse abundance in each plot was determined from live-trapping over 10 nights, simultaneously in box and non-box plots, in both July and August 1986.

A study in 1994–1997 in a coniferous forest in Lancashire, UK (11) found that red squirrels *Sciurus vulgaris* used all and bred in some nest boxes. Red squirrels used all boxes within the first three months of placement and used 16–26% of boxes for breeding each year. There was no significant difference in the use of large (18 boxes) and small nest boxes (10 boxes) by breeding females, or in the size of litters in large (2.7 young) and small (2.9) boxes. All age groups and both sexes used boxes. The study site was dominated by Scots pine *Pinus sylvestris* and Corsican pine *Pinus nigra* and contained a high density of red squirrels (3.5–4/ha in the spring). Three groups of five small (27 × 30.5 × 48 cm) and five large (32 × 35.5 × 56 cm) timber nest boxes were attached to pine trees a height of 5–8 m in February 1994. Boxes were 50 m apart and filled with hay. In 1995, eight additional large boxes were added. Boxes were waterproofed and had a 7.5-cm-diameter entrance. Boxes were checked monthly from summer 1994 to summer 1997.

A study in 1994–1996 in a forest site Victoria, Australia (12) found that nest boxes were used by feathertail gliders *Acrobates pygmaeus* and agile antechinus *Antechinus agilis*. Out of 40 nest boxes, feathertail gliders used nine (23%) and agile antechinus used one or two (3–5%). In total, 57 individual feathertail gliders and two agile antechinus used boxes. In January 1994, forty nest boxes were installed in a 7-ha forest area dominated by eucalyptus. Boxes were 50 m apart, had a 15-mm-wide slit as the entrance and were attached to tree trunks at approximately 4.5 m above ground. Nest boxes were checked approximately every two months, between July 1995 and May 1997. Inspections took place during daylight hours and all animals encountered were captured, individually marked and returned to the box.

A study in 1990–1993 in a rainforest in New South Wales, Australia (13) found that nest boxes were used by eastern pygmy-possums *Cercartetus nanus*. Over the first 16 months, the average monthly capture rate of eastern pygmy-possums was 33.5/100 nest box checks. Twenty-one months after the study commenced, part of the area was cleared and the average monthly capture rate dropped to 7.8/100 nest box checks. Ninety-eight individual pygmy-possums were caught in boxes over the study. The study was conducted in a 4-ha early regrowth rainforest plot at 1,200 m altitude. Between 28 and 55 nest boxes (the quantity changing through the study) were attached to tree trunks, 1.5–2.0 m above ground and 10–20 m apart. Boxes were made from 18-mm-wide pine wood, and were 17 × 17 cm and 25 cm tall, with a 1.5-cm-wide opening across the front under the lid. In February 1992, 1.4 ha of the study area was cleared by bulldozing and burning. Boxes were checked at least monthly, between June 1990 and December 1992, and in April 1993.

A replicated, randomized, controlled study in 1992–1998 in a forest in Washington, USA (14) found that artificial breeding sites were used by northern flying squirrels *Glaucomys sabrinus* but did not increase their abundance. Average northern flying squirrel abundance in sites with artificial dens (0.51–0.80 squirrels/ha) was not significantly higher than in sites without artificial dens (0.42–0.48 squirrels/ha). During 11 inspections of the 256 dens, a total of 349 northern flying squirrels, 201 Douglas' squirrels *Tamiasciurus douglasii* and 16 Townsend's chipmunk *Tamias townsendii* were detected. By the end of the study 74–80% of next boxes and 34–50% of artificial cavities were used. In 1992, 16 nest boxes (20 × 22 cm across and 22 cm tall, with a 3.8 × 3.8-cm entrance) and 16 artificial cavities (10 ×15 cm across and 18–33 cm tall with a 3.8 × 3.8 cm or 4.5-cm-diameter entrance) were added to eight of 16 Douglas-fir *Pseudotsuga menziesii* stands. Forest stands were 13 ha and located in four areas (≤4 km apart). Each area had two stands with supplementary dens and two stands without supplementary dens (each ≥ 80 m apart). Supplementary dens were 6 m high and were inspected once in summer and once in winter, from summer 1993 to summer 1998. Flying squirrels were trapped during 49,152 trap nights in 1997–1998, with two Tomahawk live traps at each of 64 samplings stations, in each stand.

A replicated study in 1996–2000 in three forest plantations and one native forest in Queensland, Australia (15) found that nest boxes were used by feathertail gliders *Acrobates pygmaeus*, sugar gliders *Petaurus breviceps*, squirrel gliders *Petaurus norfolcensis* and yellow-footed marsupial mice *Antechinus flavipes* at three of four sites. Between 0 and 40% of nest boxes were occupied at each check within each of the three plantations. No boxes were used in the native forest. Out of 96 boxes, feathertail gliders used 16 (17%), sugar gliders used 10 (10%), squirrel gliders used four (4%) and yellow-footed marsupial mice used one (1%). The study was conducted in three 2–18-year-old eucalyptus plantations (1.2–1.5 ha) and one native forest dominated by >30 year-old eucalyptus (1.8 ha). At each site, 24 boxes were attached to trees, 3 m or 6 m above ground and 2–25 m apart. Nest boxes (40 cm long, 20 cm wide, ≤18.5 cm deep) were made from laminated plywood and had a 15–20 mm wide slot at the bottom. Boxes were checked 5–9 times between April 1996 and November 2000.

A replicated study in 1998–2002 of two *Eucalyptus regnans*-forests in Victoria, Australia (16) found that nest boxes were used by four arboreal marsupial species, with large high boxes used more than smaller or lower boxes. No statistical analyses were performed. Leadbeater's possum *Gymnobelideus leadbeateri*, mountain brushtail possum *Trichosurus cunninghami*, common ringtail possum *Pseudocheirus peregrinus* and eastern pygmy possum *Cercartetus nanus* were recorded. There were 38 records of presence of these species in large high boxes, 16 in small high boxes, 10 in large low boxes and 18 in small low boxes. In each of two forests, 12 locations were selected. Each had four trees in a 20 × 20 m square. At each location, a large high, large low, small high and small low box was installed in October–November 1998, one on each tree. Large and small box volumes were 0.038 m^3 and 0.019 m^3 respectively. High and low boxes were set at 8 m and 3 m height respectively. Boxes were checked 10 times to January 2002. Mammal occupancy was determined by animal presence, or hairs left on sticky devices.

A replicated, randomized, paired sites, controlled, before-and-after study in 1996–1999 in three forest sites in British Columbia, Canada (17) found that nest boxes were used by northern flying squirrels *Glaucomys sabrinus* and Douglas squirrels *Tamiasciurus douglasii* but did not increase their abundance or body mass. Northern flying squirrels occupied 68–83% of boxes with Douglas squirrels occupying 0–29%. However, two years after boxes were erected, the abundance and body mass of northern flying squirrels did not differ significantly between plots with nest boxes (abundance: 9.8/ha; body mass: 134 g) and plots without nest boxes (abundance: 7.7/ha; body mass: 128 g). At the same time, the abundance and body mass of Douglas squirrels also did not differ significantly between plots with nest boxes (abundance: 15.1/ha; body mass: 198 g) and plots without nest boxes (abundance: 20.1/ha; body mass: 207 g). In February–March 1997, thirty nest boxes (12.8 × 13.6 × 15.5 cm), 100 m apart in a 5×6 grid and 5.5 m above ground, were mounted in each of three 30-ha plots. Three other 30-ha plots had no nest boxes. In each plot, squirrels were trapped every 5–6 weeks during the snow-free period, from June 1996 to March 1999, using 80 baited Tomahawk live traps, at 40-m intervals in an 8×10 grid.

A replicated study in 1993–1994 in 20 forest sites in Victoria, Australia (18) found that nest boxes were used by common brushtail possums *Trichosurus vulpecula* and common ringtail possums *Pseudocheirus peregrinus*. Over one year, common brushtail possums were detected in 43% (52) and common ringtail possums in 33% (40) of the available 120 nest boxes. The average occupancy rate of nest boxes per monthly survey was 9% for common brushtail possums and 10% for common ringtail possums. In July 2003, 120 nest boxes were installed in 20 randomly selected (from 44) forest fragments (<2 ha) within a 183-km² study area. Boxes were of two designs (12 or 25-mm-wide plywood; 30 × 30 x 27.5 or 30 cm high), had a 10-cm diameter entrance hole and were attached to tree trunks approximately 4 m above the ground. Nest boxes were installed 50 m apart, on either side of a 100-m transect crossing the centre of each fragment. Nest box monitoring commenced eight weeks after installation and each box was inspected monthly over one year.

A replicated study in 2002–2003 in four forest sites in New South Wales, Australia (19) found that nest boxes were used by eastern pygmy-possums *Cercartetus nanus* and brown antechinus *Antechinus stuartii*. Five individual pygmy-possums (three of which were encountered twice) at one site and five brown antechinus were detected over 264 nest box inspections. Additionally, nesting materials characteristic of pygmy-possums was detected in eight nest boxes at the one site and brown antechinus in 11 nest boxes across the sites. The study was conducted in four 1-ha sites within a 2,000-ha forest reserve. In July-November 2002, forty nest boxes were attached to tree trunks, 1–2 m above the ground. Boxes had a 15-mm-wide entry slot and were placed 10–20 m apart. Boxes were checked eight times, with visits in alternate months in 2002 and then monthly.

A controlled, before-and-after study in 1985–1989 and 2000–2003 in a forest site in Lithuania (20) found that after more nest boxes were provided, common dormouse *Muscardinus avellanarius* density approximately doubled. Dormouse density was higher when there were 16 boxes/ha (0.9–3.0 dormice/ha) than when there were 4 boxes/ha (0.3–1.5 dormice/ha). Dormouse density did not increase in an area where next box provision remained at 4 boxes/ha (after: 0.6–0.9 individuals/ha; before: 0.7–1.3 individuals/ha). The study was conducted in 60 ha of a 40–50-year-old forest. In 1985–1999 wooden nest boxes (12 × 12 × 24 cm) were installed in a 50 × 50 m grid (276 boxes, 4 boxes/ha). In 2001, eighty-five additional nest boxes were added to a 6.25-ha section of the forest to form a 25 × 25 m grid (increasing box density to 16 boxes/ha). Boxes were inspected twice each month from April until October in 1985–1989 and 2000–2003.

A replicated study in 2005–2007 in five eucalyptus plantation sites in New South Wales and Queensland, Australia (21) found nest boxes were used by five marsupial species with different frequencies, depending on box type. Feathertail gliders *Acrobates pygmaeus* used 15 of 45 available small rear-entry boxes, 10 large slit-entrance boxes and nine wedge-shaped boxes, but did not use any medium rear-entry boxes. Squirrel gliders *Petaurus norfolcensis* used 18 of 45 medium rear-entry boxes and three large slit-entry boxes. Yellow-footed antechinus *Antechinus flavipes* used two large slit-entry boxes and one medium rear-entry boxes. Brown antechinus *Antechinus stuartii* used three small rear-entry boxes and brush-tailed phascogales *Phascogale tapoatafa* used one large slit-entry box. Nest boxes were of four types, small rear-entry boxes (height×width×depth: 23×14×14 cm, 25-mm-diameter entrance), large slit-entrance boxes (48×28×18.5 cm, 1.5×15 cm entrance on the side), wedge-shaped boxes (19×16×12.5–5 cm, 2×16 cm entrance at the base) and medium rear-entry boxes (40×14.5×14 cm, 45-mm-diameter entrance). They were installed in February–March 2005 and March 2006, 3 m above ground, in 45 plots. Each plot had one of each box type (180 boxes in total). Boxes were surveyed five times over 22 months.

A study in 1993–2005 of restored sites within bauxite mined areas in the jarrah *Eucalyptus marginata* forest of Western Australia, Australia (22) found that nest boxes within restoration areas were used by western pygmy possums *Cercartetus concinnus*, mardo *Antechinus flavipes* and brush-tailed phascogale *Phascogale tapoatafa*. Western pygmy possum used nest boxes placed in 8–10-year-old restoration sites. Mardo and brush-tailed phascogale also used nest boxes and possibly bred in them (no further details provided). Mined areas were revegetated using various techniques. In 1993–1994, mammal nest boxes were placed in a range of sites. Control of non-native red foxes *Vulpes vulpes* was also carried out for several years from 1994. Nest box designs and monitoring protocols are not described.

A study in 2003–2007 in a forest reserve in Eastern Cape, South Africa (23) found that nest boxes were used by woodland dormice *Graphiurus murinus* and Mozambique thicket rats *Grammomys cometes*. Out of 70 nest boxes, at least 49 (70%) were occupied by dormice and seven (10%) by thicket rats. Dormouse nest box occupation was lowest during winter (3% of boxes) and peaked in spring (8%) and summer (9% of boxes). Over one year, at least 66 dormice used between one and 16 next boxes (average 4). More adult females (17) than adult males (11) used nest boxes, but they were used by similar numbers of adults (30) and juveniles (36). Between March 2003 and January 2006, seventy wooden nest boxes (11.5 × 13 × 12 cm) were erected across a 2.5-ha area. Boxes had a 3-cm-diameter entrance hole facing the tree trunk. Boxes were installed 1.1–2.4 m above the ground, in trees with an average trunk diameter at nest box height of 90 cm. Boxes were monitored 57 times (average 4.4 times/month) between June 2006 and June 2007. Captured dormice were individually marked to determine recaptures.

A study in 2003–2006 of 16 woodland fragments in Queensland, Australia (24) found that 20% of nest boxes were used by squirrel gliders *Petaurus norfolcensis*. In total, 11 out of 56 nest boxes were occupied at least once by squirrel gliders, with presence detected 15 times out of 318 box visits. No squirrel gliders were found in boxes until ≥18 months after placement. Four of the boxes were occupied by five female gliders with young. In 16 woodland remnants (from <50 ha to >1,000 ha in extent), 56 nest boxes were erected in September–December 2003. Boxes were 40 cm high, 25 cm wide and 18 cm deep. They were installed ≥3 m above the ground. There were 2–6 boxes/site, with the number dependent on site size. Boxes were checked at six-month intervals from summer 2003 to summer 2006.

A study in 2008–2011 in a forest area in North Carolina, USA (25) found that nest boxes were used by northern flying squirrels *Glaucomys sabrinus*. Sixteen northern flying squirrels were caught at nest boxes. The study was conducted in a forest area dominated by eastern hemlock *Tsuga canadensis*. The number of nest boxes used was not detailed. Nest boxes measured 30 × 18 × 15 cm, had a 5 × 5-cm entrance, and were attached 3.6 m up the trunks of trees using nails and wire. They were monitored in winters of 2008 to 2011 and in spring 2009. Captured flying squirrels were individually tagged.

A study in 2004–2005 in a forest reserve in Nagano Prefecture, Japan (26) found nest boxes were used by Japanese dormouse *Glirulus japonicus*. Of 200 nest boxes, at least 127 (64%) were occupied by dormice. Thirty-nine individuals used the nest boxes (total 82 captures), 23 males and 16 females. The number of dormice captured in nest boxes peaked in August 2004 and June 2005 (14 captured/month) and October in both years (10–13). Pup-rearing was observed twice in nest boxes. The average diameter at breast height of trees with used nest boxes (33 cm) was smaller than unused boxes (51 cm). In early 2004, two hundred nest boxes were installed at equal distances across a 3.8-ha area of dense deciduous forest. Nest boxes were constructed from 12-mm-wide pinewood boards with a 35 x 35 mm square entrance at one side. Boxes were attached to trees with a diameter at breast height <40 cm, at a height of 1.0–1.2 m. Boxes were checked 2–4 times/month (total 76 times) between April 2004 and October 2005. Captured dormice were individually marked. Nest boxes were considered occupied when either dormice were present or when nesting materials were found.

A replicated study in 2003–2014 in one urban and two rural forest sites in New South Wales and Queensland, Australia (27) found that nest boxes were used by six species of arboreal marsupial. Within the rural landscapes nest boxes were occupied by sugar gliders *Petaurus breviceps* (29% of available boxes, use affected by design), brown antechinus *Antechinus stuartii* (23%, use unaffected by design), mountain brushtail possums *Trichosurus caninus* (1%) and feathertail gliders *Acrobates pygmaeus* (1%). Within an urban landscape, nest boxes were occupied by common brushtail possum *Trichosurus* sp. (20% of available boxes), common ringtail possum *Pseudocheirus peregrinus* (4%), and sugar gliders (4%). Use of some nest boxes influenced by design (see original paper for details). All boxes accessible to squirrel gliders *Petaurus norfolcensis* at two sites were used by them over a 10-year period (6–21 adults/year in boxes; total 61 individuals). Nest boxes of five different types (11–42 × 15–29 × 26–45 cm, 3.5–21-cm diameter entrance) were installed 3–6 m above ground. In the rural landscape, five boxes in each of 32 plots (25 x 25 m; ≥ 200 m apart) were installed across nine sites (>1 km apart). At the urban site a total of 188 boxes were installed across 20 sites. Boxes were erected in 2003–2007 and inspected three times in 2008–2009 at the rural sites and once in August 2010 at the urban site. In 2005–2009, 16 additional boxes were installed or adapted for squirrel gliders across two sites and were inspected usually once/year in 2005–2014.

A study in 2003–2016 in a coniferous forest plantation in Dumfries and Galloway, UK (28) found that pine martens *Martes martes* occupied and, in most years, bred in den boxes. Each year, 30–70% of available den boxes were occupied by pine martens. Martens used 5–20% of den boxes for breeding, in 10 of the 12 years monitored. The study was conducted in an 800-km² forest into which 12 martens were reintroduced in 1980–1981. Fifty den boxes (55 cm high, 51 cm wide, 24 cm deep) were fitted to trees at approximately 4 m high. Ten boxes were installed in 2003 and 40 in 2013. Boxes were made of wood, had two entrances and had 10 cm depth of softwood shavings inside the chamber. Boxes were checked for martens, signs of use by martens and marten kits, once/year in 2004–2016 (excluding 2013).

A study in 2010–2013 of planted and remnant woodland patches at 30 sites in New South Wales, Australia (29) found that nest boxes were used by five native and one non-native mammal species. Use of boxes was detected for yellow-footed antechinus *Antechinus flavipes* (two detections), sugar glider *Petaurus breviceps* (two detections), common brushtail possum *Trichosurus vulpecula* (52 detections), common ringtail possum *Pseudocheirus peregrinus* (eight detections) and lesser long-eared bat *Nyctophilus geoffroyi* (four detections). The introduced black rat *Rattus rattus* was also detected on 24 occasions. One each of five nest box designs was placed at 30 sites. Sites comprised seven connected woodland plantations, nine isolated woodland plantations (>70 m from native vegetation), eight connected remnant woodlands, and six isolated remnant woodlands (>70m from native vegetation). Boxes were erected in February 2010 and checked in October 2010,

December–January of 2010–2011, October 2011 and December–January of 2012–2013. Mammals were identified from live animals or from signs, such as faeces.

A study in 2010–2013 in a eucalypt forest in New South Wales, Australia (30) found that nest boxes were used by a range of native and non-native mammal species. Yellow-footed antechinus *Antechinus flavipes* were found in 12–14% of nest boxes, common brushtail possum *Trichosurus vulpecula* in 11–13%, and common ringtail possum *Pseudocheirus peregrinus* in 3–7%. Brush tailed phascogale *Phascogale tapoatafa*, squirrel glider *Petaurus norfolcensis*, and sugar glider *Petaurus breviceps* were all found in <1% of nest boxes. The non-native black rat *Rattus rattus* was found in 4–14% of boxes and the house mouse *Mus musculus* in 0–2% of boxes. On an unspecified date, 587 nest boxes were installed in a woodland. Animal presence, or signs of presence, were recorded during six surveys in 2010–2013.

(1) Stuewer F.W. (1948) Artificial dens for raccoons. *The Journal of Wildlife Management*, 12, 296–301.

(2) Burger G.V. (1969) Response of gray squirrels to nest boxes at Remington Farms, Maryland. *The Journal of Wildlife Management*, 33, 796–801.

(3) Pederson J.C. & Heggen A.W. (1978) Use of artificial nest boxes by Abert's squirrels. *The Southwestern Naturalist*, 23, 700–702.

(4) Nixon C.M. & Donohoe R.W. (1979) Squirrel nest boxes: are they effective in young hardwood stands? *Wildlife Society Bulletin*, 7, 283–284.

(5) McComb W.C. & Noble R.E. (1981) Nest-box and natural-cavity use in three mid-south forest habitats. *The Journal of Wildlife Management*, 93–101.

(6) Fowler L. J. & Dimmick R.W. (1983) Wildlife use of nest boxes in eastern Tennessee. *Wildlife Society Bulletin*, 11, 178–181.

(7) Gano R.D. & Mosher J.A. (1983) Artificial cavity construction: An alternative to nest boxes. *Wildlife Society Bulletin*, 11, 74–76.

(8) Menkhorst P.W. (1984) Use of nest boxes by forest vertebrates in Gippsland: acceptance, preference and demand. *Wildlife Research*, 11, 255–264.

(9) Wardell-Johnson G. (1986) Use of nest boxes by mardos, *Antechinus flavipes leucogaster*, in regenerating karri forest in South West Australia. *Australian Wildlife Research*, 13, 407–417.

(10) Morris P.A., Bright P.W. & Woods D. (1990) Use of nestboxes by the dormouse *Muscardinus avellanarius*. *Biological Conservation*, 51, 1–13.

(11) Shuttleworth C.M. (1999) The use of nest boxes by the red squirrel *Sciurus vulgaris* in a coniferous habitat. *Mammal Review*, 29, 61–66.

(12) Ward S.J. (2000) The efficacy of nestboxes versus spotlighting for detecting feathertail gliders. *Wildlife Research*, 27, 75–79, https://doi.org/10.1071/wr99018

(13) Bladon R.V., Dickman C.R. & Hume I.D. (2002) Effects of habitat fragmentation on the demography, movements and social organisation of the eastern pygmy-possum (*Cercartetus nanus*) in northern New South Wales. *Wildlife Research*, 29, 105–116, https://doi.org/10.1071/wr01024

(14) Carey A.B. (2002) Response of northern flying squirrels to supplementary dens. *Wildlife Society Bulletin*, 30, 547–556.

(15) Smith G.C. & Agnew G. (2002) The value of 'bat boxes' for attracting hollow-dependent fauna to farm forestry plantations in southeast Queensland. *Ecological Management & Restoration*, 3, 37–46, https://doi.org/10.1046/j.1442-8903.2002.00088.x

(16) Lindenmayer D.B., MacGregor C.I., Cunningham R.B., Incoll R.D., Crane M., Rawlins D. & Michael D.R. (2003) The use of nest boxes by arboreal marsupials in the forests of the Central Highlands of Victoria. *Wildlife Research*, 30, 259–264, https://doi.org/10.1071/wr02047

(17) Ransome D.B. & Sullivan T.P. (2004) Effects of food and den-site supplementation on populations of *Glaucomys sabrinus* and *Tamiasciurus douglasii*. *Journal of Mammalogy*, 85, 206–215, https://doi.org/10.1644/bos-118

(18) Harper M.J., McCarthy M.A. & van der Ree R. (2005) The use of nest boxes in urban natural vegetation remnants by vertebrate fauna. *Wildlife Research*, 32, 509–516, https://doi.org/10.1071/wr04106

(19) Harris J.M. & Goldingay R.L. (2005) Detection of the eastern pygmy-possum *Cercartetus nanus* (Marsupialia: Burramyidae) at Barren Grounds Nature Reserve, New South Wales. *Australian Mammalogy*, 27, 85–88, https://doi.org/10.1071/am05085

(20) Juškaitis R. (2005) The influence of high nestbox density on the common dormouse *Muscardinus avellanarius* population. *Acta Theriologica*, 50, 43–50, https://doi.org/10.1007/bf03192617

(21) Goldingay R.L., Grimson M.J. & Smith G.C. (2007) Do feathertail gliders show a preference for nest box design? *Wildlife Research*, 34, 484–490, https://doi.org/10.1071/wr06174

(22) Nichols O.G. & Grant C.D. (2007) Vertebrate fauna recolonization of restored bauxite mines key findings from almost 30 years of monitoring and research. *Restoration Ecology*, 15, S116–S126, https://doi.org/10.1111/j.1526-100x.2007.00299.x

(23) Madikiza Z.J., Bertolino S., Baxter R.M. & San E.D.L. (2010) Nest box use by woodland dormice (*Graphiurus murinus*): the influence of life cycle and nest

box placement. *European Journal of Wildlife Research*, 56, 735–743, https://doi.org/10.1007/s10344-010-0369-x

(24) Ball T., Goldingay R.L. & Wake J. (2011) Den trees, hollow-bearing trees and nest boxes: management of squirrel glider (*Petaurus norfolcensis*) nest sites in tropical Australian woodland. *Australian Mammalogy*, 33, 106–116, https://doi.org/10.1071/am10050

(25) Kelly C.A., Diggins C.A. and Lawrence A.J. (2013) Crossing structures reconnect federally endangered flying squirrel populations divided for 20 years by road barrier. *Wildlife Society Bulletin*, 37, 375–379, https://doi.org/10.1002/wsb.249

(26) Nakamura-Kojo Y., Kojo N., Ootsuka T., Minami M. & Tamate H.B. (2014) Influence of tree resources on nest box use by the Japanese dormouse *Glirulus japonicus*. *Mammal study*, 39, 17–26, https://doi.org/10.3106/041.039.0104

(27) Goldingay R.L., Rueegger N.N., Grimson M.J. & Taylor B.D. (2015) Specific nest box designs can improve habitat restoration for cavity-dependent arboreal mammals. *Restoration Ecology*, 23, 482–490, https://doi.org/10.1111/rec.12208

(28) Croose E., Birks J.D. & Martin J. (2016) Den boxes as a tool for pine marten *Martes martes* conservation and population monitoring in a commercial forest in Scotland. *Conservation Evidence*, 13, 57–61.

(29) Lindenmayer D., Crane M., Blanchard W., Okada S. & Montague-Drake R. (2016) Do nest boxes in restored woodlands promote the conservation of hollow-dependent fauna? *Restoration Ecology*, 24, 244–251, https://doi.org/10.1111/rec.12306

(30) Lindenmayer D.B., Crane M., Evans M.C., Maron M., Gibbons P., Bekessy S. & Blanchard W. (2017) The anatomy of a failed offset. *Biological conservation*, 210, 286–292, https://doi.org/10.1016/j.biocon.2017.04.022

13.17. Provide more small artificial breeding sites rather than fewer large sites

https://www.conservationevidence.com/actions/2595

- **One study** evaluated the effects on mammals of providing more small artificial breeding sites rather than fewer larger sites. This study was in Spain[1].

COMMUNITY RESPONSE (0 STUDIES)

POPULATION RESPONSE (1 STUDY)

- **Abundance (1 study):** A replicated, controlled study in Spain[1] found that smaller artificial warrens supported higher rabbit densities than did larger artificial warrens.

BEHAVIOUR (0 STUDIES)

Background

When providing artificial breeding sites for colonial mammals, there may be a trade-off between providing large sites, which may support larger, more-resilient populations at each site, or a greater number of small sites, which may increase the chance of at least some sites surviving threats such as predation or disease. The size of the overall population may also be influenced if the density of animals occupying these sites differs between different sized sites.

A replicated, controlled study in 2002–2005 of two grassland and scrubland plots at a site in Andalucia, Spain (1) found that providing smaller artificial warrens for wild rabbits *Oryctolagus cuniculus* supported higher rabbit densities than did larger artificial warrens. Rabbit density was higher in small artificial warrens (4–13 rabbits/12 m² plot) than it was in large artificial warrens (11–24 rabbits/48 m² plot). Two plots (4 ha each, 2 km apart) were fenced to exclude terrestrial predators. Each plot had 18 artificial warrens, comprising 12 small and six large warrens. Warrens were skeletons of wooden pallets covered by earth and branches. Large warrens (48 m²) were the size of four small warrens (12 m²). In autumn 2002, five rabbits were released into each small warren, and 20 rabbits were released into each large warren. Rabbits were surveyed by live-trapping, three times, from November 2004 to May 2005.

(1) Rouco C., Villafuerte R., Castro F. & Ferreras P. (2011) Effect of artificial warren size on a restocked European wild rabbit population. *Animal Conservation*, 14, 117–123, https://doi.org/10.1111/j.1469-1795.2010.00401.x

14. Species management

Background

Most of the chapters in this book are aimed at minimizing threats, but there are also some interventions which aim specifically to increase population numbers by increasing reproductive rates and by introducing individuals. This chapter describes interventions that can be used to increase population size by translocating wild mammals from one area to another, by breeding or rearing mammals in captivity (ex-situ conservation) to release back into the wild or by enhancing resources available for mammals in ways that can be used to address multiple threats (such as by providing artificial dens or nest boxes).

14.1. Cease/reduce payments to cull mammals

https://www.conservationevidence.com/actions/2349

- **One study** evaluated the effects of ceasing or reducing payments to cull mammals. This study was in Sweden and Norway[1].

COMMUNITY RESPONSE (0 STUDIES)

POPULATION RESPONSE (1 STUDY)

- **Survival (1 study):** A before-and-after study in Sweden and Norway[1] found that fewer brown bears were reported killed after the removal of financial hunting incentives.

BEHAVIOUR (0 STUDIES)

    https://doi.org/10.11647/OBP.0234.14

Background

Financial incentives for hunting particular species of mammal may be awarded for a variety of reasons, including agricultural protection, disease control and human safety. Whilst the intention of making such payments is to increase hunting of focal species, hunter motivations are varied (e.g. Gigliotti & Metcalf 2016) and may include more than financial reward. Hence, removal of payments may or may not have the desired consequence of reducing hunting pressure on species.

Gigliotti L.M. & Metcalf E.C. (2016) Motivations of female black hills deer hunters. *Human Dimensions of Wildlife*, 21, 371–378.

A before-and-after study in 1888–1898 in Sweden and a before-and-after study in 1925–1935 in Norway (1) found that after the removal of financial hunting incentives fewer brown bears *Ursus arctos* were reported killed. In both Sweden and Norway, fewer bears were reported killed during the five years after the removal of financial hunting incentives (Sweden: average 14 bears/county/year; Norway: average 1 bear/county/year) than during the five years before the removal of financial hunting incentives (Sweden: average 25 bears/county/year; Norway: average 3 bears/county/year). Financial incentives to cull bears were eliminated in 1893 in Sweden and in 1930 in Norway. Additionally, in 1930, bear hunting on someone else's property was banned in Norway. Numbers of bears killed were obtained from national harvesting records.

(1) Swenson J.E., Wabakken P., Sandegren F., Bjärvall A., Franzén R. & Söderberg A. (1995) The near extinction and recovery of brown bears in Scandinavia in relation to the bear management policies of Norway and Sweden. *Wildlife Biology*, 1, 11–25.

14.2. Temporarily hold females and offspring in fenced area to increase survival of young

https://www.conservationevidence.com/actions/2351

• We found no studies that evaluated the effects on mammals of temporarily holding females and offspring in a fenced area to increase survival of young.

'We found no studies' means that we have not yet found any studies that have directly evaluated this intervention during our systematic journal and report searches. Therefore, we have no evidence to indicate whether or not the intervention has any desirable or harmful effects.

Background

Survival of new-born mammals can be low, due to a variety of factors including predation. Capturing pregnant females and temporarily holding them and their new-born offspring in fenced areas within their native range (short-term or 'maternal penning'), for the first few weeks of life when young are most vulnerable to predation, may result in increased survival of young. This could help to slow decline, maintain or increase population size.

14.3. Rehabilitate injured, sick or weak mammals

https://www.conservationevidence.com/actions/2352

• **Thirteen studies** evaluated the effects of rehabilitating injured, sick or weak mammals. Four studies were in the UK[3,4,5,8], three were in Spain[6,9,13], two were in Argentina[10,12] and one each was in Uganda[1], Australia[2], the USA[7] and Brazil[11].

COMMUNITY RESPONSE (0 STUDIES)

POPULATION RESPONSE (12 STUDIES)

• **Survival (11 studies):** Five studies, in the UK[3,4,5,8] and Spain[9], found that varying proportions of European hedgehogs released after being rehabilitated in captivity survived during post-release monitoring periods, which ranged from two weeks[3] to 136 days[9]. Five studies, in Australia[2], Spain[6,13], the USA[7] and Brazil[11], found that four koalas[2], an Iberian lynx[6], a gray wolf[7], a puma[11] and two brown bears[13] released following rehabilitation in captivity survived for varying durations during monitoring periods, which ranged in length from three months[6] to up to seven years[13]. A study in Argentina[10] found that over half of released rehabilitated and captive-reared giant anteaters survived for at least six months.

- **Condition (2 studies):** A study in Uganda[1] found that a snare wound in a white rhinoceros healed after treatment and rehabilitation. A study in the UK[3] found that two of three rehabilitated European hedgehogs lost 12–36% of their body weight after release into the wild.

BEHAVIOUR (1 STUDY)

- **Behaviour change (1 study):** A controlled study in Argentina[12] found that released wild-born rehabilitated giant anteaters were more nocturnal in their activity patterns than captive-bred individuals.

Background

Mammals that are injured, sick or found in a weak condition are sometimes taken in by wildlife rehabilitators, to be treated and released back into the wild. Often, this is done more for animal welfare reasons than for species conservation though, for rare species, release of such animals may provide opportunities for choosing where to augment populations. The success of such programmes can be difficult to judge, without benchmark data for survival of wild-reared mammals. It is also important to note that the majority of studies summarised below have very small sample sizes, and that unsuccessful attempts are less likely to have been reported.

A study in 1965 in a grassland site in West Nile District, Uganda (1) found that after rehabilitation, a snare wound in a white rhinoceros *Ceratotherium simum simum* healed. One day after an operation to retrieve a deeply embedded snare from a leg, the adult female white rhinoceros was walking and grazing. Three weeks after the operation, the wound appeared nearly healed and, after six weeks, the rhinoceros was not limping anymore. Five months after the operation, the rhinoceros produced a calf. In July 1965, a white rhinoceros found limping due to a snare wound was immobilised and the snare was cut out with a hacksaw. The wound was swabbed with alcohol, smeared with intramammary penicillin and dusted with penicillin powder. A rough bandage was applied and, during the operation, the rhinoceros was injected with dimethylchlortetracycline.

A study in 1988–1989 in a woodland site in Queensland, Australia (2) found that four injured and rehabilitated koalas *Phascolarctos cinereus* each survived for between at least 20 days and four months after release. Two males moved 2.8 and 3.5 km and left the study area within one month. One settled 6 km from the release site (duration not stated). The other could not be relocated after last being recorded 1.4 km from the release site. Two females moved 0.9 and 1.3 km in 30 days. One female was recaptured after two months (suffering from disease). The other was recaptured after four months (due to collar-induced injuries). Four koalas, rehabilitated after minor road accident injuries, were released in September–November 1988 at adjacent localities (precise spacing not stated). Koalas were monitored daily by radio-tacking for 30 days after release, then twice weekly.

A study in 1989 in a forest and grassland site in Yorkshire, UK (3) found that three of four European hedgehogs *Erinaceus europaeus* that had been treated for injuries and released back into the wild survived over two weeks, but two of the three surviving hedgehogs lost weight. Three of four released hedgehogs survived for at least two weeks in the wild, built nests, and established home ranges (total area 6–17 ha). The other hedgehog (a male) died three days after release. After two weeks, two of the three surviving hedgehogs had lost significant body weight (12–36%). Two female and two male hedgehogs were released in June 1989 following treatment in captivity for injuries. Hedgehogs were radio-tracked for 15 nights after release and were located at least once every hour throughout the night until they nested. Hedgehogs were captured and weighed at release and every 1–2 nights throughout the study.

A study in 1991 in a farmland site in Suffolk, UK (4) found that over one third of rehabilitated European hedgehogs *Erinaceus europaeus* survived more than seven weeks after release into the wild. At least three out of eight (38%) rehabilitated hedgehogs survived over seven weeks post-release, though one then drowned and one was killed in a road accident. Contact was lost with four animals, but authors report that they were probably still alive at least five weeks after release. One hedgehog died due to illness within two weeks. Eight hedgehogs, rehabilitated after being found injured, ill or underweight, were released in a mosaic of pasture, hay meadow and arable land in July 1991. Animals were radio-tagged and followed nightly during the first three weeks post-release and sporadically until the eighth week post-release.

A study in 1993 in pasture on a farm in Devon, UK (5) found that 40% of rehabilitated juvenile European hedgehogs *Erinaceus europaeus* survived for at least nine weeks after release back into the wild. Of 10 hedgehogs monitored, four were still alive at the end of the nine-week monitoring period, three had been predated by European badgers *Meles meles*, two had been killed on roads and one sick animal had been euthanized. Two further animals survived for at least three and four weeks before losing their radio transmitters. Twelve hedgehogs (6 male, 6 female) were released on or shortly after 2 April 1993. They were wild-born, but had been taken into captivity at a wildlife hospital as underweight juveniles the previous year. Hedgehogs weighed 82–312 g when taken into captivity and 560–1,106 g at time of release. Survival and movements were monitored by radio-tracking.

A study in 1991–1992 in a shrubland and grassland site in Sierra Morena, Spain (6) found that a rehabilitated Iberian lynx *Lynx pardinus* survived at least three months after release back into the wild. The lynx was still alive at least 93 days after release, and radio-collar fixes suggested it had established a 220 ha territory. On 6 July 1991, a wounded male Iberian lynx kitten (approximately four months old, weighing 2 kg) was brought into captivity with superficial wounds and a fractured femur. The wounds were treated and the animal was kept in a small cage with padded walls. After 43 days, it was moved to a 5 × 5-m outdoor enclosure where it was fed European rabbits *Oryctolagus cuniculus* for 112 days. After this, the animal (weight 4.9 kg) was fitted with a radio-collar and moved to a 1-ha enclosure with natural vegetation and wild rabbits. After 83 days in this enclosure, on 2 March 1992, the animal (weight 6.0 kg) was released in a pine stand, 9 km from where it was originally found. It was monitored daily until the radio-collar fell off.

A study in 1995–1999 in a forest and wetland site in Wisconsin, USA (7) found that a gray wolf *Canis lupus* treated for a leg injury subsequently survived in the wild for at least 4.5 years. The young adult (>1 year) male wolf sustained torn ligaments and an elbow dislocation to a front leg, following capture in a leg-hold trap on 21 May 1995. The dislocation was repaired using artificial ligaments. The wolf was transferred to a holding pen, but escaped on 23 May 1995. Roadkill deer were supplied for six months following the animal's escape. The wolf was monitored primarily by locating tracks, and was still alive on 24 September 1999. The escape site was a 36-km^2 wildlife area, enclosed in a 3-m high deer-proof fence. No other wolves were present at the time of escape though two subsequently entered and the three were observed travelling together.

A controlled study in 2004 in suburban gardens in Bristol, UK (8) found that most rehabilitated European hedgehogs *Erinaceus europaeus* survived over eight weeks after release back into the wild. The probability of rehabilitated hedgehogs surviving more than eight weeks after release into the wild was 73%. However, over the same period, resident wild hedgehogs in the same study area had a survival probability of 95%. Body weight decline in rehabilitated hedgehogs (13%) was similar to resident hedgehogs (5%). However, the night range of rehabilitated hedgehogs (0.58 km^2) was smaller than that of resident hedgehogs (1.67 km^2). Between May and June 2004, twenty rehabilitated hedgehogs were released, one each in 20 suburban gardens. Food was provided during the first week. Rehabilitated hedgehogs and 20 wild hedgehogs inhabiting the same gardens were radio-tracked over eight weeks. Hedgehogs were weighed every 10 days. No details about the rehabilitation are provided.

A study in 2006–2008 in four forest and farmland sites in a protected area near Barcelona, Spain (9) found that more than half of rehabilitated European hedgehogs *Erinaceus europaeus* released back into the wild survived over 20 days and one hedgehog survived for at least four months. Ten of 15 released hedgehogs survived for at least 9–136 days in the wild before their radio-tags were lost. Eight of them survived for at least 22–58 days, and one survived for at least four months. The other five hedgehogs died within two months of release due to predation (two hedgehogs), accidents (two hedgehogs) or unknown causes (one hedgehog). In 2006–2008, seven male and eight female rehabilitated hedgehogs were released across four sites in Collserola Natural Park. No details about rehabilitation are provided, but all individuals were considered healthy at the time of release. The released hedgehogs were radio-tagged and their locations were recorded 9–42 times over 5–136 days between July 2006 and June 2008.

A study in 2007–2014 in a grassland reserve in Corrientes Province, Argentina (10; same experimental set-up as 12) found that over half of released rehabilitated and captive-reared giant anteaters *Myrmecophaga tridactyla*, some of which were kept in holding pens and provided with supplementary food, survived for at least six months. At least 18 of 31 released giant anteaters survived for a minimum of six months. Long-term survival and the fate of the other 13 anteaters is not reported. In 2007–2013, thirty-one giant anteaters (18 males, 13 females; 1–8 years old) were released into a 124-km^2 private reserve. Hunting within the reserve was prohibited and livestock were absent. Three anteaters were wild-born but rehabilitated in captivity from injuries, 22 were wild-born but captive-reared and six were from zoos (origin not stated). Of the 18 surviving anteaters, six had been released after a short period in a 0.5-ha pen at the release site and 12 after 7–30 days in a 7-ha pen. Supplementary food was provided for several weeks after release. In 2007–2014, thirteen anteaters were tracked for less than six months, and 18 were tracked for 6–46 months.

A study in 2009–2012 in a forest area in São Paulo, Brazil (11) found that a rehabilitated puma *Puma concolor* released back into the wild survived for 14 months. Fourteen months after release, the rehabilitated puma was run over and found dead by a highway. The puma was healthy and the death resulted from the collision. A young male puma (approximately 12 months old) was rescued in September 2009 after being hit by a vehicle. It was kept and treated in a recovery enclosure (15 × 3 × 3 m). After 542 days, the puma had fully recovered and was transferred to a pre-release enclosure (35 × 30 × 5 m) in a forested mountainous area, 28 km from where it had been hit. It was radio-tagged and released after 34 days in the pre-release enclosure. The puma was tracked every 1–3 days from an ultra-light aircraft between February 2011 and April 2012.

A controlled study in 2007–2012 in a grassland reserve in Corrientes, Argentina (12; same experimental set-up as 10) found that wild-born rehabilitated giant anteaters *Myrmecophaga tridactyla* released into the wild were more nocturnal in their activity patterns than captive-bred individuals. Wild-born rehabilitated giant anteaters were proportionally more active at night than captive-bred animals (70% vs 43% of activity records were at night). During 2007–2012, four wild-born and three captive-bred adult giant anteaters were released into a 124-km^2 private reserve. Wild-born animals were rehabilitated after being injured by hunters or in road accidents. Six anteaters (all wild-born and two captive-bred anteaters) were released after spending a short period of time in a 0.5 ha acclimatisation pen. The remaining 12 anteaters spent 7–30 days in a 7 ha holding pen at the release site prior to release. Supplementary food was provided in the holding pen and for several weeks after anteaters were released. Each of the seven anteaters was fitted with a radio-transmitter and tracked for 1–2 x 24 h periods/month in 2007 and 2011. The released anteaters were further monitored using 14 baited camera traps for an average of 336 days/ trap during 2008–2012.

A study in 2008–2013 in two forested, mountainous areas of north-west Spain (13) found that after treating three young female brown bears *Ursus arctos* for injuries and releasing them back in to the wild, one was recaptured 21 days after release and two survived for at least 4–7 years. One cub was recaptured 21 days following release after repeatedly entering villages during the day. The other cub was monitored for 239 days, then seen seven years after release. One female sub-adult was monitored for 292 days, then seen four years after release with a dependent cub. The two bears remaining in the wild both established home ranges (90% of cub's home range: 182 ha; 90% of sub-adult's home range: 2,816 ha). In 2008–2013, three young bears were taken into captivity for 41–145 days to be treated for injuries and were then released to one of two sites, 3–14 km from where they were captured. One was monitored daily by radio-tracking for 239 days and two were monitored hourly by GPS for 21 and 292 days until they were recaptured, or the collar was lost.

(1) Spinage C.A. & Fairrie R.D. (1966) Removal of a snare from a white rhinoceros in the West Nile White Rhino Sanctuary. *African Journal of Ecology*, 4, 149–151.

(2) Ellis W.A.H., White N.A., Kunst N.D. & Carrick F.N. (1990) Response of koalas (*Phascolarctos cinereus*) to re-introduction to the wild after rehabilitation. *Australian Wildlife Research*, 17, 421–426.

(3) Morris P.A., Munn S. & Craig-Wood S. (1992) The effects of releasing captive hedgehogs (*Erinaceus europaeus*) into the wild. *Field Studies*, 8, 89–99.

(4) Morris P.A., Meakin K. & Sharafi S. (1993) The behaviour and survival of rehabilitated hedgehogs (*Erinaceus europaeus*). *Animal Welfare*, 2, 53–66.

(5) Morris P.A. & Warwick H. (1994) A study of rehabilitated juvenile hedgehogs after release into the wild. *Animal Welfare*, 3, 163–177.

(6) Rodriguez A., Barrios L. & Delibes M. (1995) Experimental release of an Iberian lynx (*Lynx pardinus*). *Biodiversity & Conservation*, 4, 382–394.

(7) Thiel R.P. (2000) Successful release of a wild wolf, *Canis lupus*, following treatment of a leg injury. *The Canadian Field-Naturalist*, 114, 319–319.

(8) Molony S.E., Dowding C.V., Baker P.J., Cuthill I.C. & Harris S. (2006) The effect of translocation and temporary captivity on wildlife rehabilitation success: an experimental study using European hedgehogs (*Erinaceus europaeus*). *Biological Conservation*, 130, 530–537, https://doi.org/10.1016/j.biocon.2006.01.015

(9) Cahill S., Llimona F., Tenés A., Carles S. & Cabañeros L. (2011) Radioseguimiento post recuperación de erizos europeos (Erinaceus europaeus Linnaeus, 1758) en el Parque Natural de la Sierra de Collserola (Barcelona). *Galemys*, 23, 63–72.

(10) Di Blanco Y.E., Jiménez Pérez I. & Di Bitetti M.S. (2015) Habitat selection in reintroduced giant anteaters: the critical role of conservation areas. *Journal of Mammalogy*, 96, 1024–1035, https://doi.org/10.1093/jmammal/gyv107

(11) Adania C.H., de Carvalho W.D., Rosalino L.M., de Cassio Pereira J. & Crawshaw P.G. (2017) First soft-release of a relocated puma in South America. *Mammal Research*, 62, 121–128, https://doi.org/10.1007/s13364-016-0302-0

(12) Di Blanco Y.E., Spørring K.L. & Di Bitetti M.S. (2017) Daily activity pattern of reintroduced giant anteaters (*Myrmecophaga tridactyla*): effects of

seasonality and experience. *Mammalia*, 81, 11–21, https://doi.org/10.1515/mammalia-2015-0088

(13) Penteriani, V., del Mar Delgado2, M., López-Bao, J.V., García, P.V., Monrós, J.S., Álvarez, E.V., Corominas, T.S. & Vázquez, V.M. (2017) Patterns of movement of released female brown bears in the Cantabrian Mountains, northwestern Spain. *Ursus*, 28, 165–170, https://doi.org/10.2192/ursu-d-16-00012.1

14.4. Hand-rear orphaned or abandoned young in captivity

https://www.conservationevidence.com/actions/2358

- **Six studies** evaluated the effects of hand-rearing orphaned mammals. Two were in the USA[3,4,], one each was in Australia[1], South Africa[2] and India[6] and one was in six countries across North America, Europe and Asia[5].

COMMUNITY RESPONSE (0 STUDIES)

POPULATION RESPONSE (5 STUDIES)

- **Reproductive success (1 study):** One study in India[6] found that three hand-reared orphaned or abandoned greater one-horned rhinoceroses gave birth in the wild.

- **Survival (5 studies):** Five studies (including one controlled and one replicated) in Australia[1], the USA[3,4], India[6] and in six countries across North America, Europe and Asia[5], found that some hand-reared orphaned or abandoned ringtail possums[1], white-tailed deer[3], sea otters[4], bears[5] and greater one-horned rhinoceroses[6] survived for periods of time after release.

BEHAVIOUR (1 STUDY)

- **Behaviour change (1 study):** A study in South Africa[2] found that a hand-reared, orphaned serval established a home range upon release.

Background

Young mammals believed to be orphaned or abandoned are sometimes taken in by wildlife rehabilitators, to be reared and released back into the wild. Often, this is done more for animal welfare reasons than for species conservation though for rare species, release of such animals may provide opportunities for choosing where to augment populations. Success of such programmes can be difficult to judge, without benchmark data for survival of wild-reared mammals.

This intervention includes studies where mammals are hand-reared. See also *Place captive young with captive foster parents* and *Place orphaned or abandoned wild young with captive foster parents*.

A controlled study in 1990–1994 in a park in New South Wales, Australia (1) found that ringtail possums *Pseudocheirus peregrinus* released following hand-rearing, or relocated from elsewhere, survived for a shorter time than did resident possums. The average survival of released possums was 101 days and for resident possums was 182 days. There was no difference in survival between hand-reared or relocated possums. Deaths were mostly due to predation by mammals, reptiles and birds. For possums for which their fate was known, predation accounted for 98% of released and 81% of resident animals. Possums were monitored in a 4-km² park, adjoining a suburban area. Released possums (112) included hand-reared orphaned animals (81) and those relocated from potentially dangerous situations (21). Resident possums (41) were wild animals that had not been moved or held in captivity. Possums were monitored by radio-tracking ≥twice/week.

A study in 1998–1999 in KwaZulu-Natal, South Africa (2) found that a hand-reared, orphaned, female serval *Felis serval* established a home range upon release. The serval settled in intensive farmland, suggesting elevated habituation to humans. It established a 6-km² home range. The core area of this range was 1.5 km from the release point. The serval was moved 3 km away, following poultry depredation, but returned within six days. Two wild servals (1 male, 1 female) were orphaned after birth and hand-reared for an unknown period. In October 1998, they were placed in a holding pen and were released on 14 December 1998 (with continued access to the holding pen). Radio-telemetry was used to monitor activity. The male serval disappeared after release and no movement data were collected. Precise duration of monitoring of the female was not reported, but spanned at least seven weeks.

A study in 2000–2002 in a forest reserve in Missouri, USA (3) found that less than one third of orphaned and captive-reared white-tailed deer *Odocoileus virginianus* fawns released into the wild survived for more than one year. Twelve of 42 (29%) captive-reared white-tailed deer fawns survived more than one year after release. The other 30 fawns died (22 within 30 days of release) due to predation, accidents, poaching or legal harvesting. Forty-two orphaned fawns were rehabilitated in a wildlife rescue centre and two private residences. Sick or injured fawns received medical treatment. Fawns were released at >10 weeks old into an 8,700-ha forest reserve. Twenty-three fawns (13 males, 10 females) were released in September and October 2000. Nineteen (10 male, nine female) were released between August and September 2001 after two weeks in a 0.8-ha holding pen at the release site. All 42 fawns were fitted with radio-collars and located daily for 14 days post-release, then 3–4 times/week for four months, and weekly for one year in 2000–2002.

A study in 1986–2000 in an aquarium in California, USA (4) found that approximately one-third of rehabilitated sea otter *Enhydra lutris* pups released back into the wild survived for at least one year. Eight of 26 (31%) rehabilitated sea otter pups reared in captivity survived for at least one year after release. The other pups died (16 pups; 11 of which died within one month of release) or had to be permanently returned to captivity (two pups). In 1986–2000, twenty-six stranded new-born sea otter pups were brought into captivity and rehabilitated. Pups were raised primarily in isolation (60–80% of their time during rehabilitation) but were introduced to other sea otters at 9–18 weeks old. Before release, pups were implanted with a radio-transmitter and individually tagged. After release in 1987–2000, rehabilitated otters were monitored daily from shore during the first month and then twice weekly for up to 12 months.

A replicated study in 1991–2012 of 12 programs in the USA, Canada, Romania, Greece, South Korea and India (5) found that following release, approximately half of orphaned and captive-reared American black bears *Ursus americanus*, Asiatic black bears *Ursus thibetanus* and brown bears *Ursus arctos* survived over one year. Of 141 known mortalities, 54% occurred during the first year after release when bears were 1 to 2-years old and at least two bears lived for more than 10 years in the wild. Average annual survival rates for released captive-reared bears were 73% for American black bear, 75% for brown bear and 87% for Asiatic black bear. A minority of all American (6.1%) and Asiatic black bears (9.7%) released demonstrated persistent problem behaviours and required removal, but none were reported for brown bears. Captive-reared females from all species reproduced in the wild. Orphaned American black bears were released in the USA and Canada (424 individuals, 7 programs), Asian black bears released in India and South Korea (62 individuals, 2 programs) and brown bears were released in Romania, Canada and Greece (64 individuals, 3 programs). Cubs were <1 year old when taken into captivity and were kept for 2–14 months. All bears were released (aged 11–23 months) in areas with suitable habitat. Bears were ear-tagged and/or equipped with telemetry collars. Collared bears were monitored until the collar dropped or malfunctioned. Overall, 30% of bears were not observed after release and so are not included in survival estimates.

A study in 2006–2013 in a grassland reserve in Assam, India (6) found that most orphaned or abandoned greater one-horned rhinoceros *Rhinoceros unicornis* calves survived for at least 6 or 7 years after release and gave birth in the wild. Three of four orphaned or abandoned female rhinoceroses were still alive 6–7 years after release into the wild, and all three gave birth to calves in 2013. The fourth animal died eight months after release, in October 2008. Four female rhinoceroses aged 1–5 months old were rescued in Kaziranga National Park, and hand-reared at the Centre for Wildlife Rehabilitation and Conservation. In January and February 2006–2008, at two or three years of age, the calves were moved to the 519-km² Manas National Park, and held in a 600-acre fenced enclosure before release (further details not provided).

(1) Augee M.L., Smith B. & Rose S. (1996) Survival of wild and hand-reared ringtail possums (*Pseudocheirus peregrinus*) in bushland near Sydney. *Wildlife Research*, 23, 99–108.

(2) Perrin M.A. (2002) Space use by a reintroduced serval in Mount Currie Nature Reserve. *South African Journal of Wildlife Research*, 32, 79–86.

(3) Beringer J., Mabry P., Meyer T., Wallendorf M. & Eddleman W.R. (2004) Post-release survival of rehabilitated white-tailed deer fawns in Missouri. *Wildlife Society Bulletin*, 32, 732–738, https://doi.org/10.2193/0091-7648(2004)032[0732:psorwd]2.0.co;2

(4) Nicholson T.E., Mayer K.A., Staedler M.M. & Johnson A.B. (2007) Effects of rearing methods on survival of released free-ranging juvenile southern sea otters. *Biological Conservation*, 138, 313–320, https://doi.org/10.1016/j.biocon.2007.04.026

(5) Beecham J.J., De Gabriel Hernando M., Karamanlidis A.A., Beausoleil R.A., Burguess K., Jeong D-H., Binks M., Bereczky L., Ashraf N.V.K., Skripova K., Rhodin L., Auger J. & Lee B-K. (2015) Management implications for releasing orphaned, captive-reared bears back to the wild. *The Journal of Wildlife Management*, 79, 1327–1336, https://doi.org/10.1002/jwmg.941

(6) Dutta D.K. & Mahanta R. (2015) A study on the behavior and colonization of translocated greater one-horned rhinos *Rhinoceros unicornis* (Mammalia: Perissodactyla: Rhinocerotidae) during 90 days from their release at Manas National Park, Assam India. *Journal of Threatened Taxa*, 7, 6864–6877, https://doi.org/10.11609/jott.o4024.6864-77

14.5. Place orphaned or abandoned wild young with wild foster parents

https://www.conservationevidence.com/actions/2343

- **Three studies** evaluated the effects of placing orphaned or abandoned wild young with wild foster parents. One study was in the USA[1], one was in South Africa[2] and one was in Botswana[3].

COMMUNITY RESPONSE (0 STUDIES)

POPULATION RESPONSE (3 STUDIES)

- **Survival (3 studies):** Two studies (one controlled) in the USA[1] and Botswana[3], found that orphaned young black bears[1] and African wild dogs[3] had greater[1] or equal[3] survival compared to animals released alone[1] or young of wild mammals with their biological parents[3]. A study in South Africa[2] found that an orphaned cheetah cub was not accepted by a family of cheetahs.

BEHAVIOUR (0 STUDIES)

Background

Young mammals believed to be orphaned or abandoned are sometimes taken in by wildlife rehabilitators, to be reared and released back into the wild. Often, this is done more for animal welfare reasons than for species conservation though for rare species, release of such animals may provide opportunities for choosing where to augment populations. An alternative to captive rearing may be to attempt to foster young into existing wild families. If this can be achieved, it may improve their ability to find food in the wild and reduce the extent to which they become imprinted on humans and could, thus, improve the prospects of longer-term survival in the wild. However, the success of such programmes can be difficult to judge, without benchmark data for survival of wild-reared mammals.

See also *Place orphaned or abandoned wild young with captive foster parents.*

A controlled study in 1973–1983 in temperate forests in Idaho and Pennsylvania, USA (1) found that orphaned black bears *Ursus americanus* released to wild females with cubs had higher short-term survival than did orphaned bears released alone. Ten days after release, 23 of 45 (51%) orphaned bears placed with females with cubs were seen to be in good condition, but only five of 39 (13%) cases in which orphans were released in the wild alone were deemed successful. In 1973–1983, twenty-nine cubs were released directly into dens of females with young, 11 cubs were released after chasing females and causing their young to climb trees and five cubs were placed with female bears and their young that were caught in culvert traps and then released. In seven cases, females were immobilized while the cubs were introduced. Thirty-nine orphaned bear cubs were held in captivity before being release alone into the wild. Reintroductions were regarded as successful if orphaned bears were observed with the foster mother at least 10 days after reintroduction or, for solo introductions, if animals survived for at least 30 days and did not become a nuisance to humans. Survey methods were unclear.

A study in 1994–1998 in a savannah reserve in North West province, South Africa (2) found that when an orphaned female cheetah *Acinonyx jubatus* cub was put in a holding pen with a family of cheetahs, the orphaned female was not accepted by the group and was removed after two weeks. The orphaned female was prevented from accessing food by male cubs and the adult female was hostile towards her, although did not cause physical harm. The orphaned female cub was fed separately as a result and was relocated to a captive breeding facility after two weeks. An 8-month-old orphaned female cub was placed in a holding pen with one adult female and three 18-month-old dependent male cubs in a 60,000 ha game reserve. The orphaned female cheetah had been captured on a farm, the family group were from a rehabilitation facility.

A study in 2000 and 2003 at three savannah sites in Botsawana (3) found that orphaned African wild dog *Lycaon pictus* pups released in the vicinity of wild dog packs were readily adopted into the pack and had survival rates similar to those of wild pups. A six-week-old pup was adopted into a pack of 24 adults and yearlings in August, and survived to at least October, but not to the year end. Four 10-week-old pups were adopted into a pack of seven adults and eight pups in August. Two pups survived at least to the year end. Three 10-week-old pups were adopted into a pack of three adults and four pups in August but did not survive to the year end. Where pups died before the year end, no pups born into those packs survived either. One orphaned pup was adopted within 24 hours of capture, the others after three weeks of quarantine. Four pups required moving to re-join their adoptive pack, which moved 7 km during the first night following interactions with lions *Panthera leo*.

(1) Alt G.L. & Beecham J.J. (1984) Reintroduction of orphaned black bear cubs into the wild. *Wildlife Society Bulletin*, 12, 169–174.

(2) Hofmeyr M. & van Dyk G. (1998) *Cheetah introductions to two north west parks: case studies from Pilanesberg National Park and Madikwe Game Reserve.* Proceedings of a Symposium on Cheetahs as Game Ranch Animals, Onderstepoort, 23 & 24 October 1998, 60–71.

(3) McNutt J.W., Parker M.N., Swarner M.J. & Gusset M. (2008) Adoption as a conservation tool for endangered African wild dogs (*Lycaon pictus*). *South African Journal of Wildlife Research*, 38, 109–112, https://doi. org/10.3957/0379-4369-38.2.109

14.6. Place orphaned or abandoned wild young with captive foster parents

https://www.conservationevidence.com/actions/2364

- **Two studies** evaluated the effects of placing orphaned or abandoned wild young with captive foster parents. One study was in Canada[1] and one was in the USA[2].

COMMUNITY RESPONSE (0 STUDIES)

POPULATION RESPONSE (1 STUDY)

- **Survival (1 study):** A controlled study in the USA[2] found that stranded sea otter pups reared in captivity by foster mothers had higher post-release survival than did unfostered pups reared mostly alone, and similar survival to wild pups.

BEHAVIOUR (2 STUDIES)

- **Behaviour change (2 studies):** A study in Canada[1] found that a captive white-tailed deer adopted a wild orphaned fawn. A controlled study in the USA[2] found that stranded sea otter pups reared in captivity by foster mothers began foraging earlier than did unfostered pups reared mostly alone.

Background

Young mammals believed to be orphaned or abandoned are sometimes taken in by wildlife rehabilitators, to be reared and released back into the wild. Often, this is done more for animal welfare reasons than for species conservation though for rare species, release of such animals may provide opportunities for choosing where to augment populations. If such mammals can be fostered in captivity by parents of the same species, it may reduce the extent to which they become imprinted on humans and could improve the prospects of post-release survival in the wild. However, the success of such programmes can be difficult to judge, without benchmark data for survival of wild-reared mammals.

See also *Hand-rear orphaned or abandoned young in captivity* and *Place captive young with captive foster parents*.

A study in 1993 in a captive facility in New Brunswick, Canada (1) found that a captive white-tailed deer *Odocoileus virginianus* adopted a wild orphaned fawn. The fawn was around one week old when rescued and was initially hand-fed. After five days, a captive white-tailed deer doe gave birth to a stillborn fawn. The following day, the orphaned fawn was placed with the doe. It was initially ignored, and hand-feeding continued. One day later, the hide of the stillborn fawn was wrapped around the orphaned fawn. The doe proceeded to lick the hide and nursed the fawn thereafter, even after the hide became detached after five hours, due to vigorous licking. The study took place in a captive research facility to which the orphaned fawn was delivered on 9 June 1993. Attachment of the hide, and adoption by the doe took place on 15 June 1993.

A controlled study in 1986–2004 at an aquarium and coastal site in California, USA (2) found that stranded sea otter *Enhydra lutris* pups reared in captivity by foster mothers began foraging earlier and had greater survival in the wild than unfostered pups, and similar survival to wild pups. Fostered sea otter pups began foraging independently on live prey at younger ages (average 8–19 weeks old) than unfostered pups reared mostly alone (average 11–22 weeks old). A greater proportion of fostered pups survived at least one year after release (5 of 7 pups; 71%) than unfostered pups (8 of 26 pups; 31%), and survival was similar to wild pups (9 of 12 pups; 75%). In 2001–2003, seven stranded sea otter pups were brought into captivity and reared with adult female sea otters. In 1986–2000, twenty-six stranded sea otter pups were reared in captivity without foster mothers (mostly alone). All pups were rehabilitated at the same aquarium. Before release, pups were implanted with radio-transmitters and individually tagged. After release in 1987–2004, the rehabilitated otters were monitored daily during the first month and then twice weekly for up to 12 months. Twelve wild juvenile male sea otter pups were observed during a field study prior to 2003 (date not reported).

(1) Greaves T.A. & Duffy M.S. (1994) Adoption of a white-tailed deer, *Odocoileus virginianus*, fawn by a captive doe. *The Canadian Field-Naturalist*, 108, 239.

(2) Nicholson T.E., Mayer K.A., Staedler M.M. & Johnson A.B. (2007) Effects of rearing methods on survival of released free-ranging juvenile southern sea otters. *Biological Conservation*, 138, 313–320, https://doi.org/10.1016/j.biocon.2007.04.026

14.7. Provide supplementary food to increase reproduction/survival

https://www.conservationevidence.com/actions/2367

- **Twenty-four studies** evaluated the effects on mammals of providing supplementary food to increase reproduction/survival. Nine studies were in the USA[1,2,3,8,11,12,16,17,20], two were in Canada[5,13], two were in South Africa[10,22], two were in Poland[4,24], and one each was in Sweden[6], the Netherlands[7], Swaziland[9], Spain[14], Portugal[15], Slovenia[18], Austria[23], Norway and Sweden[19] and one was across North America and Europe[21].

COMMUNITY RESPONSE (0 STUDIES)

POPULATION RESPONSE (18 STUDIES)

- **Abundance (8 studies):** Four of eight studies (incuding four controlled, two site comparisons and five before-and-after studies) in the USA[1,2], Canada[5,13], South Africa[10,22], Poland[4] and Austria[23] found that supplementary feeding increased the abundance or density of bank voles[4], red squirrels[5], striped mice[10], brown hyena[22] and black-backed jackals[22]. One study found a temporary increased in prairie vole abundance[1]. The other three studies found supplementary feeding not to increase abundance or density of white-footed mice[2], northern flying squirrels[13], Douglas squirrels[13] or Eurasian otters[23].

- **Reproduction (8 studies):** Four of five controlled studies (three also replicated) in the USA[1], South Africa[10], Norway and Sweden[19], Sweden[6] and Spain[14], found that supplementary food increased the proportion of striped mice that were breeding[10], the number of arctic fox litters[6,19] and the size of prairie vole litters[1]. However, there was no increase in the number of arctic fox cubs in each litter[6] or the proportion of female Iberian lynx breeding[14]. One of two replicated studies (one site comparison and one controlled), in the Netherlands[7] and the USA[16], found that supplementary feeding increased the number of young wild boar produced and recruited in to the population[7]. The other study found that the number of mule deer produced/adult female did not increase[16]. A review of studies across North America and Europe found that supplementary feeding increased ungulate reproductive rates in five of eight relevant studies[21].

- **Survival** (**9 studies**): Four of eight studies (including seven controlled studies and two before-and-after studies) in the USA[3,8,11,16,17], Canada[13], Poland[4] and Spain[14], found that supplementary feeding increased survival of mule deer[3], bank voles[4], northern flying squirrels[13] and eastern cottontail rabbits[17]. Five studies found no increase in survival for white-tailed deer[8], Douglas squirrels[13], mule deer[16], Rocky Mountain bighorn sheep lambs[11] or Iberian lynx[14]. A review of studies across North America and Europe found that supplementary feeding increased ungulate survival in four out of seven relevant studies[21].

- **Condition** (**4 studies**): One of three studies (including two controlled and two before-and-after studies) in Poland[4], the USA[12], and Canada[13], found that supplementary food lead to weight gain or weight recovery in bank voles[4]. One study found no body mass increase with supplementary feeding in northern flying squirrels and Douglas squirrels[13].The third study found mixed results, with supplementary feeding increasing weight gains in some cotton rats, depending on their sex, weight and the time of year[12]. A review of studies from across North America and Europe found that different proportions of studies found supplementary feeding to improve a range of measures of ungulate condition[21].

BEHAVIOUR (6 STUDIES)

- **Use** (**2 studies**): A replicated, controlled study in Sweden[6] found that supplementary food increased occupancy of Arctic fox dens. A replicated study in Portugal[15] found that artificial feeding stations were used by European rabbits.

- **Behaviour** (**4 studies**): Two of three replicated studies (two also controlled), in Swaziland[9], Slovenia[18] and the USA[20], found that supplementary feeding led to reduced home range sizes or shorter movements of red deer[18] and elk[20]. The third study found home ranges and movement distances to be similar between fed and unfed multimammate mice[9]. One replicated study in Poland[24] found that supplementary feeding of ungulates altered brown bear behaviour.

Background

Many mammals have long gained a proportion of their diet as a direct result of human activities (Oro *et al.* 2013). Many of these are cases where by-products of production, harvesting or consumption are exploited. However, in some cases, food is provided specifically for mammals. This is often to increase survival and condition of hunted animals, such as deer. In some other cases, food may be provided to aid the conservation status of rare species. Some studies are less directly conservation-motivated but are included here if the findings can help to inform conservation actions.

Studies that provide supplementary food as part of translocation or reintroduction programmes are discussed in: *Provide supplementary food during/after release of translocated mammals* and *Provide supplementary food during/after release of captive-bred mammals*.

Oro D., Genovart M., Tavecchia G., Fowler M.S. & Martínez-Abraín A. (2013) Ecological and evolutionary implications of food subsidies from humans. *Ecology Letters*, 16, 1501–1514.

A controlled study in 1975–1976 in a grassland site in Illinois, USA (1) found that where supplementary food was provided, prairie vole *Microtus ochrogaster* numbers were temporarily higher and litter size was larger than in an area with no supplementary food. Voles reached higher densities in the food supplemented area (135 voles/ha in April 1976) than in the area with no supplementary feeding (90 voles/ha in October 1975). However, 16–18 months after supplementary feeding commenced, vole numbers were similar in fed and unfed areas (<5/ha). Voles in the fed area had a longer life expectancy and were more likely to breed in winter than voles in the unfed area (data expressed as model results). Average litter size was larger in the fed area (5.1) than in the unfed area (4.3).A 1.5-ha abandoned pasture was divided into two live-trapping grids of 0.80 and 0.55 ha, separated by a 10-m-wide mown strip. On the 0.55-ha grid, 210 feeding stations (200-ml bottles, filled with rabbit pellets, replenished as required) were placed 5 m apart. No supplementary food was provided on the other grid. Voles were surveyed using 60 wooden traps in the supplementary feeding grid and 72 in the unfed grid. Every three weeks from May 1975 (supplementary food grid) and August 1975 (unfed grid) to November 1976, traps were set for three days and checked twice daily. Traps were baited for two days before setting.

A before-and-after study in 1975–1976 in a woodland in Illinois, USA (2) found that supplementary feeding did not increase white-footed mouse *Peromyscus leucopus* densities. Monthly densities varied seasonally but were not higher in supplementary feeding plots than in plots without supplementary food provision. After supplementary feeding commenced, the highest numbers were 5–9 months later, with 20–26 mice/ha in supplementary feeding plots and 22–29 mice/ha in plots without supplementary food. Four plots, 0.36 ha each, were established within a 9-ha live trap grid, with 20-m trap intervals. Traps were operated across the grid over three days/month, from January 1975 to July 1976. Additional trapping took place fortnightly on grid points within and immediately surrounding plots, from March–December 1976. Supplementary feeding, using mouse chow at 10-m intervals, commenced in two plots in January 1976. No food was provided in the other two plots.

A controlled study in 1984 on three areas of a predominantly grassland site in Colorado, USA (3) found that supplementary feeding of mule deer *Odocoileus hemionus hemionus* increased overwinter survival. Mortality was lowest for deer provided with as much supplementary food as they could consume (24%), intermediate for deer given fixed quantities of supplementary food (33%) and highest for deer not provided with supplementary food (53%). Three study areas (≥5 km apart, 660–1,000 ha extent) were monitored. Supplementary food consisted of wheat middlings, brewer's dry grain, cottonseed hulls and alfalfa formed into wafers. It was provided daily, from 7 January to 10 April 1984, in equal or greater quantities than deer consumed in one study area and at 0.9 kg/deer/day in another study area. No supplementary food was provided in the third area. Biweekly aerial deer counts were conducted from 27 January 1984 and mortality was assessed by ground surveys for carcases, during 1–15 June 1984, of randomly selected sample plots from each study area.

A before-and-after study in 1966–1969 and 1973–1974 on a forested island in Lake Beldany, Poland (4) found that when supplementary food was provided, the abundance, body weight and survival of bank voles *Clethrionomys glareolus* was higher. Annual peak vole abundance was higher in years when food was provided (835–1,068 individuals) than when no food was provided (157–368 individuals). The average body weight of young voles (3–9 weeks old) was higher in years when food was provided (17.2 g) than when they were not fed (13.9 g). The survival of individuals to autumn in the year they were born was higher in years when food was provided (49%) than when voles were not fed (8–42%). Voles were live-trapped every six weeks from spring to autumn 1966–1969 and 1973–1974, in five 10–14-day trapping sessions/ year. Two to five traps baited with oats were set at each of 159 trapping locations and checked twice daily. From spring 1973 to autumn 1974, a total of 159 boxes with 3 kg of oats each were distributed 15 m apart across the 4-ha island, next to trapping sites. Boxes were replaced when half the oats had been consumed, but were removed during trapping.

A replicated, paired sites, controlled, before-and-after study in 1983–1986 in four mixed spruce and pine forest sites in British Columbia, Canada (5) found that providing supplementary food increased the abundance of red squirrels *Tamiasciurus hudsonicus*. After two years, squirrel abundance in sites with supplementary food was higher (41–53 squirrels/site) than in unfed sites (9–15 squirrels/site). One year after supplementary feeding ceased, squirrel numbers declined in previously fed sites (23–31 squirrels/site) but not in unfed sites (11–12 squirrels/ site). A 9-ha grid, with 100 stations at 30 m intervals, was established in each of four forest sites (two each in two forests). Sunflower seeds (83–90 kg/month) were provided in cans nailed to trees distributed across two sites (50 cans/site), from September 1983 to September 1985. No food was provided at the other two sites. From June 1983 to June 1986, squirrels were captured and measured using one Tomahawk live trap at alternate stations. Traps were set for two days, every 3–4 weeks in summer (April–September) and 4–10 weeks in winter (October–March). Cans were refilled after each trapping period.

A replicated, controlled study in 1979–1990 in four mountainous grassland areas in northern Sweden (6) found that providing supplementary food increased occupancy of Arctic fox *Alopex lagopus* dens and the number of fox litters born, but not the numbers of cubs in each litter. Where supplementary food was provided, a higher proportion of dens were occupied (35%) than where no supplementary food was supplied (6%). Over five years, 17 of 65 dens (26%) where food was provided contained a litter while only three of 103 dens (3%) where no food was provided contained a litter. However, there was no significant difference in average litter size (supplementary food: 5.2 cubs; no food: 5.7 cubs). During January–April of 1985–1989, reindeer *Rangifer tarandus* and moose *Alces alces* meat was placed 50–200 m from 168 dens which showed signs of Arctic fox activity. In some cases, meat was buried in the snow. About 50–100 kg of meat/den/year was provided. Dens were surveyed for presence of foxes and offspring in June–August of 1979–1990.

A replicated, site comparison study in 1988–1992 of forest and heathland across nine management areas in the Netherlands (7) found that when supplementary feed was provided, wild boar *Sus scrofa* annual population recruitment rates were higher. No statistical analyses were performed. In seven areas, where boar were fed, annual recruitment (number of piglets >2 months old/ adult female) averaged 2.2–2.5, compared to 0.0–2.5 at a site where supplementary feeding ceased in the year before the study began. At a further site, where supplementary feeding ceased two years into the study, recruitment averaged 2.0–2.4 over those first two years and 1.5–1.7 in the subsequent three years. Recruitment data were obtained from nine boar management areas, based on spring counts at feeding locations.

A controlled study in 1986–1989 of a forested area in Wisconsin, USA (8) found that supplementary feeding of white-tailed deer *Odocoileus virginianus* did not increase their overall survival. The average annual survival of winter-fed deer (78%) or summer-fed deer (53%) did not differ significantly from that of unfed deer (64%). Summer-and winter-fed deer had higher over-winter survival during a single severe winter only (summer-fed: 96%; winter-fed: 100%; not fed: 79%), but not during other periods. From October 1986 to July 1989, deer were fed shelled corn or commercial deer food from mid-April to mid-December (summer-feeding— 53 deer), 1 December to 30 April (winter-feeding— 66 deer) or were not fed (48 deer). All deer, except 24 that were winter-fed, occupied a 15 × 30-km area. No deer was winter-fed and summer-fed in the same year. Survival was monitored through radio-tracking. Deer use of feeders was determined by direct observations.

A replicated, controlled study in 1995–1996 in a grassland in Middleveld, Swaziland (9) found that multimammate mice *Mastomys natalensis* provided with supplementary food had similar home range sizes and distance between captures to unfed mice. The average home ranges of 66 multimammate mice provided with supplementary food (600–923 m²) did not differ significantly from those of nine unfed mice (838–960 m²). Similarly, average distances between captures of mice provided with supplementary food (20–21 m) did not differ significantly from those of unfed mice (25–28 m). In May 1995, three 100 × 100-m plots were established in a natural grassland. Supplementary food (4 kg of rolled oats and 4 kg of rabbit pellets) was provided monthly, from July 1995 to May 1996, in two plots. No supplementary food was added to the third plot. From June 1995 to May 1996 mice were surveyed monthly using 100 Elliot and Sherman live traps/plot. Traps were set 10 m apart, on three consecutive nights/month. Mice were individually toe-clipped and weighed when captured. Only individuals captured at least five times were used to calculate home range sizes.

A controlled study in 1995 on a grassland in KwaZulu-Natal, South Africa (10) found that providing supplementary food increased striped mouse *Rhabdomys pumilio* density and the proportion of the population that was breeding. Three to six months after feeding began, there were more striped mice in the plot with supplementary food (30) than in the plot with no supplementary food (21). Over the same time period, a higher proportion of adult mice were reproductively active in the plot with supplementary food (85%) than in the plot with no supplementary food (38%). In one of two plots (>60 m apart) 25 trays, each with 1 kg of oat seeds, were filled weekly. The second plot had no supplementary food. In each plot, mice were monitored at 49 stations, in a 7 × 7 grid, at 10-m intervals. Each station was surveyed for two consecutive nights/month with one baited and insulated Elliot or Sherman live trap, from January–June 1995.

A randomized, replicated, controlled study in 1991–1995 in two mountain ranges in Colorado, USA (11) found that supplementary winter feeding of Rocky Mountain bighorn sheep *Ovis canadensis canadensis* did not increase lamb survival. Average annual recruitment did not differ between herds provided with food (0.5–0.7 lambs/adult female) and herds where no food was provided (0.6–0.7 lambs/adult female). Adult bighorn females of four herds were captured in February–March 1991–1995 and were marked and radio-collared. Between 1991 and 1995 the herds were either fed from mid-December for 8–10 weeks with 2 kg/individual/day of alfalfa hay and 1 kg/individual/day of apple pulp, or not given any supplementary food. Each year, one herd under each feeding regime was additionally medicated for lungworm using fenbendazole, while the other was not medicated. Treatments were rotated annually under a predetermined, randomly selected scheme. Lamb survival for 11–18 marked adult females/herd was assessed every two weeks between May and October the following year.

A replicated, controlled study in 1990–1992 in a forest reserve in Kansas, USA (12) found that cotton rats *Sigmodon hispidus* had different growth rates after the provision of supplementary food, depending on their size and sex, and the time of year. In winter, the growth rate of small cotton rats provided with supplementary food was significantly higher than that of small rats not provided with food, but the opposite was true for larger rats. In spring, males on supplemented grids grew faster than males on control grids, but the opposite was true in females. In summer, there was no difference in growth rates between supplemented and non-supplemented grids. In autumn, males were the same as in winter, but larger females grew faster with supplementary food (data presented as model results). Additionally, seven reproductive cotton rat females had a higher growth rate when provided with food (2.5 g/day) than did 14 non-supplemented females (2.0 g/day). Seven litters born to females on food supplemented grids had higher growth rates in their first month of life (1.4 g/day) than 23 litters born on non-supplemented grids (0.94 g/day). Between June 1990 and May 1992, supplementary food was distributed along two out of four trapping grids. Food (50 g each of sorghum seeds, millet seeds and commercial rabbit chow) was provided in cans that were refilled every two weeks. Grids contained 64–99 trapping stations, 15 m apart, each with two Sherman traps baited with scratch grain. Traps were set for three consecutive days/month, and checked twice daily. Rats were individually marked and weighed when captured. In June 1991, one of the food supplemented and one of the non-supplemented grids were switched.

A replicated, randomized, paired sites, controlled, before-and-after study in 1996–1999 in three forest sites in British Columbia, Canada (13) found that supplementary feeding did not alter the abundance and body mass of northern flying squirrels *Glaucomys sabrinus* and Douglas squirrels *Tamiasciurus douglasii*, but it did increase survival of northern flying squirrels. Between June 1997 and April 1999, the survival rate of northern flying squirrels was higher in plots with supplementary feeding (0.93) than without supplementary feeding (0.79). Survival did not significantly differ between plots before feeding began (plots to be fed = 0.84; control plots = 0.92). The survival of Douglas squirrels was similar between fed (0.72) and unfed (0.80) plots. The abundance and body mass of squirrels did not differ significantly between plots with supplementary food (northern flying squirrel abundance: 11.8/ha; body mass: 131 g; Douglas squirrel abundance: 14.2/ha; body mass: 200 g) and plots without supplementary food (northern flying squirrel abundance: 7.7/ha; body mass: 128 g; Douglas squirrel abundance: 20.1/ha; body mass: 207 g). From April 1997 to May 1998 and from September 1998 to April 1999, supplementary food was provided at 90 feeding stations, 60 m apart in a 9×10 grid, in each of three 30-ha forest plots. Stations were filled with 7 kg of sunflower seeds at 5–6-week intervals or when seed was depleted. Three other 30-ha plots had no feeding stations. In each plot, squirrels were trapped every 5–6 weeks (when snow-free), from June 1996 to March 1999, using 80 baited Tomahawk live traps, at 40-m intervals in an 8×10 grid.

A replicated, controlled study in 1985–2008 in two shrubland areas in southern Spain (14) found that supplementary feeding did not increase the breeding rate of Iberian lynx *Lynx pardinus* or survival of offspring. The proportion of female lynx that reproduced in areas where supplementary food was provided (66%) did not differ significantly from that in areas where it was not (83%). Similarly, survival of lynx offspring did not significantly differ (supplementary food: 100%; no supplementary food: 88%). In 2002–2008, six lynx breeding territories were each supplied, throughout the year, with live domestic rabbits at approximately three feeding stations. An unspecified number of other territories were not supplied with rabbits. Fifteen adult female lynx were fitted with radio-collars and were monitored in 1985–2007. Data on breeding were obtained in March–May of 1993–2008, by tracking females to locate dens. Lynx were also monitored by sightings, camera-trapping, and radio-tracking.

A replicated study in 2007–2009 in six agroforestry sites in Alentejo and Algarve, Portugal (15) found that European rabbits *Oryctolagus cuniculus* used most available artificial feeding stations. Rabbits used almost 70% of 48 feeding stations surveyed. Rabbit numbers were higher in areas where a higher proportion of feeding stations was used (data presented as a correlation). Over the course of the study, which included providing artificial shelters and waterholes, the number of rabbit latrines increased from 16 to 25 latrines/km (no statistical analysis conducted). Between July and September in 2008 and 2009, wheat, oat and alfalfa were made available through 120 artificial feeding stations in six agroforestry. Each station was protected by a fence, aimed at excluding large animals. However, 60% of feeding stations were destroyed by deer or wild boar, so data for 48 feeding stations were analysed. These were surveyed monthly and considered to be used if rabbit droppings were detected. Rabbit abundance was estimated based on the number of latrines/km counted along paths at each site.

A replicated, randomized, controlled study in 2001–2006 in eight forest, grassland and shrubland sites in Utah, USA (16) found that providing supplementary food over winter did not increase mule deer *Odocoileus hemionus* survival or reproductive success. The average annual survival of deer with supplementary feeding (80%) did not differ significantly from that of deer without supplementary feeding (73%). Similarly, the average reproductive success of deer with supplementary feeding (0.58 fawns/female deer) did not differ significantly from that of deer without supplementary feeding (0.57 fawns/female deer). In 2001, eight sites known to host winter concentrations of mule deer were randomly selected. Supplementary food (corn, alfalfa and protein pellets, 0.9 kg/deer/day) was provided over winter (December–March 2001–2005) at four sites. No food was provided at the other four sites. Sites with and without supplementary food were >3 km apart. Fifty-two female mule deer receiving supplementary food and 38 that were not fed were radio-collared between January and March 2001–2005. They were monitored 2–3 times/week, from May 2002 to January 2006.

A replicated, randomized, controlled study in 2009–2010 in 23 mixed wetland, scrubland, and wasteland sites in New Hampshire, USA (17) found that supplementary feeding increased survival of eastern cottontail rabbits *Sylvilagus floridanus*. After two months, rabbit survival in sites where supplementary food was provided was higher (9 of 15 animals; 60%) than in sites where no food was provided (5 of 13 animals; 38%). In November 2009–March 2010, twenty-eight rabbits were trapped and fitted with radio-collars and ear tags. Between December 2009 and March 2010, commercial rabbit food was provided every three days (450 g) at some sites and no food was provided at other sites. The number of sites where food was provided is unclear.

A replicated study in 1997–2003 in forest, meadows and farmland in a mountain range in central and southern Slovenia (18) found that in areas where supplementary food was provided, the home-range of red deer *Cervus elaphus* was smaller. Red deer had smaller home ranges in areas where more supplementary feeding occurred (data expressed as model results). Between 1997 and 2003, twenty-five adult female and 17 adult male red deer were caught across a 2,100 km² study area. Deer were radio-collared and released, and were relocated at least once a week, during all daylight hours, for at least one year. Annual home range size was estimated for each individual for each full year that it was monitored (total = 73 deer-years from 42 animals). Information on the location of supplementary feeding sites, and the type and quantity of food provided, was collected from a national register of feeding sites and used to model deer home-ranges alongside other relevant variables.

A replicated, controlled study in 1999–2011 in 10 tundra sites in Norway and Sweden (19) found that the number of artic fox *Vulpes lagopus* litters increased after supplementary winter feeding at den sites, along with control of red foxes *Vulpes vulpes*. At two sites where an average of 11–13.5 dens were fed, both the number of active arctic fox dens in winter, and the number of litters produced in summer, increased more than at sites where no feeding or a low level of feeding was undertaken (data reported as statistical model results). During winter 1999–2011, commercial dog food or remains from slaughtered reindeer *Rangifer tarandus* was provided to a large number of arctic fox dens (11–13.5) at two sites, where red foxes were also intensively culled in winter. At four other sites, low numbers of arctic fox dens (1–3) were provided with food, and low numbers of red foxes were culled (0–7). At the remaining four sites, no food was provided and no red foxes were culled (3 sites) or intensive culling was conducted (92 animals, 1 site). The number of arctic fox litters was counted in known arctic fox dens during July and August 1999–2011.

A replicated, controlled study in 2007–2013 in four forested mountain areas in Wyoming, USA (20) found that elk *Cervus canadensis* provided with supplementary food migrated shorter distances and spent less time on their summer feeding grounds than unfed elk. Elk provided with supplementary food in winter migrated shorter distances (35.4 km) than did unfed elk (54.6 km). Fed elk arrived at their summer range an average of five days later and left 10 days earlier than did unfed elk. More fed elk used stopover sites on spring (56% of elk) and autumn (49% of elk) migration than non-fed elk (48% and 42% of individuals). Two hundred and nineteen adult female elk were caught and fitted with GPS radio-collars between January and March 2007–2011 at 18 sites where supplementary food was provided and at four sites with no supplementary food. Sites were located in four mountain areas within elk winter ranges. Supplementary feeding began when elk started to congregate at feeding sites and ceased once most elk had departed. GPS locations were taken from the elk every 30–60 minutes, for 1–2 years. Fed and unfed elk were monitored for 164 and 116 elk-years, respectively. The precise number of fed and unfed elk monitored is not detailed.

A review of evidence within studies looking at effects of feeding wild ungulates in North America (48 studies), Fennoscandia (25 studies) and elsewhere in Europe (28 studies) (21) found that supplementary feeding increased ungulate survival, reproductive rates or condition in varying proportions of studies. Ungulate survival rates increased in four out of seven relevant studies. The reproductive rate increased in five of eight relevant studies. Birth mass increased in one of three relevant studies. Loss of mass in winter was reduced or winter condition improved in five of seven relevant studies. Autumn mass increased in three of 11 relevant studies. Autumn mass or condition of offspring was improved in four of six relevant studies. Carrying capacity was increased in all three relevant studies. The review reported evidence from 101 studies that met predefined criteria from an initial list of 232 papers and reports.

A before-and-after, site comparison study in 2007–2013 of a conservation park and a game park in South Africa (22) found that when carrion was provided at a vulture feeding station, there were more brown hyaena *Hyaena brunnea* and black-backed jackal *Canis mesomelas* scats in that area. At the vulture station site, there were more hyaena scats in the final year of carrion provision (5.0 scats/km) than before carrion provision (2.6 scats/km) and over the two years after carrion provision ceased (1.5–2.0 scats/km). Scat counts remained more stable over this period at a site without a vulture feeding station (3.2–4.3 scats/km). Similarly, there were more jackal scats at the vulture feeding station in the final year of carrion provision (3.3 scats/km) than before (0.5 scats/km) or over two years after (1.5–2.0 scats/km) carrion provision. Scat counts remained low (0.2–1.4 scats/km) at a site without a vulture feeding station. A vulture restaurant was operated at a conservation park from March 2008 to August 2011. Predator density at this park, and on a game park where carrion was not provided, was monitored by annual scat transects from 2007–2013.

A site-comparison study in 2011 along two rivers in Austria (23) found that on a river stocked with fish for angling, densities of resident adult Eurasian otters *Lutra lutra* were not higher than those on an unstocked river. No statistical analyses were performed. Resident adult otter density on the stocked river (0.23 otters/km) was similar to that on the unstocked river (0.22 otters/km). However, including juvenile and non-resident otters, a slightly higher density was found on the stocked river (0.37 otters/km) than on the unstocked river (0.33 otters/km). Two river stretches, with similar hydromorphology, were studied. One (21.5 km long) was stocked with fish from a hatchery in April–September each year. The other (18.3 km long) was not stocked. Otter spraints were collected daily for five days during three visits from February–April 2011. Individual otters were identified by genetic analysis of faeces. Forty-eight faeces were successfully used to genetically identify individuals from the stocked river and 33 from the unstocked river.

A replicated study in 2008–2015 in a mountain forest and grassland site in the northeast Carpathians, Poland (24) found that supplementary feeding of ungulates altered brown bear *Ursus arctos* behaviour. Bears encountered feeding sites more frequently (GPS-tracked bears: 0.15 sites/km; snow-tracked bears: 0.93 sites/km) than expected at random (0.05 sites/km). From 2008–2010, a complete inventory of 212 ungulate feeding sites in the 1,500 km² study area was compiled through interviews with land managers and field inspections. Feeding occurred regularly, often year-round but especially in autumn and winter, and usually in the same location for decades. In spring and autumn 2008–2009 and 2014–2015, nine bears were captured and fitted with GPS collars. Bear locations were recorded every 30 minutes for five days at the start of each month, and used to create 49 GPS-tracks (average 34 km long). From December–March 2010–2012, 40 snow tracks of unmarked bears longer than 500 m were recorded (average 6 km long). To determine what would be expected if movements were at random, for each of the 49 GPS tracks recorded, 100 random tracks were created using the same start point and number of locations, and by randomly choosing the distance travelled and angle turned between points.

(1) Cole F.R. & Batzli G.O. (1978) Influence of supplemental feeding on a vole population. *Journal of Mammalogy*, 59, 809–819.

(2) Hansen L.P. & Batzli G.O. (1979) Influence of supplemental food on local populations of *Peromyscus leucopus*. *Journal of Mammalogy*, 60, 335–342.

(3) Baker D.L. & Hobbs N.T. (1985) Emergency feeding of mule deer during winter: tests of a supplemental ration. *The Journal of Wildlife Management*, 49, 934–942.

(4) Banach K. (1986) The effect of increased food supply on the body growth rate and survival of bank voles in an island population. *Acta Theriologica*, 31, 45–54.

(5) Sullivan T.P. (1990) Responses of red squirrel (*Tamiasciurus hudsonicus*) populations to supplemental food. *Journal of Mammalogy*, 71, 579–590.

(6) Angerbjörn A., Arvidson B., Norén E. & Strömgren L. (1991) The effect of winter food on reproduction in the arctic fox, *Alopex lagopus*: a field experiment. *The Journal of Animal Ecology*, 60, 705–714.

(7) Bruinderink G.G., Hazebroek E. & Van Der Voot H. (1994) Diet and condition of wild boar, *Sus scrofu scrofu*, without supplementary feeding. *Journal of Zoology*, 233, 631–648.

(8) Lewis T.L. & Rongstad O.J. (1998) Effects of supplemental feeding on white-tailed deer, *Odocoileus virginianus*, migration and survival in northern Wisconsin. *The Canadian Field-Naturalist*, 112, 75–81.

(9) Monadjem A. & Perrin M.R. (1998) The effect of supplementary food on the home range of the multimammate mouse *Mastomys natalensis*. *South African Journal of Wildlife Research*, 28, 1–3.

(10) Perrin M.R. & Johnson S.J. (1999) The effect of supplemental food and cover availability on a population of the striped mouse. *South African Journal of Wildlife Research*, 29, 15–18.

(11) Miller M.W., Vayhinger J.E., Bowden D.C., Roush S.P., Verry T.E., Torres A.N. & Jurgens V.D. (2000) Drug treatment for lungworm in bighorn sheep: reevaluation of a 20-year-old management prescription. *The Journal of Wildlife Management*, 64, 505–512, https://doi.org/10.2307/3803248

(12) Eifler M.A., Slade N.A. & Doonan T.J. (2003) The effect of supplemental food on the growth rates of neonatal, young, and adult cotton rats (*Sigmodon hispidus*) in northeastern Kansas, USA. *Acta Oecologica*, 24, 187–193, https://doi.org/10.1016/s1146-609x(03)00084-5

(13) Ransome D.B. & Sullivan T.P. (2004) Effects of food and den-site supplementation on populations of *Glaucomys sabrinus* and *Tamiasciurus douglasii*. *Journal of Mammalogy*, 85, 206–215, https://doi.org/10.1644/bos-118

(14) López-Bao J.V., Palomares F., Rodríguez A. & Delibes M. (2010) Effects of food supplementation on home-range size, reproductive success, productivity and recruitment in a small population of Iberian lynx. *Animal Conservation*, 13, 35–42, https://doi.org/10.1111/j.1469-1795.2009.00300.x

(15) Loureiro F., Martins A.R., Santos E., Lecoq M., Emauz A., Pedroso N.M. & Hotham P. (2011) O papel do programa lince (LPN/FFI) na recuperação do habitat e presas do lince-ibérico no sul de Portugal. *Galemys*, 23: 17–25.

(16) Peterson C. & Messmer T.A. (2011) Biological consequences of winter-feeding of mule deer in developed landscapes in Northern Utah. *Wildlife Society Bulletin*, 35, 252–260, https://doi.org/10.1002/wsb.41

(17) Weidman T. & Litvaitis J.A. (2011) Can supplemental food increase winter survival of a threatened cottontail rabbit? *Biological Conservation*, 144, 2054–2058, https://doi.org/10.1016/j.biocon.2011.04.027

(18) Jerina K. (2012) Roads and supplemental feeding affect home-range size of Slovenian red deer more than natural factors. *Journal of Mammalogy*, 93, 1139–1148, https://doi.org/10.1644/11-mamm-a-136.1

(19) Angerbjörn A., Eide N.E., Dalén L., Elmhagen B., Hellström P., Ims R.A., Killengreen S., Landa A., Meijer T., Mela M. & Niemimaa J. (2013) Carnivore conservation in practice: replicated management actions on a large spatial scale. *Journal of Applied Ecology*, 50, 59–67, https://doi.org/10.1111/1365-2664.12033

(20) Jones J.D., Kauffman M.J., Monteith K.L., Scurlock B.M., Albeke S.E. & Cross P.C. (2014) Supplemental feeding alters migration of a temperate ungulate. *Ecological Applications*, 24, 1769–1779, https://doi.org/10.1890/13-2092.1

(21) Milner J.M., van Beest F.M., Schmidt K.T., Brook R.K. & Storaas T. (2014) To feed or not to feed? Evidence of the intended and unintended effects of feeding wild ungulates. *The Journal of Wildlife Management*, 78, 1322–1334, https://doi.org/10.1002/jwmg.798

(22) Yarnell R.W., Phipps W.L., Dell S., MacTavish L.M. & Scott D.M. (2014) Evidence that vulture restaurants increase the local abundance of mammalian carnivores in South Africa. *African Journal of Ecology*, 53, 287–294, https://doi.org/10.1111/aje.12178

(23) Sittenthaler M., Bayerl H., Unfer G., Kuehn R. & Parz-Gollner R. (2015) Impact of fish stocking on Eurasian otter (*Lutra lutra*) densities: A case study on two salmonid streams. *Mammalian Biology*, 80, 106–113, https://doi.org/10.1016/j.mambio.2015.01.004

(24) Selva N., Teitelbaum C.S., Sergiel A., Zwijacz-Kozica T., Zieba F., Bojarska K. & Mueller T. (2017) Supplementary ungulate feeding affects movement behaviour of brown bears. *Basic and Applied Ecology*, 24, 68–76, https://doi.org/10.1016/j.baae.2017.09.007

14.8. Provide supplementary water to increase reproduction/survival

https://www.conservationevidence.com/actions/2396

- **Six studies** evaluated the effects on mammals of providing supplementary water to increase reproduction/survival. Two studies were in Australia[2,6] and one each was in Oman[1], Portugal[4], Saudi Arabia[5] and the USA and Mexico[3].

COMMUNITY RESPONSE (0 STUDIES)

POPULATION RESPONSE (5 STUDIES)

- **Abundance (2 studies):** A replicated study in the USA and Mexico[3] found that providing supplementary water was associated with increases in desert bighorn sheep population size. A study in Oman[1] found that a released captive-bred Arabian oryx population initially provided with supplementary water and food increased over 14 years.

- **Reproduction (2 studies):** A study in Saudi Arabia[5] found that released captive-bred Arabian gazelles initially provided with supplementary water and food after release into a fenced area started breeding in the first year. A study in Australia[6] found that most female released captive-reared black-footed rock-wallabies provided with supplementary water after release into a large predator-free fenced area reproduced in the first two years.

- **Survival (2 studies):** A controlled, before-and-after study in Australia[2] found that most released captive-bred hare-wallabies provided with supplementary water, along with supplementary food and predator control, survived at least two months after release into a fenced peninsula. A study in Australia[6] found that over half of released captive-reared black-footed rock-wallabies provided with supplementary water after release into a large predator-free fenced area survived for at least two years.

BEHAVIOUR (1 STUDY)

- **Use (1 study):** A replicated study in Portugal[4] found that artificial waterholes were used by European rabbits and stone martens.

> ### Background
>
> In arid environments, artificial water sources may be provided to aid survival or population expansion for species of conservation concern (e.g. West *et al.* 2017). This may be done as part of translocation or reintroduction programmes and also for securing existing populations of threatened species.
>
> See also *Natural system modifications -Provide artificial waterholes in dry season* and *Climate change & severe weather -Provide dams/water holes during drought.*

West R., Ward M.J., Foster W.K. & Taggart D.A. (2017) Testing the potential for supplementary water to support the recovery and reintroduction of the black-footed rock-wallaby. *Wildlife Research*, 44, 269–279.

A study in 1982–1999 of a large desert area in Oman (1) found that a population of released captive-bred Arabian oryx *Oryx leucoryx* initially provided with supplementary water and food increased over 14 years, but then declined due to poaching. Oryx numbers in the wild peaked at >400 animals, 1–14 years after the release of 40 animals. Poachers (capturing live animals, especially females, for international trade) then removed at least 200 oryx over the next three years. Animals were taken back into captivity to re-establish a captive breeding program. Seventeen years after releases began, the captive population was 40, and approximately 104 remained in the wild, with a high male:female sex ratio. Arabian oryx became extinct in Oman in 1972. Founders for the initial captive herd were sourced from international collections. Forty individually marked oryx were released in 1982–1995. A sample of wild-born animals was individually marked to retain the marked proportion at 20–30%. The original released herd was provided with water and food for seven months after release. Population estimates were derived from sightings using mark-recapture analysis.

A controlled, before-and-after study in 2001 in five shrubland sites in Western Australia (2) found that most released captive-bred banded hare-wallabies *Lagostrophus fasciatus* and rufous hare-wallabies *Lagorchestes hirsutus* provided with supplementary water, along with supplementary food and predator control, survived at least two months after being released into a fenced peninsula. After 1–2 months, 10 of 16 rufous hare-wallabies and 12 of 18 banded hare-wallabies were still alive. Overall both rufous and banded hare-wallabies recaptured had similar body conditions to when they were released, although rufous hare-wallabies lost 12% of body condition while waiting for release in holding pens (data presented as a body condition index; see paper for details). Sixteen captive-bred rufous hare-wallabies and 18 captive-bred banded hare-wallabies were released at five sites in August 2001. Six rufous and nine banded-hare wallabies were placed in separate 3-ha enclosures with electrified fencing for 10–19 days before being released. Remaining animals were released directly into the wild. Supplementary water and food (kangaroo pellets, alfalfa) were made available to all hare-wallabies (those kept in holding pens and those not; duration of feeding not given). Hare-wallabies were monitored by radio tracking (once/week for 1.5 years after release) and live-trapping (at 4 and 8–9 weeks after release). Release areas were within a fenced peninsula where multiple introduced mammals were controlled or eradicated.

A replicated study in 1951–2007 in 10 desert sites in Arizona and New Mexico, USA, and the Gulf of California, Mexico (3) found that providing supplementary water at some sites was associated with increases in desert bighorn sheep *Ovis canadensis* population size. At three out of 10 sites where supplementary water was provided, it was associated with an increase in bighorn sheep populations. However, at one site, provision of water was associated with declines in sheep populations. The remaining six sites showed no association (data not presented). Data were obtained from historical records for ten sites with long-term survey and hunting information. Data included counts of bighorn sheep from both surveys and hunter harvests, and the number of watering sites provided.

A replicated study in 2009 in four agroforestry sites in Alentejo and Algarve, Portugal (4) found that artificial waterholes were used by European rabbits *Oryctolagus cuniculus* and stone martens *Martes foina*. European rabbits used four out of 16 artificial waterholes. At least one waterhole was used by stone martens (number of waterholes used by this species is not stated). In September and October 2009, sixteen artificial waterholes in four agroforestry estates dominated by cork *Quercus suber* (2–6 waterholes/estate) were monitored using camera traps. No description of the waterholes is provided. Waterholes were monitored for 7 or 14 days, using one camera trap/waterhole.

A study in 2011–2014 of a dry dwarf-scrubland site in Saudi Arabia (5) found that released captive-bred Arabian gazelles *Gazella arabica* initially provided with supplementary water and food after release into a fenced area started breeding in the year following the first releases. Seven females gave birth in August–September of the year after the first releases and all calves survived to at least the end of the year. Of 49 gazelles released over three years, 10 had died by the time of the final releases. In 2011–2014, three groups of captive-born gazelles, totalling 49 animals, were released in a 2,244-km² fenced reserve. They were moved from a wildlife research centre and held for 23 days to a few months before release in enclosures measuring 500 × 500 m. Water and food was provided for three weeks following release. Released gazelles were radio-tracked from the ground and air.

A study in 2011–2014 in a semi-arid area in South Australia (6) found that over half of released captive-reared black-footed rock-wallabies

Petrogale lateralis provided with supplementary water after being released into a large predator-free fenced area survived for at least two years and most females reproduced. Ten (five males, five females) of 16 rock-wallabies (63%) survived more than two years after being released. All five females that survived reproduced within 2–6 months of release. Over three years, 28 births from nine females were recorded. Between March 2011 and July 2012, sixteen captive-reared black-footed rock-wallabies (eight males, eight females; 1–5 years old) were released in three groups into a 97-ha fenced area. Ten of the 16 rock-wallabies were wild-born and fostered by yellow-footed rock-wallaby *Petrogale xanthopus* surrogate mothers in captivity. Introduced predators, common wallaroos *Macropus robustus* and European rabbits *Oryctolagus cuniculus* were removed from the enclosure. Supplementary water was provided in five 8-l tanks that were monitored with camera traps in 2011–2014. Rock-wallabies were fitted with radio-collars and tracked 1–7 times/week in 2011–2014. Trapping was carried out on seven occasions in 2011–2014.

(1) Spalton J.A., Lawrence M.W. & Brend S.A. (1999) Arabian oryx reintroduction in Oman: successes and setbacks. *Oryx*, 33, 168–175.

(2) Hardman B. & Moro D. (2006) Optimising reintroduction success by delayed dispersal: is the release protocol important for hare-wallabies? *Biological Conservation*, 128, 403–411, https://doi.org/10.1016/j.biocon.2005.10.006

(3) Wakeling BF., Lee R., Brown D., Thompson R., Tluczek M. & Weisenberger M. (2009) The restoration of desert bighorn sheep in the Southwest, 1951–2007: factors influencing success. *Desert Bighorn Council Transactions*, 50, 1–17.

(4) Loureiro F., Martins A.R., Santos E., Lecoq M., Emauz A., Pedroso N.M. & Hotham P. (2011) O papel do programa lince (LPN/FFI) na recuperação do habitat e presas do lince-ibérico no sul de Portugal. *Galemys*, 23, 17–25.

(5) Islam M.Z., Shah M.S. & Boug A. (2014) Re-introduction of globally threatened Arabian gazelles *Gazella arabica* (Pallas, 1766) (Mammalia: Bovidae) in fenced protected area in central Saudi Arabia. *Journal of Threatened Taxa*, 6, 6053–6060, https://doi.org/10.11609/jott.o3971.6053-60

(6) West R., Read J.L., Ward M.J., Foster W.K. & Taggart D.A. (2017) Monitoring for adaptive management in a trial reintroduction of the black-footed rock-wallaby *Petrogale lateralis*. *Oryx*, 51, 554–563, https://doi.org/10.1017/s0030605315001490

14.9. Graze herbivores on pasture, instead of sustaining with artificial foods

https://www.conservationevidence.com/actions/2398

- **One study** evaluated the effects of grazing mammalian herbivores on pasture, instead of sustaining with artificial foods. This study was in South Africa[1].

COMMUNITY RESPONSE (0 STUDIES)

POPULATION RESPONSE (1 STUDY)

- **Reproductive success (1 study)**: A site comparison study in South Africa[1] found that a population of roan antelope grazed on pasture had a higher population growth rate than populations provided solely with imported feed.

BEHAVIOUR (0 STUDIES)

Background

In highly managed populations of wild mammalian herbivores, locations of enclosures or other constraining features can determine what food is available for animals. Some populations may be maintained on food imported from elsewhere. However, making pasture available might provide a higher quality diet than can be offered with imported food and this might have positive effects on the population.

A site comparison study in 1995 of five conservation areas on a range of veld habitats in South Africa (1) found that in a population of roan antelope *Hippotragus equinus equinus* grazed on pasture, the population growth rate was higher than in populations provided solely with imported feed. The rate of increase of the pasture-fed population was higher than that of four other populations that were not pasture-fed (data presented as mean exponential rates). Population sex ratios, calving rates, population sizes and densities were not correlated with rates of population increase. Five conservation areas (each <3,000 ha) were studied. Population data were obtained in winter 1995. At one site, antelopes were grazed on pasture and, in the dry season, fed ≥0.5 kg of supplementary food/day (lucerne, antelope cubes and mineral lick). At the other four sites, antelopes solely received the supplementary feed, in varying proportions.

(1) Dörgeloh W.G., van Hoven W. & Rethman N.F.G. (1996) Population growth of roan antelope under different management systems. *South African Journal of Wildlife Research*, 26, 113–116.

Translocate Mammals

14.10. Translocate to re-establish or boost populations in native range

https://www.conservationevidence.com/actions/2397

- **Sixty-four studies** evaluated the effects of translocating mammals to re-establish or boost populations in their native range. Twenty studies were in the USA[5,8,9,11,12,13,19,20,23,26,31,32,34,35,38,42,48,51,52,56], eight in Italy[16,25,30,46,53,59,61,62], four in Canada[2,6,10,41] and South Africa[7,36,44,50], three in the Netherlands[14,33,47] and Spain[35,60,63], two in each of the USA and Canada[3,22], Zimbabwe[4,17], Sweden[15,18], Australia[28,57] and the USA and Mexico[45,49] and one in each of Uganda[1], the UK[24], Brazil[27], France[29], Portugal[39], Africa, Europe, North America[40], Botswana[43], Nepal[54], Chile[55], Slovakia[58], Ukraine, Slovakia and Poland[64] and one global study[21].

COMMUNITY RESPONSE (0 STUDIES)

POPULATION RESPONSE (62 STUDIES)

- **Abundance (22 studies):** Two studies (incuding one controlled and one before-and-after, site comparison study) in Spain[35] and Canada[41] found that translocating animals increased European rabbit[35] abundance or American badger[41] population growth rate at release sites. Fourteen studies (one replicated) in South Africa[7,44,50], the USA[9,23,20,31], the Netherlands[14], Italy[16,46,53,61], France[29] and Spain[60] found that following translocation, populations of warthogs[7], Eurasian beavers[14], red squirrels[16], roe deer[29], Alpine ibex[53], Iberian ibex[60], Cape mountain zebra[50], 22 species of grazing mammals[44], black bears[9], brown bear[61,46], bobcats[31] and most populations of river otters[23] increased. Two reviews in South Africa[36] and Australia[28] found that reintroductions (mainly through translocations) led to increasing populations for four of six species of large carnivores[36] and that over half of translocations were classified as successful[28]. One replicated study in the USA and Mexico[45] found that translocating desert bighorn sheep did not increase the population size. Two studies (one replicated) and a review in USA and Canada[3], the USA[34] and Australia[57] found that translocated American martens[34], and sea otters[3] at four of seven sites, established populations and that translocated and released captive-bred macropod species[57] established populations in 44 of 72 cases. A study in Italy[59] found that following the translocation of red deer, the density of Apennine chamois in the area almost halved. A worldwide review[21] found that translocating ungulates was more successful when larger numbers were released, and small populations grew faster if they contained more mature individuals and had an equal ratio of males and females.

- **Reproductive success (16 studies):** A controlled study in Italy[62] found that wild-caught translocated Apennine chamois reproduced in similar numbers to released captive-bred chamois. Fourteen studies (four replicated) in Canada[2], the USA[5,8,19,20,31,38], Zimbabwe[4], South Africa[7], the UK[24], Italy[46], the Netherlands[33,47] and Slovakia[58] found that translocated black and white rhinoceroses[4], warthogs[7], common dormice[24], European ground squirrels[58], cougars[19], bobcats[31], brown bears[38,46], sea otters[2], river otters[5,8,20] and some Eurasian otters[33,47] reproduced. A study in the Netherlands[14] found that translocated beavers were slow to breed.

- **Survival (39 studies):** Four of five studies (including three controlled, two replicated and one before-and-after, site comparison study) in the USA[11,19,32], Canada[41] and Chile[55] found that wild-born translocated long-haired field mice[55], female elk[11], cougars[19] and American badgers[41] had lower survival rates than non-translocated resident animals. One found that translocated Lower Keys marsh rabbits[32] had similar survival rates to non-translocated resident animals. Five of four studies (two replicated, four controlled) and two reviews in Canada[10], Canada and the USA[22], the USA[48], Italy[62], Sweden[15] and Africa, Europe, and North America[40] found that wild-born translocated swift foxes[10,22], European otters[15], black-footed ferret kits[48] and a mix of carnivores[40] had higher survival rates than released captive-bred animals. One study found that wild-born translocated Apennine chamois[62] had a similar survival rate to released captive-bred animals. Twenty of twenty-one studies (including two replicated and one before-and-after study) and a review in Nepal[54], France[29], Italy[25,30,59], Portugal[39], Ukraine, Slovakia and Poland[64], Canada[2,6], USA[5,12,13,20,26,38,56], Brazil[27], Uganda[1], South Africa[7], Zimbabwe[4,17] and Botswana[43] found that following translocation, populations of or individual mammals survived between two months and at least 25 years. The other two studies found that two of 10 translocated white rhinoceroses[1] died within three days of release and an American marten[26] population did not persist. A review in Australia[28] found that over half of translocations, for which the outcome could be determined, were classified as successful. Two of three studies (one replicated) and one review in Sweden[18], the UK[24], the Netherlands[47] and the USA and Mexico[49] found that following release of wild-caught translocated and captive-bred animals, European otters[18,47] and common dormice[24] survived three months to seven years. The review found that most black-footed ferret[49] releases were unsuccessful at maintaining a population. A replicated study in the USA[51] found that following translocation of bighorn sheep, 48–98% of their offspring survived into their first winter.

- **Condition (3 studies):** Three studies (including one replicated, controlled study) in the USA[37,52] and Italy[46] found that following translocation, populations of elk[37] had similar levels of genetic diversity to non-translocated populations, descendants of translocated swift fox[52] had genetic diversity at least as high as that of the translocated animals and brown bear[46] genetic diversity declined over time.

BEHAVIOUR (9 STUDIES)

- **Use (7 studies):** A study in Italy[53] found that following translocation, Alpine ibex used similar habitats to resident animals. Two of four studies (including one randomized, controlled study) in the USA[13,42], Netherlands[33] and Botswana[43] found that following translocation (and in one case release of some captive-bred animals), most Eurasian otters[33] settled and all three female grizzly bears[13] established ranges at their release site. The other two studies found that most nine-banded armadillos[42] and some white rhinoceroses (when released into areas already occupied by released animals)[43] dispersed from their release site. Two studies (one replicated) in Spain[60,63] found that following translocation, Iberian ibex[60] expanded their range and roe deer[63] increased their distribution six-fold.

- **Behaviour change (2 studies):** A replicated controlled study in Chile[55] found that following translocation, long-haired field mice[55] travelled two-to four-times further than non-translocated mice. A controlled study in Italy[62] found that wild-caught translocated Apennine chamois[62] moved further from the release site than released captive-bred animals.

Background

Translocations involve the intentional capture, movement and release of wild-caught mammals into the wild to re-establish a population that has been lost, or augment an existing population. This can reduce the risk of inbreeding, help safe guard small populations from extinction due to catastrophic events and/or increase the range of a species and therefore the maximum possible population. Translocations can also be used to move mammals to areas where threats have been removed, such as invasive predators on islands. However, translocations are typically expensive and may risk spreading pathogens to previously unexposed areas.

Release techniques vary considerably, from 'hard releases' involving the simple release of individuals into the wild to 'soft releases' which involve a variety of adaptation and acclimatisation techniques before release or post-release feeding and care. This action includes studies describing the effects of translocation programmes that do not provide details of specific release techniques. Studies that describe or compare specific release techniques, such as use of holding pens at release sites, or providing supplementary food, water or artificial refuges/breeding sites are described under each specific action.

This action includes studies where animals were released in groups but not studies where releases of different group sizes were compared, or where animals were released in family or social groups (including groups where social animals have been pre-conditioned together prior to release in holding pens). For those studies, see *Release translocated/captive-bred mammals in larger unrelated groups* and *Release translocated/captive-bred mammals in family/social groups*.

A study published in 1961 on savannah in a national park in Uganda (1) found that after release of 10 translocated white rhinoceros *Ceratoiherium simum cottoni*, two died within three days. One animal died one day after release and the other died three days after release. Both were adult females. One had a female calf that was taken into captivity. The remainder were all thought to have survived in the short-term, although only four of the seven were resighted by the end of the study. Ten rhinoceroses (four adult females, three half-grown males, one male calf and two female calves) were translocated to the park and released in March 1961. Duration of monitoring not stated.

A study in 1969–1978 in coastal waters close to Vancouver Island, British Columbia, Canada (2) found that a population of translocated sea otters *Enhydra lutris* persisted over nine years and reproduced. Eight and nine years after the translocation of 89 sea otters, a population of at least 67 individuals persisted within the surroundings of the translocation area. Pups (7 individuals), dependent juveniles (4 individuals) and subadult otters (10 individuals) were observed. A total of 89 sea otters were translocated in 1969, 1970 and 1972 from Alaska, USA to the Bunsby Islands along the west coast of Vancouver Island. No details about the translocation procedure are provided. Otters were counted almost daily by boat, scuba diving and aerial census in June-July 1978. Further census details are not provided.

A replicated study in 1965–1981 at seven coastal sites in Oregon, Washington, and Alaska, USA, and British Columbia, Canada (3) found that translocated sea otters *Enhydra lutris* established stable populations at four of the seven release sites. In south-eastern Alaska, where 412 sea otters were released, 479 were counted six years after the last release. In British Columbia, after 89 sea otters were released, 70 (including some pups) were seen five years after the last release. In Washington, 59 sea otters were released at two sites, with 36 (including one pup) counted across these sites 12 years later. In Oregon, 93 were released at two sites, but only one was found 10 years later. Fifty-five were released on the Pribilof Islands, Alaska, but only three were found nine years later. In 1965–1972, a total of 708 sea otters were translocated from Amchitka Island and Prince William Sound, Alaska to seven coastal sites where they had previously been extirpated. Populations were surveyed in 1971–1975 by boat and plane and from land.

A study in 1975–1981 of savannah in a national park and surrounding areas in Zimbabwe (4) found that translocated black rhinoceros *Diceros bicornis* and white rhinoceros *Ceratotherium simum* established populations and started to breed. Five out of seven translocated black rhinoceroses survived at least six years after release and at least one calf was born. Up to nine out of 10 translocated white rhinoceroses survived at least six years after release, with at least seven calves born. Together with immigrant animals, the white rhinoceros population numbered 23–25 individuals at that time, in widely dispersed locations (movements of 22–130 km from release points were recorded). Black rhinoceroses and white rhinoceroses were translocated from areas of encroaching human activities. Seven black rhinoceroses (four adult males, two adult females and a male calf) were translocated in October–December 1975. Ten white rhinoceroses (one adult male, one adult female, two sub-adult males and six sub-adult females) were translocated and released in two groups, reflecting two areas of capture, in April 1975.

A replicated study in 1982–1986 in two wetland sites in Missouri, USA (5) found that most translocated river otters *Lutra canadensis* survived for at least a year after release and reproduction occurred at both release sites from two years following releases. Of otters whose status could be confirmed one year after release, 15 of 17 were alive at one site and 10 of 14 survived at the second site. Reproduction was confirmed annually at both release sites from the second year after releases. Nineteen wild-caught otters were released at a 4,455-ha wildlife refuge in March–May 1982 and 20 were released at a 2,251-ha wildlife area in April 1983. All otters were implanted with radio-transmitters. Monitoring occurred daily for the first three weeks and then 2–4 times/week until death or transmitter failure (typically at 12–14 months).

A review of studies in 1964–1982 in Newfoundland, Canada (6) found that after translocation, 17 of 22 caribou *Rangifer tarandus* populations persisted for at least 1–20 years. Between 1964 and 1982, a total of 384 caribou were translocated to 22 sites in Newfoundland. Caribou populations at sites were resurveyed using unspecified methods in 1981–1982.

A study in 1976–1990 in a shrubland reserve in Cape Province, South Africa (7) found that translocated warthogs *Phacochoerus aethiopicus* survived, bred successfully and abundance increased over approximately 10 years. Ten to 11 years after the release of 20 warthogs, numbers of warthogs counted increased to 641. Thirteen to 14 years after release, 361 individuals were counted. Separate surveys of dead warthogs found that the population comprised a mixture of age groups, including juveniles (<1 year: 67–144 individuals), yearlings (1–2 years: 31–62 individuals) and adults (>2 years: 143–204 individuals). The majority of yearling and adult females examined (80–100%) were pregnant. In 1976–1977, twenty warthogs were introduced into a 6,493-ha reserve dominated by dense thorny scrub. Warthogs were surveyed by helicopter in 1981–1990. In 1987–1990, warthogs were shot at random from helicopters in order for carcasses to be examined and population age structure estimated.

A replicated study in 1982–1991 at two riverine sites in Pennsylvania, USA (8) found that translocated river otters *Lutra canadensis* released in areas with no existing otters settled and reproduced in the 6.5–8 years after release. Otter scats were widely found in both release areas, confirming continued otter presence. Two juveniles, live-trapped and released by hunters three years after translocations, provided evidence of breeding at one site. At the other site, four of seven otters killed by trappers, between three and seven years after translocations, were considered to be offspring of released animals. Twenty-two wild-caught otters (11 male, 11 female) were released in Pine Creek in 1983–1984 and four (two male, two female) were released in Kettle Creek in 1982. Follow-up monitoring of scats occurred in September–December 1990 (Pine Creek) and April 1991 (Kettle Creek). Additionally, carcasses were examined and trapping incidents reviewed.

A study in 1985–1993 of forest across two mountain areas in Arkansas, USA (9) found that a translocated population of black bears *Ursus americanus* grew steadily after animals were released. Following release of an estimated 254 bears, the population grew to >2,500 bears 20 years later. Litter sizes in two study areas were 1.6–2.4 and survival to one year was 40–65%. Black bears were extirpated from Arkansas sometime after 1931, apart from a small isolated population. Approximately 254 bears were released in 1958–1968 into three main areas from which bears had been lost. Released animals were wild-caught in Minnesota and in Manitoba, Canada. Bear densities were estimated in two study areas by mark-recapture at bait stations in 1985–1990. Litter sizes were estimated from bears radio-collared in 1988–1990 and monitored through to 1993.

A replicated, controlled study in 1990–1992 at two grassland sites in Alberta, Canada (10) found that translocated, wild-born swift foxes *Vulpes velox* had higher post-release survival rates than did released captive-born animals. No statistical analyses were performed. Nine months after release into the wild, 12 out of 28 (43%) wild-born translocated swift foxes were known to be alive, compared with at least two out of 27 (7%) released captive-born swift foxes. In May 1990 and 1991, a total of 27 captive-born and 28 wild-born swift foxes were released simultaneously. Wild-born animals had been captured in Wyoming, USA, 4–7 months before release and were quarantined for ≥30 days. Animals were released without prior conditioning in holding pens. Foxes were radio-collared and monitored from the ground and air, for at least nine months.

A controlled study in 1980–1990 in a large mountainous area dominated by coniferous forest in Oregon, USA (11) found that translocated female elk *Cervus canadensis* had a lower survival rate than non-translocated female elk. The average annual survival rate of translocated female elk (77% of 35 individuals) was siginificantly lower than that of non-translocated female elk resident at the release sites (92% of 35 individuals) and also appeared lower than the average annual survival rate of female elk in the whole study area (89% of 184 individuals, this result was not compared statistically to other survival rates). The study area included six national forests and eight state wildlife management districts. In 1980–1990 one hundred and eighty-four resident female elk were released at their capture site. In 1987–1990, 35 female elk were caught, radio-collared and translocated. A further 35 resident female elk were radio-collared in 1988–1989 at the translocation release site. Distances between capture and release sites of translocated elk are not given. Both non-translocated and translocated elk were located 2–4 times/month, mostly from an aircraft.

A study in 1987–1992 in a subalpine coniferous forest in Idaho, USA (12) found that approximately a quarter of translocated woodland caribou *Rangifer tarandus caribou* survived or had stayed at the release site two-four years after release. Fourteen out of 60 (23%) translocated woodland caribou survived two-four years after being released into the wild. Seven translocated caribou left the study area over the five-year study, of which six were during the first year after release. Twenty-seven caribou died during the same period (3 during the release process) and the outcome for 12 animals was unknown due to radio-collar failure. The average annual survival rate was 74%. Between 1987 and

1990, sixty woodland caribou were caught in British Columbia, Canada and released in the Selkirk Mountains, USA after 72 hours. Caribou were radio-tagged and were monitored weekly, from an aircraft, until February 1992.

A study in 1990–1993 in forests in Montana, USA (13) found that three translocated female grizzly bears *Ursus arctos horribilis* successfully established ranges around the release site and that two survived for at least three years. All three translocated bears established movement and habitat-use patterns similar to those of non-translocated bears (no data reported). Two of the three bears survived for at least three years. Three adult female bears were translocated from the border area of Canada and the USA to the Cabinet Mountains in Montana, USA. Bears were monitored by radio-tracking until their collars failed or to the end of the study period after three years.

A study in 1988–1993 of a freshwater estuary at a national park in the Netherlands (14) found that translocated Eurasian beavers *Castor fiber* increased in number, although were slow to breed. From 42 animals released over four years, the population grew to 47 two years after releases (including 27 animals ≥1 year old). Only in this final year did the number of births exceed the number of animals lost (through dispersal, death or other disappearance). Population Viability Analysis found that the population was unlikely to be viable (80% of simulated populations going extinct within 100 years) unless low breeding productivity was a temporary response to translocation. A total of 42 beavers, translocated from Germany, were released in October or November of 1988–1991. They were monitored by radio-tracking (from boat and plane) and direct observations of marked animals.

A replicated, controlled study in 1989–1993 in two rivers in southern Sweden (15; same experimental set-up as 18) found that wild-born translocated European otters *Lutra lutra* had a higher survival rate than did released captive-bred otters. One year after release, the survival rate of wild-born translocated otters (79%) was higher than that of released captive-bred otters (42%). Between 1989 and 1992, eleven wild-born otters and 25 captive-bred otters were released into two rivers in south-central Sweden. Thirty-four otters were released in one river catchment and two in the other. Wild-born otters were live-trapped along the Norwegian coast. Captive-bred otters were descendants of two captive females. All otters were around one year old when released. All except one were released between February and June. All were fitted with an implanted radio-transmitter and monitored for one year on 64% of days.

A study in 1986–1996 in a forest and heathland reserve in Lombardy, Italy (16) found that a population of translocated red squirrels *Sciurus vulgaris* increased in size over 10 years and expanded to nearby woodlands. Three to four years after eight translocated squirrels were released, the population had increased to 38–126 squirrels. By ten years after the first release, the squirrel population had further increased and colonized all five woodlands (squirrel abundance in 1996 is not given). Between December 1986 and August 1987, eight red squirrels were translocated to a 3,500-ha reserve containing 800-ha of woodland, from which the species was extirpated in the 1940s. In February 1990, squirrel nests were counted on a 70-ha plot and the population size was estimated based on a mean of 4.5 nests/squirrel. In spring 1990 and 1996, all five woodland blocks at the release site were searched for 30 min to 1 hour for dreys or typical feeding signs.

A before-and-after study in 1996 in a mixed miombo and mopane woodland reserve in the Midlands province, Zimbabwe (17) found that three translocated cape pangolins *Manis temminckii* survived at least a month after release and one established a new home range. During the sixty-five days after release, one translocated pangolin set up a home range covering 0.45 km^2. Of two adult females translocated, one returned to her original home range nine days after translocation and the other moved for 30 days (on average 1.25 km/day), without returning to the capture site or establishing a home range. One pangolin had been retrieved from a poacher and its origin and length of time in captivity were unknown. The two females were caught, radio-tagged and radio-tracked in their original capture location (for an unspecified period) before being moved and released about five and 18 km from their known home ranges within 24 hours of capture. Translocations were carried out to study effectiveness of releasing pangolins confiscated from poachers. Pangolins were monitored by radio-telemetry, and located during daytime by tracking on foot for approximately a month after release.

A study in 1989–1992 at seven lakes in boreal forest in Sweden (18; same experimental set-up as 15) found that following release of European otters *Lutra lutra* (a mix of wild-caught translocated and captive-bred animals), at least 38% survived for almost a year or longer. Fourteen otters established home ranges and were still alive when last recorded, 362–702 days after release. Eight further otters were monitored until their transmitters failed or they moved out of radio contact, 89–219 days after release. Fourteen were known to have died, 18–750 days after release. Otter origin (wild-caught or captive-bred) did not affect movement distance. In 1989–1992 thirty-six otters (11 wild-caught, translocated animals and 25 captive-bred) were released in lakes and rivers in southern Sweden. Otters were fitted with radio-transmitters. Radio-tracking was carried out at least monthly, in 1989–1992.

A study in 1989–1993 at nine temperate shrubland and coniferous woodland sites in New Mexico, USA (19) found that survival rates of translocated cougars *Puma concolor* were lower than those of resident populations, and two translocated females produced offspring. Nine of 13 cougars (69%) died within four years of translocation. Annual survival rates of translocated female (55%) and male (44%) cougars were lower than of non-translocated resident animals (86%). Two translocated females produced offspring. The main cause of mortality was from aggressive interactions with other cougars. In April 1989, one cougar was released at one site in the Cibola National Forest, New Mexico. From December 1990 to June 1991, thirteen cougars were released in eight sites in the Sangre de Christo Mountains, New Mexico. Released animals were radio-tracked by air or from the ground through to January 1993. Survival rates of translocated cougars were compared to those of 15 radio-tracked cougars that had not been translocated.

A study in 1995–1996 of a wildlife refuge with several wetland habitats in Indiana, USA (20) found that following translocation of North America river otters *Lutra canadensis*, most survived at least one year after release and breeding occurred in the second year post-release. Survival one year post-release was estimated at 71%. Three otter litters were documented in the second year after release. Confirmed mortalities were three otters killed by vehicles, one dying from research-related causes and one dying of an unknown cause. River otters were extirpated from Indiana by 1942. Twenty-five otters (15 male, 10 female) were translocated from Louisiana and released in a 3,125-ha refuge in Indiana, on 17 January 1995. Fifteen otters were radio-tracked five times/week for 16 weeks, and three times/week for up to one year. Field surveys and visual observations were also used, including to document breeding activity.

A worldwide review of 33 studies (21) found that translocating ungulates (Artiodactyla) to re-establish populations in their native range was more successful when larger numbers of animals were released, and small populations grew faster if they contained more mature individuals and had an equal ratio of males and females. All 10 translocated populations of ≥20 animals increased in number (by an average of 17%), whereas six of 23 translocated populations with ≤20 animals decreased. Small translocated populations (≤20 animals) were more likely to increase if they contained more mature individuals (females ≥3 years of age; males ≥5 years) and had an equal sex ratio (data reported as statistical model results). Analyses included 33 re-introduction studies involving nine ungulate species (including sheep, goats, elk, bison, reindeer and gazelle). Groups of 2–69 wild-caught animals were released within their native range and observed over 3–9 years (locations not reported). Studies were published (between 1959 and 1998) and unpublished (dates not reported).

A review of studies in 1989–1991 in prairie sites in Canada and the USA (22) found that following release, translocated wild-caught swift foxes *Vulpes velox* had higher survival rates than did captive-bred released swift foxes. Over an unspecified time period, 59% of wild-caught translocated swift foxes survived while three of 41 (7%) released captive-bred swift foxes survived. In 1989–1991, thirty-three wild-caught, adult foxes and 41 captive-bred foxes, born the previous year, were released in the spring. Methods used for monitoring animals were unclear from the review.

A study of projects carried out in 1976–1998 across 48 states in the USA (23) found that following translocations, river otter *Lutra canadensis* populations and ranges expanded in most states. Of 21 states with reintroduction programs, 15 reported having growing river otter populations, one reported a stable population and three reported stable to growing populations. Two states reported that it was too soon into their programs to judge population trends. Evidence of reproduction was reported from 18 states (82% of states with reintroductions), and range expansion was reported in 17 states (77%). In 1976–1998, river otter releases totalled 4,018 animals in 21 states. In six states, otters had been extirpated while in 15, reintroductions took place in parts of the state from which otters were absent. Releases involved an average of 19.6 otters/site. Information was gathered from telephone interviews in August–September 1998.

A replicated study in 1993–2002 in seven woodland sites across England, UK (24) found that following releases of some wild-born translocated but mainly captive-bred common dormice *Muscardinus avellanarius*, populations persisted for at least three months to over seven years and all reproduced. In at least three of seven releases, dormouse populations were stable or increased from 19–57 released individuals to 40–55 individuals between two and seven years later. At one site, only one individual was detected 7–8 years after the release of 52 individuals in two batches. In three populations, the number of released animals is not provided, but populations persisted for at least three months and up to at least three years after release. Animals in all seven populations bred in the wild. Releases took place in 1993–2000 into woodlands in Cambridgeshire, Nottinghamshire, Cheshire, Warwickshire, Buckinghamshire, Yorkshire and Suffolk. Monitoring continued to 2000–2002. Precise numbers and origins of dormice released are not given for all sites. Most were captive-bred, but some were wild-born translocated animals. Some dormice were kept in pre-release holding pens, sometimes for several weeks, before release. Nest boxes and supplementary food were provided at least at some sites (see paper for further details).

A replicated study in 1977–2002 in four alpine shrub and meadow sites in the Eastern Italian Alps, Italy (25) found that translocated alpine marmot *Marmota marmota* populations persisted for at least five years. At the first translocation site, 23 marmot families (28.4 family units/km^2) were counted 22 and 25 years after release. At the second site, 13 marmot family groups were counted 16 years after release (13.8 family units/km^2). After 12 more marmots were added to the second site in 2001, the population increased to 18 family units in 2002. A further two marmot populations were described as persisting for 5–7 years with 11–16 family groups (assisted by some restocking in one site). In 1977, 1983, 1995 and 1997, alpine marmots were released in four sites (150, 168, 472 and 1,005 ha respectively) in the Friulian Dolomites Natural Park. The number of individuals released is not reported. The origin of animals is not explicitly stated, but releases appear to be of translocated wild marmots. In May 1999–2002, winter burrows were located as marmots emerged from hibernation. Marmots were identified by tracks in the snow and each winter burrow was considered to be occupied by one family unit. Authors state that marmots were released in many isolated areas from the 1960s onward, but introduction was only successful in a few of them.

A study in 1989–1998 in two forest sites in Vermont, USA (26) found that after translocation of American martens *Martes americana*, the population did not persist. One to six years after introductions, there was evidence that 3–4 martens were present in the area but, after seven to eight years, there was no evidence of a marten population. In 1989–1991, a total of 115 martens (88 males, 27 females) were captured in Maine and New York State and released at two sites in southern Vermont. Forty of the martens were held in boxes at the release site for several days before release and 75 were released immediately after transport to the release site. Thirteen martens were fitted with radio-collars and monitored using telemetry until March 1991. In January–February 1990, surveys were carried out for marten tracks in the snow. In October 1994 to January 1995, January–March 1998 and the summers of 1997 and 1998, camera traps were placed at 20–285 locations to survey martens.

A study in 1994–2001 in two forest reserves in Espírito Santo, Brazil (27) found that translocated maned sloths *Bradypus torquatus* survived over 13 months and up to at least 36 months after release. All five translocated sloths survived the whole length of the post-release monitoring period (9–13 or 36 months). Two female sloths gave birth but all young were predated. Moving/resting and feeding time and daily distances travelled were not related to time since release. Between 1994 and 1999, five sloths were translocated from within or close to urban areas into two forests (500–900 ha, encompassing reserves and private forest land). Sloths were radio-collared and monitored 1–3 days/month for 9–13 months (four animals) and 36 months (one animal). Each sloth was observed from 07:00 to 17:00 h for totals of 182–509 hours. Data on activity budgets, home range size and diet were collected.

A review study of 66 translocations of 14 mammal species in Western Australia (28) found that over half of translocations, for which the outcome could be determined, were classified as successful. Out of

20 mammal translocations with a confirmed outcome, 11 (55%) were classed as successful and nine (45%) as non-successful. At the time of the review, the outcome of 46 translocations (68% of all translocations studied) remained uncertain. Species translocated were quokka *Setonix brachyurus*, black-flanked rock-wallaby *Petrogale lateralis*, tammar wallaby *Macropus eugenii*, brush-tailed bettong *Bettongia penicillata*, boodie *Bettongia lesueur*, common wallaroo *Macropus robustus*, numbat *Myrmecobius fasciatus*, southern brown bandicoot *Isoodon obesulus*, western barred bandicoot *Perameles bougainville*, western ringtail possum *Pseudocheirus occidentalis*, greater stick-nest rat *Leporillus conditor*, shark bay mouse *Pseudomys fieldi*, Thevenard Island mouse *Leggadina lakedownensis* and pebble-mound mouse *Pseudomys* sp. In 1993–2002, between 5–188 individuals of each species were translocated to different locations. Invasive mammals were controlled in some recipient sites. Two translocations included some captive-bred animals but most were translocated from wild populations. The definition of successful translocation was not stated for most species but, for others, it included measures of population increase and persistence.

A study in 1995–2002 in a mixed oak forest reserve in the south of France (29) found that following translocation in groups (alongside other associated actions), approximately half of female roe deer *Capreolus capreolus* survived over one year after release and that overall the deer population increased six years after the translocations began. Twenty-six out of 49 (53%) translocated female roe deer survived over one year post-release. Of the animals that died in the first year, 35% of mortality occurred within the first month after release. After six years the deer population had increased to 0.47 deer/km^2 compared to 0.06 deer/km^2 in the first year after translocation began. In February 1995–1997, fifty-two male and 52 female roe deer were translocated from Northern France into a 3,300-ha forest reserve in Southern France in seven release sessions. Animals were released in groups of approximately 15 individuals. They were initially placed into enclosures for 2–10 days and provided food during this time (pellets and fresh vegetables) prior to release. Forty-nine females (21 <1 year old and 28 >1 year old) were radio-tagged and were located from a vehicle once or twice each week, over one year post-release. In addition, surveys were carried out on foot (6 transects, each 5–7 km long) eight times a year in February-March

1996–2002 to estimate population growth. Deer were present in low numbers prior to translocation.

A study in 1999–2003 in a temperate forest site in northern Italy (30) found that most translocated brown bears *Ursus arctos* survived 2–3 years after release. Two to three years after release of 10 bears, at least eight were alive. In 1999–2002, ten bears (3 males, 7 females; all 3–6 years old) were captured in two sites in Slovenia and fitted with radio-collars and ear-tag transmitters. Animals were released in Adamello-Brenta Natural Park, Italy. Bears were located from the ground twice each day using radio antennae, from May 1999 to October 2003.

A study in 1988–1991 on an offshore island dominated by temperate forest in Georgia, USA (31) found that translocated bobcats *Lynx rufus* increased in numbers and reproduced in the wild. One year after the first releases, population density was 1 bobcat/10 km^2. One year after the second releases, population density was 3 bobcats/10 km^2. Over the two years after the first releases, 12 offspring were born. In September–December of 1988–1989, thirty-two bobcats fitted with radio-collars were released on Cumberland Island. Bobcats had previously become extinct on the island, in 1907. Radio signals were monitored throughout the year from the ground or from an aircraft. If females showed reduced movement, their location was visited to identify if they had given birth.

A replicated, controlled study in 2002–2004 on two islands in Florida, USA (32) found that translocated Lower Keys marsh rabbits *Sylvilagus palustris hefneri* had post-release survival rates similar to those of animals in established populations. Of rabbits whose fate was known, nine of eleven (81%) translocated to one island survived ≥5 months (two were predated) and all six (100%) translocated to another island survived ≥5 months. Eleven out of 14 (79%) caught and released at capture sites survived ≥5 months, with two predated and one dying from unknown causes. Transmitter failure curtailed monitoring of two further rabbits from these groups. Twelve rabbits, caught in 2002, were released within two hours of capture onto a nearby rabbit-free island. Seven rabbits, caught in 2004, were released onto a different rabbit-free island. In 2002, nine rabbits were also released at respective capture sites. Rabbit survival was determined by radio-tracking.

A study in 2002–2005 in two wetland areas in the Netherlands (33) found that following translocation, and release of some captive-bred animals, most Eurasian otters *Lutra lutra* settled in their release areas, where successful breeding then occurred. After three weeks, 14 of 23 otters settled within their release areas, while two died and seven moved away from release areas. Three years after the first translocations, five female otters had successfully reproduced, producing nine young. At this time, the total population was 12 otters. In 2002, fifteen wild-caught otters were released at one site. At a second site, in 2004–2005, eight animals, comprising a mix of wild-caught and captive-bred individuals, were released. Before release, animals were fitted with radio-transmitters and DNA samples were taken. Following release, otters were monitored by radio-tracking and by collection of faeces, which was analysed to identify animals individually.

A study in 2001–2003 in woodland across Peninsula Michigan, USA (34) found that translocated American martens *Martes americana* established a population. Ninety-four trapped martens had a sex ratio of 1.5 males for each female (1.9:1 considering just adults). This was not significantly different from the ratio of 2:1 which authors stated that for trapped animals, indicated that the harvest was sustainable. The age ratio was 3.3 juveniles (≤1.5 years old) for each adult (≥2.5 years old) female. This also was not significantly different from the ratio of 3:1, stated as indicating a sustainable harvest. Translocations into five areas in Peninsula Michigan, where martens had been extirpated, occurred in 1955–1957, 1968–1970 and 1979. These involved 276 martens. In 1989–1992, sixty-six martens were translocated internally within Peninsula Michigan. Marten trapping was permitted in limited areas from 2000. Sex and age data were determined for 94 martens obtained from commercial trappers in 2001–2003.

A controlled study in 1999–2002 in a shrubland site in Huelva, Spain (35) found that translocation of European rabbits *Oryctolagus cuniculus* increased rabbit abundance. Average rabbit abundance over the study was higher in translocation plots (5.0 pellets/m^2) than in non-translocation plots (1.9 pellets/m^2). The study was conducted in two 4-ha plots (≥1 km apart) in Doñana National Park. Annually, over three years, two batches of 32–34 rabbits were translocated into one plot and no translocations occurred in the other plot. The first two batches were translocated in November 1999 and February 2000. Plots were then switched such that the second and third pairs of translocations

(December 2000 and February 2001, and January and March 2002) were released into what was the non-translocation plot for the first batch. Between September 1999 and November 2002, rabbit abundance was estimated every two months by counting the number of pellets in 33 fixed-position 0.5-m diameter sampling points/plot. Wild rabbits were present in all plots prior to translocations beginning.

A review of studies conducted in 1985–2005 at 11 grassland and dry savanna sites in Eastern Cape, South Africa (36) found that reintroductions (mainly through translocations) of large carnivores led to increasing population sizes for four of six species. Twenty years after the first releases, there were 56 lions *Panthera leo* at seven sites (from 31 released), 41 cheetahs *Acinonyx jubatus* (seven sites, 40 released), 24 African wild dogs *Lycaon pictus* (two sites, 11 released) and 13 spotted hyena *Crocuta crocuta* (three sites, 11 released). There were reductions or unknown trends in two species with seven known surviving leopards *Panthera pardus* (five sites, 15 released) and an unknown number of servals *Leptailurus serval* (though known to be present — two sites, 16 released). Releases were made in 1985–2005, into 11 protected areas. Most schemes involved translocations of wild-caught animals, but at least one of seven lion reintroductions involved captive-bred animals. Monitoring methods are not specified.

A replicated, controlled study (year not provided) in six protected areas across five states in western USA (37) found that translocated elk *Cervus canadensis* populations had similar levels of genetic diversity compared to non-translocated populations. The genetic diversity (expressed as 'expected heterozygosity', He) of translocated elk populations (0.51–0.60 He) did not differ significantly from that of the source population (0.60 He). Between 1912 and 1985, five populations of elk were founded using animals translocated from source herds in Yellowstone National Park. Translocated populations had different founding histories but starting populations ranged from 12 to >150 individuals. The size of the translocated populations at the time of the research was 500–10,000 elk. In each population, 17–43 samples of skin or muscle tissue were collected from hunter-harvested elks. Tissue samples were frozen or stored in ethanol before DNA extraction. The dates of sample collection and laboratory work are not provided.

A study in 1990–2005 in a forest site in Montana and northern Idaho, USA (38) found that most translocated female brown bears *Ursus arctos* survived for at least one year after release and at least one of four reproduced in the release area. Three of the four translocated bears (75%) survived for at least one year. The fourth bear died of unknown causes. After 12 years, at least one translocated bear was alive and had produced two litters with different males. In 1990–1994, four young wild female bears were caught in southeastern British Columbia and released in the Cabinet Mountains (no more than one released each summer). Radio-satellite monitoring was carried out over 1–2 years after release. Hair samples were collected from 2000–2005 and genetic analysis was used to determine presence of translocated bears and their offspring.

A study in 2001–2003 in agricultural fields and mixed woodland in a mountain range in Fundão, Portugal (39) found that most translocated roe deer *Capreolus capreolus* survived more than two years after release. At least five out of seven translocated roe deer (71%) survived more than two years after release. One was found dead and the radio-transmitted of another stopped working. In winter 2001, fourteen adult roe deer were released into a 50-km² area. Roe deer had been absent for the area for more than a century. Seven of the 14 deer were radio-tagged. Tagged animals were located daily during summer 2002 (May–September) and winter 2002–2003 (November–March).

A review in 2008 of 49 studies in 1990–2006 of carnivore reintroductions in Africa, Europe, and North America (40) found that wild-born translocated animals had higher survival rates than did released captive-bred animals. Survival of wild-born translocated carnivores (53%) was higher than survival of captive-born animals following release (32%). The review analysed 20 reintroductions of 983 captive-bred carnivores and 29 reintroductions of 1,169 wild-caught carnivores. Post-release monitoring ranged in duration from 6 to 18 months.

A before-and-after, site comparison study in 2002–2006 in two alpine grassland sites in British Columbia, Canada (41) found that translocating American badgers *Taxidea taxus* increased the population growth rate at the recipient site, but survival was lower than in a nearby resident population. The badger population growth rate was higher at the recipient site after translocation than before and was similar to that found in a nearby non-translocated population (data reported as geometric growth rate). Ten young were born to translocated badgers. The adult annual survival rate was lower in the release site (77%) than in a nearby resident population (90%). In 2002, sixteen badgers were translocated from north-western Montana to supplement a declining population at a site in British Columbia. Translocated badgers were monitored in 2002–2006, by radio-tracking, from an aeroplane. Comparisons were made with a nearby site containing a resident badger population.

A randomized, controlled study in 2005–2006 in a plantation in Georgia, USA (42) found that most translocated nine-banded armadillos *Dasypus novemcinctus* dispersed from their release site within the first few days after release. Eleven out of 12 translocated armadillos (92%) dispersed from their release sites within the first few days (duration not specified) after release. Only six of the translocated animals were successfully relocated, of which two returned to their original capture sites, and three made long-distance movements away from their release sites. However, all 29 armadillos released at their original capture site remained near their release sites over the same period and maintained stable home ranges (3–30 ha). Between May 2005 and March 2006, forty-one armadillos were captured using long-handled dip nets and unbaited wire cage traps. Twelve armadillos were randomly selected to be translocated and the remainder were released at their capture sites. Translocated animals were released 0.7–8.1 km from their capture site. All individuals were tagged with transmitters and monitored 3–4 times/week for up to 358 days.

A study in 2001–2006 on grassland in a national park in Botswana (43) found that most translocated white rhinoceros *Ceratotherium simum* released in groups survived at least three years after release, but some dispersed away from the park when released into areas already occupied by released animals. Of 32 rhinoceroses released into the park in four batches during just over two years, five died soon after release and 21 remained in the park through to three years after the final release. Six (all females) left the park. All were from the final release. The authors suggest that this may be because suitable habitat close to the release site was already occupied by previously released animals. Rhinoceroses, sourced from protected sanctuaries, were all released from the same boma, in four batches, from November 2001 to November 2003. They were monitored by radio-tracking from a vehicle or aircraft, through to 2006.

A replicated study in 1949–2001 in South Africa (44) found that following translocations inside and outside of their historical ranges, population sizes of most of 22 species of grazing mammals increased. Following translocation, 82 out of 125 populations (66%) of 22 grazing mammals (white rhinoceros *Ceratotherium simum*, mountain zebra *Equus zebra*, plains zebra *Equus quagga*, giraffe *Giraffa camelopardalis*, African buffalo *Syncerus caffer* and seventeen species of antelope) exhibited positive growth rates (data presented as results of population growth models). Population models were based on long-term monitoring data from 178 populations relocated to 24 reserves in 1949–1978 (see original paper for details). Only translocations with five or more consecutive years of monitoring results were included (125 translocations, monitoring data duration: 5–47 years). Translocation details are not provided but authors state that most translocated populations began with fewer than 15 individuals and that most reserves contained water impoundments and lacked top predators, such as lions *Panthera leo* or spotted hyenas *Crocuta crocuta*. Seventeen of the 22 species were introduced outside of their historical range.

A replicated study in 1951–2007 in 10 desert sites in Arizona and New Mexico, USA, and the Gulf of California, Mexico (45) found that translocating desert bighorn sheep *Ovis canadensis* did not increase the population size at the release site. No bighorn sheep populations which were supplemented with translocated individuals significantly increased in size (data not presented). Between 1951 and 1990, a total of 654 bighorn sheep were released, but details of individual releases are not provided. Data were obtained from historical records for ten sites with long-term survey and hunting information. Data included counts of bighorn sheep from both surveys and hunter harvests, and bighorn sheep translocations.

A study in 1999–2008 in an area of mixed agricultural land, forest, and grassland in the Alps of northern Italy (46) found that following translocation, brown bears *Ursus arctos* bred successfully in the release area and the population increased, but genetic diversity declined. Three years after the first translocations, there were 10 bears in the area. By nine years after the first translocations, this increased to 27–31 bears.

Over this time, 35 cubs had been born. However, genetic diversity declined over time (data reported as allelic richness). In 1999–2002, nine bears were caught in Slovenia and translocated into Trentino, Italy, where the resident population had fallen to around three individuals. In 2002–2008, hair and faecal samples were collected opportunistically and along transects. Samples were also collected from bear carcasses found in the area. DNA from these samples was analysed to identify individuals and to measure genetic diversity.

A study in 2002–2008 in an area of peatland, fen, woodland, ditches and lakes in the Netherlands (47) found that after release of 30 translocated and captive-bred Eurasian otters *Lutra lutra*, at least six were still alive six years later and some had reproduced. Most dead otters recovered were killed in collisions with road vehicles. Fifty-four offspring from released otters or their descendants were detected. Between July 2002 and November 2007, thirty otters were released. Seventeen were translocated, wild-caught animals and 13 were captive-bred. A publicity campaign encouraged people to report dead otters that they found. These were then examined to establish cause of death.

A controlled study in 1999–2001 on three grassland sites in an area in South Dakota, USA (48) found that wild-born translocated black-footed ferret *Mustela nigripes* kits had higher survival rates after release than did captive-born kits released from holding pens. Thirty-day post-release survival of captive-born kits (66%) was lower than that of wild-born translocated kits at the same site (94%). Annual survival was also for lower for captive-born kits (females: 44%; males: 22%) than for wild-born kits (females: 67%; males: 43%). Annual survival at the donor site remained high (females: 80%; males: 51%) whilst survival of translocated and released kits was comparable with that at an unmanipulated colony (females: 59%; males: 28%). Eighteen wild-born ferrets were released along with 18 captive-bred ferrets at a site from which the species was then absent. Captive-born ferrets were transferred to outdoor conditioning pens, sited on prairie dog colonies, when about 90 days old and then released on 29 September and 13 October 1999. Wild-born ferrets were released the day after capture. All were born in 1999. Ferrets at the release site, the donor site for wild-born kits and an unmanipulated site were monitored by radio-tracking and by reading transponder chips.

A review of studies in 1991–2008 at 11 grassland sites in the USA and Mexico (49) found that most captive-bred (with some translocated) black-footed ferret *Mustela nigripes* releases were unsuccessful at maintaining a population, but success was higher where prey was abundant over larger areas. Of 11 reintroduction sites, populations of more than 30 adult black-footed ferrets were maintained at four sites over two years without further reintroductions. Two sites no longer contained ferrets by December 2008, and the other five sites only had small populations or were supplemented by further releases. Sites where populations were maintained tended to have more prairie dogs *Cynomys* spp., the main prey species of black-footed ferrets, covering a larger area (at least 4,300 ha) and with a higher density of animals (data presented as index of prairie dog abundance). From 1991–2008, around 2,964 captive-bred and 157 translocated wild ferrets were released at 18 sites in multiple releases. The study reports success of the 11 sites where initial releases occurred before 2003. Sites received on average over 200 ferrets over 10 years. Ferrets were monitored by annual spotlight surveys to locate, capture and uniquely mark individuals.

A study in 1987–2009 in grassland and shrubland in the Western Cape, South Africa (50) found that numbers of translocated Cape mountain zebra *Equus zebra zebra* increased four-fold over 19 years. Nineteen years after release, there were four times more Cape mountain zebras (48) than at the time of release (12). In the first 14 years after translocations, 13 foals were born. In 1987–1990, twelve Cape mountain zebras were translocated into a 3,435-ha national park dominated by renosterveld and fynbos vegetation. No translocation or monitoring details are provided. Grass availability was promoted by artificial fires at four-year intervals.

A replicated study in 2000–2007 in two mountain sites in northern Utah, USA (51) found that following translocation of bighorn sheep *Ovis canadensis*, 48–98% of young descended from these animals survived into their first winter. The average survival of bighorn sheep lambs to their first winter was 48% at one site and 55–98% at the second site. In January and February 2000–2002 and 2007, one hundred and fourteen wild-born bighorn sheep (including 92 adult females) were translocated to Mount Timpanogos (67 females, 11 males, 4 young) and Rock Canyon (25 females, 4 males, 3 young). Thirty-one individuals on Mount Timpanogos and 10 in Rock Canyon were fitted with radio-collars. Collared and uncollared females were relocated every 4–5 days from April–July 2001–2007 to count the number of young born. The number of young that survived to their first winter was determined by comparing the highest number of young observed during winter (October to March) with the number observed in the previous spring (April to July).

A study in 2003–2009 in a temperate grassland site in South Dakota, USA (52) found that translocating swift foxes *Vulpes velox* led to the establishment of a population in which genetic diversity of wild-born descendants was at least as high as that of the translocated animals. For two key measures of genetic diversity, values for descendants of translocated foxes (heterozygosity: 0.75; allelic richness: 11.2) were at least as high as those of the translocated animals (heterozygosity: 0.75–0.78; allelic richness: 7.5–8.6). In 2003–2006, one hundred and eight wild-caught swift foxes from Colorado and Wyoming were released into a national park in South Dakota from which the species had been extirpated. Four hundred DNA samples (108 from translocated foxes and 292 collected in 2004–2009 from their wild-born descendants) were analyzed for measures of genetic diversity.

A study in 1978–2004 and a controlled study in 2006–2009 in an alpine site comprising forest, rock and scree in Italy (53) found that following translocations of Alpine ibex *Capra ibex*, the population increased and translocated ibex used similar habitats to resident ibex. Twenty-three years after translocation, the estimated number of Alpine ibex (456 individuals) was higher than the number released (10 individuals). However, two years later the population declined by 75% due to a sarcoptic mange epidemic. Following further translocations, released ibex selected the same habitat resources as used by resident ibex (data presented as an ordination analysis), but translocated ibex initially occupied larger ranges and were separated from resident animals. By one year after release the home range size of translocated and resident ibex was similar, and by three years translocated animals were integrated into the resident social group. In 1978–1979, ten Alpine ibex were translocated from the Gran Paradiso National Park to the Marmolada massif in the Alps. In 2006–2007, fourteen additional male ibex were translocated to reinforce the Marmolada massif population.

All ibex translocated in 2006–2007 were radio-collared. From 2006–2009, sixty-seven resident male ibex from the established population were caught and ear-tagged and 52 were radio-collared. Translocated and established ibex were followed for 3–4 years.

A study in 1986–2011 in two reserves in western Nepal (54) found that translocated populations of the greater one-horned rhinoceros *Rhinoceros unicornis* persisted for at least 11–25 years post-release. On one reserve, there were 67 rhinoceroses in 2000, fourteen years after the first translocations, but this fell to a count of 24 rhinoceroses 11 years later. Poaching was thought to be the main cause of deaths. The second reserve had seven rhinoceroses 11 years after the translocations. Between 1986 and 2003, eighty-three rhinoceroses (38 males, 45 females) were translocated to Bardia National Park and, in 2000, four rhinoceros (three females and one male) were translocated to Suklaphanta Wildlife Reserve, which already held a single male. From 1986–2003, rhinoceros in Bardia National Park were protected by anti-poaching patrols formed of 10–15 soldiers and in 2007 a nationwide anti-poaching programme was launched. Monitoring details are not provided.

A replicated, randomized, controlled study in 2008–2009 in 10 pine plantation sites in Ñuble Province, Chile (55) found that translocated long-haired field mice *Abrothrix longipilis* travelled two-to four-times further and had lower survival than non-translocated mice. The average maximum distance travelled from the release site was longer in translocated mice (125–199 m) than in non-translocated mice (50 m). Mice released 0–100 m from their capture location had higher survival rates (20/20 survived) than mice translocated 500–1,300 m (14/18 survived). Additionally, eight of 10 mice that were translocated short distances (100 m) and nine of 10 mice which were released at their capture site returned to or stayed in their capture location, whereas mice which were translocated further (500 m = 1 of 10; 1,300 m = 0 of 10) did not return to their capture locations. From January–March 2008 and 2009, four male long-haired field mice were trapped at each of 10 sites in Quirihue and Cobquecura, using 80 baited live traps (3 × 3.5 × 9 inches) per site. Mice from each capture site were randomly allocated to one of four groups, which were released at sunset either at the capture site or 100, 500, and 1,300 m from their capture point. Each individual was radio-tagged and relocated once/day for three days after release.

A study in 2009–2012 on mixed grassland, shrub and woodland vegetation in a mountainous region in Wyoming, USA (56) found that following translocation of bighorn sheep *Ovis canadensis*, most animals survived at least 60 days after release. Sixty days after release, at least 62 of the 64 translocated sheep were alive. One sheep died, probably due to capture-induced stress, and the GPS collar on another malfunctioned after release, so it could not be tracked. In 2009–2012, seventy-seven bighorn sheep were released. Of these, 65 were GPS-collared and signals were received from 64 of the collars after release (including the one that subsequently failed). Location data were collected for 18 months after release though survival data only for the first 60 days are presented.

A review of translocations carried out in 1969–2006 in Australia (57) found that translocating wild-born and releasing captive-bred macropod species (kangaroos and allies) led to the successful establishment of populations in 44 of 72 cases. Of the established populations, 29 persisted for more than five years. Of the 28 releases considered to be failures, 17 were thought to have failed due to predation by non-native carnivores, such as red foxes *Vulpes vulpes*. Releases considered in the review included both wild-caught, translocated animals and captive-bred animals. The number of animals released ranged from one to 70 and included 20 different macropod species. Only translocations where animals were released into areas larger than 100 ha were considered for the review.

A replicated study in 2011–2014 of two grasslands in Slovakia (58) found that translocated European ground squirrels *Spermophilus citellus* bred in small numbers after four years of releases. Nine juveniles in four litters during the fourth year of releases were the first breeding evidence at one site (with 174 animals released up to then). At a second site, also during the fourth year of releases, a female with five young was the first breeding evidence (with 284 animals released up to then). Ground squirrels were translocated in 2011–2014. Some were lost to predators (e.g. red fox *Vulpes vulpes* and feral cat *Felis catus*). Heavy rain in spring 2013 and 2014 may have reduced the population at one site. Grass cutting was required to maintain suitable habitat at one site. Ground squirrels were translocated from nearby donor sites, especially airfields. Monitoring focussed on burrows as well as counting individuals, aided by individual fur clipping patterns.

A study in 1972–2011 in a grassland and rock area above the treeline in central Appenines, Italy (59) found that a population of translocated red deer *Cervus elaphus* released in groups persisted at least 24 years after release, but over the same period, the density of Apennine chamois *Rupicapra pyrenaica ornata* in the area almost halved. Red deer pellets were detected in 31–35 out of 38 (82–92%) sampling plots 23–24 years after translocation. However, authors reported that over a similar period, chamois density almost halved in the core area of their range (1984–1985: c. 38/100 ha; 2012: c. 20 individuals/100 ha). Authors found a large space (> 75%) and diet (> 90%) overlap between deer and chamois, an increase in unpalatable plant species and a reduced bite rate of adult female chamois in patches also used by deer (see paper for details). Forty-five red deer were translocated into Abruzzo, Lazio and Molise National Park in 1972 (0.5 individuals/100 ha). A further 36 deer were released in groups of 7–10 individuals (in 4 operations) in 1972–1987. In June–October 2010 and 2011, the presence/absence of groups of >5 red deer pellets was recorded in circular, 5-m radius, sampling plots, randomly placed in 38 grassland sites. Sites were located in a 65-ha mountainous area above the treeline.

A study in 2003–2007 in a mixed shrub, grassland and forest area near Madrid, Spain (60) found that following translocation, Iberian ibex *Capra pyrenaica* numbers increased and ibex expanded their range. In the first eight to 10 years after translocation began, ibex numbers increased by 23%/year on average (at release: 67 individuals; after 8–10 years: 359 individuals), by 36%/year for the next three years (after 11–13 years: 773 individuals), and by 19%/year in the following four years (after 15–17 years: 1,523 individuals). The birth rate was 0.76 calves/adult female and the area that ibex occupied increased from 2,102 ha in 2000 to 3,279 ha in 2007. In 1990–1992, sixty-seven wild-born Iberian ibex (41 females and 26 males) were translocated to a 4,890-ha national park. The translocated population was monitored between May and June in 2000, 2005, and 2007. Ibex were counted along 22 transects (average length 3.6 km) using binoculars. Transects were walked 2–3 hours after sunrise or 2–3 hours before sunset. The study area included high altitude (1,100–2,200 m) shrubland, grassland and forest areas.

A study in 1999–2012 of woodland in and around a national park in Italy (61) found that, following the start of translocations, a re-established brown bear *Ursus arctos* population increased steadily in numbers over 12 years. From 10 bears translocated to the area in 1999–2001, the population grew by 20% annually in 2002–2006, with the rate gradually falling to 16% annual growth by 2012. Breeding was first recorded in 2002, with ≥74 cubs born in ≥34 reproductive events up to 2012. At that point, there were 47 bears in the population (16 adults, 14 juveniles and 17 cubs). Ten bears (seven female, three male) were translocated from Slovenia in 1999–2001. Up to 2012, twenty-one young males had dispersed from the province (though six subsequently returned). Other documented population losses included those attributed to illegal hunting, road casualties and removal of problem bears.

A controlled study in 2008–2010 in a mountain site in the Central Apennines, Italy (62) found that wild-caught translocated Apennine chamois *Rupicapra pyrenaica ornata* survived and reproduced in similar numbers to released captive-bred chamois, but captive-born chamois remained closer to the release site. Seven of eight captive-born (88%) and seven of eight (88%) wild-caught translocated Apennine chamois survived over five months after release. Four of five captive-born (80%) and three of five wild-caught translocated (60%) female chamois reproduced in the first year after release. During the first week after release, captive-born chamois remained closer to the release site (within 1.1 km on average) than wild-caught chamois (average 1.8 km). Eight captive-born chamois (2.5–11.5 years old, five females and three males) and eight wild-caught translocated chamois (2.5–10.5 years old, five females and three males) were released into Sibillini Mountains National Park. Chamois were released in groups of one-three individuals; each group was all wild or all captive-born. Captive-born chamois were bred in large enclosures within four national parks. Translocated chamois were taken from a national park approximately 200 km away. All of the 16 released chamois were fitted with radio-collars and monitored for five months after release in 2008–2010.

A replicated study in 1971–2014 in 13 forested mountainous areas in Catalonia, Spain (63) found that translocating roe deer *Capreolus capreolus* resulted in a six-fold increase in distribution after multiple translocation events. Forty-two years after the first translocation roe deer were present in 85% of Catalonia (2013: 288 10 × 10 km squares), a six-fold increase on the area occupied compared to 23 years after the first translocation (1994: 52 10 × 10 km squares). Between 1971 and 2008, five hundred and fourty-two translocated roe deer were released in 13 areas across Catalonia. Deer were captured from the wild in France and Spain and released after 24 hours directly into protected areas. In 1971–1992, animals (46 individuals) were translocated into areas already occupied by roe deer and in 1993–2008 into areas where roe deer were currently absent (496 individuals). Distribution data were obtained from terrestrial mammal distributions atlases supplemented by traffic police reports, hunting data and sightings by volunteers.

A study in 1963–2010 in two areas of mixed broadleaf and montane forest with alpine meadows in the northern Carpathian mountains of Ukraine, Slovakia and Poland (64) found that three European bison *Bison bonasus* herds persisted >6 years after the last release of translocated individuals. Between 6–47 years after releases, around 320 free-ranging European bison survived in the three herds. Two herds (totalling about 300 individuals) resulted from 30–47-year-old translocations. The third herd (about 20 individuals) resulted from a translocation some six years earlier. The study was conducted in the Polish Bieszczady Mountains and in the Slovak Poloniny National. Bison were translocated to the Polish Bieszczady Mountains between 1963 and 1980 and to the Slovak Poloniny National in 2004. No details are provided on the number of animals translocated nor on their origin. GPS locations of bison were collected in 2001–2010 (29,382 records). No monitoring details are provided, but bison presence data included direct observations, tracks, faeces and signs of feeding. Six bison were radio-tracked in 2002–2006 (two locations recorded at least twice a week).

(1) Savidge J. (1961) The introduction of white rhinoceros into the Murchison Falls National Park, Uganda. *Oryx*, 6, 184–189.

(2) Morris R., Ellis D.V. & Emerson B.P. (1981). The British Columbia transplant of sea otters *Enhydra lutris*. *Biological Conservation*, 20, 291–295.

(3) Jameson R.J., Kenyon K.W., Johnson A.M. & Wight H.M. (1982) History and status of translocated sea otter populations in North America. *Wildlife Society Bulletin*, 10, 100–107.

(4) Booth V.R., Jones M.A. & Morris N.E. (1984) Black and white rhino introductions in north-west Zimbabwe. *Oryx*, 18, 237–240.

(5) Erickson D.W. & McCullough C.R. (1987) Fates of translocated river otters in Missouri. *Wildlife Society Bulletin*, 15, 511–517.

(6) Bergerud A.T. & Mercer W.E. (1989) Caribou introductions in eastern North America. *Wildlife Society Bulletin*, 17, 111–120.

(7) Somers M.J. & Penzhorn B.L. (1992) Reproduction in a reintroduced warthog population in the eastern Cape Province. *South African Journal of Wildlife Research*, 22, 57–60.

(8) Serfass T., Brooks R. & Rymon L. (1993) Evidence of long-term survival and reproduction by translocated river otters, *Lutra candensis*. *The Canadian Field-Naturalist*, 107, 59–63.

(9) Smith K.G. & Clark J.D. (1994) Black bears in Arkansas: characteristics of a successful translocation. *Journal of Mammalogy*, 75, 309–320.

(10) Carbyn L.N., Armbruster H.J. & Mamo C. (1994) The swift fox reintroduction program in Canada from 1983 to 1992. Pages 247–271 in: M.L. Bowles & C.J. Whelan (eds.) *Restoration of endangered species: conceptual issues, planning and implementation*. Cambridge University Press, Cambridge, UK.

(11) Stussy R.J., Edge W.D. & O'Neil T.A. (1994) Survival of resident and translocated female elk in the Cascade Mountains of Oregon. *Wildlife Society Bulletin*, 22, 242–247.

(12) Compton B.B., Zager P. & Servheen G. (1995) Survival and mortality of translocated woodland caribou. *Wildlife Society Bulletin*, 23, 490–496.

(13) Servheen C., Kasworm W.F. & Their T.J. (1995) Transplanting grizzly bears *Ursus arctos horribilis* as a management tool — results from the Cabinet Mountains, Montana, USA. *Biological Conservation*, 71, 261–268.

(14) Nolet B.A. & Baveco J.M. (1996) Development and viability of a translocated beaver *Castor fiber* population in the Netherlands. *Biological Conservation*, 75, 125–137.

(15) Sjöåsen T. (1996) Survivorship of captive-bred and wild-caught reintroduced European otters *Lutra lutra* in Sweden. *Biological Conservation*, 76, 161–165.

(16) Fornasari L., Casale P. & Wauters L. (1997) Red squirrel conservation: the assessment of a reintroduction experiment. *Italian Journal of Zoology*, 64, 163–167.

(17) Heath M.E. & Coulson I.M. (1997) Preliminary studies on relocation of Cape pangolins *Manis temminckii*. *South African Journal of Wildlife Research*, 27, 51–56.

(18) Sjöåsen T. (1997) Movements and establishment of reintroduced European otters *Lutra lutra*. *Journal of Applied Ecology*, 34, 1070–1080.

(19) Ruth T.K., Logan K.A., Swearnor L.L., Hornocker M.G. & Temple L.J. (1998) Evaluating cougar translocation in New Mexico. *The Journal of Wildlife Management*, 62, 1264–1275.

(20) Johnson S. & Berkley K. (1999) Restoring river otters in Indiana. *Wildlife Society Bulletin*, 27, 419–427.

(21) Komers P.E. & Curman G.P. (2000). The effect of demographic characteristics on the success of ungulate re-introductions. *Biological Conservation*, 93, 187–193, https://doi.org/10.1016/s0006-3207(99)00141-x

(22) Smeeton C. & Weagle K. (2000) The reintroduction of the swift fox *Vulpes velox* to South Central Saskatchewan, Canada. *Oryx*, 34, 171–179, https://doi.org/10.1017/s0030605300031161

(23) Raesly E.J. (2001) Progress and status of river otter reintroduction projects in the United States. *Wildlife Society Bulletin*, 29, 856–862.

(24) Bright P. & Morris P. (2002) Putting dormice (*Muscardinus avellanarius*) back on the map. *British Wildlife*, 14, 91–100.

(25) Borgo A. (2003) Habitat requirements of the Alpine marmot *Marmota marmota* in re-introduction areas of the Eastern Italian Alps. Formulation and validation of habitat suitability models. *Acta Theriologica*, 48, 557–569, https://doi.org/10.1007/bf03192501

(26) Moruzzi T.L., Royar K.J., Grove C., Brooks R.T., Bernier C., Thompson Jr. F.L., DeGraaf R.M. & Fuller T.K. (2003) Assessing an American Marten, *Martes americana*, reintroduction in Vermont. *The Canadian Field-Naturalist*, 117, 190–195, https://doi.org/10.22621/cfn.v117i2.681

(27) Chiarello A.G, Chivers D.J., Bassi C., Amelia M., Maciel F., Moreira L.S. & Bazzalo M. (2004) A translocation experiment for the conservation of maned sloths, *Bradypus torquatus* (Xenarthra, Bradypodidae). *Biological Conservation*, 118, 421–430, https://doi.org/10.1016/j.biocon.2003.09.019

(28) Mawson P.R. (2004) Translocations and fauna reconstruction sites: Western Shield review-February 2003. *Conservation Science Western Australia*, 5, 108–121.

(29) Calenge C., Maillard D., Invernia N. & Gaudin J.C. (2005) Reintroduction of roe deer *Capreolus capreolus* into a Mediterranean habitat: female mortality and dispersion. *Wildlife Biology*, 11, 153–161, https://doi.org/10.2981/0909-6396(2005)11[153:rordcc]2.0.co;2

(30) Preatoni D., Mustoni A., Martinoli A., Carlini E., Chiarenzi B., Chiozzini S., Van Dongen S., Wauters L.A. & Tosi G. (2005) Conservation of brown bear in the Alps: Space use and settlement behavior of reintroduced bears. *Acta Oecologica*, 28, 189–197, https://doi.org/10.1016/j.actao.2005.04.002

(31) Diefenbach D.R., Hansen L.A., Warren R. J. & Conroy M.J. (2006) Spatial organization of a reintroduced population of bobcats. *Journal of Mammalogy*, 87, 394–401, https://doi.org/10.1644/05-mamm-a-114r1.1

(32) Faulhaber C.A., Perry N.D., Silvy N.J., Lopez R.R., Frank P.A. & Peterson M.J. (2006) Reintroduction of Lower Keys marsh rabbits. *Wildlife Society Bulletin*, 34, 1198–1202, https://doi.org/10.2193/0091-7648(2006)34[1198:rolkmr]2.0.co;2

(33) Lammertsma D., Niewold F, Jansman H., Kuiters L., Koelewijn H.P., Perez Haro M., van Adrichem M., Boerwinkel M. & Bovenschen J. (2006) Herintroductie van de otter: een succesverhaal? *De Levende Natuur*, 107, 42–46.

(34) Swanson B.J., Peters R. L. & Kyle C.J. (2006) Demographic and genetic evaluation of an American marten reintroduction. *Journal of Mammalogy*, 87, 272–280, https://doi.org/10.1644/05-mamm-a-243r1.1

(35) Cabezas S. & Moreno S. (2007) An experimental study of translocation success and habitat improvement in wild rabbits. *Animal Conservation*, 10, 340–348, https://doi.org/10.1111/j.1469-1795.2007.00119.x

(36) Hayward M.W., Kerley G.I.H., Adendorff J., Moolman L.C., O'Brien J., Douglas A.S., Bissett C., Bean P., Fogarty A., Howarth D. & Slater R. (2007) The reintroduction of large carnivores to the Eastern Cape, South Africa: an assessment. *Oryx*, 41, 205–214, https://doi.org/10.1017/s0030605307001767

(37) Hicks J.F., Rachlow J.L., Rhodes Jr O.E., Williams C.L. & Waits L.P. (2007) Reintroduction and genetic structure: Rocky Mountain elk in Yellowstone and the western states. *Journal of Mammalogy*, 88, 129–138, https://doi.org/10.1644/06-mamm-a-051r1.1

(38) Kasworm W.F., Proctor M.F., Servheen C. & Paetkau D. (2007) Success of grizzly bear population augmentation in northwest Montana. *The Journal of Wildlife Management*, 71, 1261–1266, https://doi.org/10.2193/2006-266

(39) Carvalho P., Nogueira A.J., Soares A.M. & Fonseca C. (2008) Ranging behaviour of translocated roe deer in a Mediterranean habitat: seasonal and altitudinal influences on home range size and patterns of range use. *Mammalia*, 72, 89–94, https://doi.org/10.1515/mamm.2008.019

(40) Jule K.R., Leaver L.A. & Lea S.E.G. (2008) The effects of captive experience on reintroduction survival in carnivores: a review and analysis. *Biological Conservation*, 141, 355–366, https://doi.org/10.1016/j.biocon.2007.11.007

(41) Kinley T.A. & Newhouse J.A. (2008) Ecology and translocation-aided recovery of an endangered badger population. *The Journal of Wildlife Management*, 72, 113–122, https://doi.org/10.2193/2006-406

(42) Gammons D.J., Mengak M.T. & Conner L.M. (2009) Translocation of nine-banded armadillos. *Human-Wildlife Conflicts*, 3, 64–71,

(43) Støen O-G., Pitlagano M.L. & Moe S.R. (2009) Same-site multiple releases of translocated white rhinoceroses *Ceratotherium simum* may increase the risk of unwanted dispersal. *Oryx*, 43, 580–585, https://doi.org/10.1017/s0030605309990202

(44) Van Houtan K.S., Halley J.M., Van Aarde R. & Pimm S.L. (2009) Achieving success with small, translocated mammal populations. *Conservation Letters*, 2, 254–262, https://doi.org/10.1111/j.1755-263x.2009.00081.x

(45) Wakeling BF., Lee R., Brown D., Thompson R., Tluczek M. & Weisenberger M. (2009) The restoration of desert bighorn sheep in the Southwest, 1951–2007: factors influencing success. *Desert Bighorn Council Transactions*, 50, 1–17.

(46) De Barba M., Waits L.P., Garton O.E., Genovesi P., Randi E., Mustoni A. & Groff C. (2010) The power of genetic monitoring for studying demography, ecology, and genetics of a reintroduced brown bear population. *Molecular Ecology*, 19, 3938–3951, https://doi.org/10.1111/j.1365-294x.2010.04791.x

(47) Koelewijn H, Perez-Haro M., Jansman H.A.H., Boerwinkel M.C., Bovenschen J., Lammertsma D.R., Niewold F.J.J. & Kuiters A.T. (2010) The reintroduction of the Eurasian otter (*Lutra lutra*) into the Netherlands: hidden life revealed by noninvasive genetic monitoring. *Conservation Genetics*, 11, 601–614, https://doi.org/10.1007/s10592-010-0051-6

(48) Biggins D.E., Godbey J.L., Horton B.M. & Livieri T.M. (2011) Movements and survival of black-footed ferrets associated with an experimental translocation in South Dakota. *Journal of Mammalogy*, 92, 742–750, https://doi.org/10.1644/10-mamm-s-152.1

(49) Jachowski D.S., Gitzen R.A., Grenier M.B., Holmes B. & Millspaugh J.J. (2011) The importance of thinking big: Large-scale prey conservation drives black-footed ferret reintroduction success. *Biological Conservation*, 144, 1560–1566, https://doi.org/10.1016/j.biocon.2011.01.025

(50) Watson L.H., Kraaij T. & Novellie P. (2011) Management of rare ungulates in a small park: habitat use of bontebok and Cape mountain zebra in Bontebok National Park assessed by counts of dung groups. *South African Journal of Wildlife Research*, 41, 158–166, https://doi.org/10.3957/056.041.0202

(51) Whiting J.C., Bowyer R.T., Flinders J.T. & Eggett D.L. (2011) Reintroduced bighorn sheep: fitness consequences of adjusting parturition to local environments. *Journal of Mammalogy*, 92, 213–220, https://doi.org/10.1644/10-mamm-a-145.1

(52) Sasmal I., Jenks J.A., Waits L.P., Gonda M.G., Schroeder G.M. & Datta S. (2013) Genetic diversity in a reintroduced swift fox population. *Conservation Genetics*, 14, 93–102, https://doi.org/10.1007/s10592-012-0429-8

(53) Scillitani L., Darmon G., Monaco A., Cocca G., Sturaro E., Rossi L. & Ramanzin M. (2013) Habitat selection in translocated gregarious ungulate species: an interplay between sociality and ecological requirements. *The Journal of Wildlife Management*, 77, 761–769, https://doi.org/10.1002/jwmg.517

(54) Thapa K., Nepal S., Thapa G., Bhatta S.R. & Wikramanayake E. (2013) Past, present and future conservation of the greater one-horned rhinoceros *Rhinoceros unicornis* in Nepal. *Oryx*, 47, 345–351, https://doi.org/10.1017/s0030605311001670

(55) Villaseñor N.R., Escobar M.A. & Estades C.F. (2013) There is no place like home: high homing rate and increased mortality after translocation of a

small mammal. *European Journal of Wildlife Research*, 59, 749–760, https://doi.org/10.1007/s10344-013-0730-y

(56) Clapp J.G., Beck J.L. & Gerow K.G. (2014) Post-release acclimation of translocated low- elevation, non-migratory bighorn sheep. *Wildlife Society Bulletin*, 38, 657–663, https://doi.org/10.1002/wsb.441

(57) Clayton J.A., Pavey C.R., Vernes K. & Tighe M. (2014) Review and analysis of Australian macropod translocations 1969–2006. *Mammal Review*, 44, 109–123, https://doi.org/10.1111/mam.12020

(58) Löbbová D. & Hapl E. (2014) Conservation of European ground squirrel (Mammalia: Rodentia) in Slovakia: Results of current reintroduction programme. *Slovak Raptor Journal*, 8, 105–112, https://doi.org/10.2478/srj-2014-0012

(59) Lovari S., Ferretti F., Corazza M., Minder I., Troiani N., Ferrari C. & Saddi A. (2014) Unexpected consequences of reintroductions: competition between reintroduced red deer and Apennine chamois. *Animal Conservation*, 17, 359–370, https://doi.org/10.1111/acv.12103

(60) Refoyo P., Olmedo C., Polo I., Fandos P. & Muñoz B. (2015) Demographic trends of a reintroduced Iberian ibex *Capra pyrenaica victoriae* population in central Spain. *Mammalia*, 79, 139–145, https://doi.org/10.1515/mammalia-2013-0141

(61) Tosi G., Chirichella R., Zibordi F., Mustoni A., Giovannini R., Groff C., Zanind M. & Apollonio M. (2015) Brown bear reintroduction in the Southern Alps: To what extent are expectations being met? *Journal for Nature Conservation*, 26, 9–19, https://doi.org/10.1016/j.jnc.2015.03.007

(62) Bocci A., Menapace S., Alemanno S. & Lovari S. (2016) Conservation introduction of the threatened Apennine chamois *Rupicapra pyrenaica ornata*: post-release dispersal differs between wild-caught and captive founders. *Oryx*, 50, 128–133, https://doi.org/10.1017/s0030605314000039

(63) Torres R.T., Carvalho J., Fonseca C., Serrano E. & López-Martín J.M. (2016) Long-term assessment of roe deer reintroductions in North-East Spain: A case of success. *Mammalian Biology*, 81, 415–422, https://doi.org/10.1016/j.mambio.2016.05.002

(64) Ziółkowska E., Perzanowski K., Bleyhl B., Ostapowicz K. & Kuemmerle T. (2016) Understanding unexpected reintroduction outcomes: Why aren't European bison colonizing suitable habitat in the Carpathians?. *Biological Conservation*, 195, 106–117, https://doi.org/10.1016/j.biocon.2015.12.032

14.11. Translocate mammals to reduce overpopulation

https://www.conservationevidence.com/actions/2430

- **Three studies** evaluated the effects of translocating mammals to reduce overpopulation. Two studies were in the USA[1,2] and one was in Australia[3].

COMMUNITY RESPONSE (0 STUDIES)

POPULATION RESPONSE (3 STUDIES)

- **Abundance (1 study):** A before-and-after study in the USA[2] found that adult elk numbers approximately halved after the translocation of wolves to the reserve.

- **Reproductive success (1 study):** A before-and-after study in the USA[2] found that elk calf:cow ratios approximately halved after the translocation of wolves to the reserve.

- **Survival (2 studies):** A study in Australia[3] found that koalas translocated to reduce overpopulation had lower survival than individuals in the source population. A study in the USA[1] found that following translocation to reduce over-abundance, white-tailed deer had lower survival rates compared to non-translocated deer at the recipient site.

- **Occupancy/range (1 study):** A study in the USA[1] found that following translocation to reduce over-abundance at the source site, white-tailed deer had similar home range sizes compared to non-translocated deer at the recipient site.

BEHAVIOUR (0 STUDIES)

Background

Overpopulation can reduce the long-term persistence of a population, as competition for resources increases. Translocating individuals of the target species away from the area or predators into the area for example, to reduce population numbers may help reduce competition for resources and thus improve the fitness of the remaining population.

A study in 1993–1995 in a forest reserve in New York, USA (1) found that following translocation to reduce over-abundance at the source site, white-tailed deer *Odocoileus virginianus* had lower survival rates but similar home range sizes compared to non-translocated deer at the recipient site. One year after release, the annual survival rate for translocated deer (53%) was lower than that of non-translocated deer at the recipient site (75–88%). During the year after release, average home range sizes did not differ significantly between translocated deer (0.23 km²) and non-translocated deer at the recipient site (0.22 km²). In May–June 1994, seventeen female white-tailed deer were translocated from an over-populated site to a site 60 km away. In April–July of 1993–1995, twenty deer resident at the recipient site (16 females, 4 males) were captured. All deer were radio-collared. Before release, deer were held for 1–12 days in a 50-m² pen. Deer were monitored using radio-telemetry, 5–15 times/week, in April–August of 1993–1995, and less frequently at other times of the year.

A before-and-after study in 1986–2004 in a grassland and forest reserve in Wyoming, USA (2) found that adult elk *Cervus canadensis* numbers and elk calf:cow ratios approximately halved after the translocation of wolves *Canis lupus* to the reserve. Results were not subject to statistical analysis. Nine years after wolves were translocated, there were fewer adult elk (8,335) and a lower calf:cow ratio (12 calves/100 female elk) than the average before wolf translocation (adult elk: 16,664; 25 calves/100 female elk). A similar number of elk that had migrated out of the park were killed by hunters before (1,148 elk/year) and after (1,297 elk/year) wolves were translocated. Wolves were translocated into Yellowstone National Park in 1995. Between 1996 and 2004 wolf numbers increased from 21 to 106. Elk adults and calves were counted from aeroplanes annually during December–January 1986–2004. No counts were conducted during the winters of 1996 and 1997.

A study in 2007–2008 of forest sites on an island and the mainland of southeastern Australia (3) found that koalas *Phascolarctos cinereus* translocated to reduce overpopulation had higher mortality than individuals in the source population. Six of 16 koalas (38%) that were sterilized and translocated died within 12 months of release, whereas none of 13 koalas in the source population died within the same time period. In April–May 2007, sixteen koalas (eight females; eight males) were surgically sterilized and translocated from an overpopulated island to the mainland. Release sites were 10-ha forest blocks dominated by rough-barked manna gum Eucalyptus viminalis. Released koalas were radio-collared and tracked daily for one week followed by weekly for seven weeks and monthly until June 2008. Thirteen unsterilized koalas (eight females; five males) belonging to the source population were radio-collared and tracked over the same period in 2007–2008.

(1) Jones M.L., Mathews N.E. & Porter W.F. (1997) Influence of social organization on dispersal and survival of translocated female white-tailed deer. *Wildlife Society Bulletin*, 25, 272–278.

(2) White P.J. & Garrott R.A. (2005) Northern Yellowstone elk after wolf restoration. *Wildlife Society Bulletin*, 33, 942–955, https://doi.org/10.2193/0091-7648(2005)33[942:nyeawr]2.0.co;2

(3) Whisson D.A., Holland G.J. & Carlyon K. (2012) Translocation of overabundant species: Implications for translocated individuals. *The Journal of Wildlife Management*, 76, 1661–1669, https://doi.org/10.1002/jwmg.401

14.12. Translocate predators for ecosystem restoration

https://www.conservationevidence.com/actions/2431

- **Two studies** evaluated the effects of translocating predators for ecosystem restoration. These studies were in the USA[1,2].

COMMUNITY RESPONSE (0 STUDIES)

POPULATION RESPONSE (2 STUDIES)

- **Abundance (2 studies):** A before-and-after study in the USA[2] found that following reintroduction of wolves, populations of beavers and bison increased. A before-and-after study in the USA[1] found that after the translocation of wolves to the reserve, adult elk numbers approximately halved.

- **Reproductive success (1 study):** A before-and-after study in the USA[1] found that after the translocation of wolves to the reserve, elk calf:cow ratios approximately halved.

BEHAVIOUR (0 STUDIES)

Background

In areas where predators have historically been made locally extinct or populations severely reduced, often due to hunting, they may be translocated from other areas and released in an attempt to restore the ecosystem. Predators may help to reduce medium to large herbivore populations for example, and thus allow some recovery of the habitat and other species groups.

A before-and-after study in 1986–2004 in a grassland and forest reserve in Wyoming, USA (1) found that after the translocation of wolves *Canis lupus* to the reserve, adult elk *Cervus canadensis* numbers and elk calf:cow ratios approximately halved. Results were not subjected to statistical analysis. Nine years after wolves were translocated, there were fewer adult elk (8,335) and a lower calf:cow ratio (12 calves/100 female elk) than the average before wolf translocation (adult elk: 16,664; 25 calves/100 female elk). A similar number of elk that had migrated out of the park were killed by hunters before (1,148 elk/year) and after (1,297 elk/year) wolves were translocated. Wolves were translocated into Yellowstone National Park in 1995. Between 1996 and 2004 wolf numbers increased from 21 to 106. Elk adults and calves were counted from aeroplanes annually during December–January 1986–2004. No counts were conducted during the winters of 1996 and 1997.

A before-and-after study in 1990–2010 of riparian and adjacent upland habitat in a national park in Wyoming, USA (2) found that following reintroduction of wolves *Canis lupus*, populations of beavers *Castor canadensis* and bison *Bison bison* increased. There were more beaver colonies in a monitored area 13 years after wolf reintroduction began (12 colonies) than at the start of reintroduction (one colony). Average summer bison counts were higher in the decade after wolf reintroduction began (1,385 bison) than in the preceding decade (708 bison). Following the start of reintroduction in 1995–1996, wolf numbers in the study area increased to 98 in 2003, followed by a decline and substantial fluctuations. Their establishment was associated with a fall in elk *Cervus canadensis* numbers from >15,000 in the early 1990s to approximately 6,100 in 2010. Elk browsing on woody vegetation reduced, increasing resources available to beaver and bison. Beaver and bison numbers were derived from annual surveys.

(1) White P.J. & Garrott R.A. (2005) Northern Yellowstone elk after wolf restoration. *Wildlife Society Bulletin*, 33, 942–955, https://doi.org/10.2193/0091-7648(2005)33[942:nyeawr]2.0.co;2

(2) Ripple W.J. & Beschta R.L. (2012) Trophic cascades in Yellowstone: The first 15 years after wolf reintroduction. *Biological Conservation*, 145, 205–213, https://doi.org/10.1016/j.biocon.2011.11.005

14.13. Use holding pens at release site prior to release of translocated mammals

https://www.conservationevidence.com/actions/2434

- **Thirty-five studies** evaluated the effects of using holding pens at the release site prior to release of translocated mammals. Ten studies were in the USA[2,3,4,5,8,11,17,31,33,34], seven were in South Africa[9,15,16,22,24,25,27], four were in the UK[6,10,20,35], three studies were in France[12,18,21], two studies were in each of Canada[7,23], Australia[14,30] and Spain[28,32] and one was in each of Kenya[1], Zimbabwe[13], Italy[19], Ireland[26] and India[29].

COMMUNITY RESPONSE (0 STUDIES)

POPULATION RESPONSE (31 STUDIES)

- **Abundance (4 studies):** Three of four studies (two replicated, one before-and-after study) in South Africa[22], Canada[23], France[18] and Spain[32] found that following release from holding pens at release sites (in some cases with other associated actions), populations of roe deer[18], European rabbits[32] and lions[22] increased in size. The other study found that elk[23] numbers increased at two of four sites.

- **Reproductive success (10 studies):** A replicated study in the USA[8] found that translocated gray wolves[8] had similar breeding success in the first two years after release when adult family groups were released together from holding pens or when young adults were released directly into the wild. Seven of nine studies (including two replicated and one controlled study) in Kenya[1], South Africa[16,25], the USA[2,11], Italy[19], Ireland[26], Australia[30] and the UK[35] found that following release from holding pens at release sites (in some cases with other associated actions), translocated populations of roan[1], California ground squirrels[2], black-tailed prairie dogs[11], lions[25], four of four mammal populations[30], most female red squirrels[26] and some pine martens[35] reproduced successfully. Two studies found that one of two groups of Cape buffalo[16] and one pair out of 18 Eurasian badgers[19] reproduced.

- **Survival (26 studies):** Two of seven studies (five controlled, three replicated studies) in Canada[7], the USA[3,8,31], France[12,21], the UK[20] found that releasing animals from holding pens at release sites (in some cases with associated actions) resulted in higher survival for water voles[20] and female European rabbits[12] compared to those released directly into the wild. Four studies found that translocated swift foxes[7], gray wolves[8], Eurasian lynx[21] and Gunnison's prairie dogs[31] released from holding pens had similar survival rates to those released directly into the wild. One study found that translocated American martens[3] released from holding pens had lower survival than those released directly into the wild. Two of four studies (three controlled) in South Africa[24], Spain[28], and the USA[33,34] found that translocated African wild dogs[24] and European rabbits[28] that spent longer in holding pens at release sites had a higher survival rate after release. One study found mixed effects for swift foxes[33] and one found no effect of time in holding pens for San Joaquin kit foxes[34]. Eleven studies (one replicated) in Kenya[1], South Africa[9,22], the USA[5,11], France[18], Italy[19], Ireland[26], India[29], Australia[30] and the UK[35] found that after release from holding pens at release sites (in some cases with other associated actions), translocated populations or individuals survived between one month and six years, and four of four mammal populations[30] survived. Two studies in the UK[10] and South Africa[27] found that no released red squirrels[10] or rock hyraxes[27] survived over five months or 18 days respectively. One of two controlled studies (one replicated, one before-and-after) in South Africa[25] and the USA[34] found that following release from holding pens, survival of translocated lions[25] was higher than that of resident animals, whilst that of translocated San Joaquin kit foxes[34] was lower than that of resident animals. A study in Australia[14] found that translocated bridled nailtail wallabies[14] kept in holding pens prior to release into areas where predators had been controlled had similar annual survival to that of captive-bred animals.

- **Condition (1 study):** A controlled study in the UK[6] found that translocated common dormice held in pens before release gained weight after release whereas those released directly lost weight.

BEHAVIOUR (5 STUDIES)

- **Behaviour change (5 studies):** Three studies (one replicated) in the USA[4,17] and Canada[23] found that following release from holding pens, fewer translocated sea otters[4] and gray wolves[17] returned to the capture site compared to those released immediately after translocation, and elk[23] remained at all release sites. Two studies in Zimbabwe[13] and South Africa[15] found that following release from holding pens, translocated lions formed new prides.

Background

Holding pens at release sites (sometimes termed 'soft release') may be used to enable mammals to become accustomed to new surroundings before release. They are often enclosures containing natural habitat and enabling views of surrounding land. Additionally, some wild translocated mammals may display a homing instinct after release and pens may therefore be used to reduce the chance of animals returning.

The use of holding pens may be employed both for translocations of wild mammals to new sites and releases of captive-bred mammals, here we focus on the first group. See also: *Use holding pens at release site prior to release of captive-bred mammals.*

For studies that held translocated mammals in captivity away from the release site before release, see: *Hold translocated mammals in captivity before release.*

A study in 1970–1978 in a grassland and forest reserve in southeast Kenya (1) found that after being kept in a holding pen prior to release, a population of roan *Hippotragus equinus* translocated into an area outside their native range persisted and bred for more than six years. Only eight out of the original 38 translocated roan could be located 18 months after the last release. However, six years after the last translocations, roan numbers had increased to 22. From 1973–1976, at least 15 calves were born, of which one-third survived to nine months of age. Between 1970 and 1972, 38 roan were released in Shimba Hills National Reserve, where there is no evidence for their existence since at least 1885. Animals were captured in the Ithanga Hills, by funnelling them into a 2.5 acre corral using horses, trucks and a helicopter. Prior to release roan were kept in a 30-acre holding pen. Roan were monitored between June 1973 and January 1978, but no further monitoring details are provided.

A study in 1976–1978 in a pasture in California, USA (2) found that following release from holding pens at the release site, translocated California ground squirrels *Spermophilus beecheyi* established a reproductive colony. Reproduction occurred within one of the holding cages, but the number of young was not determined. At least three of the eight ground squirrels released from cages were still alive 8–13 months after release. Four wire-mesh cages (1.2 × 2.4 × 0.6 m high) were part-filled with soil, to 41 cm depth, in a 7.5-ha pasture. Cages each had four pipes (20 cm long, 10 cm diameter) leading down into the soil, as refuges. Cages were positioned in two adjacent pairs. Pairs were 46 m apart. In November 1976, one pair of wild-caught California ground squirrels was released into each cage. Squirrels were allowed to exit from two of the cages in March 1977 and from the other two in June 1977. In February–April 1978, tagged and non-tagged squirrels were observed and/or live-trapped near the cages.

A randomized, controlled study in 1975–1976 in a temperate forest in Wisconsin, USA (3) found that when using holding pens prior to releasing translocated animals, American marten *Martes americana* survival was lower than when animals were released immediately after translocation. Eight of 10 American martens released after being held in pens died within 154 days. Only one of 11 animals released immediately after translocation had died within 161 days. None of the martens reproduced in this time. Thirty days after release, martens that had been held in pens stayed closer to the release site than did those released immediately (data not reported). In January 1975–April 1976, 124 martens, captured in Ontario, Canada, were released at a forest site in Wisconsin, USA. Twenty-six animals were held in pens at the release site for seven days before release and 97 animals were released within 48 hours of being transported to the site. Individuals were randomly assigned a release method. Twenty-one of the martens were radio-collared. Their movements were monitored until June 1976.

A study in 1988–1989 in coastal waters of California, USA (4) found that after being held in pens at the release site, fewer translocated sea otters *Enhydra lutris* returned to the capture site compared to those released immediately after translocation. No statistical analyses were performed. None of 10 sea otters held in release pens returned to the capture site and all remained within 27 km for the duration of monitoring. Five of nine released immediately on arrival returned to the capture site. Nineteen sea otters (18 male, one female) were caught between May 1988 and May 1989 and were released 291 km further north. Nine were released immediately on arrival and 10 were held for 48 hours in floating pens before release. Sea otters were radio-tracked from the ground or air for 16–87 days after release.

A study in 1988–1989 in forest and swamp habitats in Florida and Georgia, USA (5) found that after being held in holding pens at the release site, more than half of translocated mountain lions *Puma concolor* survived over three months. Four out of seven translocated mountain lions survived at least 124–303 days after release. Individuals that had been in the wild >35 days established 96–930-km² home ranges. However, during the hunting season, these home ranges were abandoned. At least three mountain lions died during the study, including one that was shot. In 1988, seven mountain lions were captured in Texas and flown to Florida. They were released as a trial for evaluating the feasibility of translocating Florida panthers *Puma concolor coryi*. Animals were sterilized, radio-collared and kept in holding pens for one week before release. They were monitored six days/week for 306 days from an airplane. Before translocation, the study area (>12,000 km²) had no mountain lions but had a high abundance of deer and wild hog and a low density of humans.

A controlled study in 1992 in woodland edge in Somerset, UK (6) found that translocated common dormice *Muscardinus avellanarius* held in pens before release gained weight after release, whereas dormice released directly into the wild lost weight. The body mass of dormice released from pre-release pens increased after release by 0.12 g/day, whereas dormice released directly into the wild lost 0.14 g/day. The study was conducted along a 9-ha strip of trees and shrubs in August–September 1992. Six wild-caught dormice were placed in pre-release pens and 10 wild-caught dormice were released directly into the wild on their day of capture. Pre-release pens (0.45 m width, 0.5 m depth and 0.9 m height) were constructed from 1-cm^2 weldmesh. Nest boxes, food and water were provided. Dormice stayed in pens for eight nights before release. Dormice were monitored by radio-tracking and were recaptured and weighed 10–14 days after release.

A replicated, controlled study in 1983–1993 in three grassland sites in Alberta, Canada (7) found that translocated and captive-bred swift foxes *Vulpes velox* released after time in holding pens had similar survival rates to those released without use of holding pens. No statistical analyses were performed. At least six out of 45 (13%) swift foxes held in pens before release survived over two years post-release, compared with at least five out of 43 (12%) released without use of holding pens. In 1983–1987, forty-five translocated swift foxes were held in pens before release. Pens (3.7 × 7.3 m) were fenced for protection from cattle. Animals were placed in pens in October–November and released between the following spring and autumn. They were provided with supplementary food for 1–8 months after release. In 1987–1991, four hundred and thirty-three foxes were released without use of holding pens. Released foxes included both wild-born and captive-bred animals. All foxes released from pens and 155 of those released directly were radio-tracked, from the ground or air, for up to two years.

A replicated study in 1995–1996 in two forest sites in Idaho and Wyoming, USA, (8) found that translocated gray wolves *Canis lupus* had similar survival rates and breeding success in the first two years after release when adult family groups were released together from holding pens or when young adults were released directly into the wild. No statistical analyses were conducted. Thirty out of 35 young adult wolves released directly into the wild were still alive seven months after the last releases, and had produced up to 40 pups from 3–8 pairs. Thirty-one adult wolves released from holding pens in family groups had produced

23 pups four months after the last releases. From these 54 animals, nine had died. Six of the seven adult pairs released together from holding pens remained together, and five of these pairs established territories in the vicinity of the pens. Wolves were wild-caught from Canada in January 1995 and 1996. In Idaho, young adults were directly released in January 1995 and 1996. In Wyoming, family groups of 2–6 wolves spent 8–9 weeks in 0.4-ha chain-link holding pens before release in March 1995 and April 1996. Wolves were radio-tracked every 1–3 weeks until August 1996.

A study in 1994–1998 in a savannah reserve in North West province, South Africa (9) found that after release from holding pens in groups, approximately half of translocated cheetahs *Acinonyx jubatus* survived at least 18 months, of which half died within three years. Nine of 19 cheetahs survived 19–24 months, of which six were cubs that matured to independence, but only four cheetahs were known to still be alive at the end of the study period. Six cheetahs survived in the reserve less than one year, of which one died after a few weeks and two were removed to a captive breeding facility. The fate of four released cheetahs was unknown. In total 19 cheetahs were released into a game reserve between October 1994 and January 1998. Cheetahs were initially placed in 1 ha holding pens with electrified fencing for 4 weeks to several months. The feeding regime is not specified, but cheetahs were provided with at least one carcass on being placed in the pen and were lured from the pen with a carcass. Cheetahs were mostly rescued wild-caught animals, except for one that was habituated to humans (and had to be removed after two weeks). Cheetahs were either held in family groups (mothers with cubs) or as coalitions (of adult males). One animal/group was radio collared for monitoring.

A study in 1993–1994 on a forested peninsula in Dorset, UK (10) found that none of the translocated red squirrels *Sciurus vulgaris* released into holding pens (with supplementary food, water and nestboxes) survived over five months after release. Out of 14 translocated red squirrels, 11 (79%) survived over one week. Only three (21%) survived >3 months and none survived >4.5 months. At least half of the 14 squirrels were killed by mammalian predators. When intact carcasses were examined they showed signs of weight loss and stress (see original paper for details). Between October and November 1993, fourteen wild-born red squirrels were released into an 80-ha forest dominated by Scots pine *Pinus sylvestris*. The forest had no red squirrels but had introduced grey squirrels *Sciurus carolinensis*. Capture and release sites were similar habitats. Squirrels were kept in 1.5 × 1.5 × 1.5 m weldmesh pens surrounded by electric fencing for 3–6 days before release. Squirrels were kept individually except for two males who shared a pen. After release, squirrels continued to have access to food, water and nest boxes inside the pens and outside (20–100 m away). All squirrels were radio-tagged and located 1–3 times/day, for 10–20 days after release and thereafter every 1–2 days.

A replicated study in 1995–1997 in four grassland sites in New Mexico, USA (11) found that after release from holding pens and provision of supplementary food, translocated populations of black-tailed prairie dogs *Cynomys ludovicianusi* persisted at least two years and reproduced in the wild. The number of black-tailed prairie dogs approximately doubled during the first spring after release from holding pens in one site on one ranch where supplementary food was provided. Between the second spring and summer, after all supplementary feeding had ceased, the number of animals associated with both release sites on the same ranch doubled. Precise numbers are not reported. One hundred and one prairie dogs were translocated to two ranches (Armendaris Ranch received 71 individuals; Ladder Ranch: 30 individuals) between June 1995 and June 1997. At each ranch, prairie dogs were released into two 0.4-ha holding pens (number of individuals/holding pen is not provided). Holding pens were fenced and surrounded by electric wire. Animals at Armendaris ranch were provided with supplementary food in pens for up to year. Information on population persistence at Ladder Ranch is not provided. The time individuals were kept in the holding pens before subsequent release varied between a few days and weeks (see original paper for details).

A controlled study in 1997 in a mixed pasture and cultivated fields farmland site in northern France (12) found that keeping translocated European rabbits *Oryctolagus cuniculus* in holding pens for three days prior to release (and carrying out associated management such as supplementary feeding) increased survival rates of female, but not male rabbits immediately following release compared to rabbits released directly into the wild. During the first day after translocations, the survival rate of female rabbits released from pre-release pens was higher (100%) than that of females released directly into the wild (83%) and male rabbits released from release pens (78%). The survival rate of male rabbits released from pre-release pens (78%) was not significantly different to that of male rabbits released directly into the wild (92%). One hundred and four rabbits were translocated from Parc-du-Sausset to a 150-ha area of cultivated fields and pasture in Héric, approximately 400 km away in January 1997. Of these, roughly half were acclimatised in eight 100-m² enclosures (fence height: 1 m), for three days prior to release. Rabbits were provided with supplementary food. Survival was estimated by night-time relocation of ear-tagged rabbits using a spotlight, daily in the first week after release and twice a week until late February 1997.

A study in 1997–1998 on a savanna estate in Zimbabwe (13) found that a translocated lion *Panthera leo* family kept in a holding pen prior to release joined with immigrant lions and formed a new pride. A lioness was translocated with three cubs (one male, two female). Within 45 days, seven male lions were close by and the female mated with one of these. The male cub moved away and the pride then entailed the female and daughters with two adult male lions. A wild lioness joined the pride 1.7 months after release, but was killed by a snare after six months. After 12–13 months, the original lioness had three new cubs and her daughters each also had litters. Resident lions on the estate were eliminated in 1995. In January 1997, a lioness and three cubs were translocated from communal land to a holding pen and were released on the estate after 90 days. Lions were monitored through to May 1998 by radio-tracking and direct observation.

A study in 1996–1999 at a woodland reserve in Queensland, Australia (14) found that wild-born translocated bridled nailtail wallabies *Onychogalea fraenata* kept in holding pens prior to release into areas where predators had been controlled had similar average annual survival to that of captive-bred animals. Over four years, the average annual survival of wild-born translocated wallabies (77–80%) did not differ significantly from that of captive-bred bridled nailtail wallabies (57–92%). In 1996–1998, nine wild-born translocated and 124 captive-bred bridled nailtail wallabies were released into three sites across Idalia National Park. Ten captive-bred wallabies were held in a 10-ha enclosure within the reserve for six months before release, and 85 were bred within the 10-ha enclosure. All of the 133 released wallabies were kept in a holding pen (30-m diameter) for one week at each site before release. Mammalian predators were culled at release sites. A total of 67 wallabies (58 captive-bred, nine wild-born) were radio-tagged and tracked every 2–7 days in 1996–1998. Wallabies were live-trapped at irregular intervals with 20–35 wire cage traps in 1997–1999.

A study in 1998–2002 in a shrubland wildlife reserve in Limpopo, South Africa (15) found that after being held together in a pen for three months before release, five translocated African lions *Panthera leo* eventually formed two separate prides. Two months after release, there was aggression between two males and a female, which had sustained injuries shortly after release. Aggression continued intermittently for 10 weeks until the injured lion mated. Subsequently, over the following 3.5–4 years, two prides established territories. One pride comprised of a male and female half-siblings with an additional related female. The second pride was a looser association between a male and female sibling. Thus, inbreeding was likely to occur between mated pairs. Two male and three female wild-caught lions (from two locations) were released on 16 January 1998 into a 33,000-ha fenced reserve, after being held for three months in a 50 × 50-m pen. Lions were monitored by radio-tracking through to February 2002.

A study in 2000–2003 in a mixed karoo grassland reserve in Northern Cape Province, South Africa (16) found that following release from a holding pen in groups into a fenced reserve, one out of two translocated Cape buffalo *Syncerus caffer* groups scattered and escaped the reserve while the other formed a single herd and stayed in the reserve and bred. One month after release, a group of four buffalo had split into two solitary animals and a pair formed by one male and one female. One of the solitary animals was not seen again, the second solitary male animal was located two years after release on a neighbouring farm and released into the second group of translocated animals in May 2003. The pair escaped the reserve three times in 13 months. After the third escape, the male was moved to a different reserve and a new male introduced to form a herd with the remaining female. A second group of 10 translocated animals formed a single herd (along with the two remaining animals from the previous introduction) and over 10 months no animals died or escaped. A year after the introduction, five calves were born. Four subadult buffalo (2 male, 2 female) were placed in a holding pen in July 2000 and released in August into a fenced 12,000-ha reserve. A second group of seven adult and three subadult animals (4 male, 6 female) was placed into a holding pen in August 2002 and released into a 200 ha area in September before being completely released in October 2002. Both groups were monitored weekly with telemetry until October 2003.

A study in 1989–2002 in 25 temperate forest sites in Montana, Idaho, and Wyoming, USA (17) found that holding translocated wolves *Canis lupus* in pens at the release site before release (soft release) increased the chance of wolves not returning to their capture site relative to direct (hard) release. A lower proportion of soft-released wolves returned to their capture site (8%) than of hard-released wolves (30%). Soft-releases entailed confinement at release sites for ≥28 days after capture. Hard-releases were those occurring ≤7 days following capture. Eighty-eight wolves were translocated 74–515 km in 1989–2001 in response to livestock predation (75 wolves) or pre-emptively to avoid such conflict (13 wolves). Translocated wolves were radio-collared, and were monitored through to the end of 2002.

A study in 1995–2002 in a mixed oak forest reserve in the south of France (18) found that following translocation using holding pens prior to release and associated actions, approximately half of female roe deer *Capreolus capreolus* survived over one year after release and that overall the deer population increased six years after the translocations began. Twenty-six out of 49 (53%) translocated female roe deer survived over one year post-release. Of the animals that died in the first year, 35% of mortality occurred within the first month after release. After six years the deer population had increased to 0.47 deer/km² compared to 0.06 deer/km² in the first year after translocation began. In February 1995–1997, fifty-two male and 52 female roe deer were translocated from Northern France into a 3,300-ha forest reserve in Southern France in seven release sessions. Animals were placed into enclosures in groups of approximately 15 individuals for 2–10 days and provided with food (pellets and fresh vegetables) during this time prior to release. Forty-nine females (21 <1 year old and 28 >1 year old) were radio-tagged and were located from a vehicle once or twice each week, over one year post-release. In addition, surveys were carried out on foot (6 transects, each 5–7 km long) eight times a year in February-March 1996–2002 to estimate population growth. Deer were present in low numbers prior to translocation.

A study in 2001–2005 in a mixed forest and farmland site in northern Italy (19) found that just over half of translocated Eurasian badgers *Meles meles* released from holding pens (with supplementary food) in groups survived at least one month after release and one pair reproduced. Seven out of 12 badgers survived for 1–9 months, after which monitoring equipment stopped operating. One badger died almost immediately after release due to unknown causes. Two badgers escaped (one after the first month, the other after an unknown period). The fate of three other badgers was unknown. One pair of translocated animals reproduced in the wild 4 years after release. From March 2001 to May 2004, twelve badgers were captured at four sites in northern Italy. Badgers were fitted with radio-collars and transported 20–40 km to the release site where they were kept in a 350 m² enclosure in a wooded area in their release groups (2001: 2 individuals, 2002: 4 individuals, 2003: 2 individuals; 2004: 4 individuals) and provided supplementary food for 3–10 weeks before release. Seven of the 12 badgers were located once/week, for up to nine months after release.

A review of a study in 2001–2002 at a restored wetland in London, UK (20) found that using holding pens prior to release of translocated and captive-bred water voles *Arvicola terrestris* resulted in greater post-release survival than did releasing them directly into the wild. Voles released from pens were three times more likely to be recorded during the initial follow-up survey than were those released without use of pens (result presented as odds ratio). A total of 38 wild-caught and 109 captive-bred water voles were released in groups of 6–15 animals in May–July 2001. Prior to release, no water voles were present at the site. An unspecified number of animals were placed in an enclosure with food and shelter and allowed to burrow out at will. The remainder were released directly into the wild. Animals were monitored by live-trapping over three periods of five days, between autumn 2001 and early-summer 2002.

A controlled study in 1983–2002 in a temperate forest in Vosges massif, France (21) found that survival of translocated Eurasian lynx *Lynx lynx* that were held in captivity before release was similar between animals kept in holding pens at the release site and animals which were released directly. Four of eight animals which were kept in enclosures at the release site prior to release survived for 10–11 years, compared to six of 13 animals that survived 2–7 years after being released without holding pens. The distribution of lynx increased from 1,870 km² (six years after the first releases) to 3,160 km² (12 years later). At least two females, both of which were released without holding pens, produced litters. In 1983–1993, twenty-one adult lynx were brought to France from European zoos. The program sought wild-caught lynx for releases, however the exact origin of each animal, and the length of time that each spent in captivity, are unclear. Lynx were released at four sites in the Vosges mountains. The first eight animals were held in cages at the release site for 4–45 days prior to release, but the remainder were released immediately upon arrival. Animals were radio-tracked for 1–847 days. The presence of lynx was also established through sightings, lynx footprints, detection of faeces or hair and reports of attacks on domestic animals.

A study in 1992–2004 in a grassland reserve in KwaZuluNatal Province, South Africa (22) found that most translocated lions *Panthera leo* held in pens before release survived for more than one year and established stable home ranges and that the population grew. Of 15 lions released, all except three, which were removed after killing a tourist, survived ≥398 days post-release. Average post-release survival was ≥1,212 days. At least 95 cubs from 25 litters were documented among translocated lions and descendants over the 13-year study. Excluding cubs translocated to other sites or those still <18 months old at the end of the study, 51 of 65 cubs (78%) survived past 18 months old. Nine lions were released in May 1992, six in February 1993 and two in January 2003. Releases were into a fenced reserve (initially 176 km², then extended to 210 km²). Before release, lions were held in groups, each in an 80-m² acclimation pen, for 6–8 weeks, during which time socialization occurred and stable prides were formed. Eleven of the founder lions were monitored by radio-tracking and other animals were monitored by direct observations.

A replicated study in 1998–2004 within four largely forested areas in Ontario, Canada (23) found that following translocation elk *Cervus canadensis*, most of which had been kept in holding pens in groups, remained present at all release sites and numbers had increased at two of four sites. By 3–6 years after translocations, elk populations had grown at two sites and fallen at two. From 443 elk translocated, the population at the end of the study was estimated at 375–440 animals. Between 1998 and 2004, forty-one percent of translocated elk died. Causes of death included 10% lost to wolf predation, 5% to emaciation and 5% to being shot. Elk were translocated from a site in Alberta, Canada in 1998–2001 in nine releases. Transportation took 24–58 hours. Elk were held in pens at recipient sites for up to 16 weeks before release (some were released immediately) but the effect of holding pens was not tested. Of 443 elk released, 416 were monitored by radio-tracking. The overall population was estimated in March 2004.

A study in 1995–2005 in 12 dry savanna and temperate grassland sites in South Africa (24) found that translocated and captive-bred African wild dogs *Lycaon pictus* that spent more time in holding pens in groups had a higher survival rate after release. Wild dog families that had more time to socialise in holding pens prior to release into fenced areas had a higher survival rate than groups which spent less time in holding pens (data presented as model results). Overall, 85% of released animals and their wild-born offspring survived the first six months after release/birth. Released animals that survived their first year had a high survival rate 12–18 months (91%) and 18–24 months (92%) after release. Between 1995 and 2005, one hundred and twenty-seven wild dogs (79 wild-caught, 16 captive-bred, 16 wild-caught but captive-raised, 16 'mixed' pups) were translocated over 18 release events into 12 sites in five provinces of South Africa. Individuals were kept in pre-release pens for an average of 212 days, but groups were given between 15 and 634 days to socialise in pens prior to release. Animals were monitored for 24 months after release, and the 129 pups which they produced after release were monitored up to 12 months of age. Forty characteristics of the individual animals, release sites and methods of release were recorded, and their impact on post-release survival was tested.

A replicated, controlled study in 1999–2004 in three mixed savanna and woodland sites in KwaZulu-Natal, South Africa (25) found that after translocation to a fenced reserve with holding pens, survival of released lions *Panthera leo* was higher than that of resident lions, and that translocated animals reproduced successfully. No statistical tests were performed. After five years, a higher proportion of introduced animals survived (eight of 16 animals, 50%) than of resident animals (20 of 84 animals, 24%). Seven translocated females reproduced successfully. Between August 1999 and January 2001, sixteen lions were translocated to an enclosed reserve to improve genetic diversity. They were held at release sites in 0.5–1.0-ha pens for 4–6 weeks before release. Nine translocated lions were fitted with radio-collars. From August 1999 to December 2004, translocated animals were located at least every 10 days. Resident lions were also tracked at least every 10 days.

A study in 2005–2007 in a mixed conifer forest in Galway, Ireland (26) found that following release from holding pens (with nest boxes and supplementary food), over half of translocated red squirrels *Sciurus vulgaris* survived over eight months after release and most females reproduced during that period. At least 10 out of 19 (53%) translocated squirrels survived over eight months post-release and five out of nine translocated females (56%) were lactating five-seven months after release. In August 2006, seven juvenile squirrels were caught. At least one squirrel was still alive in the release location in two years after the original release. Two squirrels died while in the release pen or shortly afterwards. Another four squirrels died 1–2 months after release. Nineteen squirrels were translocated to a nature reserve (19 ha) in the middle of a 789-ha commercial pine plantation, 112 km from the capture site. Individuals were marked and radio-tagged. Squirrels were kept on average for 46 days in one of two pre-release enclosures (3.6 × 3.6 × 3.9 m high). Enclosures contained branches, platforms, nest boxes, and supplementary feeders (containing nuts, maize, seeds and fruit). Supplementary food (50/50 peanut/maize mix) was provided in six feeders in the nature reserve until July 2006. Twenty nest boxes were also provided Squirrels were radio-tracked in September and November 2005 and February and May 2006, and were trapped in February, May and August 2006 and observed once in October 2007.

A study in 2007 at rocky outcrops on a reserve in KwaZulu-Natal Province, South Africa (27) found that all translocated rock hyraxes *Procavia capensis* kept in a holding pen and released as a group died (or were presumed to have died) within 18 days of release. Eight of nine wild translocated hyraxes died within 18 days of release and the other was presumed to have died. The group split up and were not seen together after release. In October 2007, nine hyraxes (one juvenile, three sub-adults and five adults) were caught in baited mammal traps (900 × 310 × 320 mm) in an area where they were abundant, and moved 150 km to a 656-ha reserve where the species was nearly extinct. Hyraxes were kept together in a holding cage (1850 × 1,850 × 1850 mm) for 14 days before release. They were monitored daily for one week, and then every few days by direct observation and radio-tracking.

A replicated, randomized, controlled study published in 2010 of a grassland site in Andalucía, Spain (28) found that holding translocated wild European rabbits *Oryctolagus cuniculus* for longer in acclimation pens before release improved subsequent survival rates. A lower proportion of rabbits enclosed for six nights before release was killed by mammalian predators over the following 10 days (9%) than of rabbits enclosed for three nights before release (38%). Rabbits were translocated to a 4-ha grass field with artificial warrens. Food and water were provided. Of 181 rabbits released (average 10/warren), 38 randomly selected rabbits (2–5 in each of 15 warrens) were radio-collared. Twenty-three of these were released on the seventh day, following six nights of confinement and 14 were released on the fourth day, following three nights of confinement. The date of the study is not stated. Rabbits were monitored daily during confinement and for 10 days following release.

A study in 2008–2009 in a subtropical forest in Rajasthan, India (29) found that three translocated tigers *Panthera tigris tigris* that were kept in holding pens prior to release survived for at least 3–11 months after release and established home ranges. The annual home range of a released male was 169 km^2 and that of a female was 181 km^2. The summer home range of a later released female was 223 km^2. Home ranges overlapped by 54–99 km^2. Mating was observed between the male and each female. Of 115 recorded kills by tigers, 12 were of domestic animals. Thirty-two villages were located within the 881-km^2 reserve. Tigers had been absent since 2004. One male and one female wild-caught tiger were released on 6 and 8 July 2008, respectively. A further female was released on 27 February 2009. Tigers were held in 1-ha enclosures at release sites for 2–8 days before release. They were satellite-and radio-tracked from release until June 2009.

A study in 1998–2010 in a desert site in South Australia (30) found that after being kept in a holding pen, all four mammal populations released into an invasive-species-free fenced enclosure survived and bred. After being kept in a holding pen prior to release into a fenced enclosure, where red foxes *Vulpes vulpes*, cats *Felis catus* and rabbits *Oryctolagus cuniculus* had been eradicated, greater stick-nest rats *Leporillus conditor*, burrowing bettongs *Bettongia lesueur*, western barred bandicoots *Perameles bougainville* and greater bilbies *Macrotis lagotis* were detected for eight years, increased their distribution range within five years and produced a second generation within two years. In 1998–2005, eight wild-born greater stick-nest rats, 10 wild-born burrowing bettongs, 12 wild-born western barred bandicoots and nine captive-bred greater bilbies were translocated into a 14-km^2 invasive-species-free fenced area. Rabbits, cats and foxes were eradicated within the fenced area in 1999. Animals were kept in a 10-ha holding pen before full release after a few months. Between 2000 and 2010, tracks were surveyed annually along eight 1 km × 1 m transects.

A replicated, controlled study in 2008–2009 of grassland at two sites in Arizona, USA (31) found that following translocation of Gunnison's prairie dogs *Cynomys gunnisoni* into burrows that were topped with acclimation cages for one week, survival was not greater than that of prairie dogs released into uncaged burrows. Among prairie dogs whose identity could be established in the second year, 10% of both those released into borrows topped with acclimation cages and those released into uncaged burrows survived for at least one year. Additionally, pups were seen at both sites a year after release (39 and 37 pups at the two sites). No definite immigrants to the recipient colonies were recorded. Prairie dogs were trapped from 7 July to 5 August 2008 at one urban and one suburban site (74 and 75 prairie dogs, respectively) and moved approximately 50 km to two abandoned colonies (6 km apart) in a rural area. Approximately half at each colony was released directly into open burrows and half into borrows topped, for one week, with acclimation cages. Survival monitoring, from 10 June to 25 August 2009, entailed live-trapping, PIT-tag reading and direct observations.

A replicated, before-and-after study in 2008–2012 in 32 shrubland sites in Andalusia, Spain (32) found that following release from holding pens with artificial warrens to boost a local population, translocated European rabbit *Oryctolagus cuniculus* abundance was higher after three years. Rabbit abundance was around nine-fold higher three years after translocations (9.3 latrines/km) than before translocation (1.0 latrines/km). In autumn and winter of 2008–2009, between 75 and 90 rabbits/ha were released into artificial warrens located in 32 electric-fenced 0.5–7.7 ha plots (fencing was 0.5 m below ground and 1.7 m above ground). At the end of the 2009–2011 breeding season, small gates on the fences were opened and the rabbits were allowed to disperse into adjacent areas. Rabbit abundance was estimated by latrine counts along four 500-m transects (128 total transects) around each plot, in the summers of 2008–2009 before gates were opened and in 2012 after gates were opened. Wooden branches and artificial warrens were added within a 500-m radius of some plots and, in some, scrub was cleared to create pasture.

A controlled study in 2002–2007 on a large area of prairie in South Dakota, USA (33) found that using holding pens at release sites affected survival rates of translocated swift foxes *Vulpes velox*. A higher proportion of foxes released after 14–21 days in holding pens survived for ≥60 days post-release (76%) than of foxes held in pens for >250 days (66%) or released after 14–21 days in kennels at a field station (61%). A total of 179 foxes (85 males and 94 females; 91 adults and 88 sub-adults) were translocated in 2002–2007. Holding pens provided acclimatisation at release sites, with food provided at pens following release. Foxes released from short stays in holding pens, and those released having been held in kennels, were released in August–October. Long-stay foxes were released in mid-July. Survival was monitored by radio-tracking and visual observations at dens.

A controlled, before-and-after study in 1989–1992 on a hilly grassland and scrubland site in California, USA (34) found that the survival of translocated San Joaquin kit foxes *Vulpes macrotis mutica* kept in holding pens in pairs prior to release was lower than that of resident animals, but did not change with the length of time in holding pens. The survival of 40 translocated foxes in the first year after release (six alive, 32 dead, two unknown) was lower than that of 26 resident foxes (13 alive, 13 dead), but did not change with the length of time spent in holding pens. Eleven pups born in the holding pens and released with their parents all died within 17 days of release. Only four foxes were known to breed after release, all with resident foxes. At the end of the study (in 1992) one fox was known to be alive and 36 (out of 40) were known to have died. Causes of death were predation (20 foxes), road accidents (two foxes) and death during trapping operations (one fox). The cause of death was unknown for 13 foxes. In August and December 1988 and January 1989, and from June–October 1989, foxes were caught and translocated up to 50 km to a 19,120-ha reserve. Foxes were kept in male–female pairs in holding pens (6.1 × 3.1–6.1 × 1.8 m) for 32–354 days before release in spring and summer 1990 (12 adults, 1 pup) and 1991 (28 adults, 10 pups). Foxes were monitored by radio-tracking 4–5 days/week after release.

A study in 2015–2016 in a wooded mountain region in central Wales, UK (35) found that some translocated pine martens *Martes martes* held in pre-release pens and then provided with supplementary food and nest boxes survived and bred in the first year after release. At least four out of 10 females that had been kept in pre-release pens survived and bred the year after release. Around 10–12 months after release, 14 out of 20 martens were alive and in good condition. Twelve were within 10 km of their release site. Six martens died in the first year, two had a fungal infection two weeks after release. Authors suggest this may have been due to damp conditions in November. From September–November 2015, twenty breeding age (>3-years-old) pine martens were caught in Scotland, health checked, microchipped and fitted with a radio-collar, and in some cases a GPS logger. Martens were transported overnight to Wales, and held in individual pre-release pens (3.6 × 2.3 × 2 m) for up to seven nights. Males' pens were within 500 m of a female, but >2 km from the nearest male. Releases took place in autumn, and supplementary food was provided for 2–6 weeks after release (for as long as it continued to be taken). Den boxes were provided within 50 m of each release pen. Martens were radio-tracked until home-ranges were established, then located daily–weekly. Intensive tracking of females was carried out in March to locate breeding sites. Hair tubes and camera traps were used to monitor breeding success. A further 19 martens were released using the same procedure in September–October 2016.

(1) Sekulic R. (1978) Roan translocation in Kenya. *Oryx*, 14, 213–217.

(2) Salmon T.P. & Marsh R.E. (1981) Artificial establishment of a ground squirrel colony. The *Journal of Wildlife Management*, 45, 1016–1018.

(3) Davis M.H. (1983) Post-release movements of introduced marten. *The Journal of Wildlife Management*, 47, 59–66.

(4) Ralls K., Doroff A. & Mercure A. (1992) Movements of sea otters relocated along the California coast. *Marine Mammal Science*, 8, 178–184.

(5) Belden R.C. & Hagedorn B.W. (1993) Feasibility of translocating panthers into northern Florida. *The Journal of Wildlife Management*, 57, 388–397.

(6) Bright P.W. & Morris P.A. (1994) Animal translocation for conservation: performance of dormice in relation to release methods, origin and season. *Journal of Applied Ecology*, 31, 699–708.

(7) Carbyn L.N., Armbruster H.J. & Mamo C. (1994) The swift fox reintroduction program in Canada from 1983 to 1992. Pages 247–271 in: M.L. Bowles & C.J. Whelan (eds.) *Restoration of endangered species: conceptual issues, planning and implementation*. Cambridge University Press, Cambridge.

(8) Bangs E.E. & Fritts S.H. (1996) Reintroducing the gray wolf to Central Idaho and Yellowstone National Park. *Wildlife Society Bulletin*, 24, 402–413.

(9) Hofmeyr M. & van Dyk G. (1998) Cheetah introductions to two north west parks: case studies from Pilanesberg National Park and Madikwe Game Reserve. *Proceedings of a Symposium on Cheetahs as Game Ranch Animals, Onderstepoort*, 23 & 24 October 1998, 60–71.

(10) Kenward R.E. & Hodder K.H. (1998) Red squirrels (*Sciurus vulgaris*) released in conifer woodland: the effects of source habitat, predation and interactions with grey squirrels (*Sciurus carolinensis*). *Journal of Zoology*, 244, 23–32.

(11) Truett J.C. & Savage T. (1998) Reintroducing prairie dogs into desert grasslands. *Restoration and Management Notes*, 16, 189–195.

(12) Letty J., Marchandeau S., Clobert J. & Aubineau J. (2000) Improving translocation success: an experimental study of anti-stress treatment and release method for wild rabbits. *Animal Conservation*, 3, 211–219, https://doi.org/10.1111/j.1469-1795.2000.tb00105.x

(13) Hoare R. & Williamson J. (2001) Assisted re-establishment of a resident pride of lions from a largely itinerant population. *South African Journal of Wildlife Research*, 31, 179–182.

(14) Pople A.R., Lowry J., Lundie-Jenkins G., Clancy T.F., McCallum H.I., Sigg D., Hoolihan D. & Hamilton S. (2001) Demography of bridled nailtail wallabies translocated to the edge of their former range from captive and wild stock. *Biological Conservation*, 102, 285–299, https://doi.org/10.1016/s0006-3207(01)00101-x

(15) Kilian P.J. & Bothma, J. du P. (2003) Notes on the social dynamics and behaviour of reintroduced lions in the Welgevonden Private Game Reserve. *South African Journal of Wildlife Research*, 33, 119–124.

(16) Venter J.A. (2004) Notes on the introduction of Cape buffalo to Doornkloof Nature Reserve, Northern Cape Province, South Africa. *South African Journal of Wildlife Research*, 34, 95–99.

(17) Bradley E.H., Pletscher D.H., Bangs E.E., Kunkel K.E., Smith D.W., Mack C.M., Meier T.J., Fontaine J.A., Niemeyer C.C. & Jimenez M.D. (2005) Evaluating wolf translocation as a nonlethal method to reduce livestock conflicts in the Northwestern United States. *Conservation Biology*, 19, 1498–1508, https://doi.org/10.1111/j.1523-1739.2005.00102.x

(18) Calenge C., Maillard D., Invernia N. & Gaudin J.C. (2005) Reintroduction of roe deer *Capreolus capreolus* into a Mediterranean habitat: female mortality and dispersion. *Wildlife Biology*, 11, 153–161, https://doi.org/10.2981/0909-6396(2005)11[153:rordcc]2.0.co;2

(19) Balestrieri A., Remonti L. & Prigioni C. (2006) Reintroduction of the Eurasian badger (*Meles meles*) in a protected area of northern Italy. *Italian Journal of Zoology*, 73, 227–235, https://doi.org/10.1080/11250000600679603

(20) Mathews F., Moro D., Strachan R., Gelling M. & Buller N. (2006) Health surveillance in wildlife reintroductions. *Biological Conservation*, 131, 338–347, https://doi.org/10.1016/j.biocon.2006.04.011

(21) Vandel J.M., Stahl P., Herrenschmidt V. & Marboutin E. (2006) Reintroduction of the lynx into the Vosges mountain massif: from animal survival and movements to population development. *Biological Conservation*, 131, 370–385, https://doi.org/10.1016/j.biocon.2006.02.012

(22) Hunter L.T.B., Pretorius K., Carlisle L.C., Rickelton M., Walker C., Slotow R. & Skinner J.D. (2007) Restoring lions *Panthera leo*, to northern KwaZulu-Natal, South Africa: short-term biological and technical success but equivocal long-term conservation. *Oryx*, 41, 196–204, https://doi.org/10.1017/s003060530700172x

(23) Rosatte R., Hamr J., Young J., Filion I. & Smith H. (2007) The restoration of elk (*Cervus elaphus*) in Ontario, Canada: 1998–2005. *Restoration Ecology*, 15, 34–43, https://doi.org/10.1111/j.1526-100x.2006.00187.x

(24) Gusset M., Ryan S.J., Hofmeyr M., van Dyk G., Davies-Mostert H.T., Graf J.A., Owen C., Szykman M., Macdonald D.W., Monfort S.L., Wildt D.E., Maddock A.H., Mills M.G.L., Slotow R. & Somers M.J. (2008) Efforts going to the dogs? Evaluating attempts to re-introduce endangered wild dogs in South Africa. *Journal of Applied Ecology*, 45, 100–108.

(25) Trinkel M., Ferguson N., Reid A., Reid C., Somers M., Turelli L., Graf J., Szykman M., Cooper D., Haverman P., Kastberger G., Packer C. & Slotow R. (2008) Translocating lions into an inbred lion population in the Hluhluwe-iMfolozi Park, South Africa. *Animal Conservation*, 11, 138–143, https://doi.org/10.1111/j.1469-1795.2008.00163.x

(26) Poole A. & Lawton C. (2009) The translocation and post release settlement of red squirrels *Sciurus vulgaris* to a previously uninhabited woodland.

Biodiversity and Conservation, 18, 3205–3218, https://doi.org/10.1007/s10531-009-9637-z

(27) Wimberger K., Downs C.T., Perrin M.R. (2009) Two unsuccessful reintroduction attempts of rock hyraxes (*Procavia capensis*) into a reserve in the KwaZulu-Natal Province, South Africa. *South African Journal of Wildlife Research*, 39, 192–201, https://doi.org/10.3957/056.039.0213

(28) Rouco C., Ferreras P., Castro F. & Villafuerte R. (2010) A longer confinement period favors European wild rabbit (*Oryctolagus cuniculus*) survival during soft releases in low-cover habitats. *European Journal of Wildlife Research*, 56, 215–219, https://doi.org/10.1007/s10344-009-0305-0

(29) Sankar K., Qureshi Q., Nigam P., Malik P.K., Sinha P.R., Mehrotra R.N., Gopal R., Bhattacharjee S., Mondal K. & Gupta S. (2010) Monitoring of reintroduced tigers in Sariska Tiger Reserve, Western India: preliminary findings on home range, prey selection and food habits. *Tropical Conservation Science*, 3, 301–318, https://doi.org/10.1177/194008291000300305

(30) Moseby K.E., Read J.L., Paton D.C., Copley P., Hill B.M. & Crisp H.A. (2011) Predation determines the outcome of 10 reintroduction attempts in arid South Australia. *Biological Conservation*, 144, 2863–2872, https://doi.org/10.1016/j.biocon.2011.08.003

(31) Nelson E.J. & Theimer T.C. (2012) Translocation of Gunnison's prairie dogs from an urban and suburban colony to abandoned wildland colonies. *The Journal of Wildlife Management*, 76, 95–101, https://doi.org/10.1002/jwmg.281

(32) Guerrero-Casado J., Carpio A.J., Ruiz-Aizpurua L. & Tortosa F.S. (2013) Restocking a keystone species in a biodiversity hotspot: Recovering the European rabbit on a landscape scale. *Journal for Nature Conservation*, 21, 444–448, https://doi.org/10.1016/j.jnc.2013.07.006

(33) Sasmal I., Honness K., Bly K., McCaffery M., Kunkel K., Jenks J.A. & Phillips M. (2015) Release method evaluation for swift fox reintroduction at Bad River Ranches in South Dakota. *Restoration Ecology*, 23, 491–498, https://doi.org/10.1111/rec.12211

(34) Scrivner J.H., O'Farrell T.P., Hammer K. & Cypher B.L. (2016) Translocation of the endangered San Joaquin kit fox, *Vulpes macrotis mutica*: a retrospective assessment. *Western North American Naturalist*, 76, 90–100, https://doi.org/10.3398/064.076.0110

(35) MacPherson J.L. (2017) *Pine marten translocations: the road to recovery and beyond*. Bulletin of the Chartered Institute of Ecology and Environmental Management: Rewilding and species reintroductions, 95, 32–36.

14.14. Hold translocated mammals in captivity before release

https://www.conservationevidence.com/actions/2458

- **Fifteen studies** evaluated the effects of holding translocated mammals in captivity before release. Four studies were in the USA[3,11,12,13], two were in Australia[14,15] and one was in each of India[1], Canada[2], Switzerland[4], Croatia and Slovenia[5], the USA and Canada[6], the UK[7], France[8], Spain[9] and South Africa[10].

COMMUNITY RESPONSE (0 STUDIES)

POPULATION RESPONSE (13 STUDIES)

- **Abundance (2 studies):** Two studies (one replicated, before-and-after study) in Croatia and Slovenia[5] and the USA[13] found that following translocation, with time in captivity prior to release, Eurasian lynx[5] established an increasing population and Allegheny woodrat[13] numbers in four of six sites increased over the first two years.

- **Reproductive success (4 studies):** Four studies in Croatia and Slovenia[5], Spain[9], the USA and Canada[6] and Australia[14] found that following translocation, with time in captivity prior to release, Eurasian lynx[5] established a breeding population, and swift foxes[6], European otters[9] and red-tailed phascogales[14] reproduced.

- **Survival (10 studies):** Two studies (one controlled) in the UK[7] and USA[12] found that being held for longer in captivity before release increased survival rates of translocated European hedgehogs[7] and, along with release in spring increased the survival rate of translocated Canada lynx[12] in the first year. Four of six studies in India[1], the USA and Canada[6], the USA[11], France[8], South Africa[10] and Australia[14] found that following translocation, with time in captivity prior to release, most swift foxes[6] and greater Indian rhinoceroses[1] survived for at least 12–20 months, 48% of Eurasian lynx[8] survived for 2–11 years and red-tailed phascogales[14] survived for at least six years. The other two studies found that most kangaroo rats[11] and all rock hyraxes[10] died within 5–87 days. A replicated, controlled study in Canada[2] found that translocated swift foxes that had been held in captivity prior to release had higher post-release survival rates than did released captive-bred animals.

- **Condition (3 studies):** A randomised, controlled study in Australia[15] found that holding translocated eastern bettongs in captivity before release did not increase their body mass after release compared to animals released directly into the wild. A controlled study the UK[7] found that being held for longer in captivity before release, reduced weight loss after release in translocated European hedgehogs. A study in Spain[9] found that offspring of translocated European otters that were held in captivity before release, had similar genetic diversity to donor populations.

- **Occupancy/range (2 studies):** A study in the USA[3] found that most translocated and captive-bred mountain lions that had been held in captivity prior to release established home ranges in the release area. A study in Croatia and Slovenia[5] and review in Switzerland[4] found that following translocation, with time in captivity prior to release, the range of Eurasian lynx increased over time.

BEHAVIOUR (0 STUDIES)

Background

This intervention refers to holding translocated mammals in captivity away from the release site, before release. This may be done for a number of reasons such as logistics, to allow health checks to take place, to give captured animals time to form social groups, or to use animals in a captive breeding program. Time in captivity may be a few days, months or even a couple of years, depending on the reason for holding individuals in captivity before release.

See also: *Use holding pens at release site prior to release of translocated animals.*

A study in 1984–1986 in a national park in Uttar Pradesh, India (1) found that most translocated greater Indian rhinoceros *Rhinoceros unicornis* that had been held in captivity before release into a fenced reserve, survived over 20 months after release. Seven of eight translocated rhinoceroses were still alive at least 20 months after release into a fenced reserve, and three of these animals survived for over 31 months. One elderly female died three months after release, due to a paralysed limb. In March 1984, six rhinoceroses were captured in Assam and housed in a pen for 9–19 days (during which one individual escaped). The remaining five were transported to Dudhwa National Park, where one elderly female died before release (following abortion of a dead foetus) and four were released in April–May 1984. Four other animals captured in late March 1985 in Sauraha (Nepal) were released to Dudhwa National Park one week after capture. Survival data were collated up to December 1986.

A replicated, controlled study in 1990–1992 at two grassland sites in Alberta, Canada (2) found that translocated swift foxes *Vulpes velox* that had been held in captivity prior to release had higher post-release survival rates than did released captive-bred animals. No statistical analyses were performed. Nine months after release into the wild, 12 out of 28 (43%) wild-born translocated swift foxes were known to be alive, compared with at least two out of 27 (7%) captive-bred swift foxes. In May 1990 and 1991, a total of 28 wild-born and 27 captive-bred swift foxes were released simultaneously. Wild-born animals had been captured in Wyoming, USA, 4–7 months before release and were quarantined for 30 days. Animals were released without prior conditioning in holding pens. Foxes were radio-collared and monitored from the ground and air, for at least nine months.

A study in 1993–1995 in northern Florida, USA (3) found that most translocated and captive-bred mountain lions *Puma concolor stanleyana* that had been held in captivity prior to release established home ranges in the release area. Of 19 released mountain lions, 15 established one or more home ranges. Post-release survival periods for these 15 animals are not stated but two were killed (one illegally shot and one killed by a vehicle) and two were recaptured due to landowner concerns or concerns for their survival, 37–140 days after release. Nineteen mountain lions were released in northern Florida in 1993–1994. Ten were wild-caught and released within three months, three were caught and released after 3–8 years, and six released animals were captive-bred. Mountain lions were radio-tracked daily in February 1993–April 1993 and then for three days/week until June 1995.

A review in 1998 of translocations in 1971–1989 of Eurasian lynx *Lynx lynx* into nine temperate forest sites in Switzerland (4) found that after being held in captivity before release, the range of lynx in the release area increased over time. Ten years after the first releases, lynx occupied approximately 4,000 km². Seventeen years later, this had increased to >10,000 km², although the rate of range expansion had slowed. One-hundred and three lynx were confirmed dead following translocations, mostly from road accidents (27%) and illegal shootings (26%). In 1971–1989, at least 25 lynx were released at nine sites in the Alps and Jura mountains in Switzerland. Most were captured in the Slovakian Carpathian Mountains, kept in captivity for at least one month and then released. From 1971 to 1998, questionnaires were distributed among the public to gather reports of lynx sightings. To confirm deaths, lynx carcasses were collected over an unspecified time period. From 1983 to 1998, thirty-seven lynx were captured and fitted with radio-collars to assess range occupancy.

A study in 1973–1995 in forests across Croatia and Slovenia (5) found that following translocation, Eurasian lynx *Lynx lynx* that had been held in captivity prior to release established a breeding population and expanded in number and range. Over the six years after release of six lynx, 19 litters totalling 30 kittens were recorded. Dispersing animals reached Bosnia-Herzegovina 11 years after releases and, two years later, one reached the Julian Alps, near Italy. The population, 22 years after releases, was estimated at 140 lynx in Slovenia and Croatia. These occupied approximately 3,700 km² in Slovenia and 3,000 km² in Croatia. Hunting was permitted from five years after releases and was the greatest cause of mortality, accounting for 229 of 277 known deaths. Lynx became extinct in Croatia and Slovenia at the beginning of the twentieth century. In 1973, six wild-caught lynx (three female, three male) were caught in Slovakia, quarantined for 46 days and released in Kocevje, Slovenia. Monitoring was based on reviews of hunting data and communications with hunters, foresters and naturalists.

A study in 1994–1998 at seven temperate grassland sites along the USA–Canada border (6) found that most translocated swift foxes *Vulpes velox* that had been held in captivity prior to release and were released in social groups survived for at least one year, and some reproduced near release sites. Eleven of 18 (61%) translocated swift foxes survived at least one year after release. Of these, 60% of animals translocated as juveniles went on to reproduce, as did 33% of translocated adults. In 1994–1996 foxes were captured in Wyoming, fitted with radio-collars and held in captivity for 22–57 days. In autumn 1994–1996, animals were released in mixed-gender groups of up to three individuals that had been trapped in close proximity. Release sites were located in areas with pre-existing, but small, fox populations and with low numbers of predators and high prey availability. Foxes were monitored by visual surveys and ground-based and aerial radio-tracking.

A controlled study in 2004 in 20 suburban gardens in Bristol, UK (7) found that after being held for a period in captivity before release, translocated European hedgehogs *Erinaceus europaeus* had higher survival rates and lower body weight loss than did individuals translocated with minimum time in captivity. A higher proportion of hedgehogs translocated after over a month in captivity survived (82%) and they lost less body weight (9%) over the eight weeks following release compared to individuals translocated after less than six days in captivity (survival: 41%; reduction in body weight: 33%). Over the same period, 64–95% of non-translocated hedgehogs survived and these lost 5–10% of body weight. Between May and June 2004, forty-three hedgehogs were translocated from the Outer Hebrides, Scotland, to 10 suburban gardens in Bristol. Twenty-three had spent >1 month in captivity and 20 had spent <6 days in captivity. Food was provided during the first week after release. Translocated hedgehogs were radio-tracked over eight weeks. Over the same period, 20 free-living hedgehogs captured and released <50 m from the same set of 20 gardens together with 26 free-living hedgehogs caught and released at gardens >3 km away were monitored. Hedgehogs were weighed every 10 days.

A study in 1983–2002 in a temperate forest in Vosges massif, France (8) found that following translocation of Eurasian lynx *Lynx lynx* that had been held in captivity before release, around half survived for 2–11 years. Ten of 21 animals survived for 2–11 years after release. The distribution of lynx increased from 1,870 km² (six years after the first releases) to 3,160 km² (12 years later). At least two females produced litters. In 1983–1993, twenty-one adult lynx were brought to France from European zoos. The program sought wild-caught lynx for releases, however the exact origin of each animal, and the length of time that each spent in captivity, are unclear. Lynx were released at four sites in the Vosges Mountains. The first eight animals were held in cages at the release site for 4–45 days prior to release, but the remainder were released immediately upon arrival. Animals were radio-tracked for 1–847 days. The presence of lynx was also established through sightings, footprints, detection of faeces or hair and reports of attacks on domestic animals.

A study in 1995–2004 in three riparian and wetland sites in north-eastern Spain (9) found that following translocations of European otters *Lutra lutra* that were held in captivity before release, animals reproduced and offspring had similar genetic diversity to that of donor populations. By nine years after the first releases, at least 19 offspring had been born to translocated otters. Genetic diversity in these offspring was similar to that of the donor populations (data reported as genetic heterozygosity). In 1995–2002, forty-two otters were released into three wetland and river areas. All otters were caught in western Iberia and were quarantined before release. Blood samples were collected from 23 translocated otters. In February–March 2004, the study area was divided into eight zones, each of which was surveyed over five consecutive days. In total, 104 otter faeces and anal secretion samples were collected from release areas. Samples were genetically analysed and compared to samples from translocated otters.

A study in 2005–2006 at rocky outcrops on a reserve in KwaZulu-Natal Province, South Africa (10) found that translocated rock hyraxes *Procavia capensis* that were held in captivity before release in a social group, and provided with an artificial refuge and supplementary food after release, all died (or were presumed to have died) within 87 days of release. Eighty-seven days after the release of 17 hyraxes, none could be relocated. In July 2005, ten adult hyraxes were caught in baited mammal traps (900 × 310 × 320 mm), and held in captivity for 16 months, during which time three died. The remaining seven were released in November 2006, along with the eight juveniles and two pups born to them in captivity, to a 656-ha reserve where the species was nearly extinct. For four months prior to release, the group was housed together in an outdoor cage (5.9 × 2.5 × 3.2 m). Hyraxes were released into a hay-filled hutch which was left in place for several months, and were provided with cabbage for one week after release. Hyraxes were monitored by direct observations and by walking regular transects, daily for the first week decreasing to monthly by the end of the study.

A study in 2001 in a grassland and shrubland site in California, USA (11) found that most translocated Tipton kangaroo rats *Dipodomys nitratoides nitratoides* and Heermann's kangaroo rats *Dipodomys heermanni* ssp. that were held in captivity prior to release died within five days of release. All four Tipton kangaroo rats were predated within five days of translocation, and only one of seven Heermann's kangaroo rats survived over 45 days. Three Heermann's kangaroo rats were predated, two died as a result of aggression from other Heermann's kangaroo rats, and the fate of one was unknown. In September 2001, four juvenile Tipton kangaroo rats and three Heermann's kangaroo rats were captured and held in captivity for two months before release at a protected site in November. In December 2001, a further four Heermann's kangaroo rats were caught and translocated to the same site. All 11 animals were fitted with a radio-transmitter and ear tags, and monitored for seven days in captivity prior to release. The release site was already occupied by Heermann's kangaroo rats. Animals were released into individual artificial burrows (two 90-cm-long cardboard tubes with a chamber about 30 cm below the surface), dug 10–15 m apart and provided with seeds. Burrows were plugged with paper towels until dusk. Animals were radio-tracked every 1–8 days for 18–45 days after release.

A study in 1999–2007 in montane forest in Colorado, USA (12) found that more time in captivity and release in spring increased the survival rate of translocated Canada lynx *Lynx canadensis* in the first year. Lynx released in spring after >45 days in captivity had lower monthly mortality rates (0.4–2.8% in 2000–2006) than lynx released in spring after 21 days in captivity (1.4% in 2000) or released after 7 days but not in spring (20.5% in 1999). Overall, 117 of 218 released lynxes (53%) survived to at least 1–8 years after release. From 1999 to 2006, two hundred and eighteen lynx were translocated to a 20,684-km² mixed forest area in the San Juan Mountains, Colorado, from Canada and the USA. Lynx were held in captivity near their source location (for 3–68 days) prior to transfer to a holding facility (40 pens, 2.4 x 1.2 m with ceilings) in Colorado (100 km from release site). Time in the Colorado holding facility varied (5–137 days): release within 7 days following veterinary inspection (4 individuals in 1999); release after 3 weeks (9 individuals in 2000); release after >3 weeks in the spring (1 April-31 May; 28 individuals in 2000); release in spring after >3 weeks in captivity but excluding any juveniles or pregnant females (177 individuals in 2000–2006). Lynx were fed a diet of rabbit and commercial carnivore food while in captivity. Lynx were radio-collared and monitored weekly for the first year following release (5,324 locations recorded).

A replicated, before-and-after study in 2005–2009 in six riparian areas in Indiana, USA (13) found that following translocation of Allegheny woodrats *Neotoma magister* that were held in captivity prior to release, numbers in four out of six sites increased over the first two years. Two years after 54 woodrats were translocated to six sites, numbers had increased in four sites, but only one woodrat was recorded at each of the other two sites. At this time, there were more woodrats overall (total 67 animals) than before animals were translocated (16 animals). In 2007–2008, sixty-seven woodrats were captured in Kentucky and Tennessee. After five days, they were fitted with radio-transmitters and transported to release sites. In 2005–2006 (before translocations) and in 2007–2009 (after translocations), woodrat abundance was estimated using 35–100 live traps/site between June and August. Trapping was carried out over two consecutive nights at each site and traps were checked at dawn. All woodrats caught were fitted with ear tags.

A study in 2009–2015 in a forest and shrubland reserve in Western Australia, Australia (14) found that a translocated population of red-tailed phascogales *Phascogale calura*, some of which were held in captivity prior to release into a fenced area containing artificial nest boxes, survived and reproduced for at least six years, and spread outside the release area. At least nine of 12 translocated female red-tailed phascogales survived 8–9 months post-release and all nine reproduced in the wild. At least one female survived two years after release. From 1–6 years post-release, nest box occupancy within and outside the fenced area remained over 60%. In April 2009, twenty red-tailed phascogales were translocated to a 430-ha fenced area, within a 560-ha reserve surrounded by farmland, and released at dusk on the day of capture. Seven phascogales were released in June 2010, six weeks after capture. Animals were released into or adjacent to 22 nest boxes, alone or in pairs. From November 2010–January 2013, thirteen additional boxes were installed inside (four) and outside (nine) the fenced area. Invasive foxes *Vulpes vulpes* and cats *Felis catus* were absent from the fenced area, but the fence did not present a barrier to phascogales. Phascogales were monitored between April 2009 and March 2011 using baited Elliott live-traps (nine sessions, 5,341 trap nights) and through periodic monitoring between July 2009 and January 2015 of the nest boxes.

A randomised, controlled study in 2011–2014 in a woodland reserve in Australian Capital Territory, Australia (15) found that holding translocated eastern bettongs *Bettongia gaimardi* in captivity before release did not affect their body mass after release relative to animals released directly into the wild. Bettongs released after time in captivity were heavier at release (1.9 kg) than were those released immediately (1.7 kg) though subsequently there were no significant differences in body weight (see paper for details). In 2011–2012, thirty-two adult wild-born bettongs were captured in Tasmania and translocated to mainland Australia. Sixteen randomly selected individuals were immediately released into a fenced reserve, where invasive predators had been controlled. The remaining 16 were housed for 30 days in small enclosures (0.5–1.0 ha) before transfer to larger enclosures (2.6–9.4 ha). In total, they were held for 95–345 days before release. Bettongs were radio-tagged and were trapped and weighed periodically up to 18 months after release.

(1) Sale J.B. & Singh S. (1987) Reintroduction of greater Indian rhinoceros into Dudhwa National Park. *Oryx*, 21, 81–84.

(2) Carbyn L.N., Armbruster H.J. & Mamo C. (1994) The swift fox reintroduction program in Canada from 1983 to 1992. Pages 247–271 in: M.L. Bowles & C.J. Whelan (eds.) Restoration of endangered species: conceptual issues, planning and implementation. Cambridge University Press, Cambridge, UK.

(3) Belden R.C. & McCown J.W. (1996) *Florida panther reintroduction feasibility study. Final Report.* Study Number: 7507. Florida Game and Fresh Water Fish Commission.

(4) Breitenmoser U., Breitenmoser-Wursten C. & Capt S. (1998) Re-introduction and present status of the lynx in Switzerland. *Hystrix* 10, 17–30.

(5) Cop J. & Frkovic A. (1998) The reintroduction of the lynx in Slovenia and its present status in Slovenia and Croatia. *Hystrix*, 10, 65–76.

(6) Moehrenschlager A. & Macdonald D.W. (2003) Movement and survival parameters of translocated and resident swift foxes *Vulpes velox*. *Animal Conservation*, 6, 199–206, https://doi.org/10.1017/s1367943003251

(7) Molony S.E., Dowding C.V., Baker P.J., Cuthill I.C. & Harris S. (2006) The effect of translocation and temporary captivity on wildlife rehabilitation success: an experimental study using European hedgehogs (*Erinaceus europaeus*). *Biological Conservation*, 130, 530–537, https://doi.org/10.1016/j.biocon.2006.01.015

(8) Vandel J.M., Stahl P., Herrenschmidt V. & Marboutin E. (2006) Reintroduction of the lynx into the Vosges mountain massif: from animal survival and movements to population development. *Biological Conservation*, 131, 370–385, https://doi.org/10.1016/j.biocon.2006.02.012

(9) Ferrando A., Lecis R., Domingo-roura X. & Ponsà M. (2008) Genetic diversity and individual identification of reintroduced otters (*Lutra lutra*) in north-eastern Spain by DNA genotyping of spraints. *Conservation Genetics*, 9, 129–139, https://doi.org/10.1007/s10592-007-9315-1

(10) Wimberger K., Downs C.T., Perrin M.R. (2009) Two unsuccessful reintroduction attempts of rock hyraxes (*Procavia capensis*) into a reserve in the KwaZulu-Natal Province, South Africa. *South African Journal of Wildlife Research*, 39, 192–201, https://doi.org/10.3957/056.039.0213

(11) Germano D.J. (2010) Survivorship of translocated kangaroo rats in the San Joaquin Valley, California. *California Fish and Game*, 96, 82–89.

(12) Devineau, O., Shenk, T.M., Doherty Jr, P.F., White, G.C. and Kahn, R.H. (2011) Assessing release protocols for Canada lynx reintroduction in Colorado. *The Journal of Wildlife Management*, 75, 623–630, https://doi.org/10.1002/jwmg.89

(13) Smyser T.J., Johnson S.A., Page L.K., Hudson C.M. & Rhodes O.E. (2013) Use of experimental translocations of Allegheny woodrat to decipher causal agents of decline. *Conservation Biology*, 27, 752–762, https://doi.org/10.1111/cobi.12064

(14) Short J. & Hide A. (2015) Successful reintroduction of red-tailed phascogale to Wadderin Sanctuary in the eastern wheatbelt of Western Australia. *Australian Mammalogy*, 37, 234–244, https://doi.org/10.1071/am15002

(15) Batson W.G., Gordon I.J., Fletcher D.B. & Manning A.D. (2016) The effect of pre-release captivity on post-release performance in reintroduced eastern bettongs *Bettongia gaimardi*. *Oryx*, 50, 664–673, https://doi.org/10.1017/s0030605315000496

14.15. Use tranquilizers to reduce stress during translocation

https://www.conservationevidence.com/actions/2465

- **One study** evaluated the effects on mammals of using tranquilizers to reduce stress during translocation. This study was in France[1].

COMMUNITY RESPONSE (0 STUDIES)

POPULATION RESPONSE (1 STUDY)

- **Survival (1 study):** A controlled study in France[1] found that using tranquilizers to reduce stress during translocation did not increase post-release survival of European rabbits.

BEHAVIOUR (0 STUDIES)

Background

Translocation of mammals can cause elevated stress levels. This may affect post-release survival (e.g. Beringer *et al.* 2002). Tranquilizers may be administered during the translocation process in order to reduce stress to captured mammals.

Beringer J., Hansen L.P., Demand J.A., Sartwell J., Wallendorf M. & Mange R. (2002) Efficacy of translocation to control urban deer in Missouri: costs, efficiency, and outcome. *Wildlife Society Bulletin*, 30, 767–774.

A controlled study in 1997 on a farmland site in northern France (1) found that using tranquilizers to reduce stress during translocation did not increase post-release survival of European rabbits *Oryctolagus cuniculus*. The re-sighting rate of rabbits that had been tranquilized over seven weeks after release did not differ significantly from that of non-tranquilized rabbits over the same period (data reported as statistical model results). In January 1997, a total of 104 rabbits were translocated from Parc-du-Sausset to an area of cultivated fields and pasture in Héric, 400 km away. Of these, approximately half were tranquillized just after capture using two intra-muscular injections of carazolol (0.1 mg/kg). Roughly half the tranquilized and half the non-tranquilized rabbits were acclimatised in 100-m^2 enclosures for three days prior to release. Survival was estimated from nocturnal spotlight re-sighting sessions conducted every evening during the first week following release. Thereafter, monitoring was reduced to twice/week for a further six weeks, until late-February.

(1) Letty J., Marchandeau S., Clobert J. & Aubineau J. (2000) Improving translocation success: an experimental study of anti-stress treatment and release method for wild rabbits. *Animal Conservation*, 3, 211–219, https://doi.org/10.1111/j.1469-1795.2000.tb00105.x

14.16. Airborne translocation of mammals using parachutes

https://www.conservationevidence.com/actions/2466

- **One study** evaluated the effects of airborne translocation of mammals using parachutes. This study was in the USA[1].

COMMUNITY RESPONSE (0 STUDIES)

POPULATION RESPONSE (1 STUDY)

- **Survival (1 study):** A study in the USA[1] found that at least some North American beavers translocated using parachutes established territories and survived over one year after release.

BEHAVIOUR (0 STUDIES)

Background

Translocating animals into remote terrain can be logistically challenging. Holding animals for several days while moving across ground can cause them stress and, potentially, illness or mortality. Dropping animals from an airplane means that they can be held captive for shorter periods, though it may be harder to choose the precise release location. Parachutes, combined with a container that opens upon landing, can be used in aerial drops.

A study in 1948–1949 in a forest in Idaho, USA (1) found that at least some North American beavers *Castor canadensis* translocated using parachutes established territories and survived over one year after release. Seventy-six beavers were dropped from an airplane over the translocation area using parachutes. All but one survived the drop. After one year, an unspecified number of beavers had built dams and constructed houses. In the autumn of 1948, seventy-six beavers were parachuted into a remote forest area. Animals were dropped in pairs, inside wooden boxes (76 × 40 × 30 cm), using 7.3-m rayon parachutes of war surplus stock. Boxes consisted of two sections fitted together as a suitcase, with 2.5-cm ventilation holes. A system of ropes snapped the box open with the collapse of the parachute. The system had been tested on an old male beaver named 'Geronimo'. Observations were made of the surviving beavers in late 1949 (details not reported).

(1) Heter E.W. (1950) Transplanting beavers by airplane and parachute. *The Journal of Wildlife Management*, 14, 143–147.

14.17. Release translocated mammals into fenced areas

https://www.conservationevidence.com/actions/2467

- **Twenty-four studies** evaluated the effects of releasing translocated mammals into fenced areas. Nine studies were in Australia[5,11,15,19–24], six studies were in South Africa[6,7,8,10,12,16], two studies were in the USA[1,3] and one study was in each of India[2], China[4], Spain[9,18], Hungary[13], Namibia and South Africa[14] and France[17].

COMMUNITY RESPONSE (0 STUDIES)

POPULATION RESPONSE (22 STUDIES)

- **Abundance (5 studies):** Five studies (one replicated) in the USA[1,3], Australia[5,20] and South Africa[7] found that following translocation into fenced areas, 18 African elephant populations[7], tule elk[3], brushtail possum[20] and elk and bison[1] increased in number and following eradication of invasive species a population of translocated and released captive-bred burrowing bettongs[5] increased. A replicated, controlled study in Spain[9] found that the abundance of translocated European rabbits was higher in areas fenced to exclude predators than unfenced areas.

- **Reproductive success (7 studies):** Two replicated, controlled studies in France[17] and Spain[18] found that after translocation, reproductive success of common hamsters[17] and European rabbits[18] was higher inside than outside fenced areas or warrens. Four studies (one replicated, controlled) in China[4] and South Africa[6,8,10] found that following translocation into a fenced area, Père David's deer[4], lions[10], translocated and captive-bred African wild dogs[8] and one of two groups of Cape buffalo[6] reproduced. A study in Australia[15] found that four of five mammal populations[15] released into a predator-free enclosure and one population released into a predator-reduced enclosure reproduced, whereas two populations released into an unfenced area with ongoing predator management did not survive to breed.

- **Survival (13 studies):** Two replicated, controlled studies in Spain[9] and France[17] found that after translocation, survival rates of common hamsters[17] and European rabbits[9] were higher inside than outside fenced areas or warrens. A study in Australia[15] found that four of five mammal populations[15] released into a predator-free enclosure and one population released into a predator-reduced enclosure survived, whereas two populations released into an unfenced area with ongoing predator management did not persist. Five studies in India[2], China[4], South Africa[12], Namibia and South Africa[14] and Australia[19] found that following translocation into fenced areas, most black rhinoceroses[14] and greater Indian rhinoceroses[2], Père David's deer[4], most oribi[12] and offspring of translocated golden bandicoots[19] survived for between one and 10 years. Two studies in Australia[11,24] found that only two of five translocated numbats[11] survived over seven months and western barred bandicoots[24] did not persist. A study in South Africa[8] found that translocated and captive-bred African wild dogs[8] released into fenced reserves in family groups had high survival rates. A study in Australia[21] found that following release into fenced areas, a translocated population of red-tailed phascogales[21] survived longer than a released captive-bred population. A replicated, controlled study in South Africa[10] found that after translocation to a fenced reserve with holding pens, survival of released lions[10] was higher than that of resident lions.

- **Condition (3 studies):** A replicated, before-and-after study in Australia[23] found that eastern bettongs translocated into fenced predator proof enclosures increased in body weight post-release, with and without supplementary food. A replicated study in South Africa[16] found that following translocation into fenced reserves, stress hormone levels of African elephants declined over time. A study in Australia[19] found that golden bandicoots descended from a population translocated into a fenced area free from non-native predators, maintained genetic diversity relative to the founder and source populations.

BEHAVIOUR (2 STUDIES)

- **Use (2 studies):** A site comparison study in Australia[22] found that following translocation into a predator-free fenced area, woylies developed home ranges similar in size to those of an established population outside the enclosure. A study in Hungary[13] found that one fifth of translocated European ground squirrels released into a fenced area with artificial burrows remained in the area after release.

Background

Mammals that are being translocated to a new location may be released into fenced areas. This may be done to keep them within a certain area (e.g. a game reserve), or to keep predators or other problem species out of an area to increase their chances of survival. Here fenced areas refer to those that are large enough to cover the home ranges of the target species. Studies that use smaller holding or pre-release pens before releasing translocated mammals into the wild are covered in *Use holding pens at release site prior to release of translocated mammals*.

See also: *Release captive-bred mammals into fenced areas.*

A study in 1970–1973 in two grassland and forest sites in South Dakota, USA (1) found that following translocation into fenced areas, elk *Cervus canadensis* and bison *Bison bison* increased in numbers. Three years after the onset of translocations, there were more elk (214) and bison (109) than were released over that time (elk: 165; bison: 95). Additionally, over the same period, 55 elk and 22 bison were harvested by hunters. The study was conducted in two 4,000-ha game ranges. Both game ranges were enclosed by woven wire fences, approximately 2 m high. In 1970–1973, one hundred and sixty-five elk and 95 bison (origin not stated) were released across both sites (the number of individuals stocked into each game range is not provided). Mule deer *Odocoileus hemionus*, whitetail deer *Odocoileus virginianus* and pronghorn *Antilocapra americana* occurred naturally within the game ranges and were managed for game hunting.

A study in 1984–1986 in a national park in Uttar Pradesh, India (2) found that following translocation into a fenced reserve, most greater Indian rhinoceros *Rhinoceros unicornis* survived over 20 months after release. Seven of eight translocated rhinoceroses were still alive at least 20 months after release into a fenced reserve, and three of these animals had survived for over 31 months. One elderly female died three months after release, due to a paralysed limb. In March 1984, six rhinoceroses were captured in Assam. They were housed in a holding pen for 9–19 days (during which one individual escaped). The remaining five were transported to Dudhwa National Park, where one elderly female died before release (following abortion of a dead foetus) and four were released in April–May 1984. Four other animals captured in late March 1985 in Sauraha (Nepal) were released to Dudhwa National Park one week after capture. Survival data were collated up to December 1986.

A study in 1978–1998 in a grassland reserve in California, USA (3) found that numbers of tule elk *Cervus canadensis nannodes* translocated to a fenced reserve increased more than 50-fold over 20 years. In 1998, a translocated population of Tule elk grew to more than 500 individuals from the 10 individuals originally translocated 20 years earlier. In 1978, ten tule elk were translocated to a fenced reserve of approximately 1,000 ha. No monitoring details are provided.

A study in 1993–1997 in a grassland reserve in Hubei province, China (4) found that translocated Père David's deer *Elaphurus davidianus* released into a fenced area survived at least two years and bred. Père David's deer survived at least two years after being translocated and reproduced in the second year following relocation (numbers not provided). Deer were released in 1993 (30 individuals), 1994 (34 individuals) and 1995 (74 individuals) into a 16 km² paddock. The origin of some of the deer is unclear, but most were wild-born offspring from captive-bred animals that had been released into another reserve in China.

A study in 1993–1999 on an arid peninsula in Western Australia, Australia (5) found that following release into a fenced area where invasive species had been eradicated, a population of burrowing bettongs *Bettongia lesueur* increased. In 1999, six years after initial releases, the population was estimated at 263–301 bettongs, with 340 individuals born between 1995 and 1999. The population died out due to fox incursion in 1994, but was re-established with further releases. In 1990, a 1.6-m tall wire mesh fence (with an external overhang, an apron to prevent burrowing and two electrified wires) was erected to enclose a 12-km² peninsular, within which foxes *Vulpes vulpes* and cats *Felis catus* were eliminated by poisoning in 1991 and 1995, respectively. Outside the fence foxes were controlled by biannual aerial baiting with meat containing 1080 toxin, distributed at 10 baits/km² over 200 km². From October 1993, an additional 200 baits/month were distributed along the fence and roads across the study area. Cats were controlled by trapping and poisoning in a 100 km² buffer zone. In May 1992 and September 1993, twenty-two wild-caught bettongs were transferred to an 8-ha *in-situ* captive-breeding pen. In September 1993 and October 1995, twenty wild-caught bettongs were translocated to range freely in the reserve. From 1993–1998, one hundred and fourteen captive-bred bettongs were released. Artificial warrens and supplementary food and water were provided in 1993, but not for later releases. Eighty released bettongs were radio-tagged. From 1991–1995, European rabbits *Oryctolagus cuniculus* were controlled within the fenced area using 1080 'one shot' oats. Bettongs were monitored every three months using cage traps set over two consecutive nights, at both 100-m intervals along approximately 40 km of track, and at warrens used by radio-collared individuals.

A study in 2000–2003 in a mixed karoo grassland reserve in Northern Cape Province, South Africa (6) found that after translocated Cape buffalo *Syncerus caffer* were released into a fenced reserve in groups (after being held in a holding pen) one group scattered and escaped the reserve while the other formed a single herd and stayed in the reserve and bred. One month after release, a group of four buffalo had split into two solitary animals and a pair formed by one male and one female. One of the solitary animals was not seen again, the second solitary male animal was located two years after release on a neighbouring farm and released into the second group of translocated animals in May 2003. The pair escaped the reserve three times in 13 months. After the third escape, the male was moved to a different reserve and a new male introduced to form a herd with the remaining female. A second group of 10 translocated animals formed a single herd (along with the two remaining animals from the previous introduction) and over 10 months no animals died or escaped. A year after the introduction five calves were born. Four subadult buffalo (2 male, 2 female) were placed in a holding pen in July 2000 and released in August into a fenced 12,000-ha reserve. A second group of seven adult and three subadult animals (4 male, 6 female) was placed into a holding pen in August 2002 and released into a 200 ha area in September before being completely released in October 2002. Both groups were monitored weekly with telemetry until October 2003.

A replicated study in 1990–2001 in 18 savannah sites in South Africa (7) found that at least five years following translocation into fenced reserves, the population size of African elephants *Loxodonta africana* increased over time. The population size of translocated elephants increased at an average annual rate of 8.3%. Annual growth across recipient sites ranged from 1.7% to 16.5%. In 1990–1999, elephants were translocated into 18 fenced reserves. The number of animals translocated into each reserve ranged between 18 and 227. Translocation details and the data on numbers of animals present in 2001 were obtained through surveys of reserve owners or managers. All translocated elephants were wild-born, free-ranging animals.

A study in 1995–2005 in 12 dry savanna and temperate grassland sites in South Africa (8) found that translocated and captive-bred African wild dogs *Lycaon pictus* released into fenced reserves in family groups had high survival rates and bred successfully. Eighty-five percent of released animals and their wild-born offspring survived the first six months after release/birth, Released animals which survived their first year had a high survival rate 12–18 months (91%) and 18–24 months (92%) after release. Additionally, groups which had more time to socialise in holding pens prior to release had higher survival rates (data presented as statistical models). Between 1995 and 2005, one hundred and twenty-seven wild dogs (79 wild-caught, 16 captive-bred, 16 wild-caught but captive-raised, 16 'mixed' pups) were translocated over 18 release events into 12 sites in five provinces of South Africa. Animals were monitored for 24 months after release, and the 129 pups which they produced after release were monitored up to 12 months of age. Forty characteristics of the individual animals, release sites and methods of release were recorded, and their impact on post-release survival was tested.

A replicated, controlled study in 2002–2003 in three grassland and shrubland sites in south-west Spain (9) found that the survival of translocated rabbits *Oryctolagus cuniculus* was similar between fenced and unfenced areas but that abundance was higher in areas fenced to exclude predators. Three months after translocation, rabbit survival did not differ significantly between fenced and unfenced plots (0.57 vs 0.4). However, four months after translocation the relative abundance of rabbits was higher in fenced than in unfenced plots (data presented as log abundance). Two fenced (1 m below and 2.5 m above ground with an electric wire on top) and two unfenced translocation areas (4 ha, 18 artificial warrens each) were established in Los Melonares, Sierra Norte of Seville Natural Park. A total of 724 wild rabbits were released in similar numbers into each area. Rabbit survival was based on 45 radio-collared rabbits (19 in fenced and 26 in unfenced areas). Abundance was estimated four months after translocation through pellet counts in 10 circular plots (18 cm diameter).

A replicated, controlled study in 1999–2004 in three mixed savanna and woodland sites in KwaZulu-Natal, South Africa (10) found that after translocation to a fenced reserve with holding pens, survival of released lions *Panthera leo* was higher than that of resident lions, and translocated animals reproduced successfully. No statistical tests were performed. After five years, a higher proportion of translocated animals survived (eight of 16 animals, 50%) than of resident animals (20 of 84 animals, 24%). Seven translocated females reproduced successfully. Between August 1999 and January 2001, sixteen lions were translocated to an enclosed reserve to improve genetic diversity. They were held at release sites in 0.5–1-ha pens for 4–6 weeks before release. Nine translocated lions were fitted with radio-collars. From August 1999 to December 2004, translocated animals were located at least every 10 days. Resident lions were also tracked at least every 10 days.

A study in 2005–2006 of a savanna reserve in South Australia, Australia (11) found that following translocation and release into a fenced area, only two of five translocated numbats *Myrmecobius fasciatus* remained alive after seven months. One male was predated by a raptor 47 days after release. Two females were each carrying young four months after release, but both died three months later, probably due to raptor predation. Two males remained alive for at least 18 months after release. Five translocated numbats (three males and two females) were released in November 2005 into a 14-km² fenced area from which red foxes *Vulpes vulpes* and feral cats *Felis catus* were excluded. All animals were released on the day of capture or the following day. Animals were radio-tracked daily for three months and weekly for six further months. Methods for monitoring after that time are not detailed.

A study in 2004–2006 in a grassland reserve in KwaZulu-Natal, South Africa (12) found that following translocation into a fenced reserve, most oribi *Ourebia ourebi* survived at least one year after release. Fourteen of 15 (93%) oribi translocated into a fenced reserve survived for at least one year post-release. The other oribi (a male) died eight months after release but was old (based on horn length and wear). Four translocated females were pregnant and were observed with calves within three months of release (number not reported). Fifteen wild oribi from three populations (11 females, four males) were translocated into a 2,000-ha private game reserve in November 2004. The reserve was surrounded by a 2.1-m-high electric fence and was patrolled daily by armed guards. The grassland was managed for oribi by mowing and burning. All of the 15 oribi were ear-tagged and radio-collared. In 2005–2006, individuals were radio-tracked weekly for two months and monthly thereafter for one year.

A study in 2000 in a grassland site in central Hungary (13) found that one fifth of translocated European ground squirrels *Spermophilus citellus* released into a fenced area with artificial burrows remained in the area after release. From four to 10 days after release, 25 out of 117 ground squirrels were recaptured. The highest recapture rate came from the group released into plugged burrows in the morning (15 out of 30). The fence was designed to exclude predators from the site. From 22–24 April 2000, 117 wild-caught European ground squirrels were translocated to a fenced 40-ha protected grassland. Four 40 × 40-m grid cells were established, each containing vertical, artificial burrows (50 cm long, 4.5 cm diameter) spaced 4.5 m apart. Sixty animals were released into burrows plugged with wood caps (from which they could only exit by digging out) across two grid cells and 57 into unplugged artificial burrows in the other two grid cells. One individual was released/burrow. Approximately half the squirrels were released in the afternoon on the day of capture. Animals to be released in the morning were kept in individual wire cages (10 × 10 × 40 cm) for one night and provided with fresh apple slices prior to release. From 28 April–2 May, squirrels were recaptured with snares to record retention.

A study in 1981–2005 in reserves across Namibia and South Africa (14) found that 89% of translocated black rhinoceros *Diceros bicornis* released into fenced reserves survived over one year and 36% at least 10 years post-release. Seventy-four of 682 translocated black rhinoceroses died during the first year post-release. First-year post-release mortality was higher when animals were released into reserves occupied by other rhinoceroses (restocking, 13.4% mortality of 268 animals) than releases into new reserves (reintroduction, 7.9% mortality of 414 animals). At least 243 rhinoceroses survived at least 10 years after release. For restocking events, first-year post-release mortality was higher in rhinoceroses less than two years old (59%) than in all other age classes (9–20%), but there was no difference for reintroductions. Data on 89 reintroduction and 102 restocking events of black rhinoceroses into 81 reserves from 1981–2005 were compiled from the Namibia and South Africa Rhino Management Group reports. Animals were released in groups from 1 to 30 individuals, and reserves received up to five releases. Translocations were considered as different if the releases of individuals to the same reserve were more than 1 month apart. Deaths were detected by reserve staff. The location of reserves included in the study is not provided.

A study in 1998–2010 in a desert site in South Australia (15) found that four of five mammal populations released into a predator-free enclosure and one population released into a predator-reduced enclosure survived, increased their distribution and produced a second generation, whereas two populations released into an unfenced area with ongoing predator management did not persist. After release into a fenced enclosure where red foxes *Vulpes vulpes*, cats *Felis catus* and rabbits *Oryctolagus cuniculus* had been eradicated, greater stick-nest rats *Leporillus conditor*, burrowing bettongs *Bettongia lesueur*, western barred bandicoots *Perameles bougainville* and greater bilbies *Macrotis lagotis* were detected for eight years, increased their distribution within five years and produced a second generation within two years. Numbats *Myrmecobius fasciatus* were only detected for three years and did not produce a second generation. Burrowing bettongs released into a fenced enclosure with cats and rabbits but no foxes survived and increased their distribution over at least three years and produced a second generation within two years. Greater bilbies and burrowing bettongs released into an unfenced area with some predator management did not survive to produce a second generation or increase their distribution. In 1998–2005, five numbats, 106 greater stick-nest rats (6 captive-bred individuals), 30 burrowing bettongs, 12 western barred bandicoots and nine greater bilbies (all captive-bred) were released into a 14-km² invasive-species-free fenced area. Rabbits, cats and foxes were eradicated within the fenced area in 1999. All western barred bandicoots and greater bilbies, and some greater stick-nest rats (8 individuals) and burrowing bettongs (10 individuals) were put into a 10-ha holding pen before full release after a few months. All other animals were released directly into the larger fenced area. In 2004–2008, thirty-two greater bilbies and 15 burrowing bettongs were translocated to an unfenced area (200 km²) where invasive predators (cats and foxes) were managed with lethal controls and dingoes *Canis lupus dingo* were excluded by a fence on one side. In 2008, sixty-six burrowing bettongs were translocated to a 26 km² fenced area which contained small cat and rabbit populations as a result of previous eradication attempts. Between 2000 and 2010, animals were monitored using track counts, burrow monitoring and radio-tracking.

A replicated study in 2000–2006 in five savannah reserves in South Africa (16) found that following translocation into fenced reserves, stress hormone levels of African elephants *Loxodonta africana* declined with time since release. Average levels of stress hormones were respectively

10% and 40% lower in reserves where elephants had been released 10 and 24 years before sampling than in a reserve where elephants had been released one year before sampling. The concentrations of stress hormones levels (fecal glucocorticoid metabolites) were quantified from 1,567 fecal samples collected in 2000–2006 from elephants reintroduced to five fenced reserves. Translocated elephants had been released in 1981 in two of the reserves, in 1992 in two other reserves and in 2000 in one reserve. Samples were collected from all family groups on nearly consecutive days and efforts were made not to collect multiple samples from the same individual.

A replicated, controlled study in 2010–2011 in 10 agricultural plots in Alsace, France (17) found that survival rates and reproductive success of translocated common hamsters *Cricetus cricetus* were higher inside than outside fenced areas. Average reproductive success and weekly survival rates of translocated hamsters were higher inside (reproductive success: 0.44 litters/female; weekly survival: 89%) than outside fenced areas (reproductive success: 0.00 litters/female; weekly survival: 27%). Additionally, inside fenced areas, monthly survival was higher in wheat plots (harvested and unharvested wheat plots combined) than in alfalfa plots (61% vs 35%). The study was conducted in a 300-ha agricultural landscape, comprising small fields (ca. 0.75 ha) of multiple crops. In May 2010, a total of 14 hamsters were released in two batches into fenced plots and an equal number was released in two unfenced plots. Additionally, in May 2011, hamsters were released into two fenced plots each of harvested wheat (total 14 hamsters), unharvested wheat (total 14 hamsters) and mown alfalfa (total 14 hamsters). Animals were radio-tagged and released into artificial burrows. Fenced plots were surrounded by electrified wires located 10–100 cm above ground. Animals were located every 2–4 days in May–September by radio-tracking.

A replicated, controlled study in 2004–2006 in 16 grassland sites in Andalusia, Spain (18) found that European rabbits *Oryctolagus cuniculus* bred in artificial warrens and that reproductive success was higher in fenced than in unfenced warrens. One hundred and twenty-one rabbit kittens were detected during 222 artificial warren observations (0.54/ observation). More kittens were detected in fenced than in unfenced artificial warrens (data presented as model results). The study was conducted in sixteen 5-ha sites across two areas of Doñana National Park. Five artificial warrens in each site each consisted of a two-floor wooden structure ($15 \times 3 \times 1$ m) with 30 entrances, covered with a metallic net, ground cloth and sand. In eight sites, artificial warrens were fenced to deter terrestrial predators, with a 2-m tall metallic net that extended 0.5 m underground. In eight sites, warrens were not fenced. In each site, 5–19 rabbits/ha were released in October or November of 2004 or 2005. Rabbit reproductive success was surveyed the following year, between February and August, through observations of kittens in focal artificial warrens, using a spotting-scope.

A study in 2010–2013 at a grassland and woodland site in Western Australia, Australia (19) found that wild-born golden bandicoots *Isoodon auratus*, descended from a translocated population which had been released into a fenced area free from non-native predators, maintained genetic diversity relative to the founder and source populations and persisted for three years. For four measures of genetic diversity (allelic richness, the number of effective alleles per locus, observed heterozygosity and expected heterozygosity) there were no significant differences between descendants from translocated animals, founder animals that were translocated or source populations (see paper for details). The population size was estimated at 249 bandicoots in 2013. One hundred and sixty bandicoots were trapped on Barrow Island, which had a large population, in February 2010. They were released into a 1,100-ha enclosure free from introduced predators within 24 h of capture. Genetic material was sampled by ear punch biopsy from 57 founders in 2010 and from 67 wild-born progeny trapped in 2010–2012.

A study in 2010–2013 in a forest and shrubland reserve in Western Australia, Australia (20) found that following translocation into a predator-resistant fenced area, brushtail possums *Trichosurus vulpecula* numbers increased over the three years following release. Of five animals released in a formal translocation program, only one, a female, survived >8 months. This animal was still alive after three years. However, including survivors and progeny from four possums informally released two year earlier, there were 19 possums known to be alive three years after formal translocations. Twenty further possums were recorded over this time, of which most are presumed to have subsequently died or left the sanctuary area. Four possums caught on nearby farms were informally released within a 427-ha predator-fenced sanctuary in 2008. Five possums were translocated and released at the same site in winter 2010. Possums were monitored by radio-tracking and by 3–4 live-trapping surveys/year in 2010–2013.

A study in 2006–2015 in two forest and shrubland sites in Western Australia and Northern Territory, Australia (21) found that following release into fenced areas, a translocated population of red-tailed phascogales *Phascogale calura* survived for more than five years, but a captive-bred population survived for less than a year. A population of phascogales established from wild-caught animals survived longer (>5 years) than a population established from captive-bred animals (that had been kept in pre-release pens and given supplementary food; < 1 year). Authors suggest that the unsuccessful site may also have had a shortage of tree hollows for nesting. In July 2006 and January–February 2007, thirty-two captive-bred phascogales were released into a 26-ha fenced reserve (outside which feral cats *Felis catus* were abundant) after spending either 10 days or over four months in a pre-release pen (3×6×2 or 4.5×3×2.2 m). Eleven nest boxes were provided within 150m of the release pen, and supplementary food was provided for one week after release. In April 2009 and June 2010, twenty-seven wild-caught phascogales were released into a 430-ha fenced reserve with 22 nest boxes, but with no pre-release pen or supplementary food. From November 2010–January 2013, thirteen additional boxes were installed inside (four) and outside (nine) the fenced area at this site. Phascogales were monitored after each release using radio-collaring or Elliott live traps, and through periodic monitoring of the nest boxes.

A site comparison study in 2010–2011 of forest at two sites in Western Australia, Australia (22) found that following translocation into a predator-free, enclosed sanctuary, woylies *Bettongia penicillata* developed home ranges similar in size to those of an established population outside the enclosure. Home ranges did not differ significantly in size between woylies inside the enclosure (28–115 ha) and those in a population outside the enclosure (42–141 ha). The 423-ha sanctuary area was enclosed by a 2-m-high fence in September 2010. This was followed by an intensive cat *Felis catus* and fox *Vulpes vulpes* eradication programme. In December 2010, forty-one woylies sourced from nearby populations were released inside the fence. Eight woylies inside the fence (four male, four female) and seven from an established population 17 km to the north (five male, two female), were monitored by radio-tracking at night in March–April 2011.

A replicated, before-and-after study in 2011–2013 in two forest and grassland sites in the Australian Capital Territory, Australia (23) found that eastern bettongs *Bettongia gaimardi* translocated into fenced predator proof enclosures increased in body weight post-release, with and without supplementary food. Between twelve and 24 months post-release, the average body weight of translocated eastern bettongs (1.8 kg) increased compared to before release (1.7 kg). There was no difference in weight between bettongs fed supplementary food and those without (data not provided). In 2011–2012, sixty adult eastern bettongs were translocated from Tasmania to two predator-free fenced reserves. In one reserve bettongs (5 males, 7 females) received supplementary food at least weekly and were placed in 2.6–9.4 ha enclosures, whereas in a second reserve bettongs (8 males, 10 females) received no supplementary food and were not managed in enclosures. Supplementary food included fresh locally available produce and commercial pellets. Body weight was assessed before release and 12–24 months after release (May–November 2013). Bettongs were also monitored by radio-telemetry or camera traps and live-trapping every 3 months.

A study in 1995–2010 in a shrubland-dominated peninsula in Western Australia, Australia (24) found that a translocated population of western barred bandicoots *Perameles bougainville* released inside a predator-resistant fence did not persist. Nine years after translocations into a fenced area commenced, bandicoot numbers increased to 467, from 82 founders. However, then declined to four individuals eight months later and just one animal was recorded over the following three years. Fourteen bandicoots were translocated in 1995–1996 from an offshore island to a 17-ha enclosure, within a 1,200-ha section of a mainland peninsula, fenced to exclude foxes and feral cats. In 1997–2004, eighty-two bandicoots were released from the enclosure to the fenced peninsula. Bandicoots were monitored with cage traps at 100-m intervals over two nights during 47 trapping sessions between August 1995 and September 2010. The fence was built in 1989 and was rebuilt and repaired several times. However, it was considered to be an ineffective barrier to red foxes *Vulpes vulpes* and cats *Felis catus*, which were controlled inside the fenced area by poisoning, trapping and shooting.

(1) Cole R.S. (1974) Elk and bison management on the Oglala Sioux game range. *Journal of Range Management*, 27, 484–485.

(2) Sale J.B. & Singh S. (1987) Reintroduction of greater Indian rhinoceros into Dudhwa National Park. *Oryx*, 21, 81–84.

(3) Adess N. (1998) *Tule elk. The return of a species*. National Park Service Point Reyes National Seashore, California, USA.

(4) Jiang Z., Yu C., Feng Z., Zhang L., Xia J., Ding Y. & Lindsay N. (2000) Reintroduction and recovery of Père David's deer in China. *Wildlife Society Bulletin*, 28, 681–687.

(5) Short J. & Turner B. (2000) Reintroduction of the burrowing bettong *Bettongia lesueur* (Marsupialia: Potoroidae) to mainland Australia. *Biological Conservation*, 96, 185–196, https://doi.org/10.1016/s0006-3207(00)00067-7

(6) Venter J.A. (2004) Notes on the introduction of Cape buffalo to Doornkloof Nature Reserve, Northern Cape Province, South Africa. *South African Journal of Wildlife Research*, 34, 95–99.

(7) Slotow R., Garaï M.E., Reilly B., Page B. & Carr R.D. (2005) Population dynamics of elephants re-introduced to small fenced reserves in South Africa. *South African Journal of Wildlife Research*, 35, 23–32.

(8) Gusset M., Ryan S.J., Hofmeyr M., van Dyk G., Davies-Mostert H.T., Graf J.A., Owen C., Szykman M., Macdonald D.W., Monfort S.L., Wildt D.E., Maddock A.H., Mills M.G.L., Slotow R. & Somers M.J. (2008) Efforts going to the dogs? Evaluating attempts to re-introduce endangered wild dogs in South Africa. *Journal of Applied Ecology*, 45, 100–108, https://doi.org/10.1111/j.1365-2664.2007.01357.x

(9) Rouco C., Ferreras P., Castro F. & Villafuerte R. (2008) The effect of exclusion of terrestrial predators on short-term survival of translocated European wild rabbits. *Wildlife Research*, 35, 625–632, https://doi.org/10.1071/wr07151

(10) Trinkel M., Ferguson N., Reid A., Reid C., Somers M., Turelli L., Graf J., Szykman M., Cooper D., Haverman P., Kastberger G., Packer C. & Slotow R. (2008) Translocating lions into an inbred lion population in the Hluhluwe-iMfolozi Park, South Africa. *Animal Conservation*, 11, 138–143, https://doi.org/10.1111/j.1469-1795.2008.00163.x

(11) Bester A.J. & Rusten K. (2009) Trial translocation of the numbat (*Myrmecobius fasciatus*) into arid Australia. *Australian Mammalogy*, 31, 9–16, https://doi.org/10.1071/am08104

(12) Grey-Ross R., Downs C.T. & Kirkman K. (2009) Is use of translocation for the conservation of subpopulations of oribi *Ourebia ourebi* (Zimmermann)

effective? A case study. *African Journal of Ecology*, 47, 409–415, https://doi.org/10.1111/j.1365-2028.2008.01003.x

(13) Gedeon C.I., Váczi O., Koósz B. & Altbäcker V. (2011) Morning release into artificial burrows with retention caps facilitates success of European ground squirrel (*Spermophilus citellus*) translocations. *European Journal of Wildlife Research*, 57, 1101–1105, https://doi.org/10.1007/s10344-011-0504-3

(14) Linklater W.L., Adcock K., du Preez P., Swaisgood R.R., Law P.R., Knight M.H., Gedir J.V. & Kerley G.I. (2011) Guidelines for large herbivore translocation simplified: black rhinoceros case study. *Journal of Applied Ecology*, 48, 493–502, https://doi.org/10.1111/j.1365-2664.2011.01960.x

(15) Moseby K.E., Read J.L., Paton D.C., Copley P., Hill B.M. & Crisp H.A. (2011) Predation determines the outcome of 10 reintroduction attempts in arid South Australia. *Biological Conservation*, 144, 2863–2872, https://doi.org/10.1016/j.biocon.2011.08.003

(16) Jachowski D.S., Slotow R. & Millspaugh J.J. (2013) Delayed physiological acclimatization by African elephants following reintroduction. *Animal Conservation*, 16, 575–583, https://doi.org/10.1111/acv.12031

(17) Villemey A., Besnard A., Grandadam J. & Eidenschenck J. (2013) Testing restocking methods for an endangered species: Effects of predator exclusion and vegetation cover on common hamster (*Cricetus cricetus*) survival and reproduction. *Biological Conservation*, 158, 147–154, https://doi.org/10.1016/j.biocon.2012.08.007

(18) D'Amico M., Tablado Z., Revilla E. & Palomares F. (2014) Free housing for declining populations: Optimizing the provision of artificial breeding structures. *Journal for Nature Conservation*, 22, 369–376, https://doi.org/10.1016/j.jnc.2014.03.006

(19) Ottewell K., Dunlop J., Thomas N., Morris K., Coates D. & Byrne M. (2014) Evaluating success of translocations in maintaining genetic diversity in a threatened mammal. *Biological Conservation*, 171, 209–219, https://doi.org/10.1016/j.biocon.2014.01.012

(20) Short J. & Hide A. (2014) Successful reintroduction of the brushtail possum to Wadderin Sanctuary in the eastern wheatbelt of Western Australia. *Australian Mammalogy*, 36, 229–241, https://doi.org/10.1071/am14005

(21) Short J. & Hide A. (2015) Successful reintroduction of red-tailed phascogale to Wadderin Sanctuary in the eastern wheatbelt of Western Australia. *Australian Mammalogy*, 37, 234–244, https://doi.org/10.1071/am15002

(22) Yeatman G.J. & Wayne A.F. (2015) Seasonal home range and habitat use of a critically endangered marsupial (*Bettongia penicillata ogilbyi*) inside and outside a predator-proof sanctuary. *Australian Mammalogy*, 37, 157–163, https://doi.org/10.1071/am14022

(23) Portas T.J., Cunningham R.B., Spratt D., Devlin J., Holz P., Batson W., Owens J. & Manning A.D. (2016) Beyond morbidity and mortality in reintroduction

programmes: changing health parameters in reintroduced eastern bettongs *Bettongia gaimardi*. *Oryx*, 50, 674–683, https://doi.org/10.1017/s0030605315001283

(24) Short J. (2016) Predation by feral cats key to the failure of a long-term reintroduction of the western barred bandicoot (*Perameles bougainville*). *Wildlife Research*, 43, 38–50, https://doi.org/10.1071/wr15070

14.18. Provide supplementary food during/after release of translocated mammals

https://www.conservationevidence.com/actions/2470

- **Sixteen studies** evaluated the effects of providing supplementary food during/after release of translocated mammals. Four studies were in the UK[1,2,7,16], two were in each of the USA[3,11], France[4,5], Australia[13,14] and Argentina[12,15], and one was in each of Italy[6], Spain[8], Ireland[9] and South Africa[10].

COMMUNITY RESPONSE (0 STUDIES)

POPULATION RESPONSE (15 STUDIES)

- **Abundance (2 studies):** A controlled study in Spain[8] found that providing supplementary food during translocation did not increase European rabbit abundance. A study in France[5] found that following supplementary feeding in a holding pen prior to release, a translocated deer population increased over six years.

- **Reproductive success (4 studies):** Three studies (one replicated) in the USA[3], Italy[6] and Ireland[9] found that having been provided with supplementary food in holding pens prior to release, translocated black-tailed prairie dogs[3], a pair of Eurasian badgers[6] and most female red squirrels[9] reproduced in the wild. A study in the UK[16] found that some translocated pine martens released from holding pens and then provided with supplementary food and nest boxes bred in the first year after release.

- **Survival (10 studies):** Six of 10 studies (including one replicated and one controlled study) in the UK[2,16], France[5], Italy[6], Ireland[9],

South Africa[10], USA[3,11], Argentina[12] and Australia[13] found that at sites with supplementary food in holding pens before (and in two cases after) release, translocated populations of black-tailed prairie dogs[3], approximately half of female roe deer[5] and over half of red squirrels[9], Eurasian badgers[6], pine martens[16] and released rehabilitated or captive reared giant anteaters[12] survived for between one month and at least two years. Four studies found that at translocation release sites with provision of supplementary food, in most cases artificial refuges and in one case water, no red squirrels[2], rock hyraxes[10] or burrowing bettongs[13] survived over 2–5 months and most translocated Tipton and Heermann's kangaroo rat spp.[11] died within five days. A controlled study in France[4] found that translocated European rabbits provided with supplementary food in holding pens for three days prior to release had higher female (but not male) survival rates immediately following release compared to those released directly. A controlled study in the UK[7] found that survival of translocated and rehabilitated European hedgehogs that were provided with supplementary food after release varied with release method.

- **Condition (2 studies):** One of three studies (including one replicated, one controlled and two before-and-after studies) in the UK[1,7] and Australia[14] found that translocated common dormice gained weight after being provided with supplementary food. One found that translocated eastern bettongs[14] did not have increased body weights after provision of supplementary food in fenced enclosures prior to release. The other found that translocated and rehabilitated European hedgehogs provided with food after release all lost body mass, with effects varying with release method.

BEHAVIOUR (2 STUDIES)

- **Use (1 study):** A controlled study in Australia[13] found that supplementary feeding stations were visited by translocated burrowing bettongs.

- **Behaviour change (1 study):** A controlled study in Argentina[15] found that after being provided with supplementary food and kept in holding pens, released captive-bred giant anteaters were less nocturnal than wild-born rehabilitated and released individuals.

Background

Mammals that are translocated are especially vulnerable immediately after release. At this time, they may struggle to find natural food in an unfamiliar area. Furthermore, if the time they spend looking for food is increased, this may make them more vulnerable to predation. Hence, providing supplementary food at and after the period of release may improve longer term survival prospects.

See also: *Provide supplementary food during/after release of captive-bred mammals.*

A before-and-after study in 1991 in a woodland reserve in Somerset, UK (1) found that translocated common dormice *Muscardinus avellanarius* gained weight after being provided with supplementary food after release. Translocated common dormice lost an average 0.30 g/day before supplementary food was provided but then gained 0.20 g/day after supplementary food provision commenced. The study was conducted along a 9-ha strip of woodland and scrub. Seven dormice were translocated between 30 May and 28 June 1991. Dormice were weighed every 2–3 days up until 10–14 days after release. Six of the seven dormice were provided with supplementary food (sliced apple, sunflower seeds, fruits of trees from the study site) for 5–8 days. Dormice were caught in the morning and placed at the release site in the nest box in which they had been captured, by early afternoon of the same day.

A study in 1993–1994 on a forested peninsula in Dorset, UK (2) found that none of the translocated red squirrels *Sciurus vulgaris* provided with supplementary food and water in holding pens (with nestboxes) and once released survived over five months after release. Out of 14 translocated red squirrels, 11 (79%) survived over one week. Only three (21%) survived >3 months and none survived >4.5 months. At least half of the 14 squirrels were killed by mammalian predators. Intact carcasses examined showed signs of weight loss and stress (see original paper for details). Between October and November 1993, fourteen wild-born red squirrels were released into an 80-ha forest dominated by Scots pine *Pinus sylvestris*. The forest had no red squirrels but had introduced grey squirrels *Sciurus carolinensis*. Capture and release sites were similar habitats. Supplementary food comprised a mixture of seeds, nuts and fruit on trays and in feed hoppers. Squirrels were kept in 1.5 × 1.5 × 1.5 m weldmesh pens surrounded by electric fencing for 3–6 days before release. Squirrels were kept individually except for 2 males who shared a pen. After release, squirrels continued to have access to food, water and nest boxes inside the pens and outside (20–100 m away). All squirrels were radio-tagged and located 1–3 times/day, for 10–20 days after release and thereafter every 1–2 days.

A replicated study in 1995–1997 in four grassland sites in New Mexico, USA (3) found that translocated populations of black-tailed prairie dogs *Cynomys ludovicianusi* provided with supplementary food and kept in holding pens prior to release persisted at least two years after release and reproduced in the wild. The number of black-tailed prairie dogs approximately doubled during the first spring after release in one site on one ranch where supplementary food was provided. Between the second spring and summer, after supplementary feeding had ceased, the number of animals associated with both release sites on the same ranch doubled. Precise numbers are not reported. One hundred and one prairie dogs were translocated to two ranches (Armendaris Ranch received 71 individuals; Ladder Ranch: 30 individuals) between June 1995 and June 1997. At each ranch, prairie dogs were released into two 0.4-ha holding pens (number of individuals per holding pen is not provided). Holding pens were fenced and surrounded by electric wire. Animals at Armendaris ranch were provided with supplementary food in pens for several months up to a year. Information on population persistence at Ladder Ranch is not provided. The time individuals were kept in the holding pens before subsequent release varied between a few days, weeks and some weren't released from them at all (see original paper for details).

A controlled study in 1997 in a mixed pasture and cultivated fields farmland site in northern France (4) found that translocated European rabbits *Oryctolagus cuniculus* provided with supplementary food in holding pens for three days prior to release had higher female survival rates immediately following release compared to rabbits released directly, but male survival rates did not differ. During the first day after translocations, the survival rate of female rabbits released from pre-release pens with supplementary food was higher (100%) than that of females released directly into the wild (83%) and male rabbits released from release pens (78%). The survival rate of male rabbits released from pre-release pens with supplementary food (78%) was not significantly different to male rabbits released directly into the wild (92%). One hundred and four rabbits were translocated from Parc-du-Sausset to a 150-ha area of cultivated fields and pasture in Héric, approximately 400 km away in January 1997. Of these, roughly half were acclimatised in eight 100-m² enclosures (fence height: 1 m), for three days prior to release. Rabbits were provided supplementary food while in pens. Survival was estimated by night-time relocation of ear-tagged rabbits using a spotlight, daily in the first week after release and twice a week until late February 1997.

A study in 1995–2002 in a mixed oak forest reserve in the south of France (5) found that following supplementary feeding in a holding pen prior to release, approximately half of translocated female roe deer *Capreolus capreolus* survived over one year after release and overall the deer population increased six years after the translocations began. Twenty-six out of 49 (53%) translocated female roe deer survived over one year post-release. Of the animals that died in the first year, 35% of mortality occurred within the first month after release. After six years the deer population had increased to 0.47 deer/km² compared to 0.06 deer/km² in the first year after translocation began. In February 1995–1997, fifty-two male and 52 female roe deer were translocated from Northern France into a 3,300-ha forest reserve in Southern France in seven release sessions. Animals were placed into enclosures in groups of approximately 15 individuals for 2–10 days and provided with food (pellets and fresh vegetables) prior to release. Forty-nine females (21 <1 year old and 28 >1 year old) were radio-tagged and were located from a vehicle once or twice each week, over one year post-release. In addition, surveys were carried out on foot (6 transects, each 5–7 km long) eight times a year in February-March 1996–2002 to estimate population growth. Deer were present in low numbers prior to translocation.

A study in 2001–2005 in a mixed forest and farmland site in northern Italy (6) found that just over half of translocated Eurasian badgers *Meles meles* provided with supplementary food in holding pens (in groups) survived at least 1–9 months after release and one pair reproduced. Seven out of 12 badgers survived for 1–9 months, after which monitoring equipment stopped operating. One badger died almost immediately after release due to unknown causes. Two badgers escaped (one after the first month, the other after an unknown period). The fate of three other badgers was unknown. One pair of translocated animals reproduced in the wild four years after release. From March 2001 to May 2004, twelve badgers were captured at four sites in northern Italy. Badgers were fitted with radio-collars and transported 20–40 km to the release site where they were kept in a 350 m^2 enclosure in a wooded area in their release groups (2001: 2 individuals, 2002: 4 individuals, 2003: 2 individuals; 2004: 4 individuals) and provided supplementary food for 3–10 weeks before release. Seven of the 12 badgers were located once/week, for up to nine months after release.

A controlled study in 2004 in 20 suburban gardens in Bristol, UK (7) found that translocated and rehabilitated European hedgehogs *Erinaceus europaeus* that were provided with supplementary food after release all lost body mass and some did not survive, but the effects differed with release type. Directly translocated hedgehogs (<6 days in captivity) had a lower eight-week survival probability (41%) and a larger reduction in body mass over this time (33%) than did resident hedgehogs in release gardens (survival: 95%; body mass reduction: 5%) and hedgehogs kept in captivity prior to release (survival: 82%; body mass reduction: 9%). Over the same period, rehabilitated hedgehogs (survival: 73%; body mass reduction: 13%) and resident hedgehogs 3 km away (survival: 64%; body mass reduction: 10%) had statistically similar survival and body mass loss as directly translocated hedgehogs. Only one translocated hedgehog survived seven weeks after release. Between May and June 2004, hedgehogs were translocated to gardens in Bristol: after rehabilitation in a wildlife hospital (20 individuals, >1 month in captivity) in Scotland, directly from Scotland (20 individuals, <6 days in captivity); and from Scotland with >1 month in captivity (23 individuals). In addition, 23 free-living resident hedgehogs were captured and re-released <50 m from release gardens, and 26 free-living resident hedgehogs were captured and released >3 km from release gardens. Food was provided during the first week after release. Hedgehogs were radio-tracked over eight weeks. Hedgehogs were weighed every 10 days.

A controlled study in 1999–2002 in a shrubland site in Huelva, Spain (8) found that providing supplementary food during translocation of European rabbits *Oryctolagus cuniculus* did not increase their abundance relative to unfed translocated rabbits. Over three years, the average rabbit abundance in translocation plots where food was provided (8.9 pellets/ m²) was not significantly different than in plots where translocated rabbits were not fed (5.0 pellets/m²). The study was conducted in four 4-ha plots (1–6 km apart). Each year, in autumn, herbaceous crops (barley *Hordeum vulgare* and oats *Avena sativa*) were sown in two plots to provide supplementary feeding. Batches of 64–67 rabbits were translocated into each of two plots (one with and one without supplementary food) each winter from 1999–2000 to 2001–2002. Translocation plots were switched after the first year, such that translocations in the second and third year were into plots where no translocations were made in the first year. Between September 1999 and November 2002, rabbit abundance was estimated every two months by counting the number of pellets in 33 fixed-position 0.5-m diameter sampling points/plot. Wild rabbits were present in all plots prior to translocations beginning.

A study in 2005–2007 in a mixed conifer forest in Galway, Ireland (9) found that over half of translocated red squirrels *Sciurus vulgaris* provided with supplementary food in holding pens (with nest boxes) and after release survived over eight months after release and most females reproduced during that period. At least 10 out of 19 (53%) translocated squirrels survived over eight months post-release and five out of nine translocated females (56%) were lactating 5–7 months after release. In August 2006, seven juvenile squirrels were caught. At least one squirrel was still alive in the release location two years after the original release. Two squirrels died while in the release pen or shortly afterwards. Another four squirrels died 1–2 months after release. Ten of 13 squirrels established home ranges which contained supplementary feeding stations. Nineteen squirrels were translocated to a nature reserve (19 ha) in the middle of a 789-ha commercial pine plantation, 112 km from the capture site. Individuals were marked, radio-tagged and kept on average for 46 days in one of two pre-release enclosures (3.6 × 3.6 × 3.9 m high). Enclosures contained branches, platforms, nest boxes, and supplementary feeders (containing nuts, maize, seeds and fruit). Supplementary food (50/50 peanut/maize mix) was provided in six feeders in the nature reserve until July 2006. Twenty nest boxes were also provided Squirrels were radio-tracked in September and November 2005 and February and May 2006, and were trapped in February, May and August 2006 and observed once in October 2007.

A study in 2005–2006 at rocky outcrops on a reserve in KwaZulu-Natal Province, South Africa (10) found that translocated rock hyraxes *Procavia capensis* that were provided with food and an artificial refuge after release in a social group, having been held in captivity, all died (or were presumed to have died) within 87 days of release. Eighty-seven days after the release of 17 hyraxes, none could be relocated. In July 2005, ten adult hyraxes were caught in baited mammal traps (900 × 310 × 320 mm) in an area where they were abundant, and held in captivity for 16 months, during which time three died. The remaining seven were released in November 2006, along with the eight juveniles and two pups born to them in captivity, to a 656-ha reserve where the species was nearly extinct. For four months prior to release, the group was housed together in an outdoor cage (5.9 × 2.5 × 3.2 m). Hyraxes were released into a hay-filled hutch which was left in place for several months, and were provided with cabbage for one week after release. Hyraxes were monitored by direct observations and by walking regular transects, daily for the first week but decreasing to monthly by the end of the study.

A study in 2001 in a grassland and shrubland site in California, USA (11) found that most translocated Tipton kangaroo rats *Dipodomys nitratoides nitratoides* and Heermann's kangaroo rats *Dipodomys heermanni* ssp. provided with supplementary food within artificial burrows after release died within five days of release. All four Tipton kangaroo rats were predated within five days of translocation, and only one out of seven Heermann's kangaroo rats survived over 45 days. Three Heermann's kangaroo rats were predated, two died as a result of aggression from other Heermann's kangaroo rats, and the fate of one was unknown. In September 2001, four juvenile Tipton kangaroo rats and three Heermann's kangaroo rats were captured and held in captivity for two months before release at a protected site in November. In December 2001, a further four Heermann's kangaroo rats were caught and translocated to the same site. All 11 animals were fitted with a radio-transmitter and ear tags, and monitored for seven days in captivity prior to release. The release site was already occupied by Heermann's kangaroo rats. Animals were released into individual artificial burrows (two 90-cm-long cardboard tubes with a chamber about 30 cm below the surface), dug 10–15 m apart and provided with seeds. Burrows were plugged with paper towels until dusk. Animals were radio-tracked every 1–8 days for 18–45 days after release.

A study in 2007–2014 in a grassland reserve in Corrientes Province, Argentina (12; same study site as 15) found that over half of released rehabilitated or captive reared giant anteaters *Myrmecophaga tridactyla*, some of which were provided supplementary food and initially kept in holding pens, survived for at least six months. At least 18 of 31 (58%) released giant anteaters survived for a minimum of six months. Long term survival and the fate of the other 13 anteaters is not reported. In 2007–2013, thirty-one giant anteaters (18 males, 13 females; 1–8 years old) were released into a 124-km² private reserve. Hunting within the reserve was prohibited and livestock were absent. Three anteaters were wild-born but rehabilitated in captivity from injuries, 22 were wild-born but captive-reared and six were from zoos (origin not stated). Of the 18 surviving anteaters, six had been released after a short period in a 0.5-ha pen at the release site and 12 after 7–30 days in a 7-ha pen. Supplementary food was provided for several weeks after release. In 2007–2014, thirteen anteaters were tracked for less than six months, and 18 were tracked for 6–46 months.

A controlled study in 2013 at a desert site in South Australia, Australia (13) found that supplementary feeding stations were visited by translocated burrowing bettongs *Bettongia lesueur*, but populations did not persist. At a large release area, bettongs were detected at 52–80% of track pads at feeders compared to 0–8% of track pads sited 200 m from feeders. No bettongs were detected >42 days after the final release. At three smaller release areas, bettongs persisted for 10 and 53 days at sites where supplementary food was provided and for two days at a site where it was not provided. Bettongs were translocated and released into rabbit warrens in July–December 2013. In one area 1,266 bettongs were released. Five smaller releases, of 29–56 bettongs, were made at three further sites, 4 km apart. Oats were provided at five stations in the large release area and three stations each at two smaller release areas. From May–December 2003 feral cats *Felis catus* and foxes *Vulpes vulpes* were intensively controlled in a 500-km² area by 428 hours of shooting patrols. Bettong visitation at feeders was assessed using 10 track pads/feeder for three one-day periods, four days apart. Persistence was monitored using track counts, camera trapping, warren monitoring and live-trapping.

A replicated, before-and-after study in 2011–2013 in two forest and grassland sites in the Australian Capital Territory, Australia (14) found that translocated eastern bettongs *Bettongia gaimardi* provided with supplementary food in fenced predator proof enclosures did not have greater body weights than those without enclosures and supplementary food. Between twelve and 24 months post-release, the average body weight of translocated eastern bettongs (1.83 kg) did not differ significantly between populations with and without supplementary feeding (weight values for each individual population not provided). Overall, the average body weight of bettongs increased compared to before they were released (pre-release average weight: 1.69 kg). In 2011–2012, sixty adult eastern bettongs were translocated from Tasmania to two predator-free fenced reserves. In one reserve bettongs (5 males, 7 females) received supplementary food at least weekly and were placed in 2.6–9.4 ha enclosures, whereas in a second reserve bettongs (8 males, 10 females) received no supplementary food and were not managed in enclosures. Supplementary food included fresh locally available produce and commercial pellets. Body weight was assessed before reintroduction and 12–24 months after release (May–November 2013). Bettongs were also monitored by radio-telemetry or camera traps and live-trapping every 3 months.

A controlled study in 2007–2012 in a grassland reserve in Corrientes, Argentina (15; same study site as 12) found that after being provided with supplementary food and kept in holding pens, captive-bred giant anteaters *Myrmecophaga tridactyla* released into the wild were less nocturnal in their activity patterns than were wild-born rehabilitated and released individuals. Released captive-bred giant anteaters were proportionally less active at night than released wild-born animals (43% vs 70% of activity records were at night). During 2007–2012, three captive-bred and four wild-born adult giant anteaters were released into a 124-km² private reserve. Wild-born animals were rehabilitated after being injured by hunters or in road accidents. Six anteaters (all wild-born and two captive-bred anteaters) were released after spending a short period of time in a 0.5 ha acclimatisation pen. The remaining 12 anteaters spent 7–30 days in a 7 ha holding pen at the release site prior to release. Supplementary food was provided in the holding pen, and for several weeks after anteaters were released. Each of the seven anteaters was fitted with a radio-transmitter and tracked for one or two 24 h periods/month in 2007 and 2011. The released anteaters were further monitored using 14 baited camera traps for an average of 336 days/trap in 2008–2012.

A study in 2015–2016 in a wooded mountain region in central Wales, UK (16) found that some translocated pine martens *Martes martes* held in pre-release pens and then provided with supplementary food and nest boxes survived and bred in the first year after release. At least four out of 10 females that had been kept in pre-release pens survived and bred the year after release. Around 10–12 months after release, 14 out of 20 martens were alive and in good condition. Twelve were within 10 km of their release site. Six martens died in the first year, two had a fungal infection two weeks after release. Authors suggest this may have been due to damp conditions in November. From September–November 2015, twenty breeding age (>3-years-old) pine martens were caught in Scotland, health checked, microchipped and fitted with a radio-collar, and in some cases a GPS logger. Martens were transported overnight to Wales, and held in individual pre-release pens (3.6 × 2.3 × 2 m) for up to seven nights. Males' pens were within 500 m of a female, but >2 km from the nearest male. Releases took place in autumn, and supplementary food was provided for 2–6 weeks after release (for as long as it continued to be taken). Den boxes were provided within 50 m of each release pen. Martens were radio-tracked until home-ranges were established, then located daily–weekly. Intensive tracking of females was carried out in March to locate breeding sites. Hair tubes and camera traps were used to monitor breeding success. A further 19 martens were released using the same procedure in September–October 2016.

(1) Bright P.W. & Morris P.A. (1994) Animal translocation for conservation: performance of dormice in relation to release methods, origin and season. *Journal of Applied Ecology*, 31, 699–708.

(2) Kenward R.E. & Hodder K.H. (1998) Red squirrels (*Sciurus vulgaris*) released in conifer woodland: the effects of source habitat, predation and interactions with grey squirrels (*Sciurus carolinensis*). *Journal of Zoology*, 244, 23–32.

(3) Truett J.C. & Savage T. (1998) Reintroducing prairie dogs into desert grasslands. *Restoration and Management Notes*, 16, 189–195.

(4) Letty J., Marchandeau S., Clobert J. & Aubineau J. (2000) Improving translocation success: an experimental study of anti-stress treatment and release method for wild rabbits. *Animal Conservation*, 3, 211–219, https://doi.org/10.1111/j.1469-1795.2000.tb00105.x

(5) Calenge C., Maillard D., Invernia N. & Gaudin J.C. (2005) Reintroduction of roe deer *Capreolus capreolus* into a Mediterranean habitat: female mortality and dispersion. *Wildlife Biology*, 11, 153–161, https://doi.org/10.2981/0909-6396(2005)11[153:rordcc]2.0.co;2

(6) Balestrieri A., Remonti L. & Prigioni C. (2006) Reintroduction of the Eurasian badger (*Meles meles*) in a protected area of northern Italy. *Italian Journal of Zoology*, 73, 227–235, https://doi.org/10.1080/11250000600679603

(7) Molony S.E., Dowding C.V., Baker P.J., Cuthill I.C. & Harris S. (2006) The effect of translocation and temporary captivity on wildlife rehabilitation success: an experimental study using European hedgehogs (*Erinaceus europaeus*). *Biological Conservation*, 130, 530–537, https://doi.org/10.1016/j.biocon.2006.01.015

(8) Cabezas S. & Moreno S. (2007) An experimental study of translocation success and habitat improvement in wild rabbits. *Animal Conservation*, 10, 340–348, https://doi.org/10.1111/j.1469-1795.2007.00119.x

(9) Poole A. & Lawton C. (2009) The translocation and post release settlement of red squirrels *Sciurus vulgaris* to a previously uninhabited woodland. *Biodiversity and Conservation*, 18, 3205–3218, https://doi.org/10.1007/s10531-009-9637-z

(10) Wimberger K., Downs C.T., Perrin M.R. (2009) Two unsuccessful reintroduction attempts of rock hyraxes (*Procavia capensis*) into a reserve in the KwaZulu-Natal Province, South Africa. *South African Journal of Wildlife Research*, 39, 192–201, https://doi.org/10.3957/056.039.0213

(11) Germano D.J. (2010) Survivorship of translocated kangaroo rats in the San Joaquin Valley, California. *California Fish and Game*, 96, 82–89.

(12) Di Blanco Y.E., Jiménez Pérez I. & Di Bitetti M.S. (2015) Habitat selection in reintroduced giant anteaters: the critical role of conservation areas. *Journal of Mammalogy*, 96, 1024–1035, https://doi.org/10.1093/jmammal/gyv107

(13) Bannister H.L., Lynch C.E. & Moseby K.E. (2016) Predator swamping and supplementary feeding do not improve reintroduction success for a threatened Australian mammal, *Bettongia lesueur*. *Australian Mammalogy*, 38, 177–187, https://doi.org/10.1071/am15020

(14) Portas T.J., Cunningham R.B., Spratt D., Devlin J., Holz P., Batson W., Owens J. & Manning A.D. (2016) Beyond morbidity and mortality in reintroduction programmes: changing health parameters in reintroduced eastern

bettongs *Bettongia gaimardi*. *Oryx*, 50, 674–683, https://doi.org/10.1017/
s0030605315001283

(15) Di Blanco Y.E., Spørring K.L. & Di Bitetti M.S. (2017) Daily activity
pattern of reintroduced giant anteaters (*Myrmecophaga tridactyla*): effects of
seasonality and experience. *Mammalia*, 81, 11–21, https://doi.org/10.1515/
mammalia-2015-0088

(16) MacPherson J.L. (2017) *Pine marten translocations: the road to recovery and
beyond*. Bulletin of the Chartered Institute of Ecology and Environmental
Management: Rewilding and species reintroductions, 95, 32–36.

Captive-breeding

14.19. Breed mammals in captivity

https://www.conservationevidence.com/actions/2471

- **Three studies** evaluated the effects of breeding mammals in
  captivity. One study was across Europe[1], one was in the USA[2]
  and one was global[3].

COMMUNITY RESPONSE (0 STUDIES)

POPULATION RESPONSE (3 STUDIES)

- **Abundance (1 study):** A review of captive-breeding
  programmes across the world[3] found that the majority of 118
  captive-bred mammal populations increased.

- **Reproductive success (2 studies):** A review of a captive
  breeding programme across Europe[1] found that the number
  of European otters born in captivity tended to increase over 15
  years. A study in the USA[2] found that wild-caught Allegheny
  woodrats bred in captivity.

- **Survival (1 study):** A review of a captive breeding programme
  across Europe[1] found that the number of European otters born
  in captivity that survived tended to increase over 15 years.

BEHAVIOUR (0 STUDIES)

Background

Captive breeding involves taking wild animals into captivity and establishing and maintaining breeding populations. It tends to be undertaken when wild populations become very small or fragmented or when they are declining rapidly. Captive populations can be maintained while threats in the wild are reduced or removed and can provide an insurance policy against catastrophe in the wild. Captive breeding also potentially provides a method of increasing reproductive output beyond what would be possible in the wild. However, captive breeding can result in problems associated with inbreeding depression, removal of natural selection and adaptation to captive conditions.

The aim is usually to release captive-bred animals back to natural habitats, either to original sites once conditions are suitable, to reintroduce species to sites that were occupied in the past or to introduce species to new sites. Some captive populations may also be used for research to benefit wild populations.

Studies that investigate the effectiveness of releasing captive-bred mammals are discussed elsewhere. Those studies are not included in this section, unless specific details about captive breeding were included.

A review of a captive breeding programme in 1978–1992 across Europe (1) reported that the number of institutions successfully breeding European otters *Lutra lutra*, the number of otters born in captivity and that survived tended to increase over 15 years. These results were not tested for statistical significance. The number of institutions keeping otters remained fairly stable (23–32) from 1978 to 1989, whilst the number of captive animals born and surviving tended to increase from 1978–1983 (born: 0–20; survived: 0–18) to 1984–1989 (born: 18–46; survived: 12–38). Authors reported that until 1990, breeding was only successful in about 10 collections, but that in 1991–1992, when the number of institutions participating in the programme increased to 55, the number that successfully bred otters almost doubled. In 1992 the total captive population was 196 individuals, of which 67% was captive born, and 43 out of 50 cubs survived. In 1990, 36 otter keeping institutions (60% of those co-operating with the studbook) and in 1992 fifty five (91% included in the studbook) took part in the European breeding program for self-sustaining captive populations of otters. These institutions provided information about their captive breeding populations from 1978–1992.

A study in 2009–2011 in a captive facility in Indiana, USA (2) found that wild-caught Allegheny woodrats *Neotoma magister* bred in captivity. Over 26 months, 33 pairings resulted in copulation which produced 19 litters (58% pregnancy rate). Those litters comprised of 43 pups (26 male, 17 female), of which 40 (24 male, 16 female) survived to weaning at 45 days. Overall, eight of 12 wild-caught females produced offspring (1–5 litters) and four of six wild-caught males sired litters (1–8 litters). In 2009 a captive breeding program was established using eight wild-caught individuals collected from the seven populations in Indiana and four caught from populations in Pennsylvania. The breeding population was maintained at 12–13 animals with a female bias (8:4). Seven new wild animals replaced five in 2010–2011. Individuals were housed in wire mesh enclosures (91 x 61 x 46 cm or 76 x 46 x 91 cm) with access to the opposite sex and an external nest box (23 x 23 x 23 or 36 cm). Enclosures were at 20°C with 13 hours of light/24 hrs. Captive-reared juveniles were released into wild populations in April-July each year.

A review of captive-breeding programmes in 1970–2011 across the world (3) found that the majority of 118 captive-bred mammal populations increased in size. The average annual rate of population increase was 0.028, and only 17 populations (14%) declined (five 'endangered' or 'critically endangered' according to the IUCN Redlist). Authors reported that positive growth rates were maintained for a large majority of the populations in all IUCN categories except those of 'least concern'. However, average growth rates declined from 1970–1991 (0.054) to 1992–2011 (0.021). Authors reported that there was a slight decrease in average death rate of populations over time and either no change in average birth rate, or lower birth rates after 1989. Population growth rates did not vary with body mass, but were reported to decrease as the ratio of individuals in programs to populations increased (see original paper for details). Counts of births, deaths and end-of-year totals of individuals in captive populations recorded in studbooks (excluding regional studbooks) were published in the International Zoo Yearbook. Those published from 1970 to 2011 were used to calculate rates of population growth for 118 captive-bred populations (81 species and 37 subspecies). Only populations for which the sum of end-of-year totals was at least 250 over the time period were included.

(1) Vogt, P. (1995) The European Breeding Program (EEP) for *Lutra lutra*: its chances and problems. *Hystrix — Italian Journal of Mammalogy*, 7, 247–253.

(2) Smyser, T.J. & Swihart, R.K. (2014) Allegheny woodrat (Neotoma magister) captive propagation to promote recovery of declining populations. *Zoo Biology*, 33, 29–35, https://doi.org/10.1002/zoo.21114

(3) Alroy, J. (2015) Limits to captive breeding of mammals in zoos. *Conservation Biology*, 29, 926–931, https://doi.org/10.1111/cobi.12471

14.20. Place captive young with captive foster parents

https://www.conservationevidence.com/actions/2472

- **Two studies** evaluated the effects of placing captive young mammals with captive foster parents. One study was in the USA[1] and one was in Sweden and Norway[2].

COMMUNITY RESPONSE (0 STUDIES)

POPULATION RESPONSE (2 STUDIES)

- **Survival (2 studies):** A replicated, controlled study in the USA[1] found that most captive coyote pups placed with foster parents were successfully reared. A replicated study in Sweden and Norway[2] found that captive grey wolf pups placed with foster parents had higher survival rates than pups that stayed with their biological mother.

- **Condition (1 study):** A replicated study in Sweden and Norway[2] found that captive grey wolf pups placed with foster parents weighed less than pups that stayed with their biological mother.

BEHAVIOUR (0 STUDIES)

Background

Success of captive breeding programmes for endangered mammal species may be reduced if the biological parents are unable to rear any or all of their young. This may occur when there are more young than parents can rear, or through disease, injury or death of the parents. One option may be to place the young with captive foster parents of the same species, where such animals are available. This may reduce the risk of the young becoming imprinted on humans (which could occur if they were hand reared) and so could increase their chance of survival after release into the wild.

Studies reported on here are examples of where this action is carried out in an experimental way, but where the results could help inform actions in future programmes.

See also *Hand-rear orphaned or abandoned young in captivity*, *Place orphaned or abandoned wild young with captive foster parents* and *Place orphaned or abandoned wild young with wild foster parents*.

A replicated, controlled study (year not stated) in a captive animal facility in Utah, USA (1) found that most coyote *Canis latrans* pups placed with foster parents in captivity were successfully reared. All eight pups fostered into four litters at <1 week old survived beyond six weeks of age. Of six 3–4-week-old pups fostered into three litters, four pups in two litters survived beyond six weeks old. The two pups in the third litter died. Two attempts each to foster two 6–7-week-old pups failed, with pups dying within 24 hours. All pups born into these litters survived. The survival rate of litters fostered in their entirety when <10 days old (17 out of 19 pups surviving from four litters) was similar to that in litters not fostered (18 out of 20 pups surviving from four litters). Causes of death were not established for pups that died. Litters of eight coyote pairs were augmented by adding two additional pups, four litters were replaced completely and four litters were reared by their parents without additions. Survival was monitored to six weeks of age.

A replicated study in 2011 in six zoos in Sweden and Norway (2) found that grey wolf *Canis lupus lupus* pups placed with foster parents in captivity had higher survival rates but weighed less than pups that stayed with their biological mother. After 32 weeks, more fostered cubs survived (75%) than cubs that remained with their biological mother (65%). At 24–26 days age, fostered cubs weighed less (1,337 g) than cubs that remained with their biological mother (2,019 g). In 2011, eight pups born at zoos were removed from their biological mothers at 4–6 days of age. Pups were microchipped, to allow identification, given fluids to reduce dehydration, and transported by car or plane to new zoos. Foster pups were placed in litters containing 7–10 pups. On arrival, the tails of foster pups were rubbed in the urine of other pups so that they smelled similar. A total of 35 pups stayed with their biological mother. Cameras were placed at the den of each litter. Pups were weighed at irregular intervals and all deaths recorded.

(1) Kitchen A.M. & Knowlton F.F. (2006) Cross-fostering in coyotes: evaluation of a potential conservation and research tool for canids. *Biological Conservation*, 129, 221–255, https://doi.org/10.1016/j.biocon.2005.10.036

(2) Scharis I. & Amundin M. (2015) Cross-fostering in gray wolves (*Canis lupus lupus*). *Zoo Biology*, 34, 217–222, https://doi.org/10.1002/zoo.21208

14.21. Use artificial insemination

https://www.conservationevidence.com/actions/2473

- **Three studies** evaluated the effects on mammals of using artificial insemination. One study was in the USA[1], one was in Brazil[2] and one was in China[3].

COMMUNITY RESPONSE (0 STUDIES)

POPULATION RESPONSE (3 STUDIES)

- **Reproductive success (3 studies):** A study in the USA[1] found that following artificial insemination, fewer than half of female black-footed ferrets gave birth. A study in Brazil[2] found that following artificial insemination, a captive female Amazonian brown brocket deer gave birth. A replicated study in China[3] found that following artificial insemination, a lower proportion of captive female giant pandas became pregnant than after natural mating.

BEHAVIOUR (0 STUDIES)

> **Background**
>
> During programmes to rear endangered animals in captivity, in preparation for reintroductions into the wild, artificial insemination may be used to initiate pregnancies. The technique may be used instead of natural mating in situations such as animals being kept at different facilities or where natural mating has failed. It may also be carried out using preserved sperm for purposes of maintaining genetic diversity.
>
> Studies included here are those identified by our searches of conservation journals. It is likely that other relevant studies exist in biological journals that specialise in reproduction.

A study in 2008–2011 in two ex-situ facilities in Wyoming and Virginia, USA (1) found that following artificial insemination, fewer than half of female black- footed ferrets *Mustela nigripes* gave birth. Five out of 18 (28%) artificially inseminated female black-footed ferrets gave birth. Eight kits were born. Six of those kits subsequently went on to breed by natural mating. Kinship (a measure of relatedness within a population) was lower among these kits and their descendants than among the population as a whole. The study was conducted at the National Black-Footed Ferret Conservation Center and at the Smithsonian Conservation Biology Institute. Ferrets were managed in individual cages (1.0–3.6 × 1.3–6.0 m). Semen was collected from adult ferrets (1–6 years old) by electroejaculation and cryopreserved for 10–20 years. Females were inseminated by transabdominal injections of sperm.

A study in 2012–2013 in an ex-situ facility in São Paulo, Brazil (2) found that following artificial inseminated, a captive female Amazonian brown brocket deer *Mazama nemorivaga* gave birth. Seven months after being artificially inseminated, a female Amazonian brown brocket deer gave birth without veterinary intervention to a healthy male fawn. A captive adult pair of Amazonian brown brocket deer was kept in isolated pens in a deer research facility. Animals were exposed to natural light conditions and given similar diets. Every morning for one month, a trained examiner manually observed the female for signs of natural oestrus. Eight hours after oestrus was detected, the female was physically restrained, anesthetized and inseminated. Sperm was collected by electroejaculation. Tools and techniques used for artificial insemination were based on those from procedures carried out on sheep and other small ruminants.

A replicated study in 1996–2016 in Sichuan Province, China (3) found that following artificial insemination, a lower proportion of 78 captive female giant pandas *Ailuropoda melanoleucahela* became pregnant than after natural mating. Following artificial insemination, a lower percentage of female pandas became pregnant (19%) than following natural mating (61%). However, there was no significant difference in the litter size of females inseminated artificially or through natural mating (data reported as model results). Between 1996 and 2016, seventy-eight female pandas held in open-air enclosures at two facilities were subject to 65 attempts at artificial insemination and 150 attempts at natural mating. Natural mating was always attempted first but, in cases of excessive aggression between males and females, artificial insemination was used instead.

(1) Howard J.G., Lynch C., Santymire R.M., Marinari P.E. & Wildt D.E. (2016) Recovery of gene diversity using long-term cryopreserved spermatozoa and artificial insemination in the endangered black-footed ferret. *Animal Conservation*, 19, 102–111, https://doi.org/10.1111/acv.12229

(2) Oliveira M.E.F., dos Santos Zanetti E., Cursino M.S., Peroni E.F.C., Rola L.D., Feliciano M.A.R., Canola J.C. & Duarte J.M.B. (2016) First live offspring of Amazonian brown brocket deer (*Mazama nemorivaga*) born by artificial insemination. *European Journal of Wildlife Research*, 62, 767–770, https://doi.org/10.1007/s10344-016-1040-y

(3) Li D., Wintle N.J., Zhang G., Wang C., Luo B., Martin-Wintle M.S., Owen M. & Swaisgood R.R. (2017) Analyzing the past to understand the future: natural mating yields better reproductive rates than artificial insemination in the giant panda. *Biological Conservation*, 216, 10–17, https://doi.org/10.1016/j.biocon.2017.09.025

14.22. Clone rare species

https://www.conservationevidence.com/actions/2474

- **One study** evaluated the effects of cloning rare species. This study was in Iran[1].

COMMUNITY RESPONSE (0 STUDIES)

POPULATION RESPONSE (1 STUDY)

- **Reproductive success (1 study):** A controlled study in Iran[1] found that immature eggs of domestic sheep have potential to be used for cloning of Esfahan mouflon.

BEHAVIOUR (0 STUDIES)

Background

Cloning technology is advancing rapidly. For rare mammals, cloning provides the potential to increase reproductive output from a small number of individuals by using surrogate parents of closely related but non-threatened species.

Note that many relevant studies may be documented in journals that are not primarily conservation-related and which are, therefore, not included in our systematic searches for evidence.

A controlled study (date not stated) in Iran (1) found that immature eggs (oocytes) of domestic sheep have potential to be used for interspecies conservation cloning of Esfahan mouflon *Ovis orientalis isphahanica*. The success rate for transferring cell nuclei attached to Esfahan mouflon cells to domestic sheep oocytes (14.4%) did not significantly differ from that for transfer of nuclei attached to domestic sheep cells (22.1%). Subsequently, of 12 cloned mouflon blastocysts (early-stage cell mass which goes on to form an embryo) transferred to five domestic sheep recipients, two pregnancies resulted. In both cases live births of cloned Esfahan mouflon lambs resulted, but the lambs died soon after birth. Of 1,410 oocytes that had had their nucleus removed, 1,105 and 305 were attached to Esfahan mouflon and domestic sheep cells, respectively. Prior to transferring nuclei, donor cells were serum starved for 5 days. In vitro matured domestic sheep oocytes that had had their nucleus removed were then reconstituted with nuclei donor cells of mouflon and domestic sheep.

(1) Hajian M., Hosseini S.M., Forouzanfar M., Abedi P., Ostadhosseini S., Hosseini L., Moulavi F., Gourabi H., Shahverdi A.H., Vosough Taghi Dizaj A., Kalantari S.A., Fotouhi Z., Iranpour R., Mahyar H., Amiri-Yekta A. & Nasr-Esfahani M.H. (2011) 'Conservation cloning' of vulnerable Esfahan mouflon (*Ovis orientalis isphahanica*): in vitro and in vivo studies. *European Journal of Wildlife Research*, 57, 959–969, https://doi.org/10.1007/s10344-011-0510-5

14.23. Preserve genetic material for use in future captive breeding programs

https://www.conservationevidence.com/actions/2475

- **Two studies** evaluated the effects of preserving genetic material for use in future captive breeding programs. One study was in Mexico[1] and one was in the USA[2].

COMMUNITY RESPONSE (0 STUDIES)

POPULATION RESPONSE (2 STUDIES)

- **Survival (2 studies):** A study in Mexico[1] found that a series of non-traditional techniques, combined with natural mating, produced five aoudad embryos that could be cryogenically preserved. A study in USA[2], found that artificial insemination using preserved genetic material increased genetic diversity and lowered inbreeding in a captive black-footed ferret population.

BEHAVIOUR (0 STUDIES)

Background

Assisted reproductive technology is advancing rapidly. For rare mammals, preservation of genetic material provides potential to increase reproductive output from a small number of individuals and to retain embryos or other material for future development.

Note that many relevant studies may be documented in journals that are not primarily conservation-related and which are, therefore, not included in our systematic searches for evidence.

A study (date not stated) in a zoo in Mexico (1) found that using a series of non-traditional techniques, combined with natural mating, five embryos were produced from aoudad *Ammotragus lervia* that could be cryogenically preserved. The five embryos were obtained from just one of the three female aoudad, with the low embryo recovery rate being due to a low level of fertilization in vivo. The oestrus and superovulation of three female aoudad were synchronized. Procedures followed those used for domestic sheep combined with subcutaneous osmotic pumps for delivering the follicle-stimulating hormone. An aoudad ram was introduced for natural mating at the anticipated time of oestrous. Embryos were collected five and a half days later by incision through the abdominal wall. Embryos were cryopreserved, for use in conservation breeding programs (potentially by transferring to surrogates, such as domestic hybrids between aoudad and sheep or goats).

A controlled study in 1989–1998 and 2008–2011 in two captive facilities in Wyoming and Virginia, USA (2) found that artificial insemination using preserved genetic material increased genetic diversity and lowered measures of inbreeding in a captive population of black-footed ferrets *Mustela nigripes*. Genetic diversity of the captive population was greater when eight black-footed ferret kits (and their offspring) born as a result of artificial insemination with preserved semen were incorporated (86.5–86.8%) than when the population reproduced naturally (86.3–86.6%). Inbreeding also decreased by 6% (data reported as inbreeding coefficients). In 1989–1998, semen were collected from 16 male ferrets (1–6 years old) by electroejaculation and cryopreserved in liquid nitrogen for 10–20 years. In 2008–2011, a total of 18 female ferrets were inseminated with the thawed samples. Their eight offspring went on to produce 32 offspring and grand-offspring by natural mating. Selection of female recipients was based on the analysis of the pedigree of the captive population.

(1) López–Saucedo J., Ramón-Ugalde J.P., Barroso-Padilla J.J., Gutiérrez-Gutiérrez A.M., Fierro R. & Piña-Aguilar R.E. (2013) Superovulation, in vivo embryo recovery and cryopreservation for Aoudad (*Ammotragus lervia*) females using osmotic pumps and vitrification: a preliminary experience and its implications for conservation. *Tropical Conservation Science*, 6, 149–157, https://doi.org/10.1177/194008291300600105

(2) Howard J.G., Lynch C., Santymire R.M., Marinari P.E. & Wildt D.E. (2016) Recovery of gene diversity using long-term cryopreserved spermatozoa and artificial insemination in the endangered black-footed ferret. *Animal Conservation*, 19, 102–111, https://doi.org/10.1111/acv.12229

Release captive-bred mammals

14.24. Release captive-bred individuals to re-establish or boost populations in native range

https://www.conservationevidence.com/actions/2476

- **Thirty-one studies** evaluated the effects of releasing captive-bred mammals to establish or boost populations in their native range. Seven studies were in the USA[2,7,8,13,24,27,29], three were in Australia[11,23,28] and Italy[5,20,30], two studies were in each of Canada[1,17], Sweden[3,6], Saudi Arabia[4,25], the UK[9,10], the Netherlands[12,21] and South Africa[14,18] and one study was in each of France[15], Africa, Europe, and North America[16], Estonia[19], the USA and Mexico[22], Poland[26] and China[31].

COMMUNITY RESPONSE (0 STUDIES)

POPULATION RESPONSE (30 STUDIES)

- **Abundance (7 studies):** Five of five studies (one replicated) and two reviews in Saudi Arabia[4], Australia[11], the USA[13], South Africa[14], France[15], the Netherlands[21] and China[31] found that following release of captive-bred (or in one case captive-reared, or including translocated) animals, populations of mountain gazelles[4], Corsican red deer[15], Père David's deer[31], Eurasian otters[21] and swift foxes[13] increased. The two reviews found that following release of mainly translocated but some captive-bred large carnivores[14], populations of four of six species increased, and over half of mammal release programmes[11] were considered successful.

- **Reproductive success (5 studies):** Four studies (one replicated) in Saudi Arabia[4,25], the UK[9] and the Netherlands[12] found that released captive-bred (and in some cases some wild-born translocated) mountain gazelles[4], dormice[9] and some Eurasian otters[12] reproduced successfully and female Arabian oryx[25] reproduced successfully regardless of prior breeding experience. A controlled study in Italy[30] found that released captive-born Apennine chamois[30] reproduced in similar numbers to wild-caught translocated chamois.

- **Survival (24 studies):** Four of three controlled studies (two replicated) and two reviews in Canada[1], Canada and the USA[8], Sweden[3], Italy[30] and across the world[16] found that released captive-bred swift foxes[1,8], European otters[3] and mammals from a review of 49 studies[16] had lower post-release survival rates than did wild-born translocated animals. The other study found that released captive-born Apennine chamois[30] survived in similar numbers to wild-caught translocated chamois. Three studies (one replicated) in the USA[27,29] and Canada[17] found that released captive-born Key Largo woodrats[27], Vancouver Island marmots[17] and swift fox pups[29] had lower survival rates than wild-born, wild-living animals. One of the studies also found that Vancouver Island marmots[17] released at two years old were more likely to survive than those released as yearlings. Eleven studies (three replicated) in Italy[5,20], Sweden[6], the UK[9,10], Estonia[19], Poland[26], Saudi Arabia[4,25], Australia[23] and the USA[24] found that following the release of captive-bred (and in some cases some wild-born translocated) animals, Arabian oryx[25], populations of European otters[6,10,20], European mink[19] and mountain gazelle[4] survived for 2–11 years, roe deer[5] and over a third of brush-tailed rock-wallabies[23], black-footed ferrets[24] and brown hares[26] survived for 0.5–24 months and dormice[9] populations survived three months to over seven years. A review in Australia[11] found that release programmes for macropod species resulted in successful establishment of populations in 61% of cases and that 40% survived over five years, and another review in Australia[28] found that over half of programmes were considered successful. Two studies and a review in the USA[7], USA and Mexico[22] and South Africa[18] found that over 40% of released captive-bred American black bears[7] were killed or had to be removed, only one of 10 oribi[18] survived over two years and that most black-footed ferret[22] releases were unsuccessful at maintaining a population.

BEHAVIOUR (3 STUDIES)

- **Use (3 studies):** Two studies in the USA[2] and Australia[23] found that following release, most captive-bred and translocated mountain lions[2] that had been held in captivity prior to release and most released captive-bred brush-tailed rock-wallabies[23] established stable home ranges. A controlled study in Italy[30] found that released captive-born Apennine chamois remained closer to the release site than released wild-caught translocated chamois.

Background

Captive breeding is normally used to provide individuals which can then be released into the wild (often called 'reintroduction') to either re-establish a population that has been lost, or to augment an existing population ('restocking').

Release techniques vary considerably, from 'hard releases' involving the simple release of individuals into the wild to 'soft releases' which involve a variety of adaptation and acclimatisation techniques before release or post-release feeding and care. This action includes studies describing the effects of release programmes for captive-bred or captive-reared mammals that do not provide details of specific release techniques. Studies that describe or compare specific release techniques, such as use of holding pens at release sites, or providing supplementary food, water or artificial refuges/breeding sites are described under each specific action.

This action includes studies where animals were released in groups but not studies where releases of different group sizes were compared, or where animals were released in family or social groups (including groups where social animals have been pre-conditioned together prior to release in holding pens). For those studies, *see Release translocated/captive-bred mammals in larger unrelated groups* and *Release translocated/captive-bred mammals in family/social groups.*

A replicated, controlled study in 1990–1992 at two grassland sites in Alberta, Canada (1) found that captive-born swift foxes *Vulpes velox* had lower post-release survival rates than did translocated, wild-born animals. No statistical analyses were performed. Nine months after release into the wild, at least two out of 27 (7%) captive-born swift foxes were known to be alive, compared with twelve out of 28 (43%) wild-born translocated swift foxes. In May 1990 and 1991, a total of 27 captive-born and 28 wild-born swift foxes were released simultaneously. Wild-born animals had been captured in Wyoming, USA, 4–7 months before release and were quarantined for ≥30 days. Animals were released without prior conditioning in holding pens. Foxes were radio-collared and monitored from the ground and air, for at least nine months.

A study in 1993–1995 in northern Florida, USA (2) found that following release, most captive-bred and translocated mountain lions *Puma concolor stanleyana* that had been held in captivity prior to release established home ranges in the release area. Of 19 released mountain lions, 15 established one or more home ranges. Post-release survival periods for these 15 animals are not stated but two were killed (one illegally shot and one killed by a vehicle) and two were recaptured due to landowner concerns or concerns for their survival, 37–140 days after release. Nineteen mountain lions were released in northern Florida in 1993–1994. Six animals were captive-bred, 10 were wild-caught and released within three months and three were caught and released after 3–8 years. mountain lions were radio-tracked daily in February 1993–April 1993 and then for three days/week until June 1995.

A replicated, controlled study in 1989–1993 in two rivers in southern Sweden (3; same experimental set-up as 6) found that captive-bred European otters *Lutra lutra* released into the wild had a lower survival rate than did wild-born translocated otters. One year after release, the survival rate of captive-bred otters (42%) was lower than that of wild-born translocated otters (79%). Additionally, captive-bred otters with a shorter (5–48 day) period between separation from their mother and release to the wild had a higher survival rate (80%) than individuals with a longer (49–98 day) period (13%). Between 1989 and 1992, twenty-five captive-bred and 11 wild-born otters were released into two rivers. Thirty-four otters were released in one river catchment and two in the other. Captive-bred otters were descendants of two captive females. Wild-born otters were live-trapped along the Norwegian coast. All otters were around one year old when released. All except one were released between February and June. All were fitted with an implanted radio-transmitter and monitored for one year on 64% of days.

A study in 1991–1995 in a desert reserve in central Saudi Arabia (4) found that nearly half of captive-bred mountain gazelles *Gazella gazella* released into the wild survived more than two years, and the population bred successfully and more than doubled in size. Of a total of 71 released gazelles, 69–73% survived over one year and 58–59% survived over two years. Mortality was high in the first month after release (13% died), but the mean annual survival rate of gazelles which survived the first month was 78%. Gazelles that were over three years of age when released were more likely to die within 54 weeks of release than younger animals (54% vs 19% mortality) due to a higher rate of predation by wolves. Released females gave birth to at least 134 calves, of which at least 107 were conceived in the wild. By December 1994, the population had increased to 152–185 animals. Between January 1991 and June 1993, seventy-one captive-born mountain gazelles were released into three valleys inside a 2,000-km² reserve. The valleys were fenced to exclude domestic camels but allowed movement of gazelles. All released individuals were ear-tagged and 28 were fitted with a radio-collar. Gazelles were monitored using binoculars and a telescope on 396 days between January 1991 and June 1995. Gazelles were provided with water year-round.

A study in 1992–1993 in a mountain area dominated by deciduous forest in northern Italy (5) found that two captive-bred roe deer *Capreolus capreolus* that were released into the wild survived for at least 10 months. Both captive-bred roe deer survived over 10 months post-release (long term survival is not reported). Their average annual home range extended over 38.5 ha. In November 1992, the two captive-bred male roe deer (aged 17 months) were radio-tagged and released into the wild. The release site was within a 400-ha area with a roe deer population density of 0.2 deer/ha. The area was dominated by deciduous coppice (45%), mixed crops (21%), urbanized areas (14%) and meadows and pastures (13%). The two roe deer were radio-tracked for 10 months after release until September 1993.

A study in 1989–1992 at seven lakes in boreal forest in Sweden (6; same experimental set-up as 3) found that following release, at least 14 of 36 captive-bred or wild-born translocated European otters *Lutra lutra* survived for at least one to two years. Fourteen otters had established home ranges and were still alive when last recorded, 362–702 days after release. Eight further otters were monitored until their transmitters failed or they moved out of radio contact, 89–219 days after release. Fourteen were known to have died, 18–750 days after release. Otter origin (captive-bred or wild-caught) did not affect movement distance. In 1989–1992, thirty-six otters (25 captive-bred and 11 wild-born, translocated otters) were released in lakes and rivers in southern Sweden. Otters were fitted with radio-transmitters. Radio-tracking was carried out at least monthly, in 1989–1992.

A study in 1982–1997 in a mountain forest reserve in Tennessee, USA (7) found that at least 10 of 23 captive-bred American black bears *Ursus americanus* released into the wild were killed or had to be removed. Ten of 23 captive-bred black bears (43%) survived for an average of 172 days after release (range 4–468 days) before being killed (seven bears), euthanised after being hit by a vehicle (one bear), relocated (one bear) or returned to captivity (one bear). The fate of the 13 other released bears is not known (one tracked bear lost its radio-collar after 484 days, 12 bears were not radio-tracked or observed again after release). Twenty-three captive-bred, pen-reared black bears (11 male, 12 female; average 2.5 years old) were released in 1982–1995 at five sites in which bear hunting was prohibited in the Cherokee National Park. All bears were individually marked with ear-tags and/or tattoos. Seven were radio-collared and monitored an average of once every 18 days from an aircraft in 1983–1997.

A review of studies in 1989–1991 in prairie sites in Canada and the USA (8) found that following release, captive-bred swift foxes *Vulpes velox* had lower survival rates than did translocated, wild-caught swift foxes. Over an unspecified time period, 59% of wild-caught translocated swift foxes survived while three of 41 (7%) captive-bred swift foxes survived after release. In 1989–1991, thirty-three wild-caught, adult foxes and 41 captive-bred foxes, born the previous year, were released in the spring. Methods used for monitoring animals were unclear.

A replicated study in 1993–2002 in seven forest sites across England, UK (9) found that following releases of captive-bred (and some translocated wild-born) dormice *Muscardinus avellanarius*, populations persisted for between three months and over seven years and reproduced. In at least three of seven releases, dormouse populations were stable or increased from 19–57 released individuals to 40–55 individuals between two and seven years later. At one site, only one individual was detected 7–8 years after the release of 52 individuals in two batches. In three populations, the number of released animals is not provided, but populations persisted for at least three months and up to at least three years after release. Animals in all seven populations bred in the wild. Releases took place in 1993–2000 into woodlands in Cambridgeshire, Nottinghamshire, Cheshire, Warwickshire, Buckinghamshire, Yorkshire and Suffolk. Monitoring continued until 2000–2002. Precise numbers and origins of dormice released are not given for all sites. Most were captive-bred but some were wild-born translocated animals. Some dormice were kept in pre-release holding pens, sometimes for several weeks, before release. Nest boxes and supplementary food were provided at least at some sites. See paper for further details.

A replicated study in 1992–2000 on two rivers in Hertfordshire, UK (10) found that a population of released captive-bred European otters *Lutra lutra* persisted for over eight years after release. Eight years after release of six captive-bred otters into rivers with no otter populations, otters were still detected in the release area. Over this time, the range used by released otters expanded, but some of this may have been due to natural recolonization. At least one otter died during the study period. In October–December 1991, six captive-bred otters were released in two rivers with no known otter populations. Individuals were approximately two years old when released. The range and persistence of the populations were assessed by surveying droppings through to February 2000.

A review of 14 releases of six species of captive-bred mammals in Western Australia, Australia (11) found that where outcomes were available for release programmes, over half were regarded as successful. One out of two releases of rufous hare-wallabies *Lagorchestes hirsutus*, one out of two of dibblers *Parantechinus apicalis* and one out of four of western quolls *Dasyurus geoffroii* were classed as successful. However, the only release of banded hare-wallabies *Lagostrophus fasciatus* and one out of two releases of rufous hare-wallabies *Lagorchestes hirsutus* were classed as unsuccessful. At the time of the review, the outcomes of two releases of bilbies *Perameles lagotis*, three of western quolls, one of dibblers and three of Shark Bay mouse *Pseudomys fieldi* remained uncertain. In 1993–2002, sixteen to 149 captive-bred mammals were released per location. One translocation of Shark Bay mouse was partially sourced from wild stock. Invasive mammals were controlled at some release sites. The definition of successful reintroduction was not stated for most species but, for others, it included measures of population increase and persistence.

A study in 2002–2005 in two wetland areas in the Netherlands (12) found that following release of captive-bred animals, together with the release of some translocated individuals, over half of Eurasian otters *Lutra lutra* settled in their release areas and some successfully reproduced. After three weeks, 14 of 23 otters settled within their release areas, while two died and seven moved away from release areas. Three years after the first translocations, five female otters had successfully reproduced, producing nine young. At this time, the total population was 12 otters. In 2002, fifteen wild-caught otters were released at one site. At a second site, in 2004–2005, eight animals, comprising a mix of wild-caught and captive-bred individuals, were released. Before release, animals were fitted with radio-transmitters and DNA samples were taken. Following release, otters were monitored by radio-tracking and by collection of faeces, which was analysed to identify individuals.

A study in 1998–2005 at a prairie grassland site in Montana, USA (13) found that following releases of captive-reared swift foxes *Vulpes velox*, a population became established and grew. One year after releases finished, there were 62 animals, increasing to 93 animals two years later. From 50 to 100% of mature female swift foxes reproduced each year, producing 4–5 offspring. Five to seven years after reintroductions, adult swift fox annual survival was 60–73%, and that of young swift foxes was 69–77%. Of the 33 animals that died during the study, 26 were killed by coyotes *Canis latrans* or birds of prey. In 1998–2002, one-hundred and twenty-three captive-reared swift foxes were released in the Blackfeet Indian Reservation. In 2003–2005, twenty-three adult and 35 juvenile foxes were trapped and radio-collared. They were then tracked weekly, until 2005.

A review of studies conducted in 1985–2005 at 11 grassland and dry savanna sites in Eastern Cape, South Africa (14) found that reintroductions (mainly through translocations but including some captive-bred animals) of large carnivores led to increasing population sizes for four of six species. Twenty years after the first releases, there were 56 lions *Panthera leo* at seven sites (from 31 released), 41 cheetahs *Acinonyx jubatus* (seven sites, 40 released), 24 African wild dogs *Lycaon pictus* (two sites, 11 released) and 13 spotted hyena *Crocuta crocuta* (three sites, 11 released). There were reductions or unknown trends in two species with seven known surviving leopards *Panthera pardus* (five sites, 15 released) and an unknown number of servals *Leptailurus serval* (though known to be present — two sites, 16 released). Releases were made in 1985–2005, into 11 protected areas. Most schemes involved translocations of wild-caught animals but at least one of seven lion reintroductions involved captive-bred animals. Monitoring methods are not specified.

A replicated study in 1998–2004 of woodland at three sites in Corsica, France (15) found that captive-bred Corsican red deer *Cervus elaphus corsicanus*, released following extinction on the island, increased in number at all three sites. At one site, following two releases, four years apart, totalling 35 founders, there were 100 deer two years after the second release. At a second site, 24 founders grew to 60 animals over seven years. Twenty-seven founders released at a third site increased to 40 animals later that year. Corsican red deer became extinct on Corsica in 1970. Captive populations of deer, sourced from Sardinia, were established at three sites on Corsica from 1985 onwards, to provide animals for reintroductions. From 7, 14 and 17 founders, captive populations in enclosures grew and were artificially restricted to 35 each at two sites and 50 at the third site (each equating to 3.2 deer/ ha). Releases from the captive populations took place in February and March of 1998–2004 and the wild population was then estimated at each site later in 2004.

A review in 2008 of 49 studies in 1990–2006 of carnivore reintroductions in Africa, Europe, and North America (16) found that captive-bred animals released into the wild had lower survival than did wild-born translocated animals. Survival of captive-born carnivores following release (32%) was lower than survival of wild-born translocated animals (53%). The review analysed 20 reintroductions of 983 captive-bred carnivores and 29 reintroductions of 1,169 wild-caught carnivores. Post-release monitoring ranged in duration from 6 to 18 months.

A replicated study in 2003–2007 at two mountain sites on Vancouver Island, Canada (17) found that released captive-born Vancouver Island marmots *Marmota vancouverensis* had lower annual survival rates than wild-born marmots, and those released at two years old were more likely to survive than those released as yearlings. The average annual post-release survival rate of captive-bred marmots (61%) was lower than that of wild-born marmots (85%). Captive-bred marmots released at the age of two or more years had higher annual survival rates (77%) than those released as yearlings (60%). In 2003–2007, ninety-six captive-born Vancouver Island marmots were released at two sites. The released marmots were radio-tagged and monitored for a total of 154 marmot-years (one marmot-year represents one record/marmot/year). Wild-born marmots (number not reported) were also radio-tagged and monitored for 101 marmot-years in 2003–2007. All radio-tagged marmots were tracked from the ground or from a helicopter. Monitoring frequency is not stated.

A study in 2004–2006 at a grassland reserve in KwaZulu-Natal, South Africa (18) found that one of 10 captive-bred oribi *Ourebia ourebi* released into the wild survived more than two years. One captive-bred female oribi released into the wild survived for at least 27 months. Eight oribi died, six within one month of release and three within eight months. One oribi was taken back into captivity with a broken leg. Two of the eight animals that died were predated, two were poached, one died in cold weather and the cause of death in three cases was unknown. In April 2004, ten adult oribi (four males, six females) from a private breeding facility (9 x 1–3 ha enclosures) were fitted with radio-collars and released into two grassland sites (five animals at each) within three hours of capture. In 2004–2005, the released oribi were monitored weekly during the first month and monthly after the first three months post-release.

A study in 2000–2006 in an unspecified number of riparian sites on Hiiumaa Island, Estonia (19) found that captive-bred European mink

Mustela lutreola survived up to 39 months after release into the wild. Eighty days after release, 88 of 172 released mink had survived. After 39 months, at least one released mink was still alive. Seventy-five percent of deaths were caused by predators, including foxes, dogs *Canis lupus familiaris*, and raptors. In autumn 2000–2003, one-hundred and seventy-two captive-born mink were released at the site. Fifty-four mink were fitted with radio-collars before release and were monitored for up to five months. To monitor mink survival, animals were repeatedly trapped over 39 months.

A study in 2008 along a river in northern Italy (20) found that the release of a pair of captive-bred Eurasian otters *Lutra lutra* resulted in a population that persisted for at least 11 years. Eleven years after the introduction of a pair of Eurasian otters, signs of otter presence were detected along at least three of the 10 contiguous stretches of river that were surveyed. In 1997, a pair of captive-bred otters was released at a site in an area where the species had been extirpated in the late 1980s. In June–September 2008, otter presence was monitored along 5 km of the river, in 10 stretches, each 500 m long. Monitoring entailed searches for spraints and anal secretions. Each river stretch was surveyed 8–11 times.

A study in 2002–2008 in an area of peatland, fen, woodland, ditches and lakes in the Netherlands (21) found that following release of captive-bred and translocated wild-born Eurasian otters *Lutra lutra*, the population grew. By the end of the study (1–6 years after releases), six of the released otters were known to be still alive. Fifty-four offspring from released otters or their descendants were detected during the course of the study. Most dead otters found were killed in collisions with road vehicles. Between July 2002 and November 2007, thirty otters were released. Thirteen were captive-bred and 17 were translocated, wild-caught animals. Monitoring was mostly by genetic analysis of otter spraints. A publicity campaign encouraged people to report dead otters that they found. These were examined to establish cause of death.

A review of studies in 1991–2008 at 11 grassland sites in the USA and Mexico (22) found that most captive-bred (with some translocated) black-footed ferret *Mustela nigripes* releases were unsuccessful at maintaining a population, but success was higher where prey was abundant over larger areas. Of 11 reintroduction sites, populations of more than 30 adult black-footed ferrets were maintained at four sites over two years without further reintroductions. Two sites no longer contained ferrets by December 2008, and the other five sites only had small populations or were supplemented by further releases. Sites where populations were maintained tended to have more prairie dogs *Cynomys* spp., the main prey species of black-footed ferrets, covering a larger area (at least 4,300 ha) and with a higher density of animals (data presented as index of prairie dog abundance). From 1991–2008, around 2,964 captive-bred and 157 translocated wild ferrets were released at 18 sites in multiple releases. The study reports success of the 11 sites where initial releases occurred before 2003. Sites received on average over 200 ferrets over 10 years. Ferrets were monitored by annual spotlight surveys to locate, capture and uniquely mark individuals.

A study in 2009–2010 of a woodland area and adjacent escarpment in Victoria, Australia (23) found that most captive-bred brush-tailed rock wallabies *Petrogale penicillata* survived for at least five months after release and established stable home ranges. Four animals from five released were alive at least five months after release. One animal died two months after release, from undetermined causes. Additionally, three animals from an earlier release that were alive 11 months after release all survived to at least 16 months after release. Rock-wallabies established stable home ranges of 16.2–41.5 ha in extent, with core areas of 1.2–4.5 ha. Five captive-bred brush-tailed rock-wallabies were released in October 2009. Three from a release in November 2008 that were still alive in October 2009 were also monitored. Wallabies were monitored by radio-tracking, through October 2009 and for two weeks in March 2010.

A replicated study in 1996–1997 in three grassland sites in South Dakota, USA (24) found that over half of released captive-bred black-footed ferrets *Mustela nigripes* survived more than two weeks. At each of the three sites, 48% (12 of 25), 50% (9 of 18) and 89% (32 of 36) of captive-bred ferrets released into the wild survived for at least two weeks (long term survival is not reported). Overall, 53 out of 79 captive-bred black-footed ferrets (67%) survived more than two weeks after release into the wild. Twenty-four ferrets were killed by native predators (mostly great-horned owls *Bubo virginianus* and coyotes *Canis latrans*) and the cause of death of two others could not be determined. A total of 79 captive-bred black-footed ferrets were released across three mixed-grass prairie sites

(18–36 ferrets/site) in September–October 1996 and October–November 1997. Between 18 and 35 individuals were released at each site. Each of the 79 ferrets was radio-tagged and tracked every 5–30 min/night for two weeks post-release in 1996–1997.

A study in 1990–2007 in a desert reserve in west-central Saudi Arabia (25) found that released captive-bred female Arabian oryx *Oryx leucoryx* survived more than 10 years and successfully reproduced, regardless of prior breeding experience. Released captive-bred female oryx lived 11–12 years in the wild. Average birth rates were similar for 'experienced' females that had given birth prior to release (0.69 calves/ year) and 'inexperienced' females that had not (0.74 calves/year). Between 1990 and 1994, a total of 76 captive-bred oryx were released, of which 36 were females aged 0.5–8.9 years (numbers of experienced/ inexperienced mothers not specified). Animals were identified by collars, ear-tags or ear notches. Individuals were located at least once every two weeks until 2007.

A study in 2005–2009 in a mostly agricultural area in Maciejowice, Poland (26) found that approximately one third of released captive-bred brown hares *Lepus europaeus* survived for at least one year. Twenty-two of 60 hares (37%) survived for at least one year after release. Of those that died during the first year after release, males survived for an average of 57 days and females for an average of 64 days. Deaths were due to predation (31%), poaching (13%) and road kills (7%), with the remainder (49%) disappearing or dying of unknown causes. Seventy-eight brown hares bred in a 20-ha open-field enclosure were released in a landscape comprising cultivated fields, floodbanks, forest, orchards and meadows. The hares (at least six months old) were released in groups of 18–30 individuals in November 2005, 2006 and 2007. Sixty radio-collared hares (15–29 hares/group) were tracked 3–7 times/week for 1–2 years after release in 2005–2009.

A study in 2002–2011 of forest on two islands in Florida, USA (27) found that released captive-bred Key Largo woodrats *Neotoma floridana smalli* had a lower survival rate than did wild-born, wild-living animals. From 40 captive-bred woodrats radio-tracked for an average of 49 days, 33 (67%) deaths were recorded. From 58 wild-born, wild-living woodrats radio-tracked for an average of 80 days, ten (6%) deaths were recorded. All but one death, from both groups combined, was thought to be due to predation. Adult captive-bred woodrats were released on two islands between February 2010 and December 2011. They were located at least every second day by radio-tracking, for up to four months. Nineteen adult wild-born woodrats were radio-tracked at least three times/week from March to December 2002 and 39 were radio-tracked 2–5 times/week, from June 2005 to February 2006.

A review of translocations carried out in 1969–2006 in Australia (28) found that releasing captive-bred and wild-born translocated macropod species (kangaroos and allies) led to the successful establishment of populations in 44 of 72 cases, of which 29 survived for over five years. Of the established populations, 29 persisted for more than five years. Of the 28 releases considered to be failures, 17 were thought to have failed due to predation by non-native carnivores, such as red foxes *Vulpes vulpes*. Releases considered in the review included both wild-caught translocated animals and captive-bred animals. The number of animals released ranged from one to 70 and included 20 different macropod species. Only translocations where animals were released into areas larger than 100 ha were considered for the review.

A study in 2002–2007 on prairie in South Dakota, USA (29) found that post-release survival rates of captive-bred swift fox *Vulpes velox* pups were lower than survival rates of wild-born pups. The proportion of captive-bred pups that survived for 60 days after release (48%) was lower than the proportion of wild-born pups that survived for 60 days (100%). Forty-three pups (26 male, 17 female) born in pens to wild-caught foxes formed the captive-bred cohort. They were released in mid-July of 2003–2007. Survival was compared, using radio-telemetry and visual observations at dens, to that of 90 pups born in the wild in 2003–2007, to previously translocated and released foxes.

A controlled study in 2008–2010 in a mountain site in the Central Apennines, Italy (30) found that released captive-born Apennine chamois *Rupicapra pyrenaica ornata* survived and reproduced in similar numbers to wild-caught translocated chamois, but captive-born chamois remained closer to the release site. Seven of eight captive-born (88%) and seven of eight (88%) wild-caught translocated Apennine chamois survived over five months after release. Four of five captive-born (80%) and three of five wild-caught translocated (60%) female chamois reproduced in the first year after release. During the first week after release, captive-born chamois remained closer to the release site (within 1.1 km on average) than wild-caught chamois (average 1.8 km). Eight captive-born chamois (2.5–11.5 years old, five females and three males) and eight wild-caught translocated chamois (2.5–10.5 years old, five females and three males) were released into Sibillini Mountains National Park. Chamois were released in groups of one-three individuals; each group was all wild or all captive-born. Captive-born chamois were bred in large enclosures within four national parks. Translocated chamois were taken from a national park approximately 200 km away. All of the 16 released chamois were fitted with radio-collars and monitored for five months after release in 2008–2010.

A study in 1997–2016 in a grassland area in Jiangsu province, China (31) found that a population of released captive-bred Père David's deer *Elaphurus davidianus*, established and increased in number over time. From a total of 82 founders, the population increased to 325 animals by 18 years after the first of these founders were released. In 1998, seven deer were released into a 1,000-ha area in which there were no other Père David's deer. Between 2002 and 2016, a further 75 animals were released. Observations were made with binoculars and using a drone, to estimate the deer population size. No other details of monitoring were provided in the study.

(1) Carbyn L.N., Armbruster H.J. & Mamo C. (1994) The swift fox reintroduction program in Canada from 1983 to 1992. Pages 247–271 in: M.L. Bowles & C.J. Whelan (eds.) *Restoration of endangered species: conceptual issues, planning and implementation*. Cambridge University Press, Cambridge, UK.

(2) Belden R.C. & McCown J.W. (1996) *Florida panther reintroduction feasibility study. Final Report*. Study Number: 7507. Florida Game and Fresh Water Fish Commission.

(3) Sjöåsen T. (1996) Survivorship of captive-bred and wild-caught reintroduced European otters *Lutra lutra* in Sweden. *Biological Conservation*, 76, 161–165.

(4) Dunham K.M. (1997) Population growth of mountain gazelles *Gazella gazella* reintroduced to central Arabia. *Biological Conservation*, 81, 205–214.

(5) Pandini W. & Cesaris C. (1997) Home range and habitat use of roe deer (*Capreolus capreolus*) reared in captivity and released in the wild. *Hystrix, the Italian Journal of Mammalogy*, 9, 45–50.

(6) Sjöåsen T. (1997) Movements and establishment of reintroduced European otters *Lutra lutra*. *Journal of Applied Ecology*, 34, 1070–1080.

(7) Stiver W.H., Pelton M.R. & Scott C.D. (1997) Use of pen-reared black bears for augmentation or reintroductions. *Bears: Their Biology and Management*, 9, 145–150.

(8) Smeeton C. & Weagle K. (2000) The reintroduction of the swift fox *Vulpes velox* to South Central Saskatchewan, Canada. *Oryx*, 34, 171–179, https://doi.org/10.1017/s0030605300031161

(9) Bright P. & Morris P. (2002) Putting dormice (*Muscardinus avellanarius*) back on the map. *British Wildlife*, 14, 91–100.

(10) Copp G.H. & Roche K. (2003) Range and diet of Eurasian otters *Lutra lutra* (L.) in the catchment of the River Lee (south-east England) since re-introduction. *Aquatic Conservation: Marine and Freshwater Ecosystems*, 13, 65–76, https://doi.org/10.1002/aqc.561

(11) Mawson P.R. (2004) Translocations and fauna reconstruction sites: Western Shield review-February 2003. Conservation Science Western Australia, 5, 108–121.

(12) Lammertsma D., Niewold F, Jansman H., Kuiters L., Koelewijn H.P., Perez Haro M., van Adrichem M., Boerwinkel M. & Bovenschen J. (2006) Herintroductie van de otter: een succesverhaal? *De Levende Natuur*, 107, 42–46.

(13) Ausband D.E., Foresman K.R. (2007) Swift fox reintroductions on the Blackfeet Indian Reservation, Montana, USA. *Biological Conservation*, 136, 423–430, https://doi.org/10.1016/j.biocon.2006.12.007

(14) Hayward M.W., Kerley G.I.H., Adendorff J., Moolman L.C., O'Brien J., Douglas A.S., Bissett C., Bean P., Fogarty A., Howarth D. & Slater R. (2007) The reintroduction of large carnivores to the Eastern Cape, South Africa: an assessment. *Oryx*, 41, 205–214, https://doi.org/10.1017/s0030605307001767

(15) Kidjo N., Feracci G., Bideau E., Gonzalez G., Mattéi C., Marchand B. & Aulagnier S. (2007) Extirpation and reintroduction of the Corsican red deer *Cervus elaphus corsicanus* in Corsica. *Oryx*, 41, 488–494, https://doi.org/10.1017/s0030605307012069

(16) Jule K.R., Leaver L.A. & Lea S.E.G. (2008) The effects of captive experience on reintroduction survival in carnivores: a review and analysis. *Biological Conservation*, 141, 355–366, https://doi.org/10.1016/j.biocon.2007.11.007

(17) Aaltonen K., Bryant A.A., Hostetler J.A. & Oli M.K. (2009) Reintroducing endangered Vancouver Island marmots: survival and cause-specific mortality rates of captive-born versus wild-born individuals. *Biological Conservation*, 142, 2181–2190, https://doi.org/10.1016/j.biocon.2009.04.019

(18) Grey-Ross R., Downs C.T. & Kirkman K. (2009) Reintroduction failure of captive-bred oribi (*Ourebia ourebi*). *South African Journal of Wildlife Research*, 39, 34–38, https://doi.org/10.3957/056.039.0104

(19) Maran T., Podra M., Polma M. & Macdonald D.W. (2009) The survival of captive-born animals in restoration programmes — case study of the endangered European mink *Mustela lutreola*. *Biological Conservation*, 142, 1685–1692, https://doi.org/10.1016/j.biocon.2009.03.003

(20) Prigioni C., Smiroldo G., Remonti L. & Balestrieri, A. (2009) Distribution and diet of reintroduced otters (*Lutra lutra*) on the river Ticino (NW Italy). *Hystrix — Italian Journal of Mammalogy*, 20, 45–54.

(21) Koelewijn H, Perez-Haro M., Jansman H.A.H., Boerwinkel M.C., Bovenschen J., Lammertsma D.R., Niewold F.J.J. & Kuiters A.T. (2010) The reintroduction of the Eurasian otter (*Lutra lutra*) into the Netherlands: hidden life revealed by noninvasive genetic monitoring. *Conservation Genetics*, 11, 601–614, https://doi.org/10.1007/s10592-010-0051-6

(22) Jachowski D.S., Gitzen R.A., Grenier M.B., Holmes B. & Millspaugh J.J. (2011) The importance of thinking big: Large-scale prey conservation drives black-footed ferret reintroduction success. *Biological Conservation*, 144, 1560–1566, https://doi.org/10.1016/j.biocon.2011.01.025

(23) Molyneux J., Taggart D.A., Corrigan A. & Frey S. (2011) Home-range studies in a reintroduced brush-tailed rock-wallaby (*Petrogale penicillata*) population in the Grampians National Park, Victoria. *Australian Mammalogy*, 33, 128–134, https://doi.org/10.1071/am10039

(24) Poessel S.A., Breck S.W., Biggins D.E., Livieri T.M., Crooks K.R. & Angeloni L. (2011) Landscape features influence postrelease predation on endangered black-footed ferrets. Journal of Mammalogy, 92, 732–741, https://doi.org/10.1644/10-mamm-s-061.1

(25) Wronski T., Lerp H. & Ismail K. (2011) Reproductive biology and life history traits of Arabian oryx (*Oryx leucoryx*) founder females reintroduced to Mahazat as-Sayd, Saudi Arabia. *Mammalian Biology*, 76, 506–511, https://doi.org/10.1016/j.mambio.2011.01.004

(26) Misiorowska M. & Wasilewski M. (2012) Survival and causes of death among released brown hares (*Lepus europaeus* Pallas, 1778) in Central Poland. *Acta Theriologica*, 57, 305–312, https://doi.org/10.1007/s13364-012-0081-1

(27) McCleery R., Oli M.K., Hostetler J.A., Karmacharya B., Greene D., Winchester C., Gore J., Sneckenberger S., Castleberry S.B. & Mengak M.T. (2013) Are declines of an endangered mammal predation-driven, and can a captive-breeding and release program aid their recovery? *Journal of Zoology*, 291, 59–68, https://doi.org/10.1111/jzo.12046

(28) Clayton J.A., Pavey C.R., Vernes K. & Tighe M. (2014) Review and analysis of Australian macropod translocations 1969–2006. *Mammal Review*, 44, 109–123, https://doi.org/10.1111/mam.12020

(29) Sasmal I., Honness K., Bly K., McCaffery M., Kunkel K., Jenks J.A. & Phillips M. (2015) Release method evaluation for swift fox reintroduction at

Bad River Ranches in South Dakota. *Restoration Ecology*, 23, 491–498, https://doi.org/10.1111/rec.12211

(30) Bocci A., Menapace S., Alemanno S. & Lovari S. (2016) Conservation introduction of the threatened Apennine chamois *Rupicapra pyrenaica ornata*: post-release dispersal differs between wild-caught and captive founders. *Oryx*, 50, 128–133, https://doi.org/10.1017/s0030605314000039

(31) Yuan B.D., Wang L.B., Xie S.B., Ren Y.J., Liu B., Jia Y.Y., Shen H., Sun D.M. & Ruan H.H. (2017) Density dependence effects on demographic parameters-A case study of Père David's deer (*Elaphurus davidianus*) in captive and wild habitats. *Biology and Environment: Proceedings of the Royal Irish Academy*, 117, 139–144, https://doi.org/10.3318/bioe.2017.15

14.25. Captive rear in large enclosures prior to release

https://www.conservationevidence.com/actions/2507

- **Four studies** evaluated the effects of captive rearing mammals in large enclosures prior to release. Two studies were in the USA[1,2], one was in Mexico[3] and one was in Australia[4].

COMMUNITY RESPONSE (0 STUDIES)

POPULATION RESPONSE (3 STUDIES)

- **Reproductive success (1 study):** A study in Mexico[3] found that peninsular pronghorn taken from the wild and kept in a large enclosure bred successfully and the population increased, providing stock suitable for reintroductions.

- **Survival (2 studies):** A replicated, controlled study in USA[1] found that black-footed ferrets reared in outdoor pens had higher post-release survival rates than did ferrets raised indoors. A controlled study in Australia[4] found that Tasmanian devils reared free-range in large enclosures did not have greater post-release survival rates than animals from intensively managed captive-rearing facilities.

- **Condition (1 study):** A controlled study in Australia[4] found that Tasmanian devils reared free-range in large enclosures did not gain more body weight post-release compared to animals from intensively managed captive-rearing facilities.

BEHAVIOUR (1 STUDY)

- **Behaviour change (1 study)**: A controlled study in USA[2] found that captive-bred black-footed ferrets raised in large enclosures dispersed shorter distances post-release than did ferrets raised in small enclosures.

Background

Captive-bred mammals may take time to adapt to conditions in the wild post-release, making them especially vulnerable to predation, starvation and disease. If they are reared in large enclosures, with habitat that resembles natural conditions, they may develop more natural behaviour and be better able to find food and shelter in the wild, compared to those animals reared in smaller pens.

A replicated, controlled study in 1991–1996 at three grassland sites in South Dakota, Wyoming and Montana, USA (1) found that black-footed ferrets *Mustela nigripes* reared in outdoor pens had a higher survival rate after release than did ferrets raised indoors. Nine months after release, a higher proportion of black-footed ferrets that were reared in outdoor pens were still alive (20%) than of animals reared in indoor cages (2%). In 1991–1995, one hundred and ninety-one ferrets were reared in indoor cages and 58 were raised in outdoor pens. Pens were 18–280 m^2 and were stocked with white-tailed prairie dogs *Cynomys ludovicianus* (as food for ferrets and to dig burrows that were used by ferrets). Ferrets, implanted with Passive Integrated Transponder (PIT) tags, were released in August–November of 1991–1995 at three sites. In 1991–1996, each area was surveyed on at least three consecutive nights by 8–32 people, on foot or in vehicles. All ferrets located were individually identified using PIT tags.

A controlled study in 1992 in a grassland area in Wyoming, USA (2) found that captive-bred black-footed ferrets *Mustela nigripes* raised in large enclosures dispersed smaller distances and moved less after release than did ferrets raised in small enclosures. Black-footed ferrets raised in large enclosures had a lower average maximum dispersal distance during the first three days post-release (1.7 km) and lower average cumulative movement over any three-day period post-release (8.2 km) than ferrets raised in small enclosures (maximum dispersal distance: 5.6 km; average cumulative movement: 21.1 km). Between September and October 1992, twenty-five 16.5–18-week-old captive-bred black-footed ferrets were radio-tagged and released into a 20,596-ha area. Eight ferrets were born in cages but raised in 80-m² outdoor pens with prairie dog burrows and 17 were born and raised in indoor-1.5 m² cages. All ferrets were fed live prairie dogs. Ferrets were followed in October–November 1992.

A study in 1998–2003 at a captive breeding facility in Baja California Sur, Mexico (3) found that peninsular pronghorn *Antilocapra americana peninsularis* taken from the wild and kept in a large enclosure increased in number and provided a suitable resource for future reintroductions. Nine adult pronghorns and 16 fawns were captured in the wild, in 1998–2003, to establish the captive breeding herd. Births in captivity occurred from 2000, with 85 occurring up to 2003. There were 20 deaths. In 2003, the captive population stood at 90 animals. The captive breeding facility measured 1,400 × 1,850 m, with moveable internal divisions to manage animal separations where necessary. The founder animals were wild-caught. Fawns caught wild were bottle-fed until weaned. A different male was used for mating each year.

A controlled study in 2012–2015 on a forested island in Tasmania, Australia (4) found that Tasmanian devils *Sarcophilus harrisii* reared free-range in large enclosures did not have greater post-release survival rates and body weight gains compared to animals from intensively managed captive-rearing facilities. Survival of animals reared in free-range enclosures (eight of nine animals survived ≥825 days after release) did not differ from that of those reared in intensive captive facilities (18 of 19 survived ≥825 days after release). Free-range enclosure animals did not gain more body weight than did intensive captive facility animals over 440 days post-release (average 14% gain across all animals). Twenty-eight adult (c.1 year old) Tasmanian devils (13 females, 15 males) were released. Nine had been reared in free-range enclosures (22-ha pens) and 19 in intensive captive rearing facilities (which included zoos and hand-rearing).

(1) Biggins E., Godbey J.L., Hanebury L.R., Luce B., Marinari P.B., Matchett M.R. & Vargas A. (1998) The effect of rearing methods on survival of reintroduced black-footed ferrets. *The Journal of Wildlife Management*, 62, 643–653.

(2) Biggins D.E., Vargas A., Godbey J.L. & Anderson S.H. (1999) Influence of prerelease experience on reintroduced black-footed ferrets (*Mustela nigripes*). *Biological Conservation*, 89, 121–129.

(3) Cancino J., Sanchez-Sotomayor V. & Castellanos R. (2005) From the field: Capture, hand-raising, and captive management of peninsular pronghorn. *Wildlife Society Bulletin*, 33, 61–65, https://doi.org/10.2193/0091-7648(2005)33[61:ftfcha]2.0.co;2

(4) Rogers T., Fox S., Pemberton D. & Wise P. (2016) Sympathy for the devil: captive-management style did not influence survival, body-mass change or diet of Tasmanian devils 1 year after wild release. *Wildlife Research*, 43, 544–552, https://doi.org/10.1071/wr15221

14.26. Use holding pens at release site prior to release of captive-bred mammals

https://www.conservationevidence.com/actions/2510

- **Thirty-one studies** evaluated the effects of using holding pens at the release site prior to release of captive-bred mammals. Seven studies were in Australia[9,14,17,22,23,29,30], and in the USA[8,10,12,19,20,21,27], four were in the UK[1,2,3,15], three in Argentina[24,28,31], two in each of Israel[5,13], Saudi Arabia[7,25] and China[11,26] and one in each of Canada[4], Namibia[6], South Africa[16] and Germany[18].

COMMUNITY RESPONSE (0 STUDIES)

POPULATION RESPONSE (30 STUDIES)

- **Abundance (2 studies):** A study in Saudi Arabia[7] found that a population of captive-bred Arabian sand gazelles kept in holding pens prior to release nearly doubled in size over four years. A before-and-after study in China[26] found that following release of captive-bred animals from a pre-release enclosure into the semi-wild (free-roaming in summer, enclosed in winter and provided with food), Przewalski's horses increased in number.

- **Reproductive success (10 studies):** Eight studies (one replicated) and one review in the UK[1,2], Saudi Arabia[7,25], the USA[10,12], Israel[5,13] and Australia[17] found that following the use of holding pens prior to release (and in some cases provision of supplementary food), captive-bred Eurasian otters[1,2], Arabian sand gazelles[7], eastern-barred bandicoots[17], some swift foxes[10], some red wolves[12] and over 33% of Persian fallow deer[13] reproduced, Arabian gazelles[25] started breeding in the first year and the reproductive success of female Asiatic wild ass[5] increased over 10 years. A study in Australia[22] found that after being kept in a holding pen, all four mammal populations[22] released into an invasive-species-free fenced enclosure reproduced.

- **Survival (23 studies):** One of three studies (two controlled, one replicated) in the UK[15], Canada[4] and Australia[30] found that using holding pens prior to release of captive-bred (and some translocated) animals resulted in greater post-release survival for water voles[15] compared to animals released directly into the wild. The other two studies found similar survival rates for eastern barred bandicoots[30] and swift foxes[4] compared to animals released directly into the wild. A replicated study in the USA[27] found that captive-bred Allegheny woodrats kept in holding pens prior to release, had higher early survival rates than those not kept in holding pens, but overall survival rates tended to be lower than wild resident woodrats. Three studies in South Africa[16], USA[19] and Argentina[24] found that released captive-bred (and some translocated) African wild dogs[16], riparian brush rabbits[19] and guanacos[24] that spent longer in, and in one case in larger[24], holding pens had a higher survival rate. Three studies (one controlled) in Australia[9] and the USA[20,21] found that captive-bred animals kept in holding pens prior to release had similar (bridled nailtail wallabies)[9] or lower (black-footed ferret kits)[20] annual survival rate after release to that of wild-born translocated animals and lower (black-footed ferrets)[21] survival rates than resident animals. Ten studies (including one controlled, before-and-after study) and one review in

Saudi Arabia[7], the USA[10,12], Argentina[28], China[11], Israel[5,13], Australia[14,17,22] and Germany[18] found that following the use of holding pens prior to release of captive-bred animals (or in some cases captive-reared/rehabilitated, or with provision of supplementary food), four of four mammal populations[22], 19% of red wolves[12], Asiatic wild ass[5], Persian fallow deer[13], most Arabian sand gazelles[7], most swift foxes[10], eastern-barred bandicoots[17] and European mink[18] survived at least 1–10 years, over half of giant anteaters[28], hare-wallabies[14] and Père David's deer[11] survived for at least 1.5–6 months. Three studies in Namibia[6], the USA[8] and Australia[29] found that that following the use of holding pens prior to release of captive-bred or reared animals (some provided with nest boxes and/or supplementary food), red-tailed phascogales[29], most Mexican wolves[8] and African wild dogs[6] survived less than 6–12 months.

- **Condition (4 studies):** A randomized, controlled study in Australia[30] found that eastern barred bandicoots released after time in holding pens lost a similar proportion of body weight and recovered to a similar weight compared to bandicoots released directly. A controlled study in the UK[3] found that common dormice lost weight after being put into holding pens whereas wild translocated dormice gained weight. A controlled, before-and-after study in Australia[14] found that captive-bred rufous hare-wallabies placed in holding pens prior to release lost body condition in holding pens. A before-and-after study in Australia[23] found that captive-bred brush-tailed rock-wallabies placed in a holding pen prior to release maintained good health.

BEHAVIOUR (1 STUDY)

- **Behaviour change (1 study):** A controlled study in Argentina[31] found that after being kept in holding pens and provided with supplementary food, released captive-bred giant anteaters were less nocturnal in their activity patterns than released wild-born rehabilitated individuals.

Background

Holding pens at release sites (sometimes termed 'soft release') may be used to enable mammals to become accustomed to new surroundings before release. They are often enclosures containing natural habitat and enabling views of surrounding land. The technique may be employed both for releases of captive-bred mammals and for translocations of wild mammals to new sites, here we focus on the first group.

See also: *Use holding pens at release site prior to release of translocated mammals.*

This intervention does not include studies that solely document use of pens or enclosures used as part of captive-rearing processes if these are remote from release sites.

A replicated study in summer 1983–1984 at a riparian site in East Anglia, UK (1) found that captive-bred European otters *Lutra lutra* kept in a pre-release pen and provided supplementary food after release bred successfully. Footprints of at least one otter cub were found in the year after release. Otters settled near the release site, but ranged along 31.5 km of river over the first 100 days after release. In July 1983, three 18-month-old captive-bred otters (one male, two female) were released. Before release, they were held together in a pen at the release site, for an unspecified period of time. After release, supplementary food was provided in the pens for 12 days. The male otter was radio-tracked for 50 nights after release. Local bridges were monitored for 100 days after release for signs of otter faeces.

A study in 1983–1985 along river on the Norfolk-Suffolk border, UK (2) found that following the use of holding pens at release sites and short-term provision of supplementary food, released captive-bred Eurasian otters *Lutra lutra* stayed in their release area for at least two years and bred. Otters survived in the release area at least 28 months after release. Breeding was confirmed the summer after release and suspected again the following summer. Otters held in pens before release displayed similar activity periods, range sizes, and behaviours to those seen in wild otter populations. One male and two female otters (captive-bred and unrelated) were kept in a large pen with a pool where they had limited contact with humans from 10 to 18 months of age. In June 1983, at 18 months, they were moved to a 9 × 15-m pre-release pen, 10 m from a river bank, on a river island. After 20 days, the pen door was fixed open. Food was placed in the pen daily for 12 days after release. The male was radio-tracked from 5 July to 24 August 1983. Otter signs (especially spraints) were then monitored until 1985.

A controlled study in 1992 in a woodland reserve in Somerset, UK (3) found that captive-bred common dormice *Muscardinus avellanarius* lost weight after release into holding pens whereas wild-caught translocated dormice gained weight. The body mass of captive-bred common dormice decreased after release into holding pens by 0.23 g/day, whereas that of translocated wild-caught dormice increased by 0.12 g/day. After release from the holding pens, both captive-bred and wild-caught translocated dormice lost a small amount of weight (see original paper for details). The study was conducted along a 9-ha strip of woodland and scrub between 24 August and 30 September 1992. Eight captive-bred and six wild-caught dormice were held in a pre-release pen for eight nights, and then released into the wild. The pre-release pen (0.45 m wide, 0.5 m deep and 0.9 m high) was constructed from 1-cm^2 weldmesh and had food and water. Dormice were released in the same groups as they were found in nestboxes or in which they had been living in captivity. All individuals were weighed 10–14 days after release.

A replicated, controlled study in 1983–1993 in three grassland sites in Alberta, Canada (4) found that captive-bred and translocated swift foxes *Vulpes velox* released after time in holding pens had similar survival rates to those released without use of holding pens two years after release. No statistical analyses were performed. At least six out of 45 (13%) swift foxes held in pens before release survived over two years post-release, compared with at least five out of 43 (12%) released without use of holding pens. In 1983–1987, forty-five translocated swift foxes were held in pens before release. Pens (3.7 × 7.3 m) were fenced for protection from cattle. Animals were placed in pens in October–November and released between the following spring and fall. They were provided with supplementary food for 1–8 months after release. In 1987–1991, four hundred and thirty-three foxes were released without use of holding pens. Released foxes included both wild-born and captive-bred animals. All foxes released from pens and 155 of those released directly were radio-tracked, from the ground or air, for up to two years.

A study in 1982–1993 in a desert reserve in Israel (5) found that a released population of captive-reared Asiatic wild ass *Equus hemionus* spp. kept in holding pens prior to release persisted over 10 years, and the reproductive success of females increased over time. The number of adult females (≥3 years old) in the released herd was 14 in 1987 and 16 in 1993. The reproductive success of released females increased over time (first five years = 0.27; following 4–5 years = 0.74 foals/female/year). By 1993, sixty-six foals had been born in the wild, of which 24 were second or third generation. The reproductive success of wild-born females (0.81) was higher than released females (0.19) at the same age. From 1982–1987, fourteen adult females and 14 adult males aged two to six (except one 17-year-old animal) were released into a 200 km² nature reserve in the Negev Desert in four release events. Three females died immediately. Asses were sourced from zoos and maintained in a 2km² enclosure until the release program began. Before three releases, animals were kept in a holding pen for up to three months with food, water and shade. Animals were released directly into the wild in the final release. Wild asses were surveyed 2–3 times/week in the spring and summer by random visual searching from an off-road vehicle, tracking of spoor and monitoring of water sources. The population size of males is not reported.

A study in 1978–1990 on a savanna site in Namibia (6) found that released captive-bred or captive-reared African wild dogs *Lycaon pictus* held in a holding pen prior to release did not survive more than six months. None of 24 African wild dogs introduced at the site survived for more than six months. Causes of death included starvation, predation by lions *Panthera leo* and rabies. In 1978, 1989 and 1990, a total of 24 captive-bred wild dogs were released. In 1990, animals were held in an enclosure adjacent to the release site prior to release, and were vaccinated against rabies and canine distemper. While in the enclosure, wild dogs were fed daily and live springbok were released in the pen, so they could learn to hunt. Methods used for monitoring animals introduced in 1978 and 1989 were unclear. Animals introduced in 1990 were monitored for four months after release and, if dogs did not feed for 2–3 days, they were provided with a springbok carcass. The 1978 release was of captive-reared animals (details of whether or not they were born in captivity are not given). The 1989 and 1990 releases were of captive-bred animals.

A study in 1990–1994 in a desert reserve in southwest Saudi Arabia (7) found that most captive-bred Arabian sand gazelles *Gazella subgutturosa marica* kept in holding pens prior to release survived for at least four years, the population bred successfully and nearly doubled in size. Of the 164 sand gazelles released, 155 (95%) survived for at least four years. A total of 108 births were recorded in the wild and the number of sand gazelles increased to approximately 300 individuals over four years. In 1990–1993, a total of 135 sand gazelles were moved from captive-breeding facilities to a fenced 2,200-km^2 open desert steppe reserve. Before release, gazelles were kept in four 40 × 30-m quarantine enclosures for 2–3 months and then transferred to a 25-ha pre-release enclosure for 10–14 months. Twenty-five gazelle died within the enclosures before release. A total of 164 gazelle (98 translocated and 66 born in the enclosures) were released in five groups in 1991–1994. Radio-tagged individuals (number not reported) were monitored 1–2 times/week by ground telemetry and at least once each fortnight by air telemetry (dates not reported).

A study in 1998 in a grassland, shrubland and forest reserve in Arizona, USA (8) found that most captive-bred Mexican wolves *Canis lupus baileyi* kept in holding pens prior to release in groups and provided with supplementary food did not survive over eight months after release into the wild. Out of 11 captive-bred Mexican wolves released, six (55%) were illegally killed within eight months, three (27%) were returned to captivity and two (18%) survived in the wild for at least one year (long term survival is not reported). Three weeks after their release, three individuals from one family group killed an adult elk *Cervus canadensis*. Two females gave birth two months after release but only one pup survived. Eleven wolves in three family groups were released in March 1998. Before release, wolves were kept for two months in pre-release holding pens, where they were fed carcasses of native prey. Carcasses were provided as supplementary food for two months post-release when sufficient killing of prey was confirmed. The released wolves were fitted with radio-collars. No monitoring details are provided.

A study in 1996–1999 at a woodland reserve in Queensland, Australia (9) found that captive-bred bridled nailtail wallabies *Onychogalea fraenata* kept in holding pens where predators were controlled prior to release had similar average annual survival after release to that of wild-born translocated animals. Over four years, the average annual survival of captive-bred bridled nailtail wallabies (57–92%) did not differ significantly from that of wild-born translocated animals (77–80%). In 1996–1998, one hundred and twenty-four captive-bred and nine wild-born translocated bridled nailtail wallabies were released into three sites across Idalia National Park. Ten captive-bred wallabies were held in a 10-ha enclosure within the reserve for six months before release, and 85 were bred within the 10-ha enclosure. All of the 133 released wallabies were kept in a holding pen (30-m diameter) for one week at each site before release. Mammalian predators were culled at release sites. A total of 67 wallabies (58 captive-bred, nine wild-born) were radio-tagged and tracked every 2–7 days in 1996–1998. Wallabies were live-trapped at irregular intervals with 20–35 wire cage traps in 1997–1999.

A study in 1998–2001 on a grassland site in Montana, USA (10) found that after the release of captive-bred swift foxes *Vulpes velox* using holding pens prior to release, most animals survived for at least one to three years, and some successfully bred. One to three years after introduction, a maximum of 69 of the 76 reintroduced foxes were still alive. Over the three years after introduction, 24–29 cubs were born in the wild. In the summers of 1998–2000, a total of 76 foxes were held in pens at the release site and, after 10 days, were released. Twenty-four animals were radio-tracked in 1999–2001. Methods used in the study to determine mortality and breeding success were unclear.

A study in 1998–1999 in a grassland site in Jiangsu, China (11) found that following release of captive-bred animals after being held in pre-release pens, all Père David's deer *Elaphurus davidianus* survived for at least six weeks. Seven deer were released and all were still alive six weeks later. For 18 months prior to release, eight deer (one male, three female, and four immature animals) were held in a fenced enclosure. Seven deer were released into Dafeng Reserve in November 1998. One female was fitted with a radio-collar to enable location of the group. From November 1998 to April 1999, released deer were located at least three times/week.

A study in 1987–1994 in a grassland site in North Carolina, USA (12) found that following release of captive-bred animals, some of which were kept in holding pens and then provided supplementary food, 12 of 63 red wolves *Canis lupus rufus* survived for at least seven years, and some successfully reproduced. Seven years after wolves were first reintroduced, 12 of 63 translocated animals were still alive. By the same time, at least 66 pups had been born. Between October 1987 and December 1994, sixty-three captive-bred wolves were released. Twenty-nine wolves were held in pens (225 m²) on site before release (duration: 14 days-49 months), and thirty-four animals were released on arrival at the site. An unspecified number of wolves were fitted with radio-collars. From October 1987 to December 1994, wolves were radio-tracked from the ground and from an aeroplane. Monitoring frequency was not specified. Supplementary food (deer carcasses) was provided at release sites for 1–2 months after release from the ninth release onwards.

A study in 1996–2001 of a wooded valley in a reserve in the Galilee region, Israel (13) found that most captive-bred Persian fallow deer *Dama mesopotamica* kept in holding pens prior to release survived for at least five years and over one-third of females observed 1–3 years after release reproduced. Sixty of 74 (81%) captive-bred deer (13 males, 47 females) survived for at least five years post-release. Six of 15 females observed 1–3 years after release had fawns with them. A total of 124 captive-bred Persian fallow deer were released into the wild in groups of 10–19 deer in the spring and autumn during each of five years in 1996–2000. The deer were held in an 11-ha enclosure for three months before release. Seventy-four deer (57 females, 17 males) were fitted with radio-collars. Released deer were monitored for five years post-release through radio-tracking, video and direct observation.

A controlled, before-and-after study in 2001 in five shrubland sites in Western Australia, Australia (14) found that captive-bred banded hare-wallabies *Lagostrophus fasciatus* and rufous hare-wallabies *Lagorchestes hirsutus*, some of which were placed in holding pens prior to release into a fenced peninsula (with predator controls, supplementary food and water), survived at least two months after being released, although rufous hare-wallabies lost body condition while awaiting release in holding pens. After 1–2 months, 10 of 16 rufous hare-wallabies and 12 of 18 banded hare-wallabies were still alive. Overall both rufous and banded hare-wallabies recaptured had similar body conditions to when they were released regardless of whether they were initially put in holding pens, although rufous hare-wallabies lost 12% of body condition while waiting for release in holding pens (data presented as a body condition index; see paper for details). Sixteen captive-bred rufous hare-wallabies and 18 captive-born banded hare-wallabies were released at five sites in August 2001. Six rufous hare-wallabies and nine banded-hare wallabies were placed in separate 3-ha enclosures with electrified fencing for 10–19 days before release. Remaining animals were released directly into the wild. Supplementary food (kangaroo pellets, alfalfa) and water were made available to all hare-wallabies (those kept in holding pens and those not; feeding duration not given). Hare-wallabies were monitored by radio tracking (once/week for 1.5 years after release) and live-trapping (at 4 and 8–9 weeks after release). Release areas were within a fenced peninsula where multiple introduced mammals were controlled or eradicated.

A review of a study in 2001–2002 at a restored wetland in London, UK (15) found that using holding pens prior to release of captive-bred and translocated water voles *Arvicola terrestris* resulted in greater post-release survival than did releasing them directly into the wild. Voles released from pens were three times more likely to be recorded during the initial follow-up survey than were those released without use of pens (result presented as odds ratio). A total of 109 captive-bred and 38 wild-caught water voles were released in groups of 6–15 animals in May–July 2001. Prior to release, no water voles were present at the site. An unspecified number of animals were placed in an enclosure with food and shelter and allowed to burrow out at will. The remainder were released directly into the wild. Animals were monitored by live-trapping over three periods of five days, between autumn 2001 and early-summer 2002.

A study in 1995–2005 in 12 dry savanna and temperate grassland sites in South Africa (16) found that captive-bred and translocated African wild dogs *Lycaon pictus* which spent more time in holding pens had a higher survival rate after release. Wild dog families that had more time to socialise in holding pens prior to release into fenced areas had a higher survival rate than groups which spent less time in holding pens (data presented as model results). Overall, 85% of released animals and their wild-born offspring survived the first six months after release/ birth, Released animals that survived their first year had a high survival rate 12–18 months (91%) and 18–24 months (92%) after release. Between 1995 and 2005, one hundred and twenty-seven wild dogs (79 wild-caught, 16 captive-bred, 16 wild-caught but captive-raised, 16 'mixed' pups) were translocated over 18 release events into 12 sites in five provinces of South Africa. Individuals were kept in pre-release pens for an average of 212 days, but groups were given between 15 and 634 days to socialise in pens prior to release. Animals were monitored for 24 months after release, and the 129 pups which they produced after release were monitored up to 12 months of age. Forty characteristics of the individual animals, release sites and methods of release were recorded, and their impact on post-release survival was tested.

A review of eight studies in 1989–2005 in eight grassland and woodland sites in Victoria, Australia (17) found that in one study, released captive-bred eastern-barred bandicoots *Perameles gunnii*, some of which were placed in a holding pen prior to release, survived at least one year and bred. Captive-bred bandicoots, some of which were released into a holding pen prior to release into the wild survived at least one year and both pouch young and wild-born adults were observed. In total 22 captive-bred bandicoots were released into a 585 ha fenced predator-free enclosure in 2004–2005. Initially four animals were placed in a 1 ha holding pen prior to release. The remaining released animals were not placed in a holding pen prior to release. Bandicoots were released in stages in each site. Red fox *Vulpes vulpes* were controlled. Bandicoots were monitored by live-trapping but frequency and methods are not detailed.

A study in 2006–2008 in nine areas around rivers in south-west Germany (18) found that most captive-bred European mink *Mustela lutreola* kept in holding pens prior to release survived at least one year after release. Of 48 captive-bred animals released, 36 were still alive after 12 months. All animals were microchipped and 33 were fitted with radio-transmitters. For two weeks before release, mink were kept in enclosures measuring 5 × 2 m, containing small trees, branches, and small streams. In May 2006–August 2007, forty-eight animals were released. They were radio-tracked twice each day, in April 2006–May 2008. Animals not bearing transmitters were surveyed using live traps.

A study in 2001–2005 of riparian scrub at a site in California, USA (19) found that captive-bred riparian brush rabbits *Sylvilagus bachmani riparius* kept longer in holding pens at the release site before release had greater survival rates than those kept in pens for shorter times. Survival increased with duration held in soft-release pens prior to release, especially for smaller animals (result presented as model coefficient). Survival increased with time since release, with four-week post-release survival (71%) being lower than average four-weekly survival over the following eight weeks (89%). Wild rabbits taken into a captive breeding program produced 476 offspring from November 2001 to July 2005. Of these, 325 were released, in July 2002–July 2005, to unoccupied habitat within the species' historic range. They were held in soft-release pens (0.3–0.4 ha) and released after 2–20 days. Survival was monitored by radio-tracking, at least twice weekly.

A controlled study in 1999–2001 on three grassland sites in an area in South Dakota, USA (20) found that captive-born black-footed ferret *Mustela nigripes* kits initially kept in holding pens had lower survival rates after release than did wild-born translocated kits. Thirty-day post-release survival of captive-born kits (66%) was lower than that of wild-born translocated kits at the same site (94%). Annual survival was also lower for captive-born kits (females: 44%; males: 22%) than for wild-born kits (females: 67%; males: 43%). Annual survival at the donor site remained high (females: 80%; males: 51%) whilst survival of translocated and released kits was comparable with that at an unmanipulated colony (females: 59%; males: 28%). Eighteen captive-bred ferrets were released along with 18 wild-born ferrets at a site from which the species was then absent. Captive-born ferrets were transferred to outdoor conditioning pens, sited on prairie dog colonies, when about 90 days old and then released on 29 September and 13 October 1999. Wild-born ferrets were released the day after capture. All were born in 1999. Ferrets at the release site, the donor site for wild-born kits and an unmanipulated site were monitored by radio-tracking and by reading transponder chips.

A study in 1991 at a grassland site in Wyoming, USA (21) found that released captive-born black-footed ferrets *Mustela nigripes* kept in holding pens in the release site (where predators had been controlled) had higher post-release mortality than did resident wild ferrets. The estimated one-month survival rate for captive-born released ferrets

(49%) was lower than that for free-ranging wild ferrets at their ancestral site (93%). Of animals known to have died, five were predated by coyotes *Canis latrans*, one by a badger *Taxidea taxus*, one by a golden eagle *Aquila chrysaetos* and two died of starvation. Black-footed ferrets were extirpated in the wild in 1985–1986. Thirty-seven captive-bred ferrets were released in September–November 1991, when 4–6 months old, onto a white-tailed prairie dog *Cynomys leucurus* colony. Before releases, 66 coyotes and 63 badgers were removed from the site. Ferrets spent two weeks in acclimatisation cages at the reintroduction site before release. Dead prairie dogs were provided in the cage for 10 days post-release. Ferrets were monitored by radio-tracking for ≤42 days after release.

A study in 1998–2010 in a desert site in South Australia (22) found that after being kept in a holding pen, all four mammal populations released into an invasive-species-free fenced enclosure survived for eight years and bred. After being kept in a holding pen prior to release into a fenced enclosure where red foxes *Vulpes vulpes*, cats *Felis catus* and rabbits *Oryctolagus cuniculus* had been eradicated, greater stick-nest rats *Leporillus conditor*, burrowing bettongs *Bettongia lesueur*, western barred bandicoots *Perameles bougainville* and greater bilbies *Macrotis lagotis* were detected for eight years, increased their distribution range within five years and produced a second generation within two years. In 1998–2005, nine captive-bred greater bilbies, eight wild-born greater stick-nest rats, 10 wild-born burrowing bettongs, and 12 wild-born western barred bandicoots were translocated into a 14-km² invasive-species-free fenced area. Rabbits, cats and foxes were eradicated within the fenced area in 1999. Animals were released into a 10-ha holding pen before full release after a few months. Between 2000 and 2010, tracks were surveyed annually along eight 1 km × 1 m transects.

A before-and-after study in 2007–2010 of a primarily woodland and shrubland site in Victoria, Australia (23) found that captive-bred brush-tailed rock-wallabies *Petrogale penicillata* placed in a holding pen prior to release exhibited stress levels consistent with maintaining good health. Stress index values measured from blood samples of released animals, were not significantly different to those of animals held in captivity before release. For both groups, the levels indicated lower levels of stress-induced cellular damage than the animals were able to mitigate. Of 41 captive-born wallabies, 24 (aged 1.1–4.3 years) were selected, following health examinations, for transfer to a 1.3-ha pre-release enclosure. They were kept in this enclosure for 3–17 months. Shelter was provided in the enclosure but animals foraged on natural foods, except during trapping procedures. Twenty-one were then released between November 2008 and October 2010. Samples were taken from 11 that were subsequently recaptured, up to October 2010.

A study in 2007–2012 in a forest and grassland reserve in Córdoba, Argentina (24) found that captive-bred guanacos *Lama guanicoe* kept for 38–184 days in large holding pens before release had higher post-release survival than guanacos kept for 3–15 days in small holding pens. Of 25 guanacos kept for 38–184 days in large holding pens before release, 24 (96%) survived the first month of which 19 (79%) survived over one year after release. Of 113 guanacos kept for 3–15 days in small holding pens before release, only 24 (21%) survived the first month of which 17 (71%) survived over one year after release. In 2011 and 2012, twenty-five captive-bred guanacos were kept in a 20,000-m^2 holding pen for 38–184 days before release into a 24,774-ha national park. In 2007, 113 captive-bred guanacos were kept in a 1,200-m^2 holding pen and fed with alfalfa for 3–15 days before release into the same national park. Guanacos were marked and 42 individuals (6 in 2011 and 36 in 2007) were radio-tagged. Animals were monitored 2–3 times for 4–5 days during the first month post-release and 1–2 times each month for 2–3 days up to one year post-release.

A study in 2011–2014 of a dry dwarf-scrubland site in Saudi Arabia (25) found that captive-bred Arabian gazelles *Gazella arabica* kept in holding pens prior to release into a fenced reserve started breeding in the year following the first releases. Seven females gave birth in August–September of the year after the first releases and all calves survived to the year end at least. Of 49 gazelles released over three years, 10 had died by the time of the final releases. In 2011–2014, three groups of captive-born gazelles, totalling 49 animals, were released in a 2,244-km^2 fenced reserve. They were moved from a wildlife research centre and kept for 23 days to a few months in holding pens (500 × 500 m) prior to release at the reserve. Water and food were provided for three weeks following release. Released gazelles were radio-tracked from the ground and air.

A before-and-after study in 1985–2003 on a nature reserve in Xinjiang, China (26) found that following release of captive-bred animals from a pre-release enclosure into the semi-wild (free-roaming in summer, enclosed in winter and provided with food), Przewalski's horses *Equus ferus przewalskii* increased in number. The first foals were born two years after the first releases. Over the following 11 years, 107 foals were born in the semi-wild with first-year survival of 75%. At this time, released animals formed 16 groups, comprising 127 individuals. From 2001–2013, eighty-nine horses from a captive-breeding centre were held in a pre-release enclosure (20 ha) for an unspecified period of time before being released into semi-wild conditions (free-roaming except in winter, when enclosed). The founders for the captive population were sourced from zoos in Europe and North America. The release site (and adjacent areas of Mongolia) were the last refuge of Przewalski's horse, before extinction in the wild in 1969. Released animals roamed freely from spring to fall, but were kept in a coral in winter, to enable supplementary feeding and to reduce competition with domestic horse herders.

A replicated study in 2011–2012 in two forest sites in Indiana, USA (27) found that when captive-bred Allegheny woodrats *Neotoma magister* were kept in holding pens prior to release, early survival rates were higher than those not kept in holding pens, but overall survival rates of captive-bred animals tended to be lower than those of wild resident woodrats after 4–5 months. In the first 14 days after release, seven of 16 (44%) captive-bred woodrats that were not initially kept in holding pens survived, compared to nine of 13 (69%) captive-bred woodrats that were initially kept in holding pens. After 4–5 months, captive-bred woodrats not initially kept in holding pens had significantly lower survival rates (19%) than wild-born, resident woodrats (56%). The 4-5-month survival rates of captive-bred woodrats initially kept in holding pens (31%) was also lower than wild-born, resident woodrats, but not statistically significantly lower. In April–August 2011 and 2012, a total of 29 captive-bred woodrats (>90 days old) were radio-tagged and released into two unconnected wild populations. Sixteen were directly released into the wild in 2011. Thirteen were held for two weeks in wire mesh enclosures (1.2 × 2.1 × 0.6 m) with nest boxes within the release area before release in 2012. In June–August 2011 and 2012, two samples of 16 and 17 wild-born woodrats, born that year, were radio-tagged. Captive-bred and wild-born woodrats were radio-tracked 1–7 times/ week for 4–5 months after release/tagging.

A study in 2007–2014 in a grassland reserve in Corrientes Province, Argentina (28; same experimental set-up as 31) found that over half of released captive-reared or rehabilitated giant anteaters *Myrmecophaga tridactyla*, some of which were kept in holding pens and provided supplementary food, survived for at least six months. At least 18 of 31 released giant anteaters survived for a minimum of six months. Long term survival and the fate of the other 13 anteaters is not reported. In 2007–2013, thirty-one giant anteaters (18 males, 13 females; 1–8 years old) were released into a 124-km² private reserve. Hunting within the reserve was prohibited and livestock were absent. Twenty-two anteaters were wild-born but captive-reared, six were from zoos (origin not stated) and three were wild-born but rehabilitated in captivity from injuries. Of the 18 surviving anteaters, six had been released after a short period in a 0.5-ha pen at the release site and 12 after 7–30 days in a 7-ha pen. Supplementary food was provided for several weeks after release. In 2007–2014, thirteen anteaters were tracked for less than six months, and 18 were tracked for 6–46 months.

A study in 2006–2008 in a woodland and shrubland site in Northern Territory, Australia (29) found that captive-bred red-tailed phascogales *Phascogale calura* kept in pre-release pens prior to release into a fenced area with supplementary food and nest boxes survived for less than a year. Six captive-bred females survived for at least three months after release, with at least two of them carrying young. However, there were no sightings after the first year post-release, and the population is believed to have died out. Authors suggest that there may have been a shortage of tree hollows for nesting. In July 2006 and January–February 2007, thirty-two captive-bred phascogales were released into a 26-ha fenced reserve after spending either 10 days or over four months in a pre-release pen (3×6×2 or 4.5×3.0×2.2 m). Supplementary food was provided for one week after release. Feral cats were abundant outside of the fence. Eleven nest boxes were provided within 150m of the release pen. No information on monitoring is provided.

A randomized, controlled study in 2005 in a grassland and forest site in Victoria, Australia (30) found that captive-bred eastern barred bandicoots *Perameles gunnii* kept in holding pens prior to release into a fenced reserve had similar post-release survival and body weight compared to bandicoots released directly from captivity. Four out of six bandicoots (67%) released after time in holding pens survived at least 22 days after release, which was similar to the five out of six bandicoots (83%) released directly that survived this period. Maximum weight loss (released from pen: 13%; released directly: 13% loss of weight when released) and final weight 3–4 weeks after release (released from pen: 97%; released directly: 98% of weight when released) were similar. Twelve adult captive-bred bandicoots were randomly divided into two groups of six. One group was kept in a 1-ha pre-release pen (500m from the eventual release site) for one week and provided supplementary food and water and the other group was released directly from captivity. Both groups were released simultaneously into a 170-ha fenced reserve, free of invasive predators. Bandicoots were radio-tracked daily, and were trapped and weighed every 4–5 days, for one month.

A controlled study in 2007–2012 in a grassland reserve in Corrientes, Argentina (31; same experimental set-up as 28) found that after being kept in holding pens and provided with supplementary food, captive-bred giant anteaters *Myrmecophaga tridactyla* released into the wild were less nocturnal in their activity patterns than were wild-born rehabilitated individuals. Captive-bred giant anteaters were proportionally less active at night than wild-born animals (43% vs 70% of activity records were at night). During 2007–2012, three captive-bred and four wild-born adult giant anteaters were released into a 124-km² private reserve. Wild-born animals were rehabilitated after being injured by hunters or in road accidents. Six anteaters (all wild-born and two captive-bred anteaters) were released after spending a short period of time in a 0.5 ha acclimatisation pen. The remaining 12 anteaters spent 7–30 days in a 7 ha holding pen at the release site prior to release. Supplementary food was provided in the holding pen and for several weeks after anteaters were released. Each of the seven anteaters was fitted with a radio-transmitter and tracked for 1–2 x 24 h periods/month in 2007 and 2011. The released anteaters were further monitored using 14 baited camera traps for an average of 336 days/trap in 2008–2012.

(1) Wayre P. (1985) A successful reintroduction of European otters. *Oryx*, 19, 137–139.

(2) Jefferies D.J., Wayre P., Jessop R.M. & Mitchel-Jones A.J. (1986) Reinforcing the native otter *Lutra lutra* population in East Anglia: an analysis of the

behaviour and range development of the first release group. *Mammal Review*, 16, 65–79.

(3) Bright P.W. & Morris P.A. (1994) Animal translocation for conservation: performance of dormice in relation to release methods, origin and season. *Journal of Applied Ecology*, 31, 699–708.

(4) Carbyn L.N., Armbruster H.J. & Mamo C. (1994) The swift fox reintroduction program in Canada from 1983 to 1992. Pages 247–271 in: M.L. Bowles & C.J. Whelan (eds.) *Restoration of endangered species: conceptual issues, planning and implementation*. Cambridge University Press, Cambridge.

(5) Saltz D. & Rubenstein D.I. (1995) Population dynamics of a reintroduced Asiatic wild ass (*Equus hemionus*) herd. *Ecological Applications*, 5, 327–335.

(6) Scheepers J.L. & Venzke K.A.E. (1995) Attempts to reintroduce African wild dogs *Lycaon pictus* into Etosha National Park, Namibia. *South African Journal of Wildlife Research*, 25, 138–140.

(7) Haque M.N. & Smith T.R. (1996) Reintroduction of Arabian sand gazelle *Gazella subgutturosa marica* in Saudi Arabia. *Biological Conservation*, 76, 203–207.

(8) Parsons D.R. (1998) 'Green fire' returns to the Southwest: reintroduction of the Mexican wolf. *Wildlife Society Bulletin*, 26, 799–807.

(9) Pople A.R., Lowry J., Lundie-Jenkins G., Clancy T.F., McCallum H.I., Sigg D., Hoolihan D. & Hamilton S. (2001) Demography of bridled nailtail wallabies translocated to the edge of their former range from captive and wild stock. *Biological Conservation*, 102, 285–299, https://doi.org/10.1016/s0006-3207(01)00101-x

(10) Smeeton C. & Weagle K. (2001) First Swift fox, *Vulpes velox*, reintroduction in the USA: results of the first two years. *Endangered Species Update*, 18, 167–170.

(11) Hu H. & Jiang, Z. (2002) Trial release of Père David's deer *Elaphurus davidianus* in the Dafeng Reserve, China. *Oryx*, 36, 196–199, https://doi.org/10.1017/s0030605302000273

(12) Phillips M.K., Henry V.G. & Kelly B.T. (2003) Restoration of the red wolf. Pages 272–288 in D.L. Mech & Luigi Boitani (eds.) *Wolves: behavior, ecology, and conservation*. University of Chicago Press, Chicago.

(13) Bar-David S., Saltz D., Dayan T., Perelberg A. & Dolev A. (2005) Demographic models and reality in reintroductions: Persian fallow deer in Israel. *Conservation Biology*, 19, 131–138, https://doi.org/10.1111/j.1523-1739.2005.00371.x

(14) Hardman B. & Moro D. (2006) Optimising reintroduction success by delayed dispersal: is the release protocol important for hare-wallabies? *Biological Conservation*, 128, 403–411, https://doi.org/10.1016/j.biocon.2005.10.006

(15) Mathews F., Moro D., Strachan R., Gelling M. & Buller N. (2006) Health surveillance in wildlife reintroductions. *Biological Conservation*, 131, 338–347, https://doi.org/10.1016/j.biocon.2006.04.011

(16) Gusset M., Ryan S.J., Hofmeyr M., van Dyk G., Davies-Mostert H.T., Graf J.A., Owen C., Szykman M., Macdonald D.W., Monfort S.L., Wildt D.E., Maddock A.H., Mills M.G.L., Slotow R. & Somers M.J. (2008) Efforts going to the dogs? Evaluating attempts to re-introduce endangered wild dogs in South Africa. *Journal of Applied Ecology*, 45, 100–108, https://doi.org/10.1111/j.1365-2664.2007.01357.x

(17) Winnard A.L. & Coulson G. (2008) Sixteen years of eastern barred bandicoot *Perameles gunnii* reintroductions in Victoria: a review. *Pacific Conservation Biology*, 14, 34–53.

(18) Peters E., Brinkman I., Krüger F., Zwirlein S. & Klaumann, I. (2009) Reintroduction of the European mink *Mustela lutreola* in Saarland, Germany: Preliminary data on the use of space and activity as revealed by radio-tracking and live-trapping. *Endangered Species Research*, 10, 305–320.

(19) Hamilton L.P., Kelly P.A., Williams D.F., Kelt D.A. & Wittmer H.U. (2010) Factors associated with survival of reintroduced riparian brush rabbits in California. *Biological Conservation*, 143, 999–1007.

(20) Biggins D.E., Godbey J.L., Horton B.M. & Livieri T.M. (2011a) Movements and survival of black-footed ferrets associated with an experimental translocation in South Dakota. *Journal of Mammalogy*, 92, 742–750.

(21) Biggins D.E., Miller B.J., Hanebury L.R. & Powell R.A. (2011b) Mortality of Siberian polecats and black-footed ferrets released onto prairie dog colonies. *Journal of Mammalogy*, 92, 721–731.

(22) Moseby K.E., Read J.L., Paton D.C., Copley P., Hill B.M. & Crisp H.A. (2011) Predation determines the outcome of 10 reintroduction attempts in arid South Australia. *Biological Conservation*, 144, 2863–2872.

(23) Schultz D.J., Rich B.G., Rohrig W., McCarthy P.J., Mathews B., Schultz T.J., Corrigan T. & Taggart D.A. (2011) Investigations into the health of brush-tailed rock-wallabies (*Petrogale penicillata*) before and after reintroduction. *Australian Mammalogy*, 33, 235–244.

(24) Barri F.R. & Cufré M. (2014) Supervivencia de guanacos (*Lama guanicoe*) reintroducidos con y sin período de preadapatación en el parque nacional Quebrada del Condorito, Córdoba, Argentina [Survival of reintroduced guanacos (*Lama guanicoe*), with and without pre-adaptation period, in the Quebrada del Condorito national park, Córdoba, Argentina]. *Mastozoología Neotropical*, 21, 9–16.

(25) Islam M.Z., Shah M.S. & Boug A. (2014) Re-introduction of globally threatened Arabian gazelles *Gazella arabica* (Pallas, 1766) (Mammalia: Bovidae) in fenced protected area in central Saudi Arabia. *Journal of Threatened Taxa*, 6, 6053–6060.

(26) Xia C., Cao J., Zhang H., Gao X., Yang W. & Blank D. (2014) Reintroduction of Przewalski's horse (*Equus ferus przewalskii*) in Xinjiang, China: The status and experience. *Biological Conservation*, 177, 142–147.

(27) Blythe R.M., Smyser T.J., Johnson S.A. & Swihart R.K. (2015) Post-release survival of captive-reared Allegheny woodrats. *Animal Conservation*, 18, 186–195.

(28) Di Blanco Y.E., Jiménez Pérez I. & Di Bitetti M.S. (2015) Habitat selection in reintroduced giant anteaters: the critical role of conservation areas. *Journal of Mammalogy*, 96, 1024–1035.

(29) Short J. & Hide A. (2015) Successful reintroduction of red-tailed phascogale to Wadderin Sanctuary in the eastern wheatbelt of Western Australia. *Australian Mammalogy*, 37, 234–244.

(30) de Milliano J., Di Stefano J., Courtney P., Temple-Smith P. & Coulson G. (2016) Soft-release versus hard-release for reintroduction of an endangered species: an experimental comparison using eastern barred bandicoots (*Perameles gunnii*). *Wildlife Research*, 43, 1–12.

(31) Di Blanco Y.E., Spørring K.L. & Di Bitetti M.S. (2017) Daily activity pattern of reintroduced giant anteaters (*Myrmecophaga tridactyla*): effects of seasonality and experience. *Mammalia*, 81, 11–21.

14.27. Provide live natural prey to captive mammals to foster hunting behaviour before release

https://www.conservationevidence.com/actions/2518

- **Three studies** evaluated the effects of providing live natural prey to captive mammals to foster hunting behaviour before release. One study was in Spain[1], one was in the USA[2] and one was in Botswana[3].

COMMUNITY RESPONSE (0 STUDIES)

POPULATION RESPONSE (2 STUDIES)

- **Survival (2 studies):** Two studies in Spain[1] and Botswana[3] found that a rehabilitated Iberian lynx[1] and wild-born but captive-reared orphaned cheetahs and leopards[3] that were provided with live natural prey in captivity survived for between at least three months and 19 months after release.

BEHAVIOUR (1 STUDY)

- **Behaviour change** (**1 study**): A controlled study in the USA[2] found that captive-bred black-footed ferrets fed on live prairie dogs took longer to disperse after release but showed greater subsequent movements than did ferrets not fed with live prairie dogs.

Background

Predatory mammals held in captivity, either for rearing prior to release or for rehabilitation following injury or illness, may lose or not fully develop natural hunting abilities. This may reduce their chance of survival after release. Providing live prey to such animals in captivity may help them to retain or develop essential hunting skills.

A study in 1991–1992 in a shrubland and grassland site in Sierra Morena, Spain (1) found that a rehabilitated Iberian lynx *Lynx pardinus* that was provided with live natural prey to foster hunting behaviour survived at least three months after release. The lynx was still alive at least 93 days after release, and locations of the radio-collar suggested it had established a 220-ha territory. On 6 July 1991, a wounded male Iberian lynx kitten (approximately four months old, weighing 2.0 kg) was brought into captivity. The wounds were treated and after 43 days the lynx was moved to a 5 × 5 m outdoor enclosure. The lynx was initially fed dead prey but, after 15 days in the enclosure, it was given live rabbits *Oryctolagus cuniculus*. After 112 days the animal (weight = 4.9 kg) was fitted with a radio-collar and moved to a 1-ha enclosure where 100 live rabbits had been released. After 83 days in this enclosure, on 2 March 1992, the animal (weight = 6.0 kg) was released in a pine stand, 9 km from where it was originally found. It was monitored daily until the collar dropped off.

A controlled study in 1992 in a grassland area in Wyoming, USA (2) found that captive-bred black-footed ferrets *Mustela nigripes* fed on live white-tailed prairie dogs *Cynomys leucurus* took longer to disperse after release but showed greater subsequent movements than did black-footed ferrets not fed with live prairie dogs. Results were not tested for statistical significance. Black-footed ferrets fed on live prairie dogs dispersed less on average during the first three days post-release (5.6 km) than did those with no experience with live prairie dogs (7.9 km). However, they had a greater average cumulative movement over any three-day period (21.2 km) than did those without live prairie dog experience (15.6 km). Between September and October 1992, twenty-nine 16.5–18-week-old captive-bred black-footed ferrets were radio-tagged and released into a 20,596-ha site. Seventeen ferrets had been fed live white-tailed prairie dogs weekly at 13–16 weeks and 12 had no experience with live prairie dogs. All ferrets were born and raised in indoor 1.5-m² cages. Ferrets were radio-tracked in October-November 1992.

A study in 2005–2009 in three dry savannah sites in Botswana (3) found that after being provided with live prey during captive rearing, orphaned cheetah *Acinonyx jubatus* and leopard *Panthera pardus* cubs successfully hunted live prey after release and survived for between 7 months and at least 19 months. All three cheetahs survived on naturally hunted prey after release. However, they were all shot and killed within seven months of release. The leopard hunted live prey, and remained alive 19 months after release. Three 3–6-month-old, wild-born cheetahs were taken into a rearing facility in January–February 2005. They were fed 1.5–3.0 kg of meat, six days/week. This decreased as live and dead rabbits, poultry and wild prey was gradually introduced. After 16 months, they were moved to a 100-ha enclosure stocked with live prey, primarily impalas *Aepyceros melampus* and tsessebes *Damaliscus lunatus*. They were released seven months later. The leopard was kept from October 2006 (when six months old) and released after 18 months in a holding facility stocked with live prey. Animals were satellite-tracked until death for the cheetahs (seven months) and for 19 months for the leopard (to November 2009).

(1) Rodriguez A., Barrios L. & Delibes M. (1995) Experimental release of an Iberian lynx (*Lynx pardinus*). *Biodiversity & Conservation*, 4, 382–394.

(2) Biggins D.E., Vargas A., Godbey J.L. & Anderson S.H. (1999) Influence of prerelease experience on reintroduced black-footed ferrets (*Mustela nigripes*). *Biological Conservation*, 89, 121–129.

(3) Houser A., Gusset M., Bragg C., Boast L. & J. Somers (2011) Pre-release hunting training and post-release monitoring are key components in the

rehabilitation of orphaned large felids. *South African Journal of Wildlife Research*, 41, 11–20.

14.28. Train captive-bred mammals to avoid predators

https://www.conservationevidence.com/actions/2520

- **Two studies** evaluated the effects of training captive-bred mammals to avoid predators. One study was in Australia[1] and one was in the USA[2].

COMMUNITY RESPONSE (0 STUDIES)

POPULATION RESPONSE (1 STUDY)

- **Survival (1 study):** A randomized, controlled study in the USA[2] found that training captive-born juvenile black-tailed prairie dogs, by exposing them to predators, increased post-release survival.

BEHAVIOUR (1 STUDY)

- **Behaviour change (1 study):** A before-and-after study in Australia[1] found that rufous hare-wallabies could be conditioned to become wary of potential predators.

Background

Mammals raised in captivity, free of predators, may be poorly adapted if released into areas where they are likely to encounter predators. It may be possible to train captive animals to avoid predators once they are released. This intervention covers specifically training attempts on captive-bred mammals. For wild mammals, see: *Invasive and problematic species -Train mammals to avoid problematic species.*

A before-and-after study in 1992 on captive animals at a site in Australia (1) found that rufous hare-wallabies *Lagorchestes hirsutus* could be conditioned to become wary of potential predators. Hare-wallabies spent more time out of sight of a model of a fox *Vulpes vulpes* or cat

Felis catus after being subject to aversive conditioning (37–45%) than before (27–33%). Observations were made on 22 captive hare-wallabies. Training involved either a cat or fox model. One version appeared from a box at the same time as a loud noise and moved across the pen, accompanied by a recording of hare-wallaby alarm calls. The other model version jumped at hare-wallabies that approached to ≤3 m, with the animal squirted from a water pistol at the same time. Initial data collection was carried out over three nights, training (use of aversion techniques) was over three nights and subsequent behaviour in the presence of the model was measured on one night. Experiments were conducted in September–October 1992.

A randomized, controlled study in 2002–2003 on grassland at a captive facility and at a reintroduction site in New Mexico, USA (2) found that training captive-born juvenile black-tailed prairie dogs *Cynomys ludovicianus*, by exposing them to predators, enhanced post-release survival. Prairie dogs 'trained' using black-footed ferrets *Mustela nigripes*, red-tailed hawks *Buteo jamaicensis* and prairie rattlesnakes *Crotalus viridis* had greater survival one year post-release than did untrained prairie dogs (data not presented). During captive trials, only the hawk elicited fleeing behaviour. The rattlesnake caused trained juveniles to spend more time being vigilant and making alarm noises and to spend less time in shelters than untrained juveniles. In spring 2002, eighteen captive-born juvenile prairie dogs were randomly assigned to training or non-training groups. Both groups had four tests/week for two weeks. Each test involved either a predator stimulus for the training group (live ferret, live rattlesnake or stuffed red tailed hawk, each accompanied by prairie dog alarm calls) or a non-predator control for the untrained group (live desert cottontail *Sylvilagus audubonii*). Prairie dogs were then released into a vacant colony in June 2002. Post-release survival was determined by live-trapping.

(1) McLean I.G., Lundie-Jenkins G. & Jarman P.J. (1996) Teaching an endangered mammal to recognise predators. *Biological Conservation*, 75, 51–62.

(2) Shier D.M. & Owings D.H. (2006) Effects of predator training on behavior and post-release survival of captive prairie dogs (*Cynomys ludovicianus*). *Biological Conservation*, 132, 126–135.

14.29. Release captive-bred mammals into fenced areas

https://www.conservationevidence.com/actions/2521

- **Fourteen studies** evaluated the effects of releasing captive-bred mammals into fenced areas. Nine studies were in Australia[1,2,3,6,7,8,11,13,14] and one each was in Jordan[4], South Africa[5], the USA[9], Saudi Arabia[10] and Senegal[12].

COMMUNITY RESPONSE (0 STUDIES)

POPULATION RESPONSE (14 STUDIES)

- **Abundance (5 studies):** Four studies (one replicated) and a review in Australia[1,6,7], Jordan[4] and Senegal[12] found that after releasing captive-bred animals into fenced areas, a population of burrowing bettongs[1] increased, a population of Arabian oryx[4] increased six-fold in 12 years, a population of dorcas gazelle[12] almost doubled over four years, three populations of eastern barred bandicoot[6] initially increased and abundance of eastern barred bandicoots[7] increased.

- **Reproductive success (6 studies):** Four studies and a review in South Africa[5], Australia[6,14], Saudi Arabia[10] and Senegal[12] found that following release of captive-bred animals into fenced areas (in some cases with other associated management), African wild dogs[5], three populations of eastern barred bandicoot[6], dorcas gazelle[12] and most female black-footed rock-wallabies[14] reproduced, and Arabian gazelles[10] started breeding in the year following the first releases. A study in Australia[8] found that four of five mammal populations[8] released into a predator-free enclosure and one released into a predator-reduced enclosure reproduced, whereas two populations released into an unfenced area with ongoing predator management did not survive to reproduce.

- **Survival (10 studies):** A study in Australia[8] found that four of five mammal populations[8] released into a predator-free enclosure and one population released into a predator-reduced enclosure survived, whereas two populations released into an unfenced area with ongoing predator management did not. Six studies (one controlled before-and-after study and two replicated studies) in Australia[2,3,6,7,14] and the USA[9] found that following release of captive-bred animals into fenced areas (in some cases with other associated management), a burrowing bettong[2] population, three eastern barred bandicoot[6] populations and over half of black-footed rock-wallabies[14] survived between one and eight years, most captive-bred hare-wallabies[3] survived at least two months, at least half of black-footed ferrets[9] survived more than two weeks, and bandicoots[7] survived at five of seven sites up to three years after the last release. One study in Australia[11] found that following release into fenced areas, a captive-bred population of red-tailed phascogales[11] survived for less than a year. A study in South Africa[5] found that captive-bred African wild dogs[5] released into fenced reserves in family groups had high survival rates. A randomized, controlled study in Australia[13] found that captive-bred eastern barred bandicoots[13] released into a fenced reserve after time in holding pens had similar post-release survival compared to bandicoots released directly from captivity.

- **Condition (1 study):** A randomized, controlled study in Australia[13] found that captive-bred eastern barred bandicoots released into a fenced reserve after time in holding pens had similar post-release body weight compared to those released directly from captivity.

BEHAVIOUR (0 STUDIES)

Background

Captive-bred mammals may be released into fenced areas. This may be done to keep them within a certain area (e.g. a game reserve), or to keep predators or other problem species out of an area to increase their chances of survival. Here fenced areas refer to those that are large enough to cover the home ranges of the target species. Studies that use smaller holding or pre-release pens before releasing captive-bred mammals into the wild are covered in *Use holding pens at release site prior to release of captive-bred mammals.*

See also: *Release translocated mammals into fenced areas.*

A study in 1993–1999 on an arid peninsula in Western Australia, Australia (1) found that following release into a fenced area where invasive species had been eradicated, a population of burrowing bettongs *Bettongia lesueur* increased. In 1999, six years after initial releases, the population was estimated at 263–301 bettongs, with 340 individuals born between 1995 and 1999. The population died out due to fox incursion in 1994, but was re-established with further releases. In 1990, a 1.6-m tall wire mesh fence (with an external overhang, an apron to prevent burrowing and two electrified wires) was erected to enclose a 12-km² peninsular, within which foxes *Vulpes vulpes* and cats *Felis catus* were eliminated by poisoning in 1991 and 1995, respectively. Outside the fence foxes were controlled by biannual aerial baiting with meat containing 1080 toxin, distributed at 10 baits/km² over 200 km². From October 1993, an additional 200 baits/month were distributed along the fence and roads across the study area. Cats were controlled by trapping and poisoning in a 100 km² buffer zone. In May 1992 and September 1993, twenty-two wild-caught bettongs were transferred to an 8-ha *in-situ* captive-breeding pen. In September 1993 and October 1995, 20 wild-caught bettongs were translocated to range freely in the reserve. From 1993–1998, one hundred and fourteen captive-bred bettongs were released. Artificial warrens, supplementary food and water were provided in 1993, but not for later releases. Eighty released bettongs were radio-tagged. From 1991–1995, European rabbits *Oryctolagus cuniculus* were controlled within the fenced area using 1080 'one shot' oats. Bettongs were monitored every three months using cage traps set over two consecutive nights, at both 100-m intervals along approximately 40 km of track, and at warrens used by radio-collared individuals.

A study in 1998–2000 in an arid protected area in Western Australia, Australia (2) found that after releasing captive-bred burrowing bettongs *Bettongia lesueur* into a fenced area without predators, the population persisted for at least eight years. In 1992 an unspecified number of bettongs were released onto a 1,200-ha peninsula, fenced to exclude predators. In July 1998, February and August 1999, and February 2000, the population was surveyed using unspecified methods.

A controlled before-and-after study in 2001 in five shrubland sites in Western Australia, Australia (3) found that most captive-bred banded hare-wallabies *Lagostrophus fasciatus* and rufous hare-wallabies

Lagorchestes hirsutus released into a fenced peninsula (with predator control, supplementary food and water and, in some cases, holding pens prior to release), survived at least two months, although rufous hare-wallabies lost body condition while awaiting release in holding pens. After 1–2 months, 10 of 16 rufous hare-wallabies and 12 of 18 banded hare-wallabies were still alive. Overall both rufous and banded hare-wallabies recaptured had similar body conditions to when they were released, although rufous hare-wallabies lost 12% of body condition while waiting for release in holding pens (data presented as a body condition index; see paper for details). Sixteen captive-bred rufous hare-wallabies and 18 captive-bred banded hare-wallabies were released at five sites in August 2001. Six rufous and nine banded-hare wallabies were placed in separate 3-ha enclosures with electrified fencing for 10–19 days before being released. Remaining animals were released directly into the wild. Supplementary food (kangaroo pellets, alfalfa) and water were made available to all hare-wallabies (those kept in holding pens and those not; duration of feeding not given). Hare-wallabies were monitored by radio tracking (once/week for 1.5 years after release) and live-trapping (at 4 and 8–9 weeks after release). Release areas were within a fenced peninsula where multiple introduced mammals were controlled (cats *Felis catus* and goats *Capra hircus*) or eradicated (red fox *Vulpes vulpes*).

A study in 1978–1995 in a desert reserve in Jordan (4) found that following release into a fenced area, a population of captive-bred Arabian oryx *Oryx leucoryx* increased six-fold in 12 years. The herd numbered 186 animals in 1995, after being founded from 31 oryx in 1983. The project began in 1978, with 11 captive-bred founder animals (six females and five males) held in breeding pens. In 1983, thirty-one oryx were released from these pens into the 342-km² Shaumari Nature Reserve, but were fenced into a 22-km² sub-section of the reserve in 1984 to exclude domestic grazing animals. An additional three males were introduced in 1984. Release outside the fenced reserve was prevented by in influx of pastoralists displaced from a war zone. From 1997 to 2006, one hundred and five oryx were moved to other reserves to reduce overcrowding. By 2006, forty-three oryx remained in the reserve. Oryx numbers were obtained from the reserve records and independent reports.

A study in 1995–2005 in 12 dry savanna and temperate grassland sites in South Africa (5) found that translocated and captive-bred

African wild dogs *Lycaon pictus* released into fenced reserves in family groups had high survival rates and bred successfully. Eighty-five percent of released animals and their wild-born offspring survived the first six months after release/birth. Released animals which survived their first year had a high survival rate 12–18 months (91%) and 18–24 months (92%) after release. Additionally, groups which had more time to socialise in holding pens prior to release had higher survival rates (data presented as statistical models). Between 1995 and 2005, one hundred and twenty-seven wild dogs (79 wild-caught, 16 captive-bred, 16 wild-caught but captive-raised, 16 'mixed' pups) were translocated over 18 release events into 12 sites in five provinces of South Africa. Animals were monitored for 24 months after release, and the 129 pups which they produced after release were monitored up to 12 months of age. Forty characteristics of the individual animals, release sites and methods of release were recorded, and their impact on post-release survival was tested.

A review of eight studies in 1989–2005 in eight grassland and woodland sites in Victoria, Australia (6) found that three captive-bred eastern barred bandicoot *Perameles gunnii* populations that were released into fenced areas with associated management survived between 1 and 15 years, animals were breeding and populations increased in size at least initially. In two studies, bandicoots were released into fenced areas and populations increased for at least five years after releases began and there was evidence of breeding and wild-born pouch young maturing to adults. These populations subsequently declined to low numbers 12–15 years after the original releases began. A further population released into a fenced area survived at least one year and both pouch young and wild-born adults were observed. Of five studies where bandicoots were not released into a fenced area, one population survived over at least seven years, two populations were extinct after five years, and two populations declined and management ceased (due to low detection rates) after 9–10 years. Between 22 and 207 bandicoots were released into three fenced areas (100–585 ha) and 50 to 103 bandicoots were released into unfenced areas (85–500 ha) in 1989–2005. All bandicoots were captive-bred. Bandicoots were released in stages in each site. Red fox *Vulpes vulpes* were controlled in all three fenced areas and four of five unfenced areas. Supplementary food was provided in two of the fenced areas (in one for 6–10 days after release, the other was not specified). In most sites, bandicoots were monitored by live-trapping but frequency and methods are not detailed.

A replicated study in 1990–2001 in seven grassland, wetland and forest sites in Victoria, Australia (7) found that using predator-proof fencing alongside regular predator control increased abundance of captive-bred eastern barred bandicoots *Perameles gunnii* released into the wild and that bandicoots were recorded at five of seven sites up to three years after the last release. Greater amounts of predator control had a positive influence on the number of bandicoot signs found at each site (Sites with 0–2 methods of regular predator control: 0 bandicoots/site; sites with 3+ methods, including predator-proof fencing: 0.3–2 bandicoots/site). Bandicoot signs were found in five of the seven release sites (average 0.3–2 signs/quadrat) but no signs were detected in two sites. At each of seven sites (88–500 ha), 50–129 captive-bred eastern barred bandicoots were released between 1990 and 1999. Combinations of regular predator control methods were employed (e.g. poisoning, shooting, destruction of red fox *Vulpes vulpes* dens) differed between the sites (1 site: no predator control; 1 site: 2 methods used; 2 sites: 3 methods used (including 1 site with partial fencing); 3 sites: 4 methods used (including 1 site with full predator-proof fencing). Bandicoot signs (fresh diggings and scats) were collected at 10 randomly distributed 5-m^2 quadrats/site on two occasions in 2000–2001.

A study in 1998–2010 in a desert site in South Australia (8) found that four of five mammal populations released into a predator-free enclosure and one population released into a predator-reduced enclosure survived, increased their distribution and produced a second generation, whereas two populations released into an unfenced area with ongoing predator management did not persist. After release into a fenced enclosure where red foxes *Vulpes vulpes*, cats *Felis catus* and rabbits *Oryctolagus cuniculus* had been eradicated, greater stick-nest rats *Leporillus conditor*, burrowing bettongs *Bettongia lesueur*, western barred bandicoots *Perameles bougainville* and greater bilbies *Macrotis lagotis* were detected for eight years, increased their distribution within five years and reproduced within two years. Numbats *Myrmecobius fasciatus* were only detected for three years and did not produce a second generation. Burrowing bettongs released into a fenced enclosure with cats and rabbits but no foxes survived and increased their distribution over at least three years and produced a second generation within two years. Greater bilbies and burrowing bettongs released into an unfenced area with some predator management did not survive to produce a second generation or increase their distribution. In 1998–2005, five numbats, 106 greater stick-nest rats (6 captive-bred individuals), 30 burrowing bettongs, 12 western barred bandicoots and nine greater bilbies (all captive-bred) were released into a 14-km² invasive-species-free fenced area. Rabbits, cats and foxes were eradicated within the fenced area in 1999. All western barred bandicoots and greater bilbies, and some greater stick-nest rats (8 individuals) and burrowing bettongs (10 individuals) were put into a 10-ha holding pen before full release after a few months. All other animals were released directly into the larger fenced area. In 2004–2008, thirty-two greater bilbies and 15 burrowing bettongs were translocated to an unfenced area (200 km²) where invasive predators (cats and foxes) were managed with lethal controls and dingoes *Canis lupus dingo* were excluded by a fence on one side. In 2008, sixty-six burrowing bettongs were translocated to a 26 km² fenced area which contained small cat and rabbit populations as a result of previous eradication attempts. Between 2000 and 2010, animals were monitored using track counts, burrow monitoring and radio-tracking.

A replicated study in 1996–1997 in three grassland sites in South Dakota, USA (9) found that at least half of captive-bred black-footed ferrets *Mustela nigripes* released into fenced areas where predators were managed survived more than two weeks. At each of the three sites, 48% (12 of 25), 50% (9 of 18) and 89% (32 of 36) of captive-bred ferrets released into the wild survived for at least two weeks (long-term survival is not reported). Overall, twenty-four ferrets were killed by native predators (mostly great-horned owls *Bubo virginianus* and coyotes *Canis latrans*) and the cause of death of two others could not be determined. A total of 79 captive-bred black-footed ferrets were released across three mixed-grass prairie sites (18–36 ferrets/site) in September–October 1996 and October–November 1997. A 107 cm high electric fence was installed in each release site (creating 2 km² enclosures) and activated 1–2 weeks prior to ferrets being released. Ferrets were able to move in and out of the fenced areas. Low-to-moderate lethal coyote control took place for 2–3 weeks each year prior to ferrets being released. Each of the 79 ferrets was radio-tagged and tracked every 5–30 min/night for two weeks post-release in 1996–1997.

A study in 2011–2014 of a dry dwarf-scrubland site in Saudi Arabia (10) found that captive-bred Arabian gazelles *Gazella arabica* released into a fenced reserve after being kept in holding pens started breeding in the year following the first releases. Seven females gave birth in August–September of the year after the first releases and all calves survived to the year end at least. Of 49 gazelles released over three years, 10 had died by the time of the final releases. In 2011–2014, three groups of captive-born gazelles, totalling 49 animals, were released in a 2,244-km² fenced reserve. They were moved from a wildlife research centre and kept for 23 days to a few months in holding pens (500 × 500 m) prior to release at the reserve. Water and food were provided for three weeks following release. Released gazelles were radio-tracked from the ground and air.

A study in 2006–2015 in two woodland and shrubland sites in Western Australia and Northern Territory, Australia (11) found that following release into fenced areas, a captive-bred population of red-tailed phascogales *Phascogale calura* survived for less than a year, whereas a translocated population survived for more than five years. A population of phascogales established from wild-caught animals survived longer (>5 years) than a population established from captive-bred animals (which had been kept in pre-release pens and given supplementary food; < 1 year). Authors suggest that the unsuccessful site may also have had a shortage of tree hollows for nesting. In July 2006 and January–February 2007, thirty-two captive-bred phascogales were released into a 26-ha fenced reserve (outside which feral cats *Felis catus* were abundant) after spending either 10 days or over four months in a pre-release pen (3×6×2 or 4.5×3×2.2 m). Eleven nest boxes were provided within 150m of the release pen, and supplementary food was provided for one week after release. In April 2009 and June 2010, twenty-seven wild-caught phascogales were released into a 430-ha fenced reserve with 22 nest boxes, but with no pre-release pen or supplementary food. From November 2010–January 2013, thirteen additional boxes were installed inside (four) and outside (nine) the fenced area at this site. Phascogales were monitored after each release using radio-collaring or Elliott live traps, and through periodic monitoring of the nest boxes.

A study in 2009–2013 in a restored savanna site in Katané, Senegal (12) found that a population of captive-bred dorcas gazelle *Gazella dorcas neglecta* released into a fenced area reproduced successfully and almost doubled in number over four years. Over four years after release, the gazelle population increased from 26 to 50 individuals. Thirty-one births and 15 deaths were recorded. Twenty-three (nine male, 14 female) captive-bred dorcas gazelles were released into a fenced enclosure in March 2009 and a further three males were released in November 2010. The enclosure was initially 440 ha but was enlarged by 200 ha in 2010. Released gazelles shared the enclosure with scimitar-horned oryx *Oryx dammah*, mhorr gazelles *Nanger dama mhorr* and red-fronted gazelles *Eudorcas rufifrons*. Small and medium-sized animals, including predators, could pass through the enclosure fence. Natural vegetation was restored prior to the release. Dorcas gazelles were ear-tagged and monitored through direct observations twice daily during 2–3 surveys/ season from June 2009 to March 2013.

A randomized, controlled study in 2005 in a grassland and forest site in Victoria, Australia (13) found that captive-bred eastern barred bandicoots *Perameles gunnii* released into a fenced reserve after time in holding pens had similar post-release survival and body weight compared to bandicoots released directly from captivity. Four out of six bandicoots (67%) released after time in holding pens survived at least 22 days after release, which was similar to the five out of six bandicoots (83%) released directly that survived this period. Maximum weight loss (released from pen: 13%; released directly: 13% loss of weight when released) and final weight 3–4 weeks after release (released from pen: 97%; released directly: 98% of weight when released) were similar. Twelve adult captive-bred bandicoots were randomly divided into two groups of six. One group was kept in a 1-ha pre-release pen (500m from the eventual release site) for one week and provided supplementary food and water and the other group was released directly from captivity. Both groups were released simultaneously into a 170-ha fenced reserve, free of exotic predators. Bandicoots were radio-tracked daily, and were trapped and weighed every 4–5 days, for one month.

A study in 2011–2014 in a semi-arid area in South Australia, Australia (14) found that over half of captive-reared black-footed rock-wallabies *Petrogale lateralis* released into a large fenced area survived at least 20 months and most females reproduced. Ten (five males, five females) of 16 captive-raised black-footed rock-wallabies (63%) survived at least 20 months after release into a fenced area. All five females that survived reproduced within 2–6 months of release. Over three years, 28 births from nine females were recorded. Between March 2011 and July 2012, sixteen captive-reared black-footed rock-wallabies (eight males, eight females; 1–5 years old) were released into a 97-ha fenced area. The fence included a floppy overhang to deter predator entry. Ten of the 16 black-footed rock-wallabies were wild-born and fostered by yellow-footed rock-wallaby *Petrogale xanthopus* surrogate mothers in captivity. Introduced predators, common wallaroos *Macropus robustus* and European rabbits *Oryctolagus cuniculus* were removed from the enclosure by September 2012. Supplementary water was provided in five 8-l tanks that were monitored with camera traps in 2011–2014. Wallabies were fitted with radio-collars and tracked 1–7 times/week in 2011–2014. Trapping was carried out on seven occasions in 2011–2014.

(1) Short J. & Turner B. (2000) Reintroduction of the burrowing bettong *Bettongia lesueur* (Marsupialia: Potoroidae) to mainland Australia. *Biological Conservation*, 96, 185–196.

(2) Parsons B.C., Short J.C. & Calver M.C. (2002) Evidence for male-biased dispersal in a reintroduced population of burrowing bettongs *Bettongia lesueur* at Heirisson Prong, Western Australia. *Australian Mammalogy*, 24, 219–224.

(3) Hardman B. & Moro D. (2006) Optimising reintroduction success by delayed dispersal: is the release protocol important for hare-wallabies? *Biological Conservation*, 128, 403–411.

(4) Harding L.E., Abu-Eid O.F., Hamidan N. & al Sha'lan A. (2007) Reintroduction of the Arabian oryx *Oryx leucoryx* in Jordan: war and redemption. *Oryx*, 41, 478–487.

(5) Gusset M., Ryan S.J., Hofmeyr M., van Dyk G., Davies-Mostert H.T., Graf J.A., Owen C., Szykman M., Macdonald D.W., Monfort S.L., Wildt D.E., Maddock A.H., Mills M.G.L., Slotow R. & Somers M.J. (2008) Efforts going to the dogs? Evaluating attempts to re-introduce endangered wild dogs in South Africa. *Journal of Applied Ecology*, 45, 100–108.

(6) Winnard A.L. & Coulson G. (2008) Sixteen years of eastern barred bandicoot *Perameles gunnii* reintroductions in Victoria: a review. *Pacific Conservation Biology*, 14, 34–53.

(7) Cook C.N., Morgan D.G. & Marshall D.J. (2010) Reevaluating suitable habitat for reintroductions: lessons learnt from the eastern barred bandicoot recovery program. *Animal Conservation*, 13, 184–195.

(8) Moseby K.E., Read J.L., Paton D.C., Copley P., Hill B.M. & Crisp H.A. (2011) Predation determines the outcome of 10 reintroduction attempts in arid South Australia. *Biological Conservation*, 144, 2863–2872.

(9) Poessel S.A., Breck S.W., Biggins D.E., Livieri T.M., Crooks K.R. & Angeloni L. (2011) Landscape features influence postrelease predation on endangered black-footed ferrets. *Journal of Mammalogy*, 92, 732–741.

(10) Islam M.Z., Shah M.S. & Boug A. (2014) Re-introduction of globally threatened Arabian gazelles *Gazella arabica* (Pallas, 1766) (Mammalia: Bovidae) in fenced protected area in central Saudi Arabia. *Journal of Threatened Taxa*, 6, 6053–6060.

(11) Short J. & Hide A. (2015) Successful reintroduction of red-tailed phascogale to Wadderin Sanctuary in the eastern wheatbelt of Western Australia. *Australian Mammalogy*, 37, 234–244.

(12) Abáigar T., Cano M., Djigo C.A., Gomis J., Sarr T., Youm B., Fernández-Bellon H. & Ensenyat C. (2016) Social organization and demography of reintroduced Dorcas gazelle (*Gazella dorcas neglecta*) in North Ferlo Fauna Reserve, Senegal. *Mammalia*, 80, 593–600.

(13) De Milliano, J., Di Stefano, J., Courtney, P., Temple-Smith, P. & Coulson, G. (2016). Soft-release versus hard-release for reintroduction of an endangered species: an experimental comparison using eastern barred bandicoots (*Perameles gunnii*). *Wildlife Research*, 43, 1–12.

(14) West R., Read J.L., Ward M.J., Foster W.K. & Taggart D.A. (2017) Monitoring for adaptive management in a trial reintroduction of the black-footed rock-wallaby *Petrogale lateralis*. *Oryx*, 51, 554–563.

14.30. Provide supplementary food during/after release of captive-bred mammals

https://www.conservationevidence.com/actions/2527

- **Fifteen studies** evaluated the effects of providing supplementary food during/after release of captive-bred mammals. Four studies were in Australia[2,9,10,14], two were in each of the USA[5,8], China[7,12] and Argentina[13,15], and one was in each of Poland[1], the UK[3,4], Oman[6] and Saudi Arabia[11].

COMMUNITY RESPONSE (0 STUDIES)

POPULATION RESPONSE (14 STUDIES)

- **Abundance (5 studies):** Four studies (one replicated, one before-and-after study) and one review in Poland[1], Oman[6], China[7,12] and Australia[10] found that following provision of supplementary food (and in one case water) to released captive-bred animals, populations of European bison[1] increased more than six-fold over 20 years, Arabian oryx[6] increased over 14 years, eastern-barred bandicoots[10] increased for the first five years before declining, Père David's deer[7] increased more than six-fold over 12 years and Przewalski's horses (enclosed in winter)[12] increased over 11 years.

- **Reproductive success (9 studies):** Eight studies (including two replicated and one before-and-after study) and one review in Poland[1], the UK[3,4], China[7,12], the USA[8], Australia[2,10] and Saudi Arabia[11] found that following the provision of supplementary food (and in one case water or artificial nests) after release of captive-bred animals, some from holding pens, European bison[1], European otters[3,4], Père David's deer[7], eastern-barred bandicoots[10], Przewalski's horses[12] and some captive-bred red wolves[8] successfully reproduced, Arabian gazelles[11] started breeding in the year following releases and sugar gliders[2] established a breeding population.

- **Survival (6 studies):** Four of six studies (one controlled, before-and-after study) in the UK[4], USA[5,8], Argentina[13] and Australia[9,14] found that following the provision of supplementary food (and in one case water or nest boxes) after release of captive-bred animals, many from holding pens, 19% of red wolves[8] survived for at least seven years, Eurasian otters[4] survived for at least two years, over half the giant anteaters (some rehabilitated)[13] survived for at least six months and hare-wallabies[9] survived at least two months. Two of the studies found that red-tailed phascogales[14] survived for less than a year and most Mexican wolves[5] survived less than eight months.

BEHAVIOUR (1 STUDY)

- **Behaviour change (1 study):** A controlled study in Argentina[15] found that after being provided with supplementary food and kept in holding pens, released captive-bred giant anteaters were less nocturnal in their activity patterns than released wild-born rehabilitated individuals.

Background

Mammals that are captive-bred are especially vulnerable immediately after release. At this time, they may struggle to find natural food in an unfamiliar area. Furthermore, if the time they spend looking for food is increased, this may make them more vulnerable to predation. Hence, providing supplementary food at and after the period of release may improve longer term survival prospects.

See also: *Provide supplementary food during/after release of translocated mammals.*

A study in 1952–1973 in a mixed forest site in Białowieża, Poland (1) found that captive-bred European bison *Bison bonasus* provided with supplementary food after being released into the wild bred successfully and the population increased more than six-fold over 20 years. The population increased to 253 individuals (112 males, 141 females) during 20 years in which 38 captive-bred bison were released. A total of 316 births and 67 deaths were recorded. In 1952–1972, thirty-eight captive-bred bison were released from reserves into the western Białowieża Primeval Forest (580 km² area). Supplementary food (hay) was provided each winter. Numbers of bison and the number of births and deaths in the population were counted by observers each year in 1952–1973.

A study in 1979–1981 at a young planted native forest reserve in Victoria, Australia (2) found that released, captive-bred sugar gliders *Petaurus breviceps* provided with supplementary food and artificial nest hollows appeared to establish a breeding population. In the third year after releases began, approximately 37 sugar gliders were recorded. Of 17 females caught, 10 were over one year old. All six females that were over two years old had bred. Seven of the 32 animals caught had been wild-bred in the year after the first releases. Sugar gliders were almost all located near to where artificial nest hollows were installed and 58 of 70 were either occupied or showed signs of recent occupation. On a 130-ha island of planted native forest (trees ≤17 years old), 26 captive-bred juvenile gliders (12 male, 14 female) were released in February 1979. Thirty-four (21 male, 13 female) were released in January–February 1980. Twelve (six male, six female) were released in February 1981. Seventy artificial nest hollows (boxes, hollow branches and pipes) were installed. Supplementary food was provided at release points during winters of 1979 and 1980. Gliders were surveyed in May 1981, by live-trapping, using 54 traps for up to four nights, supplemented by sightings of animals flushed from nest hollows.

A replicated study in summer 1983–1984 at a riparian site in East Anglia, UK (3) found that captive-bred European otters *Lutra lutra* provided with supplementary food after being kept in a pre-release pen bred successfully following release. Footprints of at least one otter cub were found in the year after release. Otters settled near the release site, but ranged along 32 km of river over the first 100 days after release. In July 1983, three 18-month-old captive-bred otters (one male, two female) were released. Before release, they were held together in a pen at the release site, for an unspecified period of time. After release, supplementary food was provided in the pens for 12 days. The male otter was radio-tracked for 50 nights after release. Local bridges were monitored for 100 days after release for signs of otter faeces.

A study in 1983–1985 along a river on the Norfolk-Suffolk border, UK (4) found that following the short-term provision of supplementary food after release from holding pens, captive-bred Eurasian otters *Lutra lutra* survived at the release site for at least two years and reproduced. The otters survived in the release area at least 28 months after release. Breeding was confirmed the summer after release and suspected again the following summer. On the first night, otters were fed prior to being released. They returned to feed on the second, third and fifth to seventh nights but after that food was untouched. Spraint analysis suggested they were catching fish from the fourth night. One male and two female otters (captive-bred and unrelated) were kept in a large pen with a pool where they had limited contact with humans from

10 months to 18 months of age. In June 1983, at 18 months, they were moved to a 9 × 15-m pre-release pen, 10 m from a river bank, on a river island. After 20 days, the pen door was fixed open. Food was placed in the pen daily for 12 days after release in diminishing quantities and uneaten food was cleared away. The male was radio-tracked for 50 days from 5 July 1983. Otter signs (especially spraints) were then monitored until 1985.

A study in 1998 in a grassland, shrubland and forest reserve in Arizona, USA (5) found that most captive-bred Mexican wolves *Canis lupus baileyi* provided with supplementary food after being kept in holding pens and released in groups did not survive over eight months after release into the wild. Out of 11 captive-bred Mexican wolves released, six (55%) were illegally killed within eight months, three (27%) were returned to captivity and two (18%) survived in the wild for at least one year (long-term survival not reported). Three weeks after their release, three individuals from one family group killed an adult elk *Cervus canadensis*. Two females gave birth two months after release but only one pup survived. Eleven wolves in three family groups were released in March 1998. Before release, wolves were kept for two months in pre-release holding pens, where they were fed carcasses of native prey. Carcasses were provided as supplementary food for two months post-release when sufficient killing of prey was confirmed. The released wolves were fitted with radio-collars. No monitoring details are provided.

A study in 1982–1996 of a large desert area in Oman (6) found that a reintroduced captive-bred Arabian oryx *Oryx leucoryx* population initially provided with supplementary food and water grew in number over 14 years, but then declined, due to poaching. Oryx numbers in the wild peaked at >400 animals, 1–14 years after release of 40 animals. Poachers (capturing live animals, especially females, for international trade) then removed at least 200 oryx over the next three years. Animals were taken back into captivity to re-establish a captive breeding program. Seventeen years after releases began, the captive population was 40, and approximately 104 remained in the wild, with a high male:female sex ratio. Arabian oryx became extinct in Oman in 1972. Founders for the initial captive herd were sourced from international collections. Forty individually marked oryx were released in 1982–1995. A sample of wild-born animals was individually marked to retain the marked proportion at 20–30%. The original released herd was provided with food and water for seven months after release. Population estimates were derived from sightings using mark-recapture analysis.

A replicated study in 1985–1997 in two grassland reserves in Jiangsu and Beijing, China (7) found that captive-bred Père David's deer *Elaphurus davidianus* released into the wild and provided with supplementary food in the winter bred successfully and increased in number more than six-fold over 12 years. In one reserve, numbers of Père David's deer were more than six times higher 12 years after release (127 deer) than at the time of release (20 deer). At a second reserve, numbers were more than seven times higher 11 years after release (302 deer) than at the time of release (39 deer). Average annual birth and death rates were 53% and 9% respectively at one site, and 54% and 3% at the other. Wild offspring translocated from the first site to another fenced area in China survived at least two years post-relocation and reproduced in the second year. In 1985–1987, thirty-seven captive-bred deer were released into a reserve (60 ha). In 1986, thirty-nine captive-bred deer were released into three fenced paddocks (each 100 ha) at a second reserve. In 1992–1996, twenty-one deer from one population and 134 deer from the other were moved to other sites. Supplementary food was provided in both reserves during the winter. The deer populations were monitored for 11–12 years after release in 1985–1997. Details of monitoring methods are not provided.

A study in 1987–1994 in a grassland site in North Carolina, USA (8) found that having provided supplementary food after release (after some animals were kept in holding pens), 12 of 63 captive-bred red wolves *Canis lupus rufus* survived for at least seven years, and some animals successfully reproduced. Seven years after wolves were first reintroduced, 12 of 63 translocated animals were still alive. By the same time, at least 66 pups had been born. Between October 1987 and December 1994, sixty-three captive-bred wolves were released. Twenty-nine wolves were held in pens (225 m^2) on site before release (duration: 14 days-49 months), and 34 animals were released on arrival at the site. An unspecified number of wolves were fitted with radio-collars. From October 1987 to December 1994, wolves were radio-tracked from the ground and from an aeroplane. Monitoring frequency was not specified. Supplementary food (deer carcasses) was provided for 1–2 months after release from the ninth release onwards.

A controlled, before-and-after study in 2001 in five shrubland sites in Western Australia, Australia (9) found that most captive-bred banded hare-wallabies *Lagostrophus fasciatus* and rufous hare-wallabies *Lagorchestes hirsutus* provided with supplementary food and water (and in some cases having been in holding pens) survived at least two months after being released into a fenced peninsula where predators had been controlled. After 1–2 months, 10 of 16 rufous hare-wallabies and 12 of 18 banded hare-wallabies were still alive. Overall both rufous and banded hare-wallabies recaptured had similar body conditions to when they were released, although rufous hare-wallabies lost 12% of body condition while waiting for release in holding pens (data presented as a body condition index; see paper for details). Sixteen captive-bred rufous hare-wallabies and 18 captive-bred banded hare-wallabies were released at five sites in August 2001. Six rufous and nine banded-hare wallabies were placed in separate 3-ha enclosures with electrified fencing for 10–19 days before being released. Remaining animals were released directly into the wild. Supplementary food (kangaroo pellets, alfalfa) and water were made available to all hare-wallabies (those in holding pens and those not; duration of feeding not given). Hare-wallabies were monitored by radio tracking (once per week for 1.5 years after release) and live-trapping (at 4 and 8–9 weeks after release). Release areas were within a fenced peninsula where multiple introduced mammals were controlled or eradicated.

A review of eight studies in 1989–2005 in eight grassland and woodland sites in Victoria, Australia (10) found that in two studies where captive-bred eastern-barred bandicoots *Perameles gunnii* were given supplementary food as part of a release program, the populations survived and bred in the wild, increasing for the first five years prior to declining. Two captive-bred bandicoot populations provided with supplementary food increased for at least five years after releases began and there was evidence of breeding and wild-born pouch young maturing to adults. These populations subsequently declined to low numbers 12–15 years after the original releases began. Between 174 and 207 bandicoots were released into 100–300 ha fenced predator-free enclosures in 1989–2004. Bandicoots were released in stages in each site. Supplementary food was provided in both sites (in one for 6–10 days after release, the other was not specified). Red fox *Vulpes vulpes* were controlled in both sites. Bandicoots were monitored by live-trapping but frequency and methods are not detailed.

A study in 2011–2014 of a dry dwarf-scrubland site in Saudi Arabia (11) found that captive-bred Arabian gazelles *Gazella arabica* provided supplementary food and water after release into a fenced reserve started breeding in the year following the first releases. Seven females gave birth in August–September of the year after the first releases and all calves survived to the year end at least. Of 49 gazelles released over three years, 10 had died by the time of the final releases. In 2011–2014, three groups of captive-born gazelles, totalling 49 animals, were released in a 2,244-km² fenced reserve. They were moved from a wildlife research centre and kept for 23 days to a few months in holding pens (500 × 500 m) prior to release at the reserve. Water and food was provided for three weeks following release. Released gazelles were radio-tracked from the ground and air.

A before-and-after study in 1985–2003 on a nature reserve in Xinjiang, China (12) found that following release of captive-bred Przewalski's horses *Equus ferus przewalskii* into the semi-wild (free-roaming in summer, enclosed in winter and provided with food), animals reproduced and numbers increased. The first foals were born two years after the first releases. Over the following 11 years, 107 foals were born in the semi-wild with first-year survival of 75%. At this time, released animals formed 16 groups, comprising 127 individuals. From 2001–2013, eighty-nine horses from a captive-breeding centre were held in a pre-release enclosure (20 ha) for an unspecified period of time before being released into semi-wild conditions. Released animals roamed freely from spring to fall, but were kept in a coral in winter, to enable supplementary feeding and to reduce competition with domestic horse herders. The founders for the captive population were sourced from zoos in Europe and North America. The release site (and adjacent areas of Mongolia) were the last refuge of Przewalski's horse, before extinction in the wild in 1969.

A study in 2007–2014 in a grassland reserve in Corrientes Province, Argentina (13; same experimental set-up as 15) found that over half of released captive reared or rehabilitated giant anteaters *Myrmecophaga tridactyla*, some of which were provided supplementary food and initially kept in holding pens, survived for at least six months. At least 18 of 31 (58%) released giant anteaters survived for a minimum of six months. Long term survival and the fate of the other 13 anteaters is not reported. In 2007–2013, thirty-one giant anteaters (18 males, 13 females; 1–8 years old) were released into a 124-km² private reserve. Hunting within the reserve was prohibited and livestock were absent. Twenty-two anteaters were wild-born but captive-reared, six were from zoos (origin not stated) and three were wild-born but rehabilitated in captivity from injuries. Of the 18 surviving anteaters, six had been released after a short period in a 0.5-ha pen at the release site and 12 after 7–30 days in a 7-ha pen. Supplementary food was provided for several weeks after release. In 2007–2014, thirteen anteaters were tracked for less than six months, and 18 were tracked for 6–46 months.

A study in 2006–2008 in a woodland and shrubland site in Northern Territory, Australia (14) found that captive-bred red-tailed phascogales *Phascogale calura* that were initially given supplementary food when released into a fenced area with nest boxes, having been kept in pre-release pens, survived for less than a year. Six captive-bred females survived for at least three months after release, with at least two of them carrying young. However, there were no sightings after the first year post-release, and the population is believed to have died out. Authors suggest that there may have been a shortage of tree hollows for nesting. In July 2006 and January–February 2007, thirty-two captive-bred phascogales were released into a 26-ha fenced reserve after spending either 10 days or over four months in a pre-release pen (3×6×2 or 4.5×3×2.2 m). Supplementary food was provided for one week after release. Feral cats were abundant outside of the fence. Eleven nest boxes were provided within 150m of the release pen. No information on monitoring is provided.

A controlled study in 2007–2012 in a grassland reserve in Corrientes, Argentina (15; same experimental set-up as 13) found that after being provided with supplementary food and kept in holding pens, captive-bred giant anteaters *Myrmecophaga tridactyla* released into the wild were less nocturnal in their activity patterns than were wild-born rehabilitated individuals. Captive-bred giant anteaters were proportionally less active at night (43% activity records were at night) than wild-born animals (70% of activity records). During 2007–2012, three captive-bred and four wild-born adult giant anteaters were released into a 124-km² private reserve. Wild-born animals were rehabilitated after being injured by hunters or in road accidents. Six anteaters (all wild-born and two captive-bred anteaters) were released after spending a short period of time in a 0.5 ha acclimatisation pen. The remaining 12 anteaters spent 7–30 days in a 7-ha holding pen at the release site prior to release. Supplementary food was provided in the holding pen and for several weeks after anteaters were released. Each of the seven anteaters was fitted with a radio-transmitter and tracked for 1–2 x 24 h periods/month in 2007 and 2011. The released anteaters were further monitored using 14 baited camera traps for an average of 336 days/trap in 2008–2012.

(1) Krasinski Z.A. (1978) Dynamics and structure of European bison population in Bialowieza primeval forest. *Acta Theriologica*, 23, 3–48.

(2) Suckling G.C. & Macfarlane M.A. (1983) Introduction of the sugar glider, *Petaurus breviceps*, into re-established forest of the Tower Hill State Game Reserve, Vic.. *Australian Wildlife Research*, 10, 249–258.

(3)(4) Wayre P. (1985) A successful reintroduction of European otters. *Oryx*, 19, 137–139. Jefferies D.J., Wayre P., Jessop R.M. & Mitchel-Jones A.J. (1986) Reinforcing the native otter *Lutra lutra* population in East Anglia: an analysis of the behaviour and range development of the first release group. Mammal Review, 16, 65–79.

(5) Parsons D.R. (1998) 'Green fire' returns to the Southwest: reintroduction of the Mexican wolf. *Wildlife Society Bulletin*, 26, 799–807

(6) Spalton J.A., Lawrence M.W. & Brend S.A. (1999) Arabian oryx reintroduction in Oman: successes and setbacks. *Oryx*, 33, 168–175.

(7) Jiang Z., Yu C., Feng Z., Zhang L., Xia J., Ding Y. & Lindsay N. (2000) Reintroduction and recovery of Père David's deer in China. *Wildlife Society Bulletin*, 28, 681–687.

(8) Phillips M.K., Henry V.G. & Kelly B.T. (2003) Restoration of the red wolf. Pages 272–288 in D.L. Mech & Luigi Boitani (eds.) *Wolves: behavior, ecology, and conservation*. University of Chicago Press, Chicago.

(9) Hardman B. & Moro D. (2006) Optimising reintroduction success by delayed dispersal: is the release protocol important for hare-wallabies? *Biological Conservation*, 128, 403–411.

(10) Winnard A.L. & Coulson G. (2008) Sixteen years of eastern barred bandicoot *Perameles gunnii* reintroductions in Victoria: a review. *Pacific Conservation Biology*, 14, 34–53.

(11) Islam M.Z., Shah M.S. & Boug A. (2014) Re-introduction of globally threatened Arabian gazelles *Gazella arabica* (Pallas, 1766) (Mammalia: Bovidae) in fenced protected area in central Saudi Arabia. *Journal of Threatened Taxa*, 6, 6053–6060.

(12) Xia C., Cao J., Zhang H., Gao X., Yang W. & Blank D. (2014) Reintroduction of Przewalski's horse (*Equus ferus przewalskii*) in Xinjiang, China: The status and experience. *Biological Conservation*, 177, 142–147.

(13) Di Blanco Y.E., Jiménez Pérez I. & Di Bitetti M.S. (2015) Habitat selection in reintroduced giant anteaters: the critical role of conservation areas. *Journal of Mammalogy*, 96, 1024–1035.

(14) Short J. & Hide A. (2015) Successful reintroduction of red-tailed phascogale to Wadderin Sanctuary in the eastern wheatbelt of Western Australia. *Australian Mammalogy*, 37, 234–244.

(15) Di Blanco Y.E., Spørring K.L. & Di Bitetti M.S. (2017) Daily activity pattern of reintroduced giant anteaters (*Myrmecophaga tridactyla*): effects of seasonality and experience. *Mammalia*, 81, 11–21.

Release captive-bred/translocated mammals

14.31. Release translocated/captive-bred mammals in areas with invasive/problematic species eradication/control

https://www.conservationevidence.com/actions/2469

- **Twenty-two studies** evaluated the effects of releasing translocated or captive-bred mammals in areas with eradication or control of invasive or problematic species. Sixteen studies were in Australia[1–7,9,11,14,17–22], four were in the USA[10,12,13,16], and one in the UK[8,15].

COMMUNITY RESPONSE (0 STUDIES)

POPULATION RESPONSE (21 STUDIES)

- **Abundance (4 studies):** A replicated study in Australia[9] found that increasing amounts of regular predator control increased population numbers of released captive-bred eastern barred bandicoots. Two studies in Australia[1,4] found that following eradication[1] or control[4] of invasive species, a population of translocated and released captive-bred burrowing bettongs[1] increased and a population of translocated western barred bandicoots[4] increased over four years. A study in Australia[14] found that following the release of captive-bred bridled nailtail wallabies and subsequent predator controls, numbers increased over a three years, but remained low compared to the total number released.

- **Reproductive success (2 studies):** A study in Australia[11] found that four of five captive-bred mammal populations released into a predator-free enclosure and one population released into a predator-reduced enclosure produced a second generation, whereas two populations released into an unfenced area with ongoing predator management did not survive to reproduce. A study in Australia[22] found that most female captive-reared black-footed rock-wallabies released into a large predator-free fenced area reproduced.

- **Survival (18 studies):** Ten studies (one controlled, three replicated, two before-and-after studies) in Australia[3,4,5,6,9,17,18,22], and the UK[8,15] found that following the eradication/control of invasive species (and in some cases release into a fenced area), a translocated population of woylies[3], western barred bandicoots[4] and red-tailed phascogales[18] survived over four years, released captive-bred eastern barred bandicoots[9] survived up to three years at five of seven sites, offspring of translocated golden bandicoots[17] survived three years, over half of released captive-reared black-footed rock-wallabies[22] survived over two years, captive-bred water voles[8] survived for at least 20 months[15] or over 11 months at over half of release sites, most released captive-bred hare-wallabies[6] survived at least two months, most captive-bred eastern barred bandicoots[5,20] survived for over three weeks. A replicated study in Australia[19] found that after the control of invasive species, four translocated populations of burrowing bettongs died out within four months. A review of studies in Australia[7] found that in seven studies where red fox control was carried out before or after the release of captive-bred eastern-barred bandicoots, survival varied. A study in Australia[11] found that four of five captive-bred mammal populations released into a predator-free enclosure and one population released into a predator-reduced enclosure survived, whereas two populations released into an unfenced area with ongoing predator management did not. A study in Australia[2] found that captive-bred bridled nailtail wallabies released from holding pens in areas where predators had been controlled had similar annual survival rates to that of wild-born translocated animals. Two studies (one replicated) in the USA[10,12] found that where predators were managed, at least half of released captive-bred black-footed ferrets survived more than two weeks[12], but that post-release mortality was higher than resident wild ferrets[10]. A before-and-after study in the USA[13] found following the onset of translocations of black bears away from an elk calving site, survival of the offspring of translocated elk increased.

- **Condition (2 studies):** A study Australia[17] found that wild-born golden bandicoots, descended from a translocated population released into a predator-free enclosure, maintained genetic diversity relative to the founder and source populations. A replicated, before-and-after study in Australia[21] found that one to two years after release into predator-free fenced reserves, translocated eastern bettongs weighed more and had improved nutritional status compared to before release.

BEHAVIOUR (1 STUDY)

- **Behaviour change (1 study):** A replicated, before-and-after study in the USA[16] found that translocated Utah prairie dogs released after the control of native predators into an area with artificial burrows showed low site fidelity and different pre- and post-release behaviour.

Background

Mammals are sometimes wild-caught and translocated, or bred in captivity and released to areas where invasive predators or problematic native species have been eradicated or controlled, to re-establish populations that have been lost, or augment an existing population. Alternatively, ongoing predator control may be undertaken during and after releases. This action includes studies describing or comparing the effects of projects that release mammals after the eradication or control of invasive or problematic species, and studies where the problematic species has been controlled shortly after the release of the species of concern. However, it does not include such projects undertaken on islands, those are discussed under *Release translocated/captive-bred mammals to islands without invasive predators*.

A study in 1993–1999 on an arid peninsula in Western Australia, Australia (1) found that following eradication of invasive species from a fenced area, a released population of burrowing bettongs *Bettongia lesueur* increased. In 1999, six years after initial releases, the population was estimated at 263–301 bettongs, with 340 individuals born between 1995 and 1999. The population died out due to fox incursion in 1994, but was re-established with further releases. In 1990, a 1.6-m tall wire mesh fence (with an external overhang, an apron to prevent burrowing and two electrified wires) was erected to enclose a 12-km² peninsular, within which foxes *Vulpes vulpes* and cats *Felis catus* were eliminated by poisoning in 1991 and 1995, respectively. Outside the fence foxes were controlled by biannual aerial baiting with meat containing 1080 toxin, distributed at 10 baits/km² over 200 km². From October 1993, an additional 200 baits/month were distributed along the fence and roads across the study area. Cats were controlled by trapping and poisoning in a 100 km² buffer zone. In May 1992 and September 1993, twenty-two wild-caught bettongs were transferred to an 8-ha *in-situ* captive-breeding pen. In September 1993 and October 1995, twenty wild-caught bettongs were translocated to range freely in the reserve. From 1993–1998, one hundred and fourteen captive-bred bettongs were released.

Artificial warrens and supplementary food and water were provided in 1993, but not for later releases. Eighty released bettongs were radio-tagged. From 1991–1995, European rabbits *Oryctolagus cuniculus* were controlled within the fenced area using 1080 'one shot' oats. Bettongs were monitored every three months using cage traps set over two consecutive nights, at both 100-m intervals along approximately 40 km of track, and at warrens used by radio-collared individuals.

A study in 1996–1999 at a woodland reserve in Queensland, Australia (2) found that captive-bred bridled nailtail wallabies *Onychogalea fraenata* released from holding pens in areas where mammalian predators had been controlled had similar annual survival rates to that of wild-born translocated animals. Over four years, the average annual survival of released captive-bred bridled nailtail wallabies (57–92%) did not differ significantly from that of wild-born translocated animals (77–80%). In 1996–1998, one hundred and twenty-four captive-bred and nine wild-born translocated bridled nailtail wallabies were released into three sites across Idalia National Park. Ten captive-bred wallabies were held in a 10-ha enclosure within the reserve for six months before release, and 85 were bred within the 10-ha enclosure. All of the 133 released wallabies were kept in a holding pen (30-m diameter) for one week at each site before release. Mammalian predators were culled at release sites. A total of 67 wallabies (58 captive-bred, nine wild-born) were radio-tagged and tracked every 2–7 days in 1996–1998. Wallabies were live-trapped at irregular intervals with 20–35 wire cage traps in 1997–1999.

A study in 1992–1996 in a forest reserve in Western Australia, Australia (3) found that following baiting with poison to control red foxes *Vulpes vulpes*, a translocated population of woylies *Bettongia penicillata* persisted over four years. Four years after translocation into a site where red foxes were controlled, eight woylies were captured in one part of the site and 59 in another part. Foxes were controlled using poisoned baits started in 1985 in one part of the Boyagin Nature Reserve (4,780 ha) and in 1989 in another part of the reserve. Baits (1080-poison meat baits or intact fowl eggs) were deployed monthly. Forty woylies (28 female, 12 male) were translocated to the reserve in 1992. No further details of the translocation are provided. Woylies were live-trapped over 150 trap nights in each part of the reserve in 1996, using baited wire cage traps set at 100-m intervals. Traps were set at dusk and cleared each morning.

A study in 1995–1999 on an arid peninsula in Western Australia, Australia (4) found that following control of invasive species, a translocated population of western barred bandicoots *Perameles bougainville* persisted and increased in numbers over four years. Six out of 14 translocated western barred bandicoots (43%) survived over one month after release into a predator-free enclosure. From 51 bandicoots then released from this enclosure, the population increased to an estimated 130 individuals by two years after releases commenced. In 1995–1996, fourteen bandicoots were trapped in Dorre Island and released into a 17-ha enclosure. Invasive predators were unable to enter the enclosure and European rabbits *Oryctolagus cuniculus* and Gould's monitors *Varanus gouldii* were controlled by trapping. In 1997 and 1999, bandicoots were released from this enclosure into the larger study area, a 12-km² mainland peninsula. This was fenced to exclude alien predators, though was occasionally accessed by foxes *Vulpes vulpes* and cats *Felis catus*. Bandicoots were monitored by radio-tracking within the predator-free enclosure. Following release, they were live-trapped at three-month intervals, over 2–4 nights, on a 50-m grid.

A study in 2001 in a grassy woodland site in Melbourne, Australia (5) found that following control of red foxes *Vulpes vulpes,* and release of captive-bred animals, most eastern barred bandicoots *Perameles gunnii* survived for at least five weeks. After five weeks, seven of 10 released bandicoots were known to be alive. Despite control, red foxes were recorded in all monitoring locations. In May 2001, poison-laced baits were buried at 28 locations, 180 m apart, in an effort to control red foxes. In July 2001, ten captive-bred eastern barred bandicoots were released into a 400-ha reserve. To monitor bandicoot survival, 180 live traps, baited with oats, peanut butter and honey, were distributed over a 9-ha area. Trapping was carried out on seven occasions over a five-week period, with traps set for two consecutive days each time and with two to four days between trapping. Twenty-nine 1-m² pads, covered in sand, were placed close to vehicle tracks and the presence of fox prints was recorded every weekday, in March–August 2001.

A controlled, before-and-after study in 2001 in five shrubland sites in Western Australia, Australia (6) found that following control of introduced mammals, most captive-bred banded hare-wallabies *Lagostrophus fasciatus* and rufous hare-wallabies *Lagorchestes hirsutus* survived at least two months after being released into a fenced peninsula (some from holding pens and all with supplementary food and water provided). After 1–2 months, 10 of 16 rufous hare-wallabies and 12 of 18 banded hare-wallabies were still alive. Overall both rufous and banded hare-wallabies recaptured had similar body conditions to when they were released, although rufous hare-wallabies lost 12% of their body condition while waiting for release in holding pens (data presented as a body condition index; see paper for details). Sixteen captive-bred rufous hare-wallabies and 18 captive-bred banded hare-wallabies were released at five sites in August 2001. Six rufous hare-wallabies and nine banded-hare wallabies were placed in separate 3-ha enclosures with electrified fencing for 10–19 days before being released. Remaining animals were released directly into the wild. Supplementary food (kangaroo pellets, alfalfa) and water were made available to all hare-wallabies (those kept in holding pens and those not; feeding duration not given). Hare-wallabies were monitored by radio tracking (once/week for 1.5 years after release) and live-trapping (at 4 and 8–9 weeks after release). Release areas were within a fenced peninsula where multiple introduced mammals were controlled (cats *Felis catus* and goats *Capra hircus*) or eradicated (red fox *Vulpes vulpes*).

A review of eight studies in 1989–2005 in eight grassland and woodland sites in Victoria, Australia (7) found that in seven studies where red fox *Vulpes vulpes* control was carried out before or after the release of captive-bred eastern-barred bandicoots *Perameles gunnii*, survival rates of populations varied. In sites with fox control, two bandicoot populations increased for at least five years after releases began and there was evidence of breeding and wild-born pouch young maturing to adults. These populations subsequently declined to low numbers 12–15 years after the original releases began. A further population survived at least one year and both pouch young and wild-born adults were observed. However, two populations went extinct after five years, and two populations declined and management ceased (due to low detection rates) after 9–10 years. In a site without proactive fox control, released bandicoots survived and bred for at least seven years with the population comprising 74% wild-born offspring two years after releases began. Between 22 and 207 bandicoots were released into sites (85–585 ha) with fox control and 85 bandicoots were released a site with no proactive fox management (200 ha) in 1989–2005. Captive-bred bandicoots were released in stages in each site. Red fox *Vulpes vulpes* were controlled by shooting, use of 1080 poison bait, or a combination thereof before and/or after releases. In two sites with fox control, invasive European rabbits *Oryctolagus cuniculus* were also culled. Supplementary food was provided in two sites with fox management (in one for 6–10 days after release, the other was not specified). In most sites, bandicoots were monitored by live-trapping but frequency and methods are not detailed.

A replicated study in 2005–2008 at 12 riverside sites in the Upper Thames region, UK (8) found following American mink *Neovison vison* control, captive-bred water voles *Arvicola terrestris* survived over 11 months at more than half of release sites. Water voles persisted over 11 months at seven out of 12 sites (58%). Voles were released at 12 sites where previous populations had been eradicated due to mink predation. Sites were >5 km apart and comprised suitable riparian habitat on which mink control took place. Either 44 or 45 voles were released at each site, in early May of 2005–2007. Release sites had 20–22 predator-proof release pens. Pens were 120 × 120 cm cross section, 60 cm high and buried 15–20 cm into the ground. Food and water was provided for seven days but most voles burrowed out of pens within 2–3 days. Voles were monitored monthly for five months post-release, using live traps, 15 m apart along each site, over four days. Sites were checked for vole signs in the April after release.

A replicated study in 1990–2001 in seven grassland, wetland and forest sites in Victoria, Australia (9) found that increasing amounts of regular predator control increased population numbers of released captive-bred eastern barred bandicoots *Perameles gunnii*, and bandicoots were recorded at five of seven sites up to three years after the last release. Greater amounts of predator control had a positive influence on the number of bandicoot signs found at each site (Sites with 0–2 methods of regular predator control: 0 bandicoots/site; sites with 3+ methods: 0.3–2 bandicoots/site). Bandicoot signs were found in five of the seven release sites (average 0.3–2 signs/quadrat) but no signs were detected in two sites. At each of seven sites (88–500 ha), 50–129 captive-bred eastern barred bandicoots were released between 1990 and 1999. Combinations of regular predator control methods employed (e.g. poisoning, shooting, destruction of red fox *Vulpes vulpes* dens) differed between the sites (1 site: no predator control; 1 site: 2 methods used; 2 sites: 3 methods used (including 1 site with partial fencing); 3 sites: 4 methods used (including 1 site with full predator-proof fencing). Bandicoot signs (fresh diggings and scats) were collected at 10 randomly distributed 5-m^2 quadrats/site on two occasions in 2000–2001.

A study in 1991 at a grassland site in Wyoming, USA (10) found that following predator management, captive-born black-footed ferrets *Mustela nigripes* released from holding pens had higher post-release mortality than did resident wild ferrets. The estimated one-month survival rate for captive-born released ferrets (49%) was lower than that for free-ranging wild ferrets at their ancestral site (93%). Of animals known to have died, five were predated by coyotes *Canis latrans*, one by a badger *Taxidea taxus*, one by a golden eagle *Aquila chrysaetos* and two died of starvation. Black-footed ferrets were extirpated in the wild in 1985–1986. Thirty-seven captive-bred ferrets were released in September–November 1991, when 4–6 months old, onto a white-tailed prairie dog *Cynomys leucurus* colony. Before releases, 66 coyotes and 63 badgers were removed from the site. Ferrets spent two weeks in acclimatisation cages at the reintroduction site before release. Dead prairie dogs were provided in the cage for 10 days post-release. Ferrets were monitored by radio-tracking for ≤42 days after release.

A study in 1998–2010 in a desert site in South Australia (11) found that four of five captive-bred mammal populations released into a predator-free enclosure and one population released into a predator-reduced enclosure survived, increased their distribution and produced a second generation, whereas two populations released into an unfenced area with ongoing predator management did not persist. After release into a fenced enclosure where red foxes *Vulpes vulpes*, cats *Felis catus* and rabbits *Oryctolagus cuniculus* had been eradicated, greater stick-nest rats *Leporillus conditor*, burrowing bettongs *Bettongia lesueur*, western barred bandicoots *Perameles bougainville* and greater bilbies *Macrotis lagotis* were detected for eight years, increased their distribution within five years and produced a second generation within two years, but numbats *Myrmecobius fasciatus* were only detected for three years and did not produce a second generation. Burrowing bettongs released into a fenced enclosure with cats and rabbits but no foxes survived and increased their distribution over at least three years and produced a second generation within two years. Greater bilbies and burrowing bettongs released into an unfenced area with some predator management did not survive to produce a second generation or increase their distribution. In 1998–2005, five numbats, 106 greater stick-nest rats (6 captive-bred individuals), 30 burrowing bettongs, 12 western barred bandicoots and nine greater bilbies (all captive-bred) were released into a 14-km² invasive-species-free fenced area. Rabbits, cats and foxes were eradicated within the fenced area in 1999. All western barred bandicoots and greater bilbies, and some greater stick-nest rats (8 individuals) and burrowing bettongs (10 individuals) were put into a 10-ha holding pen before full release after a few months. All other animals were released directly into the larger fenced area. In 2004–2008, thirty-two greater bilbies and 15 burrowing bettongs were translocated to an unfenced area (200 km²) where invasive predators (cats and foxes) were managed with lethal controls and dingoes *Canis lupus dingo* were excluded by a fence on one side. In 2008, sixty-six burrowing bettongs were released into a 26 km² fenced area which contained small cat and rabbit populations as a result of previous eradication attempts. Between 2000 and 2010, animals were monitored using track counts, burrow monitoring and radio-tracking.

A replicated study in 1996–1997 in three grassland sites in South Dakota, USA (12) found that at least half of captive-bred black-footed ferrets *Mustela nigripes* released into an area where predators were managed survived more than two weeks. At each of the three sites, 48% (12 of 25), 50% (9 of 18) and 89% (32 of 36) of captive-bred ferrets released into the wild survived for at least two weeks (long term survival is not reported). Overall, twenty-four ferrets were killed by native predators (mostly great-horned owls *Bubo virginianus* and coyotes *Canis latrans*) and the cause of death of two others could not be determined. A total of 79 captive-bred black-footed ferrets were released across three mixed-grass prairie sites (18–36 ferrets/site) in September–October 1996 and October–November 1997. Low-to-moderate lethal coyote control took place for 2–3 weeks each year prior to ferrets being released. A 107 cm high electric fencing was installed in each release site (creating 2 km² enclosures) and activated 1–2 weeks prior to ferrets being released. Ferrets were able to move in and out of the fenced areas. Each of the

79 ferrets was radio-tagged and tracked every 5–30 min/night for two weeks post-release in 1996–1997.

A before-and-after study in 2006–2008 in a temperate forest area in Tennessee and North Carolina, USA (13) found following the onset of translocations of black bears *Ursus americanus* away from an elk *Cervus canadensis* calving site, survival of the offspring of translocated elk increased. A higher proportion of elk calves survived their first year during bear translocations (69%) than before (59%). In 2001–2002, fifty-two elk were translocated to the Great Smoky Mountains National Park. Calf survival was monitored in 2001–2006 in a previous study that indicated that black bears predated nine out of 13 elk calves killed by predators. In 2006–2008, forty-nine black bears were relocated >40 km away from the elk calving area. In 2006–2008, forty-nine elk births were documented from which 42 recently-born calves were radio-collared. Calf survival was monitored by radio-tracking and visual observation.

A study in 2001–2008 in a forest reserve in Queensland, Australia (14) found that following the release of captive-bred bridled nailtail wallabies *Onychogalea fraenata* and subsequent predator controls, numbers increased over a three-year period, but remained low compared to the total number released. Three years after the last release event, the estimated bridled nailtail wallaby population (31 individuals) was higher than at the time of the last release (15 individuals) but was lower than the total number that had been released (166 individuals). In 2001–2005, groups of 1–20 captive-bred wallabies were released on 14 occasions into a 565-ha private forest reserve. Ninety-seven wallabies were kept in two 50 × 50-m predator-proof holding pens for one week before release. Sixty-nine wallabies infested with parasites were treated before release. Predator control was carried out in 2004–2008. Wallabies were trapped in a 2-km^2 area with 5–45 wire cage traps during 7–22 nights on eight occasions in 2005–2008.

A before-and-after study in 2006–2010 in a river catchment in Herefordshire, UK (15) found that alongside control of invasive American mink *Neovison vison*, a released captive-bred water vole *Arvicola amphibius* population persisted for at least 20 months. Following releases of water voles over three years along a river where American mink were being controlled, the population persisted through to 20 months after the final release. At this time, voles occupied 13.3 km of river and authors reported that numbers remained fairly constant. Between March 2006 and February 2010, one hundred and fifteen mink were captured. Mink control entailed use of 44–114 mink rafts along 63–203 km of river within the catchment. Seven hundred captive-bred water voles were released, along the main channel of the River Dore, in August–September of 2006–2008. Voles were released from boxes in groups of up to six animals/box. Boxes were ≥25 m apart. Food was provided daily until voles vacated boxes (typically within three days). Vole signs (food stores, feedings signs and faeces) were monitored annually, each April or May, in 2007–2010.

A replicated, before-and-after study in 2010–2011 in two grassland sites in Utah, USA (16) found that translocated Utah prairie dogs *Cynomys parvidens* released after the control of native predators into an area with artificial burrows showed low release site fidelity and different pre-and post-release behaviour. After translocation in both family groups and groups of unrelated individuals, prairie dogs spent more time being vigilant (48%) than they had done before translocation (22%). Only 50 out of 779 were still present at the release sites two months after release. In July 2010 and 2011, three hundred and seventy-nine and 400 prairie dogs were caught on a golf course using baited Tomahawk wire box-traps. Individuals were marked with hair dye and ear tags and released the same day at two sites with artificial burrow systems, with up to 10 animals/burrow. Each site had four release areas at least 200 m apart, each containing five burrows, 4 m apart. Each burrow consisted of a 30 × 45 × 30 cm box, buried 1.8m deep, and with two entrances (10-cm diameter and 4-m long) made from plastic tubing. Extra holes were left in the box and tubing to allow burrow expansion. Burrow entrances were protected from predators by mesh cages. At each site, two release areas were used for family groups and two were used for non-related groups. Predator removal of coyote *Canis latrans* and badgers *Taxidea taxus* was conducted for several weeks before and after prairie dog release. In September 2010 and 2011, prairie dogs were trapped, using 100 traps/site, during two sessions of four days each to determine site retention.

A study in 2010–2013 at a grassland and woodland site in Western Australia, Australia (17) found that wild-born golden bandicoots *Isoodon auratus*, descended from a translocated population which had been released into a predator-free enclosure, maintained genetic diversity relative to the founder and source populations and persisted for three years. For four measures of genetic diversity (allelic richness, the number of effective alleles per locus, observed heterozygosity and expected heterozygosity) there were no significant differences between descendants from translocated animals, founder animals that were translocated or source populations (see paper for details). The population size was estimated at 249 bandicoots in 2013. One hundred and sixty bandicoots were trapped on Barrow Island, which has a large population, in February 2010. They were released into a 1,100-ha enclosure free from introduced predators within 24 h of capture. Genetic material was sampled by ear punch biopsy from 57 founders in 2010 and from 67 wild-born progeny trapped in 2010–2012.

A study in 2010–2014 in a woodland and shrubland site in Western Australia, Australia (18) found that following the control of invasive red foxes *Vulpes vulpes* and provision of nest boxes, a translocated population of red-tailed phascogales *Phascogale calura* survived for more than four years. Four years after the first release at least 16 phascogales were present at the site, and 90% of 30 nest boxes showed signs of use. In May 2010, twenty wild-caught phascogales were released into a 389-ha unfenced reserve, and a further 10 were released in May 2011. Poison baiting was used to control foxes on the reserve until 2012, but was suspended due to a possible positive effect on feral cats. In May 2014, phascogales were monitored using Elliott live traps (400 trap nights), and nest box checks.

A replicated study in 2013 at a desert site in South Australia, Australia (19) found that four translocated populations of burrowing bettongs *Bettongia lesueur* released after controlling invasive foxes *Vulpes vulpes* and cats *Felis catus* died out within four months. There was no significant difference in post-release survival for a large release (bettongs last recorded 42 days after the final release) and three smaller releases (bettongs persisted 41–53 days after releases). At the three smaller release areas, bettongs persisted for 53 days at the site where fewer predator tracks were recorded and for 2–10 days at two sites where more predator tracks were recorded. A total of 1,492 bettongs were translocated and released into rabbit warrens. At one 250-ha site, 1,266 bettongs were released in July–October 2013. In October–December

2013, five releases of 29–56 bettongs were made at three smaller sites, 4 km apart. From May–December 2003 feral cats *Felis catus* and foxes *Vulpes vulpes* were intensively controlled in a 500-km² area by 428 hours of shooting patrols. Bettong survival was monitored using track counts, camera trapping, warren monitoring and live-trapping.

A replicated study in 2005 in a grassland and forest site in Victoria, Australia (20) found that most captive-bred eastern barred bandicoots *Perameles gunnii* translocated into a fenced reserve where invasive predators had been eradicated survived more than 22 days after release. Nine out of 12 captive-bred bandicoots survived at least 22–26 days after release, when their radio transmitters fell off. Two individuals died within three weeks of release (one was predated by a native eastern quoll *Dasyurus viverrinus* and one was injured during trapping). The twelfth individual was returned to captivity after losing 21% of its body weight in 10 days. The nine bandicoots which survived had lost 7–19% of their body weight 6–8 days after release, but recovered to 97–98% of their pre-release weight by day 22–26. Twelve captive-bred bandicoots were released into a 170-ha fenced reserve, free of invasive predators. Six of the 12 were kept in a 1-ha pre-release pen for one week and provided with supplementary food and water. Bandicoots were radio-tracked daily, and were trapped and weighed every 4–5 days, for one month.

A replicated, before-and-after study in 2011–2013 in two forest and grassland sites in the Australian Capital Territory, Australia (21) found that one to two years after release into predator-free fenced reserves, translocated eastern bettongs *Bettongia gaimardi* weighed more and had improved nutritional status. Translocated eastern bettongs weighed more (1.8 kg) one to two years after release than before they were released (1.7 kg). Various blood characteristics changed after release, suggesting that translocated bettongs had improved nutritional status (see original paper for details). Comprehensive health assessments were completed on 30 bettongs captured in Tasmania before release (July-October 2011 and April-September 2012) and 12–24 months after release (May–November 2013) into two predator-free reserves. In one reserve, bettongs (8 males, 10 females) received no supplementary food and the population was unmanaged. In the second reserve, bettongs (5 males, 7 females) were housed in small groups in 2.6–9.4-ha enclosures and provided supplementary food.

A study in 2011–2014 in a semi-arid area in South Australia (22) found that over half of captive-reared black-footed rock-wallabies *Petrogale lateralis* released into a large predator-free fenced area survived for at least two years and most females reproduced. Ten (five males, five females) of 16 rock-wallabies (63%) survived more than two years after being released. All five females that survived reproduced within 2–6 months of release. Over three years, 28 births from nine females were recorded. Between March 2011 and July 2012, sixteen captive-reared black-footed rock-wallabies (eight males, eight females; 1–5 years old) were released in three groups into a 97-ha fenced area. Ten of the 16 rock-wallabies were wild-born and fostered by yellow-footed rock-wallaby *Petrogale xanthopus* surrogate mothers in captivity. Introduced predators, common wallaroos *Macropus robustus* and European rabbits *Oryctolagus cuniculus* were removed from the enclosure. Supplementary water was provided in five 8-l tanks that were monitored with camera traps in 2011–2014. Rock-wallabies were fitted with radio-collars and tracked 1–7 times/week in 2011–2014. Trapping was carried out on seven occasions in 2011–2014.

(1) Short J. & Turner B. (2000) Reintroduction of the burrowing bettong *Bettongia lesueur* (Marsupialia: Potoroidae) to mainland Australia. *Biological Conservation*, 96, 185–196.

(2) Pople A.R., Lowry J., Lundie-Jenkins G., Clancy T.F., McCallum H.I., Sigg D., Hoolihan D. & Hamilton S. (2001) Demography of bridled nailtail wallabies translocated to the edge of their former range from captive and wild stock. *Biological Conservation*, 102, 285–299.

(3) Kinnear J.E., Sumner N.R. & Onus M.L. (2002) The red fox in Australia—an exotic predator turned biocontrol agent. *Biological Conservation*, 108, 335–359.

(4) Richards J.D. & Short J. (2003) Reintroduction and establishment of the western barred bandicoot *Perameles bougainville* (Marsupialia: Peramelidae) at Shark Bay, Western Australia. *Biological Conservation*, 109, 181–195.

(5) Long K., Robley A.J. & Lovett K. (2005) Immediate post-release survival of eastern barred bandicoots Perameles gunnii at Woodlands Historic Park, Victoria, with reference to fox activity. *Australian Mammalogy*, 27, 17–25.

(6) Hardman B. & Moro D. (2006) Optimising reintroduction success by delayed dispersal: is the release protocol important for hare-wallabies? *Biological Conservation*, 128, 403–411.

(7) Winnard A.L. & Coulson G. (2008) Sixteen years of eastern barred bandicoot *Perameles gunnii* reintroductions in Victoria: a review. *Pacific Conservation Biology*, 14, 34–53.

(8) Moorhouse T.P., Gelling M. & Macdonald D.W. (2009) Effects of habitat quality upon reintroduction success in water voles: evidence from a replicated experiment. *Biological Conservation*, 142, 53–60.

(9) Cook C.N., Morgan D.G. & Marshall D.J. (2010) Reevaluating suitable habitat for reintroductions: lessons learnt from the eastern barred bandicoot recovery program. *Animal Conservation*, 13, 184–195.

(10) Biggins D.E., Miller B.J., Hanebury L.R. & Powell R.A. (2011b) Mortality of Siberian polecats and black-footed ferrets released onto prairie dog colonies. *Journal of Mammalogy*, 92, 721–731.

(11) Moseby K.E., Read J.L., Paton D.C., Copley P., Hill B.M. & Crisp H.A. (2011) Predation determines the outcome of 10 reintroduction attempts in arid South Australia. *Biological Conservation*, 144, 2863–2872.

(12) Poessel S.A., Breck S.W., Biggins D.E., Livieri T.M., Crooks K.R. & Angeloni L. (2011) Landscape features influence postrelease predation on endangered black-footed ferrets. *Journal of Mammalogy*, 92, 732–741.

(13) Yarkovich J., Clark J.D. & Murrow J.L. (2011) Effects of black bear relocation on elk calf recruitment at Great Smoky Mountains National Park. *The Journal of Wildlife Management*, 75, 1145–1154.

(14) Kingsley L., Goldizen A. & Fisher D.O. (2012) Establishment of an endangered species on a private nature refuge: what can we learn from reintroductions of the bridled nailtail wallaby *Onychogalea fraenata*? *Oryx*, 46, 240–248.

(15) Reynolds J.C., Richardson S.M., Rodgers B.J. & Rodgers O.R. (2013) Effective control of non- native American mink by strategic trapping in a river catchment in mainland Britain. *The Journal of Wildlife Management*, 77, 545–554.

(16) Curtis R., Frey S.N. & Brown N.L. (2014) The effect of coterie relocation on release-site retention and behavior of Utah prairie dogs. *The Journal of Wildlife Management*, 78, 1069–1077.

(17) Ottewell K., Dunlop J., Thomas N., Morris K., Coates D. & Byrne M. (2014) Evaluating success of translocations in maintaining genetic diversity in a threatened mammal. *Biological Conservation*, 171, 209–219.

(18) Short J. & Hide A. (2015) Successful reintroduction of red-tailed phascogale to Wadderin Sanctuary in the eastern wheatbelt of Western Australia. *Australian Mammalogy*, 37, 234–244.

(19) Bannister H.L., Lynch C.E. & Moseby K.E. (2016) Predator swamping and supplementary feeding do not improve reintroduction success for a

threatened Australian mammal, *Bettongia lesueur. Australian Mammalogy*, 38, 177–187.

(20) de Milliano J., Di Stefano J., Courtney P., Temple-Smith P. & Coulson G. (2016) Soft-release versus hard-release for reintroduction of an endangered species: an experimental comparison using eastern barred bandicoots (*Perameles gunnii*). *Wildlife Research*, 43, 1–12.

(21) Portas T.J., Cunningham R.B., Spratt D., Devlin J., Holz P., Batson W., Owens J. & Manning A.D. (2016) Beyond morbidity and mortality in reintroduction programmes: changing health parameters in reintroduced eastern bettongs *Bettongia gaimardi. Oryx*, 50, 674–683.

(22) West R., Read J.L., Ward M.J., Foster W.K. & Taggart D.A. (2017) Monitoring for adaptive management in a trial reintroduction of the black-footed rock-wallaby *Petrogale lateralis. Oryx*, 51, 554–563.

14.32. Release translocated/captive-bred mammals to islands without invasive predators

https://www.conservationevidence.com/actions/2464

- **Six studies** evaluated the effects of releasing translocated or captive-bred mammals to islands without invasive predators. The six studies were in Australia[1,2,3,4,5,6].

COMMUNITY RESPONSE (0 STUDIES)

POPULATION RESPONSE (7 STUDIES)

- **Abundance (2 studies):** A study in Australia[5] found that following release of captive-bred dibblers on to an island free of introduced predators, numbers increased. A replicated study in Australia[1] found that following release of captive-bred and wild-born brush-tailed bettong onto islands free of foxes or cats, numbers increased on two of four islands.

- **Reproductive success (3 studies):** A study in Australia[4] found that captive-bred proserpine rock-wallabies released on an island without introduced predators established a breeding population. Two studies in Australia[3,5] found that following release on to islands without invasive predators, captive-bred rufous hare-wallabies[3] and captive-bred dibblers[5] reproduced.

- **Survival (3 studies):** A review of 28 translocation studies in Australia[2] found that 67% of marsupial populations translocated to islands without predators survived more than five years, compared to 0% translocated to islands with predators and 20% translocated to the mainland. A study in Australia[3] found that most captive-bred rufous hare-wallabies released on an island without non-native predators survived more than a year. A replicated study in Australia[6] found that wild-born golden bandicoots descended from translocated populations released onto two predator-free islands persisted for 2–3 years.

- **Condition (1 study):** A replicated study in Australia[6] found that wild-born golden bandicoots descended from translocated populations that had been released onto two predator-free islands, maintained genetic diversity relative to founder and source populations.

BEHAVIOUR (0 STUDIES)

Background

Mammals are sometimes wild-caught and translocated or bred in captivity and released to islands that are free of invasive predators to give them the best chance of establishing breeding populations and persisting. These could either be islands that have never had non-native predators introduced to them or those from which non-native predators have been eradicated.

See also: *Release translocated/captive-bred mammals in areas with invasive/problematic species eradication/control.*

A replicated study in 1979–1984 of shrubland and grassland on five islands in South Australia, Australia (1) found that captive-bred and wild-born brush-tailed bettong *Bettongia penicillata* populations released onto islands free of foxes *Vulpes vulpes*, rabbits *Oryctolagus cuniculus* or cats *Felis catus* increased in number on two of the four islands on which they were released and monitored. On one island, seven founders increased to ≥53 animals in four years. On a second island, 10 founders increased to 12 animals (five born on the island), 14 months later. Forty released on a third island declined to one after two years. Six released on a fourth island were predated by dogs *Canis lupus familiaris* after an unspecified period. On a fifth island, where 11 were released, animals persisted for up to 12 months, but were not formally monitored. Releases were of captive-bred animals, except those on the second island, which were wild-bred offspring from the population established on the first island. Releases were made in 1979–1983 and were monitored, primarily by live-trapping, up to April 1984. The results of this study are also included in (2).

A review of 28 translocation studies in 1905–1990 on islands and mainland Australia (2) found that eight of 12 marsupial populations translocated to islands without predators survived more than five years, none of six populations translocated to islands with predators survived and two of 10 translocations to the mainland survived more than five years. One of 12 populations of marsupials translocated to islands with no predators recorded survived at least 1–5 years, four survived 6–20 years and four survived >20 years (outcome of three translocations unknown). Five of six populations of marsupials translocated to islands with predators survived <1 year and one population survived 1–5 years. Three of 10 populations of marsupials translocated to the mainland survived <1 year, four survived 1–5 years and two survived 6–20 years (outcome of 1 translocation unknown). Translocations took place in 1905–1988 and included: banded hare-wallaby *Lagostrophus fasciatus*, black-flanked rock-wallaby *Petrogale lateralis*, bridled nail-tail wallaby *Onychogalea fraenata*, brush-tailed bettong ('woylie') *Bettongia penicillata*, brush-tailed rock-wallaby *Petrogale penicillata*, burrowing wallaby *Bettongia lesueur*, parma wallaby *Macropus parma*, quokka *Setonix brachyurus*, red-bellied pademelon *Thylogale billardierii*, rufous hare-wallaby *Lagorchestes hirsutus*, tammar wallaby *Macropus eugenii*, and western grey kangaroo *Macropus fuliginosus*. Predators were recorded as limiting factors in six island studies and were controlled in two mainland studies. Numbers of translocated animals ranged from 4–113, except for quokkas, of which 673 were translocated (see original paper for details).

A study in 1998–2001 on an offshore island dominated by grassland in Western Australia, Australia (3) found that following release on an island without non-native predators, most captive-bred rufous hare-wallabies ('mala') *Lagorchestes hirsutus* survived over one year after release and some reproduced. Twenty-four (80%) of 30 rufous hare-wallabies survived at least one year after release. Rufous hare-wallabies were still present on the island three years post-release and animals had reproduced in the wild. In June 1998, thirty captive-bred rufous hare-wallabies from a captive colony were released on to a 520-ha predator-free island, part of the Montebello Islands Conservation Park. Animals were transported in 5 × 3 m holding pens and were ear-tagged and fitted with a radio-collar before release. Hare-wallabies were released within 20 hours of capture and fruit, alfalfa and water were made available to them immediately after release. They were monitored every two days for 10 days and intermittently for up to three years post-release.

A study in 1998–2002 on an offshore island in Queensland, Australia (4) found that captive-bred proserpine rock-wallabies *Petrogale persephone* released on an island without introduced predators established a breeding population. No statistical tests were carried out and no data on population size are provided. Four rock-wallabies were born on the island, 3–4 years after the translocation of 27 animals commenced. However, nine rock-wallaby deaths were recorded over the study period (33% of all animals released). Between 1998 and 2002, twenty-seven rock-wallabies were translocated from the Queensland mainland to Hayman Island. Feral goats *Capra hircus* were eradicated before the release. Released individuals were radio-tracked over three-day periods at three-week intervals in 1998–1999, over one day every month in 2000 and over one day every two months in 2001. Remote video surveillance was used occasionally in 2001 to confirm breeding.

A study in 1998–2001 on an offshore predator-free island dominated by shrubland in Western Australia, Australia (5) found that following release on to an island free of introduced predators and rodents, captive-bred dibblers *Parantechinus apicalis* reproduced and numbers increased. Three years after the first release, more dibblers were confirmed to be alive on the island (67 animals) than in the first year of releases (26 animals). After three years, the proportion of females showing signs of recent reproduction (90%) was higher than after one year (20%). Of animals released in the first year, 10 of 26 survived for at least 12 months. Between 1998 and 2000, eighty-eight captive-bred dibblers were released on an 11-ha offshore island, free of introduced predators and rodents. All dibblers were individually marked and one-third was fitted with radio-collars. Twenty-five dibblers were radio-tracked for two weeks. For three to four nights, on 10 occasions from November 1998 to October 2001, up to 100 live traps were set across the island. New animals caught were marked to enable individual identification and females were examined for signs of recent breeding.

A replicated study in 2010–2013 on two islands in Western Australia, Australia (6) found that wild-born golden bandicoots *Isoodon auratus*, descended from translocated populations which had been released onto two predator-free islands, maintained genetic diversity relative to founder and source populations and persisted for 2–3 years. For four measures of genetic diversity (allelic richness, the number of effective alleles/locus, observed heterozygosity and expected heterozygosity) there were no significant differences between descendants from translocated animals, founder animals that were translocated or source populations (see paper for details). On the larger island, the population size was estimated to be 280 animals in 2013. No estimate is provided for the smaller island. Bandicoots were trapped on Barrow Island, which has a large population, in February 2010 (165 animals) and July 2011 (92 animals). Within 24 h of capture they were released on two other islands (1,020 and 261 ha) where non-native predators had been eradicated or had never been recorded. Genetic material was sampled by ear punch biopsy from 38 and 49 founders in 2010 and 2011, and from 44 and 39 wild-born offspring in 2010–2012.

(1) Delroy L.B., Earl J., Radbone I., Robinson A.C. & Hewett M. (1986) The breeding and re-establishment of the brush-tailed bettong, *Bettongia penicillata*, in South Australia. *Australian Wildlife Research*, 13, 387–396.

(2) Short J., Bradshaw S.D., Giles J., Prince R.I.T. & Wilson G.R. (1992) Reintroduction of macropods (Marsupialia: Macropodoidea) in Australia — a review. *Biological Conservation*, 62, 189–204.

(3) Langford D. & Burbidge A.A. (2001) Translocation of mala (*Lagorchestes hirsutus*) from the Tanami desert, Northern Territory to Trimouille Island, Western Australia. *Australian Mammalogy*, 23, 37–46.

(4) Johnson P.M., Nolan B.J. & Schaper D.N. (2003) Introduction of the Proserpine rock-wallaby *Petrogale persephone* from the Queensland mainland to nearby Hayman Island. *Australian Mammalogy*, 25, 61–71.

(5) Moro D. (2003) Translocation of captive-bred dibblers *Parantechinus apicalis* (Marsupialia: Dasyuridae) to Escape Island, Western Australia. *Biological Conservation*, 111, 305–315.

(6) Ottewell K., Dunlop J., Thomas N., Morris K., Coates D. & Byrne M. (2014) Evaluating success of translocations in maintaining genetic diversity in a threatened mammal. *Biological Conservation*, 171, 209–219.

14.33. Release translocated/captive-bred mammals in family/social groups

https://www.conservationevidence.com/actions/2463

- **Twenty-six studies** evaluated the effects of releasing translocated or captive-bred mammals in family or social groups. Eleven were in the USA[1,2,4,5,7,8,10,14,16,21,24], seven were in South Africa[6a,6b,12,17,19,20a,20b] and one was in each of Poland[3], Zimbabwe[9], along the USA–Canada border[11], Russia[13], Italy[15], Canada[18], China[22] and India[23].

COMMUNITY RESPONSE (0 STUDIES)

POPULATION RESPONSE (22 STUDIES)

- **Abundance (4 studies):** A study in the USA[1] found that a translocated population of Rocky Mountain bighorn sheep released in groups increased at a similar rate to that of a population newly established through natural recolonization. A replicated, controlled study in the USA[14] found that after translocating black-tailed prairie dogs in social groups to areas with artificial burrows, colonies increased in size over four years. A replicated study in Canada[18] found that following translocation of elk, most of which had been kept in holding pens in groups, numbers increased at two of four sites. A study in the USA[10] found that following the release of captive-reared bighorn sheep in groups, the overall population declined over 14 years.

- **Reproductive success (11 studies):** A study in the USA[10] found that captive-reared bighorn sheep released in groups had similar population recruitment rates compared to wild-reared sheep. A replicated, paired study in the USA[16] found that black-tailed prairie dogs translocated as family groups had higher reproductive success than those translocated in non-family groups. A replicated study in the USA[4] found that translocated gray wolves had similar breeding success when adult family groups were released together from holding pens or when young adults were released directly into the wild. Six of eight studies (one replicated) in Poland[3], Russia[13], South Africa[6b,12,17,19], the USA[8] and the USA–Canada border[11] found that when translocated and/or captive-bred animals were released in social or family groups, cheetahs[6b], European bison[13], lions[17], African wild dogs[19], most European beavers[3] and some swift foxes[11] reproduced successfully. One study found that one of two translocated Cape buffalo[12] groups released after being held in a holding pen formed a single herd and reproduced, while the other scattered and escaped the reserve. One study found that no Gunnison's prairie dogs[8] reproduced during the first year.

- **Survival (19 studies):** One of three studies (one controlled, before-and-after) in the USA[2,10,24] found that when translocated or captive-bred animals were released in family or social groups, captive-reared bighorn sheep[10] had similar survival compared to wild-reared sheep, whereas two found lower survival compared to wild white-tailed deer[2] and San Joaquin kit foxes[24]. Three replicated studies (one controlled, one paired) in the USA[4,5,16] found that when translocated as a social or family group, black-tailed prairie dogs[16] had higher and white-tailed deer[5] and gray wolves[4] had similar survival rates to those translocated as unrelated groups[5,16] or individuals[4]. Ten studies (one replicated) in Poland[3], Russia[13], Italy[15], South Africa[6a,6b,17], the USA[8], USA–Canada border[11], China[22] and India[23] found that when translocated and/or captive-bred animals were released in social or family groups, a population of Przewalski's horses[22] and European bison[13] persisted 5–11 years, lions[17], most swift foxes[11] and European beavers[3] and half or more cheetahs[6a,6b] survived for at least one year, and one-horned rhinoceroses[23] and over half of Gunnison's prairie dogs[8] and Eurasian badgers[15] survived at least 1–6 months. Three studies in the USA[7] and South Africa[20a,20b] found that when translocated or captive-bred animals were released in family or social groups (some provided with artificial refuges and/or supplementary food), most Mexican wolves[7] did not survive over eight months and all rock hyraxes[20a,20b] died within 90 days. A study in South Africa[19] found that translocated and captive-bred African wild dogs released in family groups into fenced reserves had high survival rates.

- **Condition (1 study):** A study in China[22] found that following the release of captive-bred Przewalski's horses in groups, the population had a lower genetic diversity than two captive populations.

BEHAVIOUR (4 STUDIES)

- **Behaviour change (4 studies):** Two replicated, controlled (one before-and-after) studies in the USA[5,21] found that when translocated as a social or family group, white-tailed deer[5] had similar average dispersal distances and Utah prairie dogs[21] had similar release site fidelity and post-release behaviour compared to those translocated as unrelated groups. One found that deer translocated together did not stay together, whether they had previously been part of the same social group or not. A study in Zimbabwe[9] found that a translocated lion family joined with immigrant lions and formed a new pride. A study in South Africa[17] found that translocated lions that were released in groups that had already been socialised and formed into prides, established stable home ranges.

Background

Mammals are sometimes wild-caught and translocated or bred in captivity and released to re-establish populations that have been lost, or to augment an existing population. This action includes studies describing or comparing the effects of translocating or releasing mammals in family or social groups. This includes releasing known family or social groups and releasing captive-bred social animals in groups. It also includes releasing groups of animals or coalitions, including pairs that were captured or housed and then released together with the intention of forming a social group/pair, even if the animals did not know each other prior to capture.

See also: *Release translocated/captive-bred mammals in larger unrelated groups.*

A study in 1960–1985 of forest and grassland across a mountain range in Montana, USA (1) found that a translocated population of Rocky Mountain bighorn sheep *Ovis canadensis* released in groups increased at rate similar to that of a population newly established through natural recolonization. Following translocation of 37 adult sheep and 30 lambs, the population reached 54 sheep and 43 lambs seven years later, though was estimated at 31 sheep and 12 lambs the following year. A naturally recolonized population increased from 30 sheep at establishment to 77 sheep and 49 lambs 22 years later (the same year that the population peaked in the translocated population) though declined to 33 sheep and 15 lambs the following year. Sheep populations were studied in a 3,000-km² study area. The translocated population (released in 1976) was surveyed seven times between 1976 and 1985. The recolonized population (established in 1958–1960 and occupying a separate part of the study area) was surveyed 11 times between 1960 and 1985. Surveys were carried out on the ground or by helicopter, usually on winter ranges. Weather frequently hampered surveys of the translocated population.

A study in 1984–1987 in two shrubland ranches in Texas, USA (2) found that most captive-bred white-tailed deer *Odocoileus virginianus* released in groups that had been reared together died within one year of release, whereas all monitored wild deer survived at least one year. Eight out of 13 (62%) captive-bred white-tailed deer died within one year post-release but all 20 wild deer survived. Thirteen captive-bred white-tailed male deer (average age: 1.7 years) were released into two ranches (extending over 25,900 ha and 15,379 ha) in January 1987. Additionally, 20 wild male deer were caught and released. In 1984–1986, ten captive-bred deer were removed from their mothers at 2–4 days old and bottle-raised by humans. Three others were raised by their mothers until four months old. After removal from their mothers, captive-bred deer were kept in 1.2-ha pens. All deer were ear-tagged and fitted a radio-collar. Deer were radio-tracked after release, on average every 25 days, from an airplane. A two-month hunting season was in place on both ranches during 1987.

A replicated study in 1975–1985 in a river basin in north-eastern Poland (3) found that most translocated and captive-bred European beavers *Castor fiber* released in pairs or family groups survived over one year after release and reproduced in the wild. Ten years after the release of 168 Europeans beavers (74 pairs or families), 108 were found to be established in 64 families. Reproduction was detected in nine of 16 areas where releases occurred and by the end of 1985, forty-four new colonies had established in the reintroduction areas. The average reproduction rate of captive-bred beavers was higher (2.1 kits/litter) than wild-born beavers (1.8 kits/litter; results were not statistically compared). Twenty-two translocated beavers (14%) died during the first year in the wild. In total, 51 beavers died or were lost following translocation. In 1975–1985, a total of 168 European beavers (74 pairs) were released into 16 regions within the Vistula river basin. Release sites had abundant willow *Salix* spp. and alder *Alnus* spp. thicket. Beavers were released in small populations of two to 11 pairs (usually 4 pairs), 2–20 km apart. Eleven individuals were captive-born and the remainder were caught in the wild and translocated. Beavers were monitored annually.

A replicated study in 1995–1996 in two forest sites in Idaho and Wyoming, USA, (4) found that translocated gray wolves *Canis lupus* had similar survival rates and breeding success in the first two years after release when adult family groups were released together from holding pens or when young adults were released directly into the wild. No statistical analyses were conducted. Thirty out of 35 young adult wolves released directly into the wild were still alive seven months after the last releases, and had produced up to 40 pups from 3–8 pairs. Thirty-one adult wolves released from holding pens in family groups had produced 23 pups four months after the last releases. From these 54 animals, nine had died. Six of the seven adult pairs released together from holding pens remained together, and five of these pairs established territories in the vicinity of the pens. Wolves were wild-caught from Canada in January 1995 and 1996. In Idaho, young adults were directly released in January 1995 and 1996. In Wyoming, family groups of 2–6 wolves spent 8–9 weeks in 0.4-ha chain-link holding pens before release in March 1995 and April 1996. Wolves were radio-tracked every 1–3 weeks until August 1996.

A replicated controlled study in 1993–1995 in a mixed hardwood and conifer forest reserve in New York, USA (5) found that white-tailed deer *Odocoileus virginianus* translocated as a social group did not differ in survival or average dispersal distance compared to deer translocated as an unrelated group and deer translocated together did not stay together, regardless of whether they had previously been part of the same social group or not. Survival rates in the first year after release were similar for translocated deer from the same social group (6/12 individuals, 50%) as for those from unrelated social groups (3/5 individuals, 60%). Survival rates of translocated deer were lower than resident deer in 1993–1995 (75–88%). Deer released together did not remain together regardless of whether they had originated from the same social group or not. The average dispersal distance of deer translocated as a social group (24 km) was similar to those translocated in a group of unrelated deer (22 km). Between May-June 1994, seventeen female white-tailed deer were caught and translocated 60 km from one hardwood and coniferous forest to another (1,133 ha). Twelve were translocated from the same social group (released in groups of 1–5 animals) and five were unrelated animals (released in a group of 3 animals or individually). Each deer was ear-tagged and radio-collared. Resident deer were radio-tracked 5–15 times/week in the source forest April-August 1993–1995 and translocated deer were radio-tracked in the destination forest 1–15 times/week in May-August 1994 and 1995, every few months in September-December 1994 and 1–8 times/month in January-March 1995.

A study in 1994–1998 in a savannah reserve in North West province, South Africa (6a) found that after being kept in groups (some family groups, some unrelated groups) in holding pens, approximately half of translocated cheetahs *Acinonyx jubatus* survived at least 18 months, of which half died within three years. Nine of 19 cheetahs survived 19–24 months, of which six were cubs that matured to independence, but only four cheetahs were known to still be alive at the end of the study period. Six cheetahs survived in the reserve less than one year, of which one died after a few weeks and two were removed to a captive breeding facility. The fate of four released cheetahs was unknown. In total 19 cheetahs were released into a game reserve between October 1994 and January 1998. Cheetahs were initially placed in 1 ha holding pens with electrified fencing for 4 weeks to several months. Cheetahs were mostly rescued wild-caught animals, except for one that was habituated to humans (and had to be removed after 2 weeks). Cheetahs were either held in family groups (mothers with cubs) or as coalitions (of adult males). One animal/group was radio collared for monitoring.

A study in 1981–1998 in a savannah reserve in North West province, South Africa (6b) found that following the release of rehabilitated and captive-bred cheetahs *Acinonyx jubatus* in groups (family and unrelated) and individually, most adults survived at least one year and animals bred in the wild. Most rehabilitated adult females (3 of 4) and all rehabilitated adult males (4 of 4) survived at least one year. Two rehabilitated adult females produced a second litter within two years of release. Three of 10 cubs released survived to independence, including a female who then raised her own litter of cubs to independence. The total population numbered 17 cheetahs one year after the end of a five year release program, compared to 18 animals released. An earlier release in the same National Park found that captive-bred cheetahs had bred successfully but most animals were subsequently removed to protect ungulate populations. Between 1995 and 1997, eighteen cheetahs (4 adult males, 4 adult females and 10 dependent cubs) were introduced to a National Park (55, 000 ha) from a rehabilitation facility (it is unclear whether the animals were wild caught, captive bred or reared in captivity). Cheetahs were released in family groups (mothers with cubs), in unrelated groups (of males) or individually. In 1981–1982, seven cheetahs were released from a captive-breeding facility and after a period of time (not specified), seven cheetahs were removed leaving three males in a group behind. Individuals were monitored by radio-tracking.

A study in 1998 in a grassland, shrubland and forest reserve in Arizona, USA (7) found that most captive-bred Mexican wolves *Canis lupus baileyi* released in family groups (initially into holding pens and provided with supplementary food) did not survive over eight months after release into the wild. Out of 11 captive-bred Mexican wolves released, six (55%) were illegally killed within eight months, three (27%) were returned to captivity and two (18%) survived in the wild for at least one year. Three weeks after their release, three individuals from one family group killed an adult elk *Cervus canadensis*. Two females gave birth two months after release but only one pup survived. Eleven wolves in three family groups were released in March 1998. Before release, wolves were kept for two months in pre-release holding pens, where they were fed carcasses of native prey. Carcasses were provided as supplementary food for two months post-release when sufficient killing of prey was confirmed. The released wolves were fitted with radio-collars. No monitoring details are provided.

A study in 1997 in one desert grassland site in New Mexico, USA (8) found that over half of the translocated Gunnison's prairie dogs *Cynomys gunnisonii* released in family groups survived at least six months, but none reproduced during the first year. Thirty-six out of 60 (60%) translocated prairie dogs survived the first summer after being released into the wild, but no young were born during this period. In spring 1997 sixty prairie dogs (30 male, 30 female) were translocated to a 3.5 ha area in a former prairie dog colony site. Individuals were released with family members or near neighbours, into the existing burrows of a former prairie dog colony. Prairie dogs were monitored during summer and autumn 1997 but monitoring details are not provided.

A study in 1997–1998 on a savanna estate in Zimbabwe (9) found that a translocated lion *Panthera leo* family kept in a holding pen prior to release joined with immigrant lions and formed a new pride. A lioness was translocated with three cubs (one male, two female). Within 45 days, seven male lions were close by and the female mated with one of these. The male cub moved away and the pride then comprised the female and daughters with two adult male lions. A wild lioness joined the pride 1.7 months after release, but was killed by a snare after six months. After 12–13 months, the original lioness had three new cubs and her daughters each also had litters. Resident lions on the estate were eliminated in 1995. In January 1997, a lioness and three cubs were translocated from communal land to a holding pen and were released on the estate after 90 days. Lions were monitored through to May 1998 by radio-tracking and direct observation.

A study in 1985–1998 in a shrub-dominated mountain area in California, USA (10) found that captive-reared bighorn sheep *Ovis canadensis* released into the wild in groups had similar survival and population recruitment rates compared to wild-reared sheep, but the overall population declined over 14 years. Captive-reared released and wild-reared bighorn sheep had similar average annual survival (captive-reared: 80%; wild-reared: 81%) and recruitment rates (captive-reared: 0.14 lambs/adult female; wild-reared: 0.14 lambs/adult female). However, despite releases, the overall population at the study site declined over 14 years from an estimated 40 sheep in 1985 to 22 sheep in 1998. In 1985–1998, seventy-four captive-reared bighorn sheep were released at three sites in a 70-km² area. Captive-reared sheep included 49 captive-born and 25 wild-born lambs brought into captivity at 1–5 months of age. Captive-reared sheep were released in 33 groups of 1–6 animals, mostly when one year old. Water was provided at the release site for 3–20 days post-release. Released sheep were ear-tagged and radio-collared and monitored at least once/week during each of 14 years in 1985–1998. Survival and reproduction were compared with those of 43 wild-reared sheep radio-tracked in the study area during the same time period.

A study in 1994–1998 at seven temperate grassland sites along the USA–Canada border (11) found that most translocated swift foxes *Vulpes velox*, which had been held in captivity prior to release and were released in social groups, survived for at least one year, and some reproduced near release sites. Eleven of 18 (61%) translocated swift foxes survived at least one year after release. Of these, 60% of animals translocated as juveniles went on to reproduce, as did 33% of translocated adults. In 1994–1996, foxes were captured in Wyoming, USA, and were fitted with radio-collars while being held in captivity for 22–57 days. In autumn 1994–1996, animals were released in mixed-gender groups of up to three individuals which had been trapped in close proximity. Release sites were located in areas with pre-existing, but small, fox populations and with low numbers of predators and high prey availability. Foxes were monitored by visual surveys and ground-based and aerial radio-tracking.

A study in 2000–2003 in a mixed karoo grassland reserve in Northern Cape Province, South Africa (12) found that one out of two translocated Cape buffalo *Syncerus caffer* groups released into a fenced reserve (after being held in a holding pen) formed a single herd, stayed in the reserve and reproduced, while the other scattered and escaped the reserve. One group of 10 translocated animals formed a single herd (along with the two remaining animals from the previous introduction) and over 10 months no animals died or escaped. A year after the introduction, five calves were born. One month after release, a second group of four buffalo had split into two solitary animals and a pair formed by one male and one female. One of the solitary animals was not seen again, the second solitary male animal was located two years after release on a neighbouring farm and released into the second group of translocated animals in May 2003. The pair escaped the reserve three times in 13 months. After the third escape, the male was moved to a different reserve and a new male introduced to form a herd with the remaining female. Four subadult buffalos (2 male, 2 female) were placed in a holding pen in July 2000 and released in August into a fenced 12,000-ha reserve. A second group of seven adult and three subadult animals (4 male, 6 female) was placed into a holding pen in August 2002 and released into a 200 ha area in September before being completely released in October 2002. Both groups were monitored weekly with telemetry until October 2003.

A study in 1996–2002 of forest in a national park in Oryol Oblast, Russia (13) found that a population of captive-bred European bison *Bison bonasus* released in groups persisted five to six years post-release and bred in the wild. The first calf was born in the second year after releases began and after six years, 30 calves had been born. The total population numbered 68 individuals (6–36 individuals/group) after six years. Sixty-five captive-bred bison were released in four groups in 1996–2001. Bison were monitored by visual observations and tracking.

A replicated, controlled study in 1999–2003 on a grassland site in Montana, USA (14) found that after translocating black-tailed prairie dogs *Cynomys ludovicianus* in social groups to areas with artificial burrows, colonies increased in size over four years. Six colonies receiving translocated prairie dogs grew more in area over four years (total growth 72 ha, 924% of pre-translocation area) than did 20 similar-sized colonies, which did not receive translocated prairie dogs (total growth 27 ha, 93% increase). Two active colonies (with existing prairie dog populations at the start of the study) that each received 120 prairie dogs increased more over four years (total increase 37 ha, 971% of pre-translocation area) than did two active colonies each receiving 60 prairie dogs (total growth 31 ha, 768%). An inactive colony that received no prairie dogs remained inactive. In June–July 1999, prairie dogs were released into pre-existing burrows (up to eight prairie dogs/burrow) or drilled holes (8 cm diameter × 60 cm deep, 45° below horizontal, up to two prairie dogs/hole, 30 holes/site). Colony size was measured four years later. Nine experimental colonies, three each occupying areas of 0 ha (inactive), 0.1–2.0 ha and 2.0–6.6 ha, were studied. In each size class, translocations to the three colonies were of 0, 60 and 120 prairie dogs. Growth-rates of 20 non-supplemented colonies were also monitored.

A study in 2001–2005 in a mixed forest and farmland site in northern Italy (15) found that just over half of translocated Eurasian badgers *Meles meles* released in groups into holding pens with supplementary food survived at least one month after release. Seven out of 12 badgers survived for 1–9 months, after which monitoring equipment stopped operating. One badger died almost immediately after release due to unknown causes. Two badgers escaped (one after the first month, the other after unknown period). The fate of three other badgers was unknown. One pair of translocated animals reproduced in the wild four years after release. From March 2001 to May 2004, twelve badgers were captured at four sites in northern Italy. Badgers were fitted with radio-collars and transported 20–40 km to the release site where they were kept in a 350 m^2 enclosure in a wooded area in their release groups (2001: 2 individuals, 2002: 4 individuals, 2003: 2 individuals; 2004: 4 individuals) and provided supplementary food for 3–10 weeks before release. Seven of the 12 badgers were located once/week, for up to nine months after release.

A replicated, paired study in 2001–2003 in 10 grassland sites in New Mexico, USA (16) found that black-tailed prairie dogs *Cynomys ludovicianus* translocated as family groups had higher survival and reproductive success than black-tailed prairie dogs translocated in non-family groups. Prairie dogs translocated as a family had higher post-release survival to the following spring (39–62%) and higher reproductive success (2.2–3.9 pups/female) than did those translocated as non-family groups (survival: 7–19%; reproductive success: 0.2–3.4 pups/female). Ten sites in Vermejo Park Ranch, Colfax County, from which prairie-dogs were absent but which were within the historical range, were selected. Four hundred and eighty-four wild-caught black-tailed prairie dogs were translocated in family groups into five sites (87–100/site) and 489 were translocated as non-family groups into five sites (88–103/site). Translocations took place in June–August of 2001 and2002. Survival and reproductive success were measured by trapping marked animals during the spring in the year after release (in May–July 2002 and May-June 2003).

A study in 1992–2004 in a grassland reserve in KwaZuluNatal Province, South Africa (17) found that translocated lions *Panthera leo* that were released in groups that had already been socialised and formed into prides, established stable home ranges, reproduced successfully and survived at least a year. Of 15 lions released, all except three, which were removed for killing a tourist, survived ≥398 days post-release. Average post-release survival was ≥1,212 days. At least 95 cubs from 25 litters were documented from the population over the 13-year study. Excluding cubs translocated to other sites or those still <18 months old at the end of the study, 51 of 65 cubs (78%) reached 18 months of age. Seven lions were released in May 1992, six in February 1993 and two in January 2003. Releases were into a fenced reserve (initially 176 km², then extended to 210 km²). Before release, lions were held in groups, each in an 80-m² acclimation pen, for 6–8 weeks. During this time, socialization occurred and stable prides were formed. Eleven of the founder lions were radio-tracked and other animals were monitored by direct observations.

A replicated study in 1998–2004 within four largely forested areas in Ontario, Canada (18) found that following translocation elk *Cervus canadensis*, most of which had been kept in holding pens in groups, remained present at all recipient sites and numbers increased at two of them. By 3–6 years after translocations, elk populations had increased at two sites and decreased at two. From 443 elk translocated, the population at the end of the study was estimated at 375–440 animals. Between 1998 and 2004, forty-one percent of translocated elk died. Causes of death included 10% lost to wolf predation, 5% to emaciation and 5% were shot. Elk were translocated from a site in Alberta, Canada in 1998–2001 in nine releases. Transportation took 24–58 hours. Elk were held in pens at recipient sites for up to 16 weeks before release (some were released immediately) but the effect of holding pens was not tested. Of 443 elk released, 416 were monitored by radio-tracking. The overall population was estimated in March 2004.

A study in 1995–2005 in 12 dry savanna and temperate grassland sites in South Africa (19) found that translocated and captive-bred African wild dogs *Lycaon pictus* released in family groups into fenced reserves had high survival rates and bred successfully. Eighty-five percent of released animals and their wild-born offspring survived the first six months after release/birth. Released animals that survived their first year had a high survival rate 12–18 months (91%) and 18–24 months (92%) after release. Additionally, groups that had more time to socialise in holding pens prior to release had higher survival rates (data presented as statistical models). Between 1995 and 2005, a total of 127 wild dogs (79 wild-caught, 16 captive-bred, 16 wild-caught but captive-raised, 16 'mixed' pups) were translocated over 18 release events into 12 sites in five provinces of South Africa. Animals were monitored for 24 months after release, and the 129 pups which they produced after release were monitored up to 12 months of age. Forty characteristics of the individual animals, release sites and methods of release were recorded, and their impact on post-release survival was tested.

A study in 2007 at rocky outcrops on a reserve in KwaZulu-Natal Province, South Africa (20a) found that all translocated rock hyraxes *Procavia capensis* that were released as a group, having been kept in a holding pen, died (or were presumed to have died) within 18 days of release. Eight of nine wild translocated hyraxes died within 18 days of release and the other was presumed to have died. The group split up and were not seen together after release. In October 2007, nine hyraxes (one juvenile, three sub-adults and five adults) were caught in baited mammal traps (90 × 31 × 32 cm) in an area where they were abundant, and moved 150 km to a 656-ha reserve where the species was nearly extinct. Hyraxes were kept together in a holding cage (185 × 185 × 185 cm) for 14 days before release. They were monitored daily for one week, and then every few days by direct observation and radio-tracking.

A study in 2005–2006 at rocky outcrops on a reserve in KwaZulu-Natal Province, South Africa (20b) found that translocated rock hyraxes *Procavia capensis* that were released in a social group after being held in captivity, and were provided with an artificial refuge and supplementary food after release, all died (or were presumed to have died) within 87 days of release. Eighty-seven days after the release of 17 hyraxes, none could be relocated. In July 2005, ten adult hyraxes were caught in baited mammal traps (90 × 31 × 32 cm) in an area where they were abundant, and held in captivity for 16 months, during which time three died. The remaining seven were released in November 2006, along with the eight juveniles and two pups born to them in captivity, to a 656-ha reserve where the species was nearly extinct. For four months prior to release, the group was housed together in an outdoor cage (5.9 × 2.5 × 3.2 m). Hyraxes were released into a hay-filled hutch which was left in place for several months, and were provided with cabbage for one week after release. Hyraxes were monitored by direct observations and by walking regular transects, daily for the first week but decreasing to monthly by the end of the study.

A replicated, controlled, before-and-after study in 2010–2011 in two grassland sites in Utah, USA (21) found no differences in the release site fidelity or post-release behaviour of translocated Utah prairie dogs *Cynomys parvidens* released in family groups or in groups composed of non-related individuals. Similar numbers of prairie dogs released in family groups (24 out of 386, 6%) and in non-related groups (26 out of 393, 7%) were still present at the release sites two months after release. Additionally, the post-release behaviour did not differ between groups, but both groups behaved differently post-release than pre-release (data presented as model results). In July 2010 and 2011, three hundred and seventy-nine and 400 prairie dogs were caught on a golf course using baited Tomahawk wire box-traps. Individuals were marked with hair dye and ear tags and released the same day at two sites with artificial burrow systems, with up to 10 animals/burrow. Each site had four release areas at least 200 m apart, each containing five burrows, 4 m apart. Each burrow consisted of a 30 × 45 × 30 cm box, buried 1.8m deep, and with two entrances (10-cm diameter and 4-m long) made from plastic tubing. Burrow entrances were protected from predators by mesh cages. At each site, two release areas were used for family groups and two for non-related groups. Predator removal of coyote *Canis latrans* and badgers *Taxidea taxus* was conducted for several weeks before and after prairie dog release. In September 2010 and 2011, prairie dogs were trapped, using 100 traps/site, during two sessions of four days each to determine site retention.

A study in 2001–2012 in a desert reserve in Xinjiang province, China (22) found that following the release of captive-bred Przewalski's horses *Equus ferus przewalskii* in groups, the population persisted at least 11 years but had a lower genetic diversity than two captive populations. Over 11 years after being reintroduced, the population of Przewalski's horses increased from 27 to 99 individuals. However, reintroduced horses had a lower genetic diversity (3.3 alleles/locus) than captive horses (3.4–3.8 alleles/locus), although the result was not tested for statistical significance. In 1985–1994, two captive populations of Przewalski's horses (founded with 22 and 18 horses imported from zoos) were established at two captive breeding facilities. In 2001, twenty-seven horses (16 females, 11 males) born in captivity within the latter population were released in small groups into a 17,330-km² reserve. Details on horse surveys are not provided. In 2010–2012, faecal samples were collected from 116 captive horses (66 and 50 horses from each of the two captive populations) and 52 reintroduced horses. Genetic diversity was estimated for 10 microsatellite loci.

A study in 2008–2012 in a grassland reserve in Assam, India (23) found that translocated greater one-horned rhinoceros *Rhinoceros unicornis*, some of which were cow-calf pairs, all survived at least 90 days after release. All 18 rhinoceroses survived more than >90 days after being released. During the first day after release, rhinoceroses dispersed an average of 2.4 km from the release site. Sixteen out of 18 rhinoceroses moved in the same direction to the bank of a river. Most cow-calf pairs separated after release, but were reunited within 24 hours. Between April 2008 and March 2012, twelve adult rhinoceroses and six calves (2–3 years old) were translocated from Kaziranga National Park and Pobitora Wildlife Sanctuary to the 519-km² Manas National Park. Rhinoceroses were released in groups of 2–4, often containing cow-calf pairs. Animals were radio-collared and located three times/day over 90 days after release. Tracking was carried out by foot, elephant back, motorcycle or vehicle.

A controlled, before-and-after study in 1989–1992 on a hilly grassland and scrubland site in California, USA (24) found that the survival of translocated San Joaquin kit foxes *Vulpes macrotis mutica* kept in pairs in holding pens prior to release was lower than that of resident animals. The survival of 40 translocated foxes in the first year after release (six alive, 32 dead, two unknown) was lower than that of 26 resident foxes (13 alive, 13 died), but did not change with the length of time spent in holding pens. Eleven pups born in the holding pens and released with their parents all died within 17 days of release. Only four foxes were known to breed after release, all with resident foxes. At the end of the study (1992) one fox was known to be alive and 36 (out of 40) were known to have died. Causes of death were predation (20 foxes), road accidents (two foxes) and death during trapping operations (one fox). The cause of death was unknown for 13 foxes. In August and December 1988 and January 1989, and from June–October 1989, foxes were caught and translocated up to 50 km to a 19,120-ha reserve. Foxes were kept in male–female pairs in holding pens (6.1 × 3.1–6.1 × 1.8 m) for 32–354 days before release in spring and summer 1990 (12 adults, 1 pup) and 1991 (28 adults, 10 pups). Foxes were monitored by radio-tracking 4–5 days/week after release.

(1) Irby L.R. & Andryk T.W. (1987) Evaluation of a mountain sheep transplant in north-central Montana. *Journal of Environmental Management*, 24, 337–346.

(2) McCall T.C., Brown R.D. & DeYoung C.A. (1988) Mortality of pen-raised and wild white-tailed deer bucks. *Wildlife Society Bulletin*, 16, 380–384.

(3) Żurowski W. & Kasperczyk B. (1988) Effects of reintroduction of European beaver in the lowlands of the Vistula basin. *Acta Theriologica*, 33, 325–338.

(4) Bangs E.E. & Fritts S.H. (1996) Reintroducing the gray wolf to Central Idaho and Yellowstone National Park. *Wildlife Society Bulletin*, 24, 402–413.

(5) Jones M.L., Mathews N.E. & Porter W.F. (1997) Influence of social organization on dispersal and survival of translocated female white-tailed deer. *Wildlife Society Bulletin*, 272–278.

(6) Hofmeyr M. & van Dyk G. (1998) *Cheetah introductions to two north west parks: case studies from Pilanesberg National Park and Madikwe Game Reserve.* Proceedings of a Symposium on Cheetahs as Game Ranch Animals, Onderstepoort, 23 & 24 October 1998, 60–71.

(7) Parsons D.R. (1998) 'Green fire' returns to the Southwest: reintroduction of the Mexican wolf. *Wildlife Society Bulletin*, 26, 799–807

(8) Davidson, A. D., Parmenter, R. R., & Gosz, J. R. (1999). Responses of small mammals and vegetation to a reintroduction of Gunnison's prairie dogs. *Journal of Mammalogy*, 80, 1311–1324.

(9) Hoare R. & Williamson J. (2001) Assisted re-establishment of a resident pride of lions from a largely itinerant population. *South African Journal of Wildlife Research*, 31, 179–182.

(10) Ostermann S.D., Deforge J.R. & Edge W.D. (2001) Captive breeding and reintroduction evaluation criteria: a case study of peninsular bighorn sheep. *Conservation Biology*, 15, 749–760.

(11) Moehrenschlager A. & Macdonald D.W. (2003) Movement and survival parameters of translocated and resident swift foxes *Vulpes velox*. *Animal Conservation*, 6, 199–206.

(12) Venter J.A. (2004) Notes on the introduction of Cape buffalo to Doornkloof Nature Reserve, Northern Cape Province, South Africa. *South African Journal of Wildlife Research*, 34, 95–99.

(13) Belousova I.P., Smirnov K.A., Kaz'min V.D. & Kudrjavtsev I.V. (2005) Reintroduction of the European bison into the forest ecosystem of the Orlovskoe Poles'e National Park. *Russian Journal of Ecology*, 36, 115–119.

(14) Dullum J.A.L.D., Foresman K.R. & Matchett M.R. (2005) Efficacy of translocations for restoring populations of black-tailed prairie dogs. *Wildlife Society Bulletin*, 2005, 842–850.

(15) Balestrieri A., Remonti L. & Prigioni C. (2006) Reintroduction of the Eurasian badger (*Meles meles*) in a protected area of northern Italy. *Italian Journal of Zoology*, 73, 227–235.

(16) Shier D.M. (2006) Effect of family support on the success of translocated black-tailed prairie dogs. *Conservation Biology*, 20, 1780–1790.

(17) Hunter L.T.B., Pretorius K., Carlisle L.C., Rickelton M., Walker C., Slotow R. & Skinner J.D. (2007) Restoring lions *Panthera leo*, to northern KwaZulu-Natal, South Africa: short-term biological and technical success but equivocal long-term conservation. *Oryx*, 41, 196–204.

(18) Rosatte R., Hamr J., Young J., Filion I. & Smith H. (2007) The restoration of elk (*Cervus elaphus*) in Ontario, Canada: 1998–2005. *Restoration Ecology*, 15, 34–43.

(19) Gusset M., Ryan S.J., Hofmeyr M., van Dyk G., Davies-Mostert H.T., Graf J.A., Owen C., Szykman M., Macdonald D.W., Monfort S.L., Wildt D.E., Maddock A.H., Mills M.G.L., Slotow R. & Somers M.J. (2008) Efforts going to the dogs? Evaluating attempts to re-introduce endangered wild dogs in South Africa. *Journal of Applied Ecology*, 45, 100–108.

(20) Wimberger K., Downs C.T., Perrin M.R. (2009) Two unsuccessful reintroduction attempts of rock hyraxes (*Procavia capensis*) into a reserve in the KwaZulu-Natal Province, South Africa. *South African Journal of Wildlife Research*, 39, 192–201.

(21) Curtis R., Frey S.N. & Brown N.L. (2014) The effect of coterie relocation on release-site retention and behavior of Utah prairie dogs. *The Journal of Wildlife Management*, 78, 1069–1077.

(22) Liu G., Shafer A.B., Zimmermann W., Hu D., Wang W., Chu H., Cao J. & Zhao, C. (2014) Evaluating the reintroduction project of Przewalski's horse in China using genetic and pedigree data. *Biological Conservation*, 171, 288–298.

(23) Dutta D.K. & Mahanta R. (2015) A study on the behavior and colonization of translocated greater one-horned rhinos *Rhinoceros unicornis* (Mammalia:

Perissodactyla: Rhinocerotidae) during 90 days from their release at Manas National Park, Assam India. *Journal of Threatened Taxa*, 7, 6864–6877.

(24) Scrivner J.H., O'Farrell T.P., Hammer K. & Cypher B.L. (2016) Translocation of the endangered San Joaquin kit fox, *Vulpes macrotis mutica*: a retrospective assessment. *Western North American Naturalist*, 76, 90–100.

14.34. Release translocated/captive-bred mammals in larger unrelated groups

https://www.conservationevidence.com/actions/2462

- **Five studies** evaluated the effects of releasing translocated or captive-bred mammals in larger unrelated groups. Two studies were in South Africa[2,3], one was in Namibia and South Africa[4], one was in the USA[1] and one was in Australia[5].

COMMUNITY RESPONSE (0 STUDIES)

POPULATION RESPONSE (5 STUDIES)

- **Reproductive success (3 studies)**: A replicated, paired sites study in the USA[1] found that black-tailed prairie dogs translocated in larger groups had higher reproductive success than smaller groups. A study in South Africa[3] found that Cape buffalo translocated to a fenced reserve as a larger group formed a single herd and reproduced, whilst a smaller group separated. A study in South Africa[2] found that rehabilitated and captive-bred cheetahs released in groups (unrelated and family) and as individuals reproduced.

- **Survival (4 studies)**: A replicated, paired sites study in the USA[1] found that black-tailed prairie dogs translocated in larger groups had higher initial daily survival rate than smaller groups. Two studies (one controlled) in Namibia and South Africa[4] and Australia[5] found that releasing translocated black rhinoceroses[4] and burrowing bettongs[5] in larger groups did not increase survival. A study in South Africa[2] found that most adult rehabilitated and captive-bred cheetahs released in groups (unrelated and family) and as individuals survived at least one year.

BEHAVIOUR (2 STUDIES)

- **Behaviour change (2 studies):** A replicated, paired sites study in the USA[1] found that black-tailed prairie dogs translocated in larger groups attracted more immigrants than smaller groups. A study in South Africa[3] found that Cape buffalo translocated as a larger group formed a single herd and stayed in the fenced reserve, whilst a smaller group scattered and escaped the reserve.

Background

Mammals are sometimes wild-caught and translocated or bred in captivity and released to re-establish populations that have been lost, or to augment an existing population. This action includes studies comparing the effects of translocating or releasing mammals in larger, unrelated groups (i.e. not family or social groups), rather than in smaller groups (which might include as few as one animal). This may be done for a variety of reasons, such as increased protection against predators, greater access to potential mates or social groups and an increased chance of establishing self-sustaining breeding populations.

Studies of unrelated translocated mammals that were held together to form social groups prior to release, or unrelated captive-bred animals raised and released together are described in *Release translocated/captive-bred mammals in family/social groups*.

Studies of releases of unrelated animals that were not held together, and where the effect of group size was not tested, are described in *Translocate to re-establish or boost population in native range* and *Release captive-bred individuals to re-establish or boost population in native range*.

A replicated, paired sites study in 1990–1991 in three grassland sites in Colorado, USA (1) found that larger groups of translocated black-tailed prairie dogs *Cynomys ludovicianus* attracted more immigrants and had higher reproductive success and initial daily survival rate than smaller groups. Over one year, prairie dogs translocated in groups of 59 individuals attracted more immigrants (13.7) than those translocated in groups of 30 (4.0) or 10–11 (1.5). Reproductive success was higher in prairie dogs translocated as groups of 59 individuals (0.79 pups/animal released) than groups of 10–11 (0.28 pups/animal released), but similar to those released as groups of 30 individuals (0.62 pups/animal released). Groups of 59 prairie dogs had higher daily survival rates in the first 23–51 days after release (99.1%) than groups of 30 (98.5%) or 10 prairie dogs (97.7%) but by the second monitoring period (139–142 days later) daily survival rates were the same for all three groups sizes (99.8%). Between July and October 1990, six groups of 10–11, three of 30 and three of 59 prairie dogs were released into three experimental blocks with four plots (2–6 ha depending on group size) in each (2 containing 10–11 prairie dog groups, 1x 30 prairie dog group and 1x 59 prairie dog group, randomly assigned), within a 69-km² military area. Prairie dogs were trapped four times during one year post-release, using 1.5 traps/released individual, over four days.

A study in 1981–1998 in a savannah reserve in North West province, South Africa (2) found that following the release of rehabilitated and captive-bred cheetahs *Acinonyx jubatus* in groups (unrelated and family) and as individuals, most adults survived at least one year and animals had reproduced in the wild. Most rehabilitated adult females (3 of 4) and all rehabilitated adult males (4 of 4) survived at least one year. Two rehabilitated adult females produced a second litter within two years of release. Three of 10 cubs released survived to independence, including a female who raised a litter of cubs to independence. The total population numbered 17 cheetahs one year after the end of a five year release program, compared to 18 animals released. An earlier release in the same National Park found that captive-bred cheetahs had bred successfully but most animals were subsequently removed to protect ungulate populations. Between 1995 and 1997, eighteen cheetahs (4 adult males, 4 adult females, 10 dependent cubs) were introduced to a National Park (55, 000 ha) from a rehabilitation facility (unknown if wild-born or captive-bred). Cheetahs were released in family groups (mothers with cubs), in unrelated groups (of males) or individually. In 1981–1982, seven cheetahs were released from a captive-breeding facility and after an unspecified period of time, seven cheetahs were removed leaving a group of three males. Individuals were monitored by radio-tracking.

A study in 2000–2003 in a mixed karoo grassland reserve in Northern Cape Province, South Africa (3) found that a larger group of translocated Cape buffalo *Syncerus caffer* released into a fenced reserve (after being held in a holding pen) formed a single herd and stayed in the reserve and bred, whilst a smaller group scattered and escaped the reserve. A group of 10 translocated animals formed a single herd (with two previously released animals) and over 10 months all animals survived and remained in the reserve. A year after release, five calves were born. One month after release, a group of four buffalo had split into two solitary animals and a male-female pair. One of the solitary animals was not seen again, the second solitary male was located two years after release on a neighbouring farm and was released into the second group of translocated animals in May 2003. The pair escaped the reserve three times in 13 months. After the third escape, the male was moved to a different reserve and a new male introduced to form a herd with the remaining female. Four subadult buffalo (2 male, 2 female) were placed in a holding pen in July 2000 and released in August into a fenced 12,000-ha reserve. A second group of seven adult and three sub-adult animals (4 male, 6 female) was placed into a holding pen in August 2002 and released into a 200 ha area in September before being completely released in October 2002. Both groups were monitored weekly using radio-tracking until October 2003.

A study in 1981–2005 of 81 reserves across Namibia and South Africa (4) found that releasing translocated black rhinoceros *Diceros bicornis* in larger groups did not affect survival in the first year post-release. Seventy-four of 682 translocated black rhinoceroses died during the first year post-release, but the number of individuals released together did not affect survival in the first year (data reported as statistical result). First-year post-release mortality was higher when animals were released into reserves occupied by other rhinoceroses (restocking, 13.4% mortality of 268 animals) than releases into new reserves (reintroduction, 7.9% mortality of 414 animals). At least 243 rhinoceroses survived at least 10 years after release. For restocking events, first-year post-release mortality was higher in rhinoceroses less than two years old (59%) than in all other age classes (9–20%), but there was no difference for reintroductions. Data on 89 reintroduction and 102 restocking events of black rhinoceroses into 81 reserves from 1981–2005 were compiled from the Namibia and South Africa Rhino Management Group reports. Animals were released in groups of one to 30 individuals, and reserves received up to five releases. Translocations were considered as different if the releases of individuals to the same reserve were more than 1 month apart. Deaths were detected by reserve staff. The location of reserves included in the study is not provided.

A controlled study in 2013 at a desert site in South Australia, Australia (5) found that releasing translocated animals in a larger group, to swamp predator activities, did not promote population persistence of burrowing bettongs *Bettongia lesueur*. There was no significant difference in post-release persistence between a large release (bettongs last recorded 42 days after the final release) and three smaller releases (bettongs persisted 41–53 days after releases). A total of 1,492 bettongs were translocated between July and December 2013 and released into rabbit warrens. The large release was of 1,266 bettongs, released in July–October 2013 in a 250-ha unfenced area. Three smaller releases, of 48–56 bettongs, occurred in October 2013, at sites 4 km from the large release and from each other. Following no bettong records at two of these sites for ≥7 weeks, further releases of 29 and 39 animals were made in December 2013. From May–December 2003 feral cats *Felis catus* and foxes *Vulpes vulpes* were intensively controlled in a 500-km² area by 428 hours of shooting patrols. Bettong persistence was monitored using track counts, camera trapping, warren monitoring and live-trapping.

(1) Robinette K.W., Andelt W.F. & Burnham K.P. (1995) Effect of group size on survival of relocated prairie dogs. *The Journal of Wildlife Management*, 867–874.

(2) Hofmeyr M. & van Dyk G. (1998) *Cheetah introductions to two north west parks: case studies from Pilanesberg National Park and Madikwe Game Reserve.* Proceedings of a Symposium on Cheetahs as Game Ranch Animals, Onderstepoort, 23 & 24 October 1998, 60–71.

(3) Venter J.A. (2004) Notes on the introduction of Cape buffalo to Doornkloof Nature Reserve, Northern Cape Province, South Africa. *South African Journal of Wildlife Research*, 34, 95–99.

(4) Linklater W.L., Adcock K., du Preez P., Swaisgood R.R., Law P.R., Knight M.H., Gedir J.V. & Kerley G.I. (2011) Guidelines for large herbivore

translocation simplified: black rhinoceros case study. *Journal of Applied Ecology*, 48, 493–502.

(5) Bannister H.L., Lynch C.E. & Moseby K.E. (2016) Predator swamping and supplementary feeding do not improve reintroduction success for a threatened Australian mammal, *Bettongia lesueur*. *Australian Mammalogy*, 38, 177–187.

14.35. Release translocated/captive-bred mammals into area with artificial refuges/breeding sites

https://www.conservationevidence.com/actions/2453

- **Seventeen studies** evaluated the effects of releasing translocated or captive-bred mammals into areas with artificial refuges or breeding sites. Five studies were in the USA[4,5,9,13,16], three were in Australia[1,3,15], three were in Spain[6,12,14], two were in the UK[2,17] and one was in each of Ireland[7], South Africa[8], Hungary[10] and Slovakia, the Czech Republic and Poland[11].

COMMUNITY RESPONSE (0 STUDIES)

POPULATION RESPONSE (15 STUDIES)

- **Abundance (5 studies):** Two of three studies (two replicated, two controlled) in Spain[6,12] and the USA[16] found that translocation release sites with artificial burrows provided had higher abundances of European rabbits[12] and densities of California ground squirrels[16] compared to those without. The other study[6] found that abundance of European rabbits following translocation was similar with and without artificial burrows provided. A replicated, controlled study in the USA[4] found that after translocating black-tailed prairie dogs to areas with artificial burrows, colonies increased in size. A before-and-after study in Spain[14] found that translocating European rabbits into areas with artificial refuges to supplement existing populations did not alter rabbit abundance, although two of three populations persisted for at least three years.

- **Reproductive success (4 studies):** Three studies in Australia[1], Ireland[7] and the UK[17] found that released captive-bred sugar gliders[1], most translocated female red squirrels[7] and some translocated pine martens[17] provided with nest boxes and supplementary food reproduced. A study of 12 translocation projects in Slovakia, the Czech Republic and Poland[11] found that translocated European ground squirrels released initially into enclosures or burrows with retention caps reproduced after release, whereas those without enclosures or burrows dispersed from release sites.

- **Survival (9 studies):** Five of eight studies in Australia[1,15], the USA[5,9], UK[2,17], Ireland[7] and South Africa[8] found that at release sites with artificial refuges, and in some cases food provided, a population of captive-bred sugar gliders[1] survived at least three years, two of three populations of red-tailed phascogales[15] survived for more than four years, most translocated black bears[5] survived at least one year and over half translocated red squirrels[7] and pine martens[17] survived 8–12 months. Three studies found that at release sites with artificial refuges, food and in one case water provided, no translocated red squirrels[2] survived more than five months, all translocated rock hyraxes[8] died within three months and most translocated Tipton and Heermann's kangaroo rat spp.[9] died within five days. A randomised, replicated, controlled study in Hungary[10] found that translocated European ground squirrels released into plugged artificial burrows had higher recapture rates than those released into unplugged artificial burrows.

BEHAVIOUR (3 STUDIES)

- **Use (2 studies):** Two studies in Australia[1,3] found that released captive-bred sugar gliders used artificial nest boxes provided.
- **Behaviour change (1 study):** A replicated, before-and-after study in the USA[13] found that translocated Utah prairie dogs released into an area with artificial burrows, after the control of native predators, tended to leave the release site and spent more time being vigilant than before.

Background

Mammals that are translocated or captive-bred and released are especially vulnerable immediately after release. At this time, they may struggle to find shelter in an unfamiliar area, or there may be few suitable refuges/breeding sites available in the new area. Furthermore, if the time they spend looking for suitable shelter or breeding sites is increased, this may make them more vulnerable to predation. Hence, providing artificial refuges or breeding sites in the release area may improve longer-term survival and reproductive rates.

See also: *Habitat restoration and creation — Provide artificial refuges/ breeding sites, provide artificial dens or nest boxes on trees, provide more small artificial breeding sites rather than fewer large sites.*

A study in 1979–1981 of a young planted native forest reserve in Victoria, Australia (1) found that a population of released, captive-bred sugar gliders *Petaurus breviceps* provided with artificial nest boxes and supplementary food survived, bred and used the nest boxes. In the third year after releases began, 37 individuals were recorded. Seven animals had been wild-born in the year after release and six females >2 years old showed signs of having reproduced. Occupation by sugar gliders or signs of previous occupation were recorded in 30 of 38 boxes, all three terra-cotta pipes and in 10 of 14 artificial hollow limbs. On a 130-ha island of planted native forest (trees ≤17 years old), 72 sugar gliders were released in January or February of 1979 (26 individuals), 1980 (34 individuals) and 1981 (12 individuals). Seventy boxes, pipes or hollowed limbs (dimensions not provided) were installed on trees, 3–7 m above the ground. Supplementary food was provided at release points during winters of 1979 and 1980. Gliders and artificial nest boxes were surveyed in May 1981.

A study in 1993–1994 on a forested peninsula in Dorset, UK (2) found that none of the translocated red squirrels *Sciurus vulgaris* provided with nest boxes, supplementary food and water (in and once released from pre-release pens) survived over five months after release. Out of 14 translocated red squirrels, 11 (79%) survived over one week, three

(21%) survived >3 months and none survived >4.5 months. At least half of the 14 squirrels were killed by mammalian predators. Intact carcasses that were examined showed signs of weightloss and stress (see original paper for details). Between October and November 1993, fourteen wild-born red squirrels were released into an 80-ha forest dominated by Scots pine *Pinus sylvestris*. The forest had no red squirrels but had introduced grey squirrels *Sciurus carolinensis*. Capture and release sites were similar habitats. Squirrels were transported in wooden nest boxes filled with dry hay. Squirrels were placed with their nest boxes into 1.5 × 1.5 × 1.5 m weldmesh pens surrounded by electric fencing for 3–6 days before release. Squirrels were kept individually except for 2 males who shared a pen. Supplementary food comprised a mixture of seeds, nuts and fruit on trays and in feed hoppers. After release, squirrels continued to have access to food, water and nest boxes inside the pens and outside (20–100 m away). All squirrels were radio-tagged and located 1–3 times/day, for 10–20 days after release and thereafter every 1–2 days.

A study in 1996 of a forest in Victoria, Australia (3) found that nest boxes were used by a population of released captive-bred sugar gliders *Petaurus breviceps*. Twenty out of 67 nest boxes were occupied by sugar gliders. Additionally, 18 boxes were occupied by feral honeybees *Apis mellifera*, a potential competitor for use of boxes. Boxes used by sugar gliders were positioned higher (average 4.5 m) than boxes used by honeybees (average 3.5 m). The site was formerly logged and had subsequently been replanted. Sixty-seven boxes were inspected in July 1996. Boxes had been installed, and captive-bred sugar gliders released in 1979–1982. Boxes were 10–27 l in capacity. Fifty-three boxes were positioned 3–5 m above ground. Seven were >5 m high and seven were <3m high, including three that had fallen to the ground.

A replicated, controlled study in 1999–2003 on a grassland site in Montana, USA (4) found that after translocating black-tailed prairie dogs *Cynomys ludovicianus* in social groups to areas with artificial burrows, colonies increased in size over four years. Six colonies receiving translocated prairie dogs grew more in area over four years (total growth 72 ha, 924% of pre-translocation area) than did 20 similar-sized colonies that did not receive translocated prairie dogs (total growth 27 ha, 93% increase). Two active colonies (with existing prairie dog populations at the start of the study) that each received 120 prairie dogs increased more over four years (total increase 37 ha, 971% of pre-translocation area) than did two active colonies each receiving 60 prairie dogs (total growth 31 ha, 768%). An inactive colony that received no prairie dogs remained inactive. In June–July 1999, prairie dogs were released into pre-existing burrows (up to eight prairie dogs/burrow) or drilled holes (8 cm diameter × 60 cm deep, 45° below horizontal, up to two prairie dogs/hole, 30 holes/site). Colony size was measured four years later. Nine experimental colonies, three each occupying areas of 0 ha (inactive), 0.1–2.0 ha and 2.0–6.6 ha, were studied. In each size class, translocations to the three colonies were of 0, 60 and 120 prairie dogs. Growth-rates of 20 non-supplemented colonies were also monitored.

A study in 2000–2003 in temperate forest in a wildlife refuge in Arkansas, USA (5) found that most translocated black bears *Ursus americanus* released into man-made dens survived at least one year after release. The first-year post-release survival rate for translocated adult female bears was 62%. For those surviving >1 year after release, second-year survival was 91%. The first-year survival rate of translocated cubs was 75%. Of eight documented adult female mortalities, at least three were due to poaching. Four bears returned to their capture site. In March 2000–April 2002, twenty-three wild adult female black bears and their 54 cubs were captured in White River National Wildlife Refuge and released, 160 km away, into man-made dens at Felsenthal National Wildlife Refuge. Radio-telemetry was used track bears and gather movement data weekly, through to January 2003.

A controlled study in 1999–2002 in a shrubland site in Huelva, Spain (6) found that providing artificial warrens to translocated European rabbits *Oryctolagus cuniculus* did not increase their abundance relative to those translocated without provision of artificial warrens. Over the three-year study, average rabbit pellet density in translocation plots where warrens were provided (4.4 pellets/m^2) was not significantly different to that in plots where warrens were not provided (5.0 pellets/m^2). The study was conducted in four 4-ha square plots (1–6 km apart) in Doñana National Park. Eight artificial warrens, with internal galleries and multiple entrances, were built in each of two plots. Two batches of rabbits, each totalling 64–67 animals, were translocated into each of two plots (one with and one without warrens) each winter from 1999–2000 to 2001–2002. Translocation plots were switched after the first winter, such that translocations in the second and third winter were into plots where no translocations were made in the first winter. Between September 1999 and November 2002, rabbit abundance was estimated every two months by counting the number of pellets in 33 fixed-position 0.5-m diameter sampling points/plot. Wild rabbits were present in all plots prior to translocations beginning.

A study in 2005–2007 in a mixed conifer forest in Galway, Ireland (7) found that over half of translocated red squirrels *Sciurus vulgaris* provided with nest boxes and supplementary food (in and once released from holding pens) survived over eight months after release and most females reproduced during that period. At least 10 out of 19 (53%) translocated squirrels survived over eight months post-release and five out of nine translocated females (56%) were lactating 5–7 months after release. In August 2006, seven juvenile squirrels were caught. At least one squirrel was still alive at the release location two years after the original release. Two squirrels died while in the release pen or shortly afterwards. Another four squirrels died 1–2 months after release. Nineteen squirrels were translocated to a nature reserve (19 ha) in the middle of a 789-ha commercial pine plantation, 112 km from the capture site. Individuals were marked, radio-tagged and kept on average for 46 days in one of two pre-release enclosures (3.6 × 3.6 × 3.9 m high). Enclosures contained branches, platforms, nest boxes, and supplementary feeders (containing nuts, maize, seeds and fruit). Supplementary food (50/50 peanut/maize mix) was provided in six feeders in the nature reserve until July 2006. Twenty nest boxes were also provided. Squirrels were radio-tracked in September and November 2005 and February and May 2006, and were trapped in February, May and August 2006 and observed once in October 2007.

A study in 2005–2006 at rocky outcrops on a reserve in KwaZulu-Natal Province, South Africa (8) found that translocated rock hyraxes *Procavia capensis* that were provided with an artificial refuge and food after release in a social group, having been held in captivity, all died (or were presumed to have died) within 87 days of release. Eighty-seven days after the release of 17 hyraxes, none could be relocated. In July 2005, ten adult hyraxes were caught in baited mammal traps (900 × 310 × 320 mm) in an area where they were abundant, and held in captivity for 16 months, during which time three died. The remaining seven were released in November 2006, along with the eight juveniles and two pups born to them in captivity, to a 656-ha reserve where the species was nearly extinct. For four months prior to release, the group was housed together in an outdoor cage (5.9 × 2.5 × 3.2 m). Hyraxes were released into a hay-filled hutch which was left in place for several months, and were provided with cabbage for one week after release. Hyraxes were monitored by direct observations and by walking regular transects, daily for the first week but decreasing to monthly by the end of the study.

A study in 2001 in a grassland and shrubland site in California, USA (9) found that most Tipton kangaroo rats *Dipodomys nitratoides nitratoides* and Heermann's kangaroo rats *Dipodomys heermanni* ssp. translocated into artificial burrows provided with supplementary food died within five days of release. All four Tipton kangaroo rats were predated within five days of translocation, and only one out of seven Heermann's kangaroo rats survived over 45 days. Three Heermann's kangaroo rats were predated, two died as a result of aggression from other Heermann's kangaroo rats, and the fate of one was unknown. In September 2001, four juvenile Tipton kangaroo rats and three Heermann's kangaroo rats were captured and held in captivity for two months before release at a protected site in November. In December 2001, a further four Heermann's kangaroo rats were caught and translocated to the same site. All 11 animals were fitted with a radio-transmitter and ear tags, and monitored for seven days in captivity prior to release. The release site was already occupied by Heermann's kangaroo rats. Animals were released into individual artificial burrows (two 90-cm-long cardboard tubes with a chamber about 30 cm below the surface), dug 10–15 m apart and provided with seeds. Burrows were plugged with paper towels until dusk. Animals were radio-tracked every 1–8 days for 18–45 days after release.

A randomised, replicated, controlled study in 2000 in a grassland site in central Hungary (10) found that translocated European ground squirrels *Spermophilus citellus* released into plugged artificial burrows had higher recapture rates than did ground squirrels released into unplugged artificial burrows. From four to 10 days after release, a higher proportion of ground squirrels released into plugged artificial burrows were recaptured (19 out of 60, 32%) than squirrels released into unplugged artificial burrows (6 out of 57, 11%). The highest recapture rate came from the group released into plugged burrows in the morning (15 out of 30). From 22–24 April 2000, one hundred and seventeen wild-caught European ground squirrels were translocated to a fenced 40-ha protected grassland. Four 40 × 40-m grid cells were established, each containing vertical, artificial burrows (50 cm long, 4.5 cm diameter) spaced 4.5 m apart. Sixty animals were released into burrows plugged with wood caps (from which they could only exit by digging out) across two grid cells and 57 into unplugged artificial burrows in the other two grid cells. One individual was released/burrow. Approximately half the squirrels were released in the afternoon on the day of capture. Animals to be released in the morning were kept in individual wire cages (10 × 10 × 40 cm) for one night and provided with fresh apple slices prior to release. From 28 April–2 May, squirrels were recaptured with snares to record retention.

A study of 12 translocation projects in 1989–2010 in 14 grassland sites in Slovakia, the Czech Republic and Poland (11) found that translocated European ground squirrels *Spermophilus citellus* released initially into enclosures or burrows with retention caps ('soft-release') reproduced on site after release, but individuals released without an initial preadaptive period or support after release ('hard-release') dispersed from release sites. Translocations in which at least 23 individuals/ season were released into enclosures or capped abandoned/artificial burrows led to reproduction (results reflect statistical model outcomes). However, animals released without initial containment did not settle at release sites. The study analysed data from 12 projects, involving release of ground squirrels at 14 sites. Around 2,500 grounds squirrels were released (4–1,057 individuals/project; 4–136 individuals/release season). Animals were 'soft-released' in eleven projects, 'hard-released' in two and combined hard and soft-released in one project. Three releases involved both captive-bred and wild-bred individuals. The remainder were of wild-bred translocated animals.

A replicated, site comparison study in 2008–2012 in 32 shrubland sites in Andalusia, Spain (12) found that release sites with shelter and artificial warrens provided had higher abundances of European rabbits *Oryctolagus cuniculus* following translocation. There were more rabbit latrines at sites where artificial warrens and wooden branches were provided (1.6–7.1 latrines/km) than at sites where they were not provided (0.3–3.4 latrines/km), although the size of the effect was less when scrub coverage was high (see original paper for details). In 2008–2009, between 75 and 90 rabbits/ha were released inside 32 fenced plots (0.5–7.7 ha). Artificial warrens and wooden branches were added within a 500-m radius of some plots and, in some sites, scrubland was cleared to create pasture (number of plots/treatment not stated). Twelve plots had no wooden branches or artificial warrens (wooden pallets covered with stones, branches and earth) added. From the end of the 2009 breeding season, small gates on fences were opened and the rabbits could disperse into adjacent areas. Relative rabbit abundance was estimated by latrine counts, in four 500-m transects outside each plot, in the summers of 2008–2009 and 2012. Scrub cover was classified as low (0–30% coverage), medium (30–60%) and high (>60%).

A replicated, before-and-after study in 2010–2011 in two grassland sites in Utah, USA (13) found that translocated Utah prairie dogs *Cynomys parvidens* released into an area with artificial burrows after the control of native predators tended to leave the release site and spent more time being vigilant than before. Only 50 out of 779 (6%) were still present at the release sites two months after release. After translocation in both family groups and groups of unrelated individuals, prairie dogs spent more time being vigilant (48%) than they had done before translocation (22%). In July 2010 and 2011, prairie dogs (379 and 400) were caught on a golf course using baited Tomahawk wire box-traps. Individuals were marked with hair dye and ear tags and released the same day at two sites with artificial burrow systems, with up to 10 animals/burrow. Each site had four release areas at least 200 m apart, each containing five burrows, 4 m apart. Each burrow consisted of a 30 × 45 × 30 cm box, buried 1.8m deep, and with two entrances (10-cm diameter and 4-m long) made from plastic tubing. Extra holes were left in the box and tubing to allow burrow expansion. Burrow entrances were protected from predators by mesh cages. At each site, two release areas were used for family groups and two were used for non-related groups. Predator removal of coyote *Canis latrans* and badgers *Taxidea taxus* was conducted for several weeks before and after prairie dog release. In September 2010 and 2011, prairie dogs were trapped, using 100 traps/site, during two sessions of four days each to determine numbers remaining at the site.

A before-and-after study in 2004–2007 in three mixed pasture and scrubland sites in southwest Spain (14) found that translocating European rabbits *Oryctolagus cuniculus* into areas with artificial refuges to supplement existing populations did not alter rabbit abundance, though populations persisted at two of three sites for at least three years. Three years after artificial warrens were built and rabbits were released, rabbit abundance was not significantly different to that before warrens were built (no data reported). In two of three sites, the rabbit population persisted for at least three years, but at one site no rabbits were seen three years after release. In 2004, at three sites, 20–72 artificial warren tubes were installed. In autumn 2004, wild translocated rabbits were released at each site and, in autumn 2005, more rabbits were released at two of the sites. In total, 150–387 rabbits were released at each site. Rabbit presence was detected at two of the sites before releases of translocated animals. In June–September of 2004–2007, rabbit droppings were counted along 10–12 transects, each 500 m long.

A study in 2006–2015 in three woodland and shrubland sites in Western Australia and Northern Territory, Australia (15) found that following release into areas with artificial refuges, two translocated populations of red-tailed phascogales *Phascogale calura* survived for more than four or five years, but one captive-bred population survived for less than a year. The two populations of phascogales established from wild-caught animals survived longer (4–5 years) than one population established from captive-bred animals (which had been kept in pre-release pens and given supplementary food; < 1 year). Authors suggest that the unsuccessful site may also have had a shortage of tree hollows for nesting. In July 2006 and January–February 2007, thirty-two captive-bred phascogales were released into a 26-ha fenced reserve (outside which feral cats were abundant) after spending either 10 days or over four months in a pre-release pen (3×6×2 or 4.5×3×2.2 m). Supplementary food was provided for one week after release. In April 2009 and June 2010, twenty-seven wild-caught phascogales were released into a 430-ha fenced reserve. In May 2010 and May 2011, thirty wild-caught phascogales were released into a 389-ha unfenced reserve, where poison baiting was used to control foxes *Vulpes vulpes* until 2012, but this was suspended due to a possible positive effect on feral cats *Felis catus*. Wild-caught animals had no pre-release pen or supplementary food. Nest boxes (11–35/site) were provided in every reserve. Phascogales were monitored after each release using radio-collaring or Elliott live traps, and through periodic monitoring of the nest boxes.

A replicated, controlled study in 2011–2014 of two areas of grassland and scrubland in southern California, USA (16) found that where holes were drilled into the soil, densities of translocated California ground squirrels *Otospermophilus beecheyi* were higher than where no holes were drilled. Two years after management commenced, there were more squirrel burrows in drilled areas (43–124/subplot) than in areas that had not been drilled (11–122/subplot). Six plots each comprised a 0.8-ha circle, divided into three equal wedge-shaped subplots. Subplots were mown (in May, for two years, at 7.5–15 cm height, with cut material removed) and were either drilled with a soil auger (20 holes/subplot) or not drilled. The third subplot (data not presented here) was not mown and did not have holes drilled. Management commenced in 2011 (two plots) and 2012 (four plots). Squirrels were translocated into plots at a rate of 30–50/plot. Squirrel abundance was determined by counting squirrel burrows.

A study in 2015–2016 in a wooded mountain region in central Wales, UK (17) found that some translocated pine martens *Martes martes* held in pre-release pens and then provided with supplementary food and nest boxes survived and bred in the first year after release. At least four out of 10 females that had been kept in pre-release pens survived and bred the year after release. Around 10–12 months after release, 14 out of 20 martens were alive and in good condition. Twelve were within 10 km of their release site. Six martens died in the first year, two had a fungal infection two weeks after release. Authors suggest this may have been due to damp conditions in November. From September–November 2015, twenty breeding-age (>3-years-old) pine martens were caught in Scotland, health checked, microchipped and fitted with a radio-collar, and in some cases a GPS logger. Martens were transported overnight to Wales, and held in individual pre-release pens (3.6 × 2.3 × 2 m) for up to seven nights. Males' pens were within 500 m of a female, but >2 km from the nearest male. Releases took place in autumn, and supplementary food was provided for 2–6 weeks after release (for as long as it continued to be taken). Den boxes were provided within 50 m of each release pen. Martens were radio-tracked until home-ranges were established, then located daily–weekly. Intensive tracking of females was carried out in March to locate breeding sites. Hair tubes and camera traps were used to monitor breeding success. A further 19 martens were released using the same procedure in September–October 2016.

(1) Suckling G.C. & Macfarlane M.A. (1983) Introduction of the sugar glider, *Petaurus breviceps*, into re-established forest of the Tower Hill State Game Reserve, Vic.. *Australian Wildlife Research*, 10, 249–258.

(2) Kenward R.E. & Hodder K.H. (1998) Red squirrels (*Sciurus vulgaris*) released in conifer woodland: the effects of source habitat, predation and interactions with grey squirrels (*Sciurus carolinensis*). *Journal of Zoology*, 244, 23–32.

(3) Wood M.S. & Wallis R.L. (1998) Potential competition for nest boxes between feral honeybees and sugar gliders at Tower Hill State Game Reserve. *The Victorian Naturalist*, 115, 78–80.

(4) Dullum J.A.L.D., Foresman K.R. & Matchett M.R. (2005) Efficacy of translocations for restoring populations of black-tailed prairie dogs. *Wildlife Society Bulletin*, 2005, 842–850.

(5) Wear B.J., Eastridge R. & Clark J.D. (2005) Factors affecting settling, survival, and viability of black bears reintroduced to Felsenthal National Wildlife Refuge Arkansas. *Wildlife Society Bulletin*, 33, 1363–1374.

(6) Cabezas S. & Moreno S. (2007) An experimental study of translocation success and habitat improvement in wild rabbits. *Animal Conservation*, 10, 340–348.

(7) Poole A. & Lawton C. (2009) The translocation and post release settlement of red squirrels *Sciurus vulgaris* to a previously uninhabited woodland. *Biodiversity and Conservation*, 18, 3205–3218.

(8) Wimberger K., Downs C.T., Perrin M.R. (2009) Two unsuccessful reintroduction attempts of rock hyraxes (*Procavia capensis*) into a reserve in the KwaZulu-Natal Province, South Africa. *South African Journal of Wildlife Research*, 39, 192–201.

(9) Germano D.J. (2010) Survivorship of translocated kangaroo rats in the San Joaquin Valley, California. *California Fish and Game*, 96, 82–89.

(10) Gedeon C.I., Váczi O., Koósz B. & Altbäcker V. (2011) Morning release into artificial burrows with retention caps facilitates success of European ground squirrel (*Spermophilus citellus*) translocations. *European Journal of Wildlife Research*, 57, 1101–1105.

(11) Matějů J., Říčanová Š., Poláková S., Ambros M., Kala B., Matějů K. & Kratochvíl, L. (2012) Method of releasing and number of animals are determinants for the success of European ground squirrel (*Spermophilus citellus*) reintroductions. *European Journal of Wildlife Research*, 58, 473–482.

(12) Guerrero-Casado J., Carpio A.J., Ruiz-Aizpurua L. & Tortosa F.S. (2013) Restocking a keystone species in a biodiversity hotspot: Recovering the European rabbit on a landscape scale. *Journal for Nature Conservation*, 21, 444–448.

(13) Curtis R., Frey S.N. & Brown N.L. (2014) The effect of coterie relocation on release-site retention and behavior of Utah prairie dogs. *The Journal of Wildlife Management*, 78, 1069–1077.

(14) Guil F., Higuero R. & Moreno-Opo R. (2014) European wild rabbit (*Oryctolagus cuniculus*) restocking: effects on abundance and spatial distribution. *Wildlife Society Bulletin*, 38, 524–529.

(15) Short J. & Hide A. (2015) Successful reintroduction of red-tailed phascogale to Wadderin Sanctuary in the eastern wheatbelt of Western Australia. *Australian Mammalogy*, 37, 234–244.

(16) McCullough-Hennessy S., Deutschman D.H., Shier D.M., Nordstrom L.A., Lenihan C., Montagne J.-P., Wisinski C.L. & Swaisgood R.R. (2016) Experimental habitat restoration for conserved species using ecosystem engineers and vegetation management. *Animal Conservation*, 19, 506–514.

(17) MacPherson J.L. (2017) *Pine marten translocations: the road to recovery and beyond.* Bulletin of the Chartered Institute of Ecology and Environmental Management: Rewilding and species reintroductions, 95, 32–36.

14.36. Release translocated/captive-bred mammals at a specific time (e.g. season, day/night)

https://www.conservationevidence.com/actions/2447

- **Seven studies** evaluated the effects of releasing translocated or captive-bred mammals at a specific time (season or day/ night). Three studies were in the USA[3,5,6] and one each was in the UK[1], Canada[2], Ireland[4] and Hungary[7].

COMMUNITY RESPONSE (0 STUDIES)

POPULATION RESPONSE (7 STUDIES)

- **Survival (7 studies):** Four of five studies in the UK[1], Canada[2] and the USA[3,4,6] found that translocated common dormice[1], black bears[3] and Canadian lynx[6] and captive-bred swift foxes[2] released in a specific season had higher survival rates than those released during another season. The other study[4] found that red squirrels translocated in autumn and winter had similar survival rates. A randomised, replicated, controlled study in Hungary[7] found that translocated European ground squirrels released during the morning had higher recapture rates than those released during the afternoon. A study in the USA[5] found that most translocated kangaroo rats released at dusk in artificial burrows supplied with food died within five days of release.

- **Condition (1 study):** A study in the UK[1] found that common dormice translocated during summer lost less weight than those translocated during spring.

BEHAVIOUR (2 STUDIES)

- **Behaviour change (2 studies):** Two studies in the UK[1] and USA[3] found that common dormice translocated during spring[1] and black bears translocated during winter[3] travelled shorter distances[1] or settled closer to the release site[3] than those translocated during summer.

Background

Mammals are sometimes wild-caught and translocated or bred in captivity and released to re-establish populations that have been lost, or augment an existing population. This action includes studies describing or comparing the effects of translocation projects that release mammals at specific times, such as in specific seasons or at certain times of day or night.

A study in 1991–1992 in a woodland reserve in Somerset, UK (1) found that common dormice *Muscardinus avellanarius* translocated during spring had lower survival rates, lost more weight and travelled shorter distances than dormice translocated during summer. Overall, five of seven dormice (57%) released in spring survived the first 10 days post release compared to seven of eight (80%) dormice released in summer. Common dormice translocated in spring lost more weight (0.30 g/day) than did dormice translocated in summer (0.14 g/day). However, they moved shorter daily distances from their release site (spring translocation: 119 m/day; summer translocation: 292 m/day).

Seven dormice were translocated in spring (between 30 May and 28 June 1991) and 10 in summer (between 24 August and 30 September 1992) to a 9-ha strip of woodland and scrub. Dormice were caught during the morning, moved to the release site and placed there by early afternoon, in the nestbox in which they had been captured. Individuals were fitted with radio-transmitters and followed for 10–20 nights. Dormice were weighed until 10–14 days after release.

A replicated, controlled study in 1987–1991 in three grassland sites in Alberta, Canada (2) found that, after one year, survival of captive-bred swift foxes *Vulpes velox* released in autumn was greater than that of captive-bred swift foxes released in spring. No statistical analyses were performed. At least 10 out of 71 (14%) swift foxes released in autumn survived over one year post-release, compared with at least one out of 27 (4%) of those released in spring. Eighty-one captive-born swift foxes were released in autumn and 41 were released in spring. They were provided with supplementary food for 1–8 months. Swift foxes were radio-collared and 98 were monitored from the ground and air for over one year.

A study in 1995–1999 in a forested area of Kentucky and Tennessee, USA (3) found that black bears *Ursus americanus* translocated during winter had higher survival rates and settled closer to the release area than did bears translocated in summer. First-year post-release survival of winter-released bears (88%) was higher than that of summer-released bears (20%). Winter-released bears remained closer to release sites during the two weeks after emergence from dens (0.4–3.6 km) than did summer-released bears during the two weeks after release (1.1–15.8 km). Eight adult female bears (five with 13 cubs in total and three assumed to be pregnant) were translocated to artificial dens in a 780-km² study area in January–March 1996 and March 1997. Six adult female bears were released in June–August 1996, following two weeks in acclimation pens at release sites. Bears were radio-tracked daily on release, reducing gradually to twice/week, until December 1999. Post-release survival was calculated with emigration included within mortality.

A study in 2005–2007 in a mixed conifer forest in Galway, Ireland (4) found that red squirrels *Sciurus vulgaris* translocated in September and October had similar survival rates compared to squirrels translocated in December. The survival rate to the following May of red squirrels translocated in September and October (78%, 7/9 individuals) was not statistically different to that of squirrels released in December (50%, 5/10 individuals). In August 2006, seven juvenile squirrels were caught and at least one squirrel was still alive in the release location two years after the original release. Nineteen squirrels were translocated to a nature reserve (19 ha) in the middle of a 789-ha commercial pine plantation, 112 km from the capture site. Squirrels were kept for an average of 46 days in one of two pre-release enclosures (3.6 × 3.6 × 3.9 m high). Enclosures contained branches, platforms, nest boxes, and supplementary feeders. Food and nest boxes were also provided in the periphery of the release site. Nine squirrels were released in September or October 2005 and 10 in December 2005. Squirrels were radio-tracked in September and November 2005 and February and May 2006, and were trapped in February, May and August 2006 and observed once in October 2007.

A study in 2001 in a grassland and shrubland site in California, USA (5) found that most translocated Tipton kangaroo rats *Dipodomys nitratoides nitratoides* and Heermann's kangaroo rats *Dipodomys heermanni* ssp. released at dusk in artificial burrows supplied with food died within five days of release. All four Tipton kangaroo rats were predated within five days of translocation, and only one out of seven Heermann's kangaroo rats survived over 45 days. Three Heermann's kangaroo rats were predated, two died as a result of aggression from other kangaroo rats, and the fate of one was unknown. In September 2001, four juvenile Tipton kangaroo rats and three Heermann's kangaroo rats were captured and held in captivity for two months before release at a protected site in November. In December 2001, a further four Heermann's kangaroo rats were caught and translocated to the same site. All 11 animals were fitted with a radio-transmitter and ear tags, and monitored for seven days in captivity prior to release. The release site was already occupied by Heermann's kangaroo rats. Animals were released into individual artificial burrows (two 90-cm-long cardboard tubes with a chamber about 30 cm below the surface), dug 10–15 m apart and provided with a paper towel and seeds. Burrows were plugged with paper towels until dusk. Animals were radio-tracked every 1–8 days for 18–45 days after release.

A study in 1999–2007 in montane forest in Colorado, USA (6) found that translocated Canadian lynx *Lynx canadensis* held in captivity and released in spring had higher survival rates in the first year than those released at other times of year. Lynx released in spring after >45 days in captivity near the release location had lower monthly mortality rates (0.4–2.8% in 2000–2006) than lynx held for up to seven days in captivity near the release location (20.5% in 1999) and not released in spring. Overall, 117 of 218 released lynxes (53%) survived to at least 1–8 years after release. From 1999 to 2006, two hundred and eighteen lynx were translocated to Colorado from Canada and USA. Lynx were held in captivity near their source location (for 3–68 days) prior to transfer to a holding facility (with 40 x 2.4 x 1.2 m pens with ceilings) in Colorado (100 km from release site). Time in the Colorado holding facility varied (5–137 days): release within seven days following veterinary inspection (4 individuals in 1999); release after 3 weeks (9 individuals in 2000); release after >3 weeks in the spring (1 April-31 May; 28 individuals in 2000); release in spring after >3 weeks in captivity but excluding any juvenile females or pregnant females (177 individuals in 2000–2006). Lynx were fed a diet of rabbit and commercial carnivore food while in captivity. Lynx were monitored for the first year following release using radio-telemetry (1,878 locations/month recorded).

A randomised, replicated, controlled study in 2000 in a grassland site in central Hungary (7) found that translocated European ground squirrels *Spermophilus citellus* released during the morning had higher recapture rates than ground squirrels released during the afternoon. From four to 10 days after release, a higher proportion of ground squirrels that had been released in the morning were recaptured (18 out of 58, 29%) than those released in the afternoon (7 out of 59, 12%). The highest recapture rate came from the group released in the morning in to plugged burrows (15 out of 30, 50%). From 22–24 April 2000, one hundred and seventeen wild-caught European ground squirrels were translocated to a fenced 40-ha protected grassland. Four 40 × 40-m grid cells were established, each containing vertical, artificial burrows (50 cm long, 4.5 cm diameter) spaced 4.5 m apart. Fifty-nine animals were released into burrows in two grid cells during the afternoon on the day of capture and 58 into burrows in the other two grid cells the morning after capture. Animals to be released in the morning were kept in individual wire cages (10 × 10 × 40 cm) for one night and provided with fresh apple slices prior to release. One individual was released/

burrow. Approximately half the burrows for each release group were plugged with wood caps so that squirrels could only exit by digging out. From 28 April–2 May, squirrels were recaptured with snares.

(1) Bright P.W. & Morris P.A. (1994) Animal translocation for conservation: performance of dormice in relation to release methods, origin and season. *Journal of Applied Ecology*, 31, 699–708.

(2) Carbyn L.N., Armbruster H.J. & Mamo C. (1994) The swift fox reintroduction program in Canada from 1983 to 1992. Pages 247–271 in: M.L. Bowles & C.J. Whelan (eds.) *Restoration of endangered species: conceptual issues, planning and implementation*. Cambridge University Press, Cambridge.

(3) Eastridge R. & Clark J.D. (2001) Evaluation of 2 soft-release techniques to reintroduce black bears. *Wildlife Society Bulletin*, 29, 1163–1174.

(4) Poole A. & Lawton C. (2009) The translocation and post release settlement of red squirrels *Sciurus vulgaris* to a previously uninhabited woodland. *Biodiversity and Conservation*, 18, 3205–3218.

(5) Germano D.J. (2010) Survivorship of translocated kangaroo rats in the San Joaquin Valley, California. *California Fish and Game*, 96, 82–89.

(6) Devineau, O., Shenk, T.M., Doherty Jr, P.F., White, G.C. & Kahn, R.H. (2011) Assessing release protocols for Canada lynx reintroduction in Colorado. *The Journal of Wildlife Management*, 75.

(7) Gedeon C.I., Váczi O., Koósz B. & Altbäcker V. (2011) Morning release into artificial burrows with retention caps facilitates success of European ground squirrel (*Spermophilus citellus*) translocations. *European Journal of Wildlife Research*, 57, 1101–1105.

14.37. Release translocated/captive-bred mammals to areas outside historical range

https://www.conservationevidence.com/actions/2443

- **Seven studies** evaluated the effects of releasing translocated or captive-bred mammals to areas outside their historical range. Three studies were in Australia[2,6,7], one study was in each of Kenya[1], France[3] and South Africa[4], and one was a review of studies in Andorra, Spain and France[5].

COMMUNITY RESPONSE (0 STUDIES)

POPULATION RESPONSE (7 STUDIES)

- **Abundance (5 studies):** Three of four studies in Kenya[1], Australia[2], France[3], and South Africa[4] found that after translocating mammals to areas outside their historical range, populations increased for Alpine marmots[3], most of 22 herbivorous species[4] and bridled nailtail wallabies[2] (including captive and enclosure bred animals). A study in Kenya[1] found that a population of translocated roan persisted for more than six years but did not increase. A review of studies in Andorra, Spain and France[5] found that following translocation to areas outside their native range, alpine marmots had similar densities and family group sizes to those of populations in their native range.

- **Reproductive success (1 study):** A study in Kenya[1] found that a population of roan translocated into an area outside their native range persisted and bred for more than six years.

- **Survival (3 studies):** A study in Australia[2] found that captive-bred, translocated and enclosure born bridled nailtail wallabies released into areas outside their historical range had annual survival rates of 40–88% over four years. A study in Australia[6] found that most captive-bred Tasmanian devils released into an area outside their native range survived over four months. A study in Australia[7] found that half the captive-bred and wild-caught translocated eastern barred bandicoots released to a red fox-free island outside their historical range survived for at least two months.

BEHAVIOUR (0 STUDIES)

Background

Endangered species are sometimes translocated from other areas or bred in captivity for release into their former range. Sometimes, though, the former range remains unsuitable for the species, for example through presence of an invasive predator. In such cases, releases to sites outside the former range may be considered, if these potentially offer better conditions for persistence of the species.

A study in 1970–1978 in a grassland and forest reserve in southeast Kenya (1) found that after release of translocated roan *Hippotragus equinus* into an area outside their native range, the population persisted and bred for more than six years. Only eight out of the original 38 translocated roan could be located 18 months years after the last release. However, six years after the last translocations, roan numbers had increased to 22. From 1973–1976, at least 15 calves were born, of which one-third survived to nine months of age. Between 1970 and 1972, 38 roan were released in Shimba Hills National Reserve, where there is no evidence for their existence since at least 1885. Animals were captured in the Ithanga Hills, by funnelling them into a 2.5-acre corral using horses, trucks and a helicopter. Prior to release roan were kept in a 30-acre holding pen. Roan were monitored between June 1973 and January 1978, but no further monitoring details are provided.

A study in 1996–1999 in a woodland reserve in Queensland, Australia (2) found that translocated, captive-bred and enclosure born bridled nailtail wallabies *Onychogalea fraenata* released into areas outside their historical range had annual survival rates of 40–88% and the population increased three-fold over four years. The average annual survival of bridled nailtail wallabies varied by release group between 40 and 88%. During four years, in which 133 wallabies were released, the population increased to approximately 400 individuals. In 1996–1997, nine wild-born translocated and 39 captive-bred bridled nailtail wallabies were released in three sites across Idalia National Park. In 1997–1998, eighty-five wallabies born (from captive animals) within a 10-ha enclosure on the reserve were also released. All released wallabies were kept in a holding pen (30 m diameter) for a week at each site before release. Mammalian predators were culled at release sites. Wallabies were individually marked with ear tags. A total of 37 wallabies (9 wild-born translocated, 28 captive-bred) were radio-tagged and tracked every 2–7 days in 1996–1998. Wallabies were live-trapped at irregular intervals with 20–35 wire cage traps in 1997–1999. Vehicle spotlight surveys were carried out 3–4 times/year in 1996–1999.

A study in 1980–2007 in a mountain grassland site in the Mézenc Massif, France (3) found that after the release of translocated Alpine marmots *Marmota marmota* into a site outside their historical range, numbers increased more than four-fold over 27 years. Twenty-seven years after the onset of the translocation, marmot numbers had increased to 492, from the 108 originally released. Population growth fluctuated over time with some population declines in 1990, 1993, 1997 and 2001 (see original paper for details). In 1980, eleven marmots were translocated into a mountain area outside their historical range. This was followed by seven reinforcements (translocation dates not provided), with a total of 108 translocated individuals by 2001. Marmots were monitored discontinuously until 1988, and then annually (five times through spring to autumn). Monitoring details are not provided.

A study in 1949–2001 in South Africa (4) found that following translocations outside of the species' native ranges, population sizes of most of 22 species of herbivorous mammals increased. Following translocation, 82 out of 125 populations (66%) of 22 species of mammals (white rhinoceros *Ceratotherium simum*, mountain zebra *Equus zebra*, plains zebra *Equus quagga*, giraffe *Giraffa camelopardalis*, African buffalo *Syncerus caffer* and 17 species of antelope) had positive growth rates (data presented as results of population growth models). Seventeen of the 22 species were introduced outside of their historical range. Population models were based on long-term monitoring data from 178 populations relocated to 24 reserves in 1949–1978 (see original paper for modelling details). Only translocations with five or more consecutive years of monitoring results were included (125 translocations, monitoring data duration: 5–47 years). Translocation details are not provided but authors state that most translocated populations began with fewer than 15 individuals and that most reserves contained water impoundments and lacked top predators, such as lions *Panthera leo* or spotted hyenas *Crocuta crocuta*.

A review of studies in 1948–2003 in nine mountain grassland sites in the Pyrenees in Andorra, Spain and France (5) found that following translocation to areas outside their native range, alpine marmots *Marmota marmota* had similar densities and family group sizes to those of populations in their native range. Average marmot densities and family group sizes did not differ significantly between translocated populations (0.9 individuals/ha; 5 individuals/group) and populations within their native range (1.4 individuals/ha; 6 individuals/group). Between 1948 and 1988, around 500 alpine marmots were translocated to multiple sites across the Pyrenees in areas outside their native range.

In 1965–2003, nine marmot populations (comprising 2–14 family groups) were monitored for 1–2 years in the introduced range and 11 populations (3–50 family groups) were monitored for 1–13 years in their native range (French, German, Italian and Swiss Alps). Monitoring methods are not provided.

A study in 2012–2013 on an offshore island in Tasmania, Australia (6) found that most captive-bred Tasmanian devils *Sarcophilus harrisii* released into an area outside their native range survived over four months after release. Fourteen out of 15 captive-bred Tasmanian devils survived >4 months (122 days) after release. In November 2012, fifteen captive-bred Tasmanian devils were released onto a 9,650-ha island reserve, 12 km off the Tasmanian mainland. Seven individuals were from a captive breeding facility, where animals were raised in groups of 1–4 in 1-ha pens. Eight were from a captive breeding facility were animals were raised in groups of 20–25 in 22-ha enclosures. Animals that shared pens in captivity were released together. Supplementary wallaby meat (20 kg) was provided at two-week intervals. Tasmanian devils were monitored for 122 days through video footage obtained at feeding sites. Individuals were identified by unique markings and scars.

A study in 2012–2013 on an island with mixed forest and grassland vegetation in Victoria, Australia (7) found that, following releases of captive-bred and wild-caught translocated eastern barred bandicoots *Perameles gunnii* to a red fox *Vulpes vulpes*-free island outside of the species' historical range, half of animals survived for at least two months. Nine out of 18 released bandicoots were still alive two months after release while seven survived at least 100 days. Deaths included two to cat predation and two to disease (toxoplasmosis). Between July and September 2012, eighteen eastern barred bandicoots were released on a fox-free island outside of the historical range of the species with 9,000 ha of potentially suitable habitat. Four animals were captive-bred and 14 animals were translocated from a reintroduction site on the mainland. All were fitted with radio-transmitters and PIT-tags to allow tracking and identification of individuals. Each bandicoot was radio-tracked from the day after its release until November 2012.

(1) Sekulic R. (1978) Roan translocation in Kenya. Oryx, 14, 213–217.

(2) Pople A.R., Lowry J., Lundie-Jenkins G., Clancy T.F., McCallum H.I., Sigg D., Hoolihan D. & Hamilton S. (2001) Demography of bridled nailtail wallabies

translocated to the edge of their former range from captive and wild stock. Biological Conservation, 102, 285–299.

(3) Ramousse R., Métral J. & Le Berre M. (2009) Twenty-seventh year of the Alpine marmot introduction in the agricultural landscape of the Central Massif (France). Ethology Ecology & Evolution, 21, 243–250, https://doi.org /10.1080/08927014.2009.9522479

(4) Van Houtan K.S., Halley J.M., Van Aarde R. & Pimm S.L. (2009) Achieving success with small, translocated mammal populations. Conservation Letters, 2, 254–262.

(5) Barrio I.C., Herrero J., Bueno C.G., López B.C., Aldezabal A., Campos-Arceiz A. & García- González R. (2013) The successful introduction of the alpine marmot Marmota marmota in the Pyrenees, Iberian Peninsula, Western Europe. Mammal Review, 43, 142–155.

(6) Thalmann S., Peck S., Wise P., Potts J.M., Clarke J. & Richley, J. (2016) Translocation of a top-order carnivore: tracking the initial survival, spatial movement, home-range establishment and habitat use of Tasmanian devils on Maria Island. Australian Mammalogy, 38, 68–79.

(7) Groenewegen R., Harley D., Hill R. & Coulson G. (2017) Assisted colonisation trial of the eastern barred bandicoot (Perameles gunnii) to a fox-free island. Wildlife Research, 44, 484–496.

15. Education and awareness raising

Background

This intervention involves general information and awareness campaigns in response to a range of threats. Studies are included that measure the effect of an action that may be done to change human behaviour for the benefit of mammal populations.

It should be noted that there are many complex factors that influence human behaviour and providing education does not guarantee that behaviour will change. It may be necessary to collaborate with social scientists to design appropriate education programmes that consider the attitudes, values and social norms of the target audience.

Studies describing educational campaigns in response to specific threats are described in the chapter on that threat category.

15.1. Encourage community-based participation in land management

https://www.conservationevidence.com/actions/2395

- **Two studies** evaluated the effects of encouraging community-based participation in management of mammals to reduce mammal persecution. One study was in Pakistan[1] and one was in India[2].

COMMUNITY RESPONSE (0 STUDIES)

POPULATION RESPONSE (1 STUDY)

- **Abundance (1 study):** A study in Pakistan[1] found that involving local communities with park management was associated with an increasing population of Himalayan brown bears.

BEHAVIOUR (0 STUDIES)

OTHER (1 STUDY)

- **Human behaviour change (1 study):** A study in Namibia[2] found that fewer farmers who engaged in community-based management of land, through membership of a conservancy, removed large carnivores from their land than did non-conservancy members.

Background

When local community members are involved in management of local land resources, they may have a greater interest in ensuring long-term sustainability of that management. One potential outcome of this is a reduction in mammal persecution.

A study in 1993–2006 of a primarily mountainous grassland national park in Pakistan (1) found that involving local communities with park management was associated with an increasing population of Himalayan brown bears *Ursus arctos isabellinus*. The known population of bears in the park increased steadily from 19 in 1993 to 43 by 2006. Breeding productivity was, however, low and the increase was reported to be due in part to immigration. The paper attributes the larger population to a reduction in poaching and persecution, linked to increased community engagement in the park since its creation in 1993. This involved recognising local community grazing rights, employing local staff, supporting development projects and enabling local generation of funds from park visitors. Eighty-six bears were monitored. Ten were radio-collared. The remainder were monitored through direct observations of individually recognisable animals.

A study in 2003–2004 of farmers across a large rangeland area in Namibia (2) found that fewer farmers who engaged in community-based management of land through being members of a conservancy removed large carnivores from their land than did non-conservancy members. A lower percentage of conservancy members (57–67%) removed large carnivores compared to non-conservancy members (81–83%). Conservancies were legally protected areas, cooperatively managed by a group of land-occupiers with the goal of sharing resources among members. Some conservancy members derived income from trophy hunting of carnivores. A total of 147 farmers were surveyed from across 30,000 km² of rangeland. They comprised 76 conservancy members (44 mixed farmers, 32 livestock farmers) and 71 non-conservancy members (33 mixed farmers, 38 livestock farmers). Data were collected by face-to-face interviews or by postal questionnaires in 2003–2004.

(1) Nawaz M.A., Swenson J.E. & Zakaria V. (2008) Pragmatic management increases a flagship species, the Himalayan brown bears, in Pakistan's Deosai National Park. *Biological Conservation*, 141, 2230–2241.

(2) Schumann M., Watson L.H. & Schumann B.D. (2008) Attitudes of Namibian commercial farmers toward large carnivores: The influence of conservancy membership. *South African Journal of Wildlife Research*, 38, 123–132.

15.2. Use campaigns and public information to improve behaviour towards mammals and reduce threats

https://www.conservationevidence.com/actions/2422

- **Two studies** evaluated the effects of using campaigns and public information to improve behaviour towards mammals and reduce threats. One study was in the USA[1] and one was in Lao People's Democratic Republic[2].

COMMUNITY RESPONSE (0 STUDIES)

POPULATION RESPONSE (0 STUDIES)

BEHAVIOUR (0 STUDIES)

OTHER (2 STUDIES)

- **Human behaviour change (2 studies)**: A randomized, replicated, controlled, before-and-after study in the USA[1] found that displaying education signs did not reduce the percentage of garbage containers that were accessible to black bears. A controlled, before-and-after study in Lao People's Democratic Republic[2] found that a social marketing campaign promoting a telephone hotline increased reporting of illegal hunting.

Background

Mammals face a range of threats from humans. These may include exploitation through hunting or persecution if the mammal is perceived as a threat or a nuisance. In some cases, mammals are protected by regulations and laws but these may be difficult to enforce. Some infringements may be difficult to detect whilst, in other cases, people may be unaware of their responsibilities under such rules. Campaigns may be designed to increase compliance with laws, to encourage reporting of infringements, such as illegal hunting, or to reduce behaviours that can be a threat to mammals, such as consumption of products derived from wild mammals. These may use a variety of media and ranging from broadcasting and social media through to word of mouth.

A randomized, replicated, controlled, before-and-after study in 2007 in a residential area in Colorado, USA (1) found that displaying education signs about the danger of garbage to black bears *Ursus americanus* did not reduce the percentage of garbage containers that were not wildlife-resistant or wildlife-proof. The overall proportion of households using garbage containers that were not wildlife-resistant or wildlife-proof declined during the study. However, where signage was used, the trend in households not using wildlife-resistant garbage containers (after: 0–31%; before: 11–52%) did not differ from where signage was not used (after: 7–27%; before: 9–45%). Dumpsters were surveyed at 68 communal housing complexes. Thirty-four were randomly selected for placement of signs on dumpsters, warning of dangers of unsecured garbage to bears. Similarly, 42 construction sites were surveyed, with signage used at 22 of these. Dumpsters were surveyed in July–September 2007, for three weeks before and three weeks after installing signage. Violations were use of unsecured containers, unsecure dumpster storage outside kerbside collection times, garbage outside dumpsters and, on building sites, food waste in open dumpsters.

A controlled, before-and-after study in 2009–2010 in 57 villages in and around a protected area in Lao People's Democratic Republic (2) found that a social marketing campaign to promote a newly created telephone hotline increased reporting of illegal hunting. Villagers exposed to the social marketing campaign were significantly more likely to report illegal hunting after the campaign. The reporting rate of the villages not exposed to the campaign did not change significantly (data not reported). In 2009, a telephone hotline was set up for villagers to report illegal hunting. In 36 villages, a social marketing campaign was used to promote the hotline. Twenty-one similar villages did not receive the campaign. Surveys of both groups were conducted before and after the social marketing campaign took place.

(1) Baruch-Mordo S., Breck S.W., Wilson K.R. & Broderick J. (2011) The carrot or the stick? Evaluation of education and enforcement as management tools for human-wildlife conflicts. *PLoS ONE*, 6, e15681.

(2) Saypanya S., Hansel T., Johnson A., Bianchessi A. & Sadowsky B. (2013) Effectiveness of a social marketing strategy, coupled with law enforcement, to conserve tigers and their prey in Nam Et Phou Louey National Protected Area, Lao People's Democratic Republic. *Conservation Evidence*, 10, 57–66.

15.3. Provide education programmes to improve behaviour towards mammals and reduce threats

https://www.conservationevidence.com/actions/2423

- **Two studies** evaluated the effects of providing education programmes to improve behaviour towards mammals and reduce threats. One study was in South Africa[1] and one was in the USA[2].

COMMUNITY RESPONSE (0 STUDIES)

POPULATION RESPONSE (1 STUDY)

- **Abundance (1 study):** A before-and-after study in South Africa[1] found that educating ranchers on ways of reducing livestock losses, along with stricter hunting policies, increased leopard density.

- **Survival (1 study):** A before-and-after study in South Africa[1] found that educating ranchers on ways of reducing livestock losses, along with stricter hunting policies, reduced leopard mortalities.

BEHAVIOUR (0 STUDIES)

OTHER (1 STUDY)

- **Human behaviour change (1 study):** A replicated, controlled, before-and-after study in the USA[2] found that visiting households to educate about the danger of garbage to black bears did not increase use of wildlife-resistant dumpsters.

Background

Where human behaviour is central to the threat to a species, an education programme may be devised to address this. Such programmes may tackle a wide range of threats to mammals and be aimed to difference audiences, such as local residents, farmers or other businesses. The effects of programmes may be measured in terms of the response of target species or in terms of changes in human behaviour that directly impact the magnitude of the threat.

This intervention covers situations where awareness of ways of reducing threats to mammals is focussed on specific narrow target groups, largely through one-to-one interactions. For more widely-targeted programmes, see *Use campaigns and public information to improve behaviour towards mammals and reduce threats.*

A before-and-after study in 2002–2009 in a temperate broadleaf forest and grassland site in KwaZulu-Natal, South Africa (1) found that educating ranchers on methods for reducing livestock losses, along with implementing stricter hunting policies, increased leopard *Panthera pardus* density and reduced leopard mortalities. Four years after both livestock husbandry workshops and hunting policy changes were implemented, there were 11.2 leopards/100 km^2, compared to 7.1/100km^2 in the first year of implementation. Nine leopards were killed during the first three years after livestock husbandry workshops and hunting policy changes were implemented compared to 23 over the previous two years. In January–July 2005, workshops were held to teach improved husbandry techniques to local landowners. Before January 2005 leopards could be killed legally if they had killed livestock. After January 2005 permits were only granted if the same leopard was confirmed (using inspections and camera traps) to have killed three or more livestock within two months and if the landowner could provide evidence that they were trying to reduce attacks on livestock. Thirty-five leopards were radio-collared and monitored between April 2002 and December 2007. Camera traps were used in January–March 2005, January–March 2007, and March–May 2009 to estimate changes in the leopard population size.

A replicated, controlled, before-and-after study in 2008 of a residential area in Colorado, USA (2) found that visiting properties to educate about the danger of garbage to black bears *Ursus americanus* did not increase use of wildlife-resistant dumpsters. Where educational visits were carried out, the trend in availability of garbage to wildlife (before visits: 13–15% of households; after visits: 16–26%) did not differ from those in neighbourhoods that were not visited (before visits: 9–15% of households; after: 16–17%). Similarly, there was no difference in use of bear-resistant containers between neighbourhoods that were visited (before visits: 11–17% of households; after: 16–23%) or not visited (before visits: 14–19% of households; after: 17–18%). In two neighbourhoods, 91% and 87% of residences were visited and residents were spoken to or had educational material delivered. Two further neighbourhoods, did not receiving any visits. Household garbage disposal facilities were surveyed in July–September 2008, before and after visits. Garbage was regarded as accessible if placed outside containers, or in non-bear-resistant containers.

(1) Balme G.A., Slotow R. & Hunter L.T.B. (2009) Impact of conservation interventions on the dynamics and persistence of a persecuted leopard (Panthera pardus) population. *Biological Conservation*, 142, 2681–2690.

(2) Baruch-Mordo S., Breck S.W., Wilson K.R. & Broderick J. (2011) The carrot or the stick? Evaluation of education and enforcement as management tools for human-wildlife conflicts. *PLoS ONE*, 6, e15681.

15.4. Provide science-based films, radio programmes, or books about mammals to improve behaviour towards mammals and reduce threats

https://www.conservationevidence.com/actions/2424

- We found no studies that evaluated the effects on mammals of providing science-based films, radio programmes, or books about mammals to improve behaviour towards mammals and reduce threats.

'We found no studies' means that we have not yet found any studies that have directly evaluated this intervention during our systematic journal and report searches. Therefore, we have no evidence to indicate whether or not the intervention has any desirable or harmful effects.

Background

There are different types of media that can be used to inform people and raise their awareness about threats to mammals and their conservation. Environmental education campaigns frequently use film, sound or print media to present either factual information or fictional stories that have parallels to environmental issues. It can be difficult to assess the impact of such initiatives, as it is harder to assess people's subsequent behaviours than it is to measure their stated attitudes towards mammals after exposure to media.

15.5. Train and support local staff to help reduce persecution of mammals

https://www.conservationevidence.com/actions/2425

- **One study** evaluated the effects of training and supporting local staff to help reduce persecution of mammals. This study was in Kenya[1].

COMMUNITY RESPONSE (0 STUDIES)

POPULATION RESPONSE (1 STUDY)

- **Survival (1 study):** A replicated, before-and-after study in Kenya[1] found that employing local tribesmen to dissuade pastoralists from killing lions and to assist with livestock protection measures, alongside compensating for livestock killed by lions, reduced lion killings by pastoralists.

BEHAVIOUR (0 STUDIES)

Background

Carnivores may be killed by farmers where they feel that their livestock are threatened. National laws or policies protecting wild mammals may be difficult to enforce at a local level. Local staff, from among the same communities as the farmers, may be able to gain more respect and to work more closely with farmers to find ways to reduce losses to predators without carrying out lethal control.

A replicated, before-and-after study in 2003–2011 in savanna grassland in four ranches in southern Kenya (1) found that employing local tribesmen to dissuade pastoralists from killing lions *Panthera leo* and to assist with livestock protection measures, alongside compensating for livestock killed by lions, reduced lion killings by pastoralists. The two schemes occurred at the same time at three group ranches, so their individual effects could not be separated. Compensation for livestock losses was estimated to reduce lion killing by 87–91% whilst additionally employing lion guardians reduced killings by 99%. The four ranches comprised a 3,500-km² study area. Compensation for verified livestock losses to lions was initiated at three of the group ranches between 2003 and 2008. Respected tribesmen, 'lion guardians', were employed to dissuade pastoralists from killing lions and to assist with livestock protection measures, such as reinforcing bomas. The scheme commenced at the four sites between 2007 and 2010. Lion mortality data, from 2003 to 2011, were collated primarily from community informants and direct interviews with lion hunters.

(1) Hazzah L., Dolrenry S., Naughton L., Edwards C.T.T., Mwebi O., Kearney F. & Frank L. (2014) Efficacy of two lion conservation programs in Maasailand, Kenya. *Conservation Biology*, 28, 851–860.

15.6. Publish data on ranger performance to motivate increased anti-poacher efforts

https://www.conservationevidence.com/actions/2426

- **One study** evaluated the effects on poaching incidents of publishing data on ranger performance to motivate increased anti-poacher efforts. This study was in Ghana[1].

COMMUNITY RESPONSE (0 STUDIES)

POPULATION RESPONSE (1 STUDY)

- **Survival (1 study):** A replicated, before-and-after, site comparison study in Ghana[1] found that when data were publishing on staff performance, poaching incidents decreased on these sites and on sites from which performance data were not published.

BEHAVIOUR (0 STUDIES)

OTHER (1 STUDY)

- **Human behaviour change (1 study):** A replicated, before-and-after, site comparison in Ghana[1] found that publishing data on staff performance lead to an increase in anti-poaching patrols.

Background

Where poaching is a threat to mammals, patrols by rangers may be carried out as a deterrent and to apprehend poachers. Ranger teams may be operating in isolated sites and motivation may be negatively impacted. Publishing metrics on team performances may encourage greater effort with patrolling and pride in ranger team activities and achievements.

A replicated, before-and-after, site comparison study in 2004–2006 within savanna and forest in seven protected areas and two national parks in Ghana (1) found that publishing data on staff performance lead to more anti-poaching patrols and that detected of poaching incidents decreased on savanna sites but not on forest sites. Staff performance was 59% higher after reporting (11.8 effective patrol days/staff/month) than before (7.4 effective patrol days/staff/month). In two parks where performance indicators were not reported, performance increased by 11% over this period (after: 10.9; before: 9.8 effective patrol days/ staff/month). In four savanna sites, the average number of detected offences related to poaching (including of mammals) was 72% lower after reporting (21 offences/patrol staff-day) than before (74 offences/ patrol staff-day). In two forest sites, the average number of offences detected after reporting (179/patrol staff-day) was not significantly different to the number before (214 offences/patrol staff-day). In two parks where performance indicators were not reported, the average number of offences detected after reporting (116 offences/patrol staff-day) was not significantly different to the number before (174 offences/ patrol staff-day). Publishing evaluation reports created an awareness of poor performance and generated performance-related competition between sites. Monitoring of patrol effort and illegal activity encounters commenced from mid-2004. Metrics were published at the end of 2005 and monitoring continued through 2006.

(1) Jachmann H. (2008) Monitoring law-enforcement performance in nine protected areas in Ghana. *Biological Conservation*, 141, 89–99.

Appendix 1
Journals (and years) searched

Journals (and years) searched and for which relevant papers have been added to the Conservation Evidence discipline-wide literature database. An asterisk indicates the journals most relevant to this synopsis.

Journal	Years Searched	Topic
Acta Chiropterologica	1999–2017	All biodiversity
Acta Herpetologica	2006–2016	All biodiversity
Acta Oecologica-International Journal of Ecology	1990–2017	All biodiversity
Acta Theriologica*	1977–2014	All biodiversity
African Bird Club Bulletin	1994–2017	All biodiversity
African Journal of Ecology*	1963–2016	All biodiversity
African Journal of Herpetology	1990–2016	All biodiversity
African Journal of Marine Science	1983–2017	All biodiversity
African Primates	1995–2012	All biodiversity
African Zoology	1979–2013	All biodiversity
Agriculture, Ecosystems & Environment*	1983–2017	All biodiversity
Ambio	1972–2011	All biodiversity
American Journal of Primatology	1981–2014	All biodiversity
American Naturalist	1867–2017	All biodiversity
Amphibia-Reptilia	1980–2012	All biodiversity
Amphibian and Reptile Conservation	1996–2012	All biodiversity
Animal Biology	2003–2013	All biodiversity
Animal Conservation*	1998–2018	All biodiversity
Annales Zoologici Fennici	1964–2013	All biodiversity

Journal	Years Searched	Topic
Annales Zoologici Societatis Zoologicae Botanicae Fennicae Vanamo	1932–1963	All biodiversity
Annual Review Ecology and Systematics	1970–2017	All biodiversity
Anthrozoos	1987–2013	All biodiversity
Apidologie	1958–2009	All biodiversity
Applied Animal Behaviour Science	1998–2014	All biodiversity
Applied Herpetology	2003–2009	All biodiversity
Applied Vegetation Science	1998–2017	All biodiversity
Aquaculture Research	1972–2008	All biodiversity
Aquatic Botany	1975–2017	All biodiversity
Aquatic Conservation: Marine and Freshwater Ecosystems	1991–2017	All biodiversity
Aquatic Ecology	1968–2016	All biodiversity
Aquatic Ecosystem Health & Management	1998–2016	All biodiversity
Aquatic Invasions	2006–2016	All biodiversity
Aquatic Living Resources	1988–2016	All biodiversity
Aquatic Mammals	1972–2017	All biodiversity
Arid Land Research and Management	1987–2013	All biodiversity
Asian Primates	2008–2012	All biodiversity
Auk	1980–2016	All biodiversity
Austral Ecology*	1977–2017	All biodiversity
Australasian Journal of Herpetology	2009–2012	All biodiversity
Australian Mammalogy*	2000–2017	All biodiversity
Avian Conservation and Ecology	2005–2016	All biodiversity
Basic and Applied Ecology*	2000–2017	All biodiversity
Behavior	1948–2013	All biodiversity
Behavior Ecology	1990–2013	All biodiversity
Bibliotheca Herpetologica	1999–2017	All biodiversity
Biocontrol	1956–2016	All biodiversity
Biocontrol Science and Technology	1991–1996	All biodiversity
Biodiversity and Conservation*	1994–2017	All biodiversity
Biological Conservation*	1981–2017	All biodiversity
Biological Control	1991–2017	All biodiversity
Biological Invasions	1999–2017	All biodiversity

Journal	Years Searched	Topic
Biology and Environment: Proceedings of the Royal Irish Academy	1993–2017	All biodiversity
Biology Letters	2005–2017	All biodiversity
Biotropica	1990–2017	All biodiversity
Bird Conservation International	1991–2016	All biodiversity
Bird Study	1980–2016	All biodiversity
Boreal Environment Research	1996–2014	All biodiversity
Bulletin of the Herpetological Society of Japan	1999–2008	All biodiversity
Canadian Journal of Fisheries and Aquatic Sciences	1901–2017	All biodiversity
Canadian Journal of Forest Research*	1971–2013	All biodiversity
Caribbean Journal of Science	1961–2013	All biodiversity
Chelonian Conservation and Biology	2006–2016	All biodiversity
Collinsorum	2012–2014	All biodiversity
Community Ecology	2000–2012	All biodiversity
Conservation Biology*	1987–2017	All biodiversity
Conservation Evidence*	2004–2018	All biodiversity
Conservation Genetics	2000–2013	All biodiversity
Conservation Letters*	2008–2017	All biodiversity
Contemporary Herpetology	1998–2009	All biodiversity
Contributions to Primatology	1974–1991	All biodiversity
Copeia	1910–2016	All biodiversity
Cunninghamia	1981–2016	All biodiversity
Current Herpetology	1964–2016	All biodiversity
Dodo	1977–2001	All biodiversity
Ecological and Environmental Anthropology	2005–2008	All biodiversity
Ecological Applications	1991–2017	All biodiversity
Ecological Indicators	2001–2007	All biodiversity
Ecological Management & Restoration	2000–2017	All biodiversity
Ecological Restoration*	1981–2016	All biodiversity
Ecology	1936–2017	All biodiversity
Ecology Letters	1998–2013	All biodiversity
Ecoscience	1994–2013	All biodiversity

Journal	Years Searched	Topic
Ecosystems	1998–2013	All biodiversity
Emu	1980–2016	All biodiversity
Endangered Species Bulletin	1966–2003	All biodiversity
Endangered Species Research	2004–2017	All biodiversity
Environmental Conservation	1974–2017	All biodiversity
Environmental Evidence	2012–2017	All biodiversity
Environmental Management*	1977–2017	All biodiversity
Environmentalist	1981–1988	All biodiversity
Ethology Ecology and Evolution	1989–2014	All biodiversity
European Journal of Soil Science	1950–2012	Soil Fertility
European Journal of Wildlife Research*	1955–2017	All biodiversity
Evolutionary Anthropology	1992–2014	All biodiversity
Evolutionary Ecology	1987–2014	All biodiversity
Evolutionary Ecology Research	1999–2014	All biodiversity
Fire Ecology	2005–2016	All biodiversity
Fisheries Management and Ecology	1994–2018	All biodiversity
Fisheries Research	1990–2018	All biodiversity
Folia Primatologica	1963–2014	All biodiversity
Folia Zoologica	1959–2013	All biodiversity
Forest Ecology and Management*	1976–2013	All biodiversity
Freshwater Biology	1975–2017	All biodiversity
Freshwater Science	1982–2017	All biodiversity
Functional Ecology	1987–2013	All biodiversity
Genetics and Molecular Research	2002–2013	All biodiversity
Geoderma	1967–2012	Soil Fertility
Gibbon Journal	2005–2011	All biodiversity
Global Change Biology	1995–2017	All biodiversity
Global Ecology and Biogeography	1991–2014	All biodiversity
Grass and Forage Science	1980–2017	All biodiversity
Herpetofauna	2003–2007	All biodiversity
Herpetologica	1936–2012	All biodiversity
Herpetological Bulletin	2000–2013	All biodiversity
Herpetological Conservation and Biology	2006–2012	All biodiversity
Herpetological Journal	2005–2012	All biodiversity

Journal	Years Searched	Topic
Herpetological Monographs	1982–2012	All biodiversity
Herpetological Review	1967–2014	All biodiversity
Herpetology Notes	2008–2014	All biodiversity
Human Wildlife Interactions*	2007–2017	All biodiversity
Hydrobiologia	2000–2017	All biodiversity
Hystrix, the Italian Journal of Mammalogy*	1986–2017	All biodiversity
Ibis	1980–2016	All biodiversity
ICES Journal of Marine Science	1990–2018	All biodiversity
iForest	2008–2016	All biodiversity
Integrative Zoology	2006–2013	All biodiversity
International Journal of Pest Management (formerly PANS Pest Articles & News Summaries 1969–1975, PANS 1976–1979 & Tropical Pest Management 1980–1992)	1969–1979	All biodiversity
International Journal of the Commons	2007–2016	All biodiversity
International Journal of Wildland Fire	1991–2016	All biodiversity
International Wader Studies	1970–1972	All biodiversity
International Zoo Yearbook	1960–2015	Management of Captive Animals
Invasive Plant Science and Management	2008–2016	All biodiversity
Israel Journal of Ecology & Evolution	1963–2013	All biodiversity
Italian Journal of Zoology	1978–2013	All biodiversity
Journal for Nature Conservation*	2002–2017	All biodiversity
Journal of Animal Ecology	1932–2017	All biodiversity
Journal of Apicultural Research	1962–2009	All biodiversity
Journal of Applied Ecology*	1964–2017	All biodiversity
Journal of Aquatic Plant Management	1962–2016	All biodiversity
Journal of Arid Environments	1993–2017	All biodiversity
Journal of Avian Biology	1980–2016	All biodiversity
Journal of Bat Conservation and Research	2000–2017	All biodiversity
Journal of Cetacean Research and Management	1999–2012	All biodiversity
Journal of Ecology	1933–2017	All biodiversity
Journal of Environmental Management	1973–2017	All biodiversity

Journal	Years Searched	Topic
Journal of Experimental Marine Biology & Ecology	1980–2016	All biodiversity
Journal of Field Ornithology	1980–2016	All biodiversity
Journal of Forest Research	1996–2017	All biodiversity
Journal of Great Lakes Research	1975–2017	All biodiversity
Journal of Herpetological Medicine and Surgery	2009–2013	All biodiversity
Journal of Herpetology	1968–2015	All biodiversity
Journal of Kansas Herpetology	2002–2011	All biodiversity
Journal of Mammalian Evolution	1993–2014	All biodiversity
Journal of Mammalogy*	1919–2017	All biodiversity
Journal of Mountain Science	2004–2016	All biodiversity
Journal of Negative Results: Ecology & Evolutionary Biology	2004–2016	All biodiversity
Journal of Ornithology	2004–2017	All biodiversity
Journal of Primatology	2012–2013	All biodiversity
Journal of Raptor Research	1966–2016	All biodiversity
Journal of Sea Research	1961–2017	All biodiversity
Journal of the Japanese Institute of Landscape Architecture	1934–2017	All biodiversity
Journal of the Marine Biological Association of the United Kingdom	1887–2006	All biodiversity
Journal of Tropical Ecology	1986–2017	All biodiversity
Journal of Vegetation Science	1990–2017	All biodiversity
Journal of Wetlands Ecology	2008–2012	All biodiversity
Journal of Wetlands Environmental Management	2012–2016	All biodiversity
Journal of Wildlife Diseases	1965–2012	All biodiversity
Journal of Zoo and Aquarium Research	2013–2016	All biodiversity
Journal of Zoology*	1966–2017	All biodiversity
Jurnal Primatologi Indonesia	2009	All biodiversity
Kansas Herpetological Society Newsletter	1977–2001	All biodiversity
Lake and Reservoir Management	1984–2016	All biodiversity
Land Degradation and Development	1989–2016	All biodiversity
Land Use Policy	1984–2012	Soil Fertility

Journal	Years Searched	Topic
Latin American Journal of Aquatic Mammals	2002–2016	All biodiversity
Lemur News	1993–2012	All biodiversity
Limnologica — Ecology and Management of Inland Waters	1999–2017	All biodiversity
Mammal Research	2001–2017	All biodiversity
Mammal Review*	1970–2017	All biodiversity
Mammal Study	2005–2017	All biodiversity
Mammalia*	1937–2017	All biodiversity
Mammalian Biology*	2002–2017	All biodiversity
Mammalian Genome	1991–2013	All biodiversity
Management of Biological Invasions	2010–2016	All biodiversity
Mangroves and Salt Marshes	1996–1999	All biodiversity
Marine Ecological Progress Series	2000–2018	All biodiversity
Marine Environmental Research	1978–2017	All biodiversity
Marine Mammal Science	1985–2017	All biodiversity
Marine Pollution Bulletin	2010–2017	All biodiversity
Mires and Peat	2006–2016	All biodiversity
Natural Areas Journal	1992–2017	All biodiversity
Neobiota	2011–2017	All biodiversity
Neotropical Primates	1993–2014	All biodiversity
New Journal of Botany	2011–2013	All biodiversity
New Zealand Journal of Zoology	1974–2017	All biodiversity
New Zealand Plant Protection	2000–2016	All biodiversity
Northwest Science	2007–2016	All biodiversity
Oecologia*	1969–2017	All biodiversity
Oikos	1949–2017	All biodiversity
Ornitologia Neotropical	1990–2018	All biodiversity
Oryx*	1950–2017	All biodiversity
Ostrich	1980–2016	All biodiversity
Pacific Conservation Biology*	1993–2017	All biodiversity
Pakistan Journal of Zoology	2004–2013	All biodiversity
Plant Ecology	1948–2007	All biodiversity
Plant Protection Quarterly	2008–2016	All biodiversity

Journal	Years Searched	Topic
Polish Journal of Ecology	2002–2013	All biodiversity
Population Ecology	1952–2013	All biodiversity
PLOS	1980–2018	Key word: bat*
Preslia	1973–2017	All biodiversity
Primate Conservation	1981–2014	All biodiversity
Primates	1957–2013	All biodiversity
Rangeland Ecology & Management (previously Journal of Range Management 1948–2004)*	1948–2016	All biodiversity
Raptors Conservation	2005–2016	All biodiversity
Regional Studies in Marine Science	2015–2017	All biodiversity
Restoration Ecology*	1993–2017	All biodiversity
Revista Chilena de Historia Natural	2000–2016	All biodiversity
Revista de Biología Tropical	1976–2013	All biodiversity
River Research and Applications	1987–2016	All biodiversity
Russian Journal of Herpetology	1994–2000	All biodiversity
Slovak Raptor Journal	2007–2016	All biodiversity
Small Ruminant Research	1988–2017	All biodiversity
Soil Biology & Biochemistry	1969–2012	Soil Fertility
South African Journal of Botany	1982–2016	All biodiversity
South African Journal of Wildlife Research*	1971–2014	All biodiversity
South American Journal of Herpetology	2006–2012	All biodiversity
Southern Forests: a journal of Forest Science	2008–2013	All biodiversity
Southwestern Naturalist	1956–2013	All biodiversity
Strix	1982–2017	All biodiversity
Systematic Reviews Centre for Evidence-Based Conservation*	2004–2017	All biodiversity
The Canadian Field-Naturalist*	1987–2017	All biodiversity
The Condor	1980–2016	All biodiversity
The Journal of Wildlife Management*	1945–2017	All biodiversity
The Open Ornithology Journal	2008–2016	All biodiversity
The Rangeland Journal	1976–2016	All biodiversity
Trends in Ecology and Evolution	1986–2017	All biodiversity
Tropical Conservation Science	2008–2014	All biodiversity

Journal	Years Searched	Topic
Tropical Ecology	1960–2014	All biodiversity
Tropical Grasslands	1967–2010	All biodiversity
Tropical Zoology	1988–2013	All biodiversity
Turkish Journal of Zoology	1996–2014	All biodiversity
Vietnamese Journal of Primatology	2007–2009	All biodiversity
Wader Study Group Bulletin	1970–1977	All biodiversity
Waterbirds	1983–2016	All biodiversity
Weed Biology and Management	2001–2016	All biodiversity
Weed Research	1961–2017	All biodiversity
West African Journal of Applied Ecology	2000–2016	All biodiversity
Western North American Naturalist	2000–2016	All biodiversity
Wetlands	1981–2016	All biodiversity
Wetlands Ecology and Management	1989–2016	All biodiversity
Wildfowl	1948–2016	All biodiversity
Wildlife Biology*	1995–2013	All biodiversity
Wildlife Monographs	1958–2013	All biodiversity
Wildlife Research*	1974–2017	All biodiversity
Wildlife Society Bulletin*	1973–2017	All biodiversity
Wilson Journal of Ornithology	1980–2016	All biodiversity
Zhurnal Obshchei Biologii	1972–2013	All biodiversity
Zoo Biology	1982–2016	All biodiversity
ZooKeys	2008–2013	All biodiversity
Zoologica Scripta	1971–2014	All biodiversity
Zoological Journal of the Linnean Society	1856–2013	All biodiversity
Zootaxa	2004–2014	All biodiversity

Index

Abrothrix longipilis (long-haired field mice) 778

Acacia seyal 127

Acinonyx jubatus (cheetah)

    deter predation using nearby people 173

    install crossing points 105–106

    install non-electric fencing 119

    provide live prey to captive mammals 906

    release captive-bred individuals 873

    release mammals in family/social groups 956

    release mammals in unrelated groups 969

    translocate predators 154, 156–157

    translocate to boost populations 771

    use guardian animals 144, 147

    use holding pens 798

    use visual deterrents 133

    wild foster parents 729

Acrobates pygmaeus (feathertail gliders)

    install rope bridges between canopies 306

    provide artificial dens 698, 700–701, 704, 706

Aepyceros melampus (impala)

    provide artificial waterholes 538–539

    provide live natural prey 906

    remove vegetation by hand/machine 662

    use prescribed burning 520

Aethomys chrysophilus (red veld rat) 283

African elephants. *See Locondonta africana*

African honeybees. *See Apis mellifera scutellata*

African savanna elephants. *See Loxodonta africana*

African wild dogs. *See Lycaon pictus*

agile antechinus. *See Antechinus agilis*

Agropyron repens (grass) 288

Agropyron spicatum (bluebunch wheatgrass)

    remove vegetation by hand/machine 657, 665

    use prescribed burning 508, 521

Ailuropoda melanoleucahela (giant pandas) 861

Akodon azarae (grass mice) 690, 694

Alaskan brown bears. *See Ursus arctos*

Alcelaphus lichtensteini (Lichtenstein's hartebeest) 513

Alces alces (moose)

    allow forest to regenerate naturally 476

    apply fertilizer to trees 470–471

    install barrier fencing along railways 355

    install barrier fencing along roads 319, 322

    install crossings over/under pipelines 400–401

    install one-way gates 313

    install overpasses over roads/railways 289, 293, 298

    install underpasses along roads 327, 329–331, 337, 340, 346, 353

    modify the roadside environment 375

    modify vegetation along railways 377

    provide food/salt lick to divert mammals 396

    provide supplementary food 737

remove competing vegetation 448

remove or control predators 587

thin trees within forest 466–467

use alternative de-icers on roads 394

use chemical repellents 391–392

use road lighting to reduce vehicle collisions 388

use wildlife refuges to reduce hunting 421

Algerian mice. *See Mus spretus*

Allegheny woodrats. *See Neotoma magister*

Alnus spp. (alder) 954

Alopex lagopus (Arctic fox) 737, 745

Alouatta guariba clamitans (brown howler monkeys) 307

alpacas. *See Vicugna pacos*

Alpine ibex. *See Capra ibex*

alpine marmot. *See Marmota marmota*

Altechinus flavipes (yellow-footed antechinus) 249

American marten. *See Martes americana*

American mink. *See Neovison vison; See Mustela vison*

American pygmy shrews. *See Sorex hoyi*

Ammospermophilus nelson (San Joaquin antelope squirrel) 93

Ammotragus lervia (aoudad) 864

Amur tigers. *See Panthera tigris altaica*

Anas spp. (duck) 159

Anderson's gerbils. *See Gerbillus allenbyi*

antbears. *See Orycteropus afer*

Antechinus agilis (agile antechinus)

provide artificial dens or nest boxes 700

use patch retention harvesting 439

Antechinus flavipes (mardo)

provide artificial dens or nest boxes 699, 701, 704, 707–708

restore former mining sites 247, 250–251

Antechinus stuartii (brown antechinus)

provide artificial dens or nest boxes 698, 703–704, 706

remove/control non-native mammals 564–565

Antechinus swainsonii (dusky antechinus)

install tunnels/culverts/underpass under roads 260

install underpasses beneath ski runs 52

antelope jackrabbit. *See Lepus alleni*

Antilocapra americana (pronghorn)

apply fertilizer to vegetation 688

captive rear in large enclosures 884

install barrier fencing and underpasses 333, 346

install mammal crossing points along fences 104

release translocated mammals into fenced areas 829

remove or control predators 586

use permeable livestock fences 102

use prescribed burning 509

Antilocapra american sonoriensis (Sonoran pronghorns) 632

aoudad. *See Ammotragus lervia*

Apennine chamois. *See Rupicapra pyrenaica ornata*

Apis mellifera (feral honeybees) 975

Apis mellifera scutellata (African honeybees) 213

Aplodontia rufa (mountain beavers)

use predator scent to deter crop damage 226–227

use repellents that taste bad 234

Apodemus agrarius (striped field mouse) 267

Apodemus sylvaticus (wood mice)

create uncultivated margins around fields 59–61

establish wild flower areas on farmland 56

exclude livestock from semi-natural habitat 91

install overpasses over roads/railways 290

install tunnels/culverts/underpass under railways 275

install tunnels/culverts/underpass under roads 260, 267

leave headlands in fields unsprayed 622–623

plant trees on farmland 73

reduce pesticide or fertilizer use 620

use prescribed burning 521

use repellent on slug pellets 78

use set-aside areas on farmland 65–67

Aquila audax (wedge-tailed eagle) 597

Aquila chrysaetos (golden eagle)

release translocated/captive-bred mammals 937

use holding pens at release site 897

Arabian gazelles. *See Gazella arabica*

Arabian oryx. *See Oryx leucoryx*

Arabian sand gazelles. *See Gazella subgutturosa marica*

Arborimus albipes (white-footed vole) 463

Arctic fox. *See Vulpes lagopus; See Alopex lagopus*

argali. *See Ovis ammon polii*

armadillos. *See Dasypus novemcinctus*

Artemesia (sagebrush) 687

Arvicola amphibius (water vole)

release translocated/captive-bred mammals 939

restore or create wetlands 683

Arvicola terrestris (water vole)

release translocated/captive-bred mammals 936

remove/control non-native mammals 564, 570

use holding pens at release site 803, 894

Arvicolinae (vole spp.)

install barrier fencing and underpasses along roads 329

install tunnels/culverts/underpass under roads 263

Arviola terrestris (water vole) 406

ash-grey mouse. *See Pseudomys albocinereus*

Asian elephants. *See Elephas maximus*

Asiatic black bears. *See Ursus thibetanus*

Asiatic lion. *See Panthera leo persica*

Asiatic wild ass. *See Equus hemionus*

Avena barbat (slender oat) 671

Avena sativa (winter oats)

install electric fencing to protect crops 163

plant crops to provide supplementary food 83

provide supplementary food 848

azure-winged magpies. *See Cyanopica cyanus*

baboons. *See Papio* sp.

badgers. *See Meles meles; See Taxidea taxus*

Baiomys taylori (northern pygmy mice) 555

banded hare-wallabies. *See Lagostrophus fasciatus*

bandicoots. *See Perameles nasuta*

bank vole. *See Clethrionomys glareolus; See Myodes glareolus*

barn owl. *See Tyto alba*

barrier fencing 259, 274, 282, 284, 288–289, 291–293, 295, 310–316, 318, 320, 323–328, 330, 334, 336, 338–343, 355, 367, 389, 402–404

install barrier fencing along railways 355

install barrier fencing along roads 315–323

install barrier fencing along waterways 404–405

install barrier fencing and underpasses along roads 323–354

barriers, use of 7, 21, 24, 40–41, 53, 91, 93, 100, 102, 104–106, 114–118, 120–124, 126–127, 129, 131, 145, 160–161, 163–167, 169–171, 186, 200, 214, 220, 222, 224, 242, 247–249, 251–252, 259, 269–270, 274, 280, 282–289, 291–298, 310–355, 367–368, 370, 372–373, 388–389, 400, 402, 404–405, 453, 455, 480, 484, 492–493, 516, 534, 540–541, 543–553, 560, 613–614, 629, 642–643, 651, 662, 692, 749, 798–799, 809, 845, 894, 912, 914, 925, 935, 937–938, 956, 975, 980

build fences around protected areas 641–642

exclude wild mammals using ditches, moats, walls or other barricades 126–128

install automatically closing gates at field entrances to prevent mammals entering 170–171

install electric fencing to protect crops from mammals 161–167

install electric fencing to reduce predation of livestock by mammals 120–126

install fences around existing culverts or underpasses under roads/ railways 283–286

install mammal crossing points along fences on farmland 103–106

install non-electric fencing to exclude predators or herbivores 114–120

install one-way gates or other structures to allow wildlife to leave roadways 310–315

protect mammals close to development areas (e.g. by fencing) 24

use electric fencing to deter mammals from energy installations or mines 251–252

use fencing/netting to reduce predation of fish stock by mammals 186–187

use fencing to exclude grazers or other problematic species 543–546

use fencing to exclude predators or other problematic species 546–553

use fencing to protect water sources for use by wild mammals 540

use livestock fences that are permeable to wildlife 102–103

use tree tubes/small fences/cages to protect individual trees 479–480

Bettongia gaimardi (eastern bettongs)
   hold translocated mammals in captivity before release 822–823
   provide supplementary food during/ after release of translocated mammals 851, 854

release translocated/captive-bred mammals 942, 945

release translocated mammals into fenced areas 839, 842

Bettongia lesueur (boodie)
   provide supplementary food 850, 853
   release captive-bred mammals into fenced areas 911, 914, 918
   release translocated/captive-bred mammals 932, 937, 941, 943, 945, 947, 971–972
   release translocated mammals into fenced areas 830, 835, 840
   translocate to re-establish or boost populations 768
   use holding pens at release site 808, 897

Bettongia penicillata (woylies)
   release translocated/captive-bred mammals 933, 946–947, 949
   release translocated mammals into fenced areas 838, 841
   remove/control non-native mammals 560–562, 568, 571
   translocate to re-establish or boost populations 768

bharal. *See Pseudois nayaur*

bighorn sheep. *See Ovis canadensis*

bilbies. *See Perameles lagotis*

bison. *See Bison bison*

Bison bison (bison)
   manage vegetation using grazing 653
   release translocated mammals into fenced areas 829
   translocate predators for ecosystem restoration 791
   treat disease in wild mammals 599
   use permeable livestock fences 102
   use prescribed burning 512

Bison bison bison (plains bison)
   treat disease in wild mammals 599
   use prescribed burning 512

Bison bonasus (European bison)
   provide supplementary food 921

release translocated/captive-bred mammals 959

translocate to re-establish or boost populations 782

biting dog lice. *See Trichodectes canis*

black bear. *See Ursus americanus*

black-footed ferrets. *See Mustela nigripes*

black rat. *See Rattus rattus*

black rhinoceros. *See Diceros bicornis*

black-tailed hares. *See Lepus californicus*

black-tailed prairie dog. *See Cynomys; See Cynomys ludovicianus*

Blarina brevicauda (northern short-tailed shrews)

exclude livestock from semi-natural habitat 91

manage vegetation using grazing 653

thin trees within forest 465

use prescribed burning 516

use selective harvesting 435

Blarina carolinensis (southeastern short-tailed shrew)

install barrier fencing and underpasses 334

leave coarse woody debris 455

leave standing deadwood 453

bobcats. *See Felis rufus; See Lynx rufus*

Bolam's mice. *See Pseudomys bolami*

bontebok. *See Damaliscus pygargus pygargus*

boodie. *See Bettongia lesueur*

Bos gaurus (gaur) 664, 666

Bos taurus (cattle) 545

bot fly. *See Cuterebra*

Bradypus torquatus (maned sloths) 767, 784

Brassica napus (oilseed rape) 66

broad-toothed rat. *See Mastacomys fuscus*

Bromus inermns (brome grass) 87

brown antechinus. *See Antechinus stuartii*

brown hare. *See Lepus capensis*

brown howler monkeys. *See Alouatta guariba clamitans*

brown hyaena. *See Hyaena brunnea*

brown rats. *See Rattus norvegicus*

Brucella abortus (brucellosis) 599

brucellosis. *See Brucella abortus*

brush mice. *See Peromyscus boylii*

brush-tailed phascogale. *See Phascogale tapoatafa*

brush-tailed rock wallabies. *See Petrogale penicillata*

Bubo virginianus (great-horned owls)

release captive-bred mammals 876, 915

release captive-bred/translocated mammals 938

buffalo. *See Syncerus caffer*

Bufo marinus (cane toads) 553–554

burning, use of

burn at specific time of year 524–526

use prescribed burning 505–524

Burramys parvus (mountain pygmy-possums) 259

bushpigs. *See Potamochoerus larvatus*

bush rat. *See Rattus fuscipes*

bushveld gerbils. *See Tatera leucogaster*

bushy-tailed wood rat. *See Neotoma cinerea*

Buteo jamaicensis (red-tailed hawks) 908

cactus mouse. *See Peromyscus eremicus*

California ground squirrel. *See Spermophilus beecheyi; See Otospermophilus beecheyi*

Callitris glaucophylla (white cypress-pine) 691

Calluna vulgaris (heather)

replant vegetation 661

use prescribed burning 519

Calomys laucha (small vesper mice) 690, 694

cane toads. *See Bufo marinus*

Canis aureus (jackals) 128

Canis dingo (dingoes)

install tunnels under roads 266

use fencing to exclude predators 551

use guardian animals to deter predators 146

use non-lethal methods 502

Canis familiaris (feral dogs) 586

Canis latrans (coyotes)

  deter mammals from human-occupied areas 48

  deter predation of livestock using shock collars 175

  dispose of livestock carcasses to deter predation 141

  install barrier fencing and underpasses 329, 331, 333, 345–346, 348

  install crossings over/under pipelines 401

  install electric fencing to reduce predation 121–124

  install non-electric fencing 116

  install overpasses over roads/railways 289

  install overpasses over waterways 403

  install tunnels under roads 263–265, 267, 270

  place captive young with captive foster parents 858

  provide artificial refuges/breeding sites 691

  provide diversionary feeding for predators 594

  release captive-bred individuals 872, 876

  release captive-bred mammals 915

  release translocated/captive-bred mammals 937–938, 940, 963, 980

  remove/control non-native plants 574

  remove or control predators 586–588

  restore or create forest 679

  use conditioned taste aversion 497

  use fencing to exclude predators 550

  use flags to reduce predation 130

  use guardian animals to deter predators 142–143

  use holding pens at release site 897

  use lights and sound to deter predation 180–181

  use loud noises to deter predation 149

  use predator scent to deter crop damage 226–227

  use repellents that taste bad 237

  use scent to deter predation 183

  use taste-aversion to reduce predation 135–138

Canis lupus (grey wolves)

  control ticks/fleas/lice in wild mammals 616

  eter predation of livestock using shock collars 176–177

  install barrier fencing and underpasses 329–331, 335–336, 342

  install crossings over/under pipelines 401

  install electric fencing 124, 126

  install overpasses over roads/railways 289, 291, 295

  install overpasses over waterways 403

  keep livestock in enclosures 161

  pay farmers to compensate for losses 112, 114

  rehabilitate injured, sick or weak mammals 719, 722

  release translocated/captive-bred mammals 954

  remove or control predators 585, 587–588

  restrict hunting of a species 411

  sterilize predators 591

  translocate mammals to reduce overpopulation 789

  translocate predators 791

  translocate predators away from livestock 152, 154

  use chemical repellents 392

  use flags to reduce predation 129–132

  use guardian animals to deter predators 145

  use holding pens at release site 797, 802, 893

  use lights and sound to deter predation 181

  use scent to deter predation 184

Canis lupus baileyi (Mexican wolves)

  provide supplementary food 923

release translocated/captive-bred mammals 957

use holding pens at release site 891

Canis lupus dingo (dingoes)

release captive-bred mammals 915

release translocated/captive-bred mammals 938

release translocated mammals 835

use non-lethal methods 501, 503

Canis lupus familiaris (dogs)

deter mammals from human-occupied areas 46–48

deter predation of livestock by mammals 172

dispose of livestock carcasses to deter predation 141

install barrier fencing and underpasses 338

install tunnels under roads 266

keep dogs indoors or in outside enclosures 29

keep domestic cats and dogs well-fed 30

pay farmers to compensate for losses 112

release captive-bred individuals 875

release translocated/captive-bred mammals 947

threat: invasive alien and other problematic species 543

threat: residential and commercial development 23

translocate predators away from livestock 153

translocate problem mammals away from residential areas 34

use chili to deter crop damage 218

use dogs to guard crops 240–241

use fire to deter crop damage 223

use guardian animals to deter predators 142–147

use non-lethal methods to deter carnivores 501

use scarecrows to deter crop damage 197

Canis lupus lupus (grey wolf) 858–859

Canis lupus rufus (red wolves)

provide supplementary food 924

use holding pens at release site 893

Canis mesomelas (jackals)

fit livestock with protective collars 179

install electric fencing to reduce predation 123

install mammal crossing points 105–106

provide supplementary food 744

use guardian animals to deter predators 147

Canis rufus (red wolf) 124

Canis simensis (Ethiopian wolves) 602

Canus latrans (coyotes) 346

Cape mountain zebra. *See Equus zebra zebra*

Capra hircus (goats)

release captive-bred mammals 912

release translocated/captive-bred mammals 935, 948

Capra ibex (Alpine ibex) 777

Capra pyrenaica (Iberian ibex) 780, 787

Capra pyrenaica hispanica (Spanish ibex) 611–612

Capra sibirica (ibex) 426

Capreolus capreolus (roe deer)

install barrier fencing along roads 322

install barrier fencing and underpasses 330, 339–340, 342

install non-electric fencing 117

install overpasses over roads/railways 289–290, 293, 295

install overpasses over waterways 403

install tunnels under roads 261

install wildlife warning reflectors 362–363

keep dogs indoors 29

provide mammals with escape routes 406

provide supplementary food 846, 853

release captive-bred individuals 869, 879

translocate to re-establish or boost populations 768, 772, 781, 784

use chemical repellents 390, 392–393

use drugs to treat parasites 609, 613

use holding pens at release site 802, 812

Caracal caracal (caracals)

fit livestock with protective collars 179

use guardian animals to deter predators 147

caracals. *See Caracal caracal*

caragana. *See Caragana arborescens*

Caragana arborescens (caragana) 232

caribou. *See Rangifer tarandus*

Carolina northern flying squirrels. *See Glaucomys sabrinus; See Glaucomys sabrinus coloratus*

Castor canadensis (North American beavers)

airborne translocation of mammals using parachutes 826

provide diversionary feeding for predators 594

provide supplementary feed 481

reduce intensity of grazing 99

translocate predators 791

use prescribed burning 516, 523

use repellents that taste bad 235

Castor fiber (Eurasian beavers)

release translocated/captive-bred mammals 954

translocate to re-establish or boost populations 762, 783

cattle. *See Bos taurus*

Central American tapir. *See Tapirus bairdii*

Ceratoiherium simum cottoni (white rhinoceros) 757

Ceratotherium simum (white rhinoceros)

provide artificial waterholes in dry season 538

rehabilitate injured, sick or weak mammals 717

release translocated/captive-bred mammals 992

translocate to re-establish or boost populations 758, 773–774, 785

Cercartetus concinnus (western pygmy possum)

provide artificial dens or nest boxes 704

restore former mining sites 249

Cercartetus nanus (eastern pygmy-possums) 700, 702–703, 709

Cercocarpus spp. (mountain mahogany) 537

Cersus elaphus (elk) 103

Cervus canadensis (elk)

apply fertilizer to vegetation 689

deter mammals from human-occupied areas 47

install barrier fencing along roads 322

install barrier fencing and underpasses 329–333, 335–336, 345–346, 349

install metal grids at field entrances 169

install overpasses over roads/railways 289, 291

install wildlife exclusion grates 369

manage vegetation using grazing by wild herbivores 653

manage vegetation using livestock grazing 649–650

provide supplementary food 743, 923

reduce intensity of grazing 95

reduce legal speed limit 371

release translocated/captive-bred mammals 939, 957, 961

release translocated mammals into fenced areas 829

restrict hunting of a species 410

restrict hunting of particular sex/ breeding age animals 417

thin trees to reduce wildfire risk 529

translocate mammals to reduce overpopulation 789

translocate predators for ecosystem restoration 791

translocate to re-establish or boost populations 761, 771

use chemical repellents 391

use holding pens at release site 804, 891

use lights and sound to deter crop damage 189

use permeable livestock fences 102

use prescribed burning 508, 510, 516–518

use repellents that taste bad 233

Cervus elaphus (red deer)

install barrier fencing and underpasses 336, 339, 342

install non-electric fencing 117

install overpasses over roads/railways 290–292, 295

install overpasses over waterways 403

install wildlife warning reflectors 363

provide diversionary feeding to reduce crop damage 193

provide mammals with escape routes from canals 406

provide supplementary food 742

release captive-bred individuals 873, 880

release translocated/captive-bred mammals 966

translocate to re-establish or boost populations 780

use chemical repellents 390, 392–393

use holding pens at release site 812

Cervus nippon (sika deer)

install electric fencing 165

use chemical repellents 390

Chaetodipus fallax (San Diego pocket mouse) 676

Chaetodipus formosus (long-tailed pocket mouse) 90

Chaetodipus spinatus (spiny pocket mouse) 676

cheetahs. *See Acinonyx jubatu*

Chenopodium quinoa (quinoa) 84

chili, use as deterrent 165, 195, 198–200, 215–220, 222, 224, 242

chipmunk. *See Tamias minimus;*
See Tamias striatus

chuditch. *See Dasyurus geoffroii*

Cirsium spp. (thistles) 61

civet. *See Civettictis civetta*

Civettictis civetta (civet) 51

clearfelling 438–439, 459, 472, 478

Clethrionomys californicus (Western red-backed vole) 463

Clethrionomys gapperi (red-backed voles)

install barrier fencing and underpasses 335

install overpasses over roads/railways 290

leave coarse woody debris in forests 454

open patches in forests 443–444

provide woody debris 54

thin trees within forest 465

use patch retention harvesting 438

use selective harvesting 434

Clethrionomys glareolus (bank voles)

create uncultivated margins around fields 60

provide supplementary food 736

use repellent on slug pellets 78

collared peccaries. *See Pecari tajacu*

collars, use of 32–35, 46–47, 128, 146, 153, 155, 174, 176–177, 248, 293–294, 319, 339–340, 342, 402–403, 407, 422, 513, 518–520, 528–529, 537, 548, 568, 585, 588–589, 591, 594, 611, 663, 669, 672, 692, 739, 742, 760–761, 767, 778–779, 789–790, 796, 798, 802, 807, 816, 821, 830, 832–833, 868, 870, 873, 877, 911, 933, 939, 955–956, 958, 964, 986, 996, 1001

use collar-mounted devices 25, 25–28, 28, 543

Columbian ground squirrel. *See Spermophilus columbianus*

Commiphora (thorny trees) 118

common brushtail possum. *See Trichosurus vulpecula;*
See Trichosurus sp.

common dunnart. *See Sminthopsis murina*

common hamsters. *See Cricetus cricetus*

common mallow. *See Malva sylvestris*

common ringtail possums. *See Pseudocheirus peregrinus*

common teasel. *See Dipsacus sylvestris*

common vole. *See Microtus arvalis*

common wallaroos. *See Macropus robustus*

common wombat. *See Vombatus ursinus*

Conepatus chinga (Molina's hog-nosed skunk) 451

Connochaetes taurinus (wildebeest)

  build fences around protected areas 642

  provide artificial waterholes 538

  replant vegetation 662

  use prescribed burning 521

Corsican pines. *See Pinus nigra*

cotton mouse. *See Peromyscus gossypinus*

cottonwood. *See Populus* spp.

cougars. *See Felis concolor*

coyotes. *See Canis latrans; See Canus latrans*

coypu. *See Myocastor coypus*

crabapple. *See Malus* spp.

crest-tailed mulgara. *See Dasycercus cristicauda*

Cricetulus kamensis (Kam dwarf hamster) 98

Cricetus cricetus (common hamsters)

  establish wild flower areas 57–58

  provide mammals with escape routes 406

  release translocated mammals 836, 841

Crocidura russula (white-toothed shrew)

  exclude livestock from semi-natural habitat 91

  remove/control non-native mammals 563

  use prescribed burning 521

Crocidura suaveolens (white-toothed shrews) 563

Crocuta crocuta (spotted hyenas)

  exclude wild mammals using barricades 127–128

install non-electric fencing 119

release captive-bred mammals 873

release translocated/captive-bred mammals 992

translocate to re-establish or boost populations 771, 774

use guardian animals to deter predators 144

use people to deter predation 173

use visual deterrents to deter predation 133

Crotalus viridis (prairie rattlesnakes) 908

Cryptotis parva (North American least shrew)

  leave coarse woody debris 455

  leave standing deadwood/snags 453

culpeo fox. *See Lycalopex culpaeus*

culverts 258–261, 263–265, 267–286, 288, 291–292, 294, 297, 316, 319, 322, 324–325, 328–337, 339–341, 343, 346–348, 350–353

Cuterebra (bot fly)

  emove vegetation using herbicides 670

  remove vegetation using herbicides 668

  use prescribed burning 510–511, 522

Cyanopica cyanus (azure-winged magpies) 586

Cynomys (black-tailed prairie dog)

  captive rear in large enclosures 883

  control ticks/fleas/lice in wild mammals 616

  provide live natural prey to captive mammals 905

  provide supplementary food 845

  release captive-bred mammals 876

  release translocated/captive-bred mammals 937, 940, 957, 959–960, 963, 968, 975, 980

  remove vegetation using herbicides 667

  replant vegetation 658, 660, 665

  train captive-bred mammals to avoid predators 908

translocate to re-establish or boost populations 776

use fencing to exclude predators 550

use holding pens at release site 799, 808, 897

use prescribed burning 515, 523

Cynomys ludovicianus (black-tailed prairie dog)

captive rear in large enclosures 883

release translocated/captive-bred mammals 959–960, 968, 975

remove vegetation by hand/machine 665

replant vegetation 660

se prescribed burning 523

train captive-bred mammals to avoid predators 908

use fencing to exclude predators 550

use prescribed burning 515

Cynomys parvidens (Utah prairie dogs)

control ticks/fleas/lice in wild mammals 616

release captive-bred/translocated mammals 940, 963, 980

remove vegetation using herbicides 667

replant vegetation 658

Dama dama (fallow deer)

install acoustic wildlife warnings 365

install overpasses over roads/railways 290

install tunnels under roads 260

install wildlife warning reflectors 359, 363

use chemical repellents 390

use repellents that taste bad 235

Damaliscus lunatus (tsessebe)

provide artificial waterholes 538

provide live natural prey to captive mammals 906

use prescribed burning 513

Damaliscus pygargus pygargus (bontebok) 653

Dama mesopotamica (Persian fallow deer) 893

Dasycercus cristicauda (crest-tailed mulgara) 569

Dasypus novemcinctus (armadillos)

install barrier fencing and underpasses 328, 334

restore or create forest 679

translocate to re-establish or boost populations 773

Dasyurus geoffroii (chuditch)

release captive-bred mammals 871

remove/control non-native mammals 566

restore former mining sites 247

translocate mammals 255

Dasyurus viverrinus (eastern quoll)

install signage to warn motorists about wildlife 384

install traffic calming structures 373

install wildlife warning reflectors 360

release captive-bred/translocated mammals 942

use fencing to exclude predators 548

deer. *See Odocoileus* spp.

deer mouse. *See Peromyscus maniculatus*

Dendrolagus lumholtzi (Lumholtz's tree-kangaroos)

install barrier fencing and underpasses 347

install rope bridges between canopies 305

install tunnels under roads 266

denning sites 39–40, 499

desert cottontail rabbits. *See Sylvilagus audubonii*

dibblers. *See Parantechinus apicalis*

Diceros bicornis (black rhinoceros)

release translocated/captive-bred mammals 970

release translocated mammals into fenced areas 834

translocate to re-establish or boost populations 758

Dicotyles tajacu (javelina) 574

Didelphis albiventris (white-eared opossums) 307

Didelphis marsupialis (opossum) 331

Didelphis virginia (opossums) 138

Didelphis virginiana (opossums)
    install barrier fencing and underpasses 328, 343
    install tunnels under roads 267, 270
    provide artificial dens or nest boxes 697
    restore or create wetlands 682

Didelphis virginianus (opossums)
    install barrier fencing and underpasses 334
    install tunnels under roads 265
    provide artificial dens or nest boxes 698

dingoes. *See Canis dingo; See Canis lupus dingo*

Diorhabda spp. (leaf beetles) 574

Dipdomys heermanni (Heermann's kangaroo rat)
    exclude livestock from semi-natural habitat 93
    hold translocated mammals in captivity 820
    provide supplementary food 849
    release translocated/captive-bred mammals 978, 987

Dipodomys ingens (giant kangaroo rat) 93

Dipodomys merriami (Merriam's kangaroo rat) 90, 95, 676

Dipodomys nitratoides nitratoides (short nosed kangaroo rat)
    exclude livestock from semi-natural habitat 93
    hold translocated mammals in captivity 820
    provide supplementary food 849
    release translocated/captive-bred mammals 978, 987

Dipodomys stephensi (Stephens' kangaroo rat)

manage vegetation using livestock grazing 650
    replant vegetation 660

Dipsacus sylvestris (common teasel) 61

diversionary feeding 41, 42, 43, 44, 158, 159, 190, 191, 192, 193, 396, 593, 594. *See also* food/feeding

dogs. *See Canis lupus familiaris*

dorcas gazelle. *See Gazella dorcas neglecta*

Douglas fir. *See Pseudotsuga menziesii*

duck. *See Anas* spp.

dusky antechinus. *See Antechinus swainsonii*

dusky hopping mouse. *See Notomys fuscus*

dusky shrew. *See Sorex monticolus*

eastern barred bandicoot. *See Perameles gunnii*

eastern bettongs. *See Bettongia gaimardi*

eastern cottontail. *See Sylvilagus floridanus*

eastern grey kangaroos. *See Macropus giganteus*

eastern hemlock. *See Tsuga canadensis*

eastern pygmy-possums. *See Cercartetus nanus*

eastern quoll. *See Dasyurus viverrinus*

eastern woodrats. *See Neotoma floridana*

Echinochloa esculenta (white millet) 83

Echinococcus multicularis (small fox tapeworm) 495

Egyptian mongooses. *See Herpestes ichneumon*

eland. *See Taurotragus oryx*

Elaphurus davidianus (Père David's deer)
    provide supplementary food during/after release 924
    release captive-bred individuals 879, 882
    release translocated mammals into fenced areas 829
    use holding pens at release site 892, 902

electric fencing 40–41, 114–115, 120–124, 126, 145, 161, 163–164, 167, 251–252, 541, 547, 549–550, 799, 845, 938, 975

elephant. *See Loxondonta africana*

Elephas maximus (Asian elephants)

drive wild animals away using domestic animals 242

install electric fencing 165

use chili to deter crop damage 217, 220

use distress calls or signals to deter crop damage 212

use fire to deter crop damage 224

use light/lasers to deter crop damage 222

use loud noises to deter crop damage 200

Eliomys quercinus (garden dormouse)

install barrier fencing and underpasses 336

use prescribed burning 521

elk. *See Cervus canadensis; See Cersus elaphus*

Enhydra lutris (sea otter)

hand-rear orphaned or abandoned young 725

place abandoned wild young with captive foster parents 731

translocate mammals away from oil spill 624

translocate to re-establish or boost populations 758, 782

use holding pens at release site 796

Equus africanus asinus (horses) 545

Equus burchelli (zebra)

provide artificial waterholes in dry season 538

replant vegetation 662

use prescribed burning 513, 521

Equus ferus caballus (horses) 545

Equus ferus przewalskii (Przewalski's horses)

provide supplementary food 926, 929

release translocated/captive-bred mammals 964

use holding pens at release site 899, 904

Equus hemionus (Asiatic wild ass) 890, 902

Equus quagga (zebra)

build fences around protected areas 642

release translocated/captive-bred mammals 992

translocate to re-establish or boost populations 774

Equus zebra (mountain zebra)

release translocated/captive-bred mammals 992

translocate to re-establish or boost populations 774

Equus zebra zebra (Cape mountain zebra)

manage vegetation using grazing 653

translocate to re-establish or boost populations 776

use prescribed burning 514

Erethizon dorsatum (porcupines) 234

Erica tetralix (heather)

replant vegetation 661

use prescribed burning 519

Erinaceus europaeus (hedgehogs)

hold translocated mammals in captivity 818, 823

install barrier fencing and underpasses 336–337, 344

install overpasses over roads/railways 292

install overpasses over waterways 403

install tunnels under railways 275, 277

install tunnels under roads 260

provide mammals with escape routes 406

provide supplementary food 847, 853

rehabilitate injured, sick or weak mammals 717–720, 722

Ethiopian wolves. *See Canis simensis*

Eucalyptus

provide artificial dens or nest boxes 702, 704

provide artificial refuges/breeding sites 691

restore former mining sites 248

restore or create forest 679–680

retain understorey vegetation 451

retain wildlife corridors in logged areas 460

translocate mammals to reduce overpopulation 790

Eudorcas rufifrons (red-fronted gazelle)

release captive-bred mammals into fenced areas 917

restore or create savannas 674

Eurasian beavers. *See Castor fiber*

Eurasian lynx. *See Lynx lynx*

European bison. *See Bison bonasus*

European ground squirrels. *See Spermophilus citellus*

European larch. *See Larix decidua*

European mink. *See Mustela lutreola*

European rabbits. *See Oryctolagus cuniculus*

fallow deer. *See Dama dama*

Fascioloides magna (trematode) 610, 613

fat-tailed dunnart. *See Sminthopsis crassicaudata*

fawn-footed melomys. *See Melomys cervinipes*

feathertail gliders. *See Acrobates pygmaeus*

Felis catus (feral cats) 4, 21

hold translocated mammals in captivity 822

install barrier fencing and underpasses 331, 334, 338

install fences around culverts or underpasses 285

keep cats indoors or in outside runs 25

keep domestic cats and dogs well-fed 30

modify culverts for accessibility 278

modify traps 579

provide supplementary food 850

reintroduce top predators 576

release captive-bred mammals 911–912, 914, 916

release captive-bred/translocated mammals 932, 934–935, 937, 941–942, 946, 971, 981

release translocated mammals 830, 833, 835, 838, 840

remove/control non-native mammals 558, 568–569, 571–572

remove or control predators 586

restore former mining sites 246

restore or create wetlands 682

sterilise non-native domestic or feral species 595

threat: invasive alien and other problematic species 543

threat: residential and commercial development 23

train captive-bred mammals 908

train mammals to avoid problematic species 597

translocate predators for ecosystem restoration 808

translocate to re-establish or boost populations 779

use collar-mounted devices to reduce predation 26–28

use conditioned taste aversion 581

use fencing to exclude predators 549, 552

use holding pens at release site 897

use reward removal 583

Felis concolor (cougars)

install barrier fencing and underpasses 327–328

translocate predators away from livestock 153, 157

Felis rufus (bobcats)

install tunnels under roads 264

restore or create forest 679

Felis serval (serval) 724

Felis silvestris (wild cat)

install barrier fencing along roads 320

install barrier fencing and underpasses along roads 336, 342 install overpasses over roads/railways 289, 294

install tunnels under railways 275

install tunnels under roads 260

install wildlife warning reflectors 363

use fencing to exclude predators 551

feral cats. *See Felis catus*

feral dogs. *See Canis familiaris*

feral honeybees. *See Apis mellifera*

fertilizer 55, 245, 469–471, 619–620, 647, 688–689

Festuca scabrella (rough fescue)

manage vegetation using livestock grazing 650

use prescribed burning 510

field voles. *See Microtus agrestis*

fire 165, 200, 218, 220, 222–224, 242, 433, 445, 506–507, 509, 514–518, 520, 522–528, 530–536, 540–541, 549, 552, 554–555, 647, 652, 670, 675, 902, 928, 965

fishers. *See Martes pennanti*

food/feeding 3, 38–44, 48–50, 63, 72, 78–79, 82–84, 86–87, 115, 119, 124–125, 129, 136–138, 146, 148–150, 158–160, 162, 164, 168, 170, 175–176, 181, 187–193, 196, 201–202, 205, 207–208, 211, 226–229, 233–236, 238, 260, 281, 312, 321, 355, 358, 365–367, 375–378, 391, 394–398, 409, 429, 438, 458, 480–481, 485–486, 493, 496–501, 506–507, 519, 533, 540–541, 544, 549–550, 552, 554, 557, 559, 566, 569, 573, 576, 581, 583, 591–594, 600, 609–613, 619, 631–632, 648–649, 651–652, 660–661, 667, 683–684, 688, 709, 720–721, 728–729, 731–750, 752–753, 757, 763, 766–768, 782, 797–800, 802–803, 806, 809–810, 813, 819, 821, 828, 830, 838–839, 842–853, 867, 871, 883, 885–891, 893–894, 898–901, 911–913, 916–917, 919–928, 933, 935–936, 940, 942, 944, 952, 957, 960, 962, 972–975, 977–978, 981–982, 985–988, 993, 999

apply fertilizer to vegetation to increase food availability 688–689

apply water to vegetation to increase food availability during drought 631–632

graze herbivores on pasture, instead of sustaining with artificial foods 752–753

plant crops to provide supplementary food for mammals 82–84

prevent mammals accessing potential wildlife food sources 39–41

provide diversionary feeding for mammals to reduce human-wildlife conflict 41–44

provide diversionary feeding for predators 593–595

provide diversionary feeding to reduce crop damage by mammals 190–193

provide diversionary feeding to reduce predation of livestock by mammals 158–160

provide food/salt lick to divert mammals from roads or railways 394–396

provide supplementary feed to reduce tree damage 480–481

provide supplementary food after fire 540–541

provide supplementary food during/ after release of captive-bred mammals 919–929

provide supplementary food to increase reproduction/survival 732–747

use negative stimuli to deter consumption of livestock feed by mammals 207–208

fox squirrels. *See Sciurus niger*

Fraxinus americana (white ash) 698

garden dormouse. *See Eliomys quercinus*

gaur. *See Bos gaurus*

Gazella arabica (Arabian gazelles)

provide supplementary food 926, 929

provide supplementary water 750–751

release captive-bred mammals 916, 919

use holding pens at release site 898, 903

Gazella dorcas neglecta (dorcas gazelle) release captive-bred mammals 917, 919

restore or create savannas 674–675

Gazella gazella (mountain gazelles) 869, 879

Gazella subgutturosa marica (Arabian sand gazelles) 891, 902

Gazella thomsonii (Thomson's gazelles) 520, 524

genet. *See Genetta genetta*

Genetta genetta (genet)
install barrier fencing and underpasses 336–337, 339–340

install ledges in culverts 280

install overpasses over roads/railways 289

install tunnels under railways 275

install tunnels under roads 261

remove or control predators 586

Gerbillus allenbyi (Anderson's gerbils) 659

giant anteaters. *See Myrmecophaga tridactyla*

giant kangaroo rat. *See Dipodomys ingens*

giant pandas. *See Ailuropoda melanoleucahela*

Giraffa camelopardalis (giraffe)
provide artificial waterholes in dry season 538–539

release translocated/captive-bred mammals 992

replant vegetation 662

translocate to re-establish or boost populations 774

giraffe. *See Giraffa camelopardalis*

Glaucomys sabrinus (Carolina northern flying squirrels)
install pole crossings for gliders/flying squirrels 301

provide artificial dens or nest boxes 701–702, 706, 709

provide supplementary food 740, 746

thin trees within forest 464, 468

Glaucomys sabrinus coloratus (Carolina northern flying squirrels) 301

Glaucomys volans (southern flying squirrels)
manage vegetation using grazing 653

provide artificial dens or nest boxes 697–698

use fencing to exclude predators 550, 552

Glirulus japonicus (Japanese dormouse) 706, 710

goats. *See Capra hircus*

golden bandicoots. *See Isoodon auratus*

golden eagle. *See Aquila chrysaetos*

golden-mantled ground squirrel. *See Spermophilus lateralis*

golden mouse. *See Ochrotomys nuttalli*

Gould's monitors. *See Varanus gouldii*

Grammomys cometes (Mozambique thicket rats) 705

Graphiurus murinus (woodland dormice) 705, 709

grass mice. *See Akodon azarae*

gray-collared chipmunk. *See Tamias cinereicollis*

gray squirrel. *See Sciurus carolinensis*

gray wolves. *See Canis lupis*

Great Basin pocket mouse. *See Perognathus parvus*

greater bilbies. *See Macrotis lagotis*

greater glider. *See Petauroides volans*

greater one-horned rhinoceros. *See Rhinoceros unicornis*

greater stick-nest rat. *See Leporillus conditor*

great-horned owls. *See Bubo virginianus*

grey fox. *See Urocyon cinereoargenteus*

grey wolf. *See Canis lupus lupus*

grey wolves. *See Canis lupus*

groundhogs. *See Marmota monax*

guanacos. *See Lama guanicoe*

guiña. *See Leopardus guigna*

Gulo gulo (wolverines) 113

Gymnobelideus leadbeateri (Leadbeater's possum) 702

hares. *See Lepus* spp.

harvesting 57, 61, 65–66, 77, 86, 237, 409–410, 413, 416, 418, 432–439, 442–445, 447–449, 454, 457–460, 464–465, 468, 472–475, 477–478, 536, 616, 636, 714, 725, 734, 771, 829, 836

harvest mouse. *See Micromys minutus*

hazel dormouse. *See Muscardinus avellanarius*

heather voles. *See Phenacomys intermedius*

hedgehogs. *See Erinaceus europaeus*

hedgerows, use of 56, 60, 62, 66–71

Heermann's kangaroo rat. *See Dipdomys heermanni*

Hemitragus jemlahicus (Himalayan tahr) 419

Herpestes ichneumon (Egyptian mongooses)

    install barrier fencing and underpasses 337, 339

    remove or control predators 586

Himalayan tahr. *See Hemitragus jemlahicus*

hippopotamus. *See Hippopotamus amphibius*

Hippopotamus amphibius (hippopotamus)

    ban exports of hunting trophies 432

    install electric fencing to protect crops 166–167

Hippotragus equinus (roan antelope)

    graze herbivores on pasture 752

    provide artificial waterholes in dry season 538

    release translocated/captive-bred mammals 991

    use holding pens at release site 795

    use prescribed burning 513

Hippotragus niger (sable antelope)

    provide artificial waterholes in dry season 538

    use prescribed burning 513

hispid cotton rats. *See Sigmodon hispidus*

hispid pocket mouse. *See Perognathus hispidus*

honey badgers. *See Mellivora capensis*

honey possum. *See Tarsipes rostratus*

Hordeum vulgare (barley)

    provide supplementary food 848

    use repellent on slug pellets 79

horses. *See Equus africanus asinus; See Equus ferus caballus*

house mice. *See Mus musculus; See Mus domesticus*

human-wildlife conflict, reduction of 11, 23, 30–32, 38–42, 44, 110–111, 114, 120, 126–128, 132, 134, 140–141, 148, 151, 158, 160–161, 168, 170–174, 178–180, 182, 185–188, 190, 194–195, 201–205, 207, 209, 213, 215–217, 221, 223, 225, 228–229, 231–232, 238–241, 243, 252, 369, 390, 483, 490, 493, 496, 499

    deter predation of livestock by herding livestock using adults instead of children 173–174

    deter predation of livestock by mammals by having people close by 172–173

    deter predation of livestock by using shock/electronic dog-training collars 174–178

    dispose of livestock carcasses to deter predation of livestock by mammals 140–141

    drive wild animals away using domestic animals of the same species 241–242

    establish deviation ponds in fish farms to reduce predation of fish stock by mammals 187

    exclude wild mammals using ditches, moats, walls or other barricades 126–128

    fit livestock with protective collars to reduce risk of predation by mammal 178–179 grow unattractive crop in buffer zone around crops 215–216

install automatically closing gates at field entrances to prevent mammals entering 170–171

install electric fencing to protect crops from mammals 161–167

install electric fencing to reduce predation of livestock by mammals 120–126

install metal grids at field entrances to prevent mammals entering 168–169

install non-electric fencing to exclude predators or herbivores 114–120

issue enforcement notices to deter use of non-bear-proof garbage dumpsters 38

keep livestock in enclosures to reduce predation by mammals 160–161

pay farmers to compensate for losses due to predators/wild herbivores 111–114

play predator calls to deter crop damage by mammals 209, 209–210

provide diversionary feeding for mammals to reduce nuisance behaviour 41–44

provide diversionary feeding to reduce crop damage by mammals 190–193

provide diversionary feeding to reduce predation of livestock by mammals 158–160

relocate local pastoralist communities 110–111

scare or otherwise deter mammals from human-occupied areas 44–50

translocate crop raiders away from crops 205–207

translocate predators away from livestock 151–157

translocate problem mammals away from residential areas to reduce human-wildlife conflict 30–37

use bees to deter crop damage by mammals 213–215

use chili to deter crop damage by mammals 216–217

use dogs to guard crops 240–241

use drones to deter crop damage by mammals 204–205

use fencing/netting to reduce predation of fish stock by mammals 186–187

use fire to deter crop damage by mammals 223–224

use flags to reduce predation of livestock by mammals 128–132

use guardian animals bonded to livestock to deter predators 141–148

use light/lasers to deter crop damage by mammals 221–222

use lights and sound to deter crop damage by mammals 188–189

use lights and sound to deter predation of livestock by mammals 180–182

use loud noises to deter crop damage by mammals 194–201

use loud noises to deter predation of livestock by mammals 148–150

use mobile phone communications to warn farmers of problematic mammals 185–186

use negative stimuli to deter consumption of livestock feed by mammals 207–208

use noise aversive conditioning to deter crop damage by mammals 201–202

use pheromones to deter crop damage 225

use pheromones to deter predation of livestock by mammals 134

use predator scent to deter crop damage by mammals 225–228

use repellents that smell bad to deter crop or property damage by mammals 238–240

use repellents that taste bad to deter crop or property damage by mammals 231, 231–237 use scarecrows to deter crop damage by mammals 194

use scent to deter predation of livestock by mammals 182–184

use 'shock collars' to deter crop damage by mammals 229–230

use target species distress calls or signals to deter crop damage by mammals 209–213

use target species scent to deter crop damage by mammals 228–229

use taste-aversion to reduce predation of livestock by mammals 134–140

use tree nets to deter wild mammals from fruit crops 171

use ultrasonic noises to deter crop damage by mammals 202–203

use visual deterrents to deter predation of livestock by mammals 132–134

use watchmen to deter crop damage by mammals 185

hunting 26, 29–30, 34, 155, 398–399, 409–411, 413–422, 424, 427–428, 430–431, 485–486, 492, 518, 585–587, 613, 636, 643, 646, 713–714, 750, 774, 781–782, 791, 796, 817, 829, 870, 904–906, 954, 997–1001

Hyaena brunnea (brown hyaena) 744

Hyaena hyaena (striped hyenas)

provide artificial waterholes in dry season 539

use guardian animals to deter predators 144

Hydropotes inermis (water deer) 267

Hypericum perforatum (St John's wort) 61

Hystrix africaeaustralis (porcupine)

build fences around protected areas 642

install electric fencing to reduce predation 124

install mammal crossing points along fences 106

Iberian hare. *See Lepus granatensis*

Iberian ibex. *See Capra pyrenaica*

Iberian lynx. *See Lynx pardinus*

ibex. *See Capra sibirica*

impala. *See Aepyceros melampus*

Irish hares. *See Lepus timidus hibernicus*

island foxes. *See Urocyon littoralis*

Isoodon auratus (golden bandicoots)

release translocated/captive-bred mammals 940, 949

release translocated mammals 837

Isoodon macrourus (northern brown bandicoots)

burn at specific time of year 525

install barrier fencing and underpasses 338, 347

install tunnels under roads 262

Isoodon obesulus (southern brown bandicoots)

install barrier fencing and underpasses 348

install tunnels under roads 266

remove/control non-native mammals 566–567

translocate to re-establish or boost populations 768

jackal. *See Canis aureus; See Canis mesomelas*

jaguar. *See Panthera onca*

Japanese dormouse. *See Glirulus japonicus*

Japanese macaques. *See Macaca fuscata*

Japanese yew trees. *See Taxus cuspidata*

javelina. *See Dicotyles tajacu*

Juniperus ashei (Ashe juniper) 536

Kam dwarf hamster. *See Cricetulus kamensis*

Keen's mouse. *See Peromyscus keeni*

Key Largo woodrats. *See Neotoma floridana smalli*

koalas. *See Phascloarctos cinereus; See Phascolarctos cinereus*

Kobus leche leche (red lechwe) 637

kudu. *See Tragelaphus strepsiceros*

lacustrine vole. *See Microtus limnophilus*

Lagorchestes hirsutus (rufous hare-wallabies)

provide supplementary food 925

provide supplementary water 749

release captive-bred mammals 871–872, 912

release captive-bred/translocated mammals 934, 947–949

train captive-bred mammals to avoid predators 907

use holding pens at release site 893

Lagostrophus fasciatus (banded hare-wallabies)

provide supplementary food 925

provide supplementary water 749

release captive-bred mammals 872, 911

release captive-bred/translocated mammals 934

release translocated/captive-bred mammals 947

use holding pens at release site 893

Lama glama (llamas) 142–143

Lama guanicoe (guanacos) 898, 903

Larix decidua (European larch) 193

Lasiorhinus latifrons (wombats) 331

Leadbeater's possum. *See Gymnobelideus leadbeateri*

leaf beetles. *See Diorhabda* spp.

Leggadina lakedownensis (short-tailed mice)

remove/control non-native mammals 559

translocate to re-establish or boost populations 768

leopard. *See Panthera pardus; See Panthera pardus fusca*

leopard cat. *See Prionailurus benalensis*

Leopardus guigna (guiña) 451

Leporillus conditor (greater stick-nest rat)

release captive-bred mammals into fenced areas 914

release captive-bred/translocated mammals 937

release translocated mammals 835

translocate to re-establish or boost populations 768

use holding pens at release site 808, 897

Leptailurus serval (servals)

release captive-bred individuals 873

translocate to re-establish or boost populations 771

Lepus spp. 261

Lepus alleni (antelope jackrabbit) 574

Lepus americanus (snowshoe hare)

apply fertilizer to trees 471

install barrier fencing and underpasses 329

install tunnels under roads 263

remove competing vegetation 449

retain undisturbed patches during thinning 441

thin trees within forest 467

use selective harvesting 434, 437

Lepus californicus (black-tailed hares)

exclude livestock from semi-natural habitat 90

remove/control non-native plants 574

use fencing to exclude grazers 544

Lepus capensis (brown hare) 292

Lepus europaeus (European hare)

create uncultivated margins 62–63

install barrier fencing and underpasses 330, 338

install non-electric fencing 117

install overpasses over roads/railways 295

install wildlife warning reflectors 363

keep dogs indoors 29

pay farmers to cover the costs 75

plant new or maintain existing hedgerows 71

plant trees on farmland 73–74

provide or retain set-aside areas 67

reduce intensity of grazing 98

reduce pesticide or fertilizer use 621

release captive-bred mammals 877, 881

use traditional breeds of livestock 107

Lepus flavigularis (Tehuantepec jackrabbits) 651–652

Lepus granatensis (Iberian hare)

install barrier fencing and underpasses 336

install overpasses over roads/railways 289

install overpasses over waterways 403

install tunnels under railways 275

remove or control predators 586

Lepus saxatilis (scrub hare)

build fences around protected areas 642

pay farmers to cover the costs 286

Lepus timidus (mountain hares)

site management by field sport practitioners 585

Lepus timidus hibernicus (Irish hares)

pay farmers to cover the costs 76

plant new or maintain existing hedgerows 70

reduce intensity of grazing 98

site management for target mammal species 414

lesser anteaters. *See Tamandua tetradactyla*

Lichtenstein's hartebeest. *See Alcelaphus lichtensteini*

Linum usitatissimum (linseed) 83

lion. *See Panthera leo*

little pocket mouse. *See Perognathus longimembris*

llama. *See Lama glama*

Locondonta africana (African elephants) 51

lodgepole chipmunk. *See Neotamias speciosus*

lodgepole pine. *See Pinus contorta*

long-eared chipmunks. *See Tamias quadrimaculatus*

long-haired field mice. *See Abrothrix longipilis*

long-haired rats. *See Rattus villosissimus*

long-nosed potoroos. *See Potorous tridactylus*

long-tailed pocket mouse. *See Chaetodipus formosus*

long-tailed vole. *See Microtus longicaudus*

long-tailed weasel. *See Mustela frenata*

Lontra canadensis (otters) 328

Lower Keys marsh rabbits. *See Sylvilagus palustris*

Loxodonta africana (African savanna elephants)

deter mammals from human-occupied areas 49

install electric fencing 163

release translocated mammals 831, 835

translocate crop raiders away from crops 206–207

use bees to deter crop damage 214

use chili to deter crop damage 217–220

use dogs to guard crops 241

use drones to deter crop damage 204

use fire to deter crop damage by mammals 223

use loud noises to deter crop damage 196–197, 199–200

use target species distress calls 210

use target species scent to deter crop damage 228–229

Loxondonta africana (elephant) 538

Lumholtz's tree-kangaroos. *See Dendrolagus lumholtzi*

Lutra canadensis (river otters) 759–760, 764–765

Lutra lutra (otters)

breed mammals in captivity 855, 857

establish deviation ponds in fish farms 187

hhold translocated mammals in captivity 823

hold translocated mammals in captivity 819

install barrier fencing and underpasses 337, 344

install ledges in culverts under roads/ railways 281

install mammal crossing points 105–106

install tunnels under roads 267

provide mammals with escape routes 407

provide supplementary food 744, 747, 922

release captive-bred mammals 868, 870–872, 875, 879–881

rovide supplementary food 928

translocate to re-establish or boost populations 762–763, 770, 775, 783, 786

use fencing/netting to reduce predation 186

use holding pens at release site 888, 901

Lycalopex culpaeus (culpeo fox) 145

Lycaon pictus (African wild dogs)

establish populations isolated from disease 614–615

install non-electric fencing 119

place orphaned wild young with wild foster parents 729

release captive-bred individuals 873

release captive-bred mammals into fenced areas 913

release translocated/captive-bred mammals 962

release translocated mammals 831

retain buffer zones around core habitat 643

translocate to re-establish or boost populations 771

use guardian animals to deter predators 144

use holding pens at release site 805, 890, 894, 902

use people to deter predation of livestock 173

use scent to deter predation 183–184

use visual deterrents to deter predation 133

lynx. *See Lynx canadensis*

Lynx canadensis (lynx)

hold translocated mammals in captivity 820

install crossings over/under pipelines 401

release translocated/captive-bred mammals 987

Lynx lynx (Eurasian lynx)

hold translocated mammals in captivity 817–818

install barrier fencing and underpasses 342

install overpasses over roads/railways 295

set hunting quotas 416

use chemical repellents 392

use holding pens at release site 803

Lynx pardinus (Iberian lynx)

install electric fencing to reduce predation 125–126

install overpasses over roads/railways 289

install tunnels under railways 275

provide live natural prey to captive mammals 905–906

provide supplementary food 740

rehabilitate injured, sick or weak mammals 718, 722

Lynx rufus (bobcats)

install barrier fencing and underpasses 327–328, 343, 346, 348

install fences around existing culverts 285

install tunnels under roads 265, 267

modify culverts 278

translocate to re-establish or boost populations 769

Macaca fuscata (Japanese macaques) 165

Macaranga siamensis trees 664

Macropus eugenii (tammar wallabies)

install acoustic wildlife warnings 366

release translocated/captive-bred mammals 947

remove/control non-native mammals 560–562

translocate to re-establish or boost populations 768

Macropus fuliginosus (Western grey kangaroo)

install barrier fencing and underpasses 348

release translocated/captive-bred mammals 947

restore former mining sites 247

Macropus giganteus (eastern grey kangaroos)

install overpasses over roads/railways 292, 296

remove/control non-native mammals 558

retain/maintain road verges 381

use target species signals to deter crop damage 211

use ultrasonic noises to deter crop damage 203

Macropus irma (western brush wallaby) 246

Macropus parma (parma wallaby) 947

Macropus robustus (common wallaroos)

provide supplementary water 751

release captive-bred mammals 918

release captive-bred/translocated mammals 943

translocate to re-establish or boost populations 768

Macropus rufogriseus (red-necked wallaby)

install barrier fencing and underpasses 338, 347

install overpasses over roads/railways 292, 296

install wildlife warning reflectors 362

Macropus rufus (red kangaroos)

fit vehicles with ultrasonic warning devices 381

install wildlife warning reflectors 362

Macrotis lagotis (greater bilbies)

release captive-bred mammals 914

release captive-bred/translocated mammals 937

release translocated mammals into fenced areas 835

train mammals to avoid problematic species 596–597

use holding pens at release site 808, 897

Malus domestica (spartan apple) 191

Malus spp. (crabapple) 191

Malva sylvestris (common mallow) 61

maned sloths. *See Bradypus torquatus*

Manis temminckii (cape pangolins) 763, 783

mardo. *See Antechinus flavipes*

Marmota marmota (alpine marmot)

release translocated/captive-bred mammals 991–992, 994

translocate to re-establish or boost populations 766, 784

Marmota monax (groundhogs)

install overpasses over roads/railways 288, 297

install tunnels under roads 267–268, 270

Marmota vancouverensis (Vancouver Island marmots) 874

Marte foina (stone marten) 280

Martes americana (American marten)

install barrier fencing and underpasses 329

install tunnels under roads 263

plant trees following clearfelling 479

remove competing vegetation 450

translocate to re-establish or boost populations 767, 770, 784

use holding pens at release site 795

Martes foina (stone martens)

install barrier fencing and underpasses 330, 337, 339

install overpasses over waterways 403

install tunnels/culverts/underpass under railways 277

install tunnels under roads 261

prevent mammals accessing wildlife food sources 40

provide supplementary water 750

remove or control predators 586

Martes martes (pine martens)

install tunnels under railways 277

provide artificial dens or nest boxes 707, 710

provide supplementary food 852

release translocated/captive-bred mammals 982

remove or control predators 585

use holding pens at release site 810

Martes pennanti (fishers)

use flags to reduce predation 130

use lights and sound to deter predation 181

masked shrews. *See Sorex cinereus*

Mastacomys fuscus (broad-toothed rat) 52

Mastomys spp. (multimammate rats)

dig trenches around culverts 283

install fences around existing culverts 286

provide supplementary food 738, 746

Mazama (red brocket)

provide mammals with escape routes 407

restore or create forest 679

use artificial insemination 860–861

use wildlife refuges 422

Mazama americana (red brocket) 407

meadow jumping mouse. *See Zapus hudsonius*

meadow voles. *See Microtus pennsylvanicus*

Meles meles (badgers)

install barrier fencing and underpasses 330, 336–337, 339–340, 344

install electric fencing 163

install ledges in culverts under roads/railways 281

install overpasses over roads/railways 290, 292, 295

install overpasses over waterways 403

install tunnels under roads 261

install wildlife warning reflectors 363

provide mammals with escape routes 406

provide supplementary food 846, 853

rehabilitate injured, sick or weak mammals 718

release translocated/captive-bred mammals 960, 966

use holding pens at release site 803, 812

use repellents that taste bad 236, 238

use vaccination programme 603–604

Mellivora capensis (honey badgers) 119

Melomys cervinipes (fawn-footed melomys) 305

Mephitis mephitis (striped skunks)

install barrier fencing and underpasses 331, 333

install tunnels under roads 262, 264–265

provide diversionary feeding 159

use taste-aversion 138

Meriones tristrami (Tristram's jird) 659

Merriam's kangaroo rat. *See Dipodomys merriami*

Mexican fox squirrels. *See Sciurus nayaritensis chiricahuae*

Mexican wolves. *See Canis lupus baileyi*

Mexican woodrat. *See Neotoma exicana*

mhorr gazelle. *See Nanger dama mhorr*

mice. *See Peromyscus* spp.

Micromys minutus (harvest mouse) 81

Microtus spp. (vole)

change type of livestock 109

provide diversionary feeding 191

reduce intensity of grazing 100

Microtus agrestis (field voles)

change type of livestock 109

reduce intensity of grazing 97, 100

Microtus arvalis (common vole)

exclude livestock from semi-natural habitat 91

install overpasses over roads/railways 290

provide mammals with escape routes 406

Microtus limnophilus (lacustrine vole) 98

Microtus longicaudus (long-tailed vole)
    clear or open patches in forests 444
    gather coarse woody debris into piles 457
    use selective harvesting 434

Microtus montanus (montane voles) 191

Microtus ochrogaster (prairie vole) 734

Microtus oeconomus (root voles)
    create or maintain corridors between habitat patches 687
    reduce intensity of grazing 98

Microtus oregoni (creeping vole) 463

Microtus pennsylvanicus (meadow voles)
    exclude livestock from semi-natural habitat 91
    fell trees in groups 473
    install barrier fencing and underpasses 335
    install overpasses over roads/railways 290, 296
    install tunnels under roads 264, 269
    leave coarse woody debris in forests 454
    restore or create wetlands 682
    use patch retention harvesting 438
    use selective harvesting 436

Microtus pinetorum (woodland vole) 653

moles. *See Talpa europaea*

Molina's hog-nosed skunk. *See Conepatus chinga*

montane voles. *See Microtus montanus*

Monterey pine. *See Pinus radiata*

moose. *See Alces alces*

mouflon. *See Ovis gmelini musimon*

mountain beavers. *See Aplodontia rufa*

mountain brushtail possum. *See Trichosurus cunninghami*

mountain gazelles. *See Gazella gazella*

mountain goats. *See Oreamnos americanus*

mountain mahogany. *See Cercocarpus* spp.

mountain pygmy-possums. *See Burramys parvus*

mountain zebra. *See Equus zebra*

Mozambique thicket rats. *See Grammomys cometes*

mufflon. *See Ovis orientalis*

mule deer. *See Odocoileus hemionus; See Odocoileus hemionus hemionus; See Odocoileush emionush emionus*

mulleins. *See Verbascum* spp.

multimammate rats. *See Mastomys* spp.

Muntiacus muntjak (red muntjac) 424

Muscardinus avellanarius (hazel dormouse)
    fell trees in groups 474
    provide supplementary food 844
    release captive-bred mammals 871, 880
    release translocated/captive-bred mammals 985
    remove vegetation by hand/machine 663
    restore or create forest 679, 681
    rrovide artificial dens or nest boxes 699, 704, 708–709
    translocate to re-establish or boost populations 766, 784
    use holding pens at release site 796, 889

Mus domesticus (house mice) 559, 570

muskrat. *See Ondatra zibethicus*

Mus musculus (house mice)
    create uncultivated margins around fields 62
    install barrier fencing and underpasses 334, 338
    provide artificial dens or nest boxes 708
    provide artificial refuges/breeding sites 691
    remove vegetation by hand/machine 659
    restore former mining sites 246, 249
    restore or create wetlands 682

Mus spretus (Algerian mice)

use fencing to exclude grazers 545–546
use prescribed burning 521
Mustela sp. (weasels) 329
Mustela ermine (stoat)
  install tunnels under roads 263
  remove or control predators 585
Mustela erminea (short-tailed weasel)
  install tunnels under railways 277
  install tunnels under roads 264
Mustela frenata (long-tailed weasel)
  install barrier fencing and underpasses 331
  install tunnels under roads 262–263
Mustela lutreola (European mink)
  release captive-bred mammals 875, 881
  use holding pens at release site 895, 903
Mustela nigripes (black-footed ferrets)
  captive rear in large enclosures 883, 885
  preserve genetic material 864
  provide live natural prey 905–906
  release captive-bred mammals 875–876, 915
  release captive-bred/translocated mammals 937–938
  train captive-bred mammals to avoid predators 908
  translocate to re-establish or boost populations 775–776
  use artificial insemination 860
  use fencing to exclude predators 550
  use holding pens at release site 896
Mustela nivalis (weasel)
  install barrier fencing and underpasses 336–337
  install tunnels under railways 275, 277
  install tunnels under roads 261
Mustela putorius (western polecat)
  install barrier fencing and underpasses 339
  install tunnels under railways 277
Mustela sibirica (Siberian weasel) 267

Mustela vison (American mink)
  install barrier fencing and underpasses 337
  remove/control non-native mammals 564, 570
Mycobacterium bovis
  treat disease in wild mammals 600
  use negative stimuli to deter consumption of livestock feed 208
Myocastor coypus (coypu) 321
Myodes gapperi (southern red-backed voles)
  fell trees in groups 473
  install overpasses over roads/railways 296
  install tunnels under roads 269
  retain wildlife corridors in logged areas 460–461
  use patch retention harvesting 439–440
  use selective harvesting 435–436
Myodes glareolus (bank vole) 59
Myrmecobius fasciatus (numbat)
  release captive-bred mammals into fenced areas 914
  release captive-bred/translocated mammals 937
  release translocated mammals 833, 835, 840
  translocate to re-establish or boost populations 768
Myrmecophaga tridactyla (giant anteaters)
  provide supplementary food 850–851, 854, 927, 929
  rehabilitate injured, sick or weak mammals 720–722
  use holding pens at release site 900–901, 904

nailtail wallabies. *See Onychogalea fraenata*
Nanger dama mhorr (mhorr gazelle)
  release captive-bred mammals into fenced areas 917
  restore or create savannas 674

Nasua nasua (South American coatis) 679

Nasuella olivacea (Western mountain coatis) 679

Neofiber alleni (round-tailed muskrat) 334

Neotamias speciosus (lodgepole chipmunk)
    burn at specific time of year 526
    use prescribed burning 515

Neotoma cinerea (bushy-tailed wood rat) 263

Neotoma exicana (Mexican woodrat) 528

Neotoma floridana (eastern woodrats)
    provide artificial dens or nest boxes 697
    provide artificial refuges/breeding sites 693
    release captive-bred mammals 877

Neotoma floridana smalli (Key Largo woodrats)
    provide artificial refuges/breeding sites 693
    release captive-bred mammals 877

Neotoma lepida (desert woodrat) 676

Neotoma magister (Allegheny woodrats)
    breed mammals in captivity 856–857
    hold translocated mammals in captivity 821
    use holding pens at release site 899

Neotoma mexicana (Mexican woodrat)
    thin trees to reduce wildfire risk 529
    use prescribed burning 519

Neovison vison (American mink)
    install tunnels under roads 262
    release captive-bred/translocated mammals 936, 939
    restrict use of rodent poisons on farmland 80

Nesoryzomys swarthi (Santiago rice rats) 565

North American beavers. *See Castor canadensis*

North American least shrew. *See Cryptotis parva*

northern brown bandicoots. *See Isoodon macrourus*

northern mountain brushtail possums. *See Trichosurus caninus*

northern pocket gophers. *See Thomomys talpoides*

northern pygmy mice. *See Baiomys taylori*

northern short-tailed shrews. *See Blarina brevicauda*

Notomys alexis (spinifex hopping mouse)
    remove/control non-native mammals 569, 572

Notomys fuscus (dusky hopping mouse) 569

nuisance behaviour 34–35, 39–42, 46, 48, 154, 158, 190, 493–494, 496, 499–500

numbat. *See Myrmecobius fasciatus*

Nyctereutes procyonoides (raccoon dog)
    install electric fencing 165
    install tunnels under roads 267
    remove or control predators 585

Nyctophilus geoffroyi (lesser long-eared bat) 707

Ochrotomys nuttalli (golden mouse)
    manage vegetation using grazing 653
    provide artificial dens or nest boxes 697

Odocoileus spp. (deer)
    install barrier fencing along roads 317
    install barrier fencing and underpasses 329–331, 335
    install one-way gates 312
    install overpasses over roads/railways 289, 291
    install wildlife exclusion grates 370
    install wildlife warning reflectors 357, 360

Odocoileus hemionus (mule deer)
    apply fertilizer to trees 470–471
    close roads in defined seasons 399
    exclude livestock from semi-natural habitat 89
    fit vehicles with ultrasonic warning devices 380

install barrier fencing along roads 318, 322

install barrier fencing and underpasses 325–326, 332–333, 345–346, 348–349

install crossings over/under pipelines 401

install fences around existing culverts 284

install metal grids at field entrances 169

install one-way gates 313

install overpasses over roads/railways 296

install overpasses over waterways 403

install signage to warn motorists 385–386

install tunnels under railways 276–277

install tunnels under roads 265–266, 270, 272

install wildlife crosswalks 367

install wildlife exclusion grates 369

install wildlife warning reflectors 358

ise road lighting to reduce vehicle collisions 388

plant crops to provide supplementary food 83

prohibit or restrict hunting 418

provide food/salt lick 395

provide mammals with escape routes 406

provide supplementary food 735, 741

reduce intensity of grazing 95

release translocated mammals 829

remove/control non-native plants 574

remove or control predators 586

remove vegetation by hand/machine 657, 663

remove vegetation using herbicides 669

restore or create forest 679

restore or create grassland 672

restore or create shrubland 677

thin trees within forest 466–467

use lights and sound to deter crop damage 189

use loud noises to deter crop damage 198

use permeable livestock fences 102

use prescribed burning 508, 510

use repellents that taste bad 232–234

use 'shock collars' to deter crop damage 230

Odocoileus hemionus hemionus (mule deer) 169, 311, 369, 735

Odocoileus virginianus (white-tailed deer)

fit vehicles with ultrasonic warning devices 381

hand-rear orphaned or abandoned young 725

install automatically closing gates 170

install barrier fencing along roads 317–318, 320, 322

install barrier fencing and underpasses 327–328, 331–333, 338, 341–342, 344–345, 348–349, 352

install crossings over/under pipelines 401

install electric fencing 124, 162

install electric fencing to protect crops 164

install metal grids at field entrances 169

install signage to warn motorists 386

install tunnels under roads 264–265, 267, 270

install wildlife exclusion grates 369

install wildlife warning reflectors 357–358, 361

manage vegetation using grazing 653

place orphaned wild young with captive foster parents 731

plant crops to provide supplementary food 83

provide supplementary food 737, 745

release translocated/captive-bred mammals 953, 955

release translocated mammals into fenced areas 829

remove or control predators 587–588

remove vegetation using herbicides 668

replant vegetation 658–659

thin trees within forest 463

translocate mammals to reduce overpopulation 789

translocate problem mammals away from residential areas 32–34

treat disease in wild mammals 599–600

use drugs to treat parasites 610

use fencing to exclude predators 551

use guardian animals to deter predators 144

use light/lasers to deter crop damage 222

use lights and sound to deter crop damage 189

use loud noises to deter crop damage 196, 198

use negative stimuli 208

use noise aversive conditioning 202

use predator scent to deter crop damage 227

use prescribed burning 511, 514

use repellents that taste bad 232–233, 237

use target species distress calls 210–211

use wildlife refuges 422

Olea europaea (olive trees) 235

Ondatra zibethicus (muskrat)

install barrier fencing and underpasses 332

install tunnels under roads 264

manage wetland water levels 685

restore or create wetlands 682

Onychogalea fraenata (nailtail wallabies)

release captive-bred/translocated mammals 933, 939, 944, 947, 991

use holding pens at release site 800, 892

Onychomys torridus (southern grasshopper mouse)

exclude livestock from semi-natural habitat 90

reduce intensity of grazing 95

opossums. *See Didelphis marsupialis*; *See Didelphis virginiana*; *See Didelphis virginianus*

Opuntia polyacantha (pricklypear cactus) 509

Oreamnos americanus (mountain goats) 326, 329

oribi. *See Ourebia ourebi*

Orycteropus afer (antbears) 124

Oryctolagus cuniculus (European rabbits)

exclude livestock from semi-natural habitat 92–93

install barrier fencing and underpasses 336

install electric fencing to protect crops 162

install non-electric fencing 117

install overpasses over roads/railways 289

install tunnels under railways 275, 277

install tunnels under roads 260–261

install wildlife warning reflectors 363

plant crops to provide supplementary food 84

provide artificial refuges/breeding sites 692–694

provide diversionary feeding to reduce crop damage 192

provide live natural prey to captive mammals 905

provide mammals with escape routes 406

provide more small artificial breeding sites 711

provide supplementary food 741, 845, 848

provide supplementary water 750–751

rehabilitate injured, sick or weak mammals 719

release captive-bred mammals into fenced areas 911, 914, 918 release captive-bred/translocated mammals 933–934, 936–937, 943, 946, 976, 979, 981, 984

release translocated mammals into fenced areas 830, 832, 835–836

remove burnt trees and branches 531

remove/control non-native mammals 559, 569, 572

remove or control predators 585

replant vegetation 661, 665

restore former mining sites 246

restore or create grassland 671–672

translocate to re-establish or boost populations 770

use fencing to exclude predators 548–549

use holding pens at release site 799, 807–808, 813, 897

use loud noises to deter crop damage 195

use pheromones to deter predation 137

use prescribed burning 511, 522

use tranquilizers to reduce stress 824

use vaccination programme 602

Oryx dammah (scimitar-horned oryx)
release captive-bred mammals 917
restore or create savannas 674

Oryx leucoryx (Arabian oryx)
pelease captive-bred mammals into fenced areas 912, 918
provide supplementary food 923
provide supplementary water 748
release captive-bred individuals 877, 881

Oryzomys palustris (rice rats)
install barrier fencing along roads 319
install barrier fencing and underpasses 334

Otospermophilus beecheyi (California ground squirrel)
release translocated/captive-bred mammals 982
replant vegetation 664

otters. *See Lutra lutra; See Lontra canadensis*

Ourebia ourebi (oribi)
airborne translocation of mammals using parachutes 833, 840
release captive-bred mammals 874, 880

overpasses 12, 268–269, 272, 274–275, 287–298, 301, 306, 309–310, 316, 330–331, 335, 340–342, 349–350, 352, 401–403

Ovis sp. (sheep)
replant vegetation 661
use prescribed burning 519

Ovis ammon polii (argali) 426

Ovis aries (sheep) 137

Ovis canadensis (bighorn sheep)
install barrier fencing along roads 322
install barrier fencing and underpasses 329, 331
install overpasses over roads/railways 289
provide supplementary food 739
provide supplementary water 750
reduce legal speed limit 371
release translocated/captive-bred mammals 953, 958
remove or control predators 586
remove trees and shrubs to recreate open areas 537
translocate animals from source populations 630
translocate to re-establish or boost populations 774, 776, 779
use drugs to treat parasites 609–610
use prescribed burning 508–509, 513

Ovis gmelini musimon (mouflon)
remove vegetation by hand/machine 661
treat disease in wild mammals 599–600
use prescribed burning 519

Ovis orientalis (mufflon)
clone rare species 862–863

use chemical repellents along roads or railways 390

Ovis vignei (urial sheep) 424

Pacific jumping mouse. *See Zapus trinotatus*

Pacific shrew. *See Sorex pacificus*

Panthera leo (lion)
exclude wild mammals using barricades 127–128
increase size of protected area 644
install non-electric fencing 119
pay farmers to compensate for losses 112–114
place abandoned wild young with wild foster parents 729
release captive-bred individuals to re-establish or boost populations 873
release translocated/captive-bred mammals 957, 961, 966, 992
release translocated mammals into fenced areas 832
train and support local staff 1003
translocate predators away from livestock 154
translocate to re-establish or boost populations 771, 774
use guardian animals bonded to livestock 144
use holding pens at release site 800–801, 804–805, 812, 890
use people to deter predation of livestock 173
use visual deterrents to deter predation 133

Panthera leo persica (Asiatic lion) 110–111

Panthera onca (jaguar)
install electric fencing 125
translocate predators away from livestock 154–155, 157
use loud noises to deter predation 150
use visual deterrents to deter predation of livestock 133

Panthera pardus (leopard)

deter predation of livestock by mammals 173
exclude or limit number of visitors 488
fit livestock with protective collars 179
install mammal crossing points along fences 105
install non-electric fencing 119
prohibit or restrict hunting 411–412
provide education programmes 1001
provide/increase anti-poaching patrols 424
provide live natural prey to captive mammals 906
release captive-bred mammals 873
translocate predators away from livestock 154–157
translocate to re-establish or boost populations 771
use guardian animals bonded to livestock to deter predators 144, 147
use visual deterrents to deter predation 133

Panthera pardus fusca (leopard) 35

Panthera tigris (tiger)
provide/increase anti-poaching patrols 425
use holding pens at release site 807
use non-lethal methods to deter carnivores 501, 503

Panthera tigris altaica (Amur tigers) 34, 37

Panthera uncia (snow leopard) 426–427

Papio sp. (baboons) 119

Parantechinus apicalis (dibblers)
release captive-bred mammals 871
release translocated/captive-bred mammals 948, 950

parma wallaby. *See Macropus parma*

pebble-mound mouse. *See Pseudomys* sp.

Pecari tajacu (collared peccaries)
install overpasses over waterways 403
provide mammals with escape routes 407

use wildlife refuges to reduce hunting 422

Perameles bougainville (western barred bandicoots)

release captive-bred mammals into fenced areas 914

release captive-bred/translocated mammals 934, 937, 943

release translocated mammals into fenced areas 835, 839, 842

translocate to re-establish or boost populations 768

use fencing to exclude predators 552–553

use holding pens at release site 808, 897

Perameles gunnii (eastern barred bandicoot)

provide supplementary food during/after release 925, 929

release captive-bred mammals into fenced areas 913–914, 917–919

release captive-bred/translocated mammals 934–936, 942–945, 993–994

use fencing to exclude predators 548, 552

use holding pens at release site 895, 900, 903–904

Perameles lagotis (bilbies) 872

Perameles nasuta (bandicoots)

install barrier fencing and underpasses 347

install tunnels under railways 274

install tunnels under roads 262

remove/control non-native mammals 564–566

Père David's deer. *See Elaphurus davidianus*

Perognathus flavus (silky pocket mouse) 95

Perognathus hispidus (hispid pocket mouse) 95

Perognathus inornatus inornatus (San Joaquin pocket mouse) 93

Perognathus longimembris (little pocket mouse) 90

Perognathus parvus (Great Basin pocket mouse) 99

Peromyscus spp. (mice)

install tunnels under roads 270

thin trees within forest 464–465, 468

Peromyscus boylii (brush mice)

remove understorey vegetation 534

use prescribed burning 517

Peromyscus eremicus (cactus mouse) 676

Peromyscus gossypinus (cotton mouse)

install barrier fencing and underpasses 334

provide artificial refuges/breeding sites 693

Peromyscus keeni (Keen's mouse) 443

Peromyscus leucopus (white-footed mouse)

manage vegetation using grazing 653

provide artificial dens or nest boxes 698

provide supplementary food 735, 745

reduce intensity of grazing 95

remove trees and shrubs 536

remove understorey vegetation 533, 535

restore or create wetlands 682

use prescribed burning 523

Peromyscus maniculatus (deer mouse)

burn at specific time of year 526

clear or open patches in forests 444

create or maintain corridors between habitat patches 686

exclude livestock from semi-natural habitat 90

fell trees in groups 473

install barrier fencing and underpasses 329, 335

install overpasses over roads/railways 290, 296

install tunnels under roads 263–264, 269

leave coarse woody debris in forests 454

reduce intensity of grazing 96, 99

remove/control non-native mammals 569

remove understorey vegetation 534

remove vegetation by hand/machine 656

restore former mining sites 245

restore or create shrubland 676

thin trees to reduce wildfire risk 528–529

thin trees within forest 463, 465

use patch retention harvesting 438

use prescribed burning 515, 517, 519

use selective harvesting 434–436

Peromyscus pectoralis (white-ankled mice) 536

Persian fallow deer. *See Dama mesopotamica*

pesticide 55, 72, 619–620

Petauroides volans (greater glider)

provide artificial dens or nest boxes 698

retain wildlife corridors in logged areas 460

Petaurus breviceps (sugar gliders)

install pole crossings 302

install rope bridges between canopies 306

provide artificial dens or nest boxes 698, 701, 706–708

provide supplementary food 921, 928

release translocated/captive-bred mammals 974–975, 983

remove/control non-native mammals 564

Petaurus norfolcensis (squirrel gliders)

install pole crossings 300–302

install rope bridges between canopies 306–309

provide artificial dens or nest boxes 701

rovide artificial dens or nest boxes 704–705, 707–708, 710

Petrogale (rock wallabies) 567

Petrogale lateralis (rock wallabies)

provide supplementary water 751

release captive-bred mammals into fenced areas 917, 919

release captive-bred/translocated mammals 943, 945, 947

remove/control non-native mammals 557

translocate to re-establish or boost populations 768

Petrogale penicillata (brush-tailed rock wallabies)

release captive-bred mammals 876, 881

release translocated/captive-bred mammals 947

use holding pens at release site 897, 903

Petrogale persephone (Proserpine rock wallabies)

elease translocated/captive-bred mammals 950

install wildlife warning reflectors 362

release translocated/captive-bred mammals 948

Petrogale rothschildi (Rothschild's rock wallaby) 560

Petrogale xanthopus (yellow-footed rock wallabies)

provide supplementary water 751

release captive-bred mammals 918

release translocated/captive-bred mammals 943

remove/control non-native mammals 568

Phacochoerus aethiopicus (warthogs)

install mammal crossing points 105

translocate to re-establish or boost populations 759

Phacochoerus africanus (warthogs)

build fences around protected areas 642

install electric fencing 124

install mammal crossing points along fences 106

Phascloarctos cinereus (koalas) 248

Phascogale calura (red-tailed phascogales)
  hold translocated mammals in captivity 821
  provide supplementary food 927
  release captive-bred mammals 916
  release captive-bred/translocated mammals 941
  release translocated/captive-bred mammals 981
  release translocated mammals 838
  use holding pens at release site 900
Phascogale tapoatafa (brush-tailed phascogale)
  install rope bridges between canopies 308
  provide artificial dens or nest boxes 704–705, 708
  restore former mining sites 247
Phascolarctos cinereus (koalas)
  install barrier fencing and underpasses 331, 334
  install rope bridges between canopies 308
  plant trees on farmland 73–74
  temporarily hold females and offspring in fenced area 717, 722
  translocate mammals to reduce overpopulation 789
Phenacomys intermedius (heather voles) 54
Picea abies (Norway spruce) 193
Picea glauca × engelmannii (interior spruce) 439
Picoides borealis (red-cockaded woodpeckers)
  remove mid-storey vegetation in forest 532
  use prescribed burning 512
pine martens. *See Martes martes*
pine vole. *See Pitymys pinetorum*
Pinus contorta (lodgepole pine)
  apply fertilizer to trees 470–471
  gather coarse woody debris into piles 457–458

  provide diversionary feeding to reduce crop damage 192
  thin trees within forest 464, 466–467, 469
  use patch retention harvesting 439
Pinus nigra (Corsican pines)
  fell trees in groups 474
  provide artificial dens or nest boxes 700
Pinus ponderosa (ponderosa pine)
  provide artificial dens or nest boxes 697
  remove trees and shrubs to recreate open areas 537
  retain undisturbed patches during thinning 441
  thin trees to reduce wildfire risk 528–529
  use prescribed burning 519
Pinus radiata (Monterey pine) 451
Pinus rigida (pitch pines) 698
Pinus sylvestris (Scots pine)
  provide artificial dens or nest boxes 700
  provide supplementary food 845
  release translocated/captive-bred mammals 975
  use holding pens at release site 799
Pinus taeda (loblolly pine)
  leave coarse woody debris 455
  leave standing deadwood/snags 452
Pitymys pinetorum (pine vole) 682
plains bison. *See Bison bison bison*
plains mouse. *See Pseudomys australis*
pocket gophers. *See Thomomys mazama*
poisoning, use of 78–79, 113–114, 553, 559–560, 564, 566, 577, 830, 840, 911, 914, 932–933, 936
  restrict use of rodent poisons on farmland with high secondary poisoning risk 79–80
  use repellent on slug pellets to reduce non-target poisoning 78–79
polar bear. *See Ursus maritimus*
Populus spp. (cottonwood) 235

Populus tremuloides (aspen)
  create or maintain corridors between habitat patches 687
  thin trees within forest 462
  use repellents that taste bad 232–233
porcupine. *See Erethizon dorsatum;*
    See Hystrix africaeaustralis;
    See Sphiggurus villosus
Potamochoerus larvatus (bushpigs)
  build fences around protected areas 642
  install electric fencing 124
Potorous tridactylus (long-nosed potoroos)
  install barrier fencing and underpasses 347
  remove/control non-native mammals 566–567
prairie rattlesnakes. *See Crotalus viridis*
prairie vole. *See Microtus ochrogaster*
predation 3, 23–26, 28–30, 42, 105, 112–113, 115, 118, 120–121, 123–126, 128–129, 131–135, 137–150, 152–156, 158–161, 172–175, 177–178, 180, 182–183, 186–187, 190, 252, 258, 264, 273, 276, 299, 303–304, 309–310, 411, 428, 520, 527, 543, 552–554, 557, 568, 573, 577, 587, 595, 684, 690–691, 711, 715, 720, 724–725, 779, 802, 805, 810–811, 833, 844, 852, 869, 877–878, 881, 883, 890, 919, 921, 936, 944, 961, 965, 974, 983, 993
predators 25, 29–30, 79, 81, 92, 111–114, 118–119, 126–128, 133–135, 138–152, 157–160, 172, 174–175, 178–180, 183, 186–187, 203, 209, 226, 264, 276, 376, 391, 522, 538, 543, 546–552, 557, 559, 575–578, 583–586, 590, 593–594, 596–597, 641, 674, 692, 711, 751, 757, 774, 779, 788, 790–791, 794, 798, 800–801, 807, 818, 822, 827–828, 832, 834–835, 837, 840, 844, 875–876, 892, 896, 901, 907–908, 910–911, 915, 917–918, 925, 931–934, 938–943, 945–949, 958, 963, 968, 973, 975, 980, 991–992, 1003
pricklypear cactus. *See Opuntia polyacantha*

Prionailurus benalensis (leopard cat) 267
Procavia capensis (rock hyraxes)
  hold translocated mammals in captivity 819, 823
  provide supplementary food 849, 853
  release translocated/captive-bred mammals 962, 966, 977, 983
  use holding pens at release site 806, 813
Procyon lotor (raccoons)
  install barrier fencing and underpasses 327–328, 331, 333, 343, 345–346
  install ledges in culverts 281
  install tunnels under roads 262–265, 267–268, 270
  install wildlife warning reflectors 363
  provide artificial dens or nest boxes 696
  restore or create forest 679
  use taste-aversion to reduce predation 138–139
pronghorn. *See Antilocapra americana*
Proserpine rock wallabies. *See Petrogale persephone*
Prosopis juliflora (velvet mesquite) 574
Prunus americana (American plum) 232
Prunus virginiana (wild chokeberry) 232
Przewalski's horses. *See Equus ferus przewalskii*
Pseudalopex culpaeus (culpeo) 451
Pseudocheirus occidentalis (western ringtail possum) 768
Pseudocheirus peregrinus (common ringtail possums)
  hand-rear orphaned or abandoned young 724, 726
  install rope bridges between canopies 306, 308
  provide artificial dens or nest boxes 702–703, 706–708
  remove/control non-native mammals 564, 566
  restore or create forest 680
Pseudocheirus perigrinus (ringtail possum) 306

Pseudois nayaur (bharal) 419

Pseudomys sp. (pebble-mound mouse) 768

Pseudomys albocinereus (ash-grey mouse) 246

Pseudomys australis (plains mouse) 569

Pseudomys bolami (Bolam's mice) 572

Pseudomys fieldi (shark bay mouse)
  release captive-bred mammals 872
  translocate to re-establish or boost populations 768

Pseudotsuga menziesii (Douglas fir)
  clear or open patches in forests 443
  fell trees in groups 473
  provide artificial dens or nest boxes 701
  provide diversionary feeding to reduce crop damage 191
  replant vegetation 657
  thin trees within forest 468
  use patch retention harvesting 439
  use prescribed burning 508
  use selective harvesting 435

Pudu puda (southern pudu) 451

puma. *See Puma concolor*

Puma concolor (puma)
  hold translocated mammals in captivity 816
  install barrier fencing and underpasses 329–330, 335, 346, 348–349
  install overpasses over roads/railways 291
  install tunnels under roads 265
  rehabilitate injured, sick or weak mammals 720
  release captive-bred mammals 868
  remove or control predators 586
  set hunting quotas 415
  translocate to re-establish or boost populations 764
  use guardian animals bonded to livestock to deter predators 145
  use holding pens at release site 796
  use loud noises to deter predation 150
  use visual deterrents to deter predation 133

Pyrenean chamois. *See Rupicapra pyrenaica pyrenaica*

Quercus spp. (oak trees)
  provide artificial dens or nest boxes 698
  provide artificial refuges/breeding sites 693
  provide supplementary water 750
  remove trees and shrubs to recreate open areas 536
  use fencing to exclude grazers 545

quokka. *See Setonix brachyurus*

rabbits. *See Sylvilagus* spp.

raccoon dog. *See Nyctereutes procyonoides*

raccoons. *See Procyon lotor*

radish. *See Raphanus sativus*

Rangifer tarandus (caribou)
  install crossings over/under pipelines 400–401
  manage vegetation using livestock grazing 651
  provide diversionary feeding for predators 594
  provide supplementary food 737, 742
  remove or control predators 585, 588
  remove vegetation by hand/machine 661
  sterilize predators 591
  translocate to re-establish or boost populations 759, 761
  use chemical repellents 391

Raphanus sativus (radish) 83

rats. *See Rattus* sp.

Rattus sp. (rats)
  install overpasses over roads/railways 292
  install tunnels under railways 275
  install tunnels under roads 260

Rattus fuscipes (bush rat)
  install tunnels under railways 274
  install tunnels under roads 260–261

install underpasses beneath ski runs 52

remove/control non-native mammals 557, 564–565

use patch retention harvesting 439

Rattus lutreolus (swamp rat)

install barrier fencing and underpasses 334

install tunnels under roads 262

Rattus norvegicus (brown rats)

install barrier fencing and underpasses 337

install tunnels under railways 277

install tunnels under roads 262, 267

remove/control non-native mammals 563

Rattus rattus (black rat)

install barrier fencing and underpasses 344

install rope bridges between canopies 306

install underpasses beneath ski runs 52

provide artificial dens or nest boxes 707–708

remove/control non-native mammals 564–565

Rattus villosissimus (long-haired rats) 551

red-backed voles. *See Clethrionomys gapperi*

red-bellied pademelon. *See Thylogale billardierii*

red brocket. *See Mazama; See Mazama americana*

red-cockaded woodpeckers. *See Picoides borealis*

red deer. *See Cervus elaphus*

red fox. *See Vulpes vulpes; See Vulpes fulva*

red-fronted gazelle. *See Eudorcas rufifrons*

red imported fire ants. *See Solenopsis invicta*

red kangaroos. *See Macropus rufus*

red lechwe. *See Kobus leche leche*

red-legged pademelons. *See Thylogale stigmatica*

red muntjac. *See Muntiacus muntjak*

red-necked wallaby. *See Macropus rufogriseus*

red squirrel. *See Tamiasciurus hudsonicus; See Sciurus vulgaris*

red-tailed hawks. *See Buteo jamaicensis*

red-tailed phascogales. *See Phascogale calura*

red veld rat. *See Aethomys chrysophilus*

red wolf. *See Canis rufus; See Canis lupus rufus*

Reithrodonromys megalotis (western harvest mouse) 95

repellents, use of 45–46, 49, 131, 178, 182, 217, 227, 231–233, 235–240, 252–253, 389–393

use chemical repellents along roads or railways 389–393

use repellent on slug pellets to reduce non-target poisoning 78–79

use repellents that smell bad ('area repellents') to deter crop or property damage by mammals 238–240

use repellents that taste bad ('contact repellents') to deter crop or property damage by mammals 231–238

use repellents to reduce cable gnawing 252–254

Rhabdomys pumilio (striped mouse) 738

Rhinoceros unicornis (greater one-horned rhinoceros)

hand-rear orphaned young in captivity 726–727

hold translocated mammals in captivity 816

legally protect habitat for mammals 638, 640

release translocated/captive-bred mammals 964, 966

release translocated mammals into fenced areas 829

translocate to re-establish or boost populations 778, 786

rice rats. *See Oryzomys palustris*

ringtail possum. *See Pseudocheirus perigrinus*

riparian brush rabbits. *See Sylvilagus bachmani*

river otters. *See Lutra canadensis*

roads 12, 168, 257–262, 264–265, 267–268, 270–273, 275–290, 292, 294, 296, 298, 300–303, 305–307, 309, 311, 315–316, 319, 323–324, 332, 337, 339, 342, 344, 347–348, 356, 359, 361–368, 370, 373, 376, 379–380, 384, 388–399, 401, 424, 456, 483–484, 489, 514, 718, 830, 911, 932

roan antelope. *See Hippotragus equinus*

rock hyraxes. *See Procavia capensis*

rock wallabies. *See Petrogale; See Petrogale lateralis*

roe deer. *See Capreolus capreolus*

root voles. *See Microtus oeconomus*

Rothschild's rock wallaby. *See Petrogale rothschildi*

rough fescue. *See Festuca scabrella*

round-tailed muskrat. *See Neofiber alleni*

rufous hare-wallabies. *See Lagorchestes hirsutus*

Rupicapra pyrenaica ornata (Apennine chamois)
  release captive-bred mammals 878, 882
  translocate to re-establish or boost populations 780–781, 787

Rupicapra pyrenaica pyrenaica (Pyrenean chamois) 416

Rusa unicolor (Sambar deer) 424

sable antelope. *See Hippotragus niger*

Saccostomus campestris (South African pouched mouse) 283

Salix spp. (willow) 954

Sambar deer. *See Rusa unicolor*

San Diego pocket mouse. *See Chaetodipus fallax*

San Joaquin antelope squirrel. *See Ammospermophilus nelson*

San Joaquin kit foxes. *See Vulpes macrotis mutica*

San Joaquin pocket mouse. *See Perognathus inornatus inornatus*

Santiago rice rats. *See Nesoryzomys swarthi*

Sarcophilus harrisii (Tasmanian devils)
  captive rear in large enclosures 884
  cull disease-infected animals 607
  release translocated/captive-bred mammals 993

Sarcophilus laniarius (Tasmanian devils)
  install signage to warn motorists 384
  install traffic calming structures 373
  install wildlife warning reflectors 360

Sarcoptes scabiei (sarcoptic mange) 611–612

scimitar-horned oryx. *See Oryx dammah*

Sciurus aberti (tassel-eared squirrels)
  provide artificial dens or nest boxes 697
  retain undisturbed patches during thinning 441
  thin trees to reduce wildfire risk 529
  use prescribed burning 519

Sciurus carolinensis (gray squirrel)
  install tunnels under roads 268, 270
  provide artificial dens or nest boxes 696–698
  provide supplementary food 845, 852
  release translocated/captive-bred mammals 975, 983
  restore or create wetlands 682
  use holding pens at release site 799, 811

Sciurus granatensis (red-tailed squirrel) 679

Sciurus nayaritensis chiricahuae (Mexican fox squirrels) 520

Sciurus niger (fox squirrels) 697

Sciurus vulgaris (red squirrel)
  install barrier fencing and underpasses 336
  provide artificial dens or nest boxes 700, 709
  provide supplementary food 844, 848, 852–853

release translocated/captive-bred mammals 974, 977, 983, 986, 989

translocate to re-establish or boost populations 763

use holding pens at release site 798, 806, 811–812

sea otter. *See Enhydra lutris*

Secale cereale (rye) 671

serval. *See Felis serval; See Leptailurus serval*

set-aside areas, use of 56, 62, 64–67, 71, 76

Setonix brachyurus (quokka)

release translocated/captive-bred mammals 947

remove/control non-native mammals 563, 570

restore former mining sites 249

translocate to re-establish or boost populations 768

shark bay mouse. *See Pseudomys fieldi*

sheep. *See Ovis aries; See Ovis* sp.

Short-beaked echidna. *See Tachyglossus aculeatus*

short nosed kangaroo rat. *See Dipodomys nitratoides nitratoides*

short-tailed mice. *See Leggadina lakedownensis*

short-tailed weasel. *See Mustela erminea*

shrews. *See Sorex* spp.

Siberian weasel. *See Mustela sibirica*

Sigmodon hispidus (hispid cotton rats)

create uncultivated margins around fields 62

install barrier fencing and underpasses 334

provide supplementary food 739, 746

provide supplementary food after fire 541

reduce intensity of grazing 95

remove trees and shrubs to recreate open areas 536

use fencing to exclude predators 549

sika deer. *See Cervus nippon*

silky pocket mouse. *See Perognathus flavus*

small fox tapeworm. *See Echinococcus multicularis*

small vesper mice. *See Calomys laucha*

Sminthopsis crassicaudata (fat-tailed dunnart)

provide artificial refuges/breeding sites 691

remove/control non-native mammals 572

Sminthopsis granulipes (white-tailed dunnart) 246

Sminthopsis macroura (stripe-faced dunnart) 572

Sminthopsis murina (Common dunnart) 338

smoky shrews. *See Sorex fumeus*

Smutsia temminckii (Temminck's ground pangolins) 411

snow leopard. *See Panthera uncia*

snowshoe hare. *See Lepus americanus*

Solanum tuberosum (potatoes) 214

Solenopsis invicta (red imported fire ants) 555

Sonoran pronghorns. *See Antilocapra american sonoriensis*

Sorex spp. (shrews)

install barrier fencing and underpasses 329

install tunnels under railways 275

install tunnels under roads 260, 263

Sorex aranaeus (common shrew)

create uncultivated margins around fields 59–61

install overpasses over roads/railways 290

plant trees on farmland 73

provide or retain set-aside areas 66

use repellent on slug pellets 78

Sorex cinereus (masked shrews)

fell trees in groups 473

thin trees within forest 465

use selective harvesting 436

Sorex fumeus (smoky shrews) 516

Sorex hoyi (American pygmy shrews)
516
Sorex longirostris (southeastern shrews)
leave coarse woody debris 455
leave standing deadwood/snags 453
use prescribed burning 516
Sorex monticolus (dusky shrew)
clear or open patches in forests 443
fell trees in groups 473
Sorex pacificus (Pacific shrew) 463
Sorex trowbridgii (Trowbridge's shrew)
clear or open patches in forests 443
thin trees within forest 463
Sorex vagrans (vagrant shrew)
clear or open patches in forests 443
install tunnels under roads 264
thin trees within forest 463
Sorghum sp. (sorghum) 214
South African pouched mouse.
See Saccostomus campestris
South American coatis. *See Nasua nasua*
southeastern short-tailed shrew.
See Blarina carolinensis
southeastern shrews. *See Sorex longirostris*
southern brown bandicoots. *See Isoodon obesulus*
southern flying squirrels. *See Glaucomys volans*
southern grasshopper mouse.
See Onychomys torridus
southern pudu. *See Pudu puda*
southern red-backed voles. *See Myodes gapperi*
Spanish ibex. *See Capra pyrenaica hispanica*
Spermophilus beecheyi (California ground squirrel)
exclude livestock from semi-natural habitat 91, 93
remove understorey vegetation 534
restore or create shrubland 676
use holding pens at release site 795
use prescribed burning 517
Spermophilus citellus (European ground squirrels)

release translocated/captive-bred mammals 978–979, 983, 988–989
release translocated mammals into fenced areas 833, 841
translocate to re-establish or boost populations 779
Spermophilus columbianus (Columbian ground squirrel) 264
Spermophilus lateralis (golden-mantled ground squirrel)
thin trees to reduce wildfire risk 528–529
use prescribed burning 519
Spermophilus tridecemlineatus (thirteen-lined ground squirrels) 245
Sphiggurus villosus (porcupines) 307
Spilogale putorius (spotted skunks) 265
spinifex hopping mouse. *See Notomys alexis*
spiny pocket mouse. *See Chaetodipus spinatus*
spotted hyenas. *See Crocuta crocuta*
spotted skunks. *See Spilogale putorius*
squirrel gliders. *See Petaurus norfolcensis*
Sri Lankan hornets. *See Vespa affinis affinis*
Stephens' kangaroo rat. *See Dipodomys stephensi*
stoat. *See Mustela ermine*
stone marten. *See Martes foina; See Marte foina*
striped field mouse. *See Apodemus agrarius*
striped hyenas. *See Hyaena hyaena*
striped mouse. *See Rhabdomys pumilio*
striped skunks. *See Mephitis mephitis*
stripe-faced dunnart. *See Sminthopsis macroura*
sugar gliders. *See Petaurus breviceps*
Sus scrofa (wild boar)
install barrier fencing and underpasses 339, 342, 612
install electric fencing to protect crops 165
install non-electric fencing 117

install overpasses over roads/railways 288–290, 292, 295

install overpasses over waterways 403

install tunnels/culverts/underpass under railways 275

install tunnels under roads 261

install wildlife warning reflectors 363

provide diversionary feeding 192–193

provide mammals with escape routes 406

provide supplementary food 737

use chemical repellents 390

swamp rat. *See Rattus lutreolus*

swamp wallaby. *See Wallabia bicolor*

swift foxes. *See Vulpes velox*

Sylvilagus spp. (rabbits) 343

Sylvilagus audubonii (desert cottontail rabbits)

install barrier fencing and underpasses 348

install tunnels under roads 265

remove/control non-native plants 574

retain dead trees after uprooting 446–447

train captive-bred mammals to avoid predators 908

Sylvilagus bachmani (riparian brush rabbits) 896

Sylvilagus floridanus (eastern cottontail rabbits)

establish long-term cover 88

provide supplementary food 742

remove vegetation using herbicides 668

use prescribed burning 510–511

Sylvilagus palustris (Lower Keys marsh rabbits)

install barrier fencing and underpasses 328

translocate to re-establish or boost populations 769

Syncerus caffer (buffalo)

build fences around protected areas 642

provide artificial waterholes 538

release translocated/captive-bred mammals 959, 970, 992

release translocated mammals into fenced areas 831

retain wildlife corridors in residential areas 51

translocate to re-establish or boost populations 774

use holding pens at release site 801

Tachyglossus aculeatus (short-beaked echidna)

install barrier fencing and underpasses 338

install overpasses over roads/railways 292

restore former mining sites 246

Talpa europaea (moles) 239–240

Tamandua tetradactyla (lesser anteaters) 254–255

Tamarix ramosissima (saltcedar) 574

Tamias amoenus (northwestern chipmunk)

clear or open patches in forests 444

fell trees in groups 473

use selective harvesting 434, 436

Tamias cinereicollis (gray-collared chipmunk)

thin trees to reduce wildfire risk 528–529

use prescribed burning 519

Tamiasciurus douglasii (Douglas' squirrels)

provide artificial dens or nest boxes 701–702, 709

provide supplementary food 740, 746

Tamiasciurus hudsonicus (red squirrels)

install barrier fencing and underpasses 329

install tunnels under roads 263

provide artificial dens or nest boxes 697

provide diversionary feeding to reduce crop damage 192

provide supplementary food 736, 745

thin trees within forest 464

Tamias minimus (chipmunk) 54

Tamias quadrimaculatus (ong-eared chipmunks)

    remove understorey vegetation 534

    use prescribed burning 517

Tamias striatus (chipmunk)

    create or maintain corridors between habitat patches 686

    install tunnels under roads 268

Tamias townsendii (Townsend's chipmunk) 701

tammar wallabies. *See Macropus eugenii*

tapir. *See Tapirus terrestris*

Tapirus bairdii (Central American tapir) 422

Tapirus terrestris (tapir) 407

Tarsipes rostratus (honey possum) 246

Tasmanian devils. *See Sarcophilus laniarius; See Sarcophilus harrisi*

tassel-eared squirrels. *See Sciurus aberti*

taste aversion 137–140, 496–497, 499

Tatera leucogaster (bushveld gerbils) 286

Taurotragus oryx (eland) 538

Taxidea taxus (badgers)

    install barrier fencing and underpasses 346

    release captive-bred/translocated mammals 937, 940, 963, 980

    translocate to re-establish or boost populations 772

    use holding pens at release site 897

Taxus cuspidata (Japanese yew trees) 233, 237

Tayassu pecari (white-lipped peccaries) 422

Tehuantepec jackrabbits. *See Lepus flavigularis*

Temminck's ground pangolins. *See Smutsia temminckii*

thirteen-lined ground squirrels. *See Spermophilus tridecemlineatus*

Thomomys mazama (pocket gophers) 234

Thomomys talpoides (northern pocket gophers) 253

Thomson's gazelles. *See Gazella thomsonii*

Thuja plicata (red cedar)

    use loud noises to deter crop damage 198

    use repellents that taste bad 234

    use 'shock collars' to deter crop damage 230

Thylogale billardierii (red-bellied pademelon) 947

Thylogale stigmatica (red-legged pademelons)

    install barrier fencing and underpasses 347

    install tunnels under roads 266

tiger. *See Panthera tigris*

Townsend's chipmunk. *See Tamias townsendii*

Tragelaphus strepsiceros (kudu)

    provide artificial waterholes 538

    replant vegetation 662

    use prescribed burning 520

translocate mammals 30–36, 38, 42, 151–157, 176, 184, 205–207, 254–255, 290, 335, 487, 492–494, 543, 547–549, 552, 561, 587–588, 591, 624–625, 629–630, 639, 654, 658, 661, 664, 666–667, 673–674, 690, 713, 727, 734, 753–783, 785–811, 814–816, 818–840, 842–853, 865–868, 870–872, 874–876, 878–879, 886–889, 891–897, 902, 910–913, 915–916, 921, 924, 929–934, 938–943, 945–994

    airborne translocation of mammals using parachutes 825–826

    hold translocated mammals in captivity before release 814–823

    provide supplementary food during/after release of translocated mammals 842–854

    release translocated/captive-bred mammals at a specific time 984–989

    release translocated/captive-bred mammals in areas with invasive/problematic species eradication/control 929–945 release translocated/captive-bred mammals in family/social groups 950–967

release translocated/captive-bred mammals in larger unrelated groups 967–972

release translocated/captive-bred mammals into area with artificial refuges/breeding sites 972–984

release translocated/captive-bred mammals to areas outside historical range 989–994

release translocated/captive-bred mammals to islands without invasive predators 945–950

release translocated mammals into fenced areas 826–842

translocate animals from source populations subject to similar climatic conditions 629–630

translocate crop raiders away from crops 205–207

translocate mammals away from site contaminated by oil spill 624–625

translocate mammals away from sites of proposed energy developments 254–255

translocate mammals that have habituated to humans 492–494

translocate mammals to reduce overpopulation 788–790

translocate predators away from livestock 151–157

translocate predators for ecosystem restoration 790–792

translocate problem mammals away from residential areas 30–37

translocate to re-establish or boost populations 753–787

use holding pens at release site prior to release of translocated mammals 792–813

use tranquilizers to reduce stress during translocation 824–825

trematode. *See Fascioloides magna*

Trichodectes canis (biting dog lice) 616

Trichosurus sp. (common brushtail possum) 706

Trichosurus caninus (northern mountain brushtail possums)

install rope bridges between canopies 308

provide artificial dens or nest boxes 698, 706

retain wildlife corridors in logged areas 460

Trichosurus cunninghami (mountain brushtail possum) 702

Trichosurus vulpecula (common brushtail possum)

install barrier fencing and underpasses 348

install rope bridges between canopies 306, 308

install tunnels under roads 266

provide artificial dens or nest boxes 703, 707–708

release translocated mammals into fenced areas 837

remove/control non-native mammals 560–562, 564–567

restore former mining sites 246–247, 249

restore or create forest 680

Trifolium subterraneum (subterranean clover) 671

Tristram's jird. *See Meriones tristrami*

Triticum aestivum (wheat) 66

Trowbridge's shrew. *See Sorex trowbridgii*

tsessebe. *See Damaliscus lunatus*

Tsuga canadensis (eastern hemlock) 706

tunnels 239, 258–261, 263–265, 268, 273–274, 276–277, 280, 284, 288, 291, 293, 295, 300, 305–307, 316, 324, 340, 353

Tyto alba (barn owls)

create uncultivated margins around fields 64

plant trees on farmland 74

provide or retain set-aside areas 67

uncultivated margins, use of 58, 68, 75

underpasses, use of 51, 257–259, 261–262, 264–277, 280, 283–292, 294, 296–299, 309, 311, 314–317, 321, 323–333, 335–354

  install barrier fencing and underpasses along roads 323–353

  install fences around existing culverts or underpasses under roads/ railways 283–288

  install tunnels/culverts/underpass under railways 273–278

  install tunnels/culverts/underpass under roads 258–273

  install underpasses beneath ski runs 51–53

urial sheep. *See Ovis vignei*

Urocyon cinereoargenteus (grey fox)

  install barrier fencing and underpasses 343, 345

  install tunnels under roads 268, 270

Urocyon littoralis (island foxes)

  modify traps 579

  remove/control non-native mammals 569

  use conditioned taste aversion 581

  use reward removal 582

Ursus americanus (black bear)

  hand-rear orphaned young 726

  install barrier fencing and underpasses 327, 329–331, 333, 335, 343, 346, 348–349

  install crossings over/under pipelines 401

  install electric fencing to protect crops 166

  install overpasses over roads/railways 289, 291, 295

  install tunnels under roads 267, 270

  install wildlife exclusion grates 370

  issue enforcement notices to deter use of non-bear-proof garbage dumpsters 39

  legally protect habitat for mammals 637

place orphaned wild young with wild foster parents 728

prevent mammals accessing potential wildlife food sources 40

provide diversionary feeding for mammals 42–43

provide diversionary feeding for predators 594

provide education programmes 1001

provide supplementary feed 481

release captive-bred mammals 870

release captive-bred/translocated mammals 939, 976, 986

remove or control predators 587

scare or otherwise deter mammals 46–48

translocate crop raiders away from crops 206

translocate mammals 493

translocate predators away from livestock 153–154

translocate problem mammals away from residential areas 34–37

translocate to re-establish or boost populations 760

use campaigns and public information 998

use conditioned taste aversion 497

use flags to reduce predation of livestock 130

use lights and sound to deter predation 181

use non-lethal methods to deter carnivores 499–500

use signs or access restrictions 484

Ursus arctos (Alaskan brown bears)

  ban private ownership of hunted mammals 413

  cease/reduce payments to cull mammals 714

  encourage community-based participation 996

  exclude or limit number of visitors to reserves 487–488

  habituate mammals to visitors 491

hand-rear abandoned young 726

install barrier fencing and underpasses 329–330, 335, 342, 346, 349–350

install non-electric fencing 118

install overpasses over roads/railways 291, 295, 297

prohibit or restrict hunting of a species 410

provide diversionary feeding 42

provide diversionary feeding to reduce predation 159

provide supplementary food 744

rehabilitate injured, sick or weak mammals 721

remove or control predators 587

translocate predators away from livestock 154

translocate problem mammals away from residential areas 32–33, 36

translocate to re-establish or boost populations 762, 769, 772, 774, 781, 783

use chemical repellents along roads or railways 392

use non-lethal methods to deter carnivores 500–501

Ursus maritimus (polar bear) 45

Ursus thibetanus (Asiatic black bears)

hand-rear abandoned young 726

install electric fencing to protect crops 163

Utah prairie dogs. *See Cynomys parvidens*

vagrant shrew. *See Sorex vagrans*

Vancouver Island marmots. *See Marmota vancouverensis*

Varanus gouldii (Gould's monitors) 934

vegetation 59, 68–69, 72, 85–87, 89, 95, 100, 108, 167, 222, 232, 244, 246–248, 261, 266, 269, 274–276, 290, 294–295, 319, 328, 331, 333, 335, 341, 343, 373–378, 395–396, 401–403, 448–452, 455, 460, 474–476, 478, 514–515, 531–534, 536, 623, 631, 633, 635, 647–650, 652–656, 658–662, 664, 666–667, 669, 673–674, 676–677, 682–683, 686–688, 693, 707, 709, 719, 776, 779, 791, 841, 917, 965, 984, 993

velvet mesquite. *See Prosopis juliflora*

Verbascum spp. (mulleins) 61

Vespa affinis affinis (Sri Lankan hornets) 212

Vicugna pacos (alpacas) 146

Vitis vinifera (common grape vines)

install non-electric fencing 117

provide diversionary feeding 192

vole. *See Microtus* spp.; *See Arvicolinae*

Vombatus ursinus (common wombat) 261, 269, 272

Vulpes fulva (red fox) 331

Vulpes lagopus (arctic fox)

provide supplementary food 742

remove or control competitors 592

Vulpes macrotis mutica (San Joaquin kit foxes)

release translocated/captive-bred mammals 964, 967

use holding pens at release site 809, 813

Vulpes velox (swift foxes)

hold translocated mammals in captivity 816, 818, 823

provide artificial refuges/breeding sites 691

release captive-bred mammals 868, 870, 872, 878, 880

release translocated/captive-bred mammals 958, 966, 986

translocate to re-establish or boost populations 760, 765, 777, 784

use holding pens at release site 797, 809, 889, 892, 902

use prescribed burning 518

Vulpes vulpes (red fox)

hold translocated mammals in captivity 822

install barrier fencing and underpasses 328, 330, 333–334, 336–337, 339–340, 344

install ledges in culverts under roads/railways 281

install mammal crossing points along fences 105

install overpasses over roads/railways 289–292, 295

install tunnels under railways 275, 277

install tunnels under roads 261, 264, 267–268, 270

install wildlife warning reflectors 363

make introduction of non-native mammals 428

provide artificial dens or nest boxes 705

provide mammals with escape routes 406

provide supplementary food 742, 850, 926

release captive-bred mammals 878

release captive-bred mammals into fenced areas 911–914

release captive-bred/translocated mammals 932–937, 941–942, 946, 971, 981, 993

release translocated mammals into fenced areas 830, 833, 835, 838, 840

remove/control non-native mammals 557–569, 571–572

remove or control competitors 592

remove or control predators 585–586

restore former mining sites 247

train captive-bred mammals to avoid predators 907

train mammals to avoid problematic species 597

translocate to re-establish or boost populations 779

treat mammals to reduce conflict caused by disease 495

use fencing to exclude predators 548–549, 552

use flags to reduce predation 130

use guardian animals to deter predators 143

use holding pens at release site 808, 895, 897

use lights and sound to deter predation 181

use vaccination programme 601–602

Wallabia bicolor (swamp wallaby)
install barrier fencing and underpasses 347

install overpasses over roads/railways 292, 296

install tunnels under railways 274

install tunnels under roads 261

warthogs. *See Phacochoerus aethiopicus; See Phacochoerus africanus*

water 11, 45, 89, 92, 106, 210, 219, 226, 229, 233, 235–237, 239–240, 243, 261–264, 267, 269–270, 278–281, 312, 318, 332, 338, 375, 402–407, 499, 502, 505, 538–540, 556, 564, 570, 600, 611, 623, 630–633, 647, 681–685, 692, 747–751, 757, 774, 793, 797–799, 803, 807, 830, 843–845, 867, 869, 886, 889–890, 893–894, 901, 908, 911–912, 917–918, 920, 923–926, 930, 933, 935–936, 939–940, 942–944, 948, 973–975, 992

water deer. *See Hydropotes inermis*

water vole. *See Arviola terrestris; See Arvicola terrestris; See Arvicola amphibius*

weasel. *See Mustela nivalis; See Mustela* sp.

wedge-tailed eagle. *See Aquila audax*

western barred bandicoots. *See Perameles bougainville*

western brush wallaby. *See Macropus irma*

western grey kangaroo. *See Macropus fuliginosus*

western harvest mouse. *See Reithrodonromys megalotis*

western jumping mouse. *See Zapus princeps*

western mountain coatis. *See Nasuella olivacea*

western polecat. *See Mustela putorius*

western pygmy possum. *See Cercartetus concinnus*

western red-backed vole.
See Clethrionomys californicus

western ringtail possum.
See Pseudocheirus occidentalis

white-ankled mice. *See Peromyscus pectoralis*

white-eared opossums. *See Didelphis albiventris*

white-footed mouse. *See Peromyscus leucopus*

white-footed vole. *See Arborimus albipes*

white rhinoceros. *See Ceratotherium simum; See Ceratoiherium simum cottoni*

white-tailed deer. *See Odocoileus virginianus*

white-tailed dunnart. *See Sminthopsis granulipes*

white-toothed shrew. *See Crocidura russula; See Crocidura suaveolens*

wild boar. *See Sus scrofa*

wild cat. *See Felis silvestris*

wildebeest. *See Connochaetes taurinus*

wildlife corridors 50–51, 459, 461

retain wildlife corridors in logged areas 459–461

retain wildlife corridors in residential areas 50–51

wolverines. *See Gulo gulo*

wombats. *See Lasiorhinus latifrons*

woodland dormice. *See Graphiurus murinus*

woodland vole. *See Microtus pinetorum*

wood mice. *See Apodemus sylvaticus*

woylies. *See Bettongia penicillata*

yellow-footed antechinus. *See Altechinus flavipes*

yellow-footed rock wallabies. *See Petrogale xanthopus*

Zapus hudsonius (meadow jumping mouse) 91

Zapus princeps (western jumping mouse) 99

Zapus trinotatus (Pacific jumping mouse) 463

Zea mays (maize) 214

zebra. *See Equus burchelli; See Equus quagga*

About the Team

Alessandra Tosi was the managing editor for this book.

Lucy Barnes performed the copy-editing, proofreading and indexing.

Anna Gatti designed the cover using InDesign. The cover was produced in InDesign using Fontin (titles) and Calibri (text body) fonts.

Luca Baffa typeset the book in InDesign. The text font is Tex Gyre Pagella; the heading font is Californian FB. Luca created all of the editions — paperback, hardback, EPUB, MOBI, PDF, HTML, and XML — the conversion is performed with open source software freely available on our GitHub page (https://github.com/OpenBookPublishers).

This book need not end here...

Share

All our books — including the one you have just read — are free to access online so that students, researchers and members of the public who can't afford a printed edition will have access to the same ideas. This title will be accessed online by hundreds of readers each month across the globe: why not share the link so that someone you know is one of them?

This book and additional content is available at:

https://doi.org/10.11647/OBP.0234

Customise

Personalise your copy of this book or design new books using OBP and third-party material. Take chapters or whole books from our published list and make a special edition, a new anthology or an illuminating coursepack. Each customised edition will be produced as a paperback and a downloadable PDF.

Find out more at:

https://www.openbookpublishers.com/section/59/1

You may also be interested in:

What Works in Conservation 2020

William J. Sutherland, Lynn V. Dicks, Silviu O. Petrovan
and Rebecca K. Smith (eds)

https://www.openbookpublishers.com/section/83/1

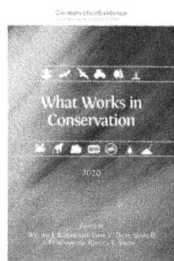

Conservation Biology in Sub-Saharan Africa

John W. Wilson and Richard B. Primack

https://doi.org/10.11647/OBP.0177

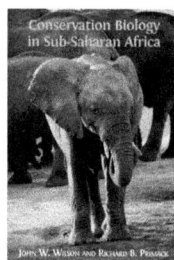

Forests and Food
Addressing Hunger and Nutrition Across Sustainable
Landscapes
Bhaskar Vira, Christoph Wildburger
and Stephanie Mansourian (eds)

https://doi.org/10.11647/OBP.0085

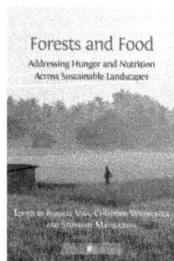

www.ingramcontent.com/pod-product-compliance
Lightning Source LLC
Chambersburg PA
CBHW060016030426

42334CB00019B/2064